Studies in Systems, Decision and Control

Volume 415

The series "Studies in Systems, Decision and Control" (SSDC) covers both new developments and advances, as well as the state of the art, in the various areas of broadly perceived systems, decision making and control–quickly, up to date and with a high quality. The intent is to cover the theory, applications, and perspectives on the state of the art and future developments relevant to systems, decision making, control, complex processes and related areas, as embedded in the fields of engineering, computer science, physics, economics, social and life sciences, as well as the paradigms and methodologies behind them. The series contains monographs, textbooks, lecture notes and edited volumes in systems, decision making and control spanning the areas of Cyber-Physical Systems, Autonomous Systems, Sensor Networks, Control Systems, Energy Systems, Automotive Systems, Biological Systems, Vehicular Networking and Connected Vehicles, Aerospace Systems, Automation, Manufacturing, Smart Grids, Nonlinear Systems, Power Systems, Robotics, Social Systems, Economic Systems and other. Of particular value to both the contributors and the readership are the short publication timeframe and the world-wide distribution and exposure which enable both a wide and rapid dissemination of research output.

Indexed by SCOPUS, DBLP, WTI Frankfurt eG, zbMATH, SCImago.

All books published in the series are submitted for consideration in Web of Science.

Peng Shi · Jovan Stefanovski · Janusz Kacprzyk
Editors

Complex Systems: Spanning Control and Computational Cybernetics: Applications

Dedicated to Professor Georgi M. Dimirovski on his Anniversary

Springer

Editors
Peng Shi
School of Electrical and Electronic Engineering
The University of Adelaide
Adelaide, SA, Australia

Jovan Stefanovski
Control and Informatics Division
JP Hydro-System "Streževo"
Bitola, North Macedonia

Janusz Kacprzyk
Systems Research Institute
Polish Academy of Sciences
Warsaw, Poland

ISSN 2198-4182 ISSN 2198-4190 (electronic)
Studies in Systems, Decision and Control
ISBN 978-3-031-00980-8 ISBN 978-3-031-00978-5 (eBook)
https://doi.org/10.1007/978-3-031-00978-5

This Springer imprint is published by the registered company Springer Nature Switzerland AG
The registered company address is: Gewerbestrasse 11, 6330 Cham, Switzerland

Dedicated to Prof. Georgi M. Dimirovski

Preface

We are honored and pleased to have an opportunity to conceive the idea of this special editorial project, and above all to obtain the support of so many top people from the broadly perceived cybernetics, systems, control, automation, etc., communities who have agreed to prepare special papers dedicated to Professor Georgi M. Dimirovski, a prominent member of these communities.

This editorial project, comprising of two volumes, is a small token of a great appreciation of the world communities in the fields mentioned above, and well beyond, to Prof. Georgi M. Dimirovski for his important research and scholarly achievements, and a lifelong service to the community, both in his home country and in the whole world. His long and illustrious career spans over many decades and it can be said without exaggeration that it reflects the tumultuous history of Europe where World War II has changed everything, implied movements of entire nations, transfers of powers, changes of borders, change of political and economic systems, etc. These all have occurred notably in Central, Eastern, and Southern Europe, notably the so-called Balkan countries. So, Professor Dimirovski was born in the former Kingdom of Greece, then lived in the former Yugoslavia, and, finally, for the last decades had lived in his home country, now known as Northern Macedonia.

Since the beginning of his academic career, he has been working on the broadly perceived systems and control, augmented with a broader cybernetics perspective, and has quickly become a driving force for the entire research community in his areas first in his home country, first in Yugoslavia and now in Northern Macedonia, and then worldwide due to his numerous influential scientific results published in prominent journals and presented at a prestigious conference. His extraordinary stature in the world's scientific and scholarly communities has been considerably amplified by his active involvement in the activities of international institutions, organizations, and societies exemplified by the European Science Foundation (ESF), Institute of Electrical and Electronics Engineering (IEEE), International Federation of Automatic Control (IFAC), Institution of Engineering and Technology (IET, formerly known as IEE), and International Academy of Systems & Cybernetics Sciences (IASCYS), just to name a few.

This great visibility of Prof. Dimirovski not only in science but also in technology and R&D as his results have also been considered very relevant for applications, has clearly implied an overwhelmingly positive reaction of the respected research and scholarly communities in his areas to this initiative to publish a special volume. This volume is dedicated to Prof. Dimirovski, with contributions by not only top researchers and scholars, his colleagues and friends, and peers but also many younger people for whom he has provided inspiration, advice, supervision, etc.

This editorial project has materialized in the present two volumes which contain relevant contributions discussing current. Challenging and promising problems are related to many aspects within the broadly perceived systems and control. The first volume is focused on more foundational aspects related to general issues in systems science and mathematical systems, various problems in control and automation, and the use of computational and artificial intelligence in the context of systems modeling and control. The second volume is concerned with a presentation of relevant applications, notably in robotics, computer networks, telecommunication, fault detection/diagnosis, as well as in biology and medicine, and economic, financial, and social systems too.

This volume, which constitutes the second volume of this editorial project, is focused on applications in various areas of science and technology that are perceived as cybernetic systems of decision and control.

Part I "New Challenges in Robotic Systems" is concerned with various aspects of robotics which constitutes a premier field at crossroads of many areas that are discussed in this volume. Novel solutions in very challenging aerospace robots and robots meant for the modeling and implementing courses of action that are characteristic of human behavior are discussed.

Jurek Z. Sąsiadek, Julius O. Adoghe, and Malik al-Isawi "Control of Unmanned Aerial Vehicle Using Vision System and Sensor Fusion for Wing Shape Deflection Measurement" presents a Deflection–Detection–Vision System (DDVS) for Unmanned Aerial Vehicles (UAV) fixed-wing for control and navigation. This technique makes it possible to measure the fixed-wing shape, deflection, and identification of the aerodynamic coefficient acting on the system by using information from the stereo camera and a strain gauge. It determines specific points to identify the shape of the wing and its deflection. The model consists of a stereo camera fixed at the top of the device with strain gauges placed at eight different points marked on the wing. Both sensors simultaneously measure the deflection in chosen locations of the wing. The performance of the DDVS and dynamic parameters are tested in a wind tunnel at speeds ranging from 10 km/h to 35 km/h, the Angles of Attack (AOA), and the roll angle ranging from 0° to 30°, respectively. In the paper, image acquisition, feature extraction, matching process, 3D reconstruction, and a stereo camera calibration are discussed. The approach presented measures the wing deflection at each selected point and identifies the maximum deflection location based on various aerodynamic conditions such as the wind speed, the AOA, and the roll angle. The drag and lift forces are obtained using the wing's surface area, and the experiment shows that less force is required for lifting as the AOA increases. The DDVS is implemented in the wind tunnel, and an extensive experiment is conducted to determine the deflection of the wing as a function of the flight parameters like the AOA, the roll angle, and the flow velocity. The results show that the integration of the strain gauge and a vision system sensor make it possible to accurately measure the wing deflections and identify the aerodynamic coefficient in comparison to simulation results, and can be used even in the most demanding environment.

Figen Özen and Dilek Bilgin Tükel ("Robotic Dance Modeling Methods") discuss the case of dance which is an example of an extraordinary complex dynamic system so that synchronized human–robot dance – maybe the most complex in comparison with even very complex systems such as those of related to climate, ecosystems, economics, organizations, social structures, and socio-political systems—is particularly difficult to deal with. A dance is an art performance that involves dedicated movements, music, and intentional interpersonal synchronization while observing the movements and interactions among herds, teams, and individual dancers. Choreography, which repeats the same movements synchronized with music is not the only aspect of dance and its positive perception. A good use of space, differentiation, and then reattaining equivalence, and dynamism are also needed. An ability of robots to respond like humans in these realms is one of the goals of this research. Two different modeling techniques are proposed for industrial robot dance modeling: Modified Laban Notation and Synchronized Petri Nets. Modified Laban Notation for Industrial Manipulators (MLIR) is designed and applied by means of a special interface. The input parameters of the manipulator are calculated using the data related to motion which are made available at the interface. The torque values are calculated and applied. Synchronized Petri Nets are then employed for the conceptualization and algorithmization of a specific dance choreography.

Oscar Castillo and Patricia Melin ("A Review of Fuzzy Metaheuristics for Optimal Design of Fuzzy Controllers in Mobile Robotics") have focused their contribution on a comprehensive review of the successfully accomplished research work that has been done in applications of *fuzzy meta-heuristics* for the optimal design of fuzzy controllers in mobile robotics. Currently, metaheuristics have become very popular as effective optimization methods in many areas of application, ranging from pattern recognition and time series prediction to robotics and control. Metaheuristic algorithms, like genetic algorithms, particle swarm optimization, ant colony optimization, and more recent ones, such as the grey wolf optimizer, firefly algorithm, and cuckoo search, and others have been applied in a plethora of problems. Furthermore, recently, fuzzy logic has been used to improve the performance of metaheuristics by dynamically adapting the involved parameters. On the other hand, the application of fuzzy logic in robot control has also gained popularity because of its advantages, like taking into account expert knowledge and providing flexible controllers that can handle uncertainty in dynamic environments. In addition, more recently, the type-2 fuzzy systems and control have achieved successful applications in robotics. In all cases, the optimal design of the fuzzy controllers requires optimization and metaheuristics have provided a very good alternative for achieving this goal. For these reasons, this review paper provides a critical appraisal overview of the existing works in this particular area of optimization techniques as well as offers a foreseeable prospect of future research trends.

Part II "New Developments in Time Series Analysis, Prediction, and Fault Detection and Control" is concerned with problems that occur in many applications in various areas of science and technology, that is with the analysis of time series, prediction, and forecasting, and some relevant issues in fault analysis and detection as well as in fault-tolerant control.

Ying Han and Kun Li ("Time Series Prediction Using Time-Series Decomposition and Multi-reservoirs Echo State Network") are concerned with the Echo State Network (ESN) which is an effective and efficient tool for time series analysis. It has a dynamic reservoir which includes input units, internal units, and output units. However, due to the randomness and non-stationarity properties of most time series, it is difficult for a single reservoir to better handle them because different scales in the time series are dealt with in a unified structure. In order to solve this, the time series decomposition (TSD) is employed to decompose the time series into different sub-sequences which can be handled by different reservoirs. Now, there are many TSD methods, such as empirical mode decomposition (EMD), ensemble empirical mode decomposition (EEMD), complementary ensemble empirical mode decomposition (CEEMD), local mean decomposition (LMD), variational mode decomposition (VMD), etc. It is shown how different TSD methods can be used to decompose the time series and then how the multi-reservoir ESNs are constructed. Finally, experimental results using several time series are presented and compared.

Adriana Villalón Falcón, Alberto Prieto Moreno, Marcos Quiñnones-Grueiro, and Orestes Llanes-Santiago ("A Proposal for Improving Remaining Useful Life Prediction in Industrial Systems: A Deep Learning Approach") discuss the problem of an accurate prediction of the remaining useful life (RUL) of engineering systems which yields valuable information which can help develop more efficient maintenance programs maximizing the equipment usage and avoiding an increase of costs due to failures. Deep learning methods have gained much popularity because of their capability to learn complex and discriminative nonlinear features that can facilitate RUL prediction. These network models are generally trained to minimize the mean square error (MSE) between RUL prediction and its true value which assigns an equal importance to the error in the beginning and at the end of a system's useful life. Since the prediction of the RUL is more critical as a system arrives at the end of its useful life, then a new performance metric for evaluating prognostic models is proposed here with the objective to establish a direct relation between the RUL prediction and maintenance planning. Moreover, a procedure is proposed to use this metric for training a recurrent neural network (RNN) to improve the network's ability to learn the relationship between the raw data and the corresponding RUL. The procedure is applied with success to the analysis of the NASA Commercial Modular Aero-Propulsion System Simulation (C-MAPSS) dataset.

Jovan Stefanovski ("Fault Diagnosis/Fault-Tolerant Control: A Survey of Results for Linear Systems Over Frequency Region in Presence of Disturbances") presents a survey of the most relevant recent research results on fault detection, fault estimation, fault-tolerant control, and fault-tolerant tracking which have appeared in the worldwide literaturę. The faults are considered to be additive inputs in respect to the plant's dynamics. Many relations and properties are formulated and analyzed. Furthermore, this survey analysis has been carried out along with a certain relevant comparison discussion of various aspects of all surveyed works. The obtained comprehensive new results are illustrated by examples.

Tarek Raïssi and his team Chaima Zamali, Jeremy Van Gorp, and Zhenhua Wang ("New Interval Observer-Based Fault Detection for Switched Systems") in two countries have focused their contributed chapter on presenting their recent synthesis design development for a new interval observer-based fault detection method for a class of discrete-time switched systems, which are subject to unknown but bounded state disturbances and measurement noise. The proposed novel technique is investigated to reduce the conservatism of gain matrices and to offer more degrees of design freedom by means of integrating weighted matrices in the structure of the fault detection observer. Using multiple quadratic Lyapunov functions with an average dwell time control condition in switched systems, novel solvable conditions are derived in terms of LMIs. Furthermore, the fault detection decision is based on residual intervals that are generated by the proposed interval observers. The efficiency of the proposed approach is highlighted through simulation results on an academic example.

Orestes Llanes-Santiago and his team Daniel Jimenez Sanchez, Marcos Quinones-Gruerro, Antonio J. Silva Neto ("A Regularized Inverse Problem Approach for Robust Condition Monitoring in Industrial Systems") in three different countries have presented a thorough investigation of a regularized inverse problem approach for industrial condition monitoring along with a case study. Condition monitoring is becoming more and more important in modern industry in order to increase the safety of industrial plants and the economic benefits. Schemes based on model inversion or system inversion represent an important branch of the available solutions in model-based condition monitoring. These techniques allow the development of detection, isolation, and successful estimation of the fault magnitude. However, most of the proposed methods do not consider the noise present in industrial control systems which significantly affects the performance of the condition monitoring systems. They do not consider either the occurrence of multiple faults. In this paper, a proposal for robust condition monitoring, formulated as the solution of a regularized inverse problem in discrete-linear time-invariant systems is presented. Single and multiple faults are reconstructed by using the vector of residuals in the presence of noise. Tikhonov regularization is used to obtain a stable solution when noise in the measurements is considered. The proposed approach is applied to a case study with rather satisfactory results.

Juan-Eduardo Velasquez-Velasquez, Rosalba Galvan-Guerra, Rafael Irriarte, and Leonid Freedman ("On Robustification Based on Continuous Integral Sliding Modes") have been focused on further exploration of robustification effects of the integral sliding modes in variable-structure control systems. They have presented a certain tutorial revisiting the fundamental principles for designing robustification schemes for conventional and switched linear time-invariant systems based on continuous integral sliding modes. State and output approaches that guarantee the exact compensation of matched uncertainties/perturbations are explained in detail. Furthermore, a constructive methodology is given to facilitate the practical design of the controller and the observer as they are involved in robustification system schemes.

Part III "Novel Applications in Human Communities and Economic Systems" dealt with the modeling, decision making, planning, and control in systems that are much more complex than the technological systems that are usually considered, namely the health care and medical diagnosis. In all these systems aspects to be taken into account include also social and economic ones too.

György Eigner, Máté´ Siket, Bence Czakó, Dániel András Drexler, Imre Rudas, Ákos Zarády, and Levente Kovács ("Model Predictive Tumour Volume Control Using Nonlinear Optimization") develop a novel nonlinear model predictive control algorithm for tumor growth regulation. Its unique feature is that the subjects of optimization are the feedback gains in the state feedback closed control loop. In the design of the controller results of the authors' previous qualitative analysis of the model to be controlled are taken into account. To realize the state feedback a discrete extended Kalman filter for the state estimation is implemented with slightly different model parameters compared to the virtual patient model to be controlled. The results show that the new algorithm proposed performs well within the predefined circumstances and it is in a position to satisfy the strict constraints.

Marjan Stoimchev and Vesna Ojleska Latkoska ("Detection of Epilepsy Using Adaptive Neuro-Fuzzy Inference System and Comparative Analysis") study the use of the Adaptive Neuro-Fuzzy Inference System (ANFIS) for the classification of EEG signals. The data consists of two types of EEG signals, i.e., epileptic patients during epilepsy and healthy patients when their eyes are open. Two algorithms for the detection of epileptic patients are proposed. In the first algorithm, the Discrete Wavelet Transform (DWT) and statistical analysis for feature extraction are used, whereas the Principal Component Analysis (PCA) is used in order to reduce the number of features in the second algorithm. The ANFIS model learns how to classify the EEG signal through a standard hybrid learning algorithm while a special form of the ANFIS model is employed which—depending on the number of inputs—splits the model into an appropriate number of substructures (sub-ANFIS models). The algorithms are evaluated in terms of training performance and classification accuracies. From the simulation results, it can be concluded that both algorithms have good potentials in the classification of the EEG signals. Moreover, a comparative analysis of the influence of the tuning parameters is performed, i.e., the influence of the different data splitting methods, the influence of the different input space partitioning methods, the usage of the different wavelet functions in the DWT, the effects of normalization, as well as the effects of using different membership functions. From the analysis, it can be seen that different combinations of input parameters differently classify the EEG signals. Lastly, a comparison of both algorithms is made in terms of training performance and classification accuracies, and it can be concluded that the algorithm that uses the PCA for feature extraction performs in some cases better than the algorithm that uses the DWT, even though the number of features is significantly reduced (from 20 to 7).

Nikolay Lomakin, Anastasia Kulachinskaya, Maxim Maramygin, and Elena Chernaya ("Improving Accuracy and Reducing Financial Risk When Forecasting Time Series of SIU0 Future Contracts Employing Neural Network with Word2vec Vector News") study some important aspects of a theoretical background of forecasting time series of financial instruments. More specifically, this work is based on the experience of using Artificial Intelligence (AI) systems for collecting and processing BigData to forecast the time series of the silver SIU0 future contracts. It is shown that the use of a neural network makes it possible to forecast the closing price of the SIU0 futures contract in a 15-min timeframe. The proposed AI system, based on the use of a neural network, improves the accuracy and reduces the financial risk. To forecast the price of the SIU0 futures contract, the system uses the parameters of Japanese candlesticks and the volume as well as «news fluctuations» from Web sites. The perceptron neural network, designed on the Deductor platform, is trained on two types of data: cost—(Pclose), and logarithmic—(ln). The proposed solution is of a great practical value due to a high forecast accuracy.

Janusz Kacprzyk, Yuriy P. Kondratenko, José M. Merigó, Jorge Hernandez Hormazabal, Gia Sirbiladze, Alexander Bozhenyuk, Eulalia Szmidt, Sławomir Zadrożny, and Jan W. Owsiński ("A Fuzzy Multistage Control Model for Stable Sustainable Agricultural Regional Development") propose a further extension of a multistage fuzzy control model of stable sustainable regional agriculture development in which an additional mechanism is added to involve a stability requirement which addresses a clear preference of the stakeholders (mainly the authorities and inhabitants) for a limited variability of crucial socioeconomic development indicators and parameters, called a stability in this context. A fuzzy dynamic programming model for solving the planning problem is presented in which many crucial aspects, notably the life quality indicators, are subject to objective, by the authorities, and subjective, by the inhabitants, evaluations which are closely related to human perception and cognitive abilities. This model is then augmented with a limitation of the variability of crucial socioeconomic development indicators and other parameters. For illustration, an example of a life quality indicator focused regional agricultural development planning is shown.

Part IV "Novel Applications in Infrastructure and Manufacturing Industry" focuses, firstly, on the use of the synergy of modern control and data science-related applications to energy, manufacturing, and production systems and, secondly, on the use of learning data-driven approaches, notably, deep neural networks, for the analysis and improvements of road systems and transportation as well as of modern virtual education systems too.

Vassil Sgurev, Mincho Hadjiski, and Nencho Deliiski ("Multicriteria Optimal Control of Industrial Thermal Processes with Distributed Parameters Under Variable Operational Conditions") discuss an intelligent system for the control of the thermal treatment process (TTP) of wood materials. A model of manufacturing with a necessity for frequent rescheduling is proposed via a combination of model-based and data-driven approaches. By using the first-principle mathematical model of TTP presented by partial differential equations in the 2D space with a suboptimal model-based control algorithm and a case-based reasoning (CBR) based approach, an explicit suboptimal control system is discussed in different operational modes. A set of virtual subspaces for feasible operational situations for a variety of objective criteria of the value assessment is obtained using the traditional "problem-decision" representation. As the search spaces are well structured, the search procedure based on the traditional k-NN algorithm is strongly simplified. In this way, a complicated computer simulation of the TTP at each time step due to the plant's parameter distribution, nonlinearity, and operational or environmental disturbances proceed in an off-line mode. A relatively small part of the calculations connected with the traditional R4–operations in the CBR, objective functions estimation, some data-based and rule-based control parameter corrections, and possible adaptation from charge to chargé are done in an online mode. The results of simulation experiments are presented and analyzed.

Nikolaj Apostolovski, Naum Trajanovski, Marko Chavdar, Tomislav Kartalov, Branislav Gerazov, and Zoran Ivanovski ("Deep Learning Based Multimodal Information Fusion for Near-Miss Event Detection in Intelligent Traffic Monitoring Systems") discuss problems related to the detection and marking of road accident "black spots" that are very important for road safety and constitute a basis for road safety improvements via changes in the construction, maintenance and operation of roads. Their detection and localization are based on accident statistics. Behind any accident statistics, there is an even larger near-miss statistics that could be employed for black spot detection, before a significant increase in accident statistics is recorded. The detection of near-miss events is difficult mainly because of their vague definition and vague manifestation. The authors propose a new approach to the detection of near-miss events based on a simultaneous occurrence of a rapid deceleration and skidding of vehicles. Though the rapid deceleration is easy to detect in videos, its occurrence alone does not always imply a near-miss event. The detection of skidding is very challenging in video due to the occlusion of wheels, and also it has a distinctive audio signature but the spatial location of the audio source is very difficult to extract from an audio-only data stream. A multi-modal algorithm for near-miss detection based on audio and video information fusion is proposed. The information extraction in both domains, audio, and video, is performed using the deep convolutional neural networks (CNN) combined with different pre- and post-processing techniques. The deceleration of vehicles is estimated using video from calibrated surveillance cameras, and vehicle positions are estimated using a CNN-based estimator and the output is corrected using content matching techniques and a Kalman predictor. Audio events are detected using the CNNs applied on Mel-frequency cepstral coefficients. In the final information fusion step, a support vector machine (SVM) classifier is used for making decision on the occurrence of potentially dangerous near-miss event.

Nurasyl Kerimbaev, Vladimir Jotsov, Aliya Akramova, and Nurdalet Nurym ("Modeling and Feedback Control for Development of Mobile Technologies in Virtual Education Environments") have focused their timely development research and implementation of mobile education technologies for virtual environments. Indeed, contemporary Information Technologies (IT) became one of the main communication means establishing the contact between the teacher and student. Modeling and feedback control are one of the main features to be explored in this direction aiming at high-quality results. The communication in real-time conditions is established using a broad variety of web-based hardware like notebooks, smartphones, and other innovation gadgets improving the education process via modern visualization and virtualization tools. For better results, the communication feedback should use Data Science methods to model and control the contemporary IT education process, especially in case of education in Control Systems. Virtual education environments are explored in this case. The usage of web-based and mobile technologies allows to improve the quality of education process and to organize more efficient understanding and active learning conditions. On the other hand, this feedback control includes a quick and efficient estimation/scoring process. The proposed complex of mobile technologies and platforms visualize different processes of communication activities in the interactive learning environments. The functional value of the mobile technologies gives an opportunity to apply visual search, voice recognition, mirror display, and so on. The elaborated feedback tools backed by mobile technologies in virtual learning environment may be broadly used in the contemporary education sphere. It is shown that improves the efficiency of the tutoring/learning process and to enhance the better upbringing process.

Igor Bimbiloski, Valentin Rakovic, and Aleksandar Risteski ("Control of Power Consumption with Integrated System of Technology, Regulation and Consumer Behavior Management") deal with environmental and social issues faced by the present world exemplified by climate change and pandemic. Their solution, even a mitigation of their negative effects, implies that science, technology, social behavior, and rules of law must act together and in an integrated way to reach the final goal. As the climate changes are mostly affected by an irrational consumption of energy, the integrated system for the management of technology, regulation, and consumer behavior can contribute to a better energy use efficiency and reduce CO_2 emission. Massive control ICT systems are currently deployed in Smart Energy Networks combining the big data management, automation, artificial intelligence support, and advanced IoT systems. To make this infrastructure more efficient, it is important to properly influence consumer behaviour and stimulate investment in green energy production. The dependency of the consumer on the smart devices and social networks can be used as a potential platform to increase the awareness of importance, and of a change in the consumption behavior. A positive regulation of green energy investments on the consumer side is also a potential tool for improvement if considered in terms of overall financial and environmental interests. The models of alignment of financial and environmental benefits use a game theory approach to find the best scenarios and to influence the consumer behavior, with a so-called "transfer of benefits" policy. This policy takes into account that the different players in the electricity market, for instance, companies and consumers, should be ruled via one system with a joint environmental benefit, not as separate players with opposing financial interest. The management of such complex systems is ruled by an advanced ICT computational structure.

Part V "Novel Applications in Computer Networks and Telecommunications" comprises papers on very up-to-date advanced applications related to computer networks and telecommunication systems, which are the backbone of the present society and are decisive for overall development of the economy and society as integrated whole.

Yuanwei Jing, Yan Zheng, Wenjuan Xu, Zanhua Li, and Kun Wang ("Double-Router TCP/AQM Network Systems: Backstepping Communication Control Design") are concerned with the problem of control of congestion communication for double-router TCP/AQM network systems. The use of the window model of the single-router TCP/AQM network system, a double-router TCP/AQM network system model is developed which follows the "priority selecting, random transferring" principle of data packet transmission. The state-space model of a non-linear system with the lower triangle form is obtained by using a class of state transformation. An active queue management congestion control algorithm for the double-router network system is proposed employing the backstepping design method. An innovated state feedback controller is proposed to make the TCP/AQM network system asymptotically stable. The simulation results are presented to show the feasibility and effectiveness of the proposed method.

Gorjan Nadzinski and Mile Stankovski ("Noise-Robust and Secure Communication Protocol for Industrial Networked Control Systems") discuss issues related to the increasing and rapid development of technology and complex systems and the emergence of the fourth industrial revolution which also influences industrial automation. Presently, industrial processes are no longer isolated entities but represent complex systems and subsystems which collect an abundance of data and are in constant communication with each other. Intelligent control, machine learning, BigData, and a rapid progress of measurement and control equipment play a significant role in this development, but communication still constitutes one of the crucial aspects. While the concept of industrial networked control systems brings about many benefits exemplified by flexibility, speed, and modularity, the key role communication implies that it is now vulnerable to both environmental and man-made interference. In the paper, a new method is presented which yields an increase in the level of security of industrial communication protocols by developing an algorithm which uses a coupling function between two dynamical systems for data encryption, and dynamical Bayesian inference for data decryption. The algorithm developed has been used to develop a communication protocol whose performance has been tested and verified in real-world experimental conditions under both the white Gaussian and colored low-frequency noise. This approach results in communication that is both cryptographically secure and noise-robust, applicable in the industrial environment, and with an increased energy efficiency.

Yang Liu, Hongyi Li, Yuanwei Jing, Xiaoping Liu, and Renquan Lu ("Distributed Adaptive NN Finite-Time Congestion Control for Multiple TCP/AWM Networks") deal with a distributed adaptive NN (neural network) congestion control problem which is is considered for the TCP/AWM networks using a practical finite-time criterion. In the first step, the TCP/AWM networks are modeled as multiple network cases. And then, under the framework of a recursion algorithm, these extend the practical finite-time criterion to multiple TCP/AWM networks. Furthermore, by a cooperative control agreement among sub-networks, the queue length of all TCP/AWM networks reaches the consensus. In addition, a simple adaptive law is designed and singular issue is avoided. Finally, simulation results are used to demonstrate the effectiveness of the proposed scheme.

Gjorgji Ilievski and Pero Latkoski ("Network Traffic Classification Using Supervised Machine Learning Algorithms in Systems with NFV Architecture") consider the concept of network functions virtualization architecture which is gaining more and more popularity and thus used in different communication network systems. The concepts of containers, virtual network functions, and application functions are implanted within the clouds and controlled with the NFV systems. Access technologies, especially the 5G, which is believed to be a crucial enabler of the IoT is also dealt with. Within such circumstances, most of the network traffic is expected to flow in the east–west direction, never leaving the cloud. The work is focused on the preparation of an experimental environment to simulate such a traffic which is analyzed by making classification of the network data flows, using a selected set of six supervised machine learning (ML) algorithms to find the algorithm with the best performance defined as a combination of the classification precision, and the time consumption which are crucial in particular from the point of view of 5G in which any packet delay introduced within the system may compromise the 5G specification calls for latency. The results obtained state that out of the 6 explored ML algorithms, the Decision Tree algorithm is the most suitable classifier that fits within the needed precision across all classes but also within the time consumption needs. Moreover. the regulatory point of view for an automated data analysis within the systems is dealt with, and the statistical features of the network flows are considered, while the payload data, the source, and destination information, as well as the network port, are excluded as the attributes for classification because the work deals with the VoIP and encrypted VoIP data used in 5G.

David Acev, Gorjan Nadzinski, Valentin Rakovic, and Aleksandar Risteski ("Manipulation of URL Addresses Using Machine Learning to Provide Better Cyber Security") are concerned with complex systems which can be in the hands of regular civilian users utilizing a combination of advanced concepts (such as control theory, artificial intelligence, and machine learning, BigData, etc.) to deliver comfort, safety, robustness, and a fast and easy communication to both individual users, households, business and industries. However, this sudden rise in complexity and opaqueness of the systems with regard to their everyday users prompts a serious rethinking of the approach to the security of IT systems. For instance, the Internet offers many points of cyber vulnerability in which regular users can be targeted. In the paper, an implementation of machine learning-based approaches for the detection of malicious URLs in order to improve cyber security is considered. The proposed approaches show promising results in exposing dangerous links and could be implemented in several points within a complex network, both locally and in different nodes within a network hierarchy, thus showing how data science and machine learning (two of the very driving forces in the rise and development of complex systems) stimulate the core of cyber security. Moreover, the same proposed approaches are used for the classification of URL addresses based on the content that each of the URLs provides to even further take advantage of the benefits of machine learning approaches.

Strahil Panev and Pero Latkoski ("Modelling of Priority Buffering Systems Applicable for Commercial Mobile Networks") discuss the Software Defined Networking

(SDN) which is a popular technology paradigm that is embedded in the basic architecture of the 5th Generation (5G) of mobile networks. Today, OpenFlow is the most common protocol used on the southbound interface. The OpenFlow switches generally involve two types of buffering mechanisms: (1) a single buffer that is used to handle both the control and user plane; (2) two different priority buffers each serving the control and user plane packets. In the work the average packet loss rate of the two different buffer design principles is analyzed, by developing an analytical proposal that incorporates the Quasi-Birth-Death (QBD) processes. The proposed numerical model is also verified via extensive computer simulations in MATLAB. The obtained results clearly show that the use of the priority buffering in the SDN switches increases the performance significantly as compared to the traditional shared buffering. When the probability of the Packet-In messages is low, the arrival rate is increasing, and as the number of the Mobile Nodes (MN) goes up, the priority buffering clearly outperforms the single buffering in most of the scenarios by nearly 99% of lower packet losses. The obtained results can be used for predicting the average packet loss rate while designing the OF-based mobile core networks.

We strongly believe that the high quality, interesting and inspiring contributions to a variety of important applications, included in this volume, will be of much interest and use for a wide research community.

We wish to thank the contributors for submitting their great works and for their support and an active participation in this editorial project. Special thanks are due to anonymous peer reviewers whose deep and constructive remarks and suggestions have helped to improve further the quality and clarity of presentations in these contributions focused on timely applications.

And last but not least, we wish to thank Dr. Tom Ditzinger, Dr. Leontina di Cecco and Mr. Holger Schaepe for their dedication and help to implement and finish this important publication project on time, while maintaining the highest publication standards.

Adelaide, Australia
Bitola, North Macedonia
Warsaw, Poland

Peng Shi
Jovan Stefanovski
Janusz Kacprzyk

Contents

Professor Georgi M. Dimirovski

The purpose of this book is to present a small token of a great appreciation of systems, control and cybernetics international communities to Prof. Georgi M. Dimirovski for his important research and scholarly achievements, and a life long service to the community. Professor Dimirovski has been, since the beginning of his long and illustrious career, a driving force for the entire research community in his areas both in his country, first in former S.F.R. of Yugoslavia, and now in Northern Macedonia, and world-wide due to decades of his involvement in the activities of our global societies and organizations, notably European Science Foundation (ESF), Institute of Electrical & Electronics Engineering (IEEE), International Federation of Automatic Control (IFAC), Institution of Engineering & Technology (IET/IEE), and International Academy of Systems & Cybernetics Sciences (IASCYS). Currently, he is one of the Vice-Presidents of IASCYS Academy.

Over the years, Prof. Dimirovski's pioneering contributions have, on the one hand, shaped the fields of the broadly perceived systems and control in their cybernetic synergy with communications and computing. On the other hand, he has always been sharing his knowledge and vision with other people, notably his young collaborators in several countries for whom he has not only been a mentor and a great teacher but also a colleague and friend, in the very European old academic legacy and tradition. It is therefore that the number of people world-wide whose life and scientific careers have been shaped thanks to Prof. Dimirovski dedicated academic work is considerably high. During the last three decades, he has paid longer- or shorter-term academic visits with seminars to universities in: Aalborg, Ankara, Belgrade, Bradford, Bochum, Brussels, Coimbra, Covilha, Duisburg, Grenoble, Hannover, Istanbul, Izmir, Linz, Lisbon, Ljubljana, London, Maribor, Nis, Ottawa (Carleton, CA), Portsmouth, Santiago de Chile, Sarajevo, Sevastopol, Sofia, Split, Valencia, Wien, Wolverhampton, and Zagreb as well as Beijing, Dalian, Harbin, Nanjing, Shanghai and Shenyang (CN), and also Kaohsiung (TW). He has been teaching to graduates on summer schools in China at universities in Dalian, Harbin, Nanjing, Shanghai, and Shenyang.

To commemorate Prof. Dimirovski's anniversary and pay tribute to him for his scientific and scholarly contributions, we have gladly dedicated our efforts to publish a special monograph with contributions by top researchers and scholars from all over the world via a selective invitation process. In the process over time this monograph grew into two volume book with more that 40 contributed chapters by more than 100 authors from 25 countries on all continenets.

* * * * * * *

At—Graduate Institute of "Dokuz Eylul" University of Izmir, Izmir, Turkey, April 2006.
Elective course on Sliding Mode Control in Mechatronics and Robotics for M.Sc. and Ph.D. students.
Relaxing after the lecture "Lyapunov Theory of Motion Stability: Theorems of Direct Method".

With the great teacher late Academician Si-Ying Zhang (May he rests in peace!): Learning theory of composite nonlinear systems and natural symmetries, while walking and talking in Russian through campus park of Northeastern University, Shenyang, Liaoning, China. Then at the beginning of an eternal friendship in July 1999 Academician Zhang blessed and encouraged my academic co-operation with both Yuanwei Jing and Jun Zhao.

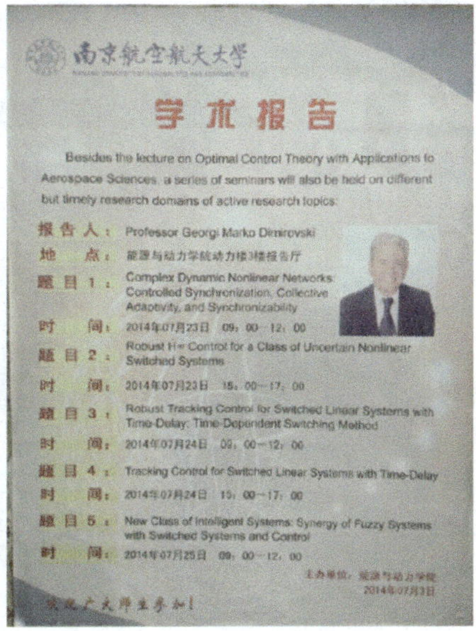

Nanjing University of Aeronautics and Astronautics—Doctoral Summer School, Nanjing, China; July 2014.
Announcements of delivered Ph.D. Course Theory of Optimal Control and Applications to Aerospace Sciences as well as Research Seminar Lectures for academic colleagues.

Research Seminar Lectures in the Institute of Mathematical Modelling of Chinese Academy of Forestry before the colleague-collaborators of his partner Prof. Yuanwei Jing and their doctoral students and research assistance; Beijing, July 2013.

On bio-diversity cultural (the Beijing Zoo) and historical (the Great Wall) expeditions along with his first colleague-partner Prof. Yuanwei Jing from Northeastern University of Shenyang, a former doctoral student of Academician Zhang (who initiated the academic cooperation in 1996), after the Beijing seminars with Prof. Jing's collaborators.

On the Natural Science expedition (Huangshan Mountain) with his second colleague-partner Prof. Jun Zhao, also a former doctoral student of Academician Zhang (who in 1999 embraced the academic cooperation), following successes of Prof. Zhao's team at Chinese Control Conference.

New Challenges in Modern Robotics Systems

Control of Unmanned Aerial Vehicle Using Vision System and Sensor Fusion for Wing Shape Deflection Measurement

Jurek Z. Sasiadek, Julius O. Adoghe, and Malik Al-Isawi

Abstract This paper presents a Deflection-Detection-Vision-System (DDVS) for unmanned aerial vehicles (UAV) fixed-wing for control and navigation. This technique allows measurement of the fixed-wing shape, deflection, and identification of the aerodynamic coefficient acting on the system, using information from the stereo camera and strain gauge. It determines specific points to identify the wing's shape and deflection. The model consists of a stereo camera fixed at the top of the device with strain gauges placed in eight different points marked on the wing. Both sensors measure the deflection in chosen locations simultaneously. The DDVS performance and dynamic parameters are tested in a wind tunnel at speeds ranging from 10 km/h to 35 km/h, angles of attack (AOA), and roll angle ranging from 0 to 30°, respectively. An image acquisition, feature extraction, matching process, 3D reconstruction, and a stereo camera calibration are presented in this paper. This approach measures the wing deflection at each selected point and identifies the maximum deflection location based on various aerodynamic conditions such as wind speed, AOA, and roll angle. The drag and lift forces were obtained using the wing's surface area, and the experiment shows that less force is required for lifting as the AOA increases. The DDVS was implemented in the wind tunnel, and extensive experiment was conducted to determine the deflection of the wing in function of flight parameters like angle of attack, roll angle and flow velocity. The results have shown that the integration of strain gauge and vision system sensor measures wing deflections and identify the aerodynamic coefficient accurately by comparing with simulation result, and it could be used even in the most demanding environment.

Keywords Control systems · Fixed-wing · Stereo camera · Strain gauges · Deflection measurement · Sensor fusion

---

J. Z. Sasiadek (✉)
Space Research Centre (CBK PAN), Warsaw, Poland
e-mail: JurekSasiadek@CUNET.CARLETON.CA

J. Z. Sasiadek · J. O. Adoghe · M. Al-Isawi
Department of Mechanical and Aerospace Engineering, Carleton University Ottawa, Ottawa, ON, Canada

P. Shi et al. (eds.), *Complex Systems: Spanning Control and Computational Cybernetics: Applications*, Studies in Systems, Decision and Control 415,
https://doi.org/10.1007/978-3-031-00978-5_1

Nomenclature

L	Length of wing (mm)
b	Width of wing (mm)
t	Thickness of wing (mm)
α	Angle of attack (degree)
β	Roll angle (degree)
V	Velocity of wind (km/h)
d	Deflection of wing (mm)
ρ	Density of air (kg/m^3)
F_L	Lift force (N)
F_D	Drag force (N)
C_L	coefficient of lift
C_D	Coefficient of drag
W	Weight of UAV (N)

1 Introduction

The modeling of a fixed-wing aircraft undergoing deformation requires a geometrical structural model combined with a reliable large motion aerodynamic model. The analyses and design of the fixed-wing configuration are important for designing unmanned aerial vehicles. This paper presents a method to identify the UAV wing's shape using two independent sensors, and the measurement from those sensors was integrated using the advanced sensor fusion method. The fusion results are more accurate and reliable than the single sensor alone, and this is done to design an advanced control system. The measurement technique is a visual method for determining aeroelastic deformation and identifying the aerodynamic properties of the fixed-wings system. The vision system is fused with strain gauge sensors for accuracy. The technique is used to determine the spatial positions of targets on a model surface from the target centroids in a series of images based on close-range photogrammetry principles. The model deformation induced by aerodynamic charge is determined from these spatial coordinates. In this case, deformation is defined as the change in the wing's shape under aerodynamic loading. A change in the design's geometry will cause variances between the acquired and the predictions based on rigid body assumptions. Therefore, the measurement of wind tunnel models' deformations is advantageous for comparing the measurement, especially in high Reynolds number facilities where dynamic pressure is typically high.

Furthermore, it is crucial to fuse both information from the strain gauge measurement with experimental vision measurements of deformation for accurate results [1, 2]. The deformation measurement includes a Zed stereo camera for vision system measurement and a set of strain gauges, considering the wing twist's change due to aerodynamic loading. A novel Deflection-Detection-Vision-System (DDVS) is used to determine the wing's deformation and control a flexible wing [3, 4]. Aerodynamic characteristics are substantial to the enhancement of aircraft performance and capacity. The present paper describes methods for obtaining fixed-wing deflection by comparing a known measurement instrument (strain gauge) to a stereo camera. This method includes carrying out testing at the wind tunnel using an airplane wing model at various wind speeds and wind directions. In [5], the attitude and angular rate of the non-cooperative space target and the trajectory of target characteristics were found using a chaser-mounted vision system. In [6], Motion estimate from far-distance measurements was established and predicted by accumulating the observed data, the motion estimate, and the dynamic model for long-term motion prediction of uncooperative space target was presented. Space robot [7, 8] studied the autonomous target capture of the space robot. The model of space robot system dynamic equations generated using one single camera. The target motion, estimate trajectory planning based on the camera. Unmanned aerial vehicles (UAVs) have been popular research for the past decade, especially in the area of military applications [9, 10]. Studies have been carried out on small UAVs because of the system's low costs and their effectiveness in special missions [11, 12]. For a system to be fully autonomous, the system should be able to make decisions at all time autonomously. This can be achieved when all the required aerodynamic parameters are identified during flight using a live system like the live vision calculation, which is implemented in this paper. The DDVS implemented in this paper measures the deflection for a fixed-wing utilizing a vision system and strain gauge.

2 Deflection Calculation with Stereo Camera

Information from the plain image is transformed into a three-dimensional (3D) image using a vision system measurement technique. The relationship between the 3D image in object space and the corresponding 2D image is achieved using a stereo camera. Considering the right and left retinal image, there is a depth perception that arises from the 3D point of view disparity. When projected under the perspective of the stereo camera, the difference of the 3D image location is expressed as;

$$d = X_l - X_r \tag{1}$$

where X_l is the position of the image from the left camera, X_r is the position of the image from the right camera, and d is the disparity.

Figure 1 shows the stereo camera optic axes. The y-axis is the axis perpendicular to the page. In the equation below, z is the image depth, f is focal length, b is the baseline.

$$\left[\frac{z}{f} = \frac{X}{x_l}\right], \left[\frac{z}{f} = \frac{x - b}{x_l}\right], \left[\frac{z}{f} = \frac{y}{y_l} = \frac{y}{yr}\right] \tag{2}$$

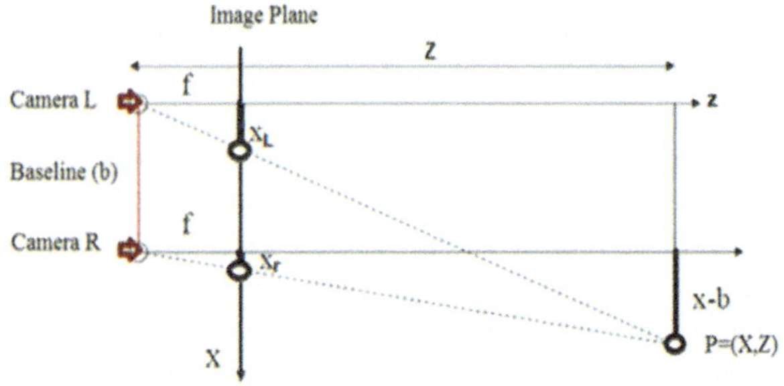

Fig. 1 Model of the stereo camera

The above equation is obtained from the triangulation, which is a result of camera calibration based on direct linear transformation (DLT). A direct linear transformation is one of the more common algorithms used in machine vision. It helps to determine the depth from disparity d and the coordinates of the 3D location for a stereo camera with parallel optical axes. The depth Z is inversely proportional to the disparity. From camera calibration, the focal length and the baseline can be determined. The corresponding point (x_r, y_r) for each (x_l, y_l) is achieved from the corresponding point distance of the images [13].

Rearrange the Eq. 2 and get:

$$DepthZ = \left[\frac{f * b}{(Xl - Xr)}\right] = \left[\frac{f * b}{d}\right] \tag{3}$$

The deflection can be found by the difference between the current depth (with load) and the depth without load.

3 Deflection Calculation Using Strain Gauge

Strain (ε) is the fractional difference in the wing dimensions due to the force applied. More specifically, the strain is characterized as a slight change in dimension of the wing. Strain can be either positive (tensile) or negative (compressive). The measured strain size is very small and is often expressed as a microstrain [14, 15]. Since the central equation for the electrical resistance R of a length of wire is

$$R = \frac{\rho \mathrm{L}}{\mathrm{A}} \tag{4}$$

where ρ is the resistivity, L the lenght and A is the cross sectional area. It follows that any change in length, and hence sectional area, will result in a change of resistance. The direct reading of the linear strain is acquired by the measurement of this resistance change with suitably calibrated equipment, and strain which may be expressed [12].

$$\text{Strain}(\varepsilon) = \frac{\Delta L}{L} \tag{5}$$

For calibration, the equation of concentrated load at the free end was used;

$$\delta_{max} = \frac{Pl^3}{3EI} \tag{6}$$

In this paper, we consider the tensile force acting on a fixed-wing. The load is uniformly distributed on a fixed-wing, which can also be seen as a flat cantilever beam. Since it is a uniform loading on the wing, the equation used in calculating the deflection of the wing during various airspeed, angle of attack, and roll angle is shown below;

$$\delta_{max} = \frac{\omega l^4}{8EI} \tag{7}$$

where E is the modulus of elasticity, I is inertia, l is length of wing, and ω load per unit length, and P is point load. The best way to calculate strain using resistance method is a Wheatstone bridge circuit. Using this circuit, we can get the total voltage difference and difference resistance.

$$V_O = \left[\frac{R_3}{R_3 + R_4} - \frac{R_2}{R_1 + R_2}\right] * V_{EX} \tag{8}$$

Null condition is satisfied when: $R_1 R_3 = R_2 R_4$. Gauge Factor $GF = (\Delta R/R)$. Knowing the GF, we can get the value of ΔR, which is known as the deflection in length. Now we are interested in determining the deflection and strain ratio for cantilever beam so we can apply it on aircraft wings to find out deflection as shown in Fig. 2.

$$\frac{V_O}{V_{EX}} = -\frac{GF * \varepsilon}{4}\left[\frac{1}{1 + GF * \frac{\varepsilon}{2}}\right] \tag{9}$$

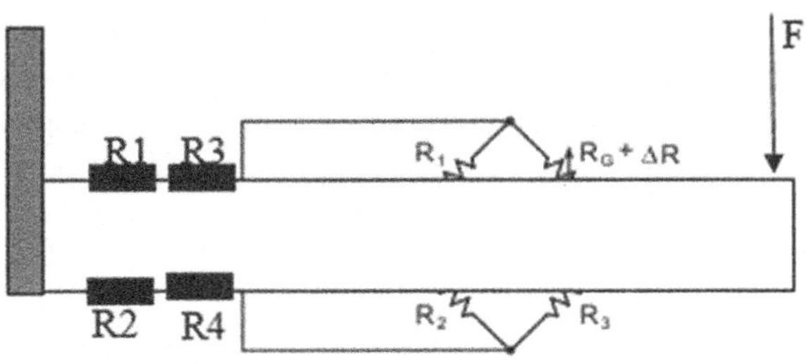

Fig. 2 Deformation due to the external force

When the four strain gauges have an equal nominal resistance i.e. $R1 = R2 = R3 = R4 = R$ then the Wheatstone bridge equation is reduced to the linear equation

$$(V_o/V_{\mathrm{ex}}) = [\Delta R/R]\alpha(F) \tag{10}$$

$$\mathrm{V_0 = a_0 + a_1 W} \tag{11}$$

where a_0, a_1 are constants related to weight. Once an unknown weight is known, an unknown mass or density easily follows.

4 Feature Matching and Tracking

Finding corresponding features from two images consists of two phases. First, using Speeded-Up Robust Features (SURF) methods to extract features from the various images, and then matching key points between the two images using Random Sample Consensus (RANSAC) [5]. The RANSAC algorithm was used to sort out the SURF algorithm results and delete the outliers [16]. The use of RANSAC and SURF method would allow to eliminate flaws in the existing matching techniques, such as high mismatch and low computational efficiency [2, 11]. The wing shape was identified by using the deflection points. The shape of the wing was determined in function of roll angle, pitch angle, and airspeed. SURF and RANSAC algorithms are the methods used for object recognition. The features are invariant to image scaling, translation, and rotation, and somewhat invariant to variations in lighting and 3D projection.

Equations (2) and (3) is used to subtract the distance of the target from the camera at different conditions. It helps to measure the deflection at a selected location on a fixed-wing. The target needs to be visible enough to acquire a perfect image. Figure 3 shows 48 targets or checkmarks marked on the black surface of fixed-wing; these points will be tracked using a stereo camera. The fixed-wing used in the wind tunnel has twenty-four checkmarks calibrated on both sides, as shown in Fig. 4. These checkmarks serve as the features camera used to measure the deflection by matching features of the image, as shown in Fig. 5. The checkmarks serve as targets placed on the model's surface, where there is a high possibility of deflection. Since the model's surface is dark, a white target is preferable indoor for the camera to detect the marks easily. The strain gauge is placed at every checkmark of the middle row to measure the deflection, and a total number of ten strain gauges were used. The marked target points are 7 cm apart, each from the tip to the fixed point of the wing. The strain gauge information was converted into deflection by using the Wheatstone bridge circuit connected to the Arduino board.

The points' position on the wing is shown in Fig. 3. The first step was to capture the image of the wind using the ZED stereo camera. The deflection was calculated for each point after extracting and matching the features using SURF and RANSAC algorithms, as shown in Fig. 5.

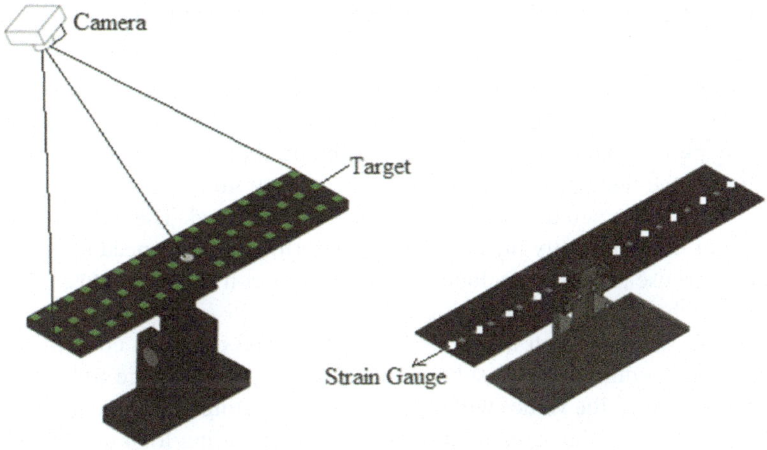

Fig. 3 Points Position (stain gauge and target points)

Fig. 4 Front and rear view of the fixed-wing in the wind tunnel

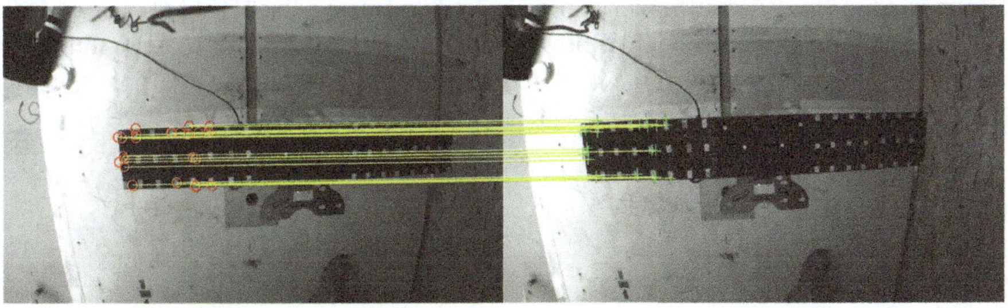

Fig. 5 Matching points procedure

5 Experiments with Camera Measurement System

Vision system captures images of the initial condition when there is no load on the wing and the memory saved. In capturing images from a live camera, the algorithm of snapping photos in the loop is needed, and the limit of that loop ranges from 1 to *N*, where *N* is the number for the last image. At every time, the Nth image will be saved in the system when under the application of wind-load. The wing's deflection measurement needs to apply the match-point algorithm, but we need to process the images first. These processes include a color image's conversation into a gray color scale and match each checkerboard's cross point.

To find out the deflection is required to subtract the cartesian coordinate of all points for image number 1, i.e., subtract coordinates of Nth image from the initial image; consequently, the values will deflect. For calculating deflection in live condition, all algorithms in the same loop of Nth image capturing loop must be considered. There would be some pause between capturing images because the deflection measuring algorithm will take time for the solution. So, in the end, the loop needs to pause the capturing algorithm to capture the images. The end loop value deflection will show on display, and those algorithms will run in a loop until its limit and show the deflection values. As a result, deflection values will show in the live conditions of cameras.

The procedure for the live camera-vision measurement system works by the camera so, the camera position is the first important step into the process and setting the camera in perfect steady condition from where the camera can cover the entire image/video. The vision system feeds the live video into the system. The system will convert the video into images at an interval of time and then use it to calculate the area using a corner detection algorithm. The pressure difference between the upper and lower surface of the wing is needed to obtain the force is required to calculate the lift force–velocity because the lift is a result of pressure. Pitot tubes are assembled at the upper surface as well as lower surface to compare the calculated pressure, which was used to achieve the lift force–velocity, as shown in Fig. 6. If that lift force value is known, then the lift coefficient can be calculated using the Eq. (13). The values of thrust force and sine product of the model's weight are required to calculate the drag force. The drag force's value is then used to calculate the drag coefficient using Eq. (16).

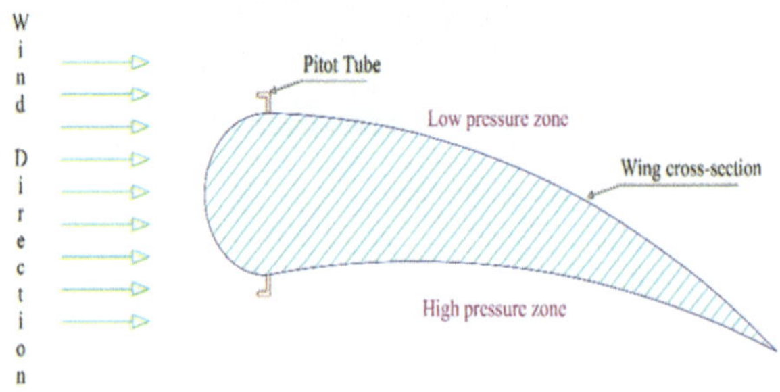

Fig. 6 Wing setup for lift force calculation

Equations used for analysis

- Lift force

$$(F_L) = \frac{1}{2} * A\rho * \left(V_{upper}^2 - V_{lower}^2\right) \tag{12}$$

- Lift co-efficient

$$(C_L) = \frac{2 * F_L}{\rho * A_L * V^2} \tag{13}$$

- Drag force during climb condition

$$(F_D) = \text{Thrust} - [W * \sin(AOA)] \tag{14}$$

- Drag force during descend condition

$$(F_D) = \text{Thrust} + [W * \sin(AOA)] \tag{15}$$

- Drag coefficient

$$(C_D) = \frac{2 * F_D}{\rho * A_D * V^2} \tag{16}$$

The velocity square difference will give the lift force magnitude from Eq. 12. Equation 13 is used to calculate the lift force's coefficient, and similarly, Eq. 16 also gives the value of the drag coefficient. For the drag force calculation, there are mainly three cases. First is during take-off (climb), which uses the sine product of the UAV's weight as subtraction. The second case during landing (decent) that sine product will be an addition in thrust as Eq. 15. The third case is during the air's stable condition when the thrust force is equal to the drag force.

6 Experimental Results and Analysis

In this paper, the experiment is to verify the Deflection-Detection-Vision-System (DDVS) that measures the deflection of a fixed-wing UAV model using the vision system, strain gauge, and simulation. The experiment took place in a static setup in the wind tunnel with various positioning of the camera, and we plan to mount the camera on the vertical stabilizer of the UAV in real-time. The platform being used in the wind tunnel is a 100 cm spam of acrylic material, which is fixed at the midpoint to allow deflection at both ends. The experiment was performed at Carleton university wind tunnel at various wind speeds 10 km/h, 20 km/h, and 30 km/h. Angles of attack and roll angles ranging from 0 to 30°, respectively. Both camera and strain gauge measurements were done simultaneously, limiting error with the time difference.

6.1 Camera Calibration

Calibration is to determine an instrument's accuracy to eliminate or reduce bias in an instrument's readings over a period. Both camera and strain gauge were calibrated in this experiment. Calibration is a key factor in vision systems as it helps to determine the relationship between the object and the location of the object in the global coordinate system. The MATLAB Toolbox [17] shows the various methods of calibrating a camera in order to determine the intrinsic parameters referring to the focal length, extrinsic parameters, which is the camera's 3D position in respect to the world coordinate system, lens distortion, and principal point. This toolbox's inputs are a series of images of a model chessboard plane that cover the calibration points. Calibrated chessboard edges are used as the reference points [11]. Using the measured homography matrix, one may use the next image's error to evaluate the error function after transferring the first image points. All of the surveyed procedures were configured and checked under the same picture noise conditions [11]. MATLAB, with a model chessboard, a camera calibration toolbox is used to get intrinsic and extrinsic parameters for the sensor, as shown in Figs. 7 and 8.

The extrinsic matrix K from the camera calibration is shown as:

$$K = \begin{bmatrix} 3380.8 & 0 & 1374 \\ 0 & 3371.6 & 978.4 \\ 0 & 0 & 1 \end{bmatrix} \tag{17}$$

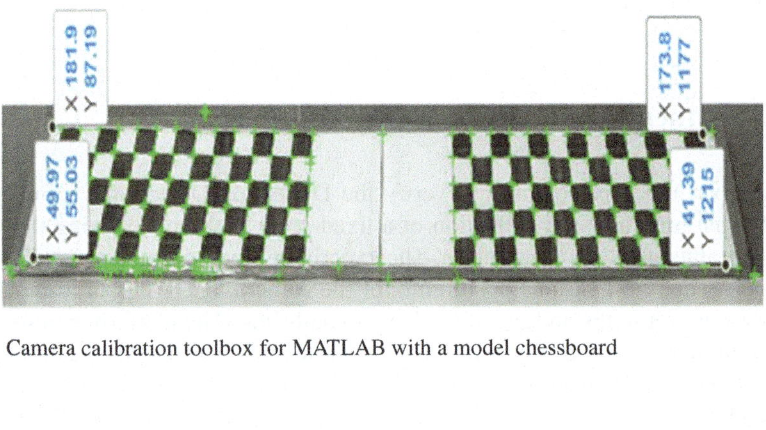

Fig. 7 Camera calibration toolbox for MATLAB with a model chessboard

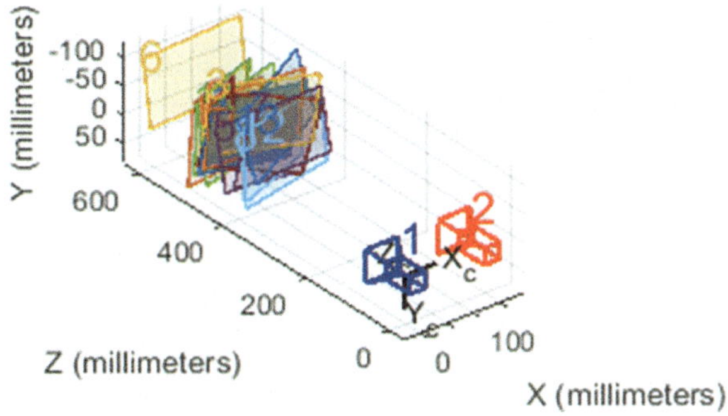

Fig. 8 Camera extrinsic parameters

6.2 *Experimental Result*

The DDVS and strain gauge starts to measure the deflection of all points in the wing at speed starting from 10 km/h to 35 km/h, AOA of 10°, and roll angle of 0°, as shown in Fig. 9. Then the deflection is measured as the roll angle is increased to 10°, as shown in Fig. 8, and there was a decrease in deflection. In Figs. 11 and 12, the deflection was measured when the roll angle is 0°, and AOA becomes 20 and 30°, and there was an increase in the deflection, respectively. The test was repeated at

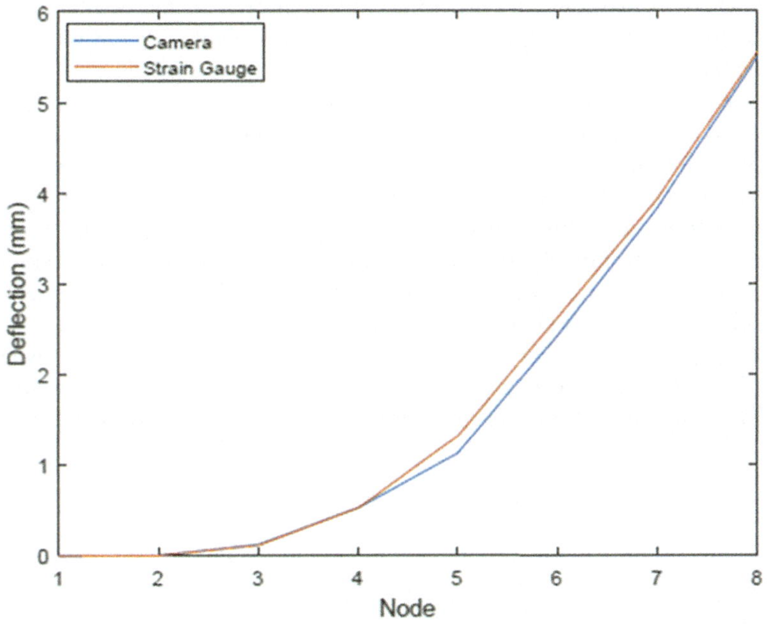

Fig. 9 Deflection versus node no. AOA = 10°, Roll angle = 0°

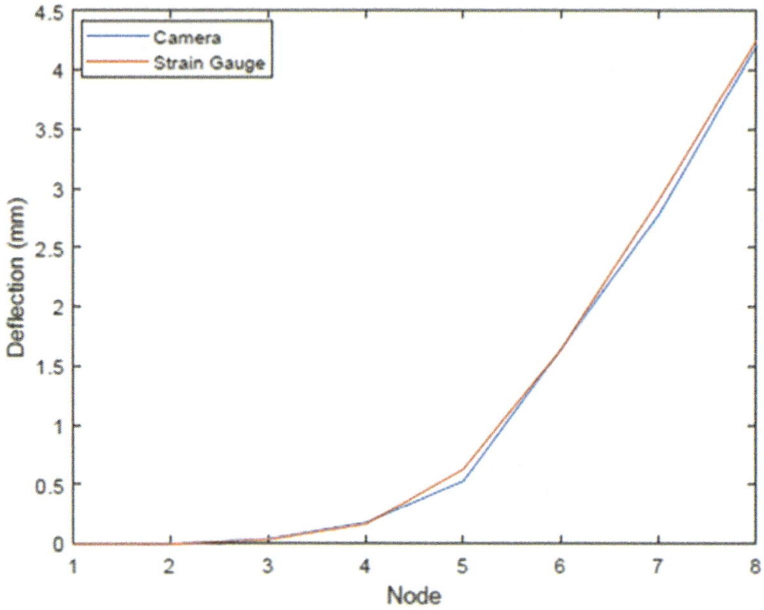

Fig. 10 Deflection versus node no., AOA = 10°, Roll angle = 10°

speed ranging from 10 km/h to 35 km/h, and roll angle 10° at AOA 30° and 20° roll angle at 30° AOA as shown in Figs. 13 and 14, respectively, and both graphs show the deflection decreases with an increase in roll angle. Nodes 1 to 8 represent the targets on the fixed wing (Figs. 9, 10, 11, 12, 13, 14).

Figures 15, 16, and 17 show the camera's deflection measurements and the strain gauge. It is noticed that the deflection was increased when the speed increased at AOA of 0 to 30°. Simultaneously, the maximum deflection occurs at a speed of 35 km/h and AOA 30°. The relationship between maximum Deflection and AOA of camera and strain gauge integration at different speeds is shown in Fig. 18. The experiment has four phases depending on time. The first phase measured the deflection for 60 s with the wind speed only. The second phase starts from 60 to 120 s. The deflection is measured when the roll angle is increased to 10°.

The third phase starts from 120 to 180 s, the deflection is determined when the speed is 35 km/s, and the roll angle is 10° and 10° AOA. The final phase starts from 180 to 240 s, and the deflection is measured when AOA increased to 20°, as shown in Fig. 19.

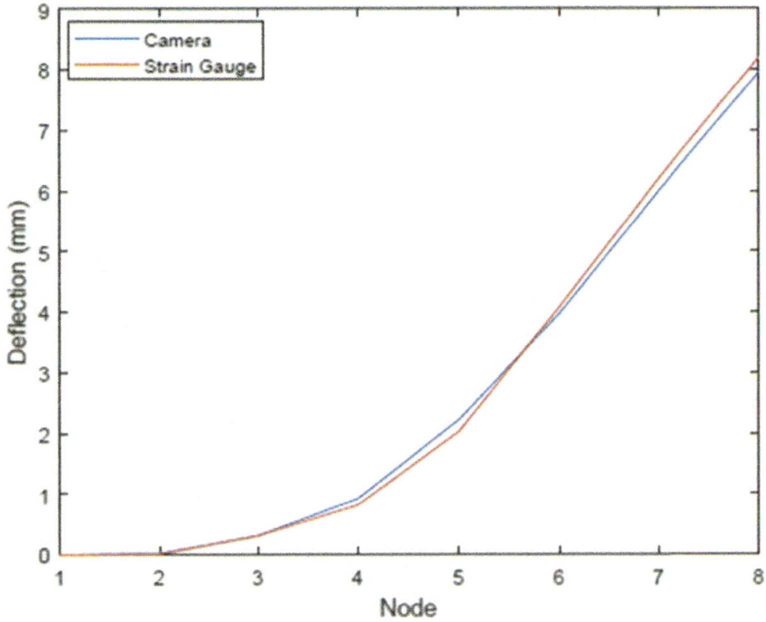

Fig. 11 Deflection versus node no. at AOA = 20°, Roll angle = 0°

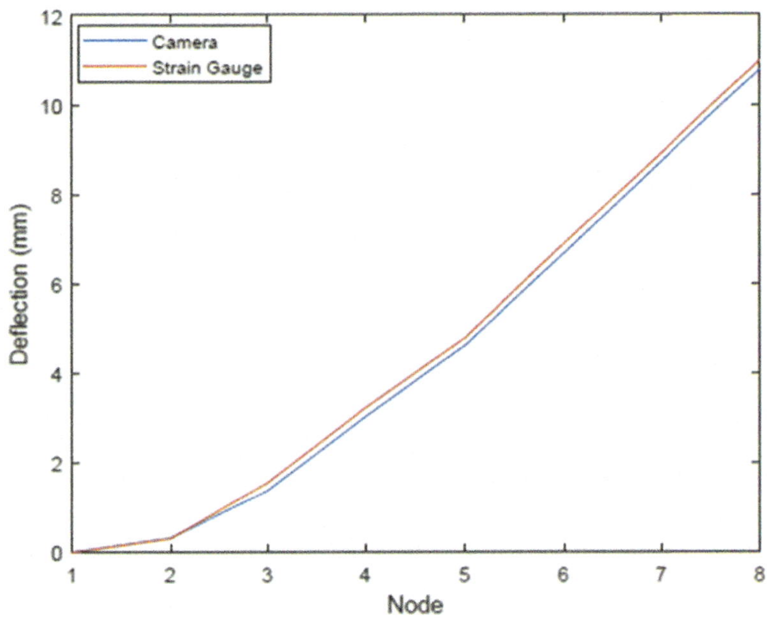

Fig. 12 Deflection versus node no. at AOA = 30°, Roll angle = 0°

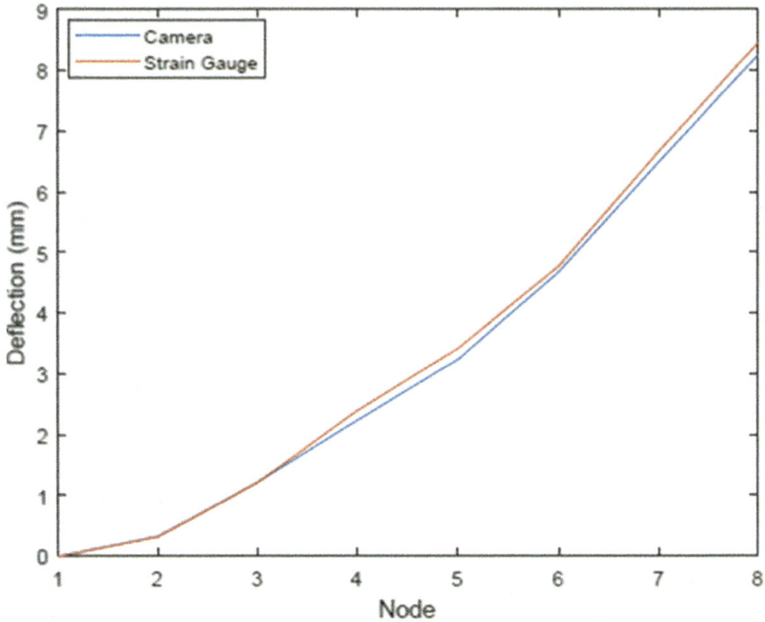

Fig. 13 Deflection versus node no. at AOA = 30°, Roll angle = 10°

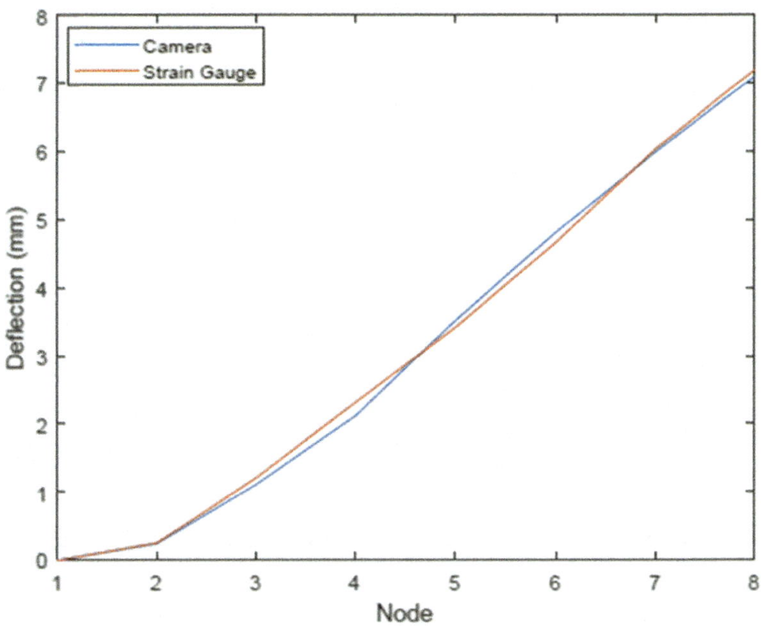

Fig. 14 Deflection versus node no. at AOA = 30°, Roll angle = 20°

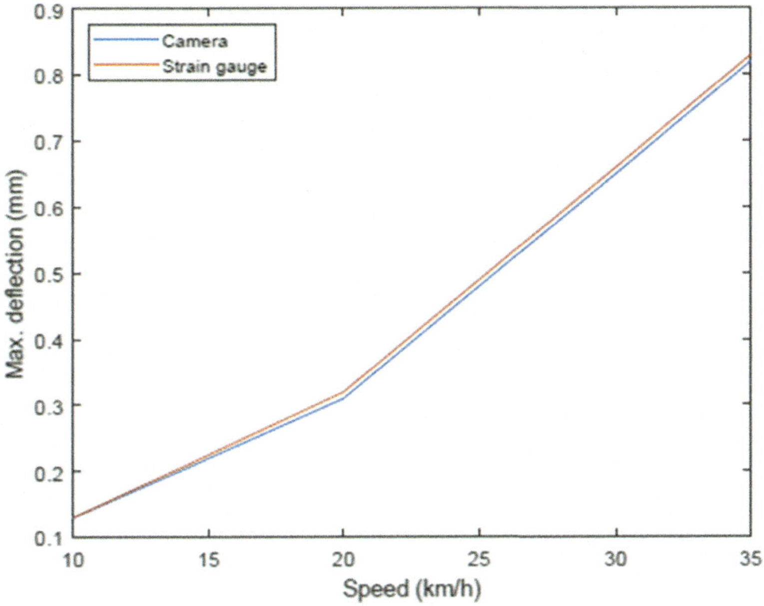

Fig. 15 Max. deflection versus different speed at AOA = 0°

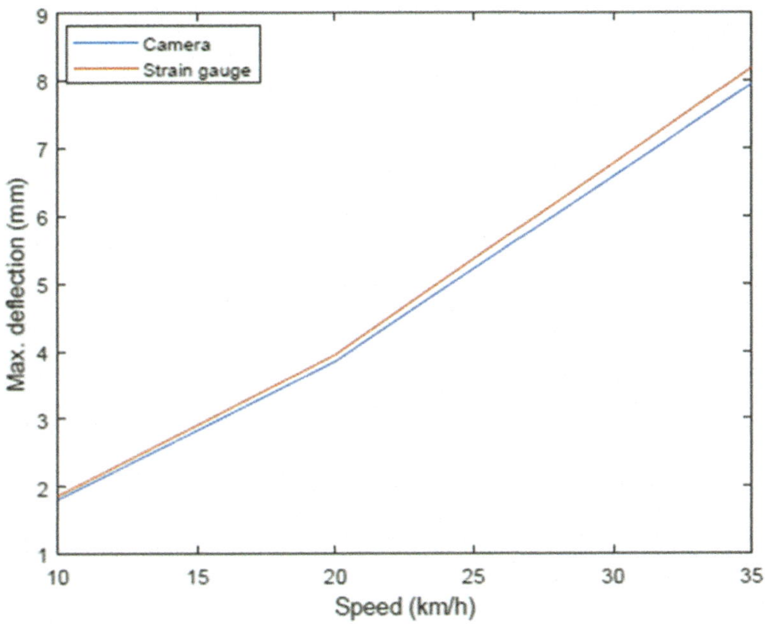

Fig. 16 Max. deflection versus different speed at AOA = 20°

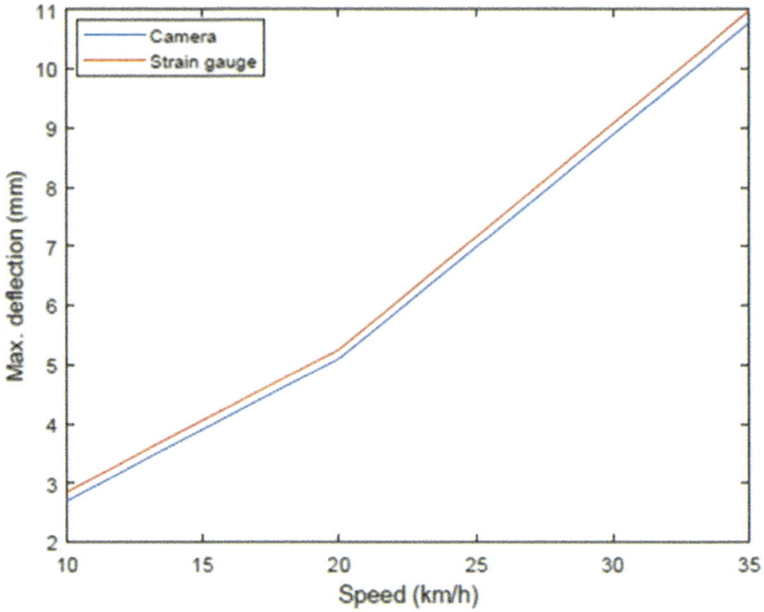

Fig. 17 Max. deflection versus different speed at AOA = 30°

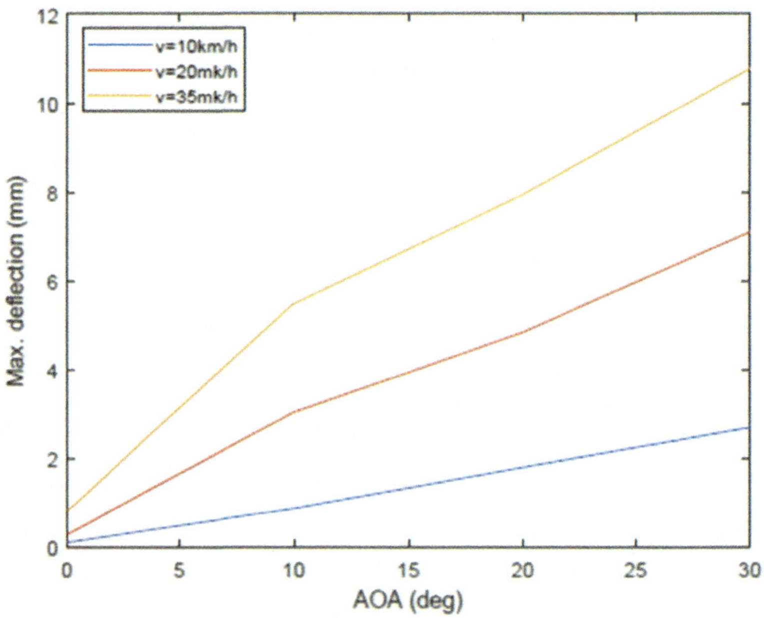

Fig. 18 Max. deflection versus AOA for the fused value

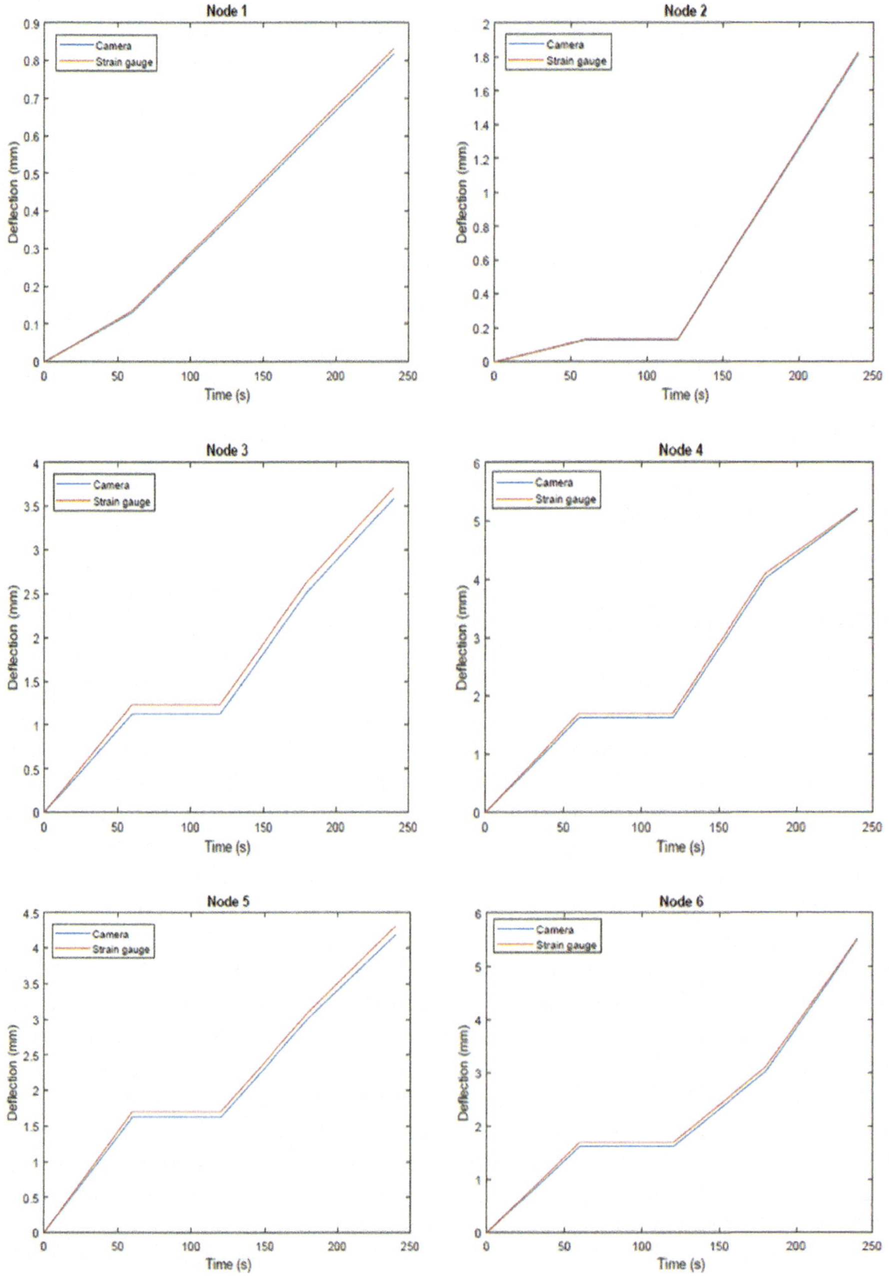

Fig. 19 Tracking points

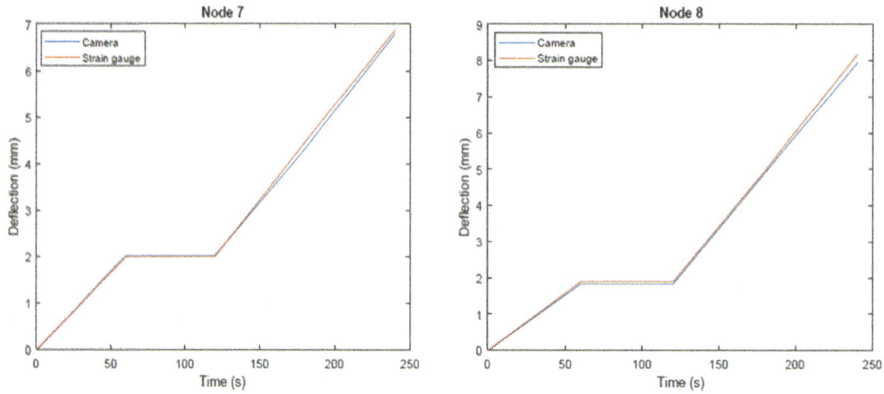

Fig. 19 (continued)

6.3 Result Verification

The fused information from the camera and strain gauge measurement compared with numerical value was used to calculate the Autodesk Fusion 360 model's fixed-wing wind load. Then developed CAD assembly model in Autodesk Fusion 360, which was converted into.step file in 12 different cases, i.e., for the angle of attack (AOA) 10, 20, 30°, and for each AOA, there are four different rolling angles 0, 10, 20, 30°. These files were input in Solidworks and for flow analysis and to obtain the aerodynamic forces. The required forces were calculated from the flow analysis and then inputted these data to Autodesk Fusion 360 to calculate the constraints, joints, materials, and forces by generating the mesh. These allow solving for the static analysis model in Fusion 360 and the deflection at each node, strain, stress, and many more aerodynamic parameters. During the inflow analysis, the assumption was made that the tunnel (computational domain) walls are adiabatic and with zero roughness. The inlet velocity of the wind in the wind tunnel is in the Z direction. The fixed-wing deflection is shown in Fig. 20. The result from the integration of camera and strain gauge and simulation measurement at different AOA and roll angles is presented in Table 1.

For accuracy and reliability of the system, an extensive experiment was conducted to determine the wing's deflection in the function of the flight parameters: angle of attack, roll angle, and flow velocity. The results from the vision system and strain gauge were fused using Unscented Kalman Filter (UKF). UKF is an advanced filtering method that is used for sensor data fusion, and its more accurate compare to Extended Kalman Filter (EKF) in several applications. The decision to use UKF is to give better performance in dealing with nonlinear systems and deal with white and color Gaussian noise. The results have shown that the integration of strain gauge and vision system sensor measures wing deflections and accurately identifies the aerodynamic coefficient by comparing with simulation results, as shown in Figs. 21 and 23. The measurement error is so minimal ash shown in Figs. 22 and 24.

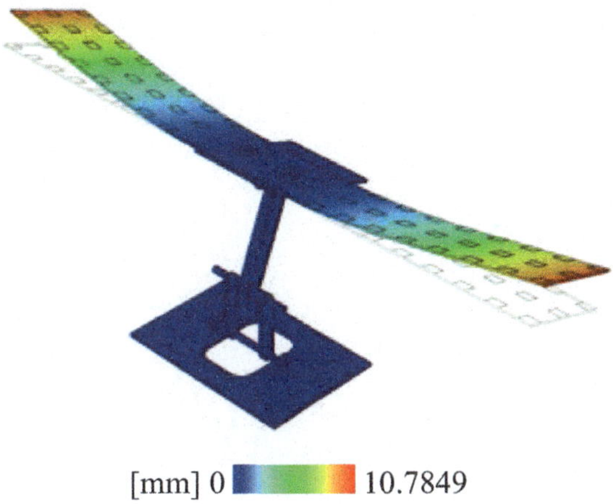

Fig. 20 Fixed-Wing Maximum deflection at AOA = 30°, Roll angle = 0°, and Wind Speed 35 km/h

Table 1 Comparison of deflection taken from camera/strain gauge and Simulation

Wind Velocity (km/h)	Angles		Simulation	Fused data	Error (%)
	Angle of attack	Rolling angle	Deflection (mm)	Deflection (mm)	
35	0	0	0.8202	0.8201	0.0100
35	10	0	5.4660	5.4941	0.5115
35	10	10	4.1810	4.2121	0.7384
35	10	20	3.6971	3.6921	0.1327
35	10	30	2.9553	3.0037	1.6213
35	20	0	8.1422	7.9549	2.3520
35	20	10	6.5320	6.7004	2.5133
35	20	20	5.8001	5.9537	2.5816
35	20	30	4.4661	4.5401	1.6321
35	30	0	10.622	10.7849	1.5290
35	30	10	8.1933	8.2549	0.7499
35	30	20	7.0072	7.1003	1.3140
35	30	30	5.2091	5.3027	1.7670

The deflection increases with an increase in AOA. Also, the deflection at each node measured by the camera is very close to the strain gauge and simulation measurement at AOA of 30°, as shown in Fig. 25.

The wing controls the airplane, and during the flying condition, the large numbers of aerodynamic forces act on it. The wing deflection plays a very crucial role because high deflection will affect the navigation of the UAV. The vision system can measure the deflection of the wing by using a stereo camera calibration. The system compares

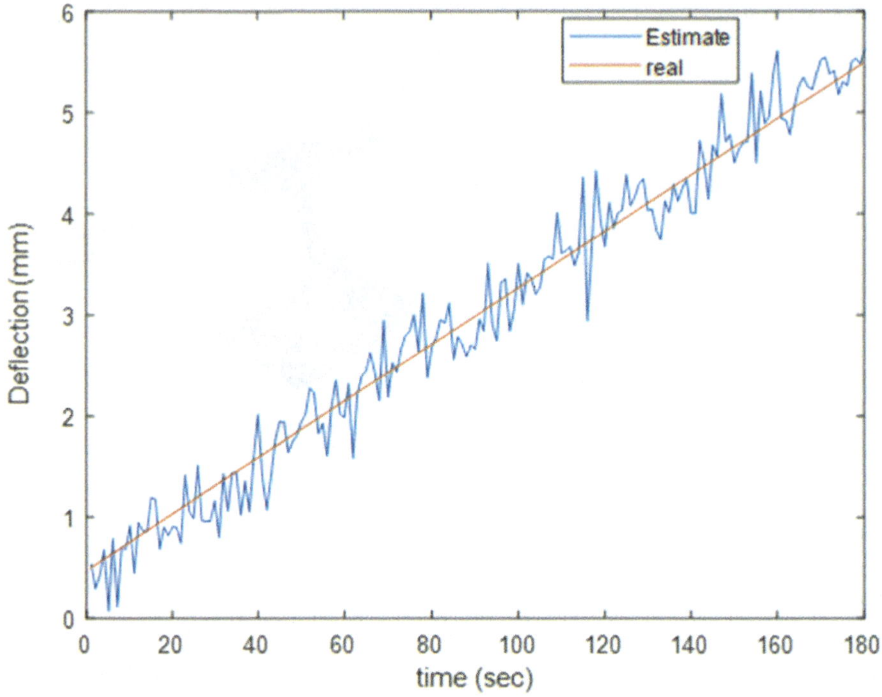

Fig. 21 UKF fusion of camera and strain gauge at AOA 10°, Roll angle 0°

the ideal results using the idealization process, by which the same conditions in the analyzing software is the same as the real-time experiment [18, 19].

6.4 The Aerodynamic Results

There is an assumption that the tunnel (computational domain) walls are adiabatic with zero roughness for the flow analysis. The input of the inlet velocity of the wind in the wind tunnel is in the Z direction to the fixed wind setup. After that, set the goals to find the various required values at the fixed-wing surface (data are required at the wing surface, especially at lower surface area). The flow distribution of the wind over the fixed-wing surface area is shown in Fig. 26. Considering the fixed-wing UAV is essential to calculate the aerodynamic force required for the fixed-wing to lift the mass of the model [18, 19]. These values were used in the live camera-vision measurement system. The original dimension of the wing used for that calculation is;

Width = 0.168 m
Length = 1.160 m
Square size = 32 × 32 mm

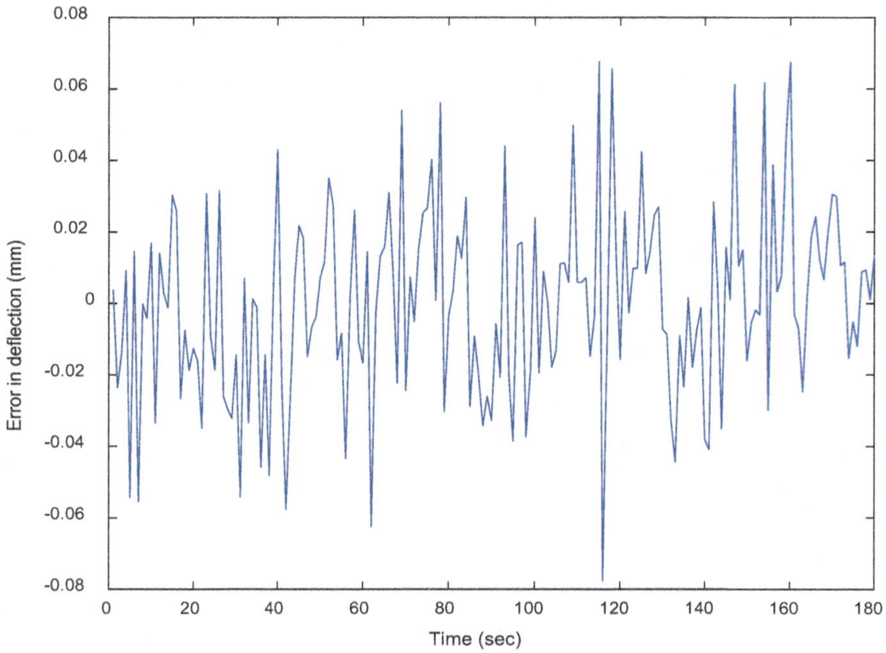

Fig. 22 Measurement error, AOA 10°, Roll angle 0°

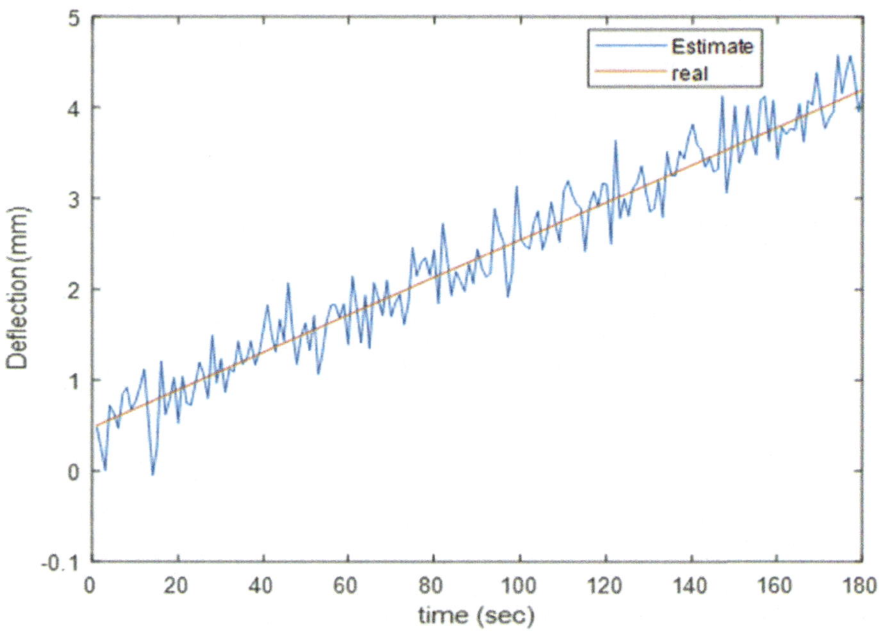

Fig. 23 UKF fusion of camera and strain gauge at AOA 10° Roll angle 10°

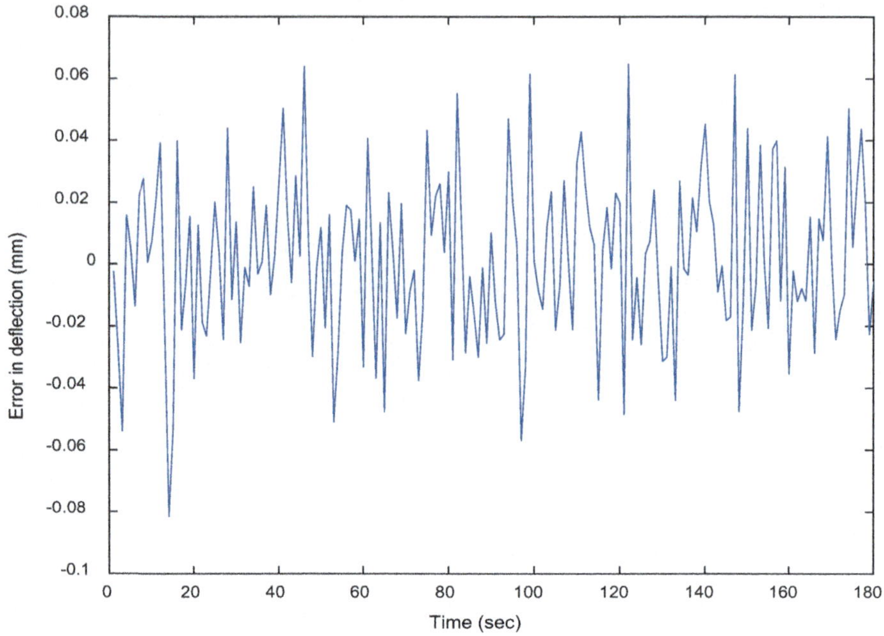

Fig. 24 Measurement error, AOA 10°, Roll angle 10°

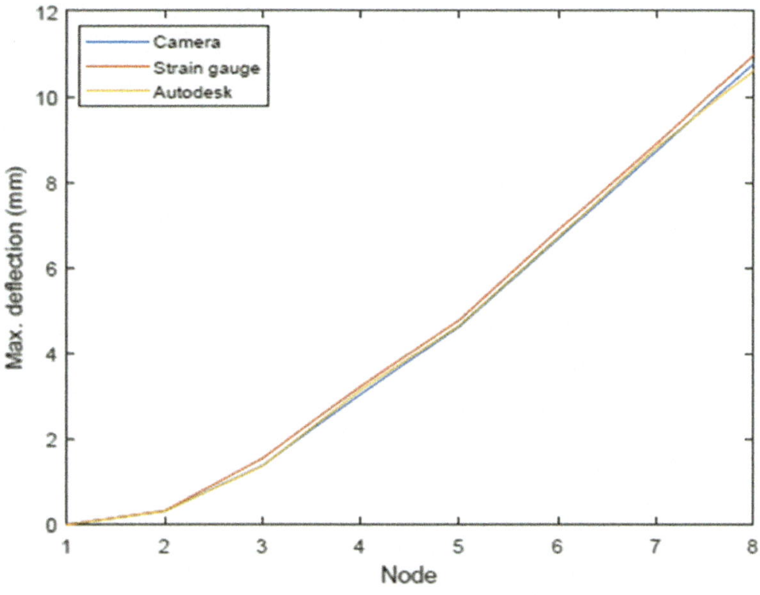

Fig. 25 Camera/Strain gauge versus Simulation deflection at AOA = 30° Roll angle = 0°

Fig. 26 Top view of wing setup during flow analysis in the computational domain

The input parameters are as follows; when the UAV is climbing, the input character is classified as 1 or write 0 = 1

Weight of Wing (N) = 12.2625
Angle of attack (Degree) = 30
Relative velocity of Airplane (m/s) = 0.3
Exit velocity of air from engine (m/s) = 2.096
density of air (kg/m^3) = 1.225
Area of the propulsive device (Engine) (m^2) = 1.44
Velocity difference at wing (m/s) = 4.612
The out parameters are as follows;
First point of Y-axis (in pixels) = 224.328598
Last point of Y-axis (in pixels) = 54.041233
First point of X-axis (in pixels) = 1222.245605
Last point of X-axis (in pixels) = 61.941708
Width (m) = 0.170287
Length (m) = 1.160304
Area (m^2) = 0.197585
Lift area (m^2) = 0.171114
Drag area (m^2) = 0.098793
Mass flowrate enter the propulsive device = 0.529200 kg/s
Mass flowrate exit the propulsive device = 3.697344 kg/s
Thrust force = 7.590873 N
Minimum Lift force require = 10.619637 N

Actual Lift force = 2.824850 N
Drag force during climb condition = 1.459623 N
Lift co-efficient = 0.285163
Drag co-efficient = 0.255211

The final output result is summarised in Table 2. Figures 27, 28 and 29 show the different drag and lift forces generated by the camera when the AOA is 10° and roll angles 0, 10, 20, and 30. With increasing roll angle, the lift and drag force values decrease due to change in the projected area of the fixed-wing. The drag force values are lesser than lift force values in all the cases, which shows the results' validity. As the AOA increases, the values of the drag and lift forces becomes higher.

Figures 30, 31 and 32 show the different deflection for roll angles 0, 10, 20, and 30 for the constant angle of attack 10°, 20°, and 30°, respectively. With increasing roll angle, the deflection values are decreasing due to changes in forces, and the deflection increases as the AOA increases. Moreover, the simulation and vision system values are compared in that graph, and the differences between both are very less.

As the AOA increases, the area will increase due to the increase in induce force, and consequently, the deflection increases. The experiment was carried out using AOA 10°, 20°, and 30°, respectively. Moreover, the software analysis output values and camera system analysis values are compared in that graph, and the difference between them is very less, as shown in Fig. 33.

The roll angle and minimum velocity to lift the fixed-wing UAV are represented in the graphs with AOA, 10°, 20°, and 30°, as shown in Figs. 34, 35, and 36. With increasing roll angles, the lift force decrease, and with respect to the decrease in force, more velocity is required to lift up the body. Therefore, more force and velocity are required during airplane take-off. In this experiment, at constant AOA of 10°, 20°, and 30° and different roll angles ranging from 0 to 30° at an interval of 10 were considered. Comparatively, the lift force's value increases with an increase of AOA so, the lift velocity at AOA 10° is higher compared to the life velocity of 30° at the same conditions.

As the AOA increases from 10 to 30°, the fixed-wing area exposed to the wind will also increase. Due to this condition, the value of the force increase also. As a result, lesser velocity will be required for lift with higher AOA, as shown in Fig. 37.

7 Conclusion

The Deflection-Detection-Vision-System (DDVS) is developed to calculate the deflection and identify the fixed-wing aerodynamic coefficients during the flight. This system employed a stereo camera to estimate the deflection for a specific point on the wing and compare it with the reading strain gauge sensors. This visual detection system is based on three main algorithms: a stereo camera calibration, visual points tracking, and deflection identification using a three-dimensional technique. The calibration for a stereo camera allows us to find intrinsic parameters. Eight

Table 2 The final output results

Wind velocity (km/h)	Angles		Camera/S.G		Solidworks		Camera/S.G		Solidworks	
	AOA	Roll angle	Drag force	Lift force	Drag force	Lift force	Co-efficient of drag	Co-efficient of lift	Co-efficient of drag	Co-efficient of lift
35	10	0	0.278	1.611	0.288	1.631	0.0432	0.2507	0.0464	0.2901
35	10	10	0.248	1.298	0.248	1.308	0.0385	0.2019	0.0389	0.2011
35	10	20	0.236	1.168	0.239	1.159	0.0367	0.1817	0.0369	0.1897
35	10	30	0.214	1.003	0.218	1.006	0.0333	0.1560	0.0335	0.1565
35	20	0	0.668	2.349	0.669	2.351	0.1039	0.3655	0.1031	0.3652
35	20	10	0.710	1.977	0.725	1.978	0.1104	0.3076	0.1117	0.3079
35	20	20	0.608	1.760	0.611	1.765	0.0946	0.2738	0.0948	0.2736
35	20	30	0.542	1.444	0.561	1.437	0.0843	0.2247	0.0853	0.2256
35	30	0	1.461	2.825	1.471	2.826	0.2273	0.4396	0.2288	0.4359
35	30	10	1.264	2.283	1.274	2.286	0.1967	0.3552	0.1945	0.3566
35	30	20	1.058	2.015	1.061	2.017	0.1646	0.3135	0.1648	0.3176
35	30	30	0.903	1.569	0.908	1.567	0.1405	0.2441	0.1401	0.2442

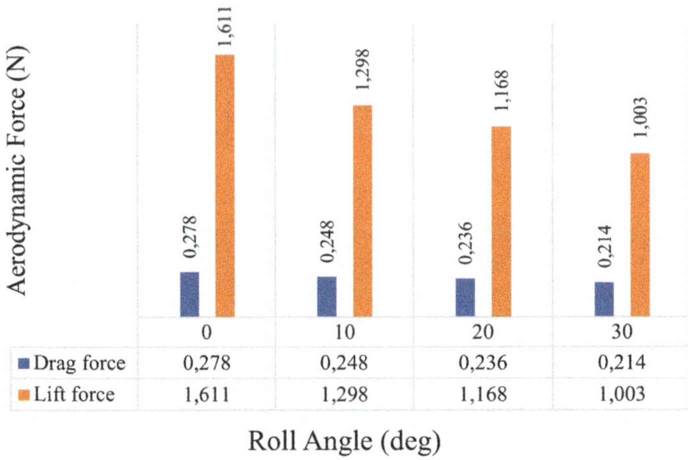

Fig. 27 Roll Angle versus Aerodynamic Forces for AOA 10°

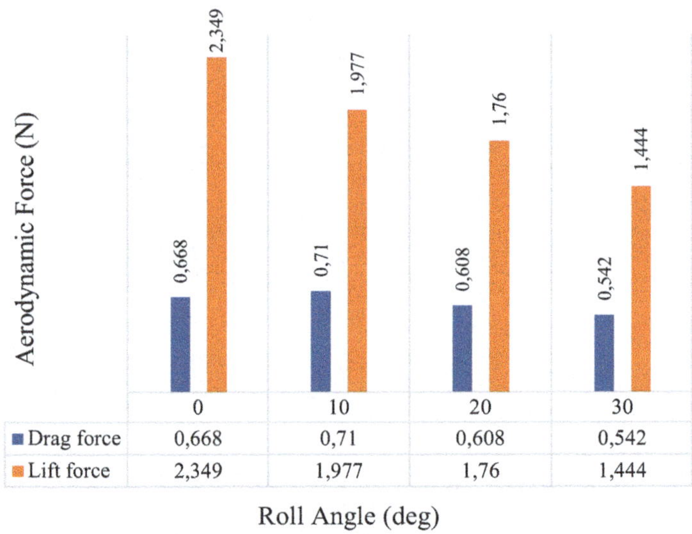

Fig. 28 Roll angle versus Aerodynamic forces for AOA 20°

selected points are used to track the wing shape visually during the flight. Extensive experiments have been conducted in a wind tunnel for different flight conditions. The experiential tests confirmed computer simulation results and measurements obtained from integrating the vision system and strain gauges are very close. When the attack angle increases, the fixed-wing deflection increases, but the deflection decreases for the same angle of attack when the roll angle increases. For the experimental setup, the maximum wing deflections occurred when the speed was 35 km/h, roll angle was

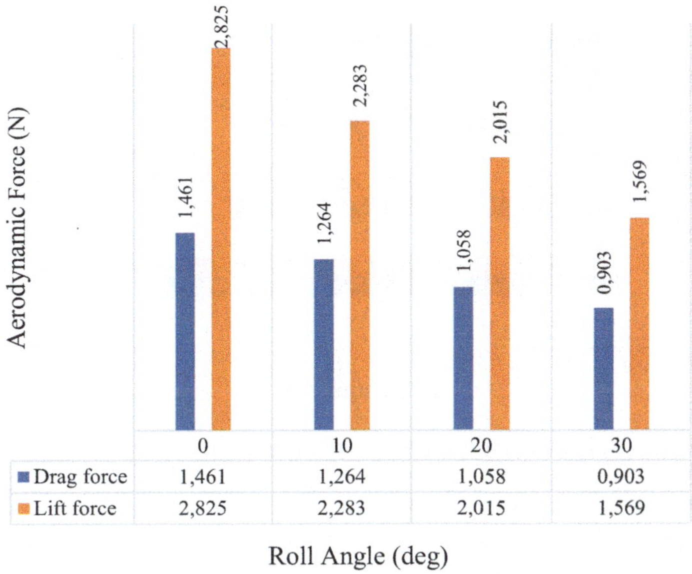

Fig. 29 Roll angle versus Aerodynamic forces for the AOA 30°

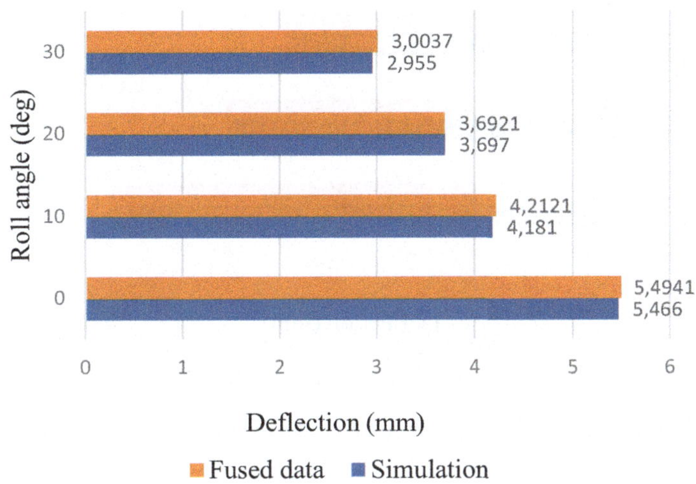

Fig. 30 Roll angle versus Deflection for AOA 10°

0°, and the AOA of 30°. In the next stage of system development, the full integration of measurements from the vision system and strain gauge sensors during flight is planned.

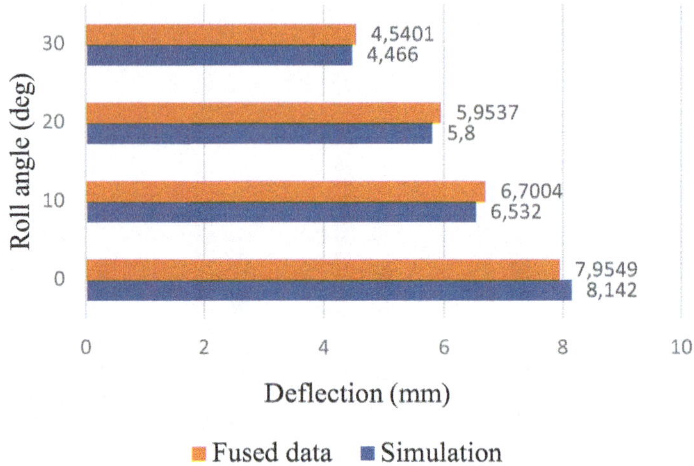

Fig. 31 Roll angle versus Deflection for AOA 20°

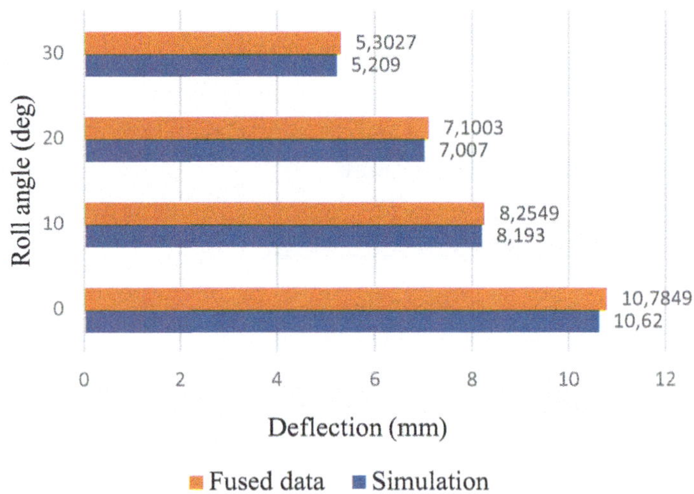

Fig. 32 Roll angle versus deflection for AOA 30°

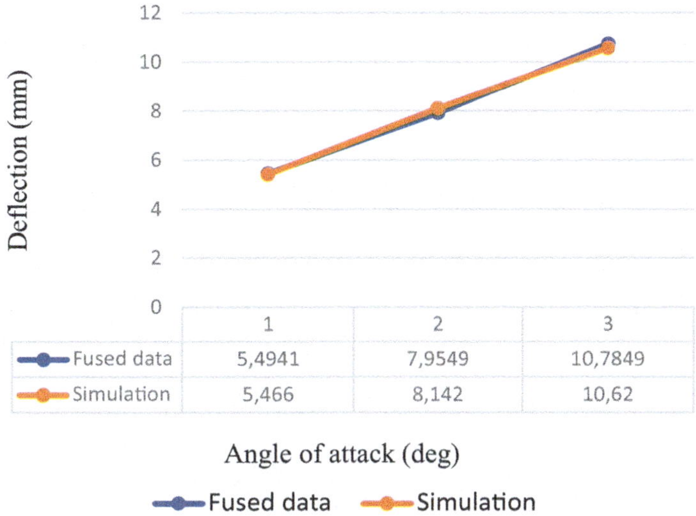

Fig. 33 Angle of attack versus deflection

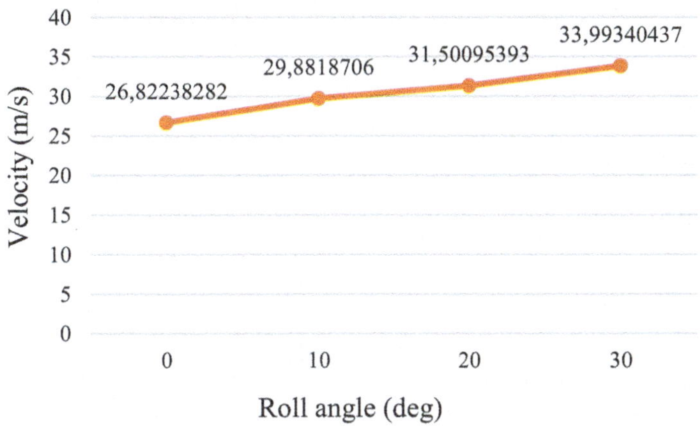

Fig. 34 Roll angle versus minimum velocity of lift for AOA 10°

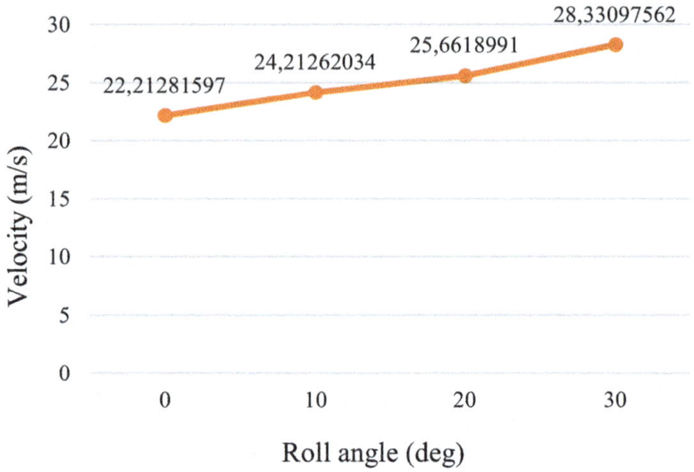

Fig. 35 Roll angle versus minimum velocity of lift for AOA 20°

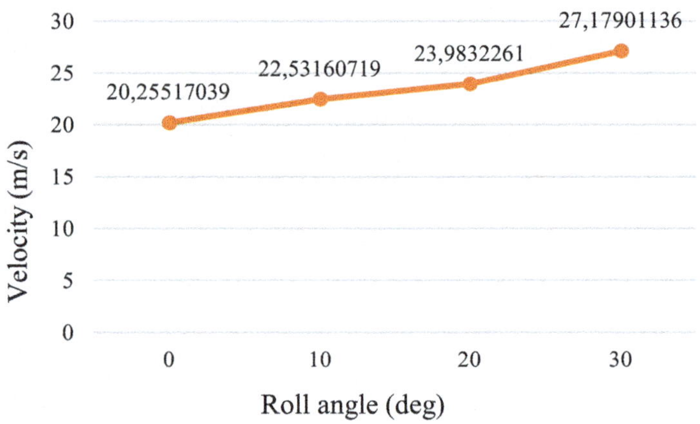

Fig. 36 Roll angle versus minimum velocity of lift for AOA 30°

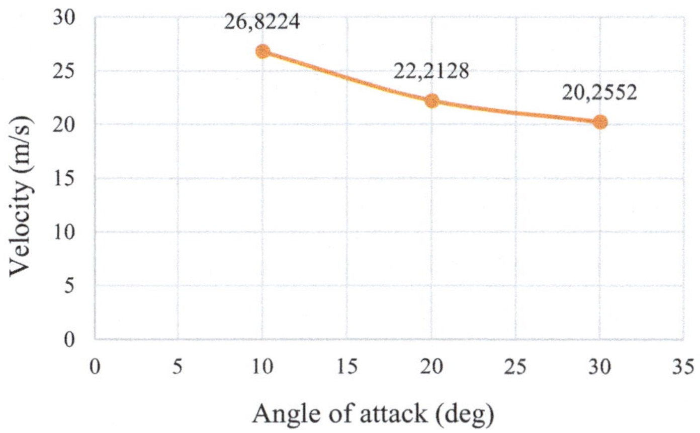

Fig. 37 Angle of attack versus minimum velocity required for lift

References

1. Burner, A.W., Tianshu, L.: Videogrammetric model deformation measurement technique, at NASA Langley research center, Hampton, Virginia 23681–2199. J. Aircraft **38**(4) (2001)
2. Burner, A.W., Wahls, R.A., Goad, W.K.: Wing twist measurements at the national transonic facility. NASA TM 110229 (1996)
3. Hooker, J.R., Burner, A.W., Valla, R.: Static aeroelastic analysis of transonic wind tunnel models using finite element methods. AIAA Paper 97–2243, June 1997
4. Al-Isawi, M.A., Sasiadek, J.Z.: Control of Flexible Wing UAV Using Stereo Camera. Springer publishers, Aerospace Robotics III (2019)
5. Biondi, G., Mauro, S., Mohtar T. et al.: Feature-based estimation of space debris angular rate via compressed sensing and Kalman filtering. In: IEEE, Metrology for Aerospace (MetroAeroSpace) (2016)
6. Hillenbrand, U., Lampariello, R.: Motion and parameter estimation of a free-rotating space object from range data for motion prediction. In: Proceedings of i-SAIRAS. 8th International Symposium on Artificial Intelligence, Robotics, and Automation in Space, Munich, Germany (2005)
7. Li, C., Liang, B., Xu, W.: Autonomous trajectory planning of free-floating robot for capturing space target. In: Proceedings of the IEEE/RSJ International Conference on Intelligent Robots and Systems, Beijing, China (2006)
8. Xu, W., Liang, B., Li, C. et al.: Autonomous target capturing of free-floating space robot: theory and experiments. In: Robotica, Vol. 27, pp. 425–445 (2009)
9. Stevens, B.L., Lewis, F.L.: Aircraft Control and Simulation. Wiley, 2nd edn (2003)
10. Teo, T.R., Jang, J.S., Tomlin, C.J.: Automated multiple UAV flight—the stanford DragonFly UAV program. 43rd IEEE Conference on Decision and Control (2004)
11. Al-Isawi, M.A., Sasiadek, J.Z.: Pose estimation for mobile robot and flying robots via visual navigation. Proceedings of the CARO3 3rd Conference on Aerospace Robotics, November 2015 and publication in Aerospace Robotics III, published by Springer (2019)
12. Patel, B.D., Srinivas, A.R.: Validation of experimental strain measurement technique and development of force transducer. Int. J. Scientific En. Res. **3**(10) (2012)
13. Watzlavick, R.L., Crowder, J.P., Wright, F.L.: Comparison of model attitude systems: active target photogrammetry, precision accelerometer, and laser interferometer. AIAA, June 1996
14. Holman, J.P.: Experimental Method for Engineers. 3rd ed, McGraw hill, pp. 44–47 (1978)

15. Viki, H.F.: NI Strain gauge tutorial, national instruments corporation (1998). Accessed 11 August 2020. https://www.scribd.com/document/12921011/ni-strain-gauge-tutorial
16. Silica, K. et al.: SURF and RANSAC: a conglomerative approach to object recognition . Int. J. Comput. Appl., IJCA J. (2015)
17. Bouguet, J.: Camera calibration toolbox for Matlab. Accessed 13 July 2020. www.vision.caltech.edu/bouguetj/calib_doc
18. Man-Vehicle Laboratory, MIT Department of Aeronautics and Astronautics. Accessed 11 August 2020. http://web.mit.edu/16.00/www/aec/flight.html
19. NASA Glenn Learning Technologies Home Page. Accessed 16 September 2020. https://www.grc.nasa.gov/WWW/K-12/airplane/guided.htm

Robotic Dance Modeling Methods

Figen Özen and Dilek Bilgin Tükel

Abstract Any art performance and dance, in particular, are examples of extraordinary complex dynamic systems, perhaps a synchronized human–robot dance being the most complex even in comparison with good examples of complex systems such as climate, ecosystems, economics, organizations, social structures, and socio-political systems. A dance is an art performance that involves dedicated movements, music, and intentional interpersonal synchronization, while observing the movements and interactions among herds, teams, and dancers. A choreography, which repeats the same movements synchronized with music is not enough for dance. Good use of space, differentiation and then reattaining equivalence, and dynamism are needed. Robots' ability to respond like humans in these realms is one of the goals of this research. Two different modeling techniques are proposed for industrial robot dance modelling: Modified Laban Notation and Synchronized Petri Nets. Modified Laban Notation for Industrial Manipulators (MLIR) is designed and applied by means of a special interface. The input parameters of the manipulator are calculated using the data related to motion, which has been made available at the interface. The torque values are calculated and applied. Second method proposed analyses to scrutinize music objects and the mathematical infrastructure. Based on the obtained data, a specific dance choreography conceptualization has been created by means of Petri nets.

F. Özen (✉)
Electrical and Electronics Engineering Department, Haliç University, Mareşal Fevzi Çakmak Cd. No: 15, Güzeltepe Mah., 34060 Eyüp, İstanbul, Turkey
e-mail: figenozen@halic.edu.tr

D. B. Tükel
R&D Department, Altınay Robot Technologies, Aydınlı Sb, Istanbul Endüstri ve Ticaret Serbest Bölgesi, Orjin Cd. 10. Sokak D: No: 3, 34953 Tuzla, İstanbul, Turkey
e-mail: dtukel@dogus.edu.tr

Software Engineering Department, Doğuş University, Esenkent, Dudullu Osb. Mah., Nato Yolu Cd 265/1, 34775 Ümraniye, İstanbul, Turkey

P. Shi et al. (eds.), *Complex Systems: Spanning Control and Computational Cybernetics: Applications*, Studies in Systems, Decision and Control 415,
https://doi.org/10.1007/978-3-031-00978-5_2

1 Introduction

Over the last decades, robot synchronization has been a popular research topic among roboticists. The ultimate aims of the researchers may vary, but they could be logically generalized as having robots work together in harmony. Researchers have used different ways of showing their results, an example of which is robot dance.

Dance is not always used for entertainment purposes in robotics, it is also used for showing how finely robot(s) can be controlled and synchronized with other robot(s).

Ramachandran and Hirstein argued that human brain is inclined to search for patterns and regularities in movements [1]. Hagendoorn proposed several 'real life' rules for complex dance choreography and improvisation, based on observations of the movements of and interactions among groups, such as swarms, flocks, teams, and dancers. His rules were: Spatial organization, alignment, clustering, levels in space, division of space, dynamics, copying or mirroring, internal differentiation, and equivalence relationships [2]. Robot dance systems are successful only when the rules of dance and human perception are considered, and therefore a researcher must extend their survey to the realm of arts and social sciences, as well.

In this chapter, we present our results on synchronization of a multi-robot dance system. The robots synchronize with music and with other robots, which means multiple synchronization. The algorithms developed here have been applied to anthropomorphic industrial robots.

The organization of the chapter is as follows: A summary of the work done on robot synchronization, Petri nets for modeling robotic tasks and robot dance modeling is given in Sect. 2. Theoretical background is provided in Sect. 3. The results of the experiments are displayed in Sect. 4, and conclusions are drawn in Sect. 5.

2 Previous Works

The subjects of robot synchronization and robot dance may be popular, however, since they are considerably new topics, there is still so much room for improvement. What follows is a short summary of the work done in robot synchronization, modeling and implementing robotic dance.

2.1 Robot Synchronization

Rodriguez-Ángeles considered the problem of having the slave robot follow the trajectory of the master robot. His system consisted of two rigid-joint robots. In his design, the unknown parameters, namely the velocity and the acceleration of the master robot, are estimated [3]. Portillo-Vélez, Cruz-Villar and Rodriguez-Ángeles added obstacle-avoidance property to the master–slave robot system. The obstacle was a single moving object. The controller was a combination of PID and optimal controllers [4]. Matsunami, Tanaka-Ishii, Frank and Matsubara constructed a team of seven Lego robots to test the cheerleading dance. One of these seven robots was assigned the role of the team-leader. The synchronization was tested for simple dance patterns and scenarios. The robots communicated through an infrared channel. They were tested in their ability to recover from fall, 70% success was obtained [5]. Mahmood and Kim simulated the leader–follower control of a group of quadcopters. They used graph theory to state the problem, applied feedback linearization to each quadcopter in the group and studied the application of PD and PID controls to satisfy the given design criteria. The heading angle of the followers were to be synchronized to the heading angle of the leader's [6]. Markus, Yskander, Agee and Jimoh proposed employing the differential flatness theory in the synchronization control of multiple single-link robots with joint flexibilities. They assumed a leader–follower architecture [7].

Pongas, Billard and Schaal investigated the synchronization of a hydraulic robot arm, holding a drumstick, with an external signal which simulated a conductor. The frequency of the music signal was varied to test the adaptation. It was found that the robot system synchronizes itself approximately in 3 s. The system was aimed to be extended to multi-robot case [8].

Mehrjerdi, Ghommam and Saad employed two-level controller in the synchronization of multi-robot system. The first level, namely the PID controller, was for individual control of the robots and the coordination was done with a Lyapunov-based controller. Graph theory was used in the solution of the coordination problem. Experiments were done for a system of 3 robots [9]. D'Ambrosio, Goodell, Lehman, Risi and Stanley used artificial neural networks to facilitate communication between robots. Each robot could communicate with a certain robot and a communication scheme, called the Hive Brain, was used. Four robots were employed in the experiment [10]. Floreano, Mitri, Magnenat and Keller studied information transfer between colonies of robots. Each robot was equipped with a neural network, simulating a short-time memory, as well [11]. Wang, Huang, Wen and Fan controlled N nonholonomic mobile robots, with some unknown parameters, using distributed adaptive control to achieve concensus tracking of the reference. The robots were not identical and only some of the robots had access to the reference information [12]. Dou and Wang treated motion synchronization as a constraint in their multiple pairs of two-link manipulator system. They applied adaptive control in parametric uncertainties [13]. They applied a force feedback PI control to achieve synchronization of a system of manipulators consisting of two rigid-link manipulators, connected through a flexible beam [14].

Steels investigated the problem of creating a common language for robots equipped with sensori-motor systems [15].

Kawai et al. studied the problem of synchronization in a humanoid robot. They concluded that the synchronization of degrees of freedom of the robot was not useful, sometimes desynchronization was needed to achieve the given task [16].

Ahmadzadeh and Masehian reviewed various approaches to synchronization in addition to other aspects of modular robotic systems. They classified messaging, locking, master control, leader–follower, hormone-based, synched internal clock, and delayed signals approaches by the researchers [17].

Iqbal and Riek studied the human–robot synchronization problem. They used the group synchronization index and time appropriateness as measures in their Synchronization Index based Anticipation algorithm, which aimed the robot to synchronize with the most synchronous dancer of dancers [18].

2.2 Dance of Robots

Apostolos, Littman, Lane, Handelman and Gelfano compared and contrasted robot dance with human dance. Theirs was one of the early works in this field [19].

Kuroki, Fujita, Ishida, Nagasaka and Yamaguchi designed new actuators to have their 28 degree-of-freedom humanoid robot to dance with grace. Using speech synthesis, they also had the robot sing and move in harmony [20].

Nakaoka et al. tackled the problem of discrepancies in the dances of a human and a humanoid, due to differences in their bodies. They focused on leg tasks and tried the method of learning from observation, they used a motion capture system for that purpose. They also dealt with the spinning problem. They used a 30 degree-of-freedom humanoid robot and tested their method using a traditional Japanese folk dance, namely, Aizu-Bandasian [21].

Landgraf, Oertel, Rhiel and Rojas built a honeybee robot equipped with cameras to study the bee dance. They tested their prototype in a beehive, to see if it was accepted by the other bees as a genuine bee. Even though the appearance and the dance were acceptable, the bees were not fooled by the artificial bee, due to differences in odor [22].

Lourenço, Urbano and Teixeira combined music analysis and robot dance. They created two Lego robots, namely a cockatoo and a leg. The dance of the cockatoo consisted of head movements, whereas the leg was built for studying the Cancan dance [23]. Santiago, Oliveira, Reis and Sousa combined a music analyzer and a 16 degree-of-freedom humanoid to study the robot dance. The music analyzer was a beat tracker, and a previously built library was used for dance movements. The robot was equipped with sonar for detecting nearby objects, to avoid obstacles [24].

LaViers, Egerstedt, Chen and Belta studied ballet movements and tried to express them for robots. They worked on the so-called grammar or the leg positions in ballet [25, 26]. Özcimder, Kong and Baillieul [27], and Baillieul and Özcimder [28] worked on the beginner's and intermediate levels of Salsa dance and tried to model human movements to translate it for humanoids. They used knot theory, finite state machine modeling and information theoretic concepts in their work. Oliveira et al. worked on humanoid robot models to translate human dance movements into robot domain. They chose Samba and Charleston dances as the examples. They extracted the spatiotemporal characteristics of the human dance movements, using a motion capture system and worked on the topological model. They analyzed their results both numerically and subjectively [29].

Meng, Tholley and Chung worked on adaptation of dance in robots according to the feedback from humans [30].

Ros, Baroni and Demiris tested a robot as a dance tutor for children. Their robot was autonomous. The assesment of child-robot interaction was done [31]. Kumra and Şahin worked on teaching a robot how to dance. They used Q-Learning algorithm of Watkins [32] on a 7 degrees-of-freedom robotic platform for two popular songs. The robot received rewards in response to its dance movements. After the learning phase, the robot selected the movements using the stored values of rewards [33]. Granados, Kinugawa, Hirata and Kosuge built a 7 degrees-of-freedom dance teaching robot and studied human–robot interaction. They used a motion capture system to draw conclusions on the center of mass displacements during dance of humans. Their aim was to have robot movements inspired by human dance. They controlled the robot, taking interaction into account. Subjects tested the robot teacher. Assessments were done [34].

Chen et al. worked on the haptic interaction in partner dance in robotics. They prepared questionnaires for subject tests for the evaluation of the robot and studied the correlation between objective and subjective measures [35].

Peng, Hu, Chao, Zhou, and Li proposed a new method, based on genetic algorithm, for dance choreography in robots. They were concerned with the aesthetics of dance and incorporated learning in their algorithm. They simulated their method on a humanoid robot using a type of Chinese folk dance [36]. Manfré, Augello, Pilato, Vella and Infantino also applied genetic algorithm to teach robot how to dance and used Hidden Markov Model for dance creation. They incorporated a music module into their system for extracting features of the music to be used in dance [37].

LaViers et al. worked on the complexities of expressive robotic systems [38].

In the collaborative artwork called OUTPUT, Cuan made choreography for dance of an industrial robot. Her team added virtual reality and App aspects to the project [39].

2.3 *Petri Nets for Robotic Tasks*

Cao and Sanderson proposed Petri nets that would perform robot tasks by processing information from sensors with fuzzy logic. The networks they proposed use three different sets of fuzzy variables. They gave an example of a robot that folds garments [40]. Kim and Yang worked on the problem of robot solving a maze and finding fire with two algorithms, which they developed by combining the concept of fuzzy logic and Petri nets [41].

Lima, Grácio, Veiga and Karlsson used Petri nets to solve the problem of having a mobile robot follow a predefined path in a competition [42]. Milutinovic and Lima developed a Petri net model for the distributed robot system [43]. Costelha and Lima proposed a general solution to the problem of task planning for robots [44]. Ziparo, Iocchi, Nardi, Palamara, and Costelha generalized the Petri net plans, creating single and multi-robot Petri net plans. They also dealt with the solution of the synchronization problem [45]. Costelha and Lima modeled the environment and robots with stochastic Petri nets, enabling robots to play football as a team [46].

Chao and Thomaz modeled the sequential operation of robots and humans with timed Petri nets. In this study, they considered multimodal interactions, as well [47].

Yasuda worked on synchronization and coordination problems in multi-robot systems and included the interaction of machines and robots in the Petri net model [48].

Losch and Roßman proposed a Petri net model that can be used in robotic production lines, including modeling at different levels [49].

Davidrajuh proposed the use of Petri nets for modeling and control of humanoid robots [50]. Furlán, Rubio, Sossa, and Ponce combined the fuzzy logic and Petri net model to plan the trajectories of humanoid robots in a closed environment [51].

Sorokin and Senkov developed a new Petri net model for modeling robots operating at risk in mining [52].

3 Theoretical Background

In this section, we provide a summary of the theoretical background needed for synchronization, modeling and application of dance in multi-robot systems.

3.1 Synchronization and Dance Using Laban Notation

3.1.1 Laban Notation

Rudolf Laban invented a notation known as Laban notation, which is used to express movements during dance. His notation can be compared to the notation of music, in the sense that they both use a staff as the medium. His motivation was to create a universal language in signs to store the information of choreography. He also tried to incorporate temporal relationships and harmony into his notation.

Laban defined the kinesphere, a dynamic sphere surrounding the moving or standing body in space and changing its size and shape as the body moves. To represent the movement, he used shape symbols and associated with them directions and shading as in Fig. 1 [53–55].

Laban notation has been used in representing robot movement [56–59].

3.1.2 Synchronization and Dance

We use music and dance modules for analyzing music and designing dance. The music module is where audio is processed, beat information is extracted and the synchronization signal is created. On the other hand, the dance module is where the

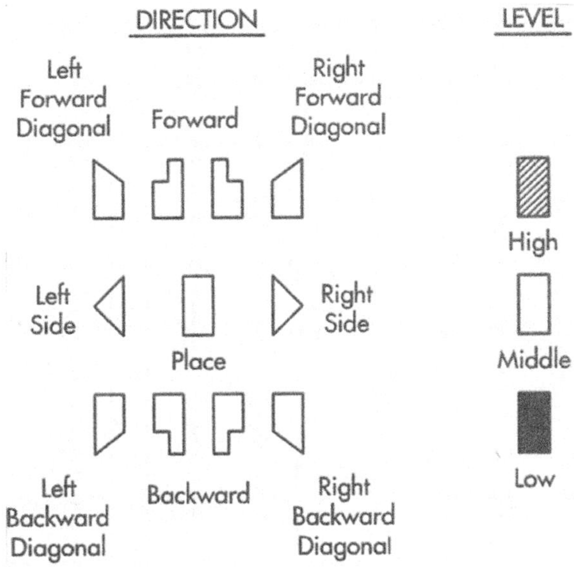

Fig. 1 Laban symbols to represent the shape and the direction of movements

choreography is designed, and the required robot trajectory is calculated. The results of each module are fed into the robot program, as in Fig. 2. The flowchart of the robot program is provided in Fig. 3. Synchronization with the music and other robots are also included.

Fig. 2 Basic building blocks of the system

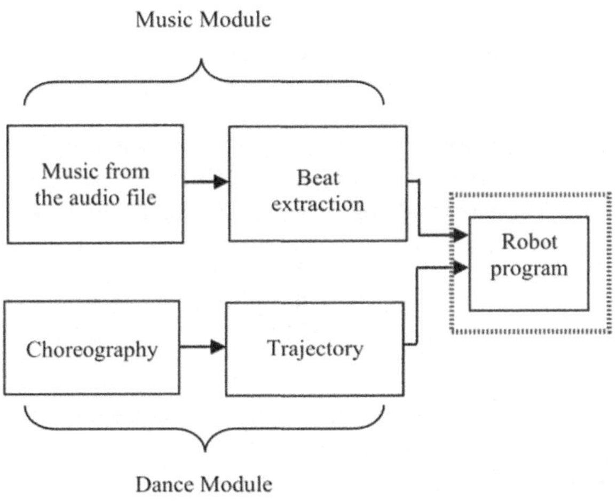

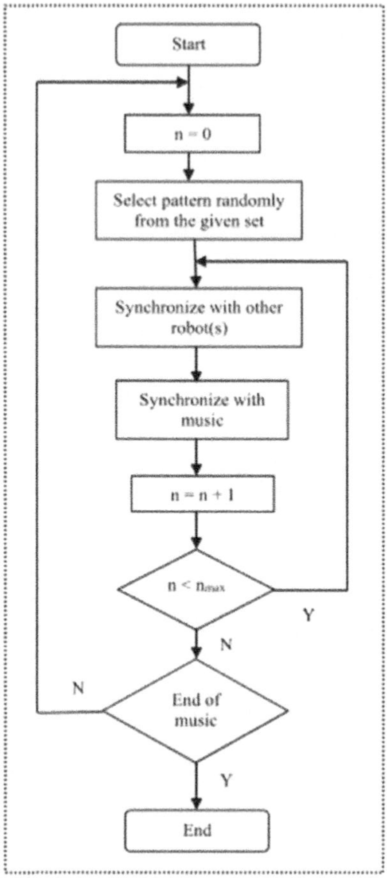

Fig. 3 Flowchart of the synchronization of multi-robotic system

We used Ellis's beat tracking algorithm in the music module. The algorithm estimates the tempo of the music and calculates the beat times, using dynamic programming. The estimation process requires a cost function to optimize, and it is defined as:

$$J(t_i) = \sum_{i=1}^{M} F_{onset}(t_i) + \beta \sum_{i=2}^{M} C\left(t_i - t_{i-1}, \tau_p\right) \quad (1)$$

where, $J(t_i)$ is used to define the cost function at the ith beat instant, t_i. $F_{onset}(t_i)$ denotes the onset strength envelope at the ith beat instant, and is calculated using Short-Time Fourier Transform and filters. $C(t_i - t_{(i-1)}, \tau_p)$ is the consistency function, which is used for comparison of the ith inter-beat interval, $t_i - t_{(i-1)}$, and the ideal beat spacing, τ_p, which is determined by target tempo. M is the number of beats, and β is a parameter to regulate the cost function. Ellis employs a consistency function, which is calculated by the negative of the square of the logarithm of inter-beat intervals divided by the ideal time spacing:

$$C(t_i - t_{i-1}, \tau) = -\left(log\left(\frac{t_i - t_{i-1}}{\tau}\right)\right)^2 \tag{2}$$

The optimum cost is calculated by:

$$J^*(t) = F_{onset}(t) + \max_{\tau=0,\ldots,t}\left\{\beta C\left(t - \tau, \tau_p\right) + J^*(\tau)\right\} \tag{3}$$

Upon optimization of the cost function, the estimation of the beat is made. This provides a sequence of beat times that correspond to the onsets in the audio signal, and composes a regular, rhythmic pattern, at the same time [60].

In the dance module, we used a modified version of the standart Laban notation. We added level and extension attributes to the classical Laban notation, to make sure that the robot movements can be represented through symbols readily [57, 61]. We also added staccato, glissando and symmetry options.

In our multi-robot dance design environment, it is possible for a choreographer to use mirroring, levels in space and division of space. In choreography, smoothness can be added using glissando, and pause using staccato. The speed of the movements is automatically determined and it is in accordance with the tempo of the music.

The contrast of high and low movements and sequencing of a composition enables the body to move in a harmonic way through space [62].

3.2 *Modeling of Robot Movement Using Petri Nets*

A Petri net is a mathematical and graphical modeling technique developed for modeling dynamic discrete-event systems and performance appraisals. From the 1960s, Petri nets have been used in the design, analysis, and control of the synchronous, asynchronous, stochastic systems [63–65].

Petri nets can be classified into two: a graphical tool and a mathematical tool. As a graphical tool, they enable the creation of a visual model of the system. As a mathematical tool, Petri nets help develop algebraic relationships of the state equations that reflect the system's behavior.

A Petri net is composed of places, transitions, and connections of places. Places are represented by circles whereas the transitions by bars or boxes as in Fig. 4.

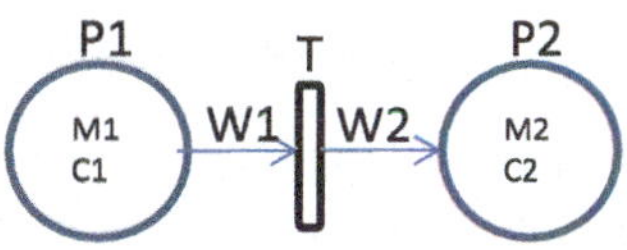

Fig. 4 Example of a Petri net consisting of two places and one transition

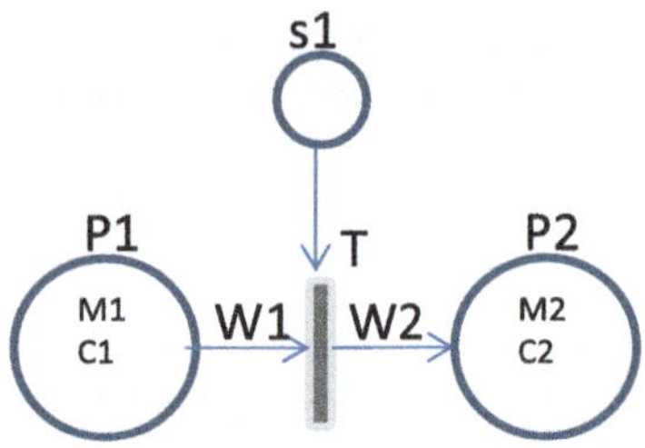

Fig. 5 Example of a synchronous Petri net

Connections and their directions are signified by arcs and arrows. Marking is represented by M and capacity by C. Accordingly, a Petri net can be defined by the 5-tuple: PN = (P, T, A, W, M_0):

where, P = {p_1, p_2, …, p_n}: places

T = {t_1, t_2, …, t_m}: transitions

$A \subseteq (P \times T) \cup (T \times P)$: connections

W: A → {1, 2, …}: weights

M_0: P → {0, 1, 2, …}: initial marking.

Marking is an important concept for Petri nets. Place capacity refers to the maximum number of markers that can be contained in the place (P). Markings (tokens) can be transferred from one place to another depending on the transition rules (T). The working rule of a Petri net is that when the transition is fired, the tokens from all connected places are multiplied by the connection weights (W) and transferred according to the capacity of the connected position.

If synchronization related transitions are added to this structure (Fig. 5) it will become applicable to more complex, namely multi-systems [66]. If we represent synchronized transitions by a dark rectangle and the synchronization signals by a small circle, firing according to the rule in synchronization transition becomes possible.

When modeling with Petri nets, actions can be represented by positions and the number of repetitions (Fig. 6), if any, can be represented by small gray circles. In general, transitions (T) consist of logical conditions and counter control.

Using sub-models to simplify the modeling is advantageous and convenient. We can use sub-models also in Petri networks and represent these models by a shaded circle and assigned name thereof. The Petri net in Fig. 7 consisting of two positions and a transition is assigned to SPETRI sub-model.

3.2.1 MIDI Music Format

MIDI (Musical Instrument Digital Interface) is a common communication protocol enabling real-time data exchange between electronic musical instruments and computers that has become an industrial standard. The fret numbers of each note

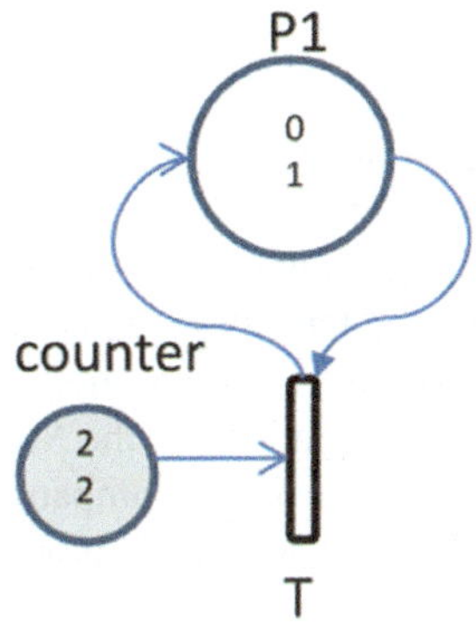

Fig. 6 Representation of repetitions by a Petri net

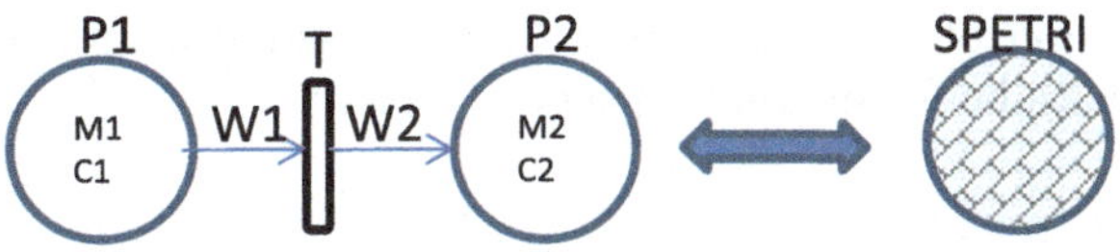

Fig. 7 Representation of a sub-model

in a MIDI file and transitions between these notes constitute the melodic line. The first data byte of the Note-on message in MIDI channel messages gives the MIDI fret numbers. These values ranging between 0 and 127 may show a sequential increase or decrease depending on the ascending or descending movement in the chromatic sequence. When the C4 fret which is given by 60 frets value is sharpened with chromatic steps the C# fret is created and the MIDI fret value becomes 61. The movement between the frets in a song toward high and low frets, therefore, the movement between MIDI fret values, is enough to create the song's melodic line and the data base to be compiled from MIDI fret numbers [67]. Using the MIDI file, music-dance synchronization and choreography can be created in the robotic system. By means of the repetitive structure of music and properties utilized in its orchestration, in this work, dance choreography has been designed by means of a Petri net.

4 Experimental Results

In this section we describe the details of our experimental work.

4.1 Robot System

The system we use in our experiments is shown in Fig. 8. The industrial robots are of 6 degrees-of freedom Mitsubishi RV-7L type. The robots are equipped with their respective CR-750D control units, and the whole system is controlled by a computer The robots are identical systems, each having a mass of 67 kg and a load capacity of 7 kg. The control units have servo drivers and motion controllers. We calculated the forward and the inverse kinematic equations of the Mitsubishi RV-7L robots, using the parameters supplied by the manufacturing company [68]. We used the kinematic equations in programming of the movements. Parameters of the robot are specified in Tables 1, 2 and 3.

We solved the forward and inverse kinematic problems of the type of the robot that we use in our experiments, using Brandstötter, Angerer, and Hofbaur's algorithm [69].

Upon calculation of the homogeneous transformation matrix to describe the movement of the end effector with respect to the base:

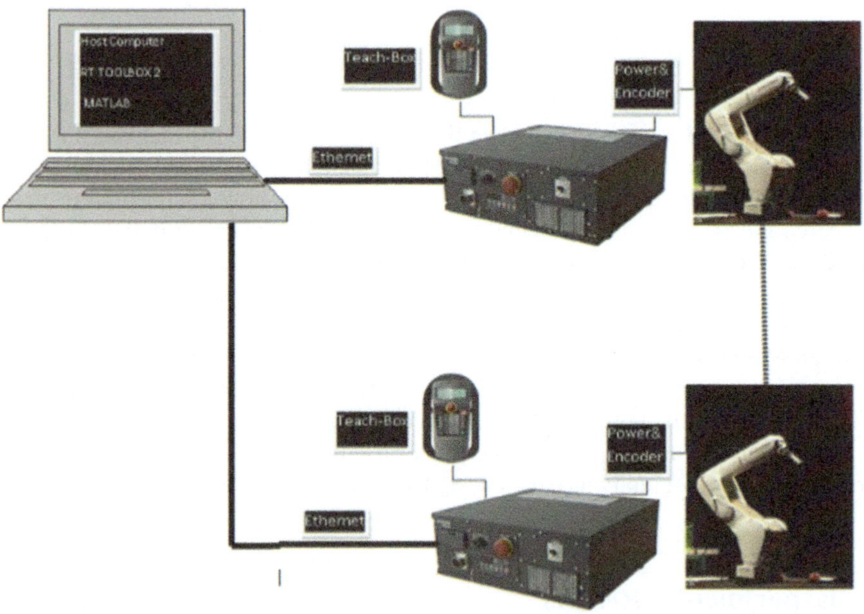

Fig. 8 The multi-robot system that we used in our experiments

Table 1 Robot parameters and explanations

Name	Representation	Rotation axis
Waist	J_1	z
Shoulder	J_2	y
Elbow	J_3	y
Wrist roll	J_4	z
Wrist pitch	J_5	y
Wrist yaw	J_6	z

Table 2 Physical parameters of the industrial robot

Type		M RV-7L	Type		M RV-7L
Degrees of freedom		6	Max. load capacity		7 kg
Max. reach radius		908 mm	Mass		67 kg
Operating range (deg)	J_1	±240	Max. speed (deg/sec)	J_1	288
	J_2	−110 +130		J_2	321
	J_3	−0 +162		J_3	360
	J_4	+200 −200		J_4	337
	J_5	−120 +120		J_5	450
	J_6	−360 +360		J_6	10,977

Table 3 Structural parameters of the industrial robot

Joint no	Joint angles	Link Length (mm)	Offsets (mm)
1	θ_1	$c_1 = 400$	$a_1 = 0$
2	θ_2	$c_2 = 435$	$a_2 = -50$
3	θ_3	$c_3 = 470$	$b = 0$
4	θ_4	$c_4 = 85$	
5	θ_5	0	
6	θ_6	0	

$$\boldsymbol{T}_0^6 = \boldsymbol{T}_0^1\boldsymbol{T}_1^2\boldsymbol{T}_2^3\boldsymbol{T}_3^4\boldsymbol{T}_4^5\boldsymbol{T}_5^6 = \begin{bmatrix} & \boldsymbol{R}(\boldsymbol{rpy}) & & x \\ & & & y \\ & & & z \\ 0 & 0 & 0 & 1 \end{bmatrix}$$

The position of the end effector is given by the following vector:

$$p = \left[x\ y\ z\ \theta_x\ \theta_y\ \theta_z\right]^T$$

where

$$
\begin{aligned}
\boldsymbol{T}_0^6(1,1) &= s_1(c_4s_6 + s_4c_5c_6) - c_1(c_{23}(s_4s_6 - c_4c_5c_6) + s_{23}s_5c_6)\\
\boldsymbol{T}_0^6(1,2) &= s_1(c_4c_6 + s_4c_5s_6) - c_1(c_{23}(s_4c_6 + c_4c_5s_6) + s_{23}s_5s_6)\\
\boldsymbol{T}_0^6(1,3) &= c_1(s_{23}c_5 + c_{23}c_4c_5) + s_1s_4s_6\\
\boldsymbol{T}_0^6(1,4) &= L_4(c_1s_{23}c_5 + c_1c_{23}c_4s_5 - s_1s_4s_5) + L_3c_1s_{23} - L_2c_1s_2 - a_2c_1c_{23}\\
\boldsymbol{T}_0^6(2,1) &= c_1(c_4s_6 + s_4c_5c_6) - s_1(c_{23}(s_4s_6 - c_4c_5c_6) + s_{23}s_5c_6)\\
\boldsymbol{T}_0^6(2,2) &= c_1(c_4c_6 + s_4c_5s_6) - s_1(c_{23}(s_4c_6 + c_4c_5s_6) + s_{23}s_5s_6)\\
\boldsymbol{T}_0^6(2,3) &= s_1(s_{23}c_5 + c_{23}c_4c_5) + c_1s_4s_5\\
\boldsymbol{T}_0^6(2,4) &= L_4(s_1s_{23}c_5 + s_1c_{23}c_4s_5 + c_1s_4s_5) + L_3c_1s_{23} + L_2s_1s_2 + a_2s_1c_{23}\\
\boldsymbol{T}_0^6(3,1) &= s_{23}(s_4s_6 - c_4c_5c_6) - c_{23}s_5c_6\\
\boldsymbol{T}_0^6(3,2) &= s_{23}(s_4c_6 + c_4c_5s_6) + c_{23}s_5s_6\\
\boldsymbol{T}_0^6(3,3) &= (c_{23}c_5 + s_{23}c_4s_5)\\
\boldsymbol{T}_0^6(3,4) &= L_1 + L_2c_2 + L_4(c_{23}s_5 - s_{23}c_4s_5) + L_3c_{23} + a_2s_1c_{23}
\end{aligned}
$$

Wrist position can be calculated as:

$$
\begin{bmatrix} x_w & y_w & z_w \end{bmatrix}^T = \begin{bmatrix} x & y & z \end{bmatrix}^T - L_4R_6^0\begin{bmatrix} 0 & 0 & 1 \end{bmatrix}^T
$$

$$
\begin{aligned}
\theta_1 &= a\tan 2(y_w, x_w)\\
\theta_2 &= a\cos\left(\frac{M^2 + L_2^2 - K^2}{2ML_2}\right) + a\tan 2(N, z_w - L_1)\\
\theta_3 &= a\cos\left(\frac{M^2 + L_2^2 - K^2}{2KL_2}\right) + +a\tan 2(a_2, L_3)\\
\theta_4 &= a\tan 2\big(\boldsymbol{T}_0^6(2,3)c_1 - \boldsymbol{T}_0^6(1,3)s_1, \boldsymbol{T}_0^6(1,3)c_{23} + \boldsymbol{T}_0^6(2,3)c_{23}s_1 - \boldsymbol{T}_0^6(3,3)s_{23}\big)\\
\theta_5 &= a\tan 2\left(\sqrt{1 - H^2}, H\right)\\
\theta_6 &= a\tan 2\big(\boldsymbol{T}_0^6(1,2)s_{23}c_1 - \boldsymbol{T}_0^6(2,2)s_1s_{23} + \boldsymbol{T}_0^6(3,2)c_{23,} - \boldsymbol{T}_0^6(1,1)s_{23}c_1\\
&\quad -\boldsymbol{T}_0^6(2,1)s_{23}s_1 - \boldsymbol{T}_0^6(3,1)c_{23}\big)
\end{aligned}
$$

where

$$N = \sqrt{x_w^2 + y_w^2}$$

$$M = \sqrt{N^2 + (z_w^2 - L_1)^2}$$

$$K = \sqrt{a_2^2 + L_3^2}$$

$$H = \left(T_0^6(1,3)s_{23}c_1 + T_0^6(2,3)s_{23}c_1 - T_0^6(3,3)s_{23}c_1\right)$$

s: sine, c: cosine.

4.2 Bolero

To implement robotic dance, Maurice Ravel's Bolero has been chosen as the music. Ravel dedicated this composition to his friend, ballerina Ida Rubinstein. Ravel described Bolero as 'a piece for orchestra without music'. This piece is made up of a single melody that repeats many times. In each repetition, new instruments are introduced to the music and the melody that has begun with a single instrument reaches a form of continuous crescendo as new instruments join [70].

The melody is inspired by fandango, a local melody in the Iberian Peninsula, where his mother was from. Ravel completed the composition upon a five-month work in November 1928. The melody that is repeated 18 times in the piece (stationary phase) creates a hypnosis effect on the audience with new instruments joining in each repetition and involves a tension phase in the final. In Ida Rubinstein's original choreography for this composition, when a young woman starts to dance in a Spanish café, it attracts other dancers' attention. As the participants increase in number, the group continuously integrates with the Bolero rhythm. At the final, the rhythm and the melody pick up a fast pace.

Bolero's musical infrastructure is influenced not only by Ravel's mother's Spanish origin but also by his father's profession, he was an engineer. Ravel's attraction to machinery and technology urged him to collect mechanical toys. Childhood experiences that shaped him is visible in the basic rhythm of Bolero, which repeats mechanically [71].

Bolero was composed for an orchestra made up of a piccolo (a small flute), two flutes, two oboes, an English horn, two clarinets, three saxophones, two bassoons, a contrabassoon, four horns, four trumpets, a tuba, three timpani, two side drums, a bass drum, a cymbal, a tam tam, a celesta, a harp and string instruments.

The melody, which is repeated twice, consists of 18 bars, the first one consists of diatonic notes and the second one consists of jazz-effect syncope (syncopated rhythm) and sub-tone notes. This melody is repeated 16 times and ends with the final section, i.e., the 17th repetition. In repetitions-periods (1) flute, (2) clarinet, (3) bassoon, (4) E flat clarinet, (5) oboe, (6) trumpet and flute, (7) tenor saxophone, (8) soprano saxophone, (9) horn, piccolo, celesta, (10) oboe, English horn and clarinet, (11) trombone, (12) wind instruments, (13) violin and wind instruments, (14–15) first and second violin and wind instruments, (16) many instruments in the orchestra, (17) all instruments in the orchestra join the melody two or three octaves higher and play with different tones.

The polytonality in the theme is visible from the four solo instruments playing three different tones in the ninth period. The C major melody is simultaneously

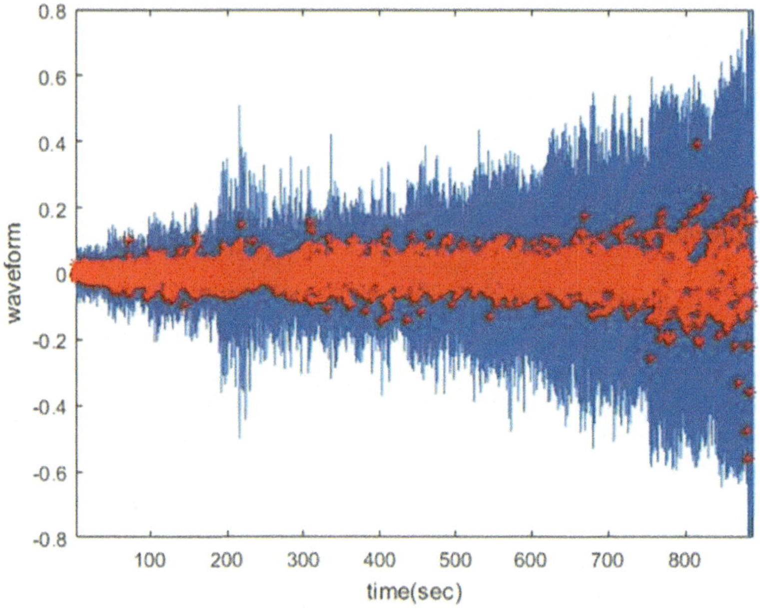

Fig. 9 Analysis of the sound waveform of Bolero

accompanied by E major and G major. A celesta (a percussion instrument resembling a small upright piano played by striking metal plates with hammers) plays the melody from the main clef. There are two piccolos accompanying the celesta, the first plays E major and the other G major. Accompanied by horn, they create this impressive blend. Wind instruments and violin together with tenor saxophone introduce variety and delicacy into the music in the 14th period [72].

The choreography of the most famous bale interpretation of Bolero was created by Maurice Béjart. In this oriental timbrical cyclical musical structure, dancers sitting on their chairs join the hypnotically repeating theme and the barefoot dancer's rhythmic movements on the round table in bunches of four, eight, twelve and sixteen and surround the table. These numbers can vary by demonstration.

In Fig. 9 the sound waveform of Bolero is shown.

4.3 *Implementation with Modified Laban Notation*

We used a modified version of Laban notation to create the dance choreography for industrial robots. We used five levels, adding extension which is represented by either vertical or horizontal bars in accordance with the dimension of the movement. We used color codes for duration and rhythm. A script written using our modified Laban notation is shorter than the one written using the original Laban notation.

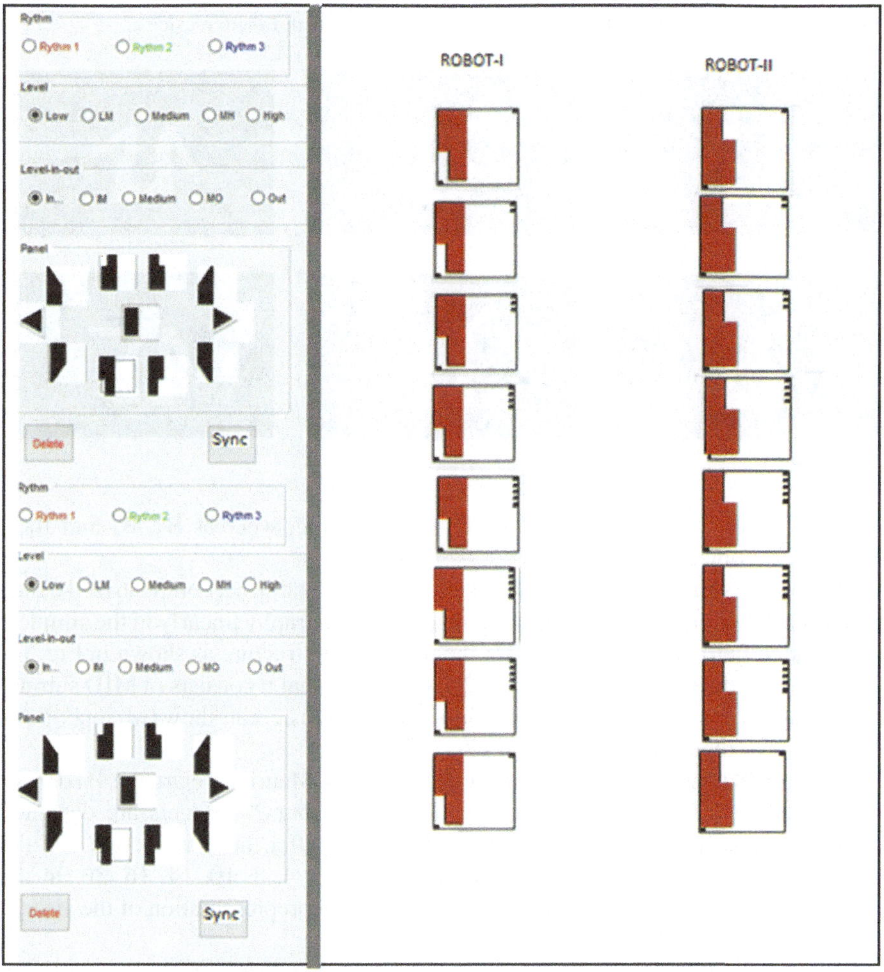

Fig. 10 User interface for dance choreography using our modified Laban notation

The user interface of the program for creating choreography using the modified Laban notation is shown in Fig. 10. The dance sequences for two robots are shown in Table 4.

4.4 Petri Net Model for Bolero

Petri nets can be used for music analysis [73, 74].

We modeled our system using the simplified rules of Petri nets. When we divide Bolero into musical objects, we can conclude that it consists of rhythmic and

Table 4 Dance sequences for two Mitsubishi RV-7FL robots with Bolero music

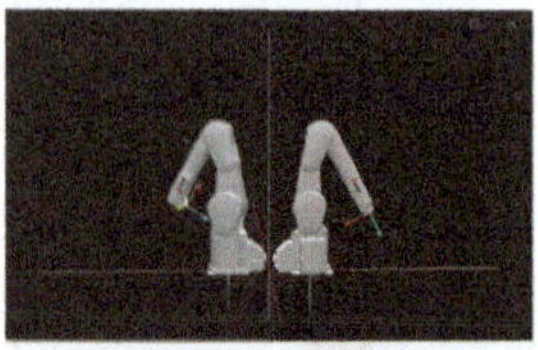	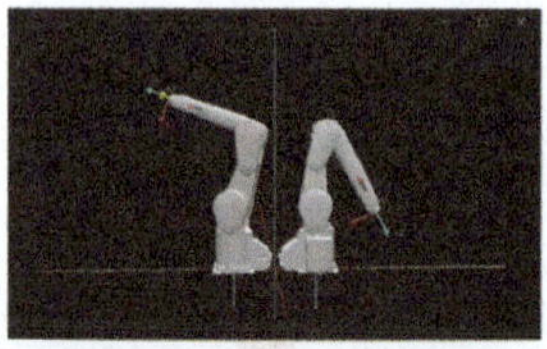	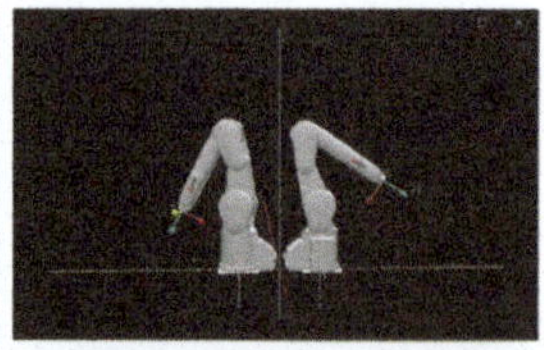
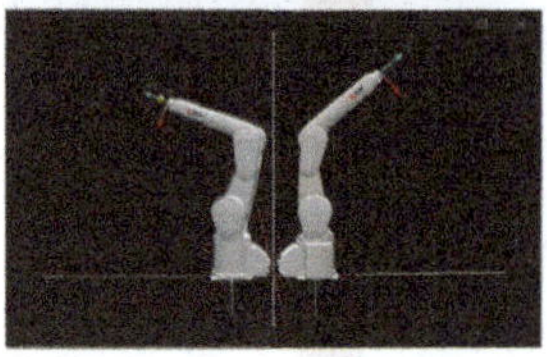	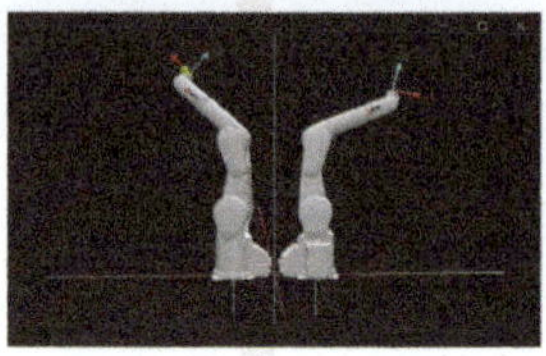	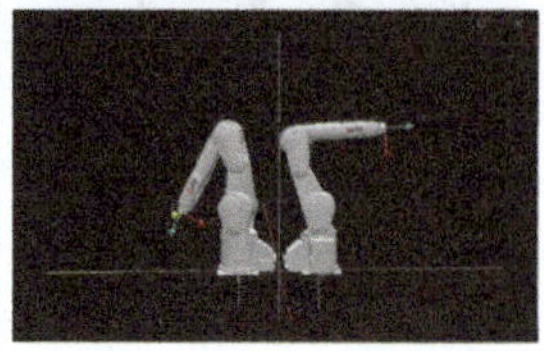

harmonic objects. In Table 5, B_1 and B_2 are rhythmic, whereas B_3, B_4 and B_{final} are harmonic objects.

The regular structure of this music piece allows to establish a computerized dance infrastructure. Modeling the principal dancer's choreography linearly in the simplest way through Petri Nets, we can obtain the three-layer structure as shown in Fig. 11.

Examining the mid-cycle net (Fig. 12), we can see that it consists of MID sub-net that repeats 4 times and MID, in accordance with the music, can show two repetitions of B_1 and two repetitions of B_2 (Fig. 13).

If we model the robot based on the choreography of Maurice Béjart, the 1st dancer is designed for all through the piece whereas the other 20 dancers are designed to participate at the beginning of each music object after the introduction and the 1st repetition. The number of dancers increases to 2, 4, 8, 10, 14, 16, 20. In the final sections, all dancers will participate. The Petri net representation of the system (Fig. 14) is shown below.

We proposed two different robot dance modeling approaches. The Laban notation model translates human choreography to robot movements. Petri net model utilizes structured musical properties. In the future, we plan to use fuzzy-Petri nets to integrate rule-based expert systems into the robotic dance system [75, 76].

Table 5 Bolero music objects

	B_1	B_1	B_2	B_2	B_1	B_3	B_4	B_{final}
Intro	4 repeats				Final			

Fig. 11 Simplified Petri net model for the principal dancer

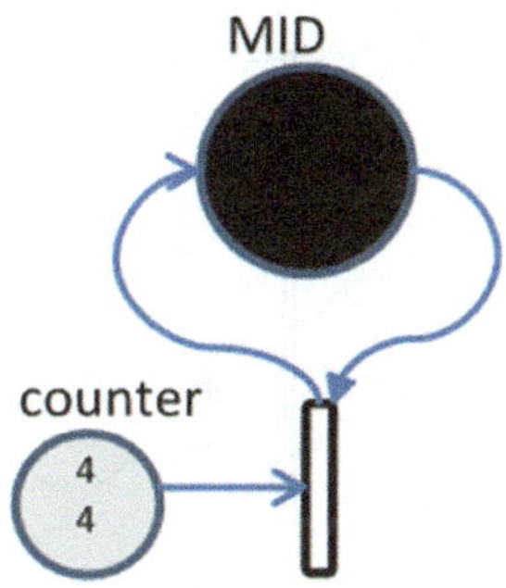

Fig. 12 Mid-cycle Petri net sub-model

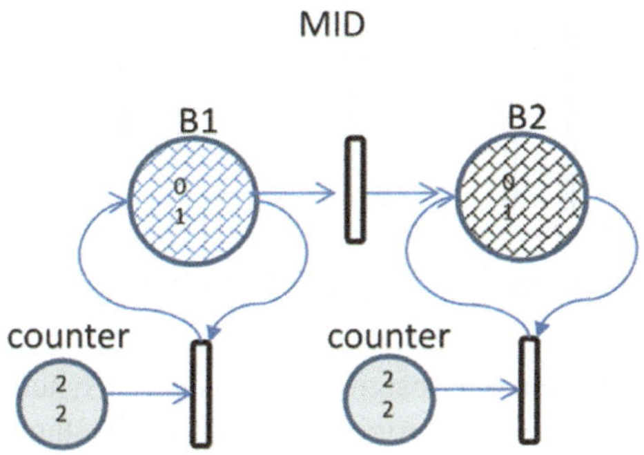

Fig. 13 Representation of Sub-Petri net mid

5 Conclusion

In this work, we presented two solutions to the modeling of robot dance. The first one was using a well-established dance notation, developed by Rudolf Laban and bearing his name. We incorporated the Laban notation and used an extended version of the classical one in our system of music analysis and dance choreography. In combination with a beat tracking algorithm, we applied our modified Laban notation to our multi-robot, synchronous dance system. Our first solution takes the audio file as the input, extracts the beats, and composes the required dance using a set of dance patterns. The synchronization of the movements is also included in the design and is determined by the inter-beat intervals. The system creates and updates synchronization according to the waiting time and the trajectory to be followed.

The second method employed Petri nets in robotic dance choreography. To ensure synchronization of dancers, a synchronization marking was added to the traditional Petri nets and synchronization of the dancers was modeled.

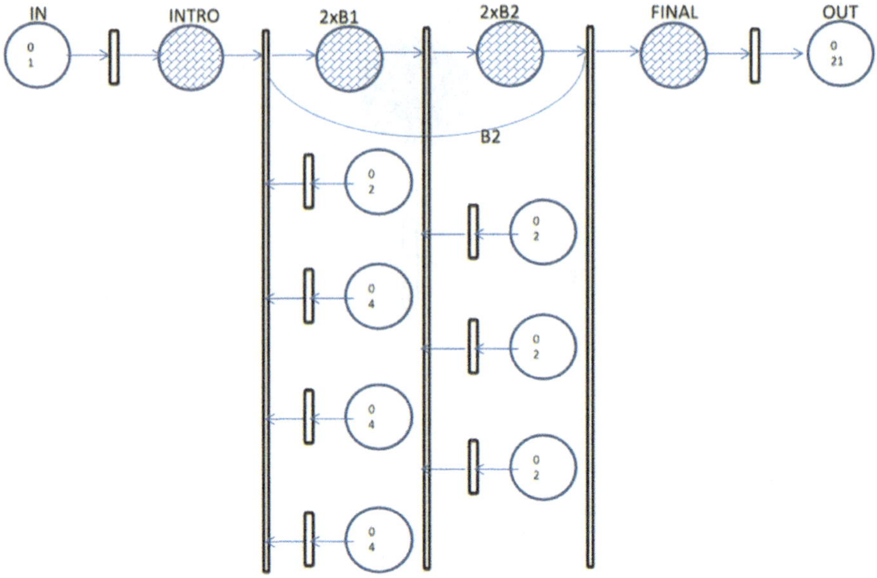

Fig. 14 The Petri net model of the dance choreography for Bolero

We used multiple robots in our designs and considered synchronization with both music and other robots. We used Maurice Ravel's Bolero as staged by Maurice Béjart, but our designs are not specific to this example and can be applied to any other music and choreography.

Acknowledgements This work was supported by Mitsubishi Electric Turkey.

References

1. Ramachandran, V.S., Hirstein, W.: The science of art: a neurological theory of aesthetic experience. J. Conscious. Stud. **6**(6–7), 15–51 (1999)
2. Hagendoorn, I.: Emergent patterns in dance improvisation and choreography. In: Minai, A.A., Bar-Yam, Y. (eds.) Unifying Themes in Complex Systems IV, pp. 183–195. Springer (2008)
3. Rodriguez-Ángeles, A.: Synchronization of mechanical systems, pp. 29–48. Ph. D. Dissertation, Technische Universiteit Eindhoven, Netherlands (2002)
4. de Portillo-Vélez, R., Cruz-Villar, C.A., Rodriguez-Ángeles, A.: On-line master/slave robot system synchronization with obstacle avoidance. Stud. Inform. Control. **21**(1), 17–26 (2012)
5. Matsunami, N., Tanaka-Ishii, K., Frank, I., Matsubara, H.: Lego mindstorms cheerleading robots. In: Nakatsu, R., Hoshino, J. (eds.) Entertainment Computing, pp. 199–206. Springer, US (2003)
6. Mahmood, A., Kim, Y.: Leader-following formation control of quadcopters with heading synchronization. Aerosp. Sci. Technol. **47**, 68–74 (2015)
7. Markus, E.D., Yskander, H., Agee, J.T., Jimoh, A.A.: Coordination control of robot manipulators using flat outputs. Robot. Auton. Syst. **83**, 169–176 (2016)

8. Pongas, D., Billard, A., Schaal, S.: Rapid synchronization and accurate phase-locking of rhytmic motor primitives. In: IEEE/RSJ International Conference on Intelligent Robots and Systems, Alberta, Canada, 2–6 August 2005
9. Mehrjerdi, H., Ghommam, J., Saad, M.: Nonlinear coordination control for a group of mobile robots using a virtual structure. In: Mechatronics, vol. 21, pp. 1147–1155. Springer (2011)
10. D'Ambrosio, D.B., Goodell, S., Lehman, J., Risi, S., Stanley, K.O.: Multirobot behaviour synchronization through direct neural network communication. In: Su, C.-Y., Rakheja, S., Liu, H. (eds.) ICIRA 2012, Part II, LNAI 7507, pp. 603–614. Springer, Berlin Heidelberg (2012)
11. Floreano, D., Mitri, S., Magnenat, S., Keller, L.: Evolutionary conditions for the emergence of communication in robots. Curr. Biol. **17**, 514–519 (2007)
12. Wang, W., Huang, J., Wen, C., Fan, H.: Distributed adaptive control for consensus tracking with application to formation control of nonholonomic mobile robots. Automatica **50**, 1254–1263 (2014)
13. Dou, H., Wang, S.: Robust adaptive motion/force control for motion synchronization of multiple uncertain two-link manipulators. Mech. Mach. Theory **67**, 77–93 (2013)
14. Dou, H., Wang, S.: A boundary control for motion synchronization of a two-manipulator system with a flexible beam. Automatica **50**, 3088–3099 (2014)
15. Steels, L.: Evolving grounded communication for robots. Trends Cogn. Sci. **7**(7), 308–312 (2003)
16. Kawai, Y., Park, J., Horii, T., Oshima, Y., Tanaka, K., Mori, H., Nagai, Y., Takuma, T., Asada, M.: Throwing skill optimization through synchronization and desynchronization of degree of freedom. In: Chen, X., Stone, P., Sucar, L.E., van der Zant, T. (eds.) RoboCup 2012: Robot Soccer World Cup XVI, pp. 178–189. Springer, Berlin Heidelberg (2013)
17. Ahmadzadeh, H., Masehian, E.: Modular robotic systems: methods and algorithms for abstraction, planning, control and synchronization. Artif. Intell. **223**, 27–64 (2015)
18. Iqbal, T., Riek, L.D.: Human coordination dynamics with heterogeneous robots in a team. In: ACM/IEEE International Conference on Human-Robot Interaction, Christchurch, New Zealand, pp. 619–620, 7–10 March 2016
19. Apostolos, M.K., Littman, M., Lane, S., Handelman, D., Gelfano, J.: Robot choreography: an artistic-scientific connection. Comput. Math. Appl. **32**(1), 1–4 (1996)
20. Kuroki, Y., Fujita, M., Ishida, T., Nagasaka, K., Yamaguchi, J.: A small biped entertainment robot exploring attarctive applications. In: Proceedings of the 2003 IEEE International Conference on Robotics & Automation, Taipei, Taiwan, pp. 471–476, September 14–19, 2003
21. Nakaoka, S., Nakazawa, A., Kanehiro, F., Kaneko, K., Morisawa, M., Hirukawa, H., Ikeuchi, K.: Learning from observation paradigm: leg task models for enabling a biped humanoid robot to imitate human dances. Int. J. Robot. Res. **26**(8), 829–844 (2007)
22. Landgraf, T., Oertel, M., Rhiel, D., Rojas, R.: A biomimetic honeybee robot for the analysis of the honeybee dance communication system. In: IEEE/RSJ International Conference on Intelligent Robots and Systems, Taipei, Taiwan, pp. 3097–3102, October 18–22, 2010
23. Lourenço, M., Urbano, P., Teixeira, C.: The first steps of robotic cancan. In: 5th Iberian Conference on Information Systems and Technologies, pp. 1–4, 16–19 June 2010
24. Santiago, C.B., Oliveira, J.L., Reis, L.P., Sousa, A.: Autonomous robot dancing synchronized to musical rhytmic stimuli. In: 6th Iberian Conference on Information Systems and Technologies, pp. 1–6, 15–18 June 2011
25. LaViers, A., Egerstedt, M., Chen, Y., Belta, C.: Automatic generation of balletic motions. In: IEEE/ACM Second International Conference on Cyber-Physical Systems, Chicago, Illinois, USA, pp. 13–21, 12–14 April 2011
26. LaViers, A., Chen, Y., Belta, C., Egerstedt, M.: Automatic sequencing of ballet poses. In: IEEE Robotics & Automation Magazine, pp. 87–95, September 2011
27. Özcimder, K., Kong, Z., Baillieul, J.: Algorithmic approaches to artistic movement. In: IEEE Conference on Decision and Control, Los Angeles, California, USA, pp. 5373–5380, December 15–17, 2014
28. Baillieul, J., Özcimder, K.: Dancing robots: The control theory of communication through movement. In: LaViers, A., Egerstedt, M. (eds.) Controls and Art, pp. 51–72. Springer, Switzerland (2014)

29. Oliveira, J.L., Naveda, L., Gouyon, F., Reis, L.P., Sousa, P., Leman, M.: A parameterizable spatiotemporal representation of popular dance styles for humanoid dancing characters. In: EURASIP Journal on Audio, Speech, and Music Processing, vol. 18 (2012)
30. Meng, Q., Tholley, I., Chung, P.W.H.: Robots learn to dance through interaction with humans. Neural Comput. Appl. **24**, 117–124 (2014)
31. Ros, R., Baroni, I., Demiris, Y.: Adaptive human-robot interaction in sensorimotor task instruction: from human to robot dance tutors. Robot. Auton. Syst. **62**, 707–720 (2014)
32. Watkins, C., Dayan, P.: Q-learning. Mach. Learn. **8**, 279–292 (1992)
33. Kumra, S., Şahin, F.: Dual flexible 7 dof arm robot learns like a child to dance using q-learning. In: System of Systems Engineering Conference, San Antonio, Texas, USA, pp. 292–297, 17–20 May 2015
34. Granados, D.F.P., Kinugawa, J., Hirata, Y., Kosuge, K.: Guiding human motions in physical human-robot interaction through COM motion control of a dance teaching robot. IEEE-RAS International Conference on Humanoid Robots (Humanoids), Cancun, Mexico, pp. 279–285, November 15–17, 2016
35. Chen, T.L., Bhattacharjee, T., McKay, J.L., Borinski, J.E., Hackney, M.E., Hing, L.T., Kemp, C.C.: Evaluation by expert dancers of a robot that performs partnered stepping via haptic interaction. PLoS One **10**(5), (2015)
36. Peng, H., Hu, H., Chao, F., Zhou, C., Li, J.: Autonomous robotic choreography creation via semi-interactive evolutionary computation. Int. J. Soc. Robot. **8**, 649–661 (2016)
37. Manfré, A., Augello, A., Pilato, G., Vella, F., Infantino, I.: Exploiting interactive genetic algorithms for creative humanoid dancing. Biol. Inspired Cogn. Arch. **17**, 12–21 (2016)
38. LaViers, A., Cuan, C., Maguire, C., Bradley, K., Brooks Mata, K., Nilles, A., Vidrin, I., Chakraborty, N., Heimerdinger, M., Huzaifa, U., McNish, R., Pakrasi, I., Zurawski, A.: Choreographic and somatic approaches for the development of expressive robotic systems. Arts **7**(2), 11 (2018)
39. Cuan, C.: OUTPUT: "Choreographed and reconfigured human and industrial robot bodies across artistic modalities." Front. Robot. AI **7**, 1–13 (2021)
40. Cao, T., Sanderson, A.C.: Modeling of sensor-based robotic task plans using fuzzy Petri nets. In: Proceedings of the Fourth International Conference on Computer Integrated Manufacturing and Automation Technology, pp. 73–80 (1994)
41. Kim, S.-Y., Yang, Y.: A self-navigating robot using Fuzzy Petri nets. Robot. Auton. Syst. **101**, 53–165 (2018)
42. Lima, P., Gracio, H., Veiga, V., Karlsson, A.: Petri nets for modeling and coordination of robotic tasks. In: SMC'98 Conference Proceedings, IEEE International Conference on Systems, Man, and Cybernetics, vol. 1, pp. 190–195 (1998)
43. Milutinovic, D., Lima, P.: Petri net models of robotic tasks. In: Proceedings 2002 IEEE International Conference on Robotics and Automation 2002, vol. 4, pp. 4059–4064 (2002)
44. Costelha, H., Lima, P.: Modelling, analysis and execution of robotic tasks using Petri nets. In: 2007 IEEE/RSJ International Conference on Intelligent Robots and Systems, pp. 1449–1454 (2007)
45. Ziparo, V.A., Iocchi, L., Nardi, D., Palamara, P.F., Costelha, H.: Petri net plans: a formal model for representation and execution of multi-robot plans. In: Proceedings of the 7th International Joint Conference on Autonomous Agents and Multiagent Systems, vol. 1, pp. 79–86 (2008)
46. Costelha, H., Lima, P.: Robot task plan representation by Petri nets: modelling, identification, analysis and execution. Auton Robot **33**, 337–360 (2012)
47. Chao, C., Thomaz, A.L.: Timing in multimodal turn-taking interactions: control and analysis using timed Petri nets. J. Hum. Robot Interact. **1**, 4–25 (2012)
48. Yasuda, G.: Modeling and distributed implementation of synchronization and coordination in multi-robot systems. Procedia Eng. **41**, 1051–1057 (2012)
49. Losch, D., Roßmann, J.: Visual programming and development of manufacturing processes based on hierarchical Petri nets. In: 3rd International Conference on Soft Computing & Machine Intelligence (ISCMI), pp. 154–158 (2016)

50. Davidrajuh, R.: Petri net based modelling and control of humanoid robots. Int. J. Simul. Syst. Sci. Technol. **17**(32), 40.1–40.9 (2016)
51. Furlán, F., Rubio, E., Sossa, H., Ponce, V.: Humanoid Robot Using Petri Nets as Tool for Decision Making. Congreso Nacional de Control Automático, Monterrey, Nuevo León, Mexico (2017)
52. Sorokin, E.V., Senkov, A.V.: Application of growing nested Petri nets for modeling robotic systems operating under risk. In: IOP Conference Series: Earth and Environmental Science vol. 87, no. 8 (2017)
53. Moore, C.-L.: The Harmonic Structure of Movement, Music, and Dance According to Rudolf Laban: An Examination of His Unpublished Writings and Drawings. Edwin Mellen Press (2009)
54. Sutil, N.S.: Laban's choreosophical model: Movement visualisation analysis and the graphic media approach to dance studies. Dance Res. **30**, 147–168 (2012)
55. Sutil, N.S.: Rudolf Laban and topological movement. Space Cult. **16**, 173–193 (2013)
56. Barakova, E., Berkel, R.V., Hiah, L., The, Y., Werts, C.: Observation scheme for interaction with embodied intelligent agents based on Laban notation. In: 2015 IEEE International Conference on Robotics and Biomimetics (ROBIO), pp. 2525–2530 (2015)
57. Özen, F., Tükel, D.B., Dimirovski, G.: Synchronized dancing of an industrial manipulator with humans on arbitrary music. Acta Polytech. Hung. **14**(2), 151–169 (2017)
58. Özen, F., Küntan, U., Tükel, D.B.: Robot-music synchronization: Self-designed dance. In: IEEE EUROCON 2017, 17th International Conference on Smart Technologies, pp. 582–587 (2017)
59. Salaris, P., Abe, N., Laumond, J.: Robot choreography: The use of the Kinetography Laban system to notate robot action and motion. IEEE Robot. Autom. Mag. **24**, 30–40 (2017)
60. Ellis, D.: Beat tracking by dynamic programming. J. New Music Res. **36**(1), 51–60 (2007)
61. Özen, F., Tükel, D.B., Tural, K.: Cooperative dancing with an industrial manipulator: Computational cybernetics complexities. In: IEEE International Conference on Systems, Man, and Cybernetics, Budapest, Hungary, Oct 9–12, 2016, pp. 1957–1962 (2016)
62. Burton, S.J., Samadani, A.-A., Gorbet, R., Kulić, D.: Laban movement analysis and affective movement generation for robots and other near-living creatures. In: Laumond, J-P., Abe, N. (eds.) Dance Notations and Robot Motion, Springer Tracts in Advanced Robotics, vol. 111, pp. 25–48 (2016)
63. Petri, C.A.: Kommunikation mit Automata. English Translation, Ph.D. Thesis, Institute for Applied Mathematics, University of Bonn (1962)
64. Reisig, W.: Petri nets an introduction. In: Brauer, W., Rozenberg, G., Salomaa, A. (eds.) Monographs in Theoretical Computer Science. Springer, Berlin (1985)
65. Wang, J.: Petri nets for dynamic event-driven system modeling. In: Fishwick, P.A. (ed.) Handbook of Dynamic System Modeling, 1st edn. Chapman and Hall/CRC (2007)
66. Kouah, S., Saidouni, D.E., Ilie, J.M.: Synchronized Petri net: a formal specification model for multi agent systems. J. Softw. **8**(3), (2013)
67. Çelik, S.: Micro-Markov: a microtonal algorithmic composition application with Markov Analysis. J. Int. Soc. Res. **9**(43), 2565–2572 (2016)
68. Mitsubishi industrial robot CR750-D/CR751-D/CR760-D controller RV-4F-D/7F-D/13F-D/20F-D/35F-D/50F-D/70F-D series standard specifications manual, BFP-A8931-S, pp. 2–44
69. Brandstötter, M., Angerer, A., Hofbaur, M.: An analytical solution of the inverse kinematics problem of industrial serial manipulators with an ortho-parallel basis and a spherical wrist. In: Proceedings of the Austrian Robotics Workshop, Linz, Austria, 22–23 May, 2014, pp. 7–11 (2014)
70. Thierens, S.: A legitimate masterpiece: Béjart Ballet dances Bolero and Light. Dance Reviews, 20 March 2014 (2014)
71. Check, J.: Perfection of the life and the work the case of Maurice Ravel. Sewanee Rev. **124**(1), 68–78 (2016)
72. Haus, G., Rodriguez, A.: Formal music representation; a case study: the model of Ravel's Bolero by Petri nets. Laboratorio di Informatica Musicale, Milano

73. Barate, A., Haus, G., Ludovico, L.A.: Real-time music composition through P-timed Petri nets. In: Georgaki, A., Kouroupetroglou, G. (eds.), Proceedings ICMC|SMC|2014, pp. 408–415,14–20 September 2014. Athens, Greece (2014)
74. Haus, G., Sametti, A.: Scoresynth: a system for the synthesis of music scores based on Petri nets and a music algebra. Computer **24**(7), 56–60 (1991)
75. Dimirovski, G.M.: Fuzzy-Petri-net reasoning supervisory controller and estimating states of Markov chain models. Comput. Inform. **24**(6), 563–576 (2005)
76. Dimirovski, G.M., Jing, Y.-W., Zhang, S.-Y.: Hybrid leader-follower and fuzzy-Petri-net traffic rate control and supervision in network systems. J. Autom. Control. **XI**(2), 1–24 (2001)

A Review of Fuzzy Metaheuristics for Optimal Design of Fuzzy Controllers in Mobile Robotics

Oscar Castillo and Patricia Melin

Abstract In this article, a review of the existing publications using fuzzy metaheuristics for optimal design of fuzzy controllers in mobile robotics is presented. Metaheuristics is an area that deals with a wide range of bio-inspired and nature-inspired optimization algorithms than can be used for diverse application areas. Fuzzy control can be thought of as a way for achieving control of dynamic systems by using fuzzy sets and fuzzy systems that are based on expert knowledge instead of traditional mathematical models. In this regard, it is natural to think that the metaheuristics area, which include techniques such as genetic algorithms and bio or nature inspired optimization techniques, will have a great impact in achieving the goals of efficiently controlling mobile robots. Actually, this review paper reveals that there have been many works in this area, and we will provide the up to date relevant statistics and analysis of the existing works. In addition, we will outline future possible trends for research on applying fuzzy metaheuristics to problems of designing fuzzy controllers in mobile robotics.

Keywords Fuzzy logic · Metaheuristics · Fuzzy control · Mobile robotics

1 Introduction

Nowadays metaheuristics, fuzzy logic and robotics have been gaining attention from industry, academia and government [1–4]. Metaheuristics can be viewed as a way for optimizing solutions to complex problems [5–9]. Fuzzy logic and fuzzy systems provide powerful methods to represent expert knowledge and solve complex problems of intelligent control, medical diagnosis, prediction and robotics [10–12]. For this reason, it is important to consider these areas in all aspects of human life [13–16]. In addition, the utilization of intelligent techniques has become also relevant for real

O. Castillo (✉) · P. Melin
Tijuana Institute of Technology, 22379 Tijuana, BC, Mexico
e-mail: ocastillo@tectijuana.mx

P. Shi et al. (eds.), *Complex Systems: Spanning Control and Computational Cybernetics: Applications*, Studies in Systems, Decision and Control 415,
https://doi.org/10.1007/978-3-031-00978-5_3

life situations, going from home appliances to autonomous robots and medical diagnosis. In this way, we are becoming aware of the interaction of both metaheuristics and fuzzy control for mobile robotics, which can become very important in the near future [14–20].

Metaheuristics and fuzzy logic are methods that belong to the Soft Computing area. Soft Computing is a recently proposed new area of Computer Science, which deals with designing and implementing systems with novel intelligent models, like fuzzy systems, neural networks, evolutionary algorithms and swarm intelligence [21–23]. Fuzzy logic was initially postulated by Zadeh and elevates bivalued logic by extending truth values to be from 0 to 1. The most important aim of fuzzy logic is to handle the inherent uncertainty in the real-world by utilizing fuzzy sets. Neural networks (NNs) are computational models that emulate the human cognitive activities by approximating real brain networks with mathematical models and by utilizing iterative training algorithms. It is now known that NNs can learn from data to build neural systems that may be utilized in diverse real-world situations going from pattern recognition to time series forecasting. Evolutionary algorithms are search techniques that emulate real natural evolution for achieving an efficient solution search to optimization problems. Evolutionary algorithms have also found numerous real applications, like in network optimization, planning, robotics, manufacturing and others. Swarm intelligence is an area formed by techniques based on collective intelligence, like particle swarm optimization, ant-based optimization, firefly algorithm, whale optimization, water cycle algorithm, fish and shark optimization, just to mention a few. These techniques have been utilized in general purpose optimization problems, as well as for optimizing fuzzy and neural models [24–28]. In addition, the appropriate mixing of the above-mentioned techniques may be utilized for constructing high performance hybrid systems, which have potential of solving real-world problems in different areas of applications, such as in control, pattern recognition, robotics, diagnosis, and others [29–31].

The contribution of this article is the review of the current works in the state of the art, where metaheuristics, fuzzy control and mobile robotics have been used jointly (in a combined fashion) in real-world problems, so that the improvements achieved by the mixing of these areas can be verified and possible future works can be postulated. In addition, regarding the scientometric analysis of the state of the art, a second contribution of the article is postulating possible future trends for research works based on mixing the concepts of metaheuristics, fuzzy control and mobile robotics.

The remaining parts of the article are organized in the following fashion. Section 2 offers an overview of existing works in the state of the art on metaheuristics, fuzzy control and mobile robotics. Section 3 presents a scientometric analysis of the existing papers in metaheuristics, fuzzy control and mobile robotics. Section 4 poses possible future research trends on the intersection of the mentioned areas. Finally, Sect. 5 postulates the conclusions and suggests future research avenues.

2 Overview of Existing Works

In this section we describe a sample of the papers that have been published in metaheuristics, fuzzy logic and mobile robotics. This sample of six papers offers a general overview of the work that has been already undertaken in these areas.

In Raiesdana et al. [32], the authors present a hybrid robust method for navigating an industrial robot in a dynamic environment that considers changing obstacles. The objectives of this work were to find the path that has the shortest distance, the minimization of energy consumption of the robot, minimizing the smoothness of the generated paths and solving the issue of dynamic obstacles.

In Tao et al. [33] a proposal for a method of global path planning is presented, which is based on a variant of the ant colony optimization to enhance speed in mobile service robot while traversing a planned path. The distribution of the initial pheromone is defined by the critical obstacle influence factor. The influence factor is put forward into the heuristic information to improve the convergence speed of the algorithm.

In Precup et al. [34] the authors propose two applications of the Grey Wolf Optimizer (GWO). First, GWO was applied to a path planning (PaPl) problem and then it was utilized for a Proportional-Integral (PI)-fuzzy controller tuning problem. Both optimization problems were solved by using the GWO algorithm and are described in detail in this article. In this work and off-line GWO-based PaPl approach for Nonholonomic Wheeled Mobile Robots in static environments is presented and very good results are obtained.

In Xu et al. [35] the authors are presenting a networked multirobot cyber-physical system that utilizes an artificial immune fuzzy optimization method to achieve distributed formation control of embedded mobile robots. The artificial immune system (AIS) algorithm is enhanced with a fuzzy system to propose a hybrid optimization approach, called AIS-fuzzy that works very well for the problem.

In Kondratenko et al. [36] is presented a hybrid multiagent method of parametric optimization of fuzzy control systems that combines the advantages of particle swarm algorithms and local search algorithms based on the elite strategy. The obtained method allows to optimize effectively various parameters of fuzzy systems, finding the global optimum of the problem to be solved, and, at the same time, has the higher convergence rate when compared to the basic particle swarm optimization method.

In Lin et al. [37] an efficient control method for navigation of mobile robots in unknown environments was presented. The proposed behavior manager switches between two behavioral controllers, wall-following behavior and toward-goal behavior, which are defined on the basis of the relationship between the mobile robot and the particular unknown environment.

The previous six papers provide an idea of the types of research works that have developed in the optimization of fuzzy controllers for mobile robots based on metaheuristics algorithms.

3 Scientometric Analysis

In this section, the Scientometric analysis for the papers in the areas of metaheuristics, fuzzy logic and mobile robots, is presented. We are presenting a sequence of plots and tables that illustrate the analysis of the data from different points of view or facets used in the analysis. We can state that this analysis aids in comprehending the nature of publications and research that has been undertaken in the mentioned areas. First, we start with Fig. 1 showing a plot of documents per year since 1991 (there are a total of 155 journal papers according to Scopus). Another important measure of the impact of the publications in these areas is the h index, which can be defined as the highest value that represents the number of times a particular paper has been cited. In this case, in Fig. 2 we show the plot for the h index, which has a value of 32 for the mentioned areas. The value of 32 means that there are 32 papers with a least 32 references or more.

Table 1 illustrates the list of the top ten most cited papers in the mentioned areas. We describe in Fig. 3 the area distribution (in percentage) of the papers that have been published in metaheuristics, fuzzy logic and mobile robots. The area with the higher percentage is Computer Science with a 34.9%.

We illustrate in Fig. 4 a plot of the citations per year, where we can notice an increasing trend.

We show in Fig. 5 a plot using bars to illustrate the number of publications of the top authors using metaheuristics, fuzzy logic and mobile robotics in their papers.

We show in Fig. 6 a plot using bars to illustrate the number of publications of the top countries in metaheuristics, fuzzy logic and mobile robotics.

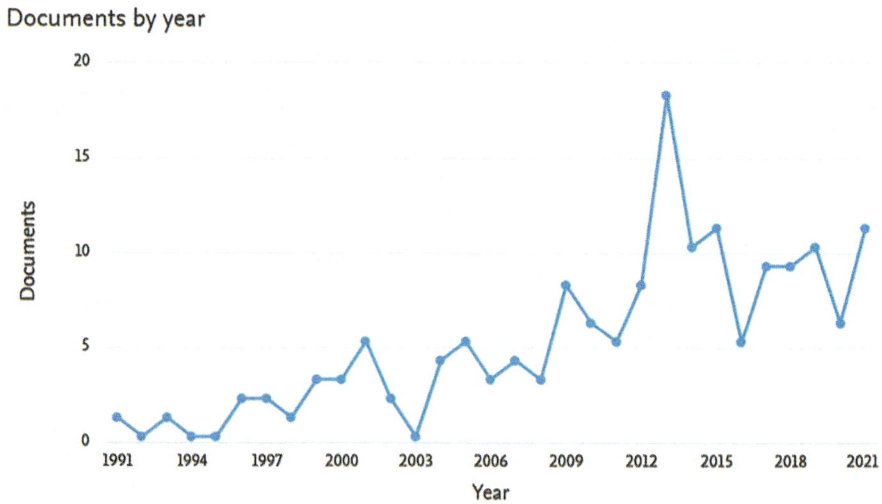

Fig. 1 Plot of documents per year for articles published since 1991

Fig. 2 Plot illustrating h index values for the papers published since 1991

We show in Fig. 7 a plot using bars to illustrate the number of publications according to the main institutions (the first one is from Mexico) in metaheuristics, fuzzy logic and mobile robotics.

In Fig. 8 we show a visualization of cluster density for the top authors based on the papers that have appear in the state of the art from 1991 to 2021. This Figure illustrates the authors that are working close together in research collaborations associated to metaheuristics, fuzzy logic and mobile robotics. We have to say that names with larger fonts represent authors that have more papers that been published.

Co-authorship is a very important for any Scientometric study and for the case of the interaction of metaheuristics, fuzzy logic and mobile robotics, this can be visualized in Fig. 9. In this case, this figure illustrates the authors grouped with respect to co-authorship in more than 3 journal papers. The plot in Fig. 9 offers an overview of the work in collaboration that has been finished to the moment in the interaction of metaheuristics, fuzzy logic and mobile robotics areas.

4 Possible Future Research Trends

Taking into account the results reported in the previous sections and the scientometric analysis, we can state that to the moment, metaheuristics like GAs, PSO, ACO and grey wolf optimization have been used more frequently. We still feel that more articles will be published in future years with these algorithms, as there numerous open problems in fuzzy control for mobile robotics that can be aimed with these techniques that have not been solved to the moment. However, there are still other

Table 1 List of the top 10 most cited papers (from a total of 155 papers)

Paper	Authors	Year of publication	Citations
Optimization of interval type-2 fuzzy logic controllers for a perturbed autonomous wheeled mobile robot using genetic algorithms [38]	Martínez, R., Castillo, O., & Aguilar, L. T	2009	285
Comparative study of bio-inspired algorithms applied to the optimization of type-1 and type-2 fuzzy controllers for an autonomous mobile robot [39]	Castillo, O., Martínez-Marroquín, R., Melin, P., Valdez, F., & Soria, J	2012	222
The RAVEN: Design and validation of a telesurgery system. International Journal of Robotics Research [40]	Lum, M. J. H., Friedman, D. C. W., Sankaranarayanan, G., King, H., Fodero II, K., Leuschke, R., Sinanan, M. N	2009	184
A particle-swarm-optimized fuzzy-neural network for voice-controlled robot systems [41]	Chatterjee, A., Pulasinghe, K., Watanabe, K., & Izumi, K	2005	168
Reinforcement ant optimized fuzzy controller for mobile-robot wall-following control [42]	Juang, C., & Hsu, C	2009	165
A new approach for dynamic fuzzy logic parameter tuning in ant colony optimization and its application in fuzzy control of a mobile robot [43]	Castillo, O., Neyoy, H., Soria, J., Melin, P., & Valdez, F	2015	138
A higher level path tracking controller for a four-wheel differentially steered mobile robot [44]	Maalouf, E., Saad, M., & Saliah, H	2006	138
Evolutionary-group-based particle-swarm-optimized fuzzy controller with application to mobile-robot navigation in unknown environments [45]	Juang, C., & Chang, Y	2011	120
Optimal design of type-2 and type-1 fuzzy tracking controllers for autonomous mobile robots under perturbed torques using a new chemical optimization paradigm [46]	Melin, P., Astudillo, L., Castillo, O., Valdez, F., & Garcia, M	2013	117

(continued)

Table 1 (continued)

Paper	Authors	Year of publication	Citations
Ant colony optimization with dynamic parameter adaptation based on interval type-2 fuzzy logic systems [47]	Olivas, F., Valdez, F., Castillo, O., Gonzalez, C. I., Martinez, G., & Melin, P	2017	104

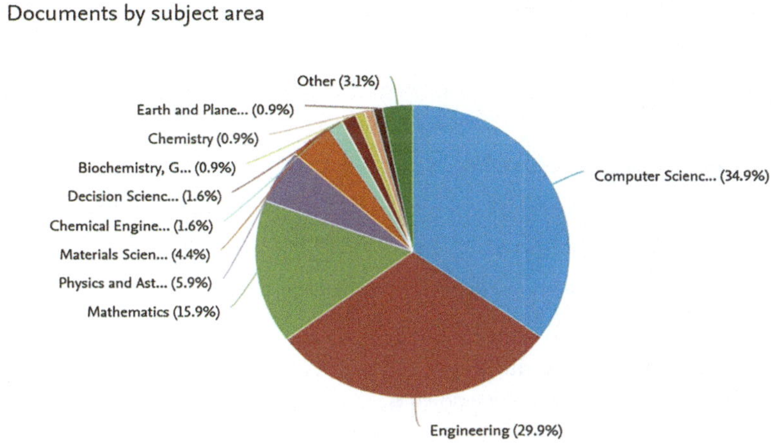

Fig. 3 Distribution of papers according to their area of application

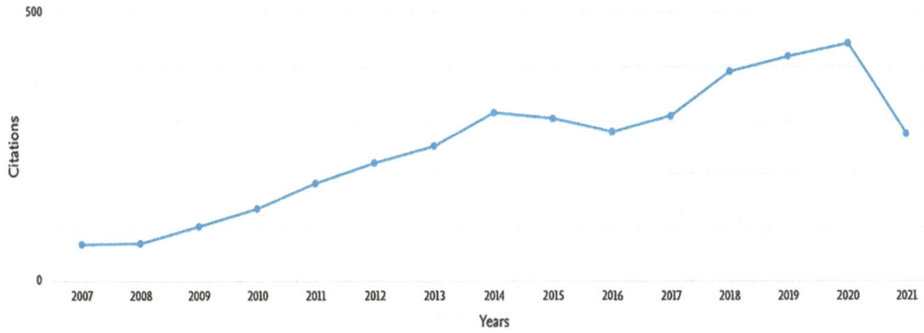

Fig. 4 Plot of the citations for published documents per year according to Scopus

types of techniques coming from soft computing that can be utilized in fuzzy control for mobile robotics. For example, recent collective intelligent and nature-inspired methods like, firefly algorithm (FA) bee colony optimization (BCO), gravitational search algorithms (GSA) and other similar methods [48–54], have not been yet applied in fuzzy control for mobile robotics problems. Also, other generalization of the initial and traditional fuzzy logic (which is called type-1) have not been considered yet, like general type-2 fuzzy systems, intuitionistic fuzzy logic, Pythagorean fuzzy

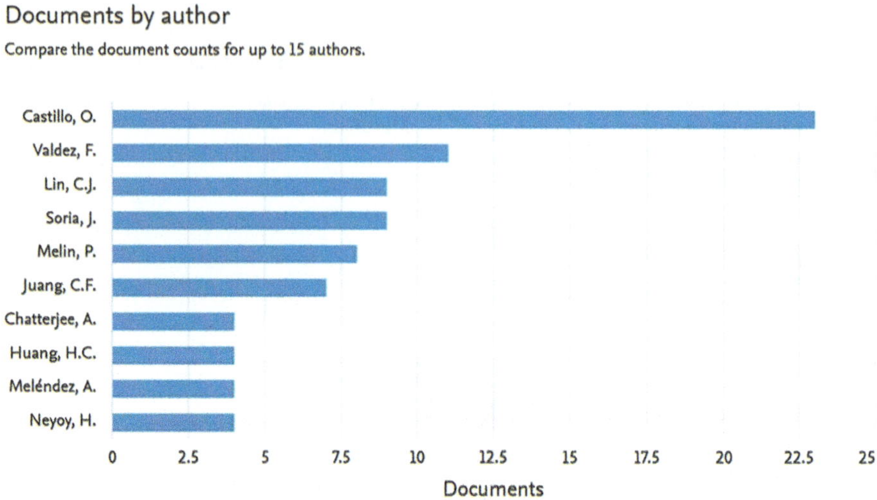

Fig. 5 Plot to illustrate the number of publications of the top authors

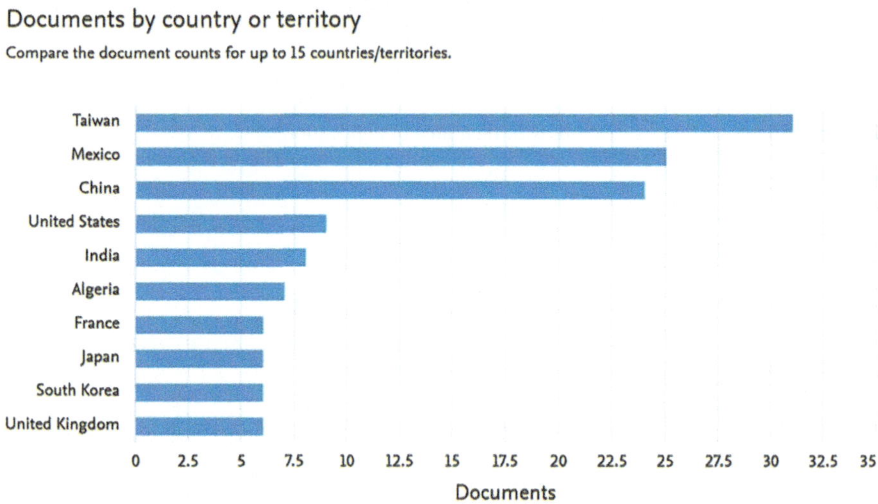

Fig. 6 Plot to illustrate the number of publications of the top countries

systems and others [55–58], have not been yet utilized in fuzzy control for mobile robotics. Finally, novel neural models, such as modular and convolutional neural models, as well as hybrid type-2 fuzzy neural systems are also to be utilized in fuzzy control for mobile robotics problems [59–64]. Finally, we have to say that there is a wide number of opportunities of future research work on applying fuzzy control for mobile robotics problems area and we envision this will continue to be a promissory area of research in the short, medium and the long terms. It is also possible, that

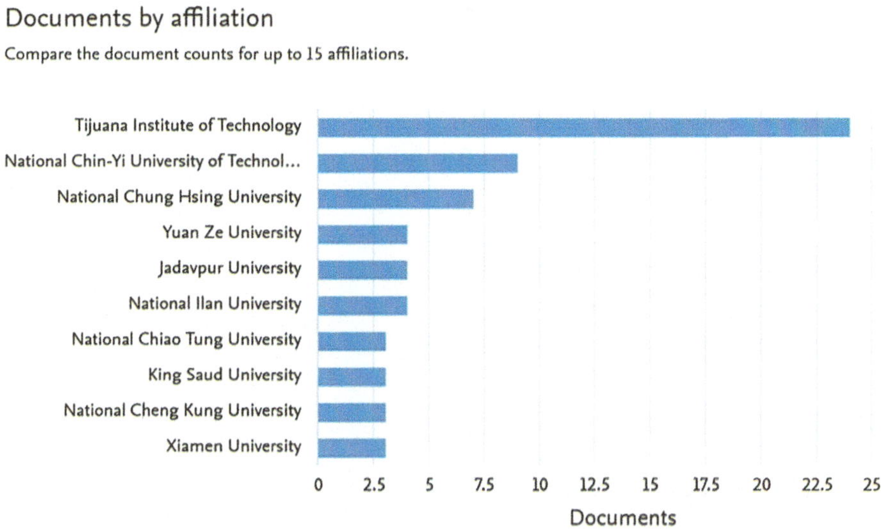

Fig. 7 Plot to illustrate the number of publications of the top institutions

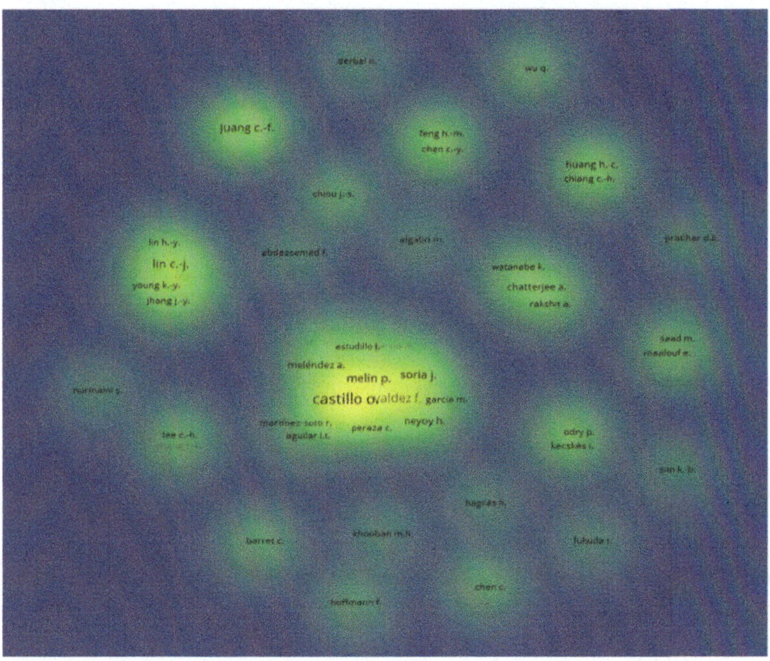

Fig. 8 Cluster density visualization of the main authors

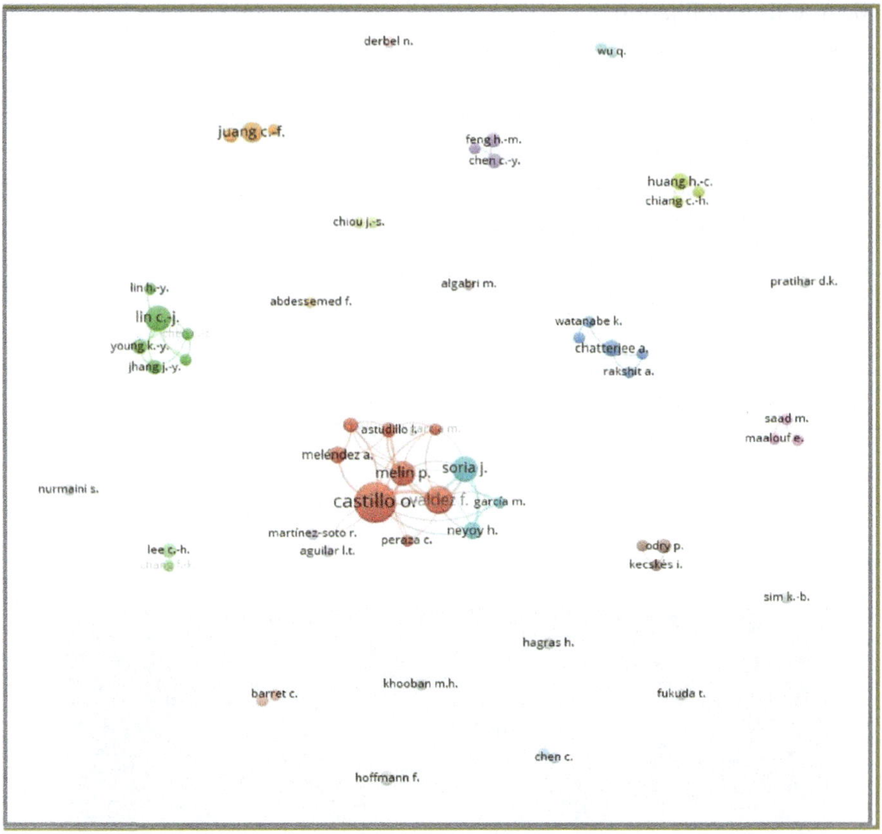

Fig. 9 Co-authorship pattern for metaheuristics, fuzzy logic and mobile robotics

newer problems will arise in fuzzy control for mobile robotics problems that may motivate more work on new theoretical constructs and methodologies.

5 Conclusion

A relatively large volume of literature on metaheuristics for optimizing fuzzy control for mobile robotics for a period of 1991–2021 recorded in Scopus (Elsevier) has been analyzed with scientometric methodologies. Several scientometric tools have been utilized in this article to comprehend the growth patterns in metaheuristics for optimizing fuzzy control for mobile robotics. Based on the analysis, the next statements have been postulated. A total of 155 articles were encountered for a period of almost 30 years. Till 2010, the growth had a slow rate. After 2010, this research area has shown an escalating growth in the number of research publications (except for 2017 and 2018). The growth rate in the literature has been noticed to be significant in the last ten years. A growth rate higher than 80% has occurred after 2010 (this is appreciated from Fig. 1). We have described in detail a sample of these articles, which illustrate the types of works already published in this area. This study can provide relevant and timely information for the academicians, and research scholars interested in scientometric analysis of metaheuristics, fuzzy control and mobile robotics. As future work, we imagine researchers doing the application of general type-2 fuzzy logic, type-3 fuzzy logic, advanced neural network models, swarm intelligence techniques or mixing of these methods, in optimal design of fuzzy controllers of mobile robots for solving real-world problems.

References

1. Castillo, O., Amador-Angulo, L.: A generalized type-2 fuzzy logic approach for dynamic parameter adaptation in bee colony optimization applied to fuzzy controller design. Inf. Sci. **460–461**, 476–496 (2018)
2. Dorigo, M.: Optimization, learning and natural algorithms. Ph.D. Thesis, Dipartimento di Elettronica, Politechico di Milano, Italy (1992)
3. Guerrero, M., Castillo, O., Garcia, M.: Fuzzy dynamic parameters adaptation in the cuckoo search algorithm using fuzzy logic. In: 2015 IEEE Congress on Evolutionary Computation (CEC), pp. 441–448. IEEE (2015)
4. Hongbo, L., Abraham, A.: A fuzzy adaptive turbulent particle swarm optimization. Int. J. Innov. Comput. Appl. **1**(1), 39–47 (2007)
5. Melin, P., Olivas, F., Castillo, O., Valdez, F., Soria, J., Garcia, J.: Optimal design of fuzzy classification systems using PSO with dynamic parameter adaptation through fuzzy logic. Elsevier Exp. Syst. Appl. **40**(8), 3196–3206 (2013)
6. Neyoy, H., Castillo, O., Soria, J.: Dynamic fuzzy logic parameter tuning for ACO and its application in TSP problems. In: Studies in Computational Intelligence, vol. 451. Springer, pp. 259–271 (2012)
7. Olivas, F., Valdez, F., Castillo, O., Melin, P.: Dynamic parameter adaptation in particle swarm optimization using interval type-2 fuzzy logic. Soft Comput. **20**(3), 1057–1070 (2016)
8. Olivas, F., Valdez, F., Castillo, O., Gonzalez, C., Martinez, G., Melin, P.: Ant colony optimization with dynamic parameter adaptation based on interval type-2 fuzzy logic systems. Appl. Soft Comput. **53**, 74–87 (2017)
9. Olivas, F., Valdez, F., Castillo, O., Melin, P.: Interval type-2 fuzzy logic for dynamic parameter adaptation in a modified gravitational search algorithm. Inf. Sci. **476**, 159–175 (2019)
10. Ochoa, P., Castillo, O., Soria, J.: Differential evolution with dynamic adaptation of parameters for the optimization of fuzzy controllers. In: Recent Advances on Hybrid Approaches for Designing Intelligent Systems, pp. 275–288. Springer International Publishing (2014)
11. Peraza, C., Valdez, F., Castillo, O.: An improved harmony search algorithm using fuzzy logic for the optimization of mathematical functions. In: Design of Intelligent Systems Based on Fuzzy Logic, Neural Networks and Nature-Inspired Optimization, pp. 605–615. Springer International Publishing (2015)
12. Perez, J., Valdez, F., Castillo, O., Melin, P., Gonzalez, C., Martinez, G.: Interval type-2 fuzzy logic for dynamic parameter adaptation in the bat algorithm. Soft Comput. 1–19 (2016)
13. Rashedi, E., Nezamabadi-Pour H., Saryazdi, S.: GSA: a gravitational search algorithm. ELSEVIER: Inf. Sci. **179**(13), 2232–2248 (2009). (Iran)
14. Shi, Y., Eberhart, R.: Fuzzy adaptive particle swarm optimization. In: Proceeding of IEEE International Conference on Evolutionary Computation, Piscataway, NJ: IEEE Service Center, Seoul, Korea, pp. 101–106 (2001)

15. Solano-Aragon, C., Castillo, O.: Optimization of Benchmark Mathematical Functions Using the Firefly Algorithm with Dynamic Parameters. In Fuzzy Logic Augmentation of Nature-Inspired Optimization Metaheuristics, pp. 81–89. Springer International Publishing (2015)
16. Sombra, A., Valdez, F., Melin, P., Castillo, O.: A new gravitational search algorithm using fuzzy logic to parameter adaptation. In: Evolutionary Computation (CEC), 2013 IEEE Congress, pp. 1068–1074. IEEE Press (2013)
17. Taher, N., Ehsan, A., Masoud, J.: A new hybrid evolutionary algorithm based on new fuzzy adaptive PSO and NM algorithms for distribution feeder reconfiguration. Elsevier Energy Convers Manag. **54**, 7–16 (2012)
18. Valdez, F., Melin, P., Castillo, O.: Evolutionary method combining particle swarm optimization and genetic algorithms using fuzzy logic for decision making. IEEE Int. Conf. Fuzzy Syst. 2114–2119 (2009)
19. Wang, B., Liang, G., Chan, L.W., Yunlong, D.: A new kind of fuzzy particle swarm optimization fuzzy_PSO algorithm. In: 1st international symposium on systems and control in aerospace and astronautics. In: ISSCAA 2006, pp. 309–311 (2006)
20. Zadeh, L.: Fuzzy sets. Inf. Control. **8**, (1965)
21. Zadeh, L.: Fuzzy logic. IEEE Comput. 83–92 (1965)
22. Zadeh, L.: The concept of a linguistic variable and its application to approximate reasoning—I. Inform. Sci. **8**, 199–249 (1975)
23. Kuntsevich, V.M.: Control Under Uncertainty: Guaranteed Results in Control and Identification Problems. Naukova Dumka, Kyiv (2006).(in Russian)
24. Kuntsevich, V.M., Gubarev, V.F., Kondratenko, Y.P., Lebedev, D.V., Lysenko, V.P. (eds.): Control Systems: Theory and Applications. Series in Automation, Control and Robotics. River Publishers (2018)
25. Leal Ramírez, C., Castillo, O., Melin, P., Rodríguez Díaz, A.: Simulation of the bird age-structured population growth based on an interval type-2 fuzzy cellular structure. Inf. Sci. **181**(3), 519–535 (2011)
26. Cázarez-Castro, N.R., Aguilar, L.T., Castillo, O.: Designing type-1 and type-2 fuzzy logic controllers via fuzzy Lyapunov synthesis for nonsmooth mechanical systems. Eng. Appl. of AI **25**(5), 971–979 (2012)
27. Castillo, O., Melin, P.: Intelligent systems with interval type-2 fuzzy logic. Int. J. Innov. Comput. Inf. Control. **4**(4), 771–783 (2008)
28. Mendez, G.M., Castillo, O.: Interval type-2 TSK fuzzy logic systems using hybrid learning algorithm, fuzzy systems, 2005. FUZZ'05. In: The 14th IEEE International Conference on, pp. 230–235
29. Melin, P., Castillo, O.: Intelligent control of complex electrochemical systems with a neuro-fuzzy-genetic approach. IEEE Trans. Ind. Electron. **48**(5), 951–955
30. Melin, P., Sánchez, D., Castillo, O.: Genetic optimization of modular neural networks with fuzzy response integration for human recognition. Inf. Sci. **197**, 1–19 (2012)
31. Melin, P., Sánchez, D.: Multi-objective optimization for modular granular neural networks applied to pattern recognition. Inf. Sci. **460–461**, 594–610 (2018)
32. Raiesdana, S.: A hybrid method for industrial robot navigation. J. Optim. Ind. Eng. **14**(1), 219–234 (2021). https://doi.org/10.22094/JOIE.2020.1863337.1629
33. Tao, Y., Gao, H., Ren, F., Chen, C., Wang, T., Xiong, H., Jiang, S.: A mobile service robot global path planning method based on ant colony optimization and fuzzy control. Appl. Sci. (Switz.) **11**(8), (2021). https://doi.org/10.3390/app11083605
34. Precup, R., Voisan, E., Petriu, E.M., Tomescu, M.L., David, R., Szedlak-Stinean, A., Roman, R.: Grey wolf optimizer-based approaches to path planning and fuzzy logic-based tracking control for mobile robots. Int. J. Comput., Commun. Control. **15**(3), (2020). https://doi.org/10.15837/IJCCC.2020.3.3844
35. Xu, S.S., Huang, H., Kung, Y., Chu, Y.: A networked multirobot cps with artificial immune fuzzy optimization for distributed formation control of embedded mobile robots. IEEE Trans. Ind. Inform. **16**(1), 414–422 (2020). https://doi.org/10.1109/TII.2019.2936045

36. Kondratenko, Y.P., Kozlov, A.V.: Parametric optimization of fuzzy control systems based on hybrid particle swarm algorithms with elite strategy. J. Autom. Inf. Sci. **51**(12), 25–45 (2019). https://doi.org/10.1615/JAutomatInfScien.v51.i12.40
37. Lin, C., Jhang, J., Young, K.: Using a type-2 neural fuzzy controller for navigation control of evolutionary robots. Sens. Mater. **31**(9), 2735–2751 (2019). https://doi.org/10.18494/SAM.2019.2343
38. Martínez, R., Castillo, O., Aguilar, L.T.: Optimization of interval type-2 fuzzy logic controllers for a perturbed autonomous wheeled mobile robot using genetic algorithms. Inf. Sci. **179**(13), 2158–2174 (2009). https://doi.org/10.1016/j.ins.2008.12.028
39. Castillo, O., Martínez-Marroquín, R., Melin, P., Valdez, F., Soria, J.: Comparative study of bio-inspired algorithms applied to the optimization of type-1 and type-2 fuzzy controllers for an autonomous mobile robot. Inf. Sci. **192**, 19–38 (2012). https://doi.org/10.1016/j.ins.2010.02.022
40. Lum, M.J.H., Friedman, D.C.W., Sankaranarayanan, G., King, H., Fodero II, K., Leuschke, R., Sinanan, M.N.: The RAVEN: design and validation of a telesurgery system. Int. J. Robot. Res. **28**(9), 1183–1197 (2009). https://doi.org/10.1177/0278364909101795
41. Chatterjee, A., Pulasinghe, K., Watanabe, K., Izumi, K.: A particle-swarm-optimized fuzzy-neural network for voice-controlled robot systems. IEEE Trans. Industr. Electron. **52**(6), 1478–1489 (2005). https://doi.org/10.1109/TIE.2005.858737
42. Juang, C., Hsu, C.: Reinforcement ant optimized fuzzy controller for mobile-robot wall-following control. IEEE Trans. Ind. Electron. **56**(10), 3931–3940 (2009). https://doi.org/10.1109/TIE.2009.2017557
43. Castillo, O., Neyoy, H., Soria, J., Melin, P., Valdez, F.: A new approach for dynamic fuzzy logic parameter tuning in ant colony optimization and its application in fuzzy control of a mobile robot. Appl. Soft Comput. J. **28**, 150–159 (2015). https://doi.org/10.1016/j.asoc.2014.12.002
44. Maalouf, E., Saad, M., Saliah, H.: A higher level path tracking controller for a four-wheel differentially steered mobile robot. Robot. Auton. Syst. **54**(1), 23–33 (2006). https://doi.org/10.1016/j.robot.2005.10.001
45. Juang, C., Chang, Y.: Evolutionary-group-based particle-swarm-optimized fuzzy controller with application to mobile-robot navigation in unknown environments. IEEE Trans. Fuzzy Syst. **19**(2), 379–392 (2011). https://doi.org/10.1109/TFUZZ.2011.2104364
46. Melin, P., Astudillo, L., Castillo, O., Valdez, F., Garcia, M.: Optimal design of type-2 and type-1 fuzzy tracking controllers for autonomous mobile robots under perturbed torques using a new chemical optimization paradigm. Expert Syst. Appl. **40**(8), 3185–3195 (2013). https://doi.org/10.1016/j.eswa.2012.12.032
47. Olivas, F., Valdez, F., Castillo, O., Gonzalez, C.I., Martinez, G., Melin, P.: Ant colony optimization with dynamic parameter adaptation based on interval type-2 fuzzy logic systems. Appl. Soft Comput. J. **53**, 74–87 (2017). https://doi.org/10.1016/j.asoc.2016.12.015
48. Olivas, F., Valdez, F., Castillo, O., Melin, P.: Dynamic parameter adaptation in particle swarm optimization using interval type-2 fuzzy logic. Soft. Comput. **20**(3), 1057–1070 (2016)
49. Olivas, F., Valdez, F., Castillo, O., Gonzalez, C.I., Martinez, G., Melin, P.: Ant colony optimization with dynamic parameter adaptation based on interval type-2 fuzzy logic systems. Appl. Soft Comput. **53**, 74–87 (2017)
50. Sanchez, D., Melin, P., Castillo, O.: Optimization of modular granular neural networks using a firefly algorithm for human recognition. Eng. Appl. AI **64**, 172–186 (2017)
51. González, B., Valdez, F., Melin, P., Prado-Arechiga, G.: Fuzzy logic in the gravitational search algorithm for the optimization of modular neural networks in pattern recognition. Expert Syst. Appl. **42**(14), 5839–5847 (2015)
52. González, B., Valdez, F., Melin, P., Prado-Arechiga, G.: Fuzzy logic in the gravitational search algorithm enhanced using fuzzy logic with dynamic alpha parameter value adaptation for the optimization of modular neural networks in echocardiogram recognition. Appl. Soft Comput. **37**, 245–254 (2015)
53. Miramontes, I., Guzman, J., Melin, P., Prado-Arechiga, G.: Optimal design of interval type-2 fuzzy heart rate level classification systems using the bird swarm algorithm. Algorithms **11**(12), 206 (2018)

54. Gonzalez, C.I., Melin, P., Castro, J.R., Castillo, O., Mendoza, O.: Optimization of interval type-2 fuzzy systems for image edge detection. Appl. Soft Comput. **47**, 631–643 (2016)
55. Castillo, O., Castro, J.R., Melin, P., Rodriguez-Diaz, A.: Application of interval type-2 fuzzy neural networks in non-linear identification and time series prediction. Soft Comput. **18**(6), 1213–1224 (2014)
56. Melin, P., Gonzalez, C.I., Castro, J.R., et al.: Edge-detection method for image processing based on generalized type-2 fuzzy logic. IEEE Trans. Fuzzy Syst. **22**(6), 1515–1525 (2014)
57. Castillo, O., Melin, P.: A review on interval type-2 fuzzy logic applications in intelligent control. Inf. Sci. **279**, 615–631 (2014)
58. Ontiveros, E., Melin, P., Castillo, O.: High order α-planes integration: a new approach to computational cost reduction of general type-2 fuzzy systems. Eng. Appl. Artif. Intell. **74**, 186–197 (2018)
59. Castillo, O., Castro, J.R., Melin, P., Rodriguez Dias, A.: Application of interval type-2 fuzzy neural networks in non-linear identification and time series prediction. Soft Comput. **18**(6), 1213–1224 (2014)
60. Sanchez, M.A., Castillo, O., Castro, J.R., Melin, P.: Fuzzy granular gravitational clustering algorithm for multivariate data. Inf. Sci. **279**, 498–511 (2014)
61. Sánchez, D., Melin, P.: Optimization of modular granular neural networks using hierarchical genetic algorithms for human recognition using the ear biometric measure. Eng. Appl. Artif. Intell. **27**, 41–56 (2014)
62. Sanchez, M.A., Castro, J.R., Castillo, O., Mendoza, O., Rodriguez-Diaz, A., Melin, P.: Fuzzy higher type information granules from an uncertainty measurement. Granul. Comput. **2**(2), 95–103 (2017)
63. Melin, P., Miramontes, I., Prado-Arechiga, G.: A hybrid model based on modular neural networks and fuzzy systems for classification of blood pressure and hypertension risk diagnosis. Expert Syst. Appl. **107**, 146–164 (2018)
64. Guzmán, J.C., Miramontes, I., Melin, P., Prado-Arechiga, G.: Optimal genetic design of type-1 and interval type-2 fuzzy systems for blood pressure level classification. Axioms **8**(1), 8 (2019)

New Developments in Time Series Analysis, Prediction, and Fault Detection and Control

Time Series Prediction Using Time-Series Decomposition and Multi-reservoirs Echo State Network

Ying Han and Kun Li

Abstract Echo State Network (ESN) is effective to do time series analysis, which has a dynamic reservoir, includes: input units, internal units and output units. However, due to the randomness and non-stationarity properties of most time series, it is difficult for single reservoir to better handle them, because different scales in the time series should be dealt with in a unified structure. In order to solve this, the time-series decomposition (TSD) is employed to decompose the time series into different sub-sequences, which can be handled by different reservoirs. Now, there are many TSD methods, such as: empirical mode decomposition (EMD), ensemble empirical mode decomposition (EEMD), complementary ensemble empirical mode decomposition (CEEMD), local mean decomposition (LMD), variational mode decomposition (VMD), etc. Different TSD methods are used to decompose the time series and then the multi-reservoirs ESN are constructed. Finally, experimental results using several time series are presented and compared.

1 Introduction

With the continuous expansion of the scale of the Internet, the network traffic data greatly increased, increasing the frequency of network congestion. It has great significance of the network traffic prediction for network management, which can realize

---

Honoring Professor Georgi M. Dimirovski for his many, Academic contributions and merits to our community.

Y. Han (✉) · K. Li
Faculty of Electrical and Control Engineering, Liaoning Technical University, Huludao 125105, China
e-mail: hyfengyan@163.com

K. Li
e-mail: phdlikun@163.com

P. Shi et al. (eds.), *Complex Systems: Spanning Control and Computational Cybernetics: Applications*, Studies in Systems, Decision and Control 415,
https://doi.org/10.1007/978-3-031-00978-5_4

network load evaluation and security early warning. Network traffic is a type of time-based data with chaotic characteristics, which can be handled by time series analysis. It is very important to establish a high precision network traffic prediction model.

At present, there are many machine learning methods can be used to establish the network traffic prediction model, such as: Least Square Support Vector Machine (LSSVM) [1], Elman Network [2], Extreme Learning Machine (ELM) [3], Finite Impulse Response Neural Network (FIRNN) [4], Radial Basis Function Neural Network (RBFNN) [5], Echo State Network (ESN) [6], etc. ESN is a type of recursive neural network, which has better computing performance to solve problems related to time series analysis. Its core part is a large scale dynamic "reservoir", including input unit, internal unit and output unit. The calculation of the model depends on the connection weights of the input layer to the "reservoir", the connection weights of the internal units of the "reservoir", the connection weights of the output layer to the internal units of the "reservoir" and the connection weights of the internal units of the "reservoir" to the output layer. All weights excepting the output weights are randomly given in the initial stage and remain unchanged in the whole training process, and only the connection weights related to the output need to be determined by training. Therefore, compared with other methods, ESN is simple and only requires little computation relatively. ESN prediction method has been successfully applied in many fields. This research adopts the network traffic prediction method based on ESN.

The network traffic data has the characteristics of fluctuation, and to a certain extent the non-linear and non-stationary characteristics will affect the prediction effects. Although the classical ESN model can better deal with the nonlinear relationship between the input and output of data series, it is difficult to accurately grasp the law of data change, which maybe leads to large prediction errors in some forecast points with dramatic changes. Time-series Decomposition (TSD) is a common method for the time series analysis [7, 8]. Empirical Mode Decomposition (EMD) [9] decomposes the complex non-stationary data into a finite number of Intrinsic Mode Functions (IMF), and each IMF component contains local characteristics of the original signal at different time scales. On the basis of EMD, Ensemble Empirical Mode Decomposition (EEMD) [10] maps the signal regions of different scales to the appropriate scales related to the white noise, by adding different random white noises to the original data sequence, in order to eliminate the phenomenon of mode aliasing. For an improved algorithm of EEMD, Complementary Ensemble Empirical Mode Decomposition (CEEMD) [11] adds a pair of opposite random white noise to the original data sequence, which can better solve the problem of balancing random error. Variational Mode Decomposition (VMD) [12] is a new non-recursive signal decomposition method, which transforms the signal decomposition problem into a constrained optimization problem. It is essentially a set of adaptive Wiener filter, in which the adopted non-recursive method will not transfer error and also will not appear the mode aliasing phenomenon, and the degree of end effect is weak.

In this research, the network traffic prediction method of multiple reservoirs ESN with different TSD methods is discussed. Firstly, the network traffic data is decomposed by TSD method, which decomposes the original data series into several sub-data series with different time scales. Then, a reservoir is constructed for each sub-data sequence. Finally, the results of multiple reservoirs are integrated for output. Compared with the single reservoir, the multiple reservoirs method can reduce the coupling of each unit and improve the processing ability of data series containing different time scales.

2 Multiple Reservoirs ESN (MRESN)

For the ESN based network traffic prediction method, multiple reservoirs can be processed separately by constructing different structural parameters, which can effectively avoid the inadequacy that a single reservoir cannot describe different time scales simultaneously. Structure of the MRESN [8] is shown in Fig. 1.

For D inputs, at the time $t + 1$, the state equation and output equation can be expressed as [8]:

$$\begin{cases} \mathbf{x}^1(t+1) = f_1(\mathbf{W}_{\text{in}}^1 \cdot \mathbf{u}^1(t+1) + \mathbf{W}_{\text{x}}^1 \cdot \mathbf{x}^1(t) + \mathbf{W}_{back}^1 \cdot \mathbf{y}(t)) \\ \mathbf{x}^2(t+1) = f_2(\mathbf{W}_{\text{in}}^2 \cdot \mathbf{u}^2(t+1) + \mathbf{W}_{\text{x}}^2 \cdot \mathbf{x}^2(t) + \mathbf{W}_{back}^2 \cdot \mathbf{y}(t)) \\ \qquad\vdots \\ \mathbf{x}^D(t+1) = f_D(\mathbf{W}_{\text{in}}^D \cdot \mathbf{u}^D(t+1) + \mathbf{W}_{\text{x}}^D \cdot \mathbf{x}^D(t) + \mathbf{W}_{back}^D \cdot \mathbf{y}(t)) \\ \mathbf{y}(t+1) = g(\mathbf{W}_{out}^1 \cdot [(\mathbf{u}^1(t+1), \mathbf{x}^1(t+1)] + \mathbf{W}_{out}^2 \cdot [(\mathbf{u}^2(t+1), \mathbf{x}^2(t+1)] \\ \qquad + \cdots + \mathbf{W}_{out}^D \cdot [(\mathbf{u}^D(t+1), \mathbf{x}^D(t+1)]) \end{cases} \tag{1}$$

where, $[\mathbf{x}^1(t+1), \mathbf{x}^2(t+1), \ldots, \mathbf{x}^D(t+1)]$ and $[\mathbf{u}^1(t+1), \mathbf{u}^2(t+1), \ldots, \mathbf{u}^D(t+1)]$ are values of D internal state units and input units at time $t + 1$; $\{\mathbf{W}_{\text{in}}^1, \mathbf{W}_{\text{x}}^1, \mathbf{W}_{\text{back}}^1, \mathbf{W}_{out}^1\}$, $\{\mathbf{W}_{\text{in}}^2, \mathbf{W}_{\text{x}}^2, \mathbf{W}_{\text{back}}^2, \mathbf{W}_{out}^2\}$, …, $\{\mathbf{W}_{\text{in}}^D, \mathbf{W}_{\text{x}}^D, \mathbf{W}_{\text{back}}^D, \mathbf{W}_{out}^D\}$ are D groups of the connection weight matrix. $\{\mathbf{W}_{\text{in}}^1, \mathbf{W}_{\text{x}}^1, \mathbf{W}_{\text{back}}^1, \mathbf{W}_{\text{in}}^2, \mathbf{W}_{\text{x}}^2, \mathbf{W}_{\text{back}}^2, \cdots, \mathbf{W}_{\text{in}}^D, \mathbf{W}_{\text{x}}^D, \mathbf{W}_{\text{back}}^D\}$ is randomly generated at the initialization stage, and $\{\mathbf{W}_{out}^1, \mathbf{W}_{out}^2, \cdots \mathbf{W}_{out}^D\}$ is got by the training of the input and output data of the system.

For the MRESN, the output of ESN can be expressed by the following linear equation,

$$\begin{aligned} g^{-1}(\mathbf{y}(t)) &= \mathbf{W}_{out}^1 \cdot [\mathbf{x}^1(t), \mathbf{u}^1(t)]^{\mathrm{T}} + \mathbf{W}_{out}^2 \cdot [\mathbf{x}^2(t), \mathbf{u}^2(t)]^{\mathrm{T}} \\ &\quad + \cdots + \mathbf{W}_{out}^D \cdot [\mathbf{x}^D(t), \mathbf{u}^D(t)]^{\mathrm{T}} \\ &= [\mathbf{W}_{out}^1, \mathbf{W}_{out}^2, \cdots, \mathbf{W}_{out}^D] \cdot \begin{bmatrix} \mathbf{q}^1(t) \\ \mathbf{q}^2(t) \\ \vdots \\ \mathbf{q}^D(t) \end{bmatrix} \end{aligned} \tag{2}$$

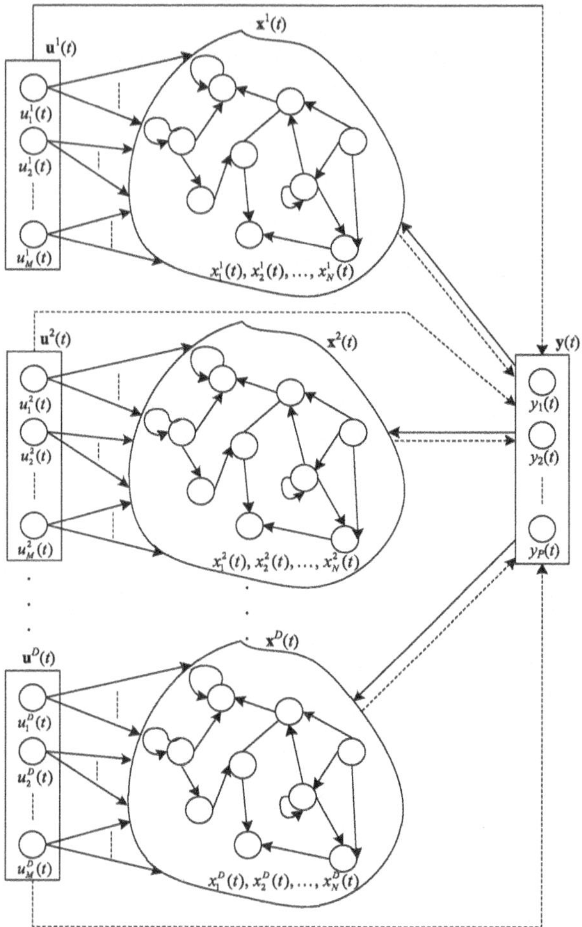

Fig. 1 Structure of the MRESN

where, $\mathbf{q}^1(t) = [\mathbf{x}^1(t),\mathbf{u}^1(t)]^{\mathrm{T}}$, $\mathbf{q}^2(t) = [\mathbf{x}^2(t),\mathbf{u}^2(t)]^{\mathrm{T}}$, …, $\mathbf{q}^D(t) = [\mathbf{x}^D(t),\mathbf{u}^D(t)]^{\mathrm{T}}$.

3 TSD Method

3.1 EMD

EMD is an adaptive signal processing method, which does not need to set any basis function, and has a high signal-to-noise ratio [9]. EMD method has obvious advantages in processing nonlinear and non-stationary signals, and can be applied to the decomposition of any type of signals in theory, especially for analyzing nonlinear and non-stationary signal sequences. It does stationary processing for the non-stationary data and decomposes the complex signal into a finite number of Intrinsic Mode functions (IMF). The decomposed IMF components contain local characteristics at different time scales of the original signal. Then, the time-spectrum diagram is obtained by Hilbert transform. Because the decomposition is based on the local characteristics of the time scale of the signal series, so it is self-adaptive.

Main steps of the decomposition for the network traffic data based on EMD method are as follows:

Step 1: The network traffic data series is represented by $x(n)$, and the upper envelope of $x(n)$ is obtained by fitting all the maximum points of it with the cubic spline interpolation function; similarly, the lower envelope of $x(n)$ is obtained by fitting all the minimum points of it with the cubic spline interpolation function. The average value of the upper and lower envelopments of $x(n)$ is denoted as $m(n)$, and a new data series $h_1(n)$ is obtained by $h_1(n) = x(n)\text{-}m(n)$.

Step 2: If $h_1(n)$ meets the conditions, then it is regarded as the first IMF component of $x(n)$, denoted as $c_1(n) = h_1(n)$; if $h_1(n)$ does not meet the conditions, it means that it is not an eigenmode function. Let $h_1(n)$ replace the original data sequence $x(n)$, and repeat **Step 1** until an IMF component that meets the conditions is obtained, denoted as $c_1(n)$.

Step 3: Calculate the residual data sequence by $x_1(n) = x(n)\text{-}c_1(n)$. $x_1(n)$ is taken as a new data sequence to be decomposed, and **Step 1** and **Step 2** are repeated to extract the second, third and until the lth IMF component, as well as the residual $r_l(n)$ of the original data sequence. The decomposition ends when the termination condition is met, and the termination condition is that the IMF component cannot be extracted from the latest data sequence. Then, the original data series $x(n)$ can be expressed as the sum of l IMF components and a residual $r_l(n)$, which is denoted by

$$x(n) = \sum_{i=1}^{l} c_i(n) + r_l(n) \tag{3}$$

Step 4: Remove the decomposed high-frequency IMF component, and sum the remaining low-frequency IMF component and the remaining residual $r_l(n)$ to reconstruct a new data series, denoted as $x'(n)$, which is denoted by

$$x'(n) = \sum_{j=l-k}^{l} c_j(n) + r_l(n) \tag{4}$$

where, k represents the number of high-frequency IMF components that are removed.

3.2 EEMD

In the decomposition process of the EMD, it is easy to produce the phenomenon of mode mixing and endpoint flying wing. In order to solve this problem, Wu-Huang et al. proposed EEMD algorithm [10] on the basis of EMD, which can effectively eliminate or weaken the end effect by adding white noise to the original signal. Main steps of the decomposition based on EEMD method are as follows:

Step 1: Add white noise to the original network traffic data series $x(n)$.

Step 2: Determine all the local minimum and maximum points of the signal, and connect them by the cubic spline interpolation curves, to form upper envelope $x_{\text{up}}(n)$ and lower envelope $x_{\text{low}}(n)$.

Step 3: Calculate the average value $m(n)$ of the upper and lower envelope, and obtain a new data sequence $h_1(n) = x(n)\text{-}m(n)$.

Step 4: Judge whether $h_1(n)$ meets the IMF judgment conditions, if not, take $h_1(n)$ as the original signal and repeat **Step 2** to **Step 3** until the ith $h_{1i}(n)$ meets the IMF conditions. Take an IMF component that meets the conditions as $c_1(n) = h_{1i}(n)$.

Step 5: Calculate the residual data sequence by $x_1(n) = x(n) - c_1(n)$.

Step 6: Repeat **Step 2** to **Step 5** until $x_k(n)$ meets the EMD screening condition and terminate the loop. At this time, $x_k(n)$ is the residual component of the original function. EMD screening condition uses the following criteria:

$$\sigma_n = \frac{|x_{\text{up}}(n) + x_{low}(n)|}{|x_{\text{up}}(n) - x_{low}(n)|} \tag{5}$$

Step 7: Repeat the above steps by p times, and the amplitude of the white noise sequence added at each time is different. The IMF component obtained by decomposing p times is averaged to get the final IMF component and residual component.

3.3 CEEMD

In EEMD, the reconstruction error is increased due to the addition of different random white noises. In order to solve this problem, CEEMD improved the EEMD by replacing the way of adding different white noises into the way of adding pairs of opposite white noises [11]. Main steps of the decomposition for the network traffic data based on CEEMD method are as follows:

Step 1: Add pairs of opposite white noises into the network traffic data series $x(n)$, as follows:

$$\begin{cases} x_i^+(n) = x(n) + w_i(n) \\ x_i^-(n) = x(n) - w_i(n) \end{cases} \tag{6}$$

where, $i = 1, 2, \ldots n$; $x_i^+(n)$ and $x_i^-(n)$ are the positive and negative data sequence after adding the pairs of opposite white noises; $w_i(n)$ is the white noise following the normal distribution.

Step 2: Handle the positive data sequence $x_i^+(n)$ firstly. Determine all the local minimum and maximum points of the signal, and connect them by the cubic spline interpolation curves, to form upper envelope $x_{\text{up}}(n)$ and lower envelope $x_{\text{low}}(n)$.

Step 3: Calculate the average value $m(n)$ of the upper and lower envelope, and obtain a new data sequence $h_1(n) = x(n)\text{-}m(n)$.

Step 4: Judge whether $h_1(n)$ meets the IMF judgment conditions, if not, take $h_1(n)$ as the original signal and repeat **Step 2** to **Step 3** until the ith $h_{1i}(n)$ meets the IMF conditions. Take an IMF component that meets the conditions as $c_1(n) = h_{1i}(n)$.

Step 5: Calculate the residual data sequence by $x_1(n) = x(n)\text{-}c_1(n)$.

Step 6: Repeat **Step 2** to **Step 5** until $x_k(n)$ meets the EMD screening condition and terminate the loop. At this time, $x_k(n)$ is the residual component of the original function.

Step 7: Repeat the above steps by p times, and the amplitude of the white noise sequence added at each time is different. The IMF component obtained by decomposing p times is averaged to get the final IMF component of the positive data sequence $x_i^+(n)$.

Step 8: For the negative data sequence $x_i^-(n)$, repeat the above **Step 2** to **Step 7** to get the final IMF component of the negative data sequence $x_i^+(n)$.

Step 9: Calculate the average value of two groups of the component of $x_i^+(n)$ and $x_i^-(n)$ obtained by **Step 7** and **Step 8**, to determine the final IMF component and residual component.

3.4 VMD

VMD decomposes the original signal into K modal components which minimizes the sum of the estimated bandwidth of each mode. So actually the signal decomposition process is the solution process of the variational problem. For an original signal $f(t)$, the model of the variational problem with constraints can be expressed by [12]

$$\begin{cases} \min\limits_{\{u_K\},\{\omega_K\}} \left\{ \sum\limits_K \left\| \partial_t [(\delta(t) + \frac{\mathrm{j}}{\pi t}) * u_K(t)] e^{-\mathrm{j}\omega_K t} \right\|_2^2 \right\} \\ s.t. \quad \sum\limits_K u_K = f \end{cases} \tag{7}$$

where, $\{u_K\} := \{u_1, \ldots, u_K\}$ and $\{\omega_K\} := \{\omega_1, \ldots, \omega_K\}$ are K modal components and center frequency of each modal component, respectively; $\delta(t)$ is an impulse function.

In order to solve the optimal solution of the above constrained problem, a Lagrange multiplicative operator λ is introduced to convert the constrained variational problem into a non-constrained variational problem, as follows:

$$
\begin{aligned}
L(\{u_K\},\{\omega_K\},\lambda) &:= \alpha\sum_K\left\|\partial_t[(\delta(t)+\frac{\mathrm{j}}{\pi t})*u_K(t)]e^{-\mathrm{j}\omega_K t}\right\|_2^2 \\
&+\left\|f(t)-\sum_K u_K(t)\right\|_2^2+\left\langle\lambda(t),f(t)-\sum_K u_K(t)\right\rangle
\end{aligned} \tag{8}
$$

where α is a quadratic penalty factor.

The Alternate Direction Method of Multipliers (ADMM) is used to solve Eq. (8) to obtain the optimal solution. Main steps are as follows:

Step 1: Initialize each modal component and center frequency, that are: $\{u_K^1\}$, $\{\omega_K^1\}$ and λ^1. Transform the variables from the time domain to the frequency domain, and let $n = 0$.

Step 2: In the non-negative frequency range, update u_K:

$$
\hat{u}_K^{n+1}(\omega) \leftarrow \frac{\hat{f}(\omega)-\sum_{i\neq K}\hat{u}_i(\omega)+\frac{\hat{\lambda}(\omega)}{2}}{1+2\alpha(\omega-\omega_K)^2} \tag{9}
$$

where, $\hat{u}_K(\omega)$, $\hat{f}(\omega)$ and $\hat{\lambda}(\omega)$ are respectively the Fourier transforms of u_K, $f(t)$ and λ.

Step 3: Update ω_K:

$$
\omega_K^{n+1} \leftarrow \frac{\int_0^\infty \omega\left|\hat{u}_K(\omega)\right|^2 \mathrm{d}\omega}{\int_0^\infty \left|\hat{u}_K(\omega)\right|^2 \mathrm{d}\omega} \tag{10}
$$

Step 4: Update λ:

$$
\hat{\lambda}^{n+1}(\omega) \leftarrow \hat{\lambda}^n(\omega)+\tau\Big(\hat{f}(\omega)-\sum_K \hat{u}_K^{n+1}(\omega)\Big) \tag{11}
$$

where τ is an iteration factor.

Step 5: Set the convergence precision $\varepsilon > 0$, there is:

$$
\frac{\sum_K\left\|\hat{u}_K^{n+1}-\hat{u}_K^n\right\|_2^2}{\left\|\hat{u}_K^n\right\|_2^2} < \varepsilon \tag{12}
$$

If Eq. (12) is satisfied, then the iteration is stopped and the result is output. Otherwise, return to **Step 2** and continue.

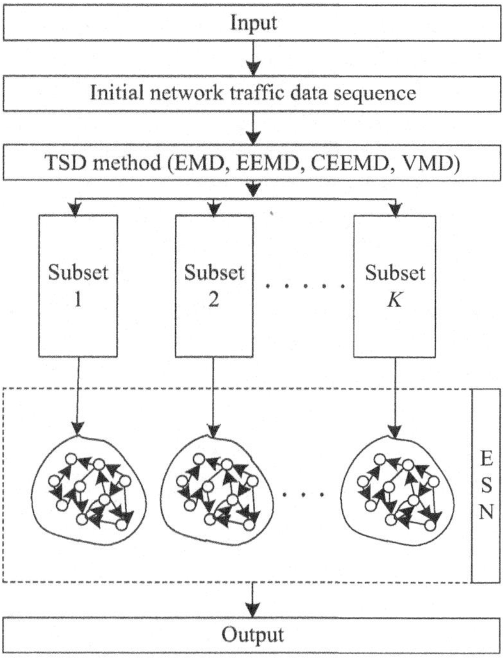

Fig. 2 Basic structure of the prediction method

4 Network Traffic Prediction Based on TSD-MRESN

In this paper, a prediction method based on TSD decomposition and multiple reservoirs ESN is discussed. The input network traffic data series was decomposed by TSD method, and K modes were obtained, corresponding to K data sub-series. The multiple reservoirs ESN model is trained by using data subsets in each mode. The final output was got by integrating the predicted results of multiple reservoirs. The basic structure of the prediction method is shown in Fig. 2.

5 Simulation Experiments

The WIDE backbone network traffic data of MAWI Working Group [http://mawi.wide.ad.jp/mawi/] are applied for simulation experiments. The "Hour" time interval was selected to carry out the experimental analysis, as shown in Table 1. Among

Table 1 Statistics information of the "Hour" network traffic dataset

Datasets	Number	Maximum (byte)	Minimum (byte)	Mean (byte)	Standard deviation	Skewness	Kurtosis
"Hour"	480	5.15E11	5.05E10	2.26E11	1.20E11	0.6547	2.3682

them, the "hour" data set contains 480 groups of data from July 1, 2018 to July 21, 2018, and the data sampling period is 1 h [8].

According to Table 1, the "Hour" network traffic dataset is shown in Fig. 3.

EMD, EEMD, CEEMD and VMD are respectively adopted to handle the "Hour" network traffic dataset, and the decomposed dataset is shown in Fig. 4.

The simulation experiments are as follows: in the "Hour" data set, the first 430 groups of data are taken as the training data, and the last 50 groups of data are taken as the test data. The parameters of each sub-reservoir are randomly selected as follows: N ∈ [10,600], IS ∈ [0.001,1], SD ∈ [0,1], SR ∈ [0.0001,0.1]. Computer configurations: Windows7, Matlab2015A, Intel Core i7 4.00 GHz CPU, 16 GB RAM. RMSE, MAE and MAPE are used to evaluate the prediction accuracy, which was defined as follows:

$$\mathrm{RMSE} = \sqrt{\frac{\sum_{i=1}^{N_{\mathrm{data}}} (y_i - y_i')^2}{N_{\mathrm{data}}}} \tag{13}$$

$$MAE = \frac{1}{N_{\mathrm{data}}} \sum_{i=1}^{N_{\mathrm{data}}} \left| y_i - y_i' \right| \tag{14}$$

$$MAPE = \frac{1}{N_{\mathrm{data}}} \sum_{i=1}^{N_{\mathrm{data}}} \left| \frac{y_i - y_i'}{y_i} \right| \times 100\% \tag{15}$$

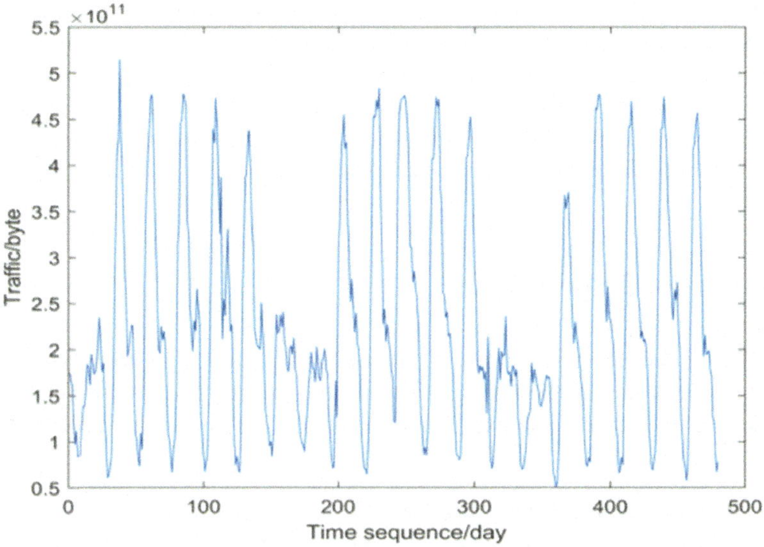

Fig. 3 "Hour" network traffic dataset

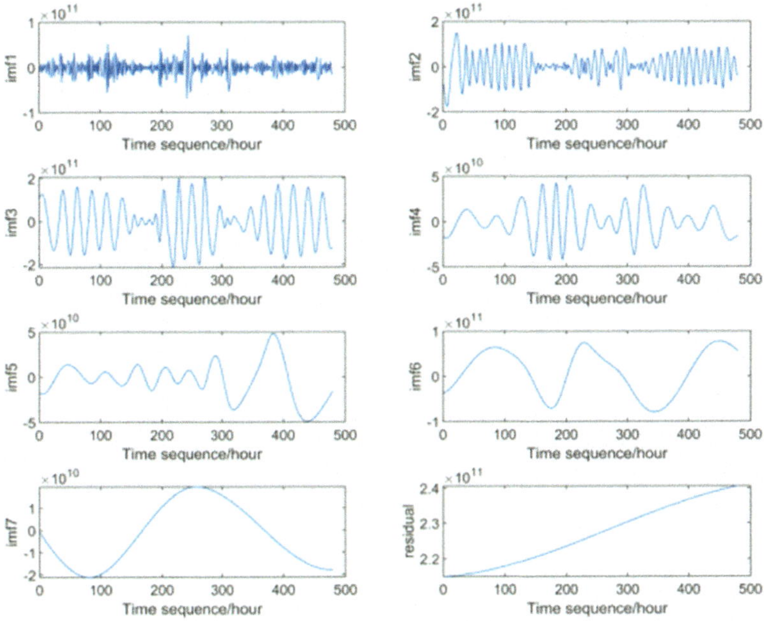

(a) EMD decomposition.

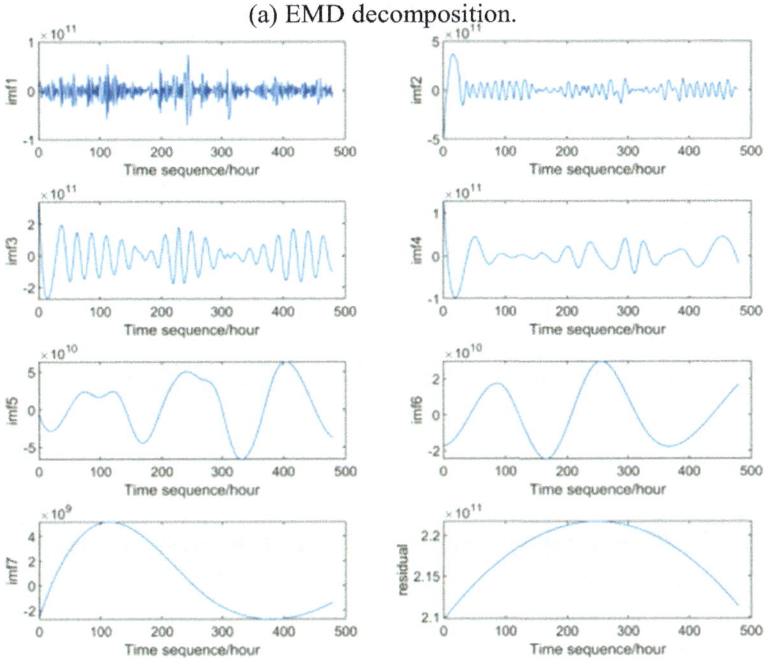

(b) EEMD decomposition.

Fig. 4 Decomposition for the "Hour" data sequence by four TSD methods

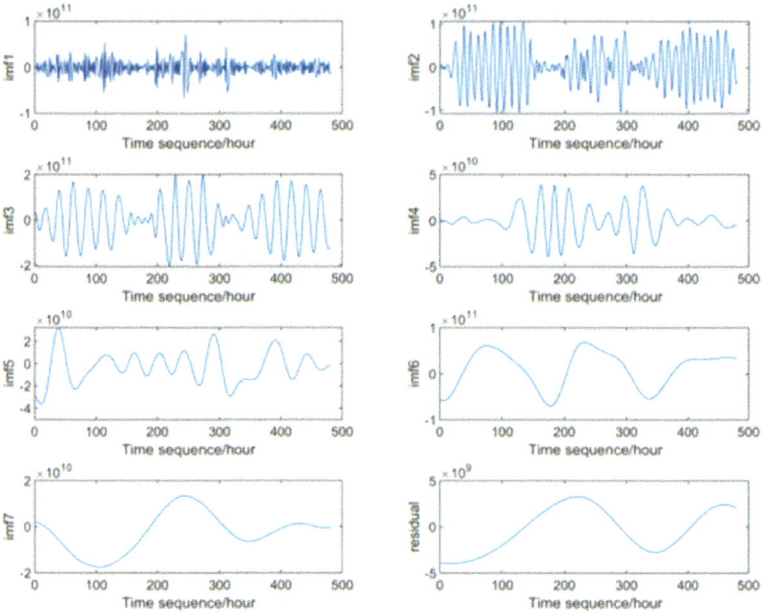

(c) CEEMD decomposition.

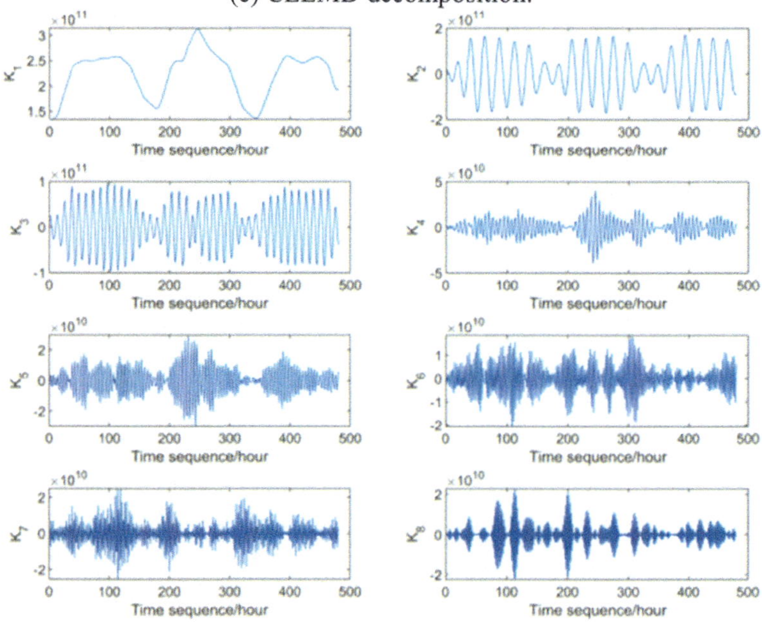

(d) VMD decomposition.

Fig. 4 (continued)

The data in all three data sets are recorded in "byte" units, and the value of them is very large, which will increase the complexity of data processing. In order to reduce the memory capacity and computational complexity required by network traffic data analysis and modeling, the logarithmic processing is carried out for all data in the original data (log2x, where x represents the data value) [13]. Since the logarithm function log2x is monotonically increasing in its domain, so taking the logarithm of the original data does not change the relative relationship of the data. Four methods, EMD-MRESN, EEMD-MRESN, CEEMD-MRESN and VMD-MRESN, are constructed by combining each TSD method with the multi-reservoir ESN, and the predicted output results of them using the test set are compared, as shown in Fig. 5.

The prediction performance of the four methods on the test set is compared by three statistical indicators: RMSE, MAE and MAPE, as shown in Table 2.

The TSD method is used to decompose the original data set into several data subsets with different scales, and then different reservoirs are built for modeling respectively. The output weights corresponding to different reservoirs are obtained through training, so as to reduce the influence of different data variation rules on the prediction performance of the model. As can be seen from Fig. 5 and Table 2, EMD-MRESN, EEMD-MRESN, CEEMD-MRESN and VMD-MRESN can all achieve good prediction performance. Among them, compared with the other three methods, VMD-MRESN method has the best predictive performance.

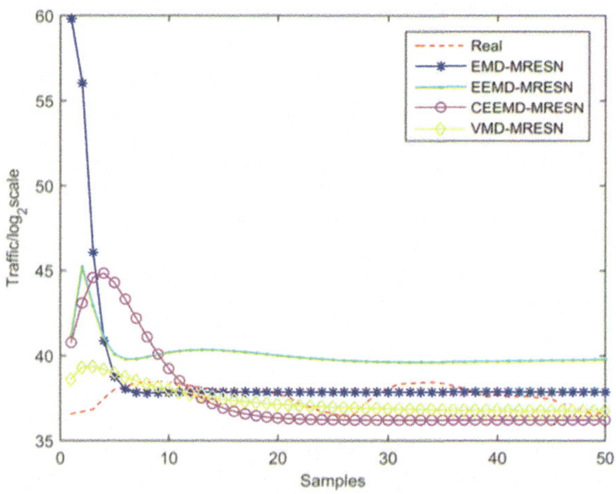

Fig. 5 Comparison of predicted output by different methods

Table 2 Comparison of the prediction performance of four methods

Method	RMSE/$\log_2{}^x$	MAE/$\log_2{}^x$	MAPE
EMD-MRESN	4.5392	1.5706	4.2532
EEMD-MRESN	2.7460	2.4535	6.5550
CEEMD-MRESN	2.5906	1.8788	4.9858
VMD-MRESN	0.9870	0.7712	2.0474

6 Conclusions

This paper discussed a network traffic prediction method based on signal decomposition and multiple reservoirs ESN. The original data were decomposed by EMD, EEMD, CEEMD and VMD, and then the structure of multiple reservoirs was constructed according to the number of data subsets, and the output of multiple reservoirs was integrated as the final output. Due to the fluctuating characteristics of the actual network traffic data, the multi-reservoir structure can effectively solve the problem of single model parameters. The stability of the model for non-linear and non-stationary data prediction could be enhanced through constructing multiple reservoirs for different data variation rules. It is found that the VMD-MRESN method has the best prediction performance by using the Wide Backbone network traffic data of MAWI Working Group in simulation experiments. Due to the random value of parameters of each sub-reservoir in its range, the structure of multi-reservoir ESN is not optimal, and the prediction performance of the model can be further improved. Therefore, the high performance swarm intelligence optimization algorithm can be combined with the multi-reservoir ESN, which can optimize the model parameters and further improve the prediction performance of the model.

References

1. Tian, Z.D., Li, S.J.: A network traffic prediction method based on IFS algorithm optimised LSSVM. Int. J. Eng. Syst. Model. Simul. **19**(4), 200–213 (2017)
2. Tian, Z.D., Li, S.J., Wang, Y.H., Wang, X.D.: A network traffic hybrid prediction model optimized by improved harmony search algorithm. Neural Netw. World **25**(6), 669–686 (2015)
3. Shi, J.M., Leau, Y.B., Li, K., Chen, H.D.: Optimal variational mode decomposition and integrated extreme learning machine for network traffic prediction. IEEE Access **2021**(9), 51818–51831 (2021)
4. Alarcon-Aquino, V., Barria, J.A.: Multiresolution FIR neural-network-based learning algorithm applied to network traffic prediction. IEEE Trans. Syst. Man Cybern. Part C **36**(2), 208–220 (2006)
5. Wei, D.: Network traffic prediction based on RBF neural network optimized by improved gravitation search algorithm. Neural Comput. Appl. **28**(8), 2303–2312 (2017)
6. Han, Y., Jing, Y.W., Dimirovski, G.M.: An improved fruit fly algorithm-unscented Kalman filter-echo state network method for time series prediction of the network traffic data with noises. Trans. Inst. Meas. Control. **42**(7), 1281–1293 (2020)
7. Li, C.S., Xiao, Z.G., Xia, X., Zou, W., Zhang, C.: A hybrid model based on synchronous optimisation for multi-step short-term wind speed forecasting. Appl. Energy **2018**(215), 131–144 (2018)
8. Han, Y., Jing, Y.W., Li, K., Dimirovski, G.M.: Network traffic prediction using variational mode decomposition and multi-reservoirs echo state network. IEEE Access **2019**(7), 138364–138377 (2019)
9. Huang, N.E., Shen, Z., Long, S.R., Wu, M.C., Shih, H.H., Zheng, Q., Yen, N.C., Tung, C.C., Liu, H.H.: The empirical mode decomposition and the Hilbert spectrum for nonlinear and non-stationary time series analysis. Proc. Math. Phys. Eng. Sci. **454**(1971), 903–995 (1998)
10. Wu, Z.H., Huang, N.E.: Ensemble empirical mode decomposition: a noise-assisted data analysis method. Adv. Adapt. Data Anal. **1**(1), 1–41 (2009)

11. Yeh, J.R., Shieh, J.S., Huang, N.E.: Complementary ensemble empirical mode decomposition: a novel noise enhanced data analysis method. Adv. Adapt. Data Anal. **2**(2), 135–156 (2010)
12. Dragomiretskiy, K., Zosso, D.: Variational mode decomposition. IEEE Trans. Signal Process. **62**(3), 531–544 (2014)
13. Fontugne, R., Abry, P., Fukuda, K., Veitch, D., Cho, K., Borgnat, P., Wendt, H.: Scaling in internet traffic: a 14 year and 3 day longitudinal study, with multiscale analyses and random projections. IEEE/ACM Trans. Netw. **25**(4), 2152–2165 (2017)

A Proposal for Improving Remaining Useful Life Prediction in Industrial Systems: A Deep Learning Approach

Adriana Villalón-Falcón, Alberto Prieto-Moreno, Marcos Quiñones-Grueiro, and Orestes Llanes-Santiago

Abstract Accurate prediction of the remaining useful life (RUL) of engineering systems provides decision-makers valuable information to apply more efficient maintenance programs maximizing the equipment usage and avoiding the increase of costs due to failures. In this area, deep learning methods have become increasingly popular because of their capability to learn complex and discriminative non-linear features that can facilitate the RUL prediction task. These network models are generally trained to minimize the mean square error (MSE) between the RUL prediction and its true value. This metric gives equal importance to the error at the beginning and at the end of a system's useful life. However, the prediction of the RUL is more critical as a system approaches the end of its useful life. In this chapter, a performance metric for evaluating prognostic models is proposed with the objective of establishing a direct relation between RUL prediction and maintenance planning. In addition, a procedure to use this metric for training a recurrent neural network (RNN) is proposed to improve the network's ability to learn the relationship between the raw data and the corresponding RUL, giving more importance to obtain accurate predictions as the system approaches to the end of its useful life. The procedure is applied to the NASA Commercial Modular Aero-Propulsion System Simulation (C-MAPSS) dataset. The satisfactory results confirm the validity of the proposal.

A. Villalón-Falcón · A. Prieto-Moreno · O. Llanes-Santiago (✉)
Universidad Tecnológica de La Habana José Antonio Echeverría, CUJAE, Marianao, La Habana, Cuba
e-mail: orestes@tesla.cujae.edu.cu

A. Villalón-Falcón
e-mail: avillalon@automatica.cujae.edu.cu

A. Prieto-Moreno
e-mail: albprieto@automatica.cujae.edu.cu

M. Quiñones-Grueiro
Institute for Software Integrated Systems, Vanderbilt University, Nashville, TN, USA
e-mail: marcos.quinones.grueiro@vanderbilt.edu

P. Shi et al. (eds.), *Complex Systems: Spanning Control and Computational Cybernetics: Applications*, Studies in Systems, Decision and Control 415,
https://doi.org/10.1007/978-3-031-00978-5_5

Keywords Remaining useful life · Prognostics · Deep learning · Performance metric · Recurrent neural network

1 Introduction

Accurate prediction of the remaining useful life (RUL) of engineering systems provides decision-makers valuable information to apply more efficient maintenance programs with the goal of reducing unplanned downtime as well as unnecessary preventative maintenance. Machine learning methods for prognostics have shown their ability to learn complex relationships obtaining accurate results for RUL prediction. Among them, deep learning methods have become increasingly popular because of their capability to learn complex and discriminative non-linear features that can facilitate the RUL prediction task [14].

Due to the mathematical nature of these models, it is hard to establish a relationship between the model parameters and the RUL, and to evaluate the effectiveness of the relationship learned by using performance metrics traditionally used for solving machine learning problems. Instead, if the objective function used to optimize the model parameters considers relevant information for decision making, it would be possible to establish the value of the obtained model for maintenance tasks.

Several metrics have been designed to measure prediction accuracy, taking into account decision-making for maintenance actions such as a time index that provides a required prediction horizon, or time required to apply a corrective action [10]. Some metrics give greater importance to the prediction error when the system is reaching the end of its useful life [7, 11, 17]. However, most literature does not report the use of these metrics during the training of prediction models. Therefore, the models proposed for RUL prediction fail to satisfactorily meet the requirements for which they were designed. Generally, the mean square error (MSE) metric is used as objective function during training [8, 12, 19]. MSE measures the prediction error by giving the same importance to the prediction of each observation irrespective of the associated time stamp. Therefore, no significant information related to RUL evolution is considered, i.e., prediction error evolution is masked, and a small error for RUL prediction near the end of useful life cannot be guaranteed.

The objective of this chapter is to propose a procedure for training neural network models, aiming to minimize the prediction error by incorporating information related to decision making for maintenance which represents the first of its contributions. A new performance metric is also presented in order to have a representation of prognostic performance that penalizes erroneous predictions near the system's end of useful life which constitutes the other contribution of the chapter. The proposed approach is validated by training and testing a deep learning model for the prediction of RUL of turbofan engines in the NASA Commercial Modular Aero-Propulsion System Simulation (CMAPSS) benchmark [11].

The chapter is organized as follows. In Sect. 2, preprocessing methods used for data preparation are presented. In Sect. 3, a literature review of performance metrics and the new performance metric are presented. In Sect. 4, the prognostic model and the proposed procedure are described. A description of the CMAPSS dataset is given in Sect. 5. Results of the application of the procedure in the CMAPSS dataset are shown in Sect. 6. Finally, some conclusions and future works are outlined.

2 Preprocessing

2.1 Data Labelling

In order to train a model for RUL prediction, a set of input and output data is required. The input data is the information from several sensors and the output data is the RUL. In this chapter, a piece-wise function (1) that represents the behavior of a system that starts to degrade after a certain time of operation, when some failure has occurred, is used to label the data [3]

$$l = \begin{cases} R_c, & \text{if} \quad 0 \leq c \leq SOF \\ EOL - c, & \text{if} \quad SOF < c \leq EOL \end{cases} \tag{1}$$

where l is the label which corresponds to the RUL in the time unit corresponding to system operation; EOL is the last time instant or end of life of the system; c is the current time instant; R_c is the initial constant value of RUL; and SOF is the start of failure which is equal to $EOL - R_c$.

2.2 Data Normalization

One of the most commonly used normalization techniques is called Z-Score, which is based on the mean and the standard deviation of the data to scale it. When working with different operating regimes, in order to ensure an equal contribution of each sensor in all regimes, it is better for the prediction task to incorporate the information of each regime in the normalization [1]. Thus, the normalization of the samples corresponding to each sensor f is performed according to Eq. (2)

$$N(x^{(r,f)}) = \frac{x^{(r,f)} - \mu^{(r,f)}}{\sigma^{(r,f)}} \tag{2}$$

where $x^{(r,f)}$ represents the data, $\mu^{(r,f)}$ and $\sigma^{(r,f)}$ are the mean and standard deviation of the data in operating regimen r, respectively.

3 Performance Metrics

Accuracy metrics quantify the similarity between the model prediction and true measured values [17]. Generally, metrics compute this similarity as the difference between the RUL predicted values (RUL^*) and true values (RUL), $\Delta = RUL^* - RUL$, also called prediction error. Accuracy metrics are created by modifying the prediction error equation, to add desired features for metrics, based on the prognostic methods capability for supporting maintenance-related decisions. The most relevant features considered in most metrics include, according to their recurrence and ordered by their importance, are:

1. *Overall performance*. Accuracy must be measured over the entire lifetime of the system, capturing the prediction error behavior. Some metrics have been used to measure the prediction error in an instant of the system degradation. RUL prediction is a continuous process, thus the evaluation of methods for this purpose requires measuring how the error changes over time. It is not enough to consider the error at specific time instants as unique evaluation measure, because the information of a single time instant is not representative of a model performance throughout the degradation process.
2. *Metric value in time units*. Accuracy value must be given in the time units of the RUL measurement (i.e. hour, day, cycle). The time unit of the prediction is key for measuring prognostic methods accuracy, allowing to establish a connection to equipment operation and maintenance planning. Some metrics provide normalized values, generally in the range [0, 1], masking the prediction time unit. Measures such as mean and median are commonly used to summarize the error made at each prediction time.
3. *Time based penalization*. Decision making is critical towards the end of life. Therefore, a penalization factor must be added to give greater importance to prediction errors made near the end of life. Usually, a function is defined to penalize the error given the time instant at which the prediction is made. A linear function (3) and a Gaussian kernel (4) have been used as penalization functions.

$$\alpha(t) = \frac{t}{\sum_{t=t_0}^{T} t} \tag{3}$$

$$\alpha(t) = e^{-\frac{(t-T)^2}{\frac{T^2}{2}}} \tag{4}$$

   where t_0 and T are the start and end of life time instants, respectively.
4. *Late prediction based penalization*. Late predictions (positive Δ) are penalized over early predictions (negative Δ), due to the impact on maintenance. Positive errors are made for predicting a higher RUL value than the true RUL, causing the system to reach the end of life before maintenance. Conversely, negative errors favor the execution of maintenance tasks before the end of the useful life of the system. Commonly, a function is defined to penalize the error. This function is used as conversion function that receives a prediction error and retrieves a value

Table 1 Features of accuracy metrics

Metric	Overall performance	Value in time units	Time penalization	Late prediction penalization
Mean Absolute Error (MAE) [18, 20]	Yes	Yes	No	No
Exponential Transformed Accuracy (ETA) [7]	No	Normalized	No	Yes
Relative Accuracy (RA) [11]	No	Normalized	No	No
Cumulative relative accuracy (CRA) [11]	Yes	Normalized	Yes	No
Mean Square Error (MSE) [2, 17]	Yes	Yes	No	No
Root Mean Square Error (RMSE) [7, 17]	Yes	Yes	No	No
Mean Absolute Percentage Error (MAPE) [17, 18]	Yes	Normalized	No	No
Sample Mean Error (SME) [2, 17]	Yes	Yes	No	No
Sample Median Error (SMeE) [2, 17]	Yes	Yes	No	No
Timeliness Weighted Error Bias (TWEB) [2, 17]	Yes	Normalized	Yes	Yes

related to its magnitude and sign. An exponential function (5) has been used as penalization function.

$$\zeta(t) = \begin{cases} e^{-\frac{\Delta}{\varphi_1}} - 1, \text{ if } \Delta < 0 \\ e^{\frac{\Delta}{\varphi_2}} - 1, \quad \text{if } \Delta \geq 0 \end{cases} \tag{5}$$

A summary of how these features are included in the most relevant accuracy metrics is presented in Table 1.

Although penalization functions presented in Eqs. (3) and (4) give a larger value as time approaches to the end of life, they do not include information about the critical moment for decision making in a particular application. It is expected that a prognostic performance metric holds some physical significance such as a time index that provides a required prediction horizon, or time required to apply a corrective action [11]. Therefore, a new accuracy metric to evaluate the performance of RUL prediction methods, that includes the first three features presented, and capture knowledge for supporting decision making, is proposed.

In order to achieve a better evaluation of the performance of RUL prediction models, this metric periodically measures the error during the degradation process and includes time-based penalization, keeping values time unit.

3.1 Proposed Performance Metric

The proposed metric is defined as follows:

Definition 1 $(Root\ Weighted\ Mean\ Squared\ Error)$ Given a set $\{E_i\} \in \mathbf{E}$ with $i = 1, 2, ..., N$ of representative systems of the same type of system **E**, the **Root Weighted Mean Squared Error** ($RWMSE$) for **E** is obtained as:

$$RWMSE = \frac{\sum_{i=1}^{N} RWMSE_i}{N} \tag{6}$$

where

$$RWMSE_i = \sqrt{\frac{\sum_{t=1}^{T} \alpha_{RUL_{i,t}} * (RUL^*_{i,t} - RUL_{i,t})^2}{\sum_{t=1}^{T} \alpha_{RUL_{i,t}}}} \tag{7}$$

is a weighted average of the prediction errors during the degradation of system i, $RUL^*_{i,t}$ and $RUL_{i,t}$ are the predicted and true RUL at instant t during the degradation of system i, respectively, T is the time frame over which the deviation is measured, $\alpha_{RUL_{i,t}}$ is the weight assigned to the RUL prediction error of system i when it is calculated for each instant t, and N is the number of systems.

Several types of functions could be used to determine $\alpha_{RUL_{i,t}}$. In this chapter, the exponential function given in Eq. (8) is selected. Considering that, as the system approaches the end of its useful life, the importance of accurately estimating the value of the RUL grows exponentially.

$$\alpha_{RUL_{i,t}} = a * e^{bx} + c \tag{8}$$

where x is defined as $RUL_{i,t} - RUL_{warning}$, and $RUL_{warning}$ is the value of RUL from which the precision of the RUL prediction is considered critical, such that the weight assigned to the error increases from that moment on. Parameters $a, c \geq 0$ such that the effects of errors do not cancel each other and each error contributes to the average. Moreover, $a = 1$ to differentiate the error weights assigned to both sides of $RUL_{warning}$. Parameter $b \in \mathbb{R}^-$ because $\alpha_{RUL_{i,t}}$ should increase as the RUL decreases. Function $\alpha_{RUL_{i,t}}$ should present a smooth shape, gradually varying as it approaches to $RUL_{i,t} = 0$. Therefore, the smooth shape is guaranteed by selecting $b = \frac{-1}{RUL_{warning}}$. In practice, $RUL_{warning}$ value can be defined by experts in the maintenance area or it can be assigned based on the knowledge acquired from the data.

Substituting the values assigned to the parameters in Eq. (8), the weight function obtained is:

$$\alpha_{RUL_{i,t}} = e^{1 - \frac{RUL_{i,t}}{RUL_{warning}}} \tag{9}$$

Figure 1 compares weights assignment using uniform distributed weights and exponential distributed weights.

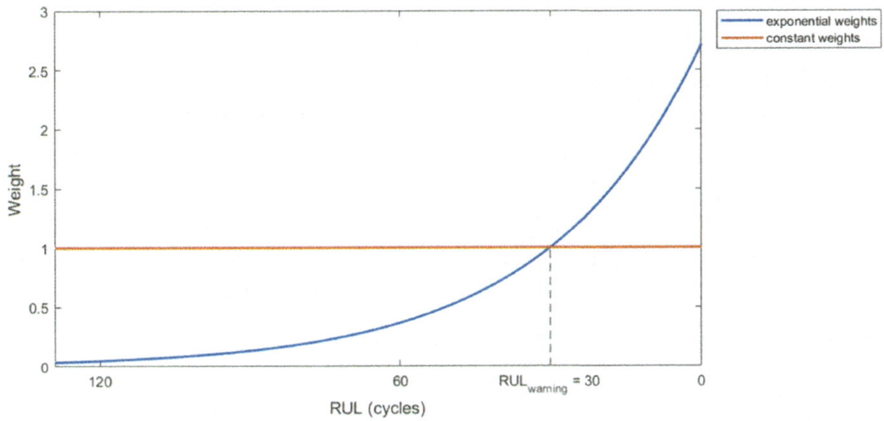

Fig. 1 Weights from exponential and constant function

4 Prognostic Model and Model Training

4.1 Prognostic Model

Neural networks are one of the most widely used data-driven methods, due to their ability to model complex nonlinear relationships among variables. Among them, deep networks have gained major attention due to their ability to automatically learn deep hidden features in data and to establish a direct relationship between raw data and the RUL, decreasing the manual feature extraction process [9, 14]. Generally, a deep network model designed for RUL prediction consists of three main parts: (1) an input layer, where sensor data is received; (2) n-hidden layers, which are selected considering the type of features to be extracted from data; and (3) an output layer, which learns the final relationship between the extracted features and the RUL.

Model selection for RUL prediction depends on the complexity of the system and the degradation patterns shown in sensor data. Convolutional neural network (CNN) [5, 8], auto-encoder (AE) [13, 15], and recurrent neural network (RNN) [12, 16] are some of the most common deep network architectures used for RUL prediction.

4.2 Model Training

In order to obtain a model to accurately predict the RUL, the deep network model should be trained with the sensor data as input data and the labels generated by the piece-wise function in Eq. (1) as output data.

Some network models, such as CNN or RNN, due to their architecture and internal functioning, require segmentation of the input data into time windows of length s.

Thus, the input of the model is a 3D tensor with shape $(T - s, s, V)$, and the output of the model is a 3D tensor with shape $(T - s, s, 1)$, where T is the overall length of a system degradation data, and V is the original feature dimension.

Algorithm 1 shows the proposed procedure for training neural networks.

Algorithm 1 Training algorithm

Parameters: $loss$: function to evaluate the fit of the model to the data, $optimizer$: method to update network parameters (weights and biases), lr: learning rate, $minibatchsize$: size of the mini-batch to update network parameters, $epochs$: number of times that the data is passed through the network
Inputs: M: training measurements, RUL: training RUL

Initialization: $w \leftarrow initialization()$
for $epoch = 1 : epochs$ **do**
  for $batch = 1 : minibatchsize$ **do**
    Obtain batch data M_b and RUL_b
    Obtain network output $RUL_b{}^* = model(M_b)$
    Obtain prediction error $l_b = loss(RUL_b, RUL_b{}^*)$
    Update network parameters $w \leftarrow optimizer(lr, l_b, w)$
  end for
end for

The loss function is one of the most important parameters of the training algorithm because it determines the computation of the prediction error, working as the objective function of the optimization algorithm. Using RWMSE metric for training neural network models implies weighting the prediction error obtained with each sample considering the evolution of the error in a degradation trajectory. Thus, batches are formed by grouping degradation trajectories that show entire degradation patterns. For training, RWMSE metric is adapted to a weighted mean squared error (WMSE) (10) to be used as a loss function.

$$WMSE = \alpha_{i,t} * (RUL_{i,t}{}^* - RUL_{i,t})^2 \tag{10}$$

5 Dataset Description

In this chapter, the NASA Commercial Modular Aero-Propulsion System Simulation (CMAPPS) dataset is used for training and validating the proposal [11]. The C-MAPPS dataset is formed by four distinct datasets. Each dataset contains a number of training engines (Engines: Training (N)) with run-to-failure information and a number of testing engines (Engines: Testing) with information terminating before a failure is observed (see Table 2). There are two failure modes: high-pressure compres-

Table 2 C-MAPSS dataset description

Data	FD001	FD002	FD003	FD004
Engines: training	100	260	100	249
Engines: test	100	259	100	248
Operating conditions	1	6	1	6
Number of failure modes	1	1	2	2

sor degradation and fan degradation. Each engine provides the following information: engine identifier, operation time (in flight cycles), 3 operating condition parameters (altitude, mach number and throttle resolver angle) showing six operating regimes, and 21 sensor signals (4 temperatures, 4 pressures, 6 speeds, and 7 additional variables).

Engines start operating with various degrees of initial wear but are considered healthy. As the number of cycles increases, the engines start to deteriorate until they can no longer function. At this point, engines are considered unhealthy. The training dataset has information collected over the entire life of the engines until failure. Unlike the training dataset, the testing dataset contains temporal data that terminates before a system failure. The objective is to predict the RUL of testing engines [11].

5.1 *Preprocessing*

5.1.1 Data Labeling

For these datasets, Eq. (1) is used to label the data. In this chapter, R_c is set to 130 since it is the value generally used in the literature [3]. Figure 2 shows the true RUL calculated for engine 17, where EOL is 276 and SOF is 146.

5.2 *Data Normalization*

In the dataset, there are three variables (altitude, mach number, and throttle resolver angle) that refer to the operating conditions of the turbines and these have a strong impact on the performance of the system. In the datasets, where the six operating regimes are present, Eq. (2) is used to normalize the data.

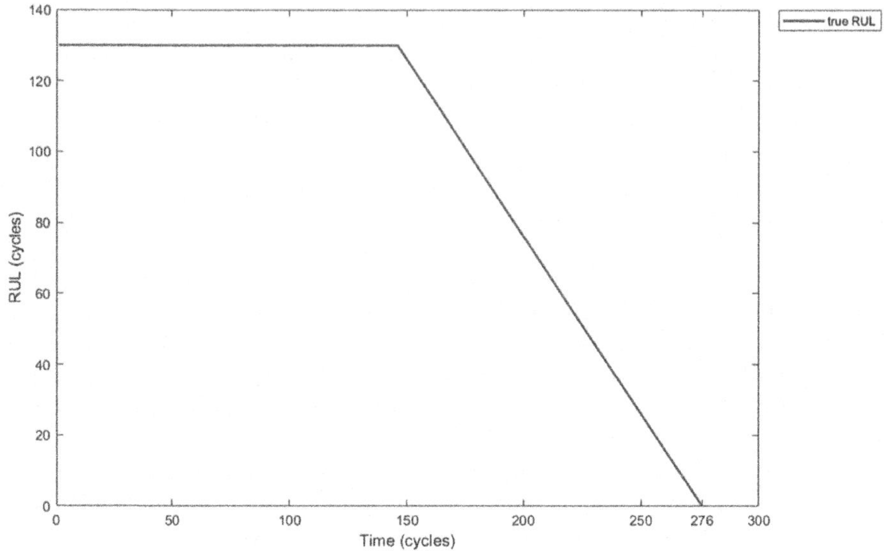

Fig. 2 RUL values from engine 17 in dataset FD001

6 Experiments and Results

In this chapter, a deep network hybrid model formed by an auto-encoder and a bidirectional long short-term memory (AE-BLSTM) networks is selected for testing the proposed procedure due to its outstanding results on prognostics [13]. The AE acts as a feature extractor to compress monitoring data and consists of three layers: (1) input layer, where the input neurons receive data corresponding to the flight cycle, 3 operating conditions and 21 sensor signals; (2) an encoding layer with 12 neurons containing a rectified linear unit (ReLU) activation function; and (3) a decoding layer with same dimension as input layer, and a sigmoid activation function. The BLSTM is a temporal modeling tool and consists of 5 layers: (1) input layer, where the input neurons receive data from the AE encoding layer; (2) a bidirectional lstm layer with 100 neurons and a hyperbolic tangent activation function; (3) a bidirectional lstm layer with 50 neurons and a hyperbolic tangent activation function; (4) a fully connected layer with 30 neurons and a ReLU activation function; (5) and an output layer with 1 neuron, and a linear activation function.

To meet the input shape requirements of BLSTM, input data are segmented into time windows of length $s = 30$ cycles. Thus, the input of the model is a tensor with shape $(T - 30, 30, 25)$, the output of autoencoder's encoding layer is a tensor with shape $(T - 30, 30, 12)$, and the output of the BLSTM is a tensor with shape $(T - 30, 30, 1)$.

The AE-BLSTM model is obtained by first training the AE model for low-dimensional representation of the data through encoding and decoding. BLSTM is appended to the encoding part of AE, and trained for learning long-range dependencies of features and mapping the learned feature representation to sample labels, RUL.

The AE model is trained with mean square error (11) as loss function to measure the error from reconstructing the data, and adaptive moment estimation (Adam) optimizer to minimize the loss function [6].

The BLSTM model is trained following Algorithm 1 using MSE (11) and WMSE (10) to demonstrate the influence of training with these loss functions on the RUL prediction. Generally, when training with MSE, the data for each batch is randomly selected from the data. However, when using WMSE, batches are formed by selecting degradation trajectories that show entire degradation patterns, instead of randomly selected samples that can be located in different trajectories, not showing a degradation pattern. RMSProp optimizer is used to update the network parameters during training [4], with a mini-batch approach to optimize the loss function by iteratively updating the network weights and biases.

$$MSE = (RUL_{i,t}{}^{*} - RUL_{i,t})^2 \tag{11}$$

Training hyper-parameters (learning rate, mini-batch size, and number of epochs) are selected according to literature or tuned by grid search. For AE model training, the mini-batch size is set to 200 samples and the number of epochs is set to 50 [13]. For BLSTM model training, the mini-batch size for MSE training is set to 200 samples and the number of epochs is set to 32.

The learning rate and the mini-batch size for WMSE training are selected by grid search, where the value range is $\{0.001, 0.01, 0.1\}$ and $\{1, 5, 10\}$, correspondingly. After a comparative analysis, the learning rate is set to 0.001 for both models. For BLSTM model training, mini-batch size for WMSE training is set to 1 trajectory.

In experiments, the training set in each dataset is used to train and validate the procedure, since the test set does not contain information of the operation of the engines until failure. This information is necessary to evaluate the prediction models throughout the degradation process. For training with WMSE loss function, the value of $RUL_{warning}$ is set to 30 cycles, considering the windows size generally used to process data for this dataset. A k-fold cross validation procedure has been used on the datasets with $k = 10$, in order to compare loss functions influence on the training process.

Since the goal is to minimize the prediction error near the end of the useful life, a first error estimation is made by obtaining the RMSE over last 30 cycles of each engine (see Table 3). The results show that the first goal is achieved by obtaining a low prediction error when training with MSE and WMSE, in which the proposed approach in this paper presents the smaller error in three of the four datasets. In all datasets, a smaller error variance is obtained with the proposal.

Now, the prediction error is calculated using RWMSE metric over the last 130 cycles of each engines to compare prognostic models based on error evolution throughout the degradation process (see Table 4). The prediction errors of each model in each partition are compared using the Wilcoxon statistical test. The significance

Table 3 RMSE (life cycles). RUL prediction performance in the last 30 cycles for MLP model when training with MSE and WMSE as loss functions

Metric	FD001		FD002		FD003		FD004	
	MSE	WMSE	MSE	WMSE	MSE	WMSE	MSE	WMSE
Average	3.65	**3.17**	5.78	**5.61**	**4.27**	4.48	6.44	**5.56**
Standard deviation	2.02	**1.52**	3.10	**2.99**	2.66	**2.55**	7.66	**3.09**

Table 4 RWMSE (life cycles). RUL prediction performance is the last 130 cycles for MLP model when training with MSE and WMSE as loss functions

Metric	FD001		FD002		FD003		FD004	
	MSE	WMSE	MSE	WMSE	MSE	WMSE	MSE	WMSE
Average	8.44	**7.09**	11.94	**10.21**	9.45	**8.75**	12.53	**9.93**
Standard deviation	3.53	**2.64**	5.11	**3.64**	5.09	**3.95**	8.16	**4.31**
Test p-value	**1.80e-03**		**2.33e-04**		0.57e-00		**2.23e-04**	

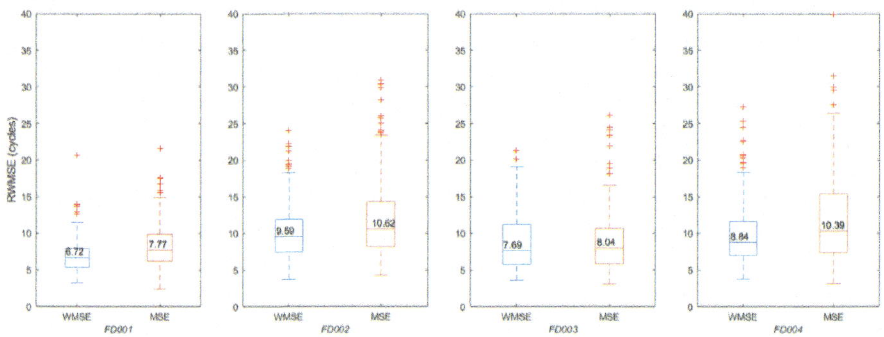

Fig. 3 RWMSE calculated over the last 130 cycles of life

level (α) of the test is 0.05. Test results show that the proposed approach in this chapter has the smaller error (see Fig. 3), with a significant difference in three of the forth datasets. With this metric, in all datasets a smaller error variance is also obtained with the proposal.

As the objective of this work is to minimize the error prediction of the RUL at the end of the useful life, Fig. 4 presents, as example, the predicted RUL for engine 17 in dataset FD001 throughout the degradation process obtained from models trained using MSE and WMSE versus the actual remaining useful life. This figure shows how with the proposed approach, the RUL predicted gets closer to the true RUL towards the end of the life of the engine.

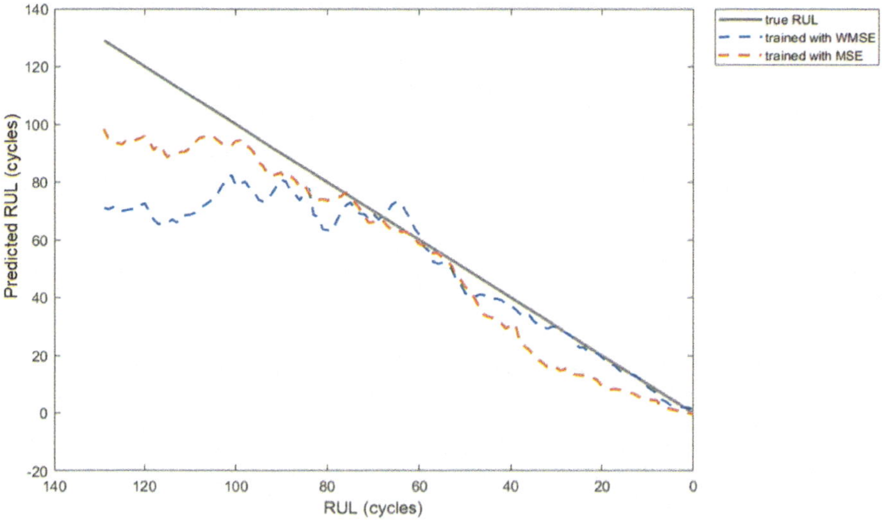

Fig. 4 RUL prediction from engine 17 in dataset FD001

The results show that the models trained with the RWMSE with $RUL_{warning} = 30$ have a small error in the last 30 cycles of the useful life of the engines. Also, the error in the RUL prediction evolves in such a way that it decreases as the engines are near the end of their useful life. In addition, maintenance actions can be planned when the RUL of the engines is in a neighborhood of 30 cycles, considering the variation of the prediction error.

7 Conclusions and Future Work

In this chapter, a procedure is proposed to improve neural network's RUL prediction accuracy. The procedure allows to minimize the prediction error as a system approaches the end of its useful life, weighting the contribution of the prediction error at each time instant to update model parameters during training. The results show a significant improvement in most of the datasets, demonstrating how it is possible to obtain a model with a lower RUL prediction error near the end of the useful life of the system by modifying the training process. To evaluate its effectiveness, an AE-BLSTM architecture was trained for predicting the RUL of turbofan engines from the C-MAPSS dataset.

The new performance metric, RWMSE, proposed to evaluate RUL prediction models, allows linking the prediction of the RUL with the planning of the maintenance tasks by considering the term $RUL_{warning}$, providing useful information for decision making. Therefore, with the prognostic model built on this basis, a significant and understandable relationship can be established between the RUL prediction and maintenance tasks. This metric is a first step for generalizing prognostic metrics with the goal of unifying their strengths and removing their limitations in evaluating how well does a model perform towards predictive maintenance.

In further research, a thorough analysis and comparison of the procedure application on several machine and deep learning models will be made to establish the limitations of the proposal.

References

1. Alberto-Olivares, M., Gonzalez-Gutierrez, A., Tovar-Arriaga, S., Gorrostieta-Hurtado, E.: Remaining useful life prediction for turbofan based on a multilayer perceptron and kalman filter. In: 2019 16th International Conference on Electrical Engineering, Computing Science and Automatic Control (CCE), pp. 1–6. IEEE (2019)
2. Di Maio, F., Turati, P., Zio, E.: Prediction capability assessment of data-driven prognostic methods for railway applications. In: Proceedings of the Third European Conference of the Prognostic and Health Management Society (2016)
3. Heimes, F.O.: Recurrent neural networks for remaining useful life estimation. In: 2008 International Conference on Prognostics and Health Management, pp. 1–6. IEEE (2008)
4. Hinton, G., Srivastava, N., Swersky, K.: Neural networks for machine learning lecture 6a overview of mini-batch gradient descent. Cited on **14**(8), (2012)
5. Huang, C.G., Huang, H.Z., Li, Y.F., Peng, W.: A novel deep convolutional neural network-bootstrap integrated method for RUL prediction of rolling bearing. J. Manuf. Syst. (2021)
6. Kingma, D.P., Ba, J.: Adam: a method for stochastic optimization. arXiv:1412.6980 (2014)
7. Lei, Y., Li, N., Guo, L., Li, N., Yan, T., Lin, J.: Machinery health prognostics: a systematic review from data acquisition to RUL prediction. Mech. Syst. Signal Process. **104**, 799–834 (2018)
8. Li, H., Zhao, W., Zhang, Y., Zio, E.: Remaining useful life prediction using multi-scale deep convolution neural network. Appl. Soft Comput. **89**, 106–113 (2020)
9. Pei, H., Si, X.S., Hu, C.H., Zheng, J.F., Li, T.M., Zhang, J.X., Pang, Z.N.: An adaptive prognostics method for fusing CDBN and diffusion process: application to bearing data. Neurocomputing **421**, 303–315 (2021)
10. Saxena, A., Celaya, J., Saha, B., Saha, S., Goebel, K.: Metrics for offline evaluation of prognostic performance. Int. J. Progn. Health Manag. **1**(1), 4–23 (2010)
11. Saxena, A., Goebel, K., Simon, D., Eklund, N.: Damage propagation modeling for aircraft engine run-to-failure simulation. In: 2008 International Conference on Prognostics and Health Management, pp. 1–9. IEEE (2008)
12. Shi, Z., Chehade, A.: A dual-LSTM framework combining change point detection and remaining useful life prediction. Reliab. Eng. Syst. Saf. **205**, 107257 (2021)
13. Song, Y., Shi, G., Chen, L., Huang, X., Xia, T.: Remaining useful life prediction of turbofan engine using hybrid model based on autoencoder and bidirectional long short-term memory. J. Shanghai Jiatong Univ. (Sci.) **23**(1), 85–94 (2018)
14. Wang, B., Lei, Y., Yan, T., Li, N., Guo, L.: Recurrent convolutional neural network: a new framework for remaining useful life prediction of machinery. Neurocomputing **379**, 117–129 (2020)
15. Wang, H., Peng, M.J., Miao, Z., Liu, Y.K., Ayodeji, A., Hao, C.: Remaining useful life prediction techniques for electric valves based on convolution auto encoder and long short term memory. ISA Trans. **108**, 333–342 (2021)
16. Wu, S., Jiang, Y., Luo, H., Yin, S.: Remaining useful life prediction for ion etching machine cooling system using deep recurrent neural network-based approaches. Control Eng. Pract. **109**, 104748 (2021)

17. Zeng, Z., Di Maio, F., Zio, E., Kang, R.: A hierarchical decision-making framework for the assessment of the prediction capability of prognostic methods. Proc. Inst. Mech. Eng., Part 0: J. Risk Reliab. **231**(1), 36–52 (2017)
18. Zhang, H., Mo, Z., Wang, J., Miao, Q.: Nonlinear-drifted fractional brownian motion with multiple hidden state variables for remaining useful life prediction of lithium-ion batteries. IEEE Trans. Reliab. **69**(2), 768–780 (2019)
19. Zheng, S., Ristovski, K., Farahat, A., Gupta, C.: Long short-term memory network for remaining useful life estimation. In: 2017 IEEE International Conference on Prognostics and Health Management (ICPHM), pp. 88–95. IEEE (2017)
20. Zhu, J., Chen, N., Peng, W.: Estimation of bearing remaining useful life based on multiscale convolutional neural network. IEEE Trans. Ind. Electron. **66**(4), 3208–3216 (2018)

Fault Diagnosis/Fault-Tolerant Control: A Survey of Results for Linear Systems Over Frequency Region in Presence of Disturbances

Jovan Stefanovski

Abstract This chapter presents a survey of recent results on the fault detection, fault estimation, fault-tolerant control, and fault-tolerant tracking, published in the most important control system journals today. The faults are considered additive inputs in respect to the plant dynamics. Although the results are given without complete proofs, the key arguments that are used in the original proofs are given. The results are illustrated by examples.

Keywords Fault-tolerant control · Fault-tolerant tracking · Fault detection · Fault estimation

1 Introduction

This chapter elaborates on the following plant, given by the following descriptor system:

$$
\begin{aligned}
E\dot{x}(t) &= Ax(t) + B_{\mathrm{f}} f(t) + B_{\mathrm{d}} d(t) + B_2 u(t)\,, \quad x(0) = x_0\,,\\
z(t) &= C_1 x(t) + D_{1\mathrm{f}} f(t) + D_{1\mathrm{d}} d(t) + D_{12} u(t)\,,\\
y(t) &= C_2 x(t) + D_{2\mathrm{f}} f(t) + D_{2\mathrm{d}} d(t) + D_{22} u(t)\,,
\end{aligned}
\tag{1}
$$

where $f(t)$ is m_{f}-dimensional input called fault,
$d(t)$ is m_{d}-dimensional input called disturbance,
$u(t)$ is m_2-dimensional input called control,
$z(t)$ is p_1-dimensional output called controlled output, and

J. Stefanovski (✉)
Control and Informatics Div., JP "Streževo", Bitola, Republic of North Macedonia
e-mail: jovanstef@t.mk

P. Shi et al. (eds.), *Complex Systems: Spanning Control and Computational Cybernetics: Applications*, Studies in Systems, Decision and Control 415,
https://doi.org/10.1007/978-3-031-00978-5_6

$y(t)$ is p_2-dimensional output called measurement.

For the fault detection (FD) (and also for the fault estimation (FE)), Sect. 3), there is no need of controlled output z in (1). Therefore, we can simplify the equation for y in (1) by writing $y(t) = Cx(t) + D_{\mathrm{f}} f(t) + D_{\mathrm{d}} d(t) + D_{\mathrm{u}} u(t)$, and re-denote matrix B_2 by B_{u}. Then we define the RMs

$$[\boldsymbol{G}_{\mathrm{f}}, \boldsymbol{G}_{\mathrm{d}}, \boldsymbol{G}_{\mathrm{u}}] = [D_{\mathrm{f}}, D_{\mathrm{d}}, D_{\mathrm{u}}] + C(sE - A)^{-1}[B_{\mathrm{f}}, B_{\mathrm{d}}, B_{\mathrm{u}}] , \tag{2}$$

while $y(t) \in \mathbb{R}^{n_{\mathrm{y}}}$, $f(t) \in \mathbb{R}^{n_{\mathrm{f}}}$, $d(t) \in \mathbb{R}^{n_{\mathrm{d}}}$, $u(t) \in \mathbb{R}^{n_{\mathrm{u}}}$.

Since the theory for descriptor systems is well-developed, see [19], it was not hard to present the results for those systems.

The notion fault diagnosis is related to all notions fault detection, fault location (somewhere called fault isolation) and fault estimation. The disturbance and fault spectrum can lie in some known frequency regions, denoted by $\mathbb{B}_{\mathrm{d}}$ and $\mathbb{B}_{\mathrm{f}}$, respectively, or their frequency region is not known. In the former case, the finite frequency region theory can be applied [20, 21]. In this chapter we apply the finite frequency region theory of [39]. The fault detection is to detect the appearance of a fault in an arbitrary channel fault. The corresponding filter is called fault detector. The fault location is to detect the channel in which a fault appears. The corresponding filter is called fault locator. The fault estimation is to estimate the magnitude of the fault in any time instance. The corresponding filter is called fault estimator.

In some cases, it is more efficient if one applies banks of fault detectors, fault locators and fault estimators. The theory of this chapter can be applied also to construct all of those banks of filters. In particular, each filter of the bank, dedicated to a particular fault, is constructed by considering that the remaining faults are disturbances. Actually, fault locators can be realized by a bank of fault detectors. In some cases of estimation of faults that appear simultaneously, it is more efficient to apply all three banks of filters. In the cases that the faults can not appear simultaneously, each filter of the bank can be constructed by considering that the remaining faults are zero. In all those cases, the filters can also be constructed by the theory of this chapter.

In Sect. 1 we present various criteria for FD. The most general one is Problem 1 in Sect. 1, which requires arbitrary dimension n_{r} of the residual vector and arbitrary frequency regions of the fault and disturbances. The problem, which generalizes the $\mathscr{H}_{/}\mathscr{H}_{\infty}$ FD problem, [22, 31], is solved in Theorem 1, using the generalized spectral factorization, [39], with sufficient conditions. Another FD criterion is formulated as Problem 1S, where a requirement is $n_{\mathrm{r}} \geq n_{\mathrm{f}}$. The problem is solved with necessary and sufficient conditions in Theorem 3. In Theorem 5, all solutions of the FD problem with the constraints $n_{\mathrm{r}} = n_{\mathrm{f}}$ and over the whole extended imaginary axis (EIA) are found. In Theorem 6, necessary and sufficient conditions for Problem 1 are given, under the assumption $\mathbb{B}_{\mathrm{f}} \subseteq \mathbb{B}_{\mathrm{d}}$. In Theorem 7, Problem 1 over the EIA is solved with sufficient conditions, under the constraint $n_{\mathrm{r}} < n_{\mathrm{f}}$.

The fault estimation in presence of disturbances results in the literature are actually results on input estimation in presence of disturbances. In this chapter we formulate three such problems and corresponding filters, see Problems 1, 2 and 3:$\mathbb{B}$ in Sect. 3, with comparison. Theorem 8 is the main result, which gives a solution to Problem 3:$\mathbb{B}$.

To formulate criteria for fault-tolerant control (FTC), the nature of the faults and disturbances must be explained at first. The disturbances appear frequently (or are persistent), and the faults appear rarely (or never). For that reason, we require almost the best performance of the closed loop system, in times when there are no faults. In other words, we require a behaviour of the closed loop system that is similar to the behaviour of the closed loop system with applied controller (called nominal controller) that is constructed under the condition that there are no faults. This requirement is formulated in [68]. However, when faults appear, for safety reasons, we allow a worse than nominal performance (for example, more spent energy), while retaining some basic system properties (for example, stability). As a nominal controller, a controller $\boldsymbol{K}_{\mathrm{d}}$, obtained by solving the (sub)optimal $\mathscr{H}_\infty$ control problem without faults, can be used. Denote by β_{d} the infimum of the latter problem.

The FTC can be solved by fault diagnosis blocks, included in the control loop, and reconfigurable controllers, known as active FTC, see page 2 of [3]. Sub-classes of the active FTC are: (a) a fault detection block is included in the control loop, and (b) a fault location block is included in the control loop. A combination of filters in (a) and (b) can be used, also, Therefore, active controllers are reconfigurable ones, with the reconfiguration made on the basis of the information obtained from a fault detector/locator. In general, a drawback of the active FTC is that a controller cannot be constructed for systems in which the fault detection/location/estimation is not reliable (see Introduction of [1]).

The active FTC apply "switching controller algorithms". Namely, in the control design stage, a nominal controller and at least fault detection/location blocks are designed. Then, in real-time, when a fault situation is detected/located, the acting controller switches with a controller that corresponds to the fault situation (see Fig. 3 in [51]). Drawbacks of this "switching controller algorithm" are that it is complex (see Introduction of [1]), that certain time for fault detection/location before another controller is computed and applied is needed, and that the variables (outputs and state) are not smooth, as functions of time (see Sect. 9.5 of [3]). One of the reasons is that the mentioned establishing of the information about the fault appearance can take time. Related is the appearance of impulses in times when a fault appears and the controller switches.

On the other hand, a passive FTC is one that is not active, and the controller is permanent, not dependant on the appearance of faults (see a comparison of passive versus active FTC, page 347 of [3]). The passive FTC controllers are designed under the criterion of a pre-defined performance and a pre-defined insensitivity to faults, of the closed loop system. The passive FTC controllers: (a) can include a fault estimator in the control loop, and (b) are constructed without fault diagnosis blocks. A passive controller in the group (b) is $\boldsymbol{K}_{\mathrm{fd}}$. It is an optimal $\mathscr{H}_\infty$ controller for the system in Fig. 13, constructed by considering that both f and d are disturbances (see Sect. 9.4.4 of [3]). Denote by β_{fd} the infimum of this problem. With the controller

$\boldsymbol{K}_{\mathrm{fd}}$, the performance of the closed-loop system without faults is worse in respect to the case when the nominal controller $\boldsymbol{K}_{\mathrm{d}}$ is applied, therefore, the requirement of [68] is not satisfied with $\boldsymbol{K}_{\mathrm{fd}}$.

There are also active controllers which include fault estimator in the control loop. The inclusion of a fault estimator in the control loop is not a surprise, because by Theorem 2 of [13] (see also Fig. 6 in [13]), each stabilizing controller can be realized with an included residual generator (see also [2, 16]). Controllers that include FE in the control loop are presented in [15, 24, 26] and Sect. 9.2 in [3], for matched actuator faults, and in [25, 61], for unmatched actuator faults.

In Sect. 5 we present three types of passive FTC: One that needs an FE (Sect. 4.1), and two that do not need an FE (Sects. 4.2 and 4.4).

In Sect. 4.1, we present a criterion for FTC (see criterion (49) below). In Sect. 4.2 we present another criterion (see criterion (66)) and a quite general FTC scheme, with included fault estimator (see Fig. 18). In Sect. 4.4, under structural Assumption 21, we find a minimal realization of the controller presented in Fig. 18, so that the information of the fault estimation is lost.

Concerning the fault-tolerant tracking (FTT) problem, our tracking controller will be a passive one. Actually, the performance will be perfect tracking with a minimal norm of the control. We use some ideas of the existing theory of tracking with disturbance rejection, in the construction of the controller. The existing works can be grouped in two global groups: The first one elaborates on the class of plants and inputs in which ideal tracking and ideal disturbance rejection is achievable, i.e. the tracking error tends to zero when $t \to \infty$, for all assumed inputs ([7], Chap. 13 of [34, 50, 57], and Sect. VII of [58]).

The second group elaborates on the plants and/or inputs in which the ideal behaviour is not achievable. The authors set some analytic criteria (for a distance to the ideal performance), and minimize the criteria ([10, 17, 27, 52, 53, 59]).

The first group is closer to our work, therefore we give more details. There are at least three classes of plants and inputs of ideal tracking with disturbance rejection:

(1) The most restrictive one (in respect to the class of inputs, but the most general in respect to the class of plants) requires a strong condition, that $\bar{\boldsymbol{Q}}^{-1}\boldsymbol{r}$ is a proper stable rational vector, where $\boldsymbol{G} = \bar{\boldsymbol{Q}}^{-1}\bar{\boldsymbol{P}}$ is a left coprime factorization of the plant transfer matrix $\boldsymbol{G}$, and $\boldsymbol{r}$ is the Laplace transform of the assumed *shape-deterministic* reference signal input $r(t)$ (see [50, 59] and references therein). An analogous condition holds for the ideal disturbance rejection.
(2) Another class pre-assumes a model of the inputs (see Theorem 2.4.1 in [34], the so called *internal model principle*). The tracking and disturbance rejection is ideal under a necessary and sufficient condition that is satisfied generically if the plant transfer matrix $\boldsymbol{G}$ is right-invertible. A drawback is that if the model changes in time, i.e. when different reference or disturbance inputs are applied in real time, or if the model of the plant is not precise, or changes in time, then the tracking with disturbance rejection can be unsatisfactory.
(3) The *perfect tracking* (see [7] and references therein) is feasible under the right-invertibility of the plant transfer matrix $\boldsymbol{G}$ and that it has no zeros in $\Re[s] \geq 0$

and infinity. The *exact disturbance decoupling* is feasible under very restrictive conditions on the disturbance dynamics of the plant (see Theorems 13.2.1 and 13.2.2 of [34]). Although the elaborated conditions on the plant are most restrictive, there are no constraints on the inputs.

In Sect. 5 of this chapter we adopt a performance criterion that is most close to the latter one (see Problem 1 in Sect. 5).

The sources of this chapter are as follows: The results of Sect. 2 are taken from the papers [36–39]. The results of Sect. 3 are taken from the papers [39] and [40]. The results of Sect. 4.1 are taken from the papers [36, 41]. The results of Sect. 4.4 are taken from the papers [42–45]. The results of Sect. 4.4 are taken from the papers [39, 46]. The results of Sect. 4 are taken from the papers [47, 48].

Besides the works [36–48], there are many works in the literature on fault diagnosis/fault-tolerant control. Without insisting on completeness, here we list some of them. Differences and improvements are explained in [36–48].

The fault detection without a frequency region constraint is elaborated in [3] (Chap. 6), [5, 8], [12] (Chap. 4), [14, 22, 23, 28–31, 55, 65, 66, 69, 70], and with a frequency region constraint, in [11, 54, 56, 62, 64].

The fault estimation without the frequency region constraint is elaborated in [4, 15, 16, 18, 26, 32, 35], and with a frequency region constraint is elaborated in [56].

The fault-tolerant control is elaborated in [1, 2, 13, 15, 16, 33, 51, 63, 68]. The approach with fault accommodation, [3] (Chap. 9), [24], is most close to the approach of this chapter, used in Sect. 4.2.

The fault-tolerant tracking is elaborated in [52, 53, 59].

Remarks on the notation. The matrices are denoted by upper-case letters, and vectors and scalars are denoted by lower-case letters. All functions of s are real rational, will be bold-faced, and if not ambiguous, without the argument. The abbreviations RM, EIA, FD, FE, FTC and FTT mean rational matrix, extended imaginary axis, fault detection, fault estimation, fault-tolerant control (or fault-tolerant controller) and fault-tolerant tracking. We denote $\overline{\mathbb{R}} = \mathbb{R} \cup \{\infty\}$, where $\mathbb{R}$ is the set of real numbers. The $\mathscr{L}_\infty$ norm of a RM $\boldsymbol{H}$ is defined by $\|\boldsymbol{H}\|_\infty = \sup_{\omega\in\mathbb{R}} \overline{\sigma}(\boldsymbol{H}(\mathrm{j}\omega))$, where by $\overline{\sigma}(H)$ we denote the maximal singular value of the complex matrix H. by $\underline{\sigma}(H)$ we denote the minimal singular value of the complex matrix H. If $\boldsymbol{H}(s)$ is a RM, by $\boldsymbol{H}^\#$ we denote the RM $\boldsymbol{H}^{\mathrm{T}}(-s)$. Given RM $\boldsymbol{H}$ is called para-hermitian if $\boldsymbol{H}^\# = \boldsymbol{H}$. The poles and zeros (including ones at infinity) of a RM are defined through its McMillan form. If a RM is without poles in $\Re[s] \geq 0$ (the closed right complex half-plane), then we say that it is stable, and we say that it is proper if it has no poles at infinity. By the superscripts ^{T} and * we denote transpose and conjugate transpose of a complex matrix. If H is a hermitian matrix, we define the inertia of H by $\mathrm{In}(H) = (m_-, m_0, m_+)$, where m_-, m_0, and m_+ are the numbers of negative, zero and positive eigenvalues of H. If $\boldsymbol{H}$ is a para-hermitian RM and $\mathbb{B} \subseteq \overline{\mathbb{R}}$, the notation $\mathrm{In}(\boldsymbol{H}) = (m_-, m_0, m_+)$ $\widetilde{\text{on}}$ $\mathrm{j}\mathbb{B}$ (or the statement that the RM $\boldsymbol{H}$ has m_- negative eigenvalues $\widetilde{\text{on}}$ $\mathrm{j}\mathbb{B}$, has m_0 zero eigenvalues $\widetilde{\text{on}}$ $\mathrm{j}\mathbb{B}$ and has m_+ positive eigenvalues $\widetilde{\text{on}}$ $\mathrm{j}\mathbb{B}$) means that $\mathrm{In}(\boldsymbol{H}(\mathrm{j}\omega)) = (m_-, m_0, m_+)$ for all $\omega \in \mathbb{B}$, except a finite number of points. Specially, the notation $\boldsymbol{H} > 0$ $\widetilde{\text{on}}$ $\mathrm{j}\mathbb{B}$ means that

$\mathrm{In}(\boldsymbol{H}) = (0,\ 0,\ m_+)\ \tilde{\text{on}}\ \mathrm{j}\mathbb{B}$, for some nonzero integer m_+. Analogously, the notation $\boldsymbol{H} \geq 0\ \tilde{\text{on}}\ \mathrm{j}\mathbb{B}$ means that $\mathrm{In}(\boldsymbol{H}) = (0,\ m_0,\ m_+)\ \tilde{\text{on}}\ \mathrm{j}\mathbb{B}$, for some integers $m_0 \geq 0$ and $m_+ \geq 0$. If $\boldsymbol{H}$ has no poles on $\mathrm{j}\mathbb{B}$, the condition $\boldsymbol{H} > 0\ \tilde{\text{on}}\ \mathrm{j}\mathbb{B}$, by the continuity argument, implies that $\boldsymbol{H}(\mathrm{j}\omega) \geq 0$ for all $\omega \in \mathbb{B}$. By $\left[\begin{array}{c|c} A & B \\ \hline C & D \end{array}\right]$ and $\left[\begin{array}{c|c} A - sE & B \\ \hline C & D \end{array}\right]$ we denote the transfer matrices $D + C(sI - A)^{-1}B$ and $D + C(sE - A)^{-1}B$. For given RMs $\boldsymbol{G} = \begin{bmatrix} \boldsymbol{G}_{11} & \boldsymbol{G}_{12} \\ \boldsymbol{G}_{21} & \boldsymbol{G}_{22} \end{bmatrix}$ and $\boldsymbol{K}$ of compatible dimensions, we denote

$$\mathscr{F}(\boldsymbol{G},\ \boldsymbol{K}) = \boldsymbol{G}_{11} + \boldsymbol{G}_{12}\boldsymbol{K}(I - \boldsymbol{G}_{22}\boldsymbol{K})^{-1}\boldsymbol{G}_{21}\,. \tag{3}$$

By $\boldsymbol{T}_{\mathrm{yr}}$ we denote the transfer matrix from r to y.

2 Fault Detection

Fault detection theory is a source of many engineering applications, not only for FTC. Before we give a formal definition for FD, some preliminary material is needed.

With respect to (2), the following identity between the Laplace transforms of y, u, f and d holds:

$$\boldsymbol{y} = \boldsymbol{G}_{\mathrm{u}}\boldsymbol{u} + \boldsymbol{G}_{\mathrm{f}}\boldsymbol{f} + \boldsymbol{G}_{\mathrm{d}}\boldsymbol{d}\ . \tag{4}$$

There exist proper and stable rational matrices (RMs) $\boldsymbol{M}$, $\boldsymbol{N}_{\mathrm{u}}$, $\boldsymbol{N}_{\mathrm{f}}$, and $\boldsymbol{N}_{\mathrm{d}}$, such that $\boldsymbol{M}$ and $[\boldsymbol{N}_{\mathrm{u}}, \boldsymbol{N}_{\mathrm{f}}, \boldsymbol{N}_{\mathrm{d}}]$ are left coprime and

$$[\boldsymbol{G}_{\mathrm{u}}, \boldsymbol{G}_{\mathrm{f}}, \boldsymbol{G}_{\mathrm{d}}] = \boldsymbol{M}^{-1}[\boldsymbol{N}_{\mathrm{u}}, \boldsymbol{N}_{\mathrm{f}}, \boldsymbol{N}_{\mathrm{d}}]\,. \tag{5}$$

For some unknown $n_{\mathrm{r}} \times n_{\mathrm{y}}$-dimensional filter transfer matrix $\boldsymbol{F}(s)$, introduce the following residual signal vector $r(t) \in \mathbb{R}^{n_{\mathrm{r}}}$ by its Laplace transform:

$$\boldsymbol{r} = \boldsymbol{F}(\boldsymbol{M}\boldsymbol{y} - \boldsymbol{N}_{\mathrm{u}}\boldsymbol{u})\,. \tag{6}$$

Replacing (4) in (6), we obtain $\boldsymbol{r} = \boldsymbol{T}_{\mathrm{rf}}\boldsymbol{f} + \boldsymbol{T}_{\mathrm{rd}}\boldsymbol{d}$, where $\boldsymbol{T}_{\mathrm{rf}} := \boldsymbol{F}\boldsymbol{N}_{\mathrm{f}}$ and $\boldsymbol{T}_{\mathrm{rd}} := \boldsymbol{F}\boldsymbol{N}_{\mathrm{d}}$.

The fault detection problem is to distinguish the appearance of a fault from the appearance of a disturbance in a reasonable short time, on the basis of the signal $r(t)$. In a perfect situation, a filter $\boldsymbol{F}$ can be constructed such that the residual doesn't respond to the disturbances, but is sensitive to the faults. In practice, the perfect situation rarely happens, so the role of the filter is to attenuate the effect of the disturbances on the residual in respect to the faults.

To formulate our problem, let $\mathbb{B}_{\mathrm{d}}$ and $\mathbb{B}_{\mathrm{f}}$ be closed subsets of the imaginary axis, symmetric with respect to $0 \in \mathbb{R}$, where the spectra of $\boldsymbol{N}_{\mathrm{f}}\boldsymbol{f}$ and $\boldsymbol{N}_{\mathrm{d}}\boldsymbol{d}$ are located, respectively.

Problem 1 *Find a possibly minimal γ and a proper and stable filter transfer matrix $\boldsymbol{F}$, where n_r is also unknown, such that the following inequalities hold:*

$$\boldsymbol{T}_{rd}\boldsymbol{T}_{rd}^* \leq \gamma^2 \boldsymbol{T}_{rf}\boldsymbol{T}_{rf}^* \,, \quad j\omega \in j\mathbb{B}_d \,, \tag{7}$$

$$\|\boldsymbol{T}_{rf}\| \geq 1 \,, \quad j\omega \in j\mathbb{B}_f \,. \tag{8}$$

Problem 1 is a generalization of the well-known $\mathscr{H}_-/\mathscr{H}_\infty$ fault detection problem.

With Problem 1, we introduce two performance indicators $\boldsymbol{\psi}_{F,\gamma}(\omega)$ and $\boldsymbol{\varphi}_F(\omega)$, of the quality of the FD filter $\boldsymbol{F}$, at all frequencies: We define

$$\boldsymbol{\psi}_{F,\gamma}(\omega) = \overline{\sigma}\left(\boldsymbol{F}(j\omega)\left[\boldsymbol{N}_d(j\omega)\boldsymbol{N}_d(j\omega)^* - \gamma^2\boldsymbol{N}_f(j\omega)\boldsymbol{N}_f(j\omega)^*\right]\boldsymbol{F}(j\omega)^*\right), \quad \omega \in \mathbb{B}_d \tag{9}$$

This performance indicator is negative for $\omega \in \mathbb{B}_d$. A more negative indicator for some $\omega \in \mathbb{B}_d$ means that the FD is better.

$$\boldsymbol{\varphi}_F(\omega) := \frac{\overline{\sigma}(\boldsymbol{F}(j\omega)\boldsymbol{N}_d(j\omega))}{\underline{\sigma}(\boldsymbol{F}(j\omega)\boldsymbol{N}_f(j\omega))} \,, \quad \omega \in \mathbb{B}_d \tag{10}$$

This performance indicator is a positive function of $\omega \in \mathbb{B}_d$. At frequencies where it is "small" the FD is "good". If the magnitudes of the disturbances and faults are scaled so that they have approximately same magnitudes, then the FD filter is "good" at some frequency ω, if $\boldsymbol{\varphi}_F(\omega)$ is small, less that γ.

Note that the same performance indicators can be applied to the fault estimation, defined in Problem 3:$\mathbb{B}$ of Sect. 3.

Denote by $\boldsymbol{\Pi}_\gamma$ the para-hermitian RM:

$$\boldsymbol{\Pi}_\gamma = \boldsymbol{N}_d\boldsymbol{N}_d^\# - \gamma^2\boldsymbol{N}_f\boldsymbol{N}_f^\# \,. \tag{11}$$

Assumption 1 There is a generalized factorization of $\boldsymbol{\Pi}_\gamma$,

$$\boldsymbol{\Pi}_\gamma = \boldsymbol{\Psi}\begin{bmatrix}\boldsymbol{\Phi}_1 & 0\\ 0 & \boldsymbol{\Phi}_2\end{bmatrix}\boldsymbol{\Psi}^\# =: \boldsymbol{\Psi}\boldsymbol{\Phi}\boldsymbol{\Psi}^\# \,, \tag{12}$$

where the RM $\boldsymbol{\Psi}$ is a nonsingular factor and $\boldsymbol{\Phi}$ is a para-hermitian polynomial matrix such that the dimension of $\boldsymbol{\Phi}_1$ is $n_y - n_r$, the dimension of $\boldsymbol{\Phi}_2$ is n_r, and $\boldsymbol{\Phi}_2$ is negative semidefinite on $j\mathbb{B}_d$.

Define the RMs $\widetilde{\boldsymbol{F}}$ by $\widetilde{\boldsymbol{F}} = [0, I_{n_r}]\boldsymbol{\Psi}^{-1}$.

Assumption 2 The RM $\widetilde{\boldsymbol{F}}\boldsymbol{N}_f$ has no zeros on $j\mathbb{B}_f$.

Theorem 1 ([36]) *Under Assumptions 1 and 2, Problem 1 is solvable by a filter with an arbitrary pole distribution.*

Sketch of the proof. The inequality (7) is satisfied with the pre-filter $\widetilde{\boldsymbol{F}}$. Indeed,

$$\begin{aligned}&\widetilde{\boldsymbol{F}}(\boldsymbol{N}_{\mathrm{d}}\boldsymbol{N}_{\mathrm{d}}^* - \gamma^2 \boldsymbol{N}_{\mathrm{f}}\boldsymbol{N}_{\mathrm{f}}^*)\widetilde{\boldsymbol{F}}^* = \widetilde{\boldsymbol{F}}\boldsymbol{\Pi}_\gamma \widetilde{\boldsymbol{F}}^* \\ &= [0, I_{n_{\mathrm{r}}}]\boldsymbol{\Psi}^{-1}\boldsymbol{\Psi}\begin{bmatrix}\boldsymbol{\Phi}_1 & 0\\ 0 & \boldsymbol{\Phi}_2\end{bmatrix}\boldsymbol{\Psi}^*\boldsymbol{\Psi}^{-*}\begin{bmatrix}0\\ I_{n_{\mathrm{r}}}\end{bmatrix} = \boldsymbol{\Phi}_2 \le 0\,, \quad \mathrm{j}\omega \in \mathrm{j}\mathbb{B}\,.\end{aligned} \tag{13}$$

Let $\widetilde{\boldsymbol{F}}$ be given by the minimal descriptor realization $\widetilde{\boldsymbol{F}} = \widetilde{D} + \widetilde{C}(s\widetilde{E} - \widetilde{A})^{-1}\widetilde{B}$. There is a matrix $\widetilde{L}$ such that the RMs $\boldsymbol{P}$ and $\widehat{\boldsymbol{F}}$, defined by

$$[\boldsymbol{P}, \widehat{\boldsymbol{F}}] = \left[\begin{array}{c|cc}\widetilde{A} - \widetilde{L}\widetilde{C} - s\widetilde{E} & -\widetilde{L} & \widetilde{B} - \widetilde{L}\widetilde{D}\\ \hline \widetilde{C} & I_{n_{\mathrm{r}}} & \widetilde{D}\end{array}\right], \quad \widehat{\boldsymbol{F}} = \boldsymbol{P}\widetilde{\boldsymbol{F}} \tag{14}$$

are proper and stable, and with an arbitrary pole distribution. Since by the definition (14), the zeros of $\widehat{\boldsymbol{F}}$ are included in the zeros of $\widetilde{\boldsymbol{F}}$, under Assumption 2, we can define $\nu_- = \|\widehat{\boldsymbol{F}}\boldsymbol{N}_{\mathrm{f}}\|_-^{\mathbb{B}_{\mathrm{f}}} \neq 0$. Then we can define the filter $\boldsymbol{F} = \widehat{\boldsymbol{F}}/\nu_-$, which is a solution. ■

Under the following assumption:

Assumption 3 The pair $(C, A - sE)$ is finite-mode detectable and impulse observable

there is a matrix H such that $(A - HC - sE)^{-1}$ is a proper and stable RM. Then the RMs in (5) can be constructed as

$$[\boldsymbol{M}, \boldsymbol{N}_{\mathrm{u}}, \boldsymbol{N}_{\mathrm{f}}, \boldsymbol{N}_{\mathrm{d}}] = \left[\begin{array}{c|cccc}\mathscr{A} - sE & -H & \mathscr{B}_{\mathrm{u}} & \mathscr{B}_{\mathrm{f}} & \mathscr{B}_{\mathrm{d}}\\ \hline C & I_{n_y} & D_{\mathrm{u}} & D_{\mathrm{f}} & D_{\mathrm{d}}\end{array}\right], \tag{15}$$

where $\mathscr{A} = A - HC$, $\mathscr{B}_{\mathrm{u}} = B_{\mathrm{u}} - HD_{\mathrm{u}}$, $\mathscr{B}_{\mathrm{f}} = B_{\mathrm{f}} - HD_{\mathrm{f}}$ and $\mathscr{B}_{\mathrm{d}} = B_{\mathrm{d}} - HD_{\mathrm{d}}$.

Then $\boldsymbol{\Pi}_\gamma$ can be rewritten as

$$\boldsymbol{\Pi}_\gamma = [C(sE - \mathscr{A})^{-1}, I]\Sigma\begin{bmatrix}(-sE^{\mathrm{T}} - \mathscr{A}^{\mathrm{T}})^{-1}C^{\mathrm{T}}\\ I\end{bmatrix} \tag{16}$$

where

$$\begin{aligned}&\Sigma = \begin{bmatrix}\mathscr{B}J_\gamma\mathscr{B}^{\mathrm{T}} & \mathscr{B}J_\gamma\mathscr{D}^{\mathrm{T}}\\ \mathscr{D}J_\gamma\mathscr{B}^{\mathrm{T}} & \mathscr{D}J_\gamma\mathscr{D}^{\mathrm{T}}\end{bmatrix},\\ &\mathscr{B} = [\mathscr{B}_{\mathrm{d}}, \mathscr{B}_{\mathrm{f}}]\,, \quad \mathscr{D} = [D_{\mathrm{d}}, D_{\mathrm{f}}]\,, \quad J_\gamma = \begin{bmatrix}I & 0\\ 0 & -\gamma^2 I\end{bmatrix}.\end{aligned}$$

The generalized symmetric factorization (12) of $\boldsymbol{\Pi}_\gamma$ is elaborated in the paper [39]. It is based on the following theorem.

Theorem 2 ([39]) *Given arbitrary real para-hermitian matrix pencil $sM - N$ ($M^{\mathrm{T}} = -M$, $N^{\mathrm{T}} = N$), there is an orthogonal matrix U such that*

$$U^{\mathrm{T}}(sM - N)U = \begin{bmatrix} 0 & 0 & -sE_1^{\mathrm{T}} - A_1^{\mathrm{T}} \\ 0 & sM_1 - N_1 & \times \\ sE_1 - A_1 & \times & \times \end{bmatrix}, \tag{17}$$

where $sE_1 - A_1$ is a regular pencil with det $E_1 \neq 0$, *by $\times$ are denoted some matrix pencils, and $sM_1 - N_1$ is a para-hermitian matrix pencil with the property that there exists a nonsingular real matrix U_1 such that*

$$U_1^{\mathrm{T}}(sM_1 - N_1)U_1 =$$

$$\begin{bmatrix} \begin{bmatrix} \omega_1 & s \\ -s & \omega_1 \end{bmatrix} & \cdots & 0 & 0 \\ \vdots & \ddots & \vdots & \vdots \\ 0 & \cdots & \begin{bmatrix} \omega_k & s \\ -s & \omega_k \end{bmatrix} & 0 \\ 0 & \cdots & 0 & -N_{k+1} \end{bmatrix}, \tag{18}$$

where $\omega_1, \ldots, \omega_k$ are nonzero real numbers, and N_{k+1} is a symmetric matrix.

The property of Theorem 2 that the ω_i's can be positive as well as negative will be used further on. For $i \in \{1, \ldots, k\}$, introduce the following identity

$$\begin{bmatrix} \omega_i & s \\ -s & \omega_i \end{bmatrix} = \begin{bmatrix} 1 & 0 \\ -\dfrac{s}{\omega_i} & 1 \end{bmatrix}^{\#} \begin{bmatrix} \dfrac{s^2 + \omega_i^2}{\omega_i} & 0 \\ 0 & \omega_i \end{bmatrix} \begin{bmatrix} 1 & 0 \\ -\dfrac{s}{\omega_i} & 1 \end{bmatrix} \tag{19}$$

For $\omega \in \mathbb{R}$, the matrix

$$\begin{bmatrix} \omega_i & \mathrm{j}\omega \\ -\mathrm{j}\omega & \omega_i \end{bmatrix} \cong \begin{bmatrix} \dfrac{\omega_i^2 - \omega^2}{\omega_i} & 0 \\ 0 & \omega_i \end{bmatrix}, \tag{20}$$

where by $\cong$ we denote matrix congruency, has the following inertia properties:

(a) If $\omega_i > 0$ and $|\omega| \leq \omega_i$, matrix (20) is positive semi-definite.
(b) If $\omega_i < 0$ and $|\omega| \leq |\omega_i|$, matrix (20) is negative semi-definite.
(c) If $\omega_i > 0$ and $|\omega| > \omega_i$, matrix (20) has inertia $(1, 0, 1)$.
(d) If $\omega_i < 0$ and $|\omega| > |\omega_i|$, matrix (20) has inertia $(1, 0, 1)$.

Example. Consider the modified F16XL system, [22]. With $n_{\mathrm{r}} = n_{\mathrm{f}} = 2$, the minimal $\gamma = 1.1768$ is given in the literature, and a filter is given. Let us apply that

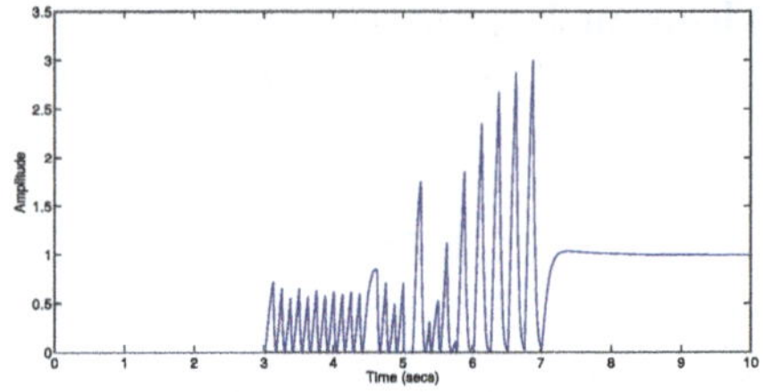

Fig. 1 Residual response $r(t)^{\mathrm{T}} r(t)$ for the filter of [22]

filter to the following test signals: $f_1(t) = 0$, $f_2(t)$ is a unit step appearing at $t = 5s$, $d_1(t) = 0$, and

$$d_2(t) = \begin{cases} 1, & 3 + i/8 < t < 3 + (i+1)/8, \quad i = 0, 2, \ldots, 30, \\ -1, & 3 + i/8 < t < 3 + (i+1)/8, \quad i = 1, 3, \ldots, 31, \\ 0, & \text{elsewhere.} \end{cases}$$

In Fig. 1, it is given the corresponding power signal $r(t)^{\mathrm{T}} r(t)$, with the filter obtained in [22]. It is seen that the fault can hardly be distinguished from the disturbance. The reason is that the minimal $\gamma = 1.1768$ is not sufficiently small.

Now we apply the algorithm for fault detection over frequency region. We take again $n_{\mathrm{r}} = 2$, but $\gamma = 0.38 < 1.1768$, and obtain that matrix pencil $sM - N$ gets the finite generalized eigenvalue $\mathrm{j}5.9929 =: \mathrm{j}\omega_1$ on the imaginary axis. To the matrix pencil $sM_1 - N_1$ we apply a nonsingular transformation matrix U_1, and obtain

$$U_1^{\mathrm{T}}(sM_1 - N_1)U_1 = \operatorname{diag}\left\{ \begin{bmatrix} \omega_1 & s \\ -s & \omega_1 \end{bmatrix}, \ 1, \ -I_2 \right\},$$

A pair of infinite generalized eigenvalues can be separated of this pencil, and obtain the pencil

$$\operatorname{diag}\left\{ \begin{bmatrix} \omega_1 & s \\ -s & \omega_1 \end{bmatrix}, \ -1 \right\}. \tag{21}$$

We observe that case (c), defined earlier, appears, since $\omega_1 > 0$, hence by the nonconstant transformation (19) and permutation, we obtain the polynomial matrix in ω:

$$\operatorname{diag}\left\{ \omega_1, \ \begin{bmatrix} \dfrac{\omega_1^2 - \omega^2}{\omega_1} & 0 \\ 0 & -1 \end{bmatrix} \right\} =: \operatorname{diag}\{\boldsymbol{\Phi}_1(\mathrm{j}\omega), \ \boldsymbol{\Phi}_2(\mathrm{j}\omega)\},$$

where $\boldsymbol{\Phi}_1(\mathrm{j}\omega) = \omega_1 > 0$ and $\boldsymbol{\Phi}_2(\mathrm{j}\omega) \leq 0$ in the interval $[\omega_1, \infty)$. As a result, we can take $\mathbb{B}_{\mathrm{d}} = [\omega_1, \infty]$.

The filter $\boldsymbol{F}$ of realization order 5 is found, and the order is slightly greater than the order of the system (which is 4).

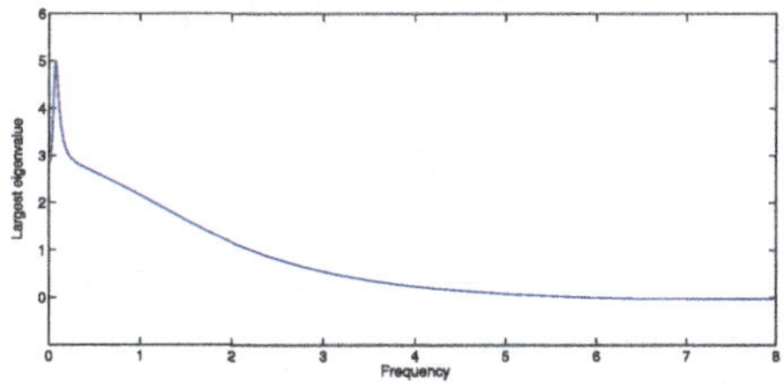

Fig. 2 Performance indicator $\boldsymbol{\psi}_{F,\gamma}(\omega)$ for our filter and minimal γ

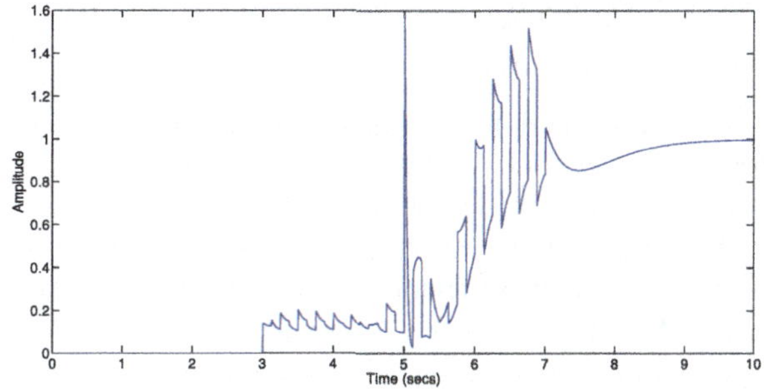

Fig. 3 Residual response $r(t)^{\mathrm{T}}r(t)$ for our filter

In Fig. 2, it is presented the frequency dependence $\boldsymbol{\psi}_{F,\gamma}(\omega)$, where the performance indicator $\boldsymbol{\psi}_{F,\gamma}(\omega)$ is defined in (9). It is seen that for $\omega > \omega_1 = 5.9929$, it is negative, as required. A waterbed effect is evident for $\omega < \omega_1$.

Since a large part of the spectrum of the disturbance in our test signal is located around $\varpi = 2\pi/0.25 = 25.2327 \in \mathbb{B}_{\mathrm{d}}$ (the period of the disturbance is 0.25), we can expect that the fault detection will be feasible.

In Fig. 3, it is given the corresponding power signal $r(t)^{\mathrm{T}}r(t)$ with our filter. It is seen that the fault indeed can be distinguished from the disturbance. In particular, an impulse of amplitude 1.6 appears at $t = 5$ in the residual power, while for $t < 5$, the amplitude of the residual power is ≤ 0.2 (the disturbance is more attenuated than in Fig. 1).

Next we elaborate on the case $n_{\mathrm{r}} \geq n_{\mathrm{f}}$, $\mathbb{B}_{\mathrm{f}} = \overline{\mathbb{R}}$ and $\mathbb{B}_{\mathrm{d}} = \overline{\mathbb{R}}$.

A solution of Problem 1 can be a RM $\boldsymbol{F}$ with $n_{\mathrm{r}} < n_{\mathrm{f}}$. In that case there is a fault vector $\boldsymbol{f}$ such that $\boldsymbol{T}_{\mathrm{rf}}\boldsymbol{f} = 0$, and the residual $\boldsymbol{r}$, by $\boldsymbol{r} = \boldsymbol{T}_{\mathrm{rf}}\boldsymbol{f} + \boldsymbol{T}_{\mathrm{rd}}\boldsymbol{d}$, is insensitive to this $\boldsymbol{f}$. Then we can introduce a strong version of Problem 1:

Problem 1S *Find a solution of Problem 1 with* $n_{\mathrm{r}} \geq n_{\mathrm{f}}$.

In the next theorem we shall present a necessary and sufficient condition for Problem 1S, but under the following assumption:

Assumption 4 The RM $[\boldsymbol{N}_{\mathrm{f}},\ \boldsymbol{N}_{\mathrm{d}}]$ has full row rank on the EIA.

Theorem 3 ([37]) *Under Assumption 4, Problem 1S with* $\mathbb{B}_{\mathrm{f}} = \overline{\mathbb{R}}$ *and* $\mathbb{B}_{\mathrm{d}} = \overline{\mathbb{R}}$ *is solvable if and only if the RM* $\boldsymbol{\Pi}_{\gamma}$ *has a constant inertia on the EIA, equal to* $(n_{\mathrm{f}} - q,\ q,\ n_{\mathrm{y}} - n_{\mathrm{f}})$*, where* q *is the rank defect of* $\boldsymbol{\Pi}_{\gamma}$.

2.1 Solvability of Problem 1 in Terms of LMI. New KYP Lemma

It can be seen that the constant inertia condition of $\mathbf{\Pi}_\gamma$ must hold also in the case $\mathbb{B}_f \neq \overline{\mathbb{R}}$ and $\mathbb{B}_d \neq \overline{\mathbb{R}}$. For this purpose, the following generalization of KYP lemma can be used.

Let $n_y \times n_d$-dimensional and $n_y \times n_f$-dimensional RMs $\mathbf{G}_d$ and $\mathbf{G}_f$, possibly unstable and improper, be given, and a frequency region $\mathbb{B}$. Define the RM $\mathbf{G} := [\mathbf{G}_d,\ \mathbf{G}_f]$ and matrix $J_\gamma = \begin{bmatrix} I_{n_d} & 0 \\ 0 & -\gamma^2 I_{n_f} \end{bmatrix}$.

In this Subsection, we pose the problem to find necessary and sufficient conditions for the RM $\mathbf{G} J_\gamma \mathbf{G}^\#$ to have constant inertia on j$\mathbb{B}$, equal to $(n_f - q,\ q,\ n_y - n_f)$, where q is the rank defect of $\mathbf{G} J_\gamma \mathbf{G}^\#$.

We derive our necessary and sufficient conditions through an algorithm, and then we formulate them in Theorem 4. The RM $\mathbf{G} J_\gamma \mathbf{G}^\#$ can be replaced by $\mathbf{\Pi}_\gamma = \mathbf{N} J_\gamma \mathbf{N}^\#$, where $\mathbf{N} = [\mathbf{N}_d,\ \mathbf{N}_f]$ is a proper and stable RM, defined in (5) and (15). Then we transform this descriptor realization into a state-space one, as

$$[\mathbf{N}_d,\ \mathbf{N}_f] = D + C(sI - A)^{-1}B = [D_d,\ D_f] + C(sI - A)^{-1}[B_d,\ B_f]\ , \tag{22}$$

where A is a stable matrix. (The same realization matrices for $[\mathbf{N}_d,\ \mathbf{N}_f]$ are taken as for $[\mathbf{G}_d,\ \mathbf{G}_f]$, for simplicity, except that matrix E does not appear in (22).)

Denote by ρ the rank of the RM $\mathbf{G}_d$. There is a full column normal rank RM $\mathbf{Q}$ with ρ columns, such that

$$\mathbf{N}_d \mathbf{N}_d^\# = \mathbf{Q}\mathbf{Q}^\#\ . \tag{23}$$

It is known that $\mathbf{Q}$ has a realization

$$\mathbf{Q} = L + C(sI - A)^{-1}K\ ,$$

for some matrices L and K, hence $[\mathbf{N}_f,\ \mathbf{Q}] = [D_f,\ L] + C(sI - A)^{-1}[B_f,\ K] =: D_1 + C(sI - A)^{-1}B_1$.

It is proved in [37] that matrix D_1 has full row rank, under Assumption 4.

Define the matrix D_0 such that its columns span a basis of $\ker(D_1)$, and let $D_1^\dagger$ be a right inverse of D_1, and define the realization

$$(A - B_1 D_1^\dagger C,\ B_1 D_0,\ -D_1^\dagger C,\ D_0) \tag{24}$$

Then we find a controllable realization of (24), which we denote by $(\mathcal{A},\ \mathcal{B},\ \mathcal{C},\ \mathcal{D})$.

Let Φ and Ψ be matrices depending on $\mathbb{B}$, constructed as shown in the paper [21]. Denote by $\otimes$ the Kronecker product, and denote $\mathcal{J}_\gamma = \mathrm{diag}\{I_{n_f},\ -\gamma^2 I_\rho\}$.

Theorem 4 ([37]) *If RM* $\boldsymbol{G}$ *has full row normal rank and has no zeros at infinity, then RM* $\boldsymbol{G}J_\gamma\boldsymbol{G}^\#$ *has constant inertia on* $\mathrm{j}\mathbb{B}$*, equal to* $(n_\mathrm{f}-q,\ q,\ n_\mathrm{y}-n_\mathrm{f})$*, if and only if there are hermitian matrices* $\mathcal{P}$ *and* $\mathcal{Q}$ *which solve the following LMIs,*

$$\begin{bmatrix} \mathcal{A}^\mathrm{T} & I \\ \mathcal{B}^\mathrm{T} & 0 \end{bmatrix} (\Phi \otimes \mathcal{P} + \Psi \otimes \mathcal{Q}) \begin{bmatrix} \mathcal{A} & \mathcal{B} \\ I & 0 \end{bmatrix} + \begin{bmatrix} \mathcal{C}^\mathrm{T} \\ \mathcal{D}^\mathrm{T} \end{bmatrix} \mathcal{J}_\gamma [\mathcal{C},\ \mathcal{D}] \leq 0\,, \quad \mathcal{Q} \geq 0\,. \tag{25}$$

2.2 All Solutions to Problem 1S with $n_\mathrm{r} = n_\mathrm{f}$, over the EIA

Here we present all solutions to Problem 1S, for the systems that satisfy three additional assumptions, besides Assumption 4. The first one is the most restrictive.

Assumption 5 $n_\mathrm{r} = n_\mathrm{f}$.

Assumption 6 The RM $\boldsymbol{\Pi}_\gamma$ is nonsingular.

Under the solvability of Problem 1S there is a factorization

$$\boldsymbol{\Pi}_\gamma = \boldsymbol{\Psi} J' \boldsymbol{\Psi}^\#\,, \quad J' = \begin{bmatrix} I_{n_\mathrm{y}-n_\mathrm{f}} & 0 & 0 \\ 0 & 0_{q\times q} & 0 \\ 0 & 0 & -I_{n_\mathrm{f}-q} \end{bmatrix} \tag{26}$$

where $\boldsymbol{\Psi}$ is a nonsingular factor RM.

Assumption 7 The RM $\boldsymbol{\Psi}$ has no poles in $\Re[s] > 0$.

Introduce the following RMs

$$\boldsymbol{\Psi}^{-1}\boldsymbol{N}_\mathrm{f} =: \begin{bmatrix} \boldsymbol{S}_1 \\ \boldsymbol{S}_2 \end{bmatrix} \begin{matrix} \}n_\mathrm{y}-n_\mathrm{f} \\ \}\quad n_\mathrm{f} \end{matrix}\,, \quad \boldsymbol{U}\boldsymbol{S}_1 + \boldsymbol{S}_2 =: \boldsymbol{H}_{\mathrm{o}U}\boldsymbol{H}_{\mathrm{i}U}\,, \tag{27}$$

where $\boldsymbol{H}_{\mathrm{i}U}$ is a square inner RM, $\boldsymbol{H}_{\mathrm{o}U}$ is a nonsingular RM without zeros in $\Re[s] \geq 0$, and $\boldsymbol{U}$ is an $n_\mathrm{f} \times (n_\mathrm{y}-n_\mathrm{f})$-dimensional RM.

Theorem 5 ([38]) *Under Assumptions 3, 4, 5, 6 and 7, all solutions to Problem 1S with* $n_\mathrm{r} = n_\mathrm{f}$ *over the EIA are given by*

$$\boldsymbol{F} = \boldsymbol{F}_3 \boldsymbol{H}_{\mathrm{o}U}^{-1} [\boldsymbol{U},\ I_{n_\mathrm{f}}] \boldsymbol{\Psi}^{-1}\,, \tag{28}$$

where $\boldsymbol{U}$ *is an arbitrary* $n_\mathrm{f} \times (n_\mathrm{y}-n_\mathrm{f})$*-dimensional RM satisfying*

$$\|\boldsymbol{U}\|_\infty \leq 1\,, \tag{29}$$

and $\boldsymbol{F}_3$ *is an arbitrary proper and stable* $n_\mathrm{f} \times n_\mathrm{f}$*-dimensional RM, satisfying:*

(a) $\boldsymbol{F}_3^*\boldsymbol{F}_3 \geq I_{n_\mathrm{f}}$*, and*

(b) $(\boldsymbol{F}_3^*\boldsymbol{F}_3)^{-1} \geq I_{n_\mathrm{f}} + \gamma^{-2}\boldsymbol{H}_{0U}^{-1}(\boldsymbol{U}\boldsymbol{U}^* - I_{n_\mathrm{f}})\boldsymbol{H}_{0U}^{-*}$,

on the EIA.

2.3 New Class of FD Filters with $n_\mathrm{r} = n_\mathrm{f}$

The following Theorem 6 and algorithm can be used to compute a class of FD filters, which solve Problem 1S.

If Problem 1S admits a solution, then, by (8), the RM $\boldsymbol{N}_\mathrm{f}$ has full column rank on $\mathrm{j}\mathbb{B}_\mathrm{f}$. We assume, more strictly, that

Assumption 8 The RM $\boldsymbol{N}_\mathrm{f}$ has full column rank on the EIA.

Under Assumption 8, there exists the factorization of $\boldsymbol{N}_\mathrm{f}$: $\boldsymbol{N}_\mathrm{f} = \boldsymbol{N}_\mathrm{co}\boldsymbol{N}_\mathrm{i}$, where $\boldsymbol{N}_\mathrm{co}$ is a stable RM that has full column rank in $\Re[s] \geq 0$ and infinity, and the RM $\boldsymbol{N}_\mathrm{i}$ is square inner. We rewrite this factorization as

$$\boldsymbol{N}_\mathrm{f} = \boldsymbol{N}_\mathrm{co}\boldsymbol{N}_\mathrm{i} = \overbrace{[\boldsymbol{N}_\mathrm{co},\ \boldsymbol{T}_\perp]}^{\boldsymbol{N}_\mathrm{o}}\begin{bmatrix} I_{n_\mathrm{f}} \\ 0_{(n_\mathrm{y}-n_\mathrm{f})\times n_\mathrm{f}} \end{bmatrix}\boldsymbol{N}_\mathrm{i}\ , \tag{30}$$

for some stable RM $\boldsymbol{T}_\perp$ such that the RM $\boldsymbol{N}_\mathrm{o}$ is nonsingular in $\Re[s] \geq 0$ and $\boldsymbol{N}_\mathrm{o}^{-1}$ is proper (if $n_\mathrm{f} = n_\mathrm{y}$ then the RM $\boldsymbol{T}_\perp$ is void).

Define the proper stable RMs $\boldsymbol{T}_{11}$ and $\boldsymbol{T}_{21}$ as

$$\boldsymbol{N}_\mathrm{o}^{-1}\boldsymbol{N}_\mathrm{d} =: \begin{bmatrix} \boldsymbol{T}_{11} \\ \boldsymbol{T}_{21} \end{bmatrix}\begin{matrix} \}\ \ n_\mathrm{f} \\ \}n_\mathrm{y} - n_\mathrm{f} \end{matrix}\ . \tag{31}$$

Re-write the identities (31) and (30) as:

$$[\boldsymbol{N}_\mathrm{d},\ \boldsymbol{N}_\mathrm{f}] = \boldsymbol{N}_\mathrm{o}\begin{bmatrix} \boldsymbol{T}_{11} & I_{n_\mathrm{f}} \\ \boldsymbol{T}_{21} & 0 \end{bmatrix}\begin{bmatrix} I_{n_\mathrm{d}} & 0 \\ 0 & \boldsymbol{N}_\mathrm{i} \end{bmatrix}. \tag{32}$$

By (32), we have:

$$\begin{aligned} \boldsymbol{\Pi}_\gamma = [\boldsymbol{N}_\mathrm{d},\ \boldsymbol{N}_\mathrm{f}]J_\gamma\begin{bmatrix} \boldsymbol{N}_\mathrm{d}^\# \\ \boldsymbol{N}_\mathrm{f}^\# \end{bmatrix} &= \boldsymbol{N}_\mathrm{o}\begin{bmatrix} \boldsymbol{T}_{11}\boldsymbol{T}_{11}^\# - \gamma^2 I & \boldsymbol{T}_{11}\boldsymbol{T}_{21}^\# \\ \boldsymbol{T}_{21}\boldsymbol{T}_{11}^\# & \boldsymbol{T}_{21}\boldsymbol{T}_{21}^\# \end{bmatrix}\boldsymbol{N}_\mathrm{o}^\# \\ &= \boldsymbol{\Psi}_\mathrm{p}\begin{bmatrix} \boldsymbol{Z} - \gamma^2 I & 0 \\ 0 & \boldsymbol{T}_{21}\boldsymbol{T}_{21}^\# \end{bmatrix}\boldsymbol{\Psi}_\mathrm{p}^\#, \end{aligned} \tag{33}$$

where

$$\boldsymbol{\Psi}_\mathrm{p} = \boldsymbol{N}_\mathrm{o}\begin{bmatrix} I & \boldsymbol{T}_{11}\boldsymbol{T}_{21}^\#(\boldsymbol{T}_{21}\boldsymbol{T}_{21}^\#)^{-1} \\ 0 & I \end{bmatrix},$$

$$\boldsymbol{Z} := \boldsymbol{T}_{11}\boldsymbol{T}_{11}^{\#} - \boldsymbol{T}_{11}\boldsymbol{T}_{21}^{\#}(\boldsymbol{T}_{21}\boldsymbol{T}_{21}^{\#})^{-1}\boldsymbol{T}_{21}\boldsymbol{T}_{11}^{\#} \,. \tag{34}$$

Assumption 9 The RM $\boldsymbol{N}_{\mathrm{d}}$ has full row rank on the EIA.

If Assumption 9 is not satisfied, but Assumption 4 is satisfied, then there is a perfect fault detector, at least in the generic case (Proposition 4 in [38]).

It is proved also in [38] that $\boldsymbol{Z}$ is a nonsingular RM on the EIA and $\boldsymbol{Z} \geq 0$ on the EIA, hence there is a biproper RM Ξ, without poles in $\Re[s] \leq 0$ and without zeros in $\Re[s] \geq 0$, such that

$$\Xi \cdot \Xi^{\#} = \boldsymbol{Z} \,. \tag{35}$$

Assumption 10 $\mathbb{B}_{\mathrm{f}} \subseteq \mathbb{B}_{\mathrm{d}}$

Assumption 10 is aimed for FD problems in which the disturbance and its spectrum location are very uncertain. In that case we have to take a very wide $\mathbb{B}_{\mathrm{d}}$.

Theorem 6 ([38]) *Under Assumptions 3, 8, 9, and 10, Problem 1S with $n_{\mathrm{r}} = n_{\mathrm{f}}$ is solvable, for some γ, if and only if*

$$\|\Xi\| \leq \gamma \,, \quad \mathrm{j}\omega \in \mathrm{j}\mathbb{B}_{\mathrm{f}} \,. \tag{36}$$

The filter satisfies the identity

$$\boldsymbol{T}_{\mathrm{rd}}\boldsymbol{T}_{\mathrm{rd}}^{*} = \gamma^2 I \,, \quad \underline{\sigma}(\boldsymbol{T}_{\mathrm{rf}}) \geq 1 \,. \tag{37}$$

An important consequence of Theorem 6 is that the optimal attenuation γ can be easily computed by solving the LMIs (25).

2.4 New Class of FD Filters with $n_{\mathrm{r}} < n_{\mathrm{f}}$, over the EIA

Assumption 11 There is a generalized symmetric factorization of $\boldsymbol{\Pi}_{\gamma}$:

$$\boldsymbol{\Pi}_{\gamma} = \boldsymbol{\Psi} \begin{bmatrix} \boldsymbol{\Phi}_1 & 0 \\ 0 & -I_{n_{\mathrm{r}}} \end{bmatrix} \boldsymbol{\Psi}^{\#} \,, \tag{38}$$

for some n_{r} such that $n_{\mathrm{f}} > n_{\mathrm{r}} \geq 1$, where the RM $\boldsymbol{\Psi}$ is a nonsingular factor with unspecified distribution of its poles and zeros, and possibly improper, and $\boldsymbol{\Phi}_1$ is a para-hermitian polynomial matrix.

The factorization (38) can be obtained as shown in Corollary 1 of [36], starting with the realization (16) of $\boldsymbol{\Pi}_{\gamma}$. Note that Assumption 11 is not very restrictive.

Theorem 7 ([38]) *Under Assumptions 3, 9 and 11, Problem 1 with* $\mathbb{B}_f = \overline{\mathbb{R}}$, $\mathbb{B}_d = \overline{\mathbb{R}}$ *and* $n_r < n_f$ *admits a solution. The solution satisfies the relations* (37) *on the* EIA.

The FD filter can be constructed analogously as in the sketch of the proof of Theorem 1.

The importance of Theorem 7 is in construction of a bank of filters, when we take $n_r = 1 < n_f$, for each filter of the bank. Moreover, there are at least two more reasons to consider filters with $n_r = 1$: (a) In general, the filters with $n_r = 1$ have the smallest minimal γ. (b) We do not need to consider filters with $n_r > 1$, i.e., with vector output $r(t)$, because in that case we have to reduce it to the scalar output defined by $r(t)^T r(t)$, in order to apply the FD decision, based on a scalar threshold.
Example. Consider the plant given by the following matrices: $E = I_7$,

$$A = \begin{bmatrix} -1.5 & -0.7 & 0.2 & 0.3 & -0.2 & -0.5 & -0.2 \\ -0.2 & -1.9 & -0.1 & 0.1 & 0.3 & 0.2 & 0.4 \\ -0.1 & 0.2 & -1.8 & -0.2 & 0.1 & -0.3 & 0.3 \\ -0.1 & 0.1 & -0.2 & -2 & 0.4 & -0.4 & 0.1 \\ 0.3 & -0.2 & 0.4 & -1 & -1.4 & 0.9 & -0.8 \\ 0.2 & -0.3 & -0.4 & 0.3 & -0.9 & -1 & -0.8 \\ 0 & -0.1 & 0.3 & 0.4 & -0.5 & 0.6 & -0.7 \end{bmatrix},$$

$$B_f = \begin{bmatrix} 1 & -0.8 \\ -0.8 & 1.8 \\ -1.4 & 1.4 \\ 1.2 & -1.8 \\ 1 & -0.8 \\ 0.6 & -1 \\ -0.6 & 0.4 \end{bmatrix}, \quad B_d = \begin{bmatrix} -0.4 & -1.1 & -1.2 & 1 \\ -1.1 & -1.3 & -1.2 & -0.9 \\ -1.8 & -1 & 0 & 1.1 \\ -0.1 & -0.5 & -1.2 & 0.8 \\ -0.2 & 1 & 0.3 & -1 \\ -1.1 & 0.3 & 0.4 & 0.4 \\ -0.2 & 0.2 & -0.4 & 1.1 \end{bmatrix},$$

$$C = \begin{bmatrix} -5 & 2 & 1 & 2.5 & 2 & -1.5 & 3 \\ 3 & 5 & 2.5 & 1.5 & -2 & 1 & 2.5 \\ 1 & 15 & 2.5 & 3 & -4 & -1.5 & 1 \end{bmatrix},$$

$$D_f = \begin{bmatrix} 2 & 1.2 \\ -2 & 1.6 \\ -1 & 2 \end{bmatrix}, \quad D_d = \begin{bmatrix} 1.4 & 0.5 & 0.7 & -0.3 \\ 1.3 & 1.6 & 1 & 0.2 \\ -0.5 & 1.1 & 2.2 & 0.4 \end{bmatrix}.$$

Test signal T1: The disturbance $d_1(t)$ is a step function with magnitude 1 appearing at $t = 3$, and the fault $f_2(t)$ is a step function with magnitude -1 appearing at $t = 7$. The remaining inputs are zero, i.e. $d_2(t) = d_3(t) = d_4(t) = f_1(t) = 0$.
Test signal T2: The disturbance $d_1(t)$ is zero for $t \leq 3$ and $d_1(t) = \sin(\omega_0(t-3))$, $t > 3$, where $\omega_0 = 16\pi$, and $f_2(t)$ is a step function with magnitude -1 appearing at $t = 5$. The remaining inputs are zero, i.e. $d_2(t) = d_3(t) = d_4(t) = f_1(t) = 0$.

The considered initial value of the plant, for all numerical simulations with the five filters, is $x(0) = 0.25[1,\ 1,\ 1,\ 1,\ 1,\ 1,\ 1]^T$.

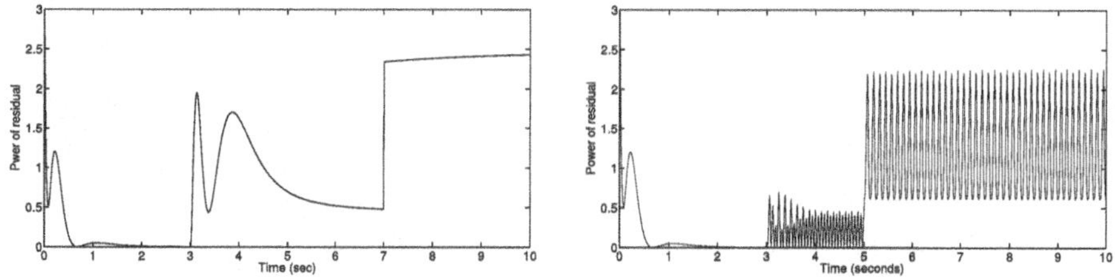

Fig. 4 Residual response $r(t)^T r(t)$ for the Jaimoukha et al. [22] filter and test signal T1 (left) and T2 (right)

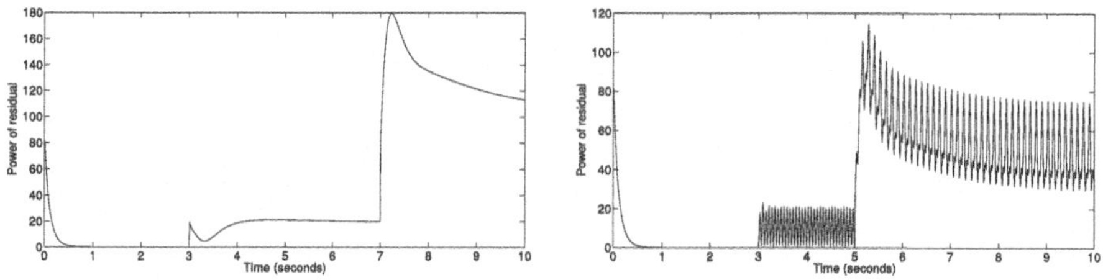

Fig. 5 Residual response $r(t)^T r(t)$ for the Liu and Zhou [31] filter and test signal T1 (left) and T2 (right)

We shall present a numerical simulation with five filters, the first two from the papers [22, 31], and the rest from Theorems 6, 7 and 1, respectively.

Properties of our filters:

The filter of Theorem 6 has a realization of order $2n = 14$.

The filter from Theorem 7 has minimal $\gamma = 1.1770 < 4.8114$ for $n_r = 1$. It is found using Corollary 1 of [36]. Using the generalized spectral factorization algorithm of [36], we obtain $\Phi_1 = \begin{bmatrix} \omega_1 & s \\ -s & \omega_1 \end{bmatrix}$, $\omega_1 = 3.8241$, and $n_r = 1$ (Assumption 11 is satisfied). Continuing with the algorithm of Theorem 7, and applying the command minreal of MATLAB, we obtain an 6-th order filter.

It is seen from Fig. 7 (left) that the FD for the test signal T1 is of a similar quality with the FD presented in Fig. 5 (left).

However, it is seen from Fig. 7 (right) that the FD for the test signal T2 is much better than that presented in Figures Fig. 4 (right) and Fig. 5 (right).

The filter from Theorem 1 is obtained by solving the problem given by (37), where the second inequality in (37) holds on $\mathbb{B} = [15.5191, \infty]$, with $\gamma = 0.3 < 1.1770$ and $n_r = 1$.

A numerical simulation of this filter with the test signals T1 and T2 is given in Fig. 9. Since $16\pi = 50.27 \in [15.5191, \infty]$, with the test signal T2, the disturbance is greatly attenuated in respect to the fault. Indeed, it is seen in Fig. 9 (right) that the disturbance sinusoid in the residual is almost zero, for $t < 5$.

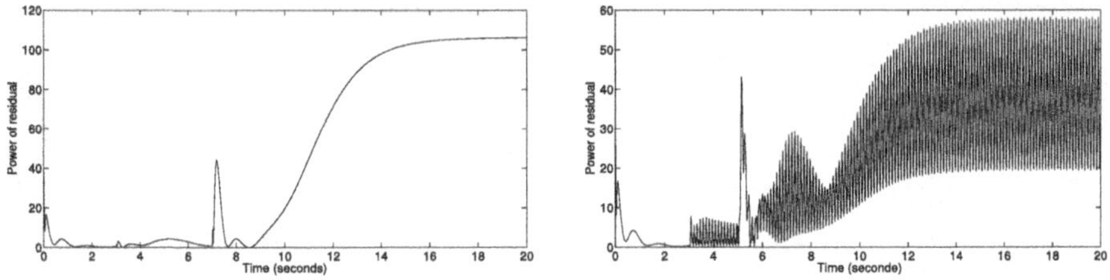

Fig. 6 Residual response $r(t)^{\mathrm{T}}r(t)$ for the filter of Theorem 6 and test signal T1 (left) and T2 (right)

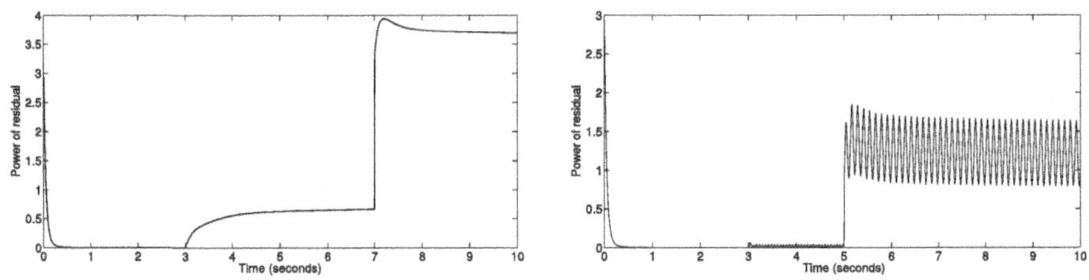

Fig. 7 Residual response $r(t)^{\mathrm{T}}r(t)$ for the filter of Theorem 7 and test signal T1 (left) and T2 (right)

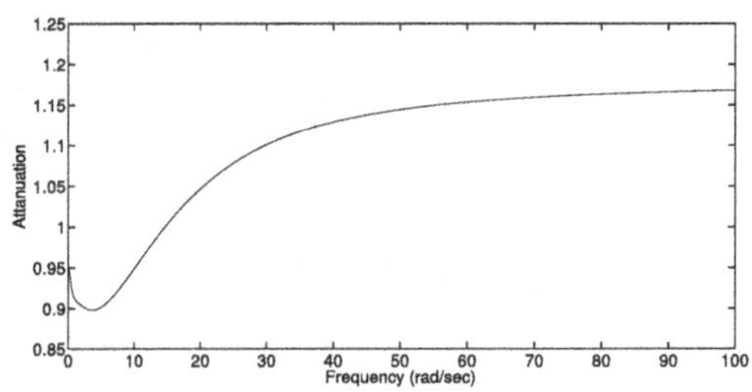

Fig. 8 Frequency dependence $\boldsymbol{\varphi}_{\boldsymbol{F}}(\omega)$ for filter of Theorem 7

From Fig. 9 (left) we see that the FD is worse than the FD with the remaining filters (for instance, one of Theorem 7, presented in Fig. 7 (left)), because the fault pattern in the residual in Fig. 9 decays fast after $t = 7$. The latter phenomenon is obvious by the fact that the spectrum of the fault is located around the frequency $\omega = 0$, and the "water-bed" effect which appears in a neighbourhood of $\omega = 0$ (see Fig. 10), where $\boldsymbol{\varphi}_{\boldsymbol{F}}(\omega)$ is the performance indicator, defined in (10)).

If the frequency region is the whole frequency axis, then the best frequency dependence $\boldsymbol{\varphi}(\omega)$ among the first four filters, in respect to both "flatness" and "smallness", is one obtained by Theorem 7 and presented in Fig. 8.

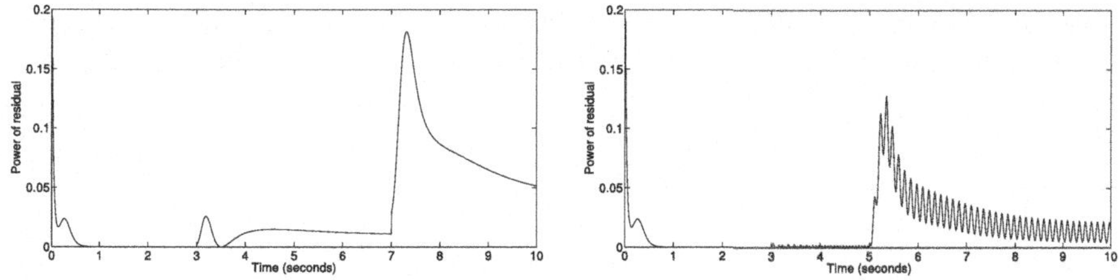

Fig. 9 Residual response $r(t)^{\mathrm{T}}r(t)$ for filter of Theorem 1 and test signal T1 (left) and T2 (right)

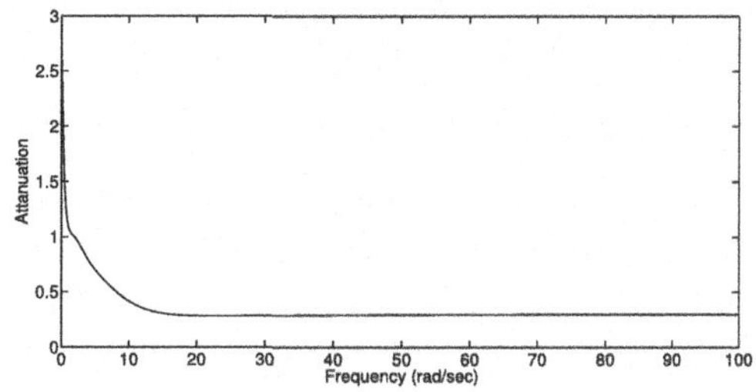

Fig. 10 Frequency dependence $\varphi_{\boldsymbol{F}}(\omega)$ for filter of Theorem 1

3 Fault Estimation

For the system given by (2), three non-equivalent fault estimation problems are:

Problem 1 *Find a possibly minimal β with $0 \le \beta < \infty$ and a proper and stable filter transfer matrix $\boldsymbol{F}$ such that $\left\| \boldsymbol{F}[\boldsymbol{N}_{\mathrm{f}},\ \boldsymbol{N}_{\mathrm{d}}] - [I_{n_{\mathrm{f}}},\ 0] \right\|_{\infty} \le \beta$.*

Problem 1 minimizes the $\mathcal{H}_\infty$ distance from the perfect FE. It is elaborated in the textbooks (see Sect. 6.5.2 of [3], Sect. 8.3.3 of [8] and Sect. 14.2.2 of [12], where in the latter reference, a slightly modified problem has been elaborated on), and it can be solved by the standard $\mathcal{H}_\infty$-filtering theory. As shown in Theorem 17.5 of [67], the filter can be constructed in an observer form.

Problem 2 *Find a possibly minimal γ with $0 \le \gamma < \infty$ and a proper and stable filter transfer matrix $\boldsymbol{F}$ such that $\|\boldsymbol{T}_{\mathrm{rd}}\|_\infty \le \gamma$, and $\boldsymbol{T}_{\mathrm{rf}} = I_{n_{\mathrm{f}}}$.*

Problem 2 can be also reduced to the standard $\mathcal{H}_\infty$ problem.

To formulate the third FE problem, we consider that the spectra of the fault and disturbance are located in some frequency regions $\mathbb{B}_{\mathrm{f}}$ and $\mathbb{B}_{\mathrm{d}}$, respectively, which are union of closed intervals in $\overline{\mathbb{R}}$, which can include the infinity point.

If we intend to apply a pre-filter to eliminate the part in $\mathbb{B}_{\mathrm{d}} \backslash \mathbb{B}_{\mathrm{f}}$ of the disturbance spectrum, where by $\backslash$ we denote the subtraction of sets, then we define $\mathbb{B} := \mathbb{B}_{\mathrm{f}} \cap \mathbb{B}_{\mathrm{d}}$, otherwise we define $\mathbb{B} := \mathbb{B}_{\mathrm{d}}$.

Problem 3:$\mathbb{B}$ *Find a possibly minimal γ with $0 < \gamma < \infty$ and a proper and stable filter transfer matrix $\boldsymbol{F}$ such that the poles of $\boldsymbol{T}_{\mathrm{rf}}$ (if any) are arbitrary assignable, and*

$$\boldsymbol{T}_{\mathrm{rd}}\boldsymbol{T}_{\mathrm{rd}}^{\#} \leq \gamma^2 \boldsymbol{T}_{\mathrm{rf}}\boldsymbol{T}_{\mathrm{rf}}^{\#}, \ \ \mathrm{j}\omega \in \mathrm{j}\mathbb{B} \tag{39}$$

$$\boldsymbol{T}_{\mathrm{rf}}(0) = I_{n_{\mathrm{f}}}\,. \tag{40}$$

We simplify the notation Problem 3:$\overline{\mathbb{R}}$ by writing Problem 3.

Note that condition (39) is a special case of condition (7) with $n_{\mathrm{r}} = n_{\mathrm{f}}$.

Next Theorem 8 is a sufficiency result on Problem 3:$\mathbb{B}$. Note that the para-hermitian RM $\boldsymbol{\Pi}_\gamma(s)$ is defined in (11).

Assumption 12 The RM $[\boldsymbol{N}_{\mathrm{d}},\ \boldsymbol{N}_{\mathrm{f}}]$ is right-invertible at $s = 0$.

Assumption 12 is not restrictive, at least in the generic case. Indeed, if the RM $[\boldsymbol{N}_{\mathrm{d}},\ \boldsymbol{N}_{\mathrm{f}}]$ is not right-invertible, it is left-invertible in the generic case. Then $f = [0,\ I_{n_{\mathrm{f}}}][\boldsymbol{N}_{\mathrm{d}},\ \boldsymbol{N}_{\mathrm{f}}]^{\dagger}(\boldsymbol{M}\boldsymbol{y} - \boldsymbol{N}_{\mathrm{u}}\boldsymbol{u})$, where $[\boldsymbol{N}_{\mathrm{d}},\ \boldsymbol{N}_{\mathrm{f}}]^{\dagger}$ is a left-inverse of $[\boldsymbol{N}_{\mathrm{d}},\ \boldsymbol{N}_{\mathrm{f}}]$, and therefore, the fault $f(t)$ can be determined without solving Problem 3.

Theorem 8 ([40]) *In addition to Assumptions 3, 12, and the full column rank of the matrix $\boldsymbol{N}_{\mathrm{f}}(0)$, assume that there is a generalized factorization of $\boldsymbol{\Pi}_\gamma$,*

$$\boldsymbol{\Pi}_\gamma = \boldsymbol{\Psi} \cdot \mathrm{diag}\{\boldsymbol{\Phi}_1,\ \boldsymbol{\Phi}_2\} \cdot \boldsymbol{\Psi}^{\#}\,, \tag{41}$$

where RM $\boldsymbol{\Psi}$ is a nonsingular factor, possibly improper and with unspecified location of poles and zeros, $\boldsymbol{\Phi}_1$ and $\boldsymbol{\Phi}_2$ are nonsingular para-hermitian $(n_{\mathrm{y}} - n_{\mathrm{f}})$- and n_{f}-dimensional polynomial matrices, such that $\boldsymbol{\Phi}_1$ is positive semi-definite on j$\mathbb{B}$, *$\boldsymbol{\Phi}_2$ is negative semi-definite on* j$\mathbb{B}$, *and* $\det(\boldsymbol{\Phi}_2(0)) \neq 0$. *Then Problem 3:$\mathbb{B}$ admits a solution $\boldsymbol{F}$.*

In order to construct a solution $\boldsymbol{F}$ to Problem 3:$\mathbb{B}$, define the RM $\widetilde{\boldsymbol{F}} = [0,\ I_{n_{\mathrm{f}}}]\boldsymbol{\Psi}^{-1}$, and let $\widetilde{\boldsymbol{F}}$ be given by the descriptor realization $\widetilde{\boldsymbol{F}} = \widetilde{D} + \widetilde{C}(s\widetilde{E} - \widetilde{A})^{-1}\widetilde{B}$ that is finite mode observable, impulse observable and finite mode controllable at $s = 0$ (the latter property means that $[\widetilde{A},\ \widetilde{B}]$ is a full row rank matrix).

Using the realizations (15) of $\boldsymbol{N}_{\mathrm{f}}$ and $\boldsymbol{N}_{\mathrm{d}}$, we obtain

$$\widetilde{\boldsymbol{F}}\boldsymbol{N}_{\mathrm{f}} = \left[\begin{array}{cc|c} \widetilde{A} - s\widetilde{E} & \widetilde{B}C & \widetilde{B}D_{\mathrm{f}} \\ 0 & \mathscr{A} - sE & \mathscr{B}_{\mathrm{f}} \\ \hline \widetilde{C} & \widetilde{D}C & \widetilde{D}D_{\mathrm{f}} \end{array}\right] =: \left[\begin{array}{c|c} \overline{A} - s\overline{E} & \overline{B}_{\mathrm{f}} \\ \hline \overline{C} & \overline{D}_{\mathrm{f}} \end{array}\right]$$

Using the fact that $\mathscr{A} - sE$ is stable and impulse-free matrix pencil, and that the pair $(\widetilde{C}, \widetilde{A} - s\widetilde{E})$ is finite mode observable and impulse observable, it can be proved that the pair $(\overline{C}, \overline{A} - s\overline{E})$ is finite mode observable and impulse observable. (The proof of impulse observability uses Theorems 1 and 2 of [19].)

By those properties of $(\overline{C}, \overline{A} - s\overline{E})$, there exists a matrix $\overline{K}$ such that matrix pencil $\overline{A} - \overline{K}\,\overline{C} - s\overline{E}$ is regular, stable, impulse-free and its finite zeros are arbitrary assignable. Define the RM$\boldsymbol{R} := I + \overline{C}(s\overline{E} - \overline{A})^{-1}\overline{K}$, and define the RM$\widehat{\boldsymbol{F}} := \boldsymbol{R}^{-1}\widetilde{\boldsymbol{F}}$.

A realization of RM $\widehat{\boldsymbol{F}}$ is given by

$$\widehat{\boldsymbol{F}} = \left[\begin{array}{cc|c} \widetilde{A} - K_1\widetilde{C} - s\widetilde{E} & (\widetilde{B} - K_1\widetilde{D})C & \widetilde{B} - K_1\widetilde{D} \\ -K_2\widetilde{C} & \mathscr{A} - K_2\widetilde{D}C - sE & -K_2\widetilde{D} \\ \hline \widetilde{C} & \widetilde{D}C & \widetilde{D} \end{array}\right], \tag{42}$$

where $\overline{K} =: \begin{bmatrix} K_1 \\ K_2 \end{bmatrix}$. Then we put

$$\boldsymbol{F} = Y\widehat{\boldsymbol{F}}, \tag{43}$$

for some nonsingular matrix Y, chosen to satisfy the requirement (40) of Problem 3:$\mathbb{B}$. It is proved in [40] that $\widehat{\boldsymbol{F}}(0)\boldsymbol{N}_{\mathrm{f}}(0)$ is a nonsingular matrix. Then we take $Y = (\widehat{\boldsymbol{F}}(0)\boldsymbol{N}_{\mathrm{f}}(0))^{-1}$.

Example. Consider the plant (91) given by $E = I_4$,

$$A = \begin{bmatrix} -1.5 & -0.7 & 0.2 & 0.3 \\ -0.2 & -1.9 & -0.1 & 0.1 \\ -0.1 & 0.2 & -1.8 & -0.2 \\ -0.1 & 0.1 & -0.2 & -2 \end{bmatrix}, \quad B_{\mathrm{f}} = \begin{bmatrix} 1 & -1.2 \\ -4 & -1.2 \\ -2.6 & 0 \\ 0.4 & -1.2 \end{bmatrix},$$

$$B_{\mathrm{d}} = \begin{bmatrix} -0.4 & -1.1 \\ -1.1 & -1.3 \\ -1.8 & -1.00 \\ -0.1 & -0.5 \end{bmatrix}, \quad C = \begin{bmatrix} -5 & 2 & 1 & 2.5 \\ 3 & 5 & 2.5 & 1.5 \\ 1 & 15 & 2.5 & 3 \end{bmatrix},$$

$$D_{\mathrm{f}} = \begin{bmatrix} 2 & 0.7 \\ -2.4 & 1 \\ -2.6 & 2.2 \end{bmatrix}, \quad D_{\mathrm{d}} = \begin{bmatrix} 1.4 & 0.5 \\ 1.3 & 1.3 \\ -0.5 & 1.1 \end{bmatrix}.$$

In this example we elaborate on the first filter of the bank, by introducing the disturbance d_3, which is equal to the fault f_2. A justification is that the minimal γ for Problem 3 with the original matrices is 2.42978924, while for the first filter in the bank, it is only 0.21839718.

To estimate f_1, we shall obtain six filters, and compare the FE by numerical simulation.

We consider the initial value of the plant $x(0) = [1,\ 1,\ 1,\ 1]^{\mathrm{T}}$, zero initial values for all six filters, and the following test signal.

The disturbance d_2 is such that $d_2(t) = 0$, $t \le 3$ and $d_2(t) = 0.3\sin(\omega_0(t-3))$, $t > 3$, where $\omega_0 = 16\pi$. The disturbances $d_1(t)$ is zero. The fault signal $f_1(t)$ is a step with magnitude 1 appearing at $t = 5$. The fault signal $f_2(t)$ $(= d_3(t))$ satisfies $f_2(t) = 0$, $t \le 3$ and $f_2(t) = \sin(\omega_0(t-3))$, $t > 3$.

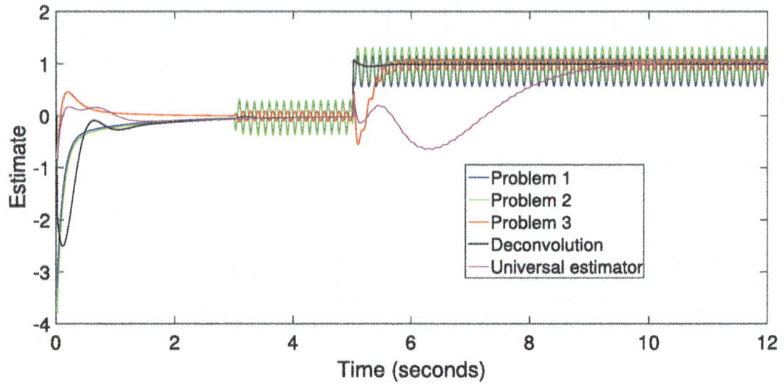

Fig. 11 $r(t)$ for the filters solving Problems 1, 2, 3 over EIA, and universal estimator

A deconvolution algorithm for FE can be obtained if we apply the standard discrete $\mathscr{H}_2$ minimization, after we discretize the model, with sampling period 0.01. We take the controlled output $z_k = u_k - f_k + \gamma(f_k - f_{k-1})$, $k = 0, 1, \ldots$, where "the control" u_k is the estimate of f_k. Then we construct a deconvolution filter of realization order 5. For the particular test signal, the best estimate is obtained for $\gamma = 1.12$.

The response of the residual of the filters solving Problems 1, 2, 3, deconvolution and universal estimator[1] to the test signal is given in Fig. 11 (in blue, green, red, black and magenta, respectively). It is seen that the disturbance pattern with a big amplitude appears in the residual for the first two filters. In the stationary regime, we see that there is an offset of $r(t)$ in respect to $f_1(t)$ for the filter solving Problem 1. We see that the amplitude of the disturbance pattern for the filter solving Problem 3 is less than that of the filters solving Problems 1 and 2. It is seen that the disturbance attenuation with the universal estimator is satisfactory, but the time for FE is longer than the FE times of the previous filters.

Finally, we obtain a filter from Theorem 8, i.e. filter (43). For that purpose, at first we find a factorization (41) of $\mathbf{\Pi}_\gamma$ with $\gamma = 0.15$. The RM $\mathbf{\Pi}_\gamma$ has two zeros on the positive EIA: j5.2798 and j36.78257. Following the generalized spectral factorization theory of [36, 39], we obtain $\omega_1 = -5.2798$ and $\omega_2 = 36.78257$. Using the inequalities $\omega_1 < 0$ and $\omega_2 > 0$, we obtain $\mathbb{B} = [36.78257, \infty]$, and $\mathbf{\Phi}_1(\mathrm{j}\omega) = \mathrm{diag}\{(\omega_1^2 - \omega^2)/\omega_1, \omega_2\} \geq 0$, $\mathbf{\Phi}_2(\mathrm{j}\omega) = (\omega_2^2 - \omega^2)/\omega_2 \leq 0$, $\omega \in \mathbb{B}$. We compute a balanced realization of the filter $\boldsymbol{F}$ solving Problem 3:$\mathbb{B}$.

Since the spectra of d_2 and d_3 are located at the frequency $16\pi \in \mathbb{B} = [36.78257, \infty]$, the attenuation of the disturbance will be at least $\gamma = 0.15$. Indeed, the response of the residual with this filter to the test signal is given in Fig. 12. The FE is reliable, because the amplitude of the disturbance pattern in the residual is very small. Unlike

[1] The universal estimator, elaborated in Theorem 3 of [40] is an FE filter that is independent of γ, such that frequency region $\mathbb{B}$ is maximal with respect to inclusion of subspaces, and $\mathbb{B}$ is specific for a given γ.

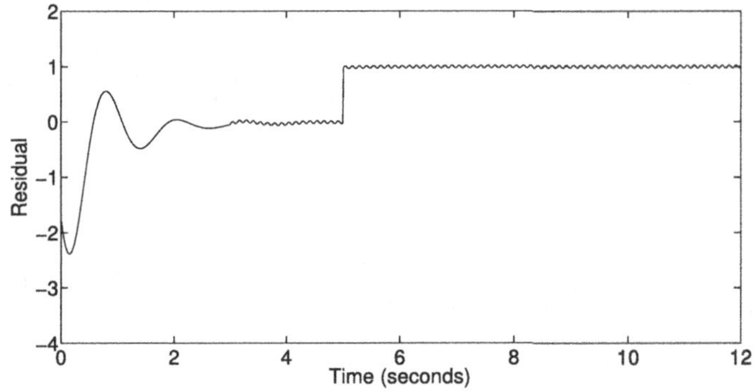

Fig. 12 Residual $r(t)$ for the filter solving Problem 3:$\mathbb{B}$

the universal estimator, for which $\mathbb{B} = [0, 5.2798] \cup [36.78257, \infty]$, the filter (43) is "dedicated" to the frequency region $\mathbb{B} = [36.78257, \infty]$.

Note that the response of the filter obtained by deconvolution is of the approximately same quality as the filter (43). A reason for the good behaviour of the deconvolution algorithm is that an estimation of $f(k)$ is not obtained, but an estimation of $f(k-1)$. The delay in the estimation of one sample period (which can be very small) makes the estimation easier.

4 Passive Fault-Tolerant Control

One of the most important control problems for systems with faults is to find a controller such that the behaviour of the closed loop system is tolerant to faults. For the case that there are no disturbances, there are various fault-tolerant control (FTC) system architectures and results in the literature, based on inclusion of an residual observer in the control loop. The problem of FTC with disturbances is much harder, because the residual signal is not zero in times when there are no faults.

The most general control system with faults and disturbances is presented in Fig. 13, and our results deal with the plant, given by the descriptor system (1).

4.1 FTC Based on Fault Reduction Effect in Respect to Disturbances

For some controller $\boldsymbol{K}$, denote by $\boldsymbol{T}_{\mathrm{zf}}$ and $\boldsymbol{T}_{\mathrm{zd}}$ the closed loop transfer matrices from f to z and from d to z, respectively. With respect to Fig. 13, we have

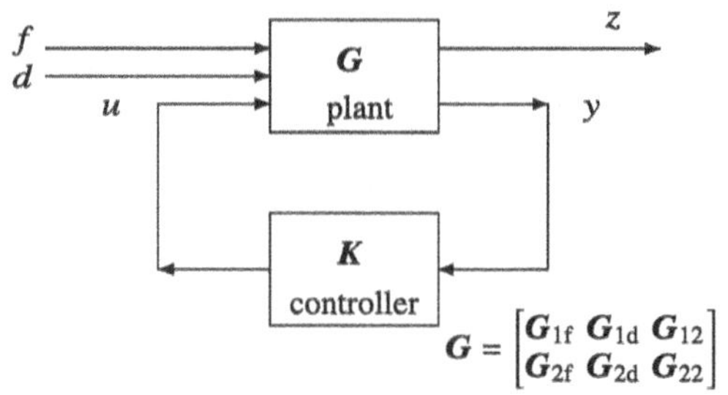

Fig. 13 Plant with controller

$$z = T_{\mathrm{zf}} f + T_{\mathrm{zd}} d \,, \tag{44}$$

under zero initial conditions, where $z(s)$, $f(s)$ and $d(s)$ are the Laplace transforms of $z(t)$, $f(t)$ and $d(t)$.

Let $\mathbb{B}$ be a given frequency region. It depends on $\mathbb{B}_{\mathrm{f}}$ and $\mathbb{B}_{\mathrm{d}}$.

Problem 1 *Find a necessary and sufficient condition for existence of a controller* $K(s)$ *such that the matrix inequality*

$$\gamma^{-2} T_{\mathrm{zf}} T_{\mathrm{zf}}^{*} < T_{\mathrm{zd}} T_{\mathrm{zd}}^{*} \ \tilde{\text{on}}\ \mathrm{j}\mathbb{B} \,, \tag{45}$$

holds, for some possibly infimal γ. *If the necessary and sufficient condition is satisfied, find all controllers such that* (45) *holds.*

If the controller $K(s)$ is stabilizing, we refer Problem 1 to as Problem 1S.

Step 1 of our FTC design algorithm is to find a controller generator RM K_{a} such that all controllers K solving Problem 1 are given in the form

$$K = \mathscr{F}(K_{\mathrm{a}}, U) \,, \tag{46}$$

where U is an RM satisfying some bound on $\mathrm{j}\mathbb{B}$.

As a second step of our algorithm, we consider an auxiliary system without faults, with the transfer matrix

$$G_{\mathrm{a}} = \begin{bmatrix} \overbrace{G_{\mathrm{a}11}}^{m_{\mathrm{d}}} & \overbrace{G_{\mathrm{a}12}}^{m_2} \\ G_{\mathrm{a}21} & G_{\mathrm{a}22} \end{bmatrix} \begin{matrix} \}p_1 \\ \}p_2 \end{matrix} \tag{47}$$

where $\mathscr{F}(G_{\mathrm{a}}, U) = \mathscr{F}(G_{\mathrm{dd}}, \mathscr{F}(K_{\mathrm{a}}, U)) = T_{\mathrm{zd}}$,

$$G_{\mathrm{dd}} := \begin{bmatrix} G_{1\mathrm{d}} & G_{12} \\ G_{2\mathrm{d}} & G_{22} \end{bmatrix} \tag{48}$$

and $U = \mathcal{F}(K_{\mathrm{a}}, U)$ as a controller (see Fig. 14).

Let the frequency region $\mathbb{B}_1 \supseteq \mathbb{B}$ be given. It also depends on $\mathbb{B}_{\mathrm{f}}$ and $\mathbb{B}_{\mathrm{d}}$.

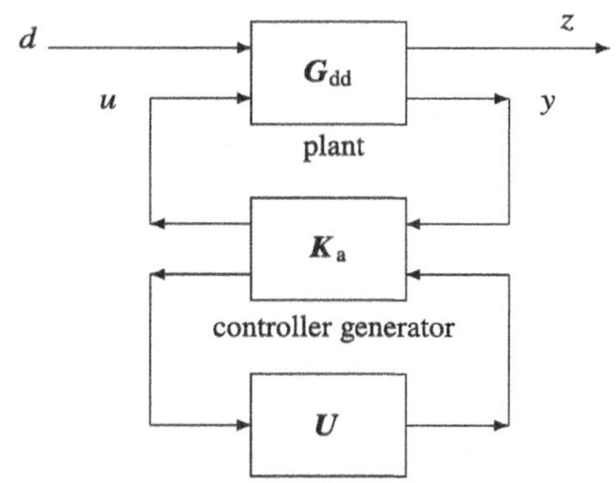

Fig. 14 Plant with parameterized by U controller and without fault

We find iteratively as small as possible $\|T_{zd}\|$ on $j\mathbb{B}_1$, denote it by β, and the corresponding U, under the mentioned constraint on U and the stability of the closed-loop system. Since the function $\max_{\omega\in\mathbb{B}_1}\|T_{zd}(j\omega)\|$ is not explicit, the gradient-based minimization algorithms cannot be applied. For that reason, we shall apply the Nelder-Mead derivative-free algorithm.

As a result of this two-step design algorithm, our controller K will be given by (46), and we will have

$$\begin{aligned} &\gamma^{-2}T_{zf}T_{zf}^{*} \le T_{zd}T_{zd}^{*}\,, \quad j\omega \in j\mathbb{B}\,,\\ &T_{zd}T_{zd}^{*} \le \beta^2 I\,, \quad j\omega \in j\mathbb{B}_1\,. \end{aligned} \tag{49}$$

In order to find all solutions to Problem 1, we proceed to formulate Step 1 of the algorithm. Denote

$$\begin{aligned} G_{11} &:= [G_{1f},\ G_{1d}] = [D_{1f},\ D_{1d}] + C_1(sE-A)^{-1}[B_f,\ B_d]\,,\\ G_{12} &:= D_{12} + C_1(sE-A)^{-1}B_2\,,\\ G_{21} &:= [G_{2f},\ G_{2d}] = [D_{2f},\ D_{2d}] + C_2(sE-A)^{-1}[B_f,\ B_d]\,,\\ G_{22} &:= D_{22} + C_2(sE-A)^{-1}B_2\,, \end{aligned}$$

and $G = \begin{bmatrix} G_{11} & G_{12} \\ G_{21} & G_{22} \end{bmatrix}$.

Under zero initial conditions $x_0 = 0$, we have

$$\begin{aligned} z &= G_{1f}f + G_{1d}d + G_{12}u\,,\\ y &= G_{2f}f + G_{2d}d + G_{22}u\,. \end{aligned} \tag{50}$$

Denote by K a (measurement feedback) $m_2 \times p_2$-dimensional transfer matrix of the controller, which satisfies $u = Ky$ (see Fig. 13). We have

$$\begin{aligned} \boldsymbol{T}_{\mathrm{zf}} &= \boldsymbol{G}_{1\mathrm{f}} + \boldsymbol{G}_{12}\boldsymbol{K}(I_{p_2} - \boldsymbol{G}_{22}\boldsymbol{K})^{-1}\boldsymbol{G}_{2\mathrm{f}} \,, \\ \boldsymbol{T}_{\mathrm{zd}} &= \boldsymbol{G}_{1\mathrm{d}} + \boldsymbol{G}_{12}\boldsymbol{K}(I_{p_2} - \boldsymbol{G}_{22}\boldsymbol{K})^{-1}\boldsymbol{G}_{2\mathrm{d}} \,, \end{aligned}$$

where $\boldsymbol{T}_{\mathrm{zf}}$ and $\boldsymbol{T}_{\mathrm{zd}}$ are the transfer matrices in (44).

Introduce the following matrix and RM

$$J_\gamma = \begin{bmatrix} I_{m_\mathrm{f}} & 0 \\ 0 & -\gamma^2 I_{m_\mathrm{d}} \end{bmatrix}, \quad \boldsymbol{G}_1 = \begin{bmatrix} \boldsymbol{G}_{11} \\ \boldsymbol{G}_{21} \end{bmatrix} = \begin{bmatrix} \boldsymbol{G}_{1\mathrm{f}} & \boldsymbol{G}_{1\mathrm{d}} \\ \boldsymbol{G}_{2\mathrm{f}} & \boldsymbol{G}_{2\mathrm{d}} \end{bmatrix}.$$

Assumption 13 The RM $\boldsymbol{G}_1 J_\gamma \boldsymbol{G}_1^*$ has at least p_2 positive eigenvalues $\widetilde{\text{on}}$ j$\mathbb{B}$.

LMI (25) can be used to check Assumption 13.

Assumption 14 The RM $\boldsymbol{G}_{12}$ is left-invertible.

Here we shall present the theory under Assumption 13. The theory under Assumption 14 is given in [41].

Since $\boldsymbol{G}_1 J_\gamma \boldsymbol{G}_1^*$ is a nonsingular RM, we can define the para-hermitian $(p_2 + m_2) \times (p_2 + m_2)$-dimensional RM:

$$\boldsymbol{\Pi}_\gamma := \begin{bmatrix} 0 & -\boldsymbol{G}_{12} \\ I_{p_2} & -\boldsymbol{G}_{22} \end{bmatrix}^\# (\boldsymbol{G}_1 J_\gamma \boldsymbol{G}_1^\#)^{-1} \begin{bmatrix} 0 & -\boldsymbol{G}_{12} \\ I_{p_2} & -\boldsymbol{G}_{22} \end{bmatrix}. \tag{51}$$

We find a $2n$-order descriptor realization of $\boldsymbol{\Pi}_\gamma$. Introduce the matrices

$$D := \begin{bmatrix} D_{1\mathrm{f}} & D_{1\mathrm{d}} \\ D_{2\mathrm{f}} & D_{2\mathrm{d}} \end{bmatrix}, \quad B := [B_\mathrm{f},\ B_\mathrm{d}], \quad C := \begin{bmatrix} C_2 \\ C_2 \end{bmatrix}. \tag{52}$$

Assume that the matrix $D_\mathrm{x} := D J_\gamma D^\mathrm{T}$ is nonsingular. We have $\boldsymbol{G}_1 = D + C(sE - A)^{-1}B$, and

$$\boldsymbol{G}_1 J_\gamma \boldsymbol{G}_1^\# = \left[\begin{array}{cc|c} A - sE & B J_\gamma B^\mathrm{T} & B J_\gamma D^\mathrm{T} \\ 0 & -A^\mathrm{T} - sE^\mathrm{T} & -C^\mathrm{T} \\ \hline C & D J_\gamma B^\mathrm{T} & D J_\gamma D^\mathrm{T} \end{array}\right] =: \left[\begin{array}{c|c} A_\mathrm{x} - sE_\mathrm{x} & B_\mathrm{x} \\ \hline C_\mathrm{x} & D_\mathrm{x} \end{array}\right]. \tag{53}$$

Note that $C_\mathrm{x} = -B_\mathrm{x}^\mathrm{T}\widetilde{J}$, $A_\mathrm{x}^\mathrm{T}\widetilde{J} = -\widetilde{J}A_\mathrm{x}$, and $E_\mathrm{x}^\mathrm{T}\widetilde{J} = \widetilde{J}E_\mathrm{x}$, where $\widetilde{J} = \begin{bmatrix} 0 & I_n \\ -I_n & 0 \end{bmatrix}$.

Introduce also the matrices

$$B_\mathrm{y} = \begin{bmatrix} 0 & -B_2 \\ 0 & 0 \end{bmatrix}, \quad D_\mathrm{y} = \begin{bmatrix} 0 & -D_{12} \\ I_{p_2} & -D_{22} \end{bmatrix}.$$

Then

$$\begin{bmatrix} 0 & -\boldsymbol{G}_{12} \\ I_{p_2} & -\boldsymbol{G}_{22} \end{bmatrix} = D_\mathrm{y} + C_\mathrm{x}(sE_\mathrm{x} - A_\mathrm{x})^{-1} B_\mathrm{y} \,. \tag{54}$$

We obtain

$$\mathbf{\Pi}_\gamma = \left[\begin{array}{c|c} A_x - B_x D_x^{-1} C_x - s E_x & B_y - B_x D_x^{-1} D_y \\ \hline -(B_y - B_x D_x^{-1} D_y)^T J & D_y^T D_x^{-1} D_y \end{array}\right]. \tag{55}$$

A necessary condition for Problem 1 is

$$\text{In}(\mathbf{\Pi}_\gamma) = (m_{2-},\ m_{20},\ p_2)\ \tilde{\text{on}}\ \text{j}\mathbb{B}\,, \tag{56}$$

where m_{2-} and m_{20} are integers such that m_{20} is the rank defect of $\mathbf{\Pi}_\gamma$ and $m_{2-} + m_{20} = m_2$. Under (56), there exists a nonsingular factor $\mathbf{\Psi}$, such that

$$\mathbf{\Pi}_\gamma = \mathbf{\Psi}^\# \mathbf{\Phi} \mathbf{\Psi}\,, \tag{57}$$

where

$$\mathbf{\Phi} = \text{diag}\{\mathbf{\Phi}_1,\ 0_{m_{20}\times m_{20}},\ -\mathbf{\Phi}_2\}\,, \tag{58}$$

where $\mathbf{\Phi}_1$ is a $p_2 \times p_2$-dimensional para-hermitian RM satisfying $\mathbf{\Phi}_1 > 0\ \tilde{\text{on}}\ \text{j}\mathbb{B}$, $\mathbf{\Phi}_2$ is a $m_{2-} \times m_{2-}$-dimensional para-hermitian RM satisfying $\mathbf{\Phi}_2 > 0\ \tilde{\text{on}}\ \text{j}\mathbb{B}$, and m_{20} is the rank defect of $\mathbf{\Pi}_\gamma$.

The factorization (57) can be found as shown in [36], starting with the realization (55).

Without loss of generality, we can consider the controller $\boldsymbol{K}$ in the form $\boldsymbol{K} = \boldsymbol{K}_2 \boldsymbol{K}_1^{-1}$, for some RMs $\boldsymbol{K}_2$ and nonsingular $\boldsymbol{K}_1$. Define RMs $\boldsymbol{U}_1$ and $\boldsymbol{U}_2$ by

$$\begin{bmatrix} \boldsymbol{U}_1 \\ \boldsymbol{U}_2 \end{bmatrix} = \mathbf{\Psi} \begin{bmatrix} \boldsymbol{K}_1 \\ \boldsymbol{K}_2 \end{bmatrix}, \quad \boldsymbol{U}_2 = \begin{bmatrix} \boldsymbol{U}_{20} \\ \boldsymbol{U}_{2-} \end{bmatrix} \begin{matrix} \}m_{20} \\ \}m_{2-} \end{matrix}. \tag{59}$$

Inequality (45) is equivalent with the following inequality

$$[\boldsymbol{K}_1^*,\ \boldsymbol{K}_2^*] \mathbf{\Pi}_\gamma \begin{bmatrix} \boldsymbol{K}_1 \\ \boldsymbol{K}_2 \end{bmatrix} > 0\ \tilde{\text{on}}\ \text{j}\mathbb{B}\,. \tag{60}$$

Combining identity (59) with the matrix inequality (60), we obtain the matrix inequality

$$\boldsymbol{U}_1^* \mathbf{\Phi}_1 \boldsymbol{U}_1 - \boldsymbol{U}_{2-}^* \mathbf{\Phi}_2 \boldsymbol{U}_{2-} > 0\ \tilde{\text{on}}\ \text{j}\mathbb{B}\,, \tag{61}$$

and the RM $\boldsymbol{U}_{20}$ remains arbitrary. Introduce the following partitions of the RMs $\mathbf{\Psi}$ and $\mathbf{\Theta} := \mathbf{\Psi}^{-1}$.

$$\mathbf{\Psi} =: \overset{\quad p_2 \quad\ m_2}{\begin{bmatrix} \mathbf{\Psi}_{11} & \mathbf{\Psi}_{12} \\ \mathbf{\Psi}_{21} & \mathbf{\Psi}_{22} \end{bmatrix}} \begin{matrix} \}p_2 \\ \}m_2 \end{matrix}, \quad \mathbf{\Theta} =: \overset{\quad p_2 \quad\ m_2}{\begin{bmatrix} \mathbf{\Theta}_{11} & \mathbf{\Theta}_{12} \\ \mathbf{\Theta}_{21} & \mathbf{\Theta}_{22} \end{bmatrix}} \begin{matrix} \}p_2 \\ \}m_2 \end{matrix}.$$

By (59), we have the two identities:

$$\boldsymbol{K}_1 = \boldsymbol{\Theta}_{11}\boldsymbol{U}_1 + \boldsymbol{\Theta}_{12}\boldsymbol{U}_2, \quad \boldsymbol{K}_2 = \boldsymbol{\Theta}_{21}\boldsymbol{U}_1 + \boldsymbol{\Theta}_{22}\boldsymbol{U}_2,$$

and the following fractional parameterization by $\boldsymbol{U}_1$ and $\boldsymbol{U}_2 = \begin{bmatrix} \boldsymbol{U}_{20} \\ \boldsymbol{U}_{2-} \end{bmatrix}$ satisfying (61), of controllers $\boldsymbol{K} = \boldsymbol{K}_2\boldsymbol{K}_1^{-1}$:

$$\boldsymbol{K} = (\boldsymbol{\Theta}_{21}\boldsymbol{U}_1 + \boldsymbol{\Theta}_{22}\boldsymbol{U}_2)(\boldsymbol{\Theta}_{11}\boldsymbol{U}_1 + \boldsymbol{\Theta}_{12}\boldsymbol{U}_2)^{-1}. \tag{62}$$

If Assumption 14 holds, the RM $\boldsymbol{\Pi}_\gamma$ is nonsingular, and $m_{20} = 0$ and the RM $\boldsymbol{U}_{20}$ is void (and $\boldsymbol{U}_2 = \boldsymbol{U}_{2-}$). Then $\boldsymbol{U}_1$ is a nonsingular RM, hence there exists the RM $\boldsymbol{U}$ defined by $\boldsymbol{U} = \boldsymbol{U}_2\boldsymbol{U}_1^{-1}$. It satisfies

$$\boldsymbol{U}^*\boldsymbol{\Phi}_2\boldsymbol{U} < \boldsymbol{\Phi}_1 \ \widetilde{\text{on}}\ \mathrm{j}\mathbb{B}, \tag{63}$$

and the controller (62) becomes

$$\boldsymbol{K} = (\boldsymbol{\Theta}_{21} + \boldsymbol{\Theta}_{22}\boldsymbol{U})(\boldsymbol{\Theta}_{11} + \boldsymbol{\Theta}_{12}\boldsymbol{U})^{-1}. \tag{64}$$

Theorem 9 ([41]) *Under Assumption 13, Problem 1 admits a solution only if the condition* (56) *holds. If in addition to Assumption 13, there exists the factorization* $\boldsymbol{\Pi}_\gamma = \boldsymbol{\Psi}^\#\boldsymbol{\Phi}\boldsymbol{\Psi}$, *i.e.,* (57), *all problem solutions are given by* (62), *where* $\boldsymbol{U}_{20}$ *is an arbitrary RM, and* $\boldsymbol{U}_1$ *and* $\boldsymbol{U}_{2-}$ *are RMs satisfying* (61) *and* $\det(\boldsymbol{\Theta}_{11}\boldsymbol{U}_1 + \boldsymbol{\Theta}_{12}\boldsymbol{U}_2) \neq 0$. *If in addition to Assumption 13 and* (57), *Assumptions 14 holds, then all problem solutions are given by* (64), *where* $\boldsymbol{U}$ *is an RM satisfying* (63) *and* $\det(\boldsymbol{\Theta}_{11} + \boldsymbol{\Theta}_{12}\boldsymbol{U}) \neq 0$.

In the case $\mathbb{B} = \overline{\mathbb{R}}$, the condition (61) becomes

$$\|\boldsymbol{U}\|_\infty < 1. \tag{65}$$

We proceed with a proof of stability of the closed-loop system. Here we elaborate on the case $\mathbb{B} = \overline{\mathbb{R}}$, and introduce three additional assumptions, in order to guarantee stability for all stable $\boldsymbol{U}$ with the constraint (65).

Assumption 15 The plant (1) is stabilizable, i.e. the matrix pencil $(A - sE,\ B_2)$ is finite mode stabilizable and impulse controllable, and the matrix pencil $(C_2,\ A - sE)$ is finite mode detectable and impulse observable.

Assumption 16 There exists a factorization $\boldsymbol{\Pi}_\gamma = \boldsymbol{\Psi}^\#\mathcal{J}\boldsymbol{\Psi}$, where the factor $\boldsymbol{\Psi}$ and its inverse are proper and stable, and $\mathcal{J} := \operatorname{diag}\{I_{p_2},\ -I_{m_2}\}$.

Assumption 17

$$\boldsymbol{G}_{12}^{*}[I,\ 0](\boldsymbol{G}_1 J_\gamma \boldsymbol{G}_1^{*})^{-1}\begin{bmatrix} I \\ 0 \end{bmatrix}\boldsymbol{G}_{12} \leq 0 \ \text{on} \ \mathrm{j}\overline{\mathbb{R}}.$$

Theorem 10 ([41]) *Under Assumptions 13, 15, 16 and 17, Problem 1S with* $\mathbb{B} = \overline{\mathbb{R}}$ *admits a solution if and only if one of the controllers* (64) *is stabilizing, where* $\boldsymbol{U}$ *ranges in the set of stable RMs satisfying* (65). *In that case, all solutions of Problem 1S are given by* (64), *where* $\boldsymbol{U}$ *is a stable RM and satisfies* (65).

To obtain $\boldsymbol{U}$, we apply the second step of the algorithm, as explained in the following.

Consider that there is a stabilizing $\boldsymbol{U}$ (see Theorem 10), which we denote by $\boldsymbol{U}_0$. To find a stabilizing $\boldsymbol{U}$, satisfying (65) such that $\|\boldsymbol{T}_{\mathrm{zd}}\|_\infty$ is minimal, where $\boldsymbol{T}_{\mathrm{zd}} = \mathscr{F}(\boldsymbol{G}_{\mathrm{a}},\ \boldsymbol{U})$ and RM $\boldsymbol{G}_{\mathrm{a}}$ is defined by (47) and (48), we apply an iterative algorithm, starting with $\boldsymbol{U}_0$. Let us group at first the coefficients of the unknown RM $\boldsymbol{U}$ in the matrix $\begin{bmatrix} A_U & B_U \\ C_U & D_U \end{bmatrix} =: P_U$, where (A_U, B_U, C_U, D_U) is a realization of $\boldsymbol{U}$. In each step of the algorithm, we find a new $\boldsymbol{U}$ such that $\|\boldsymbol{T}_{\mathrm{zd}}\|_\infty$ is decreased.

We have no guaranty that the new $\boldsymbol{U}$ will be stabilizing. However, for the case $\mathbb{B} = \overline{\mathbb{R}}$, if the distances between the successive matrices P_U are sufficiently small, which is a property of iterative algorithms, we can assure the stability. Namely, if there is a pole of $\boldsymbol{T}_{\mathrm{zd}}$ on the imaginary axis, then $\|\boldsymbol{T}_{\mathrm{zd}}\|_\infty = \infty$. Therefore, by the strategy to decrease $\|\boldsymbol{T}_{\mathrm{zd}}\|_\infty$, we avoid the crossing of the imaginary axis.

Besides the constraint on stabilizability of $\boldsymbol{U}$, we have the constraint (65). Although the set of RMs satisfying (65) is convex, the function $\|\boldsymbol{T}_{\mathrm{zd}}\|_\infty$ of $\boldsymbol{U}$ is not convex in general. Therefore, the existence of global minimum is not guaranteed.

Since the constraint (65) is not explicit, we apply algorithms for unconstrained minimization (for instance, the Nelder-Mead algorithm), so that we modify the minimizing function by introducing penalties.

In each iteration, the most complex computation is finding the $\mathscr{H}_\infty$ norms of $\boldsymbol{T}_{\mathrm{zd}}$ and $\boldsymbol{U}$.

Example. Consider the following plant, and $\mathbb{B} = \overline{\mathbb{R}}$.

$$A = \begin{bmatrix} -7.6528 & -0.0471 & -4.1759 & -1.9042 & -1.6904 \\ 0.7942 & -0.0696 & 0.2998 & 0.5543 & 0.5432 \\ 6.0399 & 2.2009 & -1.4744 & -0.0193 & 4.2586 \\ 0.9197 & -0.3532 & -0.3314 & -3.5169 & 1.2905 \\ -0.0643 & 0.0041 & -0.0530 & -0.1778 & 0 \end{bmatrix}, \quad E = \begin{bmatrix} I_4 & 0 \\ 0 & 0 \end{bmatrix}$$

$$B_{\mathrm{f}} = \begin{bmatrix} 4.0527 & 2.9827 \\ -1.5255 & -0.4934 \\ -1.5799 & 0.4476 \\ 2.9668 & 0.0748 \\ -0.2033 & 0.4067 \end{bmatrix}, \ B_{\mathrm{d}} = \begin{bmatrix} 2.1876 & 2.1058 & -0.2903 \\ -1.6820 & -0.9204 & -0.0459 \\ -1.2234 & 0.6050 & -0.8282 \\ 3.1464 & 0.6915 & 0.7043 \\ 0.4067 & -0.4067 & 0.2033 \end{bmatrix}, \ B_2 = \begin{bmatrix} -3.1026 & 8.6517 \\ 0.1800 & -0.8874 \\ 0.7204 & -2.4297 \\ -0.4479 & 2.1280 \\ -0.0000 & 0.2033 \end{bmatrix},$$

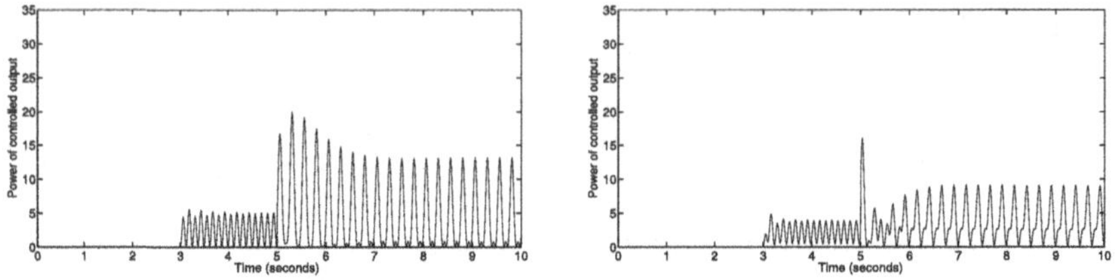

Fig. 15 $z(t)^{\mathrm{T}}z(t)$ for Test signal T1, with $\boldsymbol{K}_{\mathrm{d}}$ (left) and $\boldsymbol{K}_{\mathrm{fd}}$ (right)

$$D_{12} = \begin{bmatrix} 0.2444 & -2.7002 \\ 0.3111 & -2.1808 \end{bmatrix}, \; D_{22} = \begin{bmatrix} 0.3111 & -0.9636 \\ -0.1889 & -1.2155 \end{bmatrix}.$$

$$C_1 = \begin{bmatrix} 7.4334 & 6.7367 & -3.1013 & 3.7289 & 6.7666 \\ 2.8308 & -0.2017 & 0.0572 & -0.3951 & 1.6917 \end{bmatrix},$$

$$C_2 = \begin{bmatrix} 1.1687 & 3.3296 & -2.4898 & 1.4395 & 1.6917 \\ 1.8179 & 2.4023 & -1.2695 & 0.9500 & 1.6917 \end{bmatrix},$$

$$D_{1\mathrm{f}} = \begin{bmatrix} -4.3921 & 1.2855 \\ -0.8423 & -0.1901 \end{bmatrix}, \; D_{1\mathrm{d}} = \begin{bmatrix} -4.6237 & 1.0435 & -0.3915 \\ -0.6674 & -0.2277 & -0.3536 \end{bmatrix},$$

$$D_{2\mathrm{f}} = \begin{bmatrix} -1.0594 & 1.2442 \\ -1.3076 & 0.7405 \end{bmatrix}, \; D_{2\mathrm{d}} = \begin{bmatrix} -1.2331 & 1.3380 & -0.1365 \\ -1.2368 & 0.8418 & -0.3883 \end{bmatrix}.$$

Note that the RM $\boldsymbol{G}$ is improper and unstable.

For comparison purposes, at first we construct four other controllers:

(i) $\boldsymbol{K}_{\mathrm{d}}$, constructed under the assumption that there are no faults.
(ii) $\boldsymbol{K}_{\mathrm{fd}}$, constructed under the assumption that the faults are disturbances.
(iii) A controller based on fault detection. It is an active FTC constructed with the fault detector, elaborated on in Sect. 2.
(iv) A controller based on fault estimation. It is constructed as shown in Sect. 4.2.

All numerical simulations in this example are made for zero initial conditions of the plant and the test signal, which we denote by T1, defined as follows:

The disturbance $d_1(t)$ is zero for $t \leq 3$ and $d_1(t) = \sin(8\pi(t-3))$, $t > 3$, and $d_2(t) = d_3(t) = 0$. $f_1(t)$ is a step function with magnitude 0.3 appearing at $t = 5$, and $f_2(t)$ is a step function with magnitude 1 appearing at $t = 5$.

The response of $z(t)^{\mathrm{T}}z(t)$ for the controllers (i)—(iv) is presented in Figs. 15 and 16.

To apply the controller design algorithm of this subsection, we have to find as small as possible γ, as small as possible $\beta \geq \beta_{\mathrm{d}} = 3.5254$ and a stabilizing controller $\boldsymbol{K}$ such that (49) holds. We find a solution of Problem 1 with $\gamma = 0.075$. This γ is

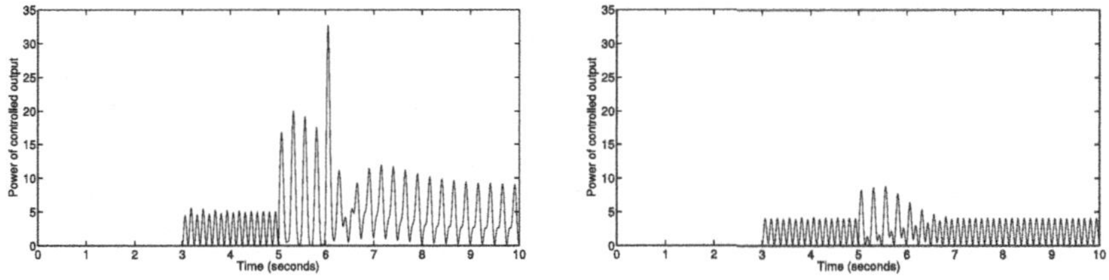

Fig. 16 $z(t)^{\mathrm{T}}z(t)$ for Test signal T1, with controller based on fault detection (left) and fault estimation (right)

extremely acceptable, but by Theorem 10, the closed loop system cannot be stabilized with $\boldsymbol{U}$ satisfying (65) (Assumption 17 is satisfied, and at least one controller is not stabilizing). Therefore, we have to increase γ.

We then consider $\gamma = 0.75$. For this γ, Assumptions 13, 15 and 16 are satisfied, but Assumption 17 is not satisfied (the RM $\boldsymbol{G}_{a22}$ has an unstable pole, equal to 112.04, and it has $\mathscr{L}_\infty$-norm equal to 3.4341). Then we find $\boldsymbol{K}_a$.

Nevertheless, for $\boldsymbol{U} = \boldsymbol{U}_0 := \begin{bmatrix} -0.2 & 0.305 \\ 0.2 & -0.1 \end{bmatrix}$, the closed loop system is stable.

For $\boldsymbol{U} = -0.9 \cdot I_2$ the closed loop system is unstable, illustrating the case that Assumption 17 is not satisfied, and there are both stabilizing and non-stabilizing RMs $\boldsymbol{U}$.

Then we apply the Nelder-Mead algorithm. To deal with the constraint (65), we modify the function $f(\boldsymbol{U}) = \|\mathscr{F}(\boldsymbol{G}_a, \boldsymbol{U})\|_\infty$, by introducing the following penalizing function

$$f(\boldsymbol{U}) = \begin{cases} \|\mathscr{F}(\boldsymbol{G}_a, \boldsymbol{U})\|_\infty, & \|\boldsymbol{U}\|_\infty < 1 \\ \|\mathscr{F}(\boldsymbol{G}_a, \boldsymbol{U}/\|1.1\boldsymbol{U}\|_\infty)\|_\infty, & \|\boldsymbol{U}\|_\infty \geq 1 \end{cases}$$

and, if in some iteration we obtain $\boldsymbol{U}$ that does not satisfy (65), we change $\boldsymbol{U}$ to $\boldsymbol{U}/\|\boldsymbol{U}\|_\infty$. We consider that $\boldsymbol{U}$ is a constant matrix satisfying (65). With the initial $\boldsymbol{U} = \boldsymbol{U}_0$, we have $\|\boldsymbol{T}_{zd}\|_\infty = 10.3291$. The value 10.3291 is too big to be taken as β. We require β that is not much bigger than $\beta_d = 3.5254$. We obtain $\boldsymbol{U} = \begin{bmatrix} 0.979 & 0.1821 \\ 0.003072 & 0.4384 \end{bmatrix}$, which satisfies $\|\boldsymbol{U}\|_\infty = 1$, and results in a stable closed loop system, with $\beta = 4.1232$.

The response of $z(t)^{\mathrm{T}}z(t)$ for the controller designed by our algorithm is presented in Fig. 17. It is seen that the latter behaviour is much better than ones presented in Figs. 15 and 16.

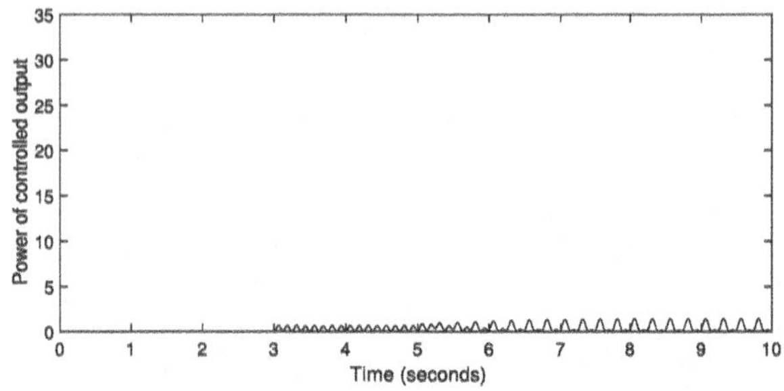

Fig. 17 $z(t)^{\mathrm{T}}z(t)$ for our controller $\boldsymbol{K}$ and Test signal T1

4.2 FTC Based on Fault Estimation

In this subsection we take $\mathbb{B} := \mathbb{B}_{\mathrm{f}} \cup \mathbb{B}_{\mathrm{d}}$. The results of this subsection are actually solutions of the following problem.

Problem 2 *Find a stabilizing controller for the plant given in Fig. 13 such that, under zero initial conditions,*

$$\|z(\mathrm{j}\omega)\| \leq \alpha\|\boldsymbol{f}(\mathrm{j}\omega) - \widehat{\boldsymbol{f}}(\mathrm{j}\omega)\| + \beta\|\boldsymbol{d}(\mathrm{j}\omega)\| \ , \tag{66}$$

for all $\omega \in \mathbb{B}$, where α and β are non-negative numbers.

As mentioned in Introduction, the disturbances appear frequently (or are persistent), and the faults appear rarely (or never). For that reason, we require almost the best performance of the closed loop system, in times when there are no faults. In other words, we require a behaviour of the closed loop system that is similar to the behaviour of the closed loop system with applied controller (called nominal controller) that is constructed under the condition that there are no faults. In this subsection, the existence of a solution of the problem to find a nominal controller is an assumption. Namely,

Assumption 18 :$\mathbb{B}$ There is a stabilizing controller $\boldsymbol{K}$, which we denote by $\boldsymbol{K}_{\mathrm{nm}}$, and a minimal $\beta \geq 0$ such that the inequality

$$\|\boldsymbol{T}_{\mathrm{zd}}\|_{\infty}^{\mathbb{B}} \leq \beta \tag{67}$$

is satisfied.

If $\mathbb{B} = \overline{\mathbb{R}}$, instead of Assumption 18:$\overline{\mathbb{R}}$, we write simply Assumption 18.

Necessary and sufficient condition for the problem given by (67) are not known yet, in the literature, except in the case $\mathbb{B} = \overline{\mathbb{R}}$, where a solution is the classical suboptimal $\mathscr{H}_\infty$ controller, Section 17 of [67], or optimal $\mathscr{H}_\infty$ controller, [42, 43]. Namely, in [21] (see also [63]), the problem (67) is solved under sufficient conditions. In [39], the stability question is not solved. In [49], the case with $\boldsymbol{G}_{12}$ and $\boldsymbol{G}_{1\mathrm{d}}$ right-invertible and left-invertible RMs is elaborated on and solved, with necessary and sufficient conditions that β in (67) is an arbitrary small number. In [44] and references therein, the disturbance is assumed a harmonic function (with unknown frequencies).

Further we show why the solutions of Problem 2 are FTC controllers. If $\widehat{f} = f$ then, with the nominal controller,

$$z = T_{z\mathrm{d}} d \,, \tag{68}$$

By minimization of α, we compensate for the faults, when the fault estimation is not ideal. As a conclusion, the behaviour of the controlled output used in the nominal regime will be approximately nominal in the both regimes, if the fault estimate is accurate, or if α is sufficiently small such that the left summand in (66) is significantly less than the right-hand one. As a consequence, the solutions of Problem 2 satisfy the requirement of [68], mentioned in Introduction.

In order to find a solution to Problem 2, at first we apply the feedback control

$$u = K_{\mathrm{nm}} y + v \,, \tag{69}$$

for some new control v, to be obtained, where K_{nm} is the nominal controller. Then, using (50) and (69), we obtain

$$z = \widehat{G}_{1\mathrm{f}} f + \widehat{G}_{1\mathrm{d}} d + \widehat{G}_{12} v \,, \tag{70}$$

where

$$\widehat{G}_{1\mathrm{f}} = G_{1\mathrm{f}} + G_{12} K_{\mathrm{nm}} (I_{p_2} - G_{22} K_{\mathrm{nm}})^{-1} G_{2\mathrm{f}} \,,$$

$$\widehat{G}_{1\mathrm{d}} = G_{1\mathrm{d}} + G_{12} K_{\mathrm{nm}} (I_{p_2} - G_{22} K_{\mathrm{nm}})^{-1} G_{2\mathrm{d}}$$

and

$$\widehat{G}_{12} = G_{12} (I_{m_2} - K_{\mathrm{nm}} G_{22})^{-1}.$$

Consider the control law

$$v = X \widehat{f} \,. \tag{71}$$

for some RM $X(s)$. Fig. 18 presents the obtained block-scheme.

We have

$$z = \widehat{G}_{1\mathrm{d}} d + (\widehat{G}_{1\mathrm{f}} + \widehat{G}_{12} X) f - \widehat{G}_{12} X (f - \widehat{f}) \,. \tag{72}$$

Of particular interest is the case that

Assumption 19 There a solution X of the equation

$$\widehat{G}_{1\mathrm{f}} + \widehat{G}_{12} X = 0 \,. \tag{73}$$

If Eq. (73) admits a solution X, instead of (72), we have

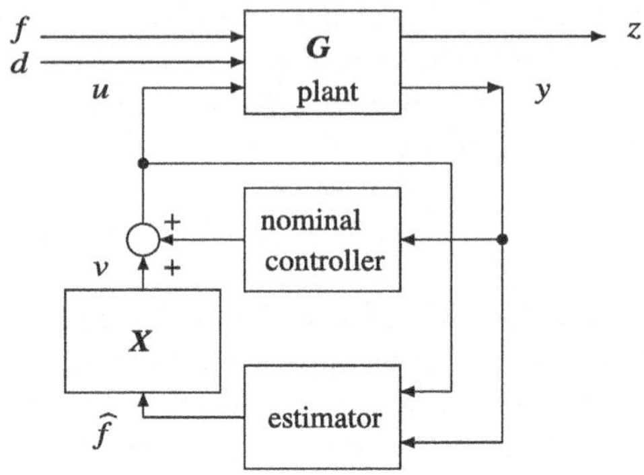

Fig. 18 Plant with fault-tolerant controller

$$z = \widehat{\boldsymbol{G}}_{1\mathrm{d}}\boldsymbol{d} + \widehat{\boldsymbol{G}}_{1\mathrm{f}}(\boldsymbol{f} - \widehat{\boldsymbol{f}})\,. \tag{74}$$

By elimination of the variables $\widehat{\boldsymbol{f}}$ and $\boldsymbol{v}$ from the Eqs. (6), (69) and (71), we obtain $\boldsymbol{u} = \boldsymbol{K}\boldsymbol{y}$, where the overall controller $\boldsymbol{K}$ is given by

$$\boldsymbol{K} = (I + \boldsymbol{X}\boldsymbol{F}\boldsymbol{N}_{\mathrm{u}})^{-1}(\boldsymbol{K}_{\mathrm{nm}} + \boldsymbol{X}\boldsymbol{F}\boldsymbol{M})\,. \tag{75}$$

The closed loop is well-posed if RM $I + \boldsymbol{X}\boldsymbol{F}\boldsymbol{N}_{\mathrm{u}}$ is nonsingular.

Identity (74) indicates why controller (75) can be a solution to Problem 2. Namely, $\alpha = \|\widehat{\boldsymbol{G}}_{1\mathrm{f}}\|_{\infty}^{\mathbb{B}}$ and β is one defined in Assumption 18:$\mathbb{B}$.

Theorem 11 ([45]) *If the controller given by the triple* $(\boldsymbol{K}_{\mathrm{nm}}, \boldsymbol{F}, \boldsymbol{X})$ *such that* $\boldsymbol{K}_{\mathrm{nm}}$ *satisfies Assumption 18:*$\mathbb{B}$*,* $\boldsymbol{F}$ *is a solution to Problem 3:*$\mathbb{B}$ *of Sect. 3,* $\boldsymbol{X}$ *satisfies Assumption 19, and such that RMs* $\boldsymbol{K}_{\mathrm{nm}}$ *and* $\boldsymbol{X}$ *are proper and stable, and RM* $I + \boldsymbol{X}\boldsymbol{F}\boldsymbol{N}_{\mathrm{u}}$ *is nonsingular, it is a solution to Problem 2.*

Sketch of the proof.

Let $\boldsymbol{V}$ and $\boldsymbol{U}$ be proper stable RMs, which are left coprime and $\boldsymbol{K}_{\mathrm{nm}} = \boldsymbol{V}^{-1}\boldsymbol{U}$. Re-write identity (75) as

$$\boldsymbol{K} = (\boldsymbol{V} + \boldsymbol{Q}\boldsymbol{N}_{\mathrm{u}})^{-1}(\boldsymbol{U} + \boldsymbol{Q}\boldsymbol{M})\,, \tag{76}$$

where $\boldsymbol{Q} = \boldsymbol{V}\boldsymbol{X}\boldsymbol{F}$. Since $\boldsymbol{V}$ and $\boldsymbol{U}$ are left coprime RMs, $\boldsymbol{M}$ and $\boldsymbol{N}_{\mathrm{u}}$ are left coprime RMs, identity (76) is a parametrization of stabilizing controllers [68], if $\boldsymbol{Q}$ is a proper stable RM. Therefore, the closed-loop system is stable and impulse-free as RM $\boldsymbol{X}\boldsymbol{F}$ is proper and stable. ■

A proper stable solution $\boldsymbol{X}$ of (73), which can be used for the fault accommodation in Theorem 11, can be found under the following assumption.

Assumption 20 Matrix pencil

$$\begin{bmatrix} sE - A & -B_2 \\ C_1 & D_{12} \end{bmatrix} \tag{77}$$

has full row rank in $\Re[s] \geq 0$ and infinity, and matrix D_{12} has full row rank, and RM $\boldsymbol{K}_{\mathrm{nm}}$ is proper and stable.

An algorithm for obtaining an $\mathscr{H}_\infty$ controller, which is proper stable, a property required in Assumption 20, can be found under sufficient conditions in [60].

Proposition 1 ([45]) *Under Assumption 20, there exists a proper stable solution* $\boldsymbol{X}$ *of the matrix equation* (73).

Example. Let be given a plant (1) with the following matrices:

$$E = I_4\,,\quad A = \begin{bmatrix} -1 & 0 & -2 & -4 \\ 1 & 2 & 0 & -1 \\ 0 & 1 & -3 & 3 \\ -2 & 1 & -4 & -3 \end{bmatrix},\ B_{\mathrm{f}} = \begin{bmatrix} 0 & 0 \\ 0 & 2 \\ 0 & -1 \\ 0 & 1 \end{bmatrix},\quad B_{\mathrm{d}} = \begin{bmatrix} 0 & -1 \\ 0 & 1 \\ 0 & -0.5 \\ 0 & 0 \end{bmatrix},\quad B_2 = \begin{bmatrix} 2 & 0 \\ 0 & -1 \\ -2 & 1 \\ -1 & 0 \end{bmatrix},$$

$$C_1 = \begin{bmatrix} 2 & 4 & -1 & 0 \\ 2 & 1 & -1 & 3 \end{bmatrix},\quad D_{1\mathrm{f}} = \begin{bmatrix} 0 & 0 \\ 0 & 0 \end{bmatrix},$$

$$D_{1\mathrm{d}} = \begin{bmatrix} 0 & 0 \\ 0 & 0 \end{bmatrix},\quad D_{12} = \begin{bmatrix} 1 & -1 \\ 0 & 2 \end{bmatrix},$$

$$C_2 = \begin{bmatrix} 2 & -2 & 0 & 1 \\ -2 & -1 & -2 & 2 \end{bmatrix},\quad D_{2\mathrm{f}} = \begin{bmatrix} 2 & 0 \\ 0 & 0 \end{bmatrix},$$

$$D_{2\mathrm{d}} = \begin{bmatrix} 0 & 0 \\ 0 & 0 \end{bmatrix},\quad D_{22} = \begin{bmatrix} 0 & 0 \\ 0 & 0 \end{bmatrix}.$$

The considered f_1 is a fault of the first sensor, while f_2 is an actuator fault. Note that the papers [15, 24, 26], Sect. 9.2 in [3], [25], and [61], deal with actuator faults only.

The matrix pencils $\begin{bmatrix} A - sI & B_{\mathrm{d}} \\ C_2 & D_{2\mathrm{d}} \end{bmatrix}$ and $\begin{bmatrix} A - sI & B_2 \\ C_1 & D_{12} \end{bmatrix}$ have no finite generalized eigenvalues in $\Re[s] \geq 0$. Using the MATLAB function hinfsyn which uses the algorithm of Sect. 17 in [67], with the option 'lmi' and $\beta = 0.0033$, we find a nominal controller $\boldsymbol{K}_{\mathrm{nm}}$. The RM $\boldsymbol{K}_{\mathrm{nm}}$ is proper and stable.

Since matrix pencil $\begin{bmatrix} A - sI & B_{\mathrm{f}} \\ C_2 & D_{2\mathrm{f}} \end{bmatrix}$ have finite generalized eigenvalues in $\Re[s] > 0$, and zeros at infinity, the FE problem is hard. Indeed, we find $\gamma = 2.24 > 1$.

Then we find $\boldsymbol{X}$, which is unique solution of (73). It is proper and stable.

All assumptions of Theorem 11 are satisfied, with $\alpha = 7.77$, $\beta = 0.0033$ and $\gamma = 2.24$. Using $\boldsymbol{K}_{\mathrm{nm}}$, $\boldsymbol{F}$ and $\boldsymbol{X}$, we construct a controller, as in Fig. 18.

We take the following test signal, with $\omega_0 = 4\pi$:

$$d_1(t) = \begin{cases} 0, & t \leq 30\,s \\ 0.3\sin(\omega_0(t-30)), & t > 30\,s \end{cases},\quad d_2(t) = \begin{cases} 0, & t \leq 30\,s \\ -0.2\sin(\omega_0(t-30)), & t > 30\,s \end{cases}$$

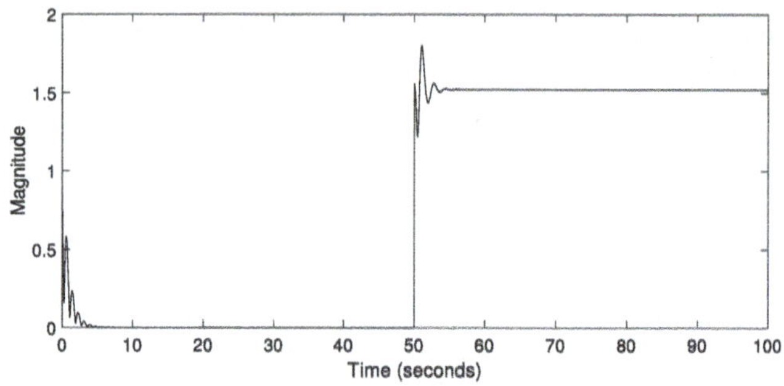

Fig. 19 Response $\|z(t)\|$ for the nominal controller

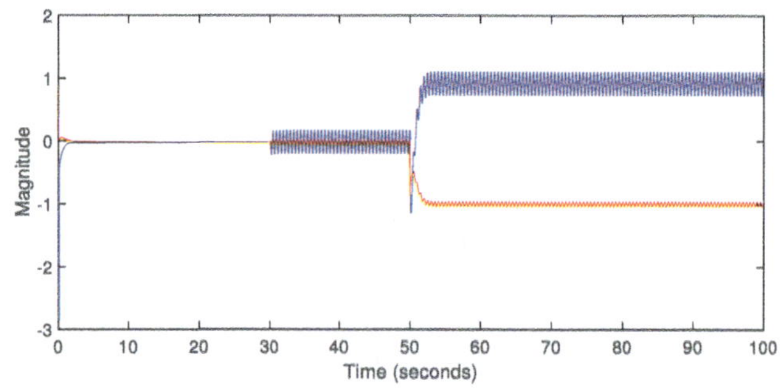

Fig. 20 Response of estimates $\widehat{f_1}(t)$ (blue) and $\widehat{f_2}(t)$ (red)

$f_1(t)$ is a step signal appearing at $t = 50\,s$ with magnitude 1, and
$f_2(t)$ is a step signal appearing at $t = 50\,s$ with magnitude -1.

Note that we have the hardest situation when the faults f_1 and f_2 appear simultaneously.

The initial values of the plant are $0.1[1, 1, 1, 1]$, and the initial values of all considered controllers are zero.

A computer simulation with several controllers, for $t \leq 100$ s, is shown in Figs. 19, 20, 21, 22 and 23. In Fig. 19, it is shown the time response of the norm of the controlled output $\|z(t)\|$ of the closed loop system with the nominal controller $\boldsymbol{K}_{\mathrm{d}}$. We see that the controlled output z is quite big (> 1.6) when the faults appear, therefore this behaviour is not acceptable. Note that $\|z(t)\|$ is approximately zero before the faults appear and when the disturbance appear (for $30\,s < t < 50\,s$) (the nominal controller is disturbance decoupling).

The estimates $\widehat{f_1}(t)$ and $\widehat{f_2}(t)$ of the faults $f_1(t)$ and $f_2(t)$ are shown in Fig. 20 in blue and red, respectively. We see that the fault estimation is not accurate, however, the contours of the fault signal can be recognized.

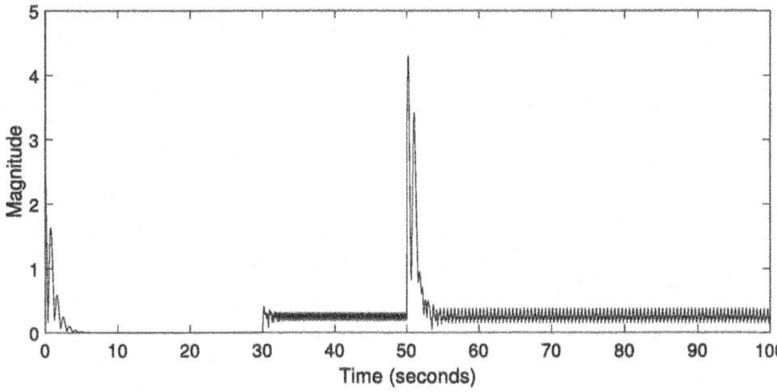

Fig. 21 Response $\|z(t)\|$ for the FTC controller

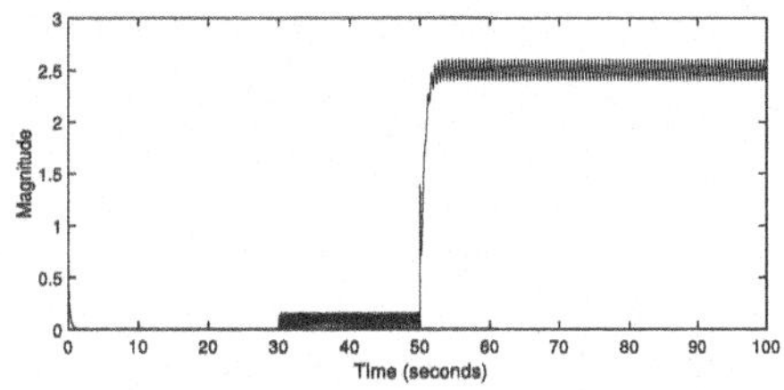

Fig. 22 Response $\|z(t)\|$ for controller $\boldsymbol{K}_{\rm fd}$

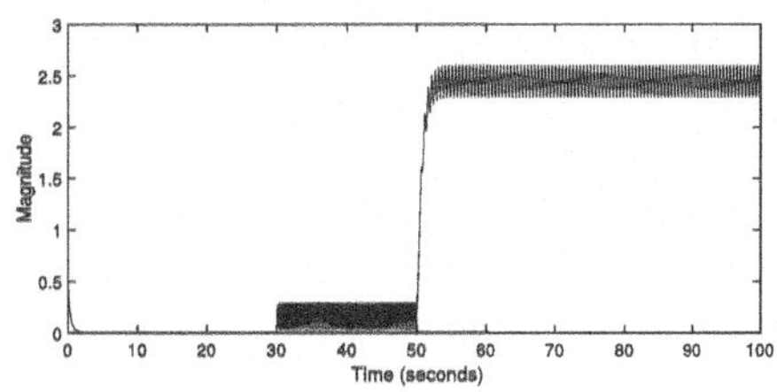

Fig. 23 Response $\|z(t)\|$ for modified controller $\boldsymbol{K}_{\rm fd}$

The behaviour of the closed loop system with the controller of Theorem 11 is illustrated in Fig. 21. We see that after the overshoot of magnitude 4.5, which appears in the time interval [50, 60], the norm of the controlled output $\|z(t)\|$ stabilizes at a magnitude 0.37, which characterizes also the fault-free behaviour. In this respect, the controller is fault-tolerant.

For comparison purposes, we present in Fig. 22 a computer simulation with the $\mathscr{H}_\infty$-controller $\boldsymbol{K}_{\rm fd}$ obtained under the assumption that both signals f and d are disturbances. The behaviour in the faulty regime ($t > 50\,s$) is not satisfactory.

Since the magnitude of the faults f_1 and f_2 are 1 and the magnitudes of the disturbances d_1 and d_2 are 0.3 and 0.2, one may think that the controller $\boldsymbol{K}_{\rm fd}$ will be improved if we scale all inputs so that they have magnitudes 1, i.e. if we multiply the first column of the matrices $B_{\rm d}$, $D_{1\rm d}$ and $D_{2\rm d}$ by 0.3 and their second column by

0.2. However, from the numerical simulation presented in Fig. 23 with the modified controller $\boldsymbol{K}_{\rm fd}$ we see that the behaviour is not better than one presented in Fig. 22.

Furthermore, our controller is better in respect to the switching fault-tolerant controllers. Indeed, when a fault is detected, the nominal controller has to be changed (switched) with the $\boldsymbol{K}_{\rm fd}$ controller, since it is optimal in the faulty regime. However, the $\boldsymbol{K}_{\rm fd}$ controller will respond with the same magnitude given in Fig. 22 after $t > 50\,s$. The reason for this behaviour is the system multivariability.

4.3 Generalization of the Result of Sect. 4.2

The assumption of Theorem 11, that RMs $\boldsymbol{K}_{\rm nm}$ and $\boldsymbol{X}$ are proper and stable, is restrictive. It guarantees the stability of the closed-loop system presented in Fig. 18, in which the fault estimate is accessible. In the following, we avoid the assumption that RMs $\boldsymbol{K}_{\rm nm}$ and $\boldsymbol{X}$ are proper and stable.

For this purpose, in the following, we avoid the inclusion of a fault estimator in the control loop, and consider a dedicated block for FE, which is needed for Problem 2. Namely, we unify all three blocks $\boldsymbol{K}_{\rm nm}$, $\boldsymbol{F}$ and $\boldsymbol{X}$, presented in Fig. 18 in a single block, which we denote by $\boldsymbol{K}$ with a minimal realization, as presented in Fig. 13. Since we need the estimate $\widehat{f}$ for monitoring the quality of the FTC, we use a dedicated estimator (6), as in a proposed scheme presented in Fig. 24.

Let $\boldsymbol{X}$ be an RM satisfying Assumption 19, which we factorize as

$$\boldsymbol{X} = \boldsymbol{U}_X \boldsymbol{V}_X^{-1}, \tag{78}$$

for some right coprime RMs $\boldsymbol{U}_X$ and $\boldsymbol{V}_X$. Note that the unstable poles of $\boldsymbol{K}_{\rm nm}$ appear as unstable zeros of $\boldsymbol{V}_X$, in the generic case.

Theorem 12 ([45]) *Under Assumption 18:𝔹, Assumption 19, where $\boldsymbol{X}$ is without poles in $s = 0$, and Assumption 3:𝔹 of Sect. 3 with fault estimator $\boldsymbol{F}_{\rm a}$ such that RM $I + \boldsymbol{U}_X \boldsymbol{F}_{\rm a} \boldsymbol{N}_{\rm u}$ is nonsingular, there exists an RM $\boldsymbol{F}$ satisfying Assumption 3:𝔹 of Sect. 3 such that the overall controller $\boldsymbol{K}$ solves Problem 2.*

Sketch of the proof.

Fig. 24 Plant with controller and dedicated fault estimator

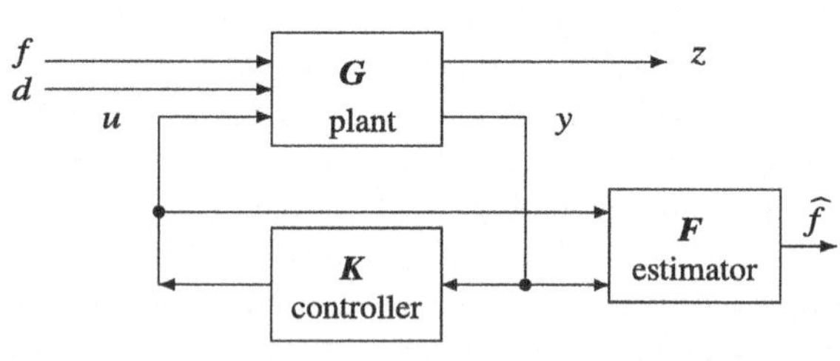

We use the parametrization of stabilizing controllers (76). As $\boldsymbol{X}$ is a solution of (73) without poles in $s = 0$, the RM $\boldsymbol{V}_X$ in (78) can be taken such that $\boldsymbol{V}_X(0) = I$. Define the proper stable RM $\boldsymbol{F} = \boldsymbol{V}_X \boldsymbol{F}_\mathrm{a}$. The closed-loop system is well-posed, stable and impulse-free because $\boldsymbol{X}\boldsymbol{F} = \boldsymbol{U}_X \boldsymbol{V}_X^{-1} \boldsymbol{F} = \boldsymbol{U}_X \boldsymbol{F}_\mathrm{a}$ is a proper and stable RM.

It remains to prove that $\boldsymbol{F}$ satisfies Assumption 3:$\mathbb{B}$ of Sect. 3. Since the validity of matrix inequality (39) with $\boldsymbol{F}_\mathrm{a}$ does not change if we left- and right-multiply it by $\boldsymbol{V}_X$ and $\boldsymbol{V}_X^{\#}$, respectively, the RM $\boldsymbol{F}$ satisfies the inequality (39). It also satisfies identity (40) because $\boldsymbol{F}(0)\boldsymbol{N}_\mathrm{f}(0) = \boldsymbol{V}_X(0)\boldsymbol{F}_\mathrm{a}(0)\boldsymbol{N}_\mathrm{f}(0) = I$. Finally, the poles of $\boldsymbol{T}_\mathrm{rf} = \boldsymbol{F}\boldsymbol{N}_\mathrm{f} = \boldsymbol{V}_X \boldsymbol{F}_\mathrm{a} \boldsymbol{N}_\mathrm{f}$ are arbitrary assignable, because the poles of $\boldsymbol{V}_X$ are arbitrary assignable. ■

A sufficient condition for existence a solution $\boldsymbol{X}$ of (73) without poles in $s = 0$ is that matrix pencil (77) has full row rank in $s = 0$ and nominal controller $\boldsymbol{K}_\mathrm{nm}$ has no poles in $s = 0$.

4.4 FTC in presence of Disturbances and Un-Estimable Faults

A necessary condition for fault estimation is that the RM $\boldsymbol{G}_{2\mathrm{f}}$ is left invertible. (Because the RM $\boldsymbol{T}_\mathrm{rf}$ has to be nonsingular, by (40).) In the case that $\boldsymbol{G}_{2\mathrm{f}}$ is not a left invertible RM, we can proceed as follows. Since this case, which always appears when the number of faults is greater than the number of measured outputs, is hard, it can be solved only if we introduce an assumption for the class of plants. Such is the following assumption.

Assumption 21

$$\boldsymbol{G}_{1\mathrm{f}} = \boldsymbol{L}\boldsymbol{G}_{2\mathrm{f}} \,, \tag{79}$$

for some proper stable RM $\boldsymbol{L}$.

Justifications for Assumption 21 are as follows:

(a) It follows by (50) that

$$z = \boldsymbol{L}\boldsymbol{y} + (\boldsymbol{G}_{1\mathrm{d}} - \boldsymbol{L}\boldsymbol{G}_{1\mathrm{d}})\boldsymbol{d} + (\boldsymbol{G}_{12} - \boldsymbol{L}\boldsymbol{G}_{22})\boldsymbol{u}$$

Since $y(t)$ and $u(t)$ are known in real time, we see by this equation that $z(t)$ is known in real time, in the absence of disturbances. The knowledge of $z(t)$ is important, because we cannot judge on the quality of our fault-tolerant controller if we do not know how "big" is $z(t)$. More important, it is hard for the controller to achieve $z(t) \to 0$ when $t \to \infty$ in absence of disturbances, if $z(t)$ cannot be reconstructed from the known $y(t)$ and $u(t)$.

(b) Assumption 21 holds if the case that $\boldsymbol{L}$ is a constant matrix, which we denote by L. It is shown in Theorem 9.2.1 of [6] that, in the absence of disturbances,

Assumption 21 is a necessary condition for almost insensitivity of $z(t)$ on the initial conditions of the plant and on the faults. Namely, Condition 4 in Theorem 9.2.1 of [6] is equivalent with existence of a constant matrix L such that $L[C_2, D_{2\mathrm{f}}] = [C_1, D_{1\mathrm{f}}]$. The latter condition implies the condition (79). The insensitivity is a highly desirable property.

(c) The case that the vector z is "included" in the vector y satisfies Assumption 21, because then $z(t) = Ly(t)$, for some constant matrix L.

(d) The important case $z = y$ satisfies Assumption 21. Then $\boldsymbol{L} = I$.

(e) Assumption 21 is automatically satisfied is RM $\boldsymbol{G}_{2\mathrm{f}}$ is nonsingular and $\boldsymbol{G}_{1\mathrm{f}}\boldsymbol{G}_{2\mathrm{f}}^{-1}$ is proper and stable RM (the stability is equivalent with that RM $\boldsymbol{G}_{2\mathrm{f}}$ has no zeros in $\Re[s] \geq 0$). Then $\boldsymbol{L} = \boldsymbol{G}_{1\mathrm{f}}\boldsymbol{G}_{2\mathrm{f}}^{-1}$.

If $\boldsymbol{L}$ is given by a realization (A_L, B_L, C_L, D_L), i.e., $\boldsymbol{L} = D_L + C_L(sI - A_L)^{-1}B_L$, for some matrices A_L, B_L, C_L and D_L, then identity (79) is an algebraic constraint on the plant matrices A, B_f, C_1, C_2, $D_{1\mathrm{f}}$ and $D_{2\mathrm{f}}$. A realization of the plant transfer matrix $\boldsymbol{G}$ can be given under various assumptions:

(a) If the realization (A_L, B_L, C_L, D_L) is fixed, we can define

$$\boldsymbol{G} = \begin{bmatrix} \boldsymbol{G}_{1\mathrm{f}} & \boldsymbol{G}_{1\mathrm{d}} & \boldsymbol{G}_{12} \\ \boldsymbol{G}_{2\mathrm{f}} & \boldsymbol{G}_{2\mathrm{d}} & \boldsymbol{G}_{22} \end{bmatrix} = \left[\begin{array}{cc|ccc} A_L & B_L C_2 & B_L D_{2\mathrm{f}} & 0 & 0 \\ 0 & A & B_\mathrm{f} & B_\mathrm{d} & B_2 \\ \hline C_L & D_L C_2 & D_L D_{2\mathrm{f}} & D_{1\mathrm{d}} & D_{12} \\ 0 & C_2 & D_{2\mathrm{f}} & D_{2\mathrm{d}} & D_{32} \end{array}\right], \tag{80}$$

and then $C_1 = [C_L, D_L C_2]$.

(b) If matrix C_1 is fixed, assume that Sylvester equation

$$A_L X - XA = B_L C_2 \tag{81}$$

admits a solution X. A sufficient condition is that the spectra of the matrices A and A_L are disjoint. Then we define

$$\boldsymbol{G} = \begin{bmatrix} \boldsymbol{G}_{1\mathrm{f}} & \boldsymbol{G}_{1\mathrm{d}} & \boldsymbol{G}_{12} \\ \boldsymbol{G}_{2\mathrm{f}} & \boldsymbol{G}_{2\mathrm{d}} & \boldsymbol{G}_{22} \end{bmatrix} = \left[\begin{array}{cc|ccc} A_L & B_L C_2 & B_L D_{2\mathrm{f}} & -XB_\mathrm{d} & -XB_2 \\ 0 & A & B_\mathrm{f} & B_\mathrm{d} & B_2 \\ \hline C_L & D_L C_2 & D_L D_{2\mathrm{f}} & D_{1\mathrm{d}} & D_{12} \\ 0 & C_2 & D_{2\mathrm{f}} & D_{2\mathrm{d}} & D_{22} \end{array}\right]$$

$$= \left[\begin{array}{cc|ccc} A_L & 0 & B_L D_{2\mathrm{f}} + XB_\mathrm{f} & 0 & 0 \\ 0 & A & B_\mathrm{f} & B_\mathrm{d} & B_2 \\ \hline C_L & C_1 & D_L D_{2\mathrm{f}} & D_{1\mathrm{d}} & D_{12} \\ 0 & C_2 & D_{2\mathrm{f}} & D_{2\mathrm{d}} & D_{22} \end{array}\right], \tag{82}$$

where matrices A_L, B_L, C_L and D_L have to satisfy the Eq. (81) and the equation

$$C_1 = D_L C_2 - C_L X\,. \tag{83}$$

Since matrix A_L is stable, both realizations (80) and (82) are stabilizable.

Introduce the following two assumptions, of which the first one is a generalization of the stabilizability of the pair (A, B_f), which is generic, and the second is generic in the case $p_2 < m_f$.

Assumption 22 Matrix pencil $[A - sI, B_f]$ has full row rank in $\Re[s] > 0$.

Assumption 23 The matrix D_{2f} is right-invertible.

In the FTC problem for plants satisfying Assumption 21, the original fault $\boldsymbol{f}$ can be replaced by some signal, denoted by $\widetilde{\boldsymbol{f}}$, of reduced dimension in respect to $\boldsymbol{f}$. This statement is precisely formulated in the following theorem.

Theorem 13 ([46]) *Given stabilizable plant* (1), *under Assumptions 21, 22 and 23, there are RM* $\widetilde{\boldsymbol{G}}_{1f}$ *and RM* $\widetilde{\boldsymbol{G}}_{2f}$, *which is invertible and without zeros in* $\Re[s] \geq 0$ *and infinity, such that*

$$\begin{aligned} z &= \widetilde{\boldsymbol{G}}_{1f}\widetilde{\boldsymbol{f}} + \boldsymbol{G}_{1d}\boldsymbol{d} + \boldsymbol{G}_{12}\boldsymbol{u} \\ y &= \widetilde{\boldsymbol{G}}_{2f}\widetilde{\boldsymbol{f}} + \boldsymbol{G}_{2d}\boldsymbol{d} + \boldsymbol{G}_{22}\boldsymbol{u} \end{aligned} \tag{84}$$

where the input $\widetilde{\boldsymbol{f}}$ *depends on the fault* $\boldsymbol{f}$ *only and has less or equal dimension in respect to* $\boldsymbol{f}$. *We have*

$$\begin{bmatrix} \widetilde{\boldsymbol{G}}_{1f} & \boldsymbol{G}_{1d} & \boldsymbol{G}_{12} \\ \widetilde{\boldsymbol{G}}_{2f} & \boldsymbol{G}_{2d} & \boldsymbol{G}_{22} \end{bmatrix} = \left[\begin{array}{c|ccc} A & \widetilde{B}_f & B_d & B_2 \\ \hline C_1 & \widetilde{D}_{1f} & D_{1d} & D_{12} \\ C_2 & \widetilde{D}_{2f} & D_{2d} & D_{22} \end{array}\right], \tag{85}$$

for some matrices $\widetilde{B}_f$, $\widetilde{D}_{1f}$ *and* $\widetilde{D}_{2f}$. *Also we have*

$$\widetilde{\boldsymbol{G}}_{1f} = \boldsymbol{L}\widetilde{\boldsymbol{G}}_{2f} .$$

For simplicity of the formulation of Theorem 14, re-denote the symbols $\widetilde{B}_f$, $\widetilde{D}_{1f}$, $\widetilde{D}_{2f}$, $\widetilde{m}_f$ and $\widetilde{\boldsymbol{G}}_{2f}$ in the realization (85) by the symbols B_f, D_{1f}, D_{2f}, m_f and $\boldsymbol{G}_{2f}$. Also, re-denote the symbol $\widetilde{\boldsymbol{f}}$ by $\boldsymbol{f}$.

Assumption 24 There is a right inverse $D_{12}^{\dagger}$ of matrix D_{12} such that matrix $A - B_2D_{12}^{\dagger}C_1$ is stable.

Assumption 24 is equivalent with the static output stabilizability.

Theorem 14 ([46]) *Given stabilizable plant* (1), *under Assumptions 21, 23 (with* $p_2 = m_f$*) and 24, the controller for the case* $D_{22} = 0$ *has a realization*

$$\boldsymbol{K} = -D_{12}^{\dagger}\left[\begin{array}{c|c} A - B_f D_{2f}^{-1} C_2 - B_2 D_{12}^{\dagger} C_{1f} & B_f - B_2 D_{12}^{\dagger} D_{1f} \\ \hline C_{1f} & D_{1f} \end{array}\right] D_{2f}^{-1} . \tag{86}$$

is stabilizing, and satisfies $\boldsymbol{T}_{zf} = 0$,

$$\boldsymbol{T}_{zd} = D_{1d} + C_{1f}(sI - A + B_f D_{2f}^{-1} C_2)^{-1}(B_d - B_f D_{2f}^{-1} D_{2d}) , \tag{87}$$

where $C_{1f} = C_1 - D_{1f} D_{2f}^{-1} C_2$.

Corollary 1 ([46])

(I) If the realization of **G** *is given with* (80), *then the controller realization in* (86) *reduces to:*

$$\boldsymbol{K} = -D_{12}^{\dagger} \left[\begin{array}{c|c} A_L & B_L \\ \hline C_L & D_L \end{array}\right] = -D_{12}^{\dagger} \boldsymbol{L} . \tag{88}$$

(II) If the realization of **G** *is given with* (82), *then the controller realization in* (86) *reduces to:*

$$\boldsymbol{K} = -D_{12}^{\dagger} \left[\begin{array}{c|c} A_L + X B_2 D_{12}^{\dagger} C_L & B_L + X B_2 D_{12}^{\dagger} D_L \\ \hline C_L & D_L \end{array}\right] . \tag{89}$$

Example. Let be given a plant with the following matrices:

$$A = \begin{bmatrix} -2 & 1 & -2 & -1 & -2 & 2 \\ -1 & -3 & -2 & 5 & 2 & -4 \\ 0 & 0 & -1 & 0 & -2 & -4 \\ 0 & 0 & 1 & 2 & 0 & -1 \\ 0 & 0 & 0 & 1 & -3 & 3 \\ 0 & 0 & -2 & 1 & -4 & -3 \end{bmatrix}, \quad B_f = \begin{bmatrix} -1 & 0 & 1 \\ -1 & -4 & -1 \\ -1 & 0 & 2 \\ 0 & 2 & -1 \\ -1 & -1 & 0 \\ 3 & 1 & -2 \end{bmatrix}, \quad B_d = \begin{bmatrix} 0 & 0 \\ 0 & 0 \\ -2 & -1 \\ 2 & 0 \\ 2 & -1 \\ 0 & -2 \end{bmatrix}, \quad B_2 = \begin{bmatrix} 0 \\ 0 \\ 2 \\ -6 \\ -4 \\ -1 \end{bmatrix},$$

$$C_1 = \begin{bmatrix} 1 & 0 & 6 & 0 & 4 & -3 \end{bmatrix}, \quad D_{1f} = \begin{bmatrix} 3 & 2 & -2 \end{bmatrix}, \quad D_{1d} = \begin{bmatrix} -6 & -2 \end{bmatrix}, \quad D_{12} = 1 ,$$

$$C_2 = \begin{bmatrix} 0 & 0 & 2 & -2 & 0 & 1 \\ 0 & 0 & -2 & -1 & -2 & 2 \end{bmatrix}, \quad D_{2f} = \begin{bmatrix} 1 & 2 & 0 \\ -1 & 0 & 1 \end{bmatrix}, \quad D_{2d} = \begin{bmatrix} -1 & 2 \\ 1 & -1 \end{bmatrix},$$

and $D_{22} = \begin{bmatrix} 0 \\ 0 \end{bmatrix}$.

The plant is unstable, but stabilizable. Assumption 21 is satisfied, with

$$\boldsymbol{L} = \left[\begin{array}{cc|cc} -2 & 1 & 0 & 1 \\ -1 & -3 & -2 & -1 \\ \hline 1 & 0 & 1 & -2 \end{array}\right]$$

The plant **G** has a realization form (80).

For a comparison purpose, we construct a nominal controller $\boldsymbol{K}_{nm}$ by solving the $\mathscr{H}_\infty$ control problem using the command hinfsyn in MATLAB, by assuming that the faults f_1, f_2 and f_3 are zero. Then we construct the FTC controller (75) by finding RMs **F** and **X**. Note that the found RM **X** is unstable, because a pole of $\boldsymbol{K}_{nm}$ and of **X** is 9.4741. Therefore, Theorem 11 can not be applied.

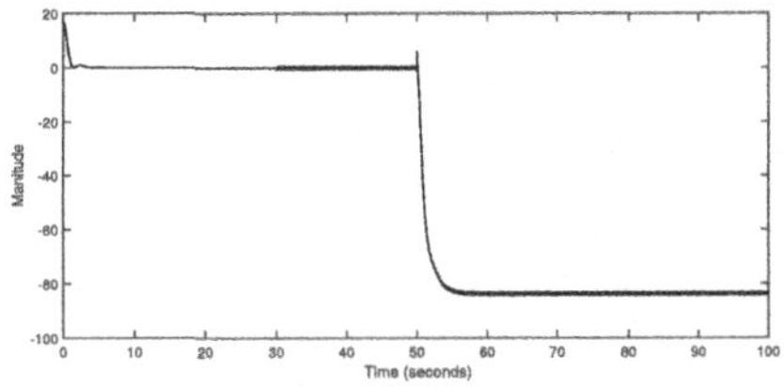

Fig. 25 Response $z(t)$ for the nominal controller $\boldsymbol{K}_{\mathrm{nm}}$

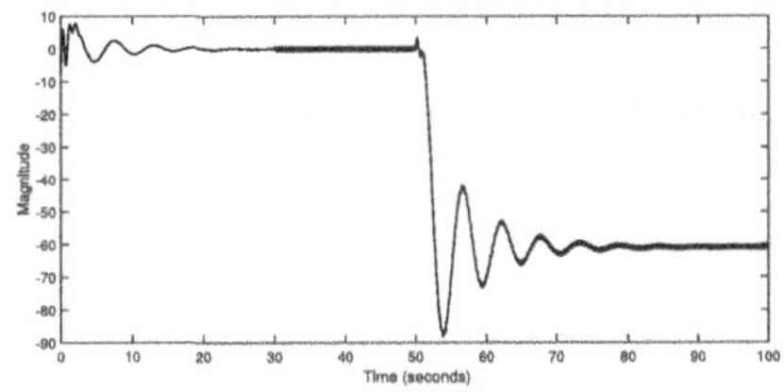

Fig. 26 Response $u(t)$ for the nominal controller $\boldsymbol{K}_{\mathrm{nm}}$

Using the realization (88), we find

$$\boldsymbol{K} = -\boldsymbol{L} \,. \tag{90}$$

In order to perform a numerical simulation, we consider the following test signal:

$$d_1(t) = \begin{cases} 0, & t \leq 30 \\ 0.2\sin(20t), & t > 30 \end{cases}, \quad d_2(t) = \begin{cases} 0, & t \leq 30 \\ -0.1\sin(20t), & t > 30 \end{cases},$$

$f_1(t)$ is a step signal appearing at $t = 50$ with magnitude 2.5, and
$f_2(t)$ is a step signal appearing at $t = 50$ with magnitude -3.
$f_3(t)$ is a step signal appearing at $t = 50$ with magnitude -2.

Note that the disturbances have less magnitudes in respect to the faults, and that we have the hardest situation when the faults appear simultaneously.

The initial values of the plant are [3, 3, 3, 3, 3, 3], and the initial values of the controllers in all simulations are zero.

A computer simulation for $t \leq 100$ s is shown in Figs. 25, 26, 27, 28, 29 and 30. In Fig. 25, it is shown the time response of the controlled output $z(t)$ of the closed loop system with the nominal controller. We see that the controlled output is quite big (> 82) in module when the faults appear, therefore this behaviour is not acceptable and the nominal controller is far from being fault-tolerant. Also, the overshoot at $t = 0$ is quite big, 16.68.

Note that in the interval $t \in (30, 50)$, the controlled output $z(t)$ is approximately a sinusoid with amplitude 0.5.

The behaviour of the closed loop system with the controller (90) is presented in Fig. 27. Good properties are: (i) that there is smaller overshoot at $t = 0$, less than 3, (ii) that the closed loop system is insensitive to the faults, and (iii) that the amplitude

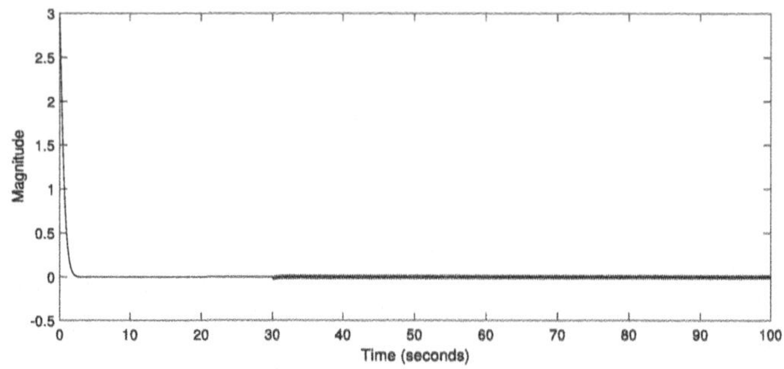

Fig. 27 Response $z(t)$ for the FTC controller $\boldsymbol{K}$

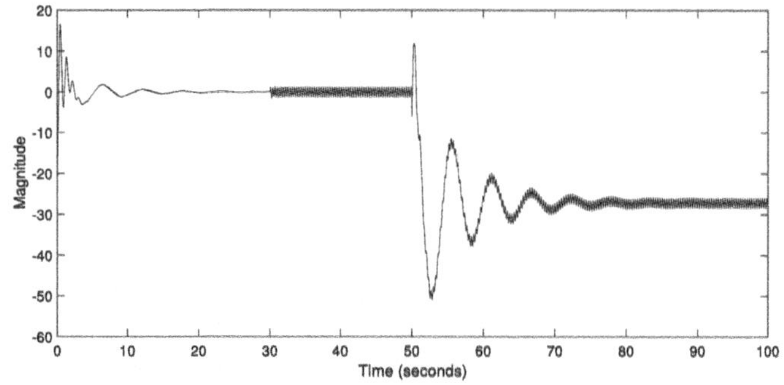

Fig. 28 Response $u(t)$ for the FTC controller $\boldsymbol{K}$

of the disturbance pattern (sinusoid) in $z(t)$ is only 0.016 for all $t > 39$ (compare with the amplitude of 0.5 in Fig. 25, in the case of applied nominal controller). In this respect, the controller (90) is fault-tolerant.

Note that although the controlled output $z(t)$ is not sensitive to faults, it is sensitive to initial conditions. By Theorem 9.2.1 of [6], the controlled output $z(t)$ will be insensitive to both initial conditions and faults if the RM $\boldsymbol{L}$ is a constant matrix.

The control magnitude in the case of FTC controller (90), which is presented in Fig. 28, is more than twice smaller in respect to the control with the nominal controller, as we see by comparing Fig. 26 with Fig. 28.

For comparison purposes, we present in Fig. 29 a computer simulation with the $\mathscr{H}_\infty$ controller obtained (using the command hinfsyn in MATLAB) under the assumption that both f and d are disturbances, denoted by $\boldsymbol{K}_{\mathrm{fd}}$, see Section 9.4.4 of [3]. We see that by comparing Figs. 27 and 29 that our controller (90) behaves better than the $\boldsymbol{K}_{\mathrm{fd}}$ controller.

We can improve the $\boldsymbol{K}_{\mathrm{fd}}$ controller by scaling the disturbance inputs, to be of the same magnitude with the fault inputs. Namely, we multiply the first columns of the matrices B_{d}, $D_{1\mathrm{f}}$ and $D_{2\mathrm{d}}$ by 0.2 and the seconds columns of those matrices by 0.1.

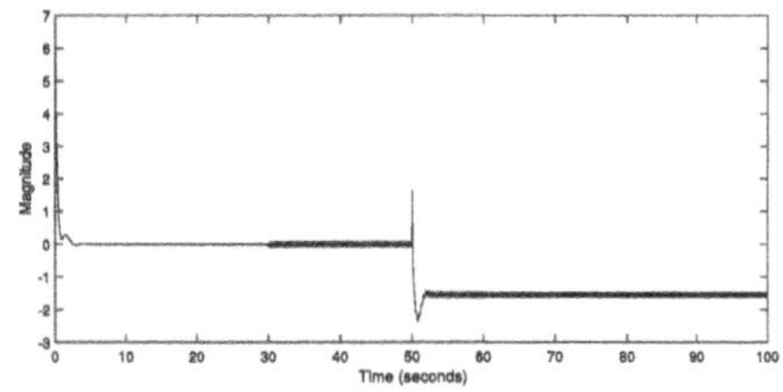

Fig. 29 Response $z(t)$ for the controller $\boldsymbol{K}_{\mathrm{fd}}$

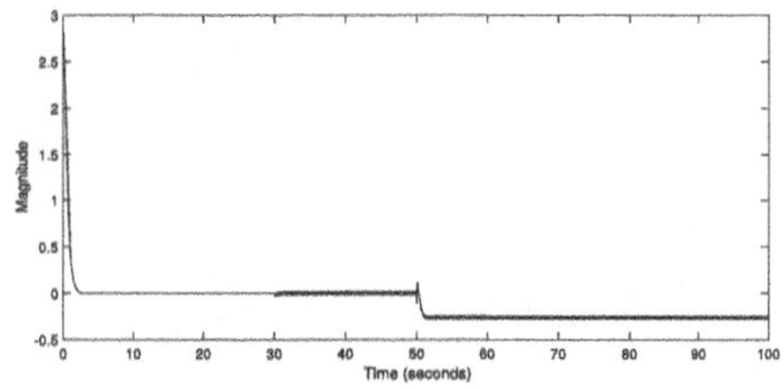

Fig. 30 Response $z(t)$ for the improved controller $\boldsymbol{K}_{\mathrm{fd}}$

From the computer simulation, presented in Fig. 30, we see that the controller $\boldsymbol{K}_{\mathrm{fd}}$ is indeed improved, in respect to one presented in Fig. 29, but it is still worse with respect to our controller (90).

Moreover, our FTC controller can be better in respect to the active fault-tolerant controller, constructed by switching at $t = 50$ between the nominal controller $\boldsymbol{K}_{\mathrm{nm}}$, which is optimal for $t < 50$, and the controller $\boldsymbol{K}_{\mathrm{fd}}$, which is optimal in the faulty regime $t > 50$. The switching criterion is based on FD.

In particular, we see by Fig. 25 for $t \in (0, 50)$ that the overshoot at $t = 0$ and the oscillations for $t \in (30, 50)$ are greater than ones given in Fig. 27. On the other hand, we see by Fig. 29 or Fig. 30 that an offset of $z(t)$ appears for $t > 50$, which is not the case in Fig. 27. The reason for this behaviour is the multivariability of the faults and disturbances.

5 Passive Fault-Tolerant Perfect Tracking

Consider a linear time-invariant plant with the following inputs and output: $f(t)$ is m_{f}-dimensional fault, $y(t)$ is p-dimensional measurement, without disturbances, and the system architecture in Fig. 31, where r is an arbitrary reference signal.

By the RMs $\boldsymbol{G}$ and $\boldsymbol{G}_{\mathrm{f}}$ is given the plant, and $\boldsymbol{X}_1$, $\boldsymbol{X}_2$ and $\boldsymbol{K}_{\mathrm{v}}$ are unknown RMs, which constitute the controller.

We take that $\boldsymbol{K}_{\mathrm{v}}$ has m_{f} rows. Denote by $e = y - r$ the tracking error. Introduce a controlled variable z by $z = \begin{bmatrix} e \\ \beta u \end{bmatrix}$, for some weighting design parameter $\beta > 0$. We pose the following problem.

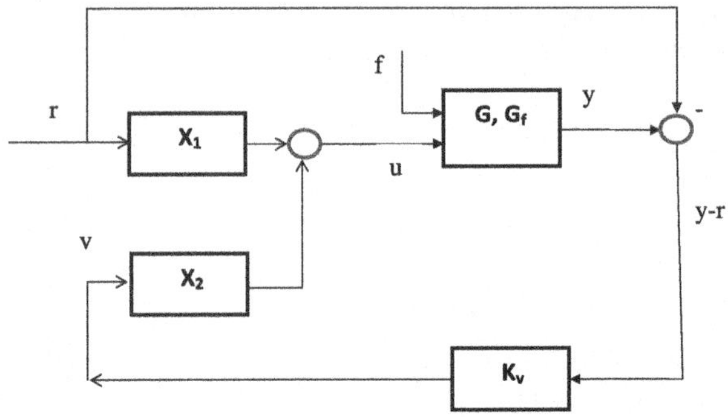

Fig. 31 Control system architecture

Problem 1 Find a controller for the system given in Fig. 31 such that

(1) the system is stable,
(2) $\boldsymbol{T}_{\mathrm{yr}} = I_p$,
(3) $\|\boldsymbol{T}_{\mathrm{ur}}\|_\infty$ is minimal, and
(4) $\|\boldsymbol{T}_{\mathrm{zf}}\|_\infty$ is minimal.

By $\boldsymbol{X}_1$ we satisfy the requirements (2) and (3) of Problem 1, and by $\boldsymbol{X}_2$ and $\boldsymbol{K}_{\mathrm{v}}$ we satisfy the requirements (1) and (4).

Consider the following system state-space plant model, without disturbances, and with measurement output:

$$\begin{aligned} \dot{x} &= Ax + Bu + B_{\mathrm{f}} f \ , \quad x(0) = x_0 \in \mathbb{R}^n \ , \\ y &= Cx + Du + D_{\mathrm{f}} f \ . \end{aligned} \tag{91}$$

Then the plant transfer matrices $\boldsymbol{G}$ and $\boldsymbol{G}_{\mathrm{f}}$ are

$$[\boldsymbol{G}, \boldsymbol{G}_{\mathrm{f}}] = \left[\begin{array}{c|cc} A & B & B_{\mathrm{f}} \\ \hline C & D & D_{\mathrm{f}} \end{array} \right].$$

It is easy to see that the following assumption is necessary for Problem 1.

Assumption 25 (1) The pair (C, A) is detectable, the pair (A, B) is stabilizable, and (2) the RM $\boldsymbol{G}$ is right-invertible and has no zeros in $\Re[s] \geq 0$ and infinity.

We search a solution $\boldsymbol{X}$ partitioned as $[\boldsymbol{X}_1, \boldsymbol{X}_2]$ of the equation $\boldsymbol{G}\boldsymbol{X} = \boldsymbol{M} = [I_p, \boldsymbol{G}_{\mathrm{f}}]$ where a realization of $\boldsymbol{M}$ is given by

$$\boldsymbol{M} = \left[\begin{array}{c|cc} A & 0 & B_{\mathrm{f}} \\ \hline C & I_p & D_{\mathrm{f}} \end{array} \right].$$

Having found $\boldsymbol{X}_1$ and $\boldsymbol{X}_2$, we find $\boldsymbol{K}_{\mathrm{v}}$ by solving the $\mathscr{H}_\infty$ control problem with the plant

$$P = \left[\begin{array}{c|c} P_{11} & P_{12} \\ \hline P_{21} & P_{22} \end{array}\right] = \left[\begin{array}{cc} G_{\rm f} & G_{\rm f} \\ 0 & \beta X_2 \\ \hline G_{\rm f} & G_{\rm f} \end{array}\right], \tag{92}$$

Namely, with the generalized plant (92), we consider that the disturbance input is $\boldsymbol{f}$, the control input is $\boldsymbol{v}$ (presented in Fig. 31), the controlled output is $\boldsymbol{z}$ and the measured output is $\boldsymbol{e}$.

Theorem 15 ([47]) (I) *Under Assumption 25, the system presented in Fig. 31 is stable.*

(II) *If in addition to Assumption 25, the* RM $\boldsymbol{G}_{\rm f}$ *has no zeros on the extended imaginary axis, then*

$$\inf_{\beta, X_2, K_{\rm v}} \|T_{\rm ef}\|_\infty = 0\,.$$

Example. Let be given the following plant

$$A = \begin{bmatrix} -0.5 & 0 & -2 & 1 \\ 2 & -0.5 & -3 & 0 \\ 1 & 0 & 0.5 & 0 \\ 1 & 1 & 0 & -3.5 \end{bmatrix}, \ B = \begin{bmatrix} 1 & -2 & 2 \\ 1 & 0 & 1 \\ 2 & -2 & 1 \\ 0 & 1 & 0 \end{bmatrix}, \ B_{\rm f} = \begin{bmatrix} 2 \\ -1 \\ 0 \\ -1 \end{bmatrix},$$

$$C = \begin{bmatrix} -1 & -1 & 0 & 1 \\ 2 & -2 & 1 & -1 \end{bmatrix}, \ D = \begin{bmatrix} 2 & -1 & 0 \\ 1 & -1 & 2 \end{bmatrix}, \ D_{\rm f} = \begin{bmatrix} 0 \\ -1 \end{bmatrix}.$$

Two eigenvalues of the matrix A are in $\Re[s] > 0$.

We compute a proper stable X_1 with minimal $\|X_1\|_\infty = 1.5327$, and a minimal $\mathscr{H}_\infty$-norm proper stable solution of $GX_2 = G_{\rm f}$, satisfying $\|X_2\|_\infty = 0.9645$.

To obtain $K_{\rm v}$, we take the design parameter $\beta = 0.1$, chosen for an acceptable magnitude of the control in the faulty regime. By solving the $\mathscr{H}_\infty$ control problem for the generalized plant (92), we obtain $K_{\rm v}$.

We consider the following test signal: (i) The first reference $r_1(t)$ is a periodic square signal with zero mean value and magnitude ± 1, starting at $t = 0$ with -1 and period 10 s, (ii) the second reference $r_2(t)$ is zero, (iii) the fault is a step appearing at 17th second, of magnitude 100. All initial states of the plant (91) are 1.

The response of the first output $y_1(t)$ is given in Fig. 32 (left). We can see that the tracking is "almost" perfect, regardless of the fault of magnitude 100. The controller enables a very large attenuation of the fault, since $\|T_{\rm ef}\|_\infty = 0.0084$.

However, the norm of the control $\|u(t)\|$ is of order 100, as we can see in Fig. 32 (right), and that is a price paid for the insensitivity of the output to the big fault.

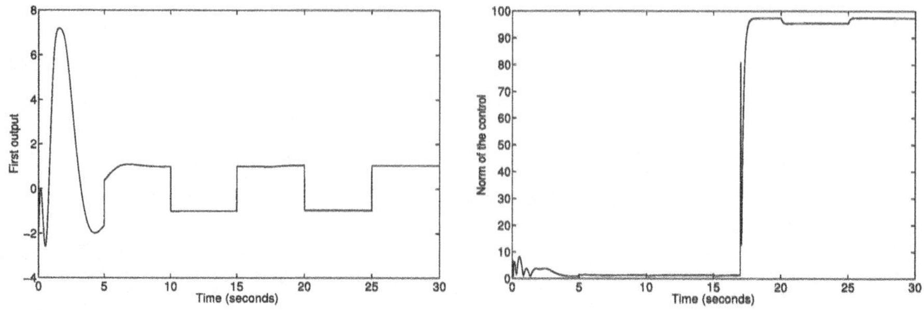

Fig. 32 Response of output $y_1(t)$ to the test signal (left) and of $\|u(t)\|$ (right)

6 Conclusions

A survey of recent results on fault diagnosis and fault-tolerant control is given in this chapter. A further research can be:

- Theorem 5 can be used to parametrize all fault estimators. A further research could be to use this parametrization in finding optimal FTC, based on fault estimation.
- To consider linear time-invariant discrete-time systems. Those systems allow time-lag in the filters or controllers, while still dealing with RMs. The time-lag can improve the FE, for instance, as we have seen in Fig. 11 (deconvolution filter).
- Theorem 7 could be applied to construct fault locators and fault estimators. In both cases a bank of filters can be found, and for each of the filters we take $n_r = 1$. The latter idea could be used for a further research.
- If we consider the problem of fault-tolerant tracking, then the necessary condition is that the plant is minimum phase. In that case, a feedforward controller have to be added in the controller structure. The feedforward controller of [48] can be used for this purpose, which solves a more general fault-tolerant tracking problem, namely, the almost fault-tolerant tracking problem, which does not require the absence of invariant zeros at infinity of the plant.
- In this chapter the faults are considered additive inputs in respect to the plant dynamics. However, the faults can appear on the plant components. Then they are not additive inputs, but functions of the state variables.

References

1. Allerhand, L.I., Shaked, U.: Robust switching-based fault-tolerant control. IEEE Trans. Autom. Control **60**(8), 2272–2276 (2015)
2. Benallouch, M., Boutayeb, M., Trinh, H.: $\mathscr{H}_\infty$ observer-based control for discrete-time one-sided Lipschitz systems with unknown inputs. SIAM J. Control Optim. **52**(6), 3751–3775 (2014)

3. Blanke, M., Kinnaert, M., Lunze, J., Staroswiecki, M.: Diagnosis and Fault-Tolerant Control, 3rd edn. Springer, Berlin, Heidelberg (2016)
4. Cerutti, S., Marchesi, C., Sparacino, G., Pillonetto, G., De Nicolao, G., Cobelli, C.: Deconvolution for Physiological Signal Analysis. https://doi.org/10.1002/9781118007747.ch8
5. Chadli, M., Abdob, A., Ding, S.X.: $\mathscr{H}_-/\mathscr{H}_\infty$ fault detection filter design for discrete-time Takagi-Sugeno fuzzy system. Automatica **49**, 1996–2005 (2013)
6. Chen, B.M.: Robust and $\mathscr{H}_\infty$ Control. Springer, London (2000)
7. Chen, B.M., Lin, Z., Liu, K.: Robust and perfect tracking of discrete-time systems. Automatica **38**, 293–299 (2002)
8. Chen, J., Patton, R.J.: Robust model-based fault diagnosis for dynamic systems. Springer Science+Business Media, New York (1999)
9. Chibani, A., Chadli, M., Shi, P., Braiek, N.B.: Fuzzy fault detection filter design for T-S fuzzy systems in finite frequency domain. IEEE Trans. Fuzzy Syst. (2017). https://doi.org/10.1109/TFUZZ.2016.2593921
10. Cirka, L., Fikar, M., Mikle, J.: Youla-Kučera parameterisation approach to LQ tracking and disturbance rejection problem. In: Proceedings of 16th IFAC World Congress (2005)
11. Di, H., Yu, C., Zhisheng, D., Qishao, W.: An iterative approach to $\mathscr{H}_-/\mathscr{H}_\infty$ fault detection for discrete-time systems in finite frequency domain. In: Proceedings of 35th Chinese Control Conference, July 27–29, Chengdu, China (2016). https://doi.org/10.1109/ChiCC.2016.7554411
12. Ding, S.X.: Model-based Fault Diagnosis Techniques: Design Schemes, Algorithms and Tools, 2nd edn. Springer, Berlin, Heidelberg (2013)
13. Ding, S.X., Yang, G., Zhang, P., Ding, E., Jeinsch, T., Weinhold, N., Schulalbers, M.: Feedback control structures, embedded residual signals and feedback control schemes with an integrated residual access. IEEE Trans. Control Syst. Technol. **18**, 352–367 (2010)
14. Frisk, E., Nyberg, M.: A minimal polynomial basis solution to residual generation for fault diagnosis in linear systems. Automatica **37**, 1417–1424 (2001)
15. Gao, Z., Ding Steven, X.: Actuator fault robust estimation and fault-tolerant control for a class of nonlinear descriptor systems. Automatica **43**, 912–920 (2007)
16. He, G., Liu, Y., Ji, J., Yu, W.Y.: Observer-based scheme for fault estimation and robust tolerant control: an LMI approach. In: Proceedings of 28th Chinese Control and Decision Conference (2016)
17. Hoover, D.N., Longchamp, R., Rosenthal, J.: Two-degree-of-freedom ℓ_2-optimal tracking with preview. Automatica **40**, 155–162 (2004)
18. Hou, M., Patton, R.J.: Input observability and input reconstruction. Automatica **34**(6), 789–794 (1998)
19. Ishihara, J.Y., Terra, M.H.: Impulse controllability and observability of rectangular descriptor systems. IEEE Trans. Autom. Control **46**(6), 991–994 (2001)
20. Iwasaki, T., Meinsma, G., Fu, M.: Generalized S-procedure and finite frequency KYP lemma. Math. Probl. Eng. **6**(2–3), 305–320 (2000)
21. Iwasaki, T., Hara, S.: Generalized KYP lemma: unified frequency domain inequalities with design applications. IEEE Trans. Autom. Control **50**(1), 41–59 (2005)
22. Jaimoukha, I.M., Zhenhai, L., Papakos, V.: A matrix factorization solution to the $\mathscr{H}_-/\mathscr{H}_\infty$ fault detection problem. Automatica **42**, 1907–1912 (2006)
23. Jaimoukha, I.M., Li, Z., Mazars, E.F.M.: Linear matrix inequality solution to the $\mathscr{H}_-/\mathscr{H}_\infty$ fault detection problem. In: Proceedings of 7th IASTED International Conference Control and Applications, May 18–20, pp. 183–188. Cancun, Mexico (2005)
24. Jiang, B., Staroswiecki, M., Cocquempot, V.: Fault accommodation for nonlinear dynamic systems. IEEE Trans. Autom. Control **51**(9) (2006)
25. Lan, J., Patton, R.J.: A decoupling approach to integrated fault-tolerant control for linear systems with unmatched non-differentiable faults. Automatica **89**, 290–299 (2018)
26. Lan, J., Patton, R.J.: A new strategy for integration of fault estimation within fault-tolerant control. Automatica **69**, 48–59 (2016)

27. Leva, A., Bascetta, L.: Designing the feedforward part of 2-d.o.f. industrial controllers for optimal tracking. In: Control Engineering Practice, vol. 15, pp. 909–921 (2007)
28. Li, W., Ding, S.: Optimal $\mathscr{H}_-/\mathscr{H}_\infty$ fault detection filter design: an iterative LMI approach. Proc. IEEE Conf. Decis. Control (2010). https://doi.org/10.1109/CDC2009.5399722
29. Li, Z., Mazars, E., Zhang, Z., Jaimoukha, I.: State space solutions to the H_-/H_∞ fault detection problem. Int. J. Robust Nonlinear Control **22**, 282–299 (2012)
30. Liu, N., Zhou, K.: Optimal robust fault detection for linear discrete time systems. J. Control Sci. Eng. (2008). https://doi.org/10.1155/2008/829456
31. Liu, N., Zhou, K.: Optimal solutions to multi-objective robust fault detection problems. In: Proceedings of 46th IEEE Conference on Decision and Control New Orleans, LA, USA, Dec 12–14, pp. 981–988 (2007)
32. Osorio-Gordillo, G.-L., Darouach, M., Astorga-Zaragoza, C.-M.: $\mathscr{H}_\infty$ dynamical observers design for linear descriptor systems. Application to state and unknown input estimation. Eur. J. Control **26**, 35–43 (2015)
33. Rodrigues, M., Hamdi, H., Braiek, N.B., Theilliol, D.: Observer-based fault-tolerant control design for a class of LPV descriptor systems. J. Franklin Inst. **351**, 3104–3125 (2014)
34. Saberi, A., Stoorvogel, A.A., Sannuti, P.: Control of Linear Systems with Regulation and Input Constraints. Springer, London (2000)
35. Shi, P., Liu, M., Zhang, L.: Fault-tolerant sliding mode observer synthesis of markovian jump systems using quantized measurements. IEEE Trans. Ind. Electron. https://doi.org/10.1109/TIE.2015.2442221
36. Stefanovski, J.: Fault detection over frequency region: generalized spectral factorization approach. IEEE Trans. Autom. Control **62**, 5296–5301 (2017)
37. Stefanovski, J.: New condition for FD, all filters, and new KYP lemma. Asian J. Control **40**(4), 1–11 (2018)
38. Stefanovski, J.D.: New class of FD filters in presence of disturbances. J. Franklin Inst. **355**, 1311–1337 (2018)
39. Stefanovski, J.: Canonical form of para-Hermitian pencils, generalized spectral factorization, and optimal control over frequency region. Int. J. Robust Nonlinear Control **23**, 1301–1323 (2013)
40. Stefanovski, J., Juricic, D.: Input estimation over frequency region in presence of disturbances. IEEE Trans. Autom. Contr. **64**, 5074–5079 (2019)
41. Stefanovski, J.: Fault-tolerant control of descriptor systems with disturbances. IEEE Trans. Autom. Control **64**, 976–988 (2018). https://doi.org/10.1109/TAC.2018.2827702
42. Stefanovski, J.: $\mathscr{H}_\infty$ problem with nonstrict inequality and all solutions: interpolation approach. SIAM J. Control Optim. **53**(4), 1734–1767 (2015)
43. Stefanovski, J.D.: Strongly (J, J')-lossless rational matrices and $\mathscr{H}_\infty$ problem. Int. J. Robust Nonlinear Control (2018). https://doi.org/10.1002/rnc.4231
44. Stefanovski, J.: General optimal attenuation of harmonic disturbance with unknown frequencies. Int. J. Control **85**(3), 260–279 (2012)
45. Stefanovski, J., Juričić, D.: Fault-tolerant control in presence of disturbances based on fault estimation. Syst. Control Lett. **138** (April 2020). https://doi.org/10.1016/j.sysconle.2020.104646
46. Stefanovski, J., Juričić, D.: FTC in presence of disturbances and un-estimable faults. Automatica **115** (May 2020)
47. Stefanovski, J.: Passive fault-tolerant perfect tracking with additive faults. Automatica **87**, 432–436 (2018)
48. Stefanovski, J.: Almost fault-tolerant tracking. Int. J. Robust Nonlinear Control **30**(6), 2219–2247 (2020)
49. Stefanovski, J., Georgijević, D.: Interpolation with constraint on frequency region and systems & control application. Syst. Control Lett. **97**, 70–82 (2016)
50. Stefanovski, J.: Simplified formula for a controller in optimal control problems. SIAM J. Control Optim. **45**(5), 2011–2034 (2007)
51. Stoustrup, J., Niemann, H.H.: Fault-tolerant feedback control using the Youla parameterization. In: European Control Conference, pp. 1970–1974. Portugal, Sept, Porto (2001)

52. Tsai, M.-C., Yang, F.-Y., Chen, C.-L.: A double-loop control structure for tracking control and disturbance attenuation. In: Proceedings of 19th World Congress IFAC (2014)
53. Wang, B., Guan, Z.-H., Yuan, F.-S.: Optimal tracking and two-channel disturbance rejection under control energy constraint. Automatica **47**, 733–738 (2011)
54. Wang, H., Yang, G.-H.: A finite frequency domain approach to fault detection observer design for linear continuous-time systems. Asian J. Control **10**(5), 1–10 (2008)
55. Wei, X., Verhaegen, M.: Robust fault detection observer for linear uncertain systems. Int. J. Control **84**(1), 197–215 (2011)
56. Wei, X., Verhaegen, M.: Robust fault detection observer and fault estimation filter design for LTI systems based on GKYP lemma. Eur. J. Control. (2010). https://doi.org/10.3166/EJC.16.366-383
57. Willems, J.L., Mareels, I.M.Y.: A rigorous solution of the infinite time interval LQ problem with constant state tracking. Syst. Control Lett. **52**, 289–296 (2004)
58. Wolovich, W.A.: Linear Multivariable Systems. Springer, New-York-Heidelberg-Berlin (1974)
59. Xie, L., Xue, D., Shiu, K., Tso, S.K.: On the two-degree-of-freedom Wiener-Hopf optimal design with tracking and disturbance rejection constraints. Automatica **36**, 1897–1904 (2000)
60. Zeren, M., Ozbay, H.: On the synthesis of stable controllers. IEEE Trans. Autom. Control **44**, 431–435 (1999)
61. Zhang, K., Jiang, B., Yan, X., Mao, Z., Polycarpou, M.P.: Fault-tolerant control for systems with unmatched actuator faults and disturbances. IEEE Trans. Autom. Control (2020). https://doi.org/10.1109/TAC.2020.2997347
62. Zhang, S., Wang, A., Ding, D., Shu, H., Hayat, T., Dobaie, A.: On design of robust fault detection filter in finite frequency domain with regional pole assignment. IEEE Trans. Circ. Syst. II Express Briefs **62**(4), 382–386 (2015)
63. Zhang, X.-N., Yang, G.-H.: Dynamic output feedback control synthesis with mixed frequency small gain specifications. Acta Autom. Sinica **34**(5), 551–557 (2008)
64. Zhang, Z., Jaimoukha, I.: Optimal state space solution to the fault detection problem at single frequency. In: Proceedings of the 18th IFAC Word Congress Milano (Italy), Aug. 28–Sept. 2, 2011, pp. 7619–7624 (2011)
65. Zhang, Z., Jaimoukha, I.: An optimal solution to an $\mathcal{H}_-/\mathcal{H}_\infty$ fault detection problem. In: 50th IEEE Conference on Decision and Control and European Control Conference (CDC-ECC) Orlando FL, USA, Dec. 12–15, 2011, pp. 903–908 (2011)
66. Zhang, Z., Jaimoukha, I.M.: An optimal solution to an $\mathcal{H}_-/\mathcal{H}_\infty$ fault detection Problem. In: Proceedings of 50th Conference Decision and Control and European Control Conference (CDC-ECC), Orlando, FL, USA, Dec. 12–15, 2011, pp. 903–908 (2011)
67. Zhou, K., Doyle, J.C., Glover, K.: Robust and Optimal Control. Prentice-Hall, Upper Saddle River, NJ (1996)
68. Zhou, K., Ren, Z.: A new controller architecture for high performance, robust, and fault-tolerant control. IEEE Trans. Autom. Contr. **46**, 1613–1618 (2001)
69. Zhu, X., Xia, Y., Wang, M., Ma, S.: $\mathscr{H}_\infty$ fault detection for discrete-time hybrid systems via a descriptor system method. Circ. Syst. Signal Process. **33**(9), 2807–2826 (2014)
70. Zhu, X., Xia, Y.: $\mathscr{H}_\infty$ descriptor fault detection filter design for T-S fuzzy discrete-time systems. J. Franklin Inst. **351**(12), 5358–5375 (2014)

New Interval Observer-Based Fault Detection for Switched Systems

Chaima Zammali, Jérémy Van Gorp, Zhenhua Wang, and Tarek Raïssi

Abstract The primary focus of this chapter is to design a new interval observer-based Fault Detection (FD) method for a class of discrete-time switched systems subject to unknown but bounded state disturbances and measurement noise. The proposed technique is investigated to reduce the conservatism of gain matrices and to offer more degrees of design freedom by integrating weighted matrices in the structure of the FD observer. Using multiple quadratic Lyapunov functions (MQLF) with an average dwell time (ADT) control condition, novel solvable conditions are derived in terms of linear matrix inequalities (LMIs). Furthermore, the FD decision is based on residual intervals generated by the proposed interval observers. The efficiency of the proposed approach is highlighted through simulation results on an academic example.

1 Introduction

With the continuous growing complexity of industrial systems and rising demand for higher performance and safety, the implementation of optimized fault detection strategies has become a priority for many industrials in order to minimize performance degradation and to avoid severe and irreversible damage to human operator and equipment [1–4]. Usually, physical systems as mechanical, embedded, power,

C. Zammali (✉) · J. Van Gorp · T. Raïssi
Conservatoire National des Arts et Métiers (CNAM), Cedric-Laetitia, 292 Rue Saint-Martin, 75141 Paris, France
e-mail: chaima.zammali@lecnam.net

J. Van Gorp
e-mail: jeremy.vangorp@lecnam.net

T. Raïssi
e-mail: tarek.raissi@cnam.fr

Z. Wang
School of Astronautics, Harbin Institute of Technology, Harbin 150001, China
e-mail: zhenhua.wang@hit.edu.cn

P. Shi et al. (eds.), *Complex Systems: Spanning Control and Computational Cybernetics: Applications*, Studies in Systems, Decision and Control 415,
https://doi.org/10.1007/978-3-031-00978-5_7

communication networks, are characterized by hybrid processes which exhibit both discrete and continuous dynamics. Switched systems represent a special class of hybrid systems [5]. They involve a finite number of subsystems and a switching law specifying the active subsystem at each time instant. They can model and control a wide range of engineering systems. Some typical examples of switched systems with greater theoretical challenges and high practical interest can be found, for instance, in power systems, computer disk drives, transmission and stepper motors, robotics and automated highways.

In the presence of unknown but bounded uncertainties coming either from external disturbances or from the mismatch between the model and the real system, interval observers have been proven to be effective in determining whether the system is faulty or not [6–8]. The major advantage of using interval FD observers consists in providing systematic residual evaluation and avoids the design of threshold generators. In the other side, interval techniques are based on the theory of positive systems, which require the nonnegativity of the error dynamics. Unfortunately, this assumption is still restrictive and may lead to more theoretical difficulty and computational complexity in searching for a gain matrix to simultaneously ensure the nonnegativity and stability of the error system. Some methods propose a coordinate transformation to relax the design conditions of interval observers. However, it is hard to merge the coordinate transformation approach with performance constraints such as disturbance attenuation performance [9]. In view of this major drawback, the main idea of this chapter is to design a new interval observer-based (TNL structure) FD method for discrete-time switched systems subject to unknown but bounded disturbances. The proposed method offers more degrees of design freedom by integrating weighted matrices in the FD observer structure.

The accuracy of fault detection is highly affected by the modeling uncertainties. Thus, one of the most essential requirements imposed on FD algorithms is the robustness against uncertainties using, for instance, H_∞ technique [10, 11]. It is worth noting that the H_∞ norm is a measurement of energy-to-energy gain. Nevertheless, the practical signals are not necessarily energy bounded but have bounded peak values. Other methods such as disturbances decoupling methods have been proposed with the aim to completely cancel the disturbances effect from the residual signals. However, it may be problematic, in some cases, because the fault impact may also be removed. Hence, appropriate criteria should be designed in order to take into account the effects of disturbances. To the best of our knowledge, the optimisation of the observer gain in order to achieve robust fault detection is still an open problem for switched systems. In this chapter, robust fault detection is developed by introducing set-membership approaches with an L_∞ analysis which describes the peak-to-peak performance index. The efficiency of the present approach is highlighted through simulation results on an academic example.

2 Problem Statement

Consider the following discrete-time switched system:

$$\begin{cases} x_{k+1} = A_q x_k + B_q u_k + D_q w_k \\ y_k = C x_k + D_v v_k + F f_k, \end{cases} \tag{1}$$

where $x \in \mathbb{R}^{n_x}$, $u \in \mathbb{R}^{n_u}$, $y \in \mathbb{R}^{n_y}$, $f \in \mathbb{R}^{n_f}$, $w \in \mathbb{R}^{n_w}$ and $v \in \mathbb{R}^{n_v}$ are respectively the state vector, the input, the output, the sensor fault, the state disturbances and the measurement noise. $\mathbb{R}^n$ denotes the n dimensional Euclidean space. The known matrices A_q, B_q, C, D_q, D_v and F are given with appropriate dimensions. The index q specifies, at each discrete instant k, the subsystem that is currently followed. $q \in \mathcal{I} = \overline{1, N}$, $N \in \mathbb{Z}_+$, N is the number of linear subsystems. $\mathcal{I}$ denotes the set of non-negative integers $\{1, ..., N\}$. The switching signal is assumed to be known.

Assumption 1 Assume that the state disturbances and the measurement noises are unknown but bounded with a priori known bounds such that

$$\underline{w} \leq w \leq \overline{w}, \qquad \underline{v} \leq v \leq \overline{v}$$

where $\underline{w}, \overline{w} \in \mathbb{R}^{n_w}$ and $\underline{v}, \overline{v} \in \mathbb{R}^{n_v}$.

Assumption 2 The pairs (A_q, C) are detectable, $\forall q = 1, \ldots, N$.

In the following, the goal is to design residual framers based on robust FD interval observers for discrete-time linear switched systems subject to sensor faults. An L_∞ criterion is developed in this chapter in order to compute the observer gains and to take into account the presence of state disturbances and measurement noises in the design of the robust FD procedure.

3 Interval Observer Design: TNL Structure

The proposed FD interval observer for system (1) is given by

$$\begin{cases} \overline{\xi}_{k+1} = \overline{T}_q A_q \overline{x}_k + \overline{T}_q B_q u_k + \overline{L}_q (y_k - C\overline{x}_k) + \overline{\Delta} \\ \overline{x}_k = \overline{\xi}_k + \overline{N}_q y_k \\ \underline{\xi}_{k+1} = \underline{T}_q A_q \underline{x}_k + \underline{T}_q B_q u_k + \underline{L}_q (y_k - C\underline{x}_k) + \underline{\Delta} \\ \underline{x}_k = \underline{\xi}_k + \underline{N}_q y_k \\ \overline{y}_k = C^+ \overline{x}_k - C^- \underline{x}_k + D_v{}^+ \overline{v} - D_v{}^- \underline{v} \\ \underline{y}_k = C^+ \underline{x}_k - C^- \overline{x}_k + D_v{}^+ \underline{v} - D_v{}^- \overline{v} \\ \overline{r}_k = \overline{y}_k - y_k \\ \underline{r}_k = \underline{y}_k - y_k \end{cases} \tag{2}$$

where $\overline{\xi}_k$, $\underline{\xi}_k \in \mathbb{R}^{n_x}$ are intermediate variables, $\overline{x}_k$, $\underline{x}_k \in \mathbb{R}^{n_x}$ are the estimated upper and lower bounds of x_k respectively. $\overline{\Delta}$ and $\underline{\Delta}$ are given by:

$$\begin{cases} \overline{\Delta} = (\overline{T}_q D_q)^+ \overline{w} - (\overline{T}_q D_q)^- \underline{w} + (\overline{L}_q D_v)^+ \overline{v} - (\overline{L}_q D_v)^- \underline{v} + (\overline{N}_q D_v)^+ \overline{v} - (\overline{N}_q D_v)^- \underline{v} \\ \underline{\Delta} = (\underline{T}_q D_q)^+ \underline{w} - (\underline{T}_q D_q)^- \overline{w} + (\underline{L}_q D_v)^+ \underline{v} - (\underline{L}_q D_v)^- \overline{v} + (\underline{N}_q D_v)^+ \underline{v} - (\underline{N}_q D_v)^- \overline{v}. \end{cases}$$

In (2), $\overline{L}_q \in \mathbb{R}^{n_x \times n_y}$ and $\underline{L}_q \in \mathbb{R}^{n_x \times n_y}$ are the observer gains. $\overline{T}_q \in \mathbb{R}^{n_x \times n_x}$, $\underline{T}_q \in \mathbb{R}^{n_x \times n_x}$, $\overline{N}_q \in \mathbb{R}^{n_x \times n_y}$ and $\underline{N}_q \in \mathbb{R}^{n_x \times n_y}$ are constant matrices that should be designed to satisfy

$$\overline{T}_q + \overline{N}_q C = I_{n_x} \tag{3}$$

$$\underline{T}_q + \underline{N}_q C = I_{n_x}. \tag{4}$$

Lemma 1 *[12] Given matrices $A \in \mathbb{R}^{a \times b}$, $B \in \mathbb{R}^{b \times c}$ and $C \in \mathbb{R}^{a \times c}$, if* $\text{rank}(B) = c$, *then the general solution of the following equation $AB = C$ is given by*

$$A = CB^\dagger + S(I - BB^\dagger),$$

where $B^\dagger$ represents the pseudo-inverse of B, defined by $B^\dagger = B^T(BB^T)^{-1}$, and $S \in \mathbb{R}^{a \times b}$ is an arbitrary matrix.

Based on Lemma 1, the general solutions of (3) and (4) are given by

$$[\overline{T}_q \; \overline{N}_q] = \begin{bmatrix} I_{n_x} \\ C \end{bmatrix}^\dagger + \overline{S}_q \left(I_{n_x+n_y} - \begin{bmatrix} I_{n_x} \\ C \end{bmatrix} \begin{bmatrix} I_{n_x} \\ C \end{bmatrix}^\dagger \right), \tag{5}$$

$$[\underline{T}_q \; \underline{N}_q] = \begin{bmatrix} I_{n_x} \\ C \end{bmatrix}^\dagger + \underline{S}_q \left(I_{n_x+n_y} - \begin{bmatrix} I_{n_x} \\ C \end{bmatrix} \begin{bmatrix} I_{n_x} \\ C \end{bmatrix}^\dagger \right), \tag{6}$$

where $\overline{S}_q$, $\underline{S}_q \in \mathbb{R}^{n_x \times (n_x+n_y)}$ for $q = 1, \ldots, N$ are arbitrary matrices which are designed such that all matrices $\overline{T}_q$, $\underline{T}_q$ are of full rank.

Under the new structure of the proposed FD interval observer (2), the objective is to compute the observer gains $\overline{L}_q$ and $\underline{L}_q$ that minimize the effect of state disturbances and measurement noise on the upper and lower bounds of the residual vectors $\overline{r}_k$, $\underline{r}_k$, respectively. Let $\overline{e}_k = \overline{x}_k - x_k$ and $\underline{e}_k = x_k - \underline{x}_k$ be the upper and the lower estimation errors. By combining (1), (3) and (4), x_{k+1} can be written in two different ways:

$$\begin{aligned} x_{k+1} &= (\overline{T}_q + \overline{N}_q C)x_{k+1} \\ &= \overline{T}_q x_{k+1} + \overline{N}_q(y_{k+1} - D_v v_{k+1} - F f_{k+1}) \\ &= \overline{T}_q A_q x_k + \overline{T}_q B_q u_k + \overline{T}_q D_q w_k + \overline{N}_q y_{k+1} - \overline{N}_q D_v v_{k+1} - \overline{N}_q F f_{k+1}, \end{aligned} \tag{7}$$

$$\begin{aligned} x_{k+1} &= (\underline{T}_q + \underline{N}_q C)x_{k+1} \\ &= \underline{T}_q x_{k+1} + \underline{N}_q (y_{k+1} - D_v v_{k+1} - F f_{k+1}) \\ &= \underline{T}_q A_q x_k + \underline{T}_q B_q u_k + \underline{T}_q D_q w_k + \underline{N}_q y_{k+1} - \underline{N}_q D_v v_{k+1} - \underline{N}_q F f_{k+1}. \end{aligned} \tag{8}$$

Then, the dynamics of the upper and lower errors are given by:

$$\begin{cases} \overline{e}_{k+1} = (\overline{T}_q A_q - \overline{L}_q C)\overline{e}_k + \overline{\Delta} + \overline{L}_q D_v v_k + \overline{N}_q D_v v_{k+1} \\ \qquad -\overline{T}_q D_q w_k + \overline{L}_q F f_k + \overline{N}_q F f_{k+1} \\ \underline{e}_{k+1} = (\underline{T}_q A_q - \underline{L}_q C)\underline{e}_k - \underline{\Delta} - \underline{L}_q D_v v_k - \underline{N}_q D_v v_{k+1} \\ \qquad +\underline{T}_q D_q w_k - \underline{L}_q F f_k - \underline{N}_q F f_{k+1}. \end{cases} \tag{9}$$

We introduce

$$\overline{d}_k = \begin{bmatrix} \overline{\Delta} - \overline{T}_q D_q w_k \\ D_v v_k \\ D_v v_{k+1} \end{bmatrix}, \quad \underline{d}_k = \begin{bmatrix} -\underline{\Delta} + \underline{T}_q D_q w_k \\ -D_v v_k \\ -D_v v_{k+1} \end{bmatrix}, \quad \tilde{f}_k = \begin{bmatrix} f_k \\ f_{k+1} \end{bmatrix}.$$

Accordingly, the error dynamics in (9) can be rewritten as

$$\begin{cases} \overline{e}_{k+1} = (\overline{T}_q A_q - \overline{L}_q C)\overline{e}_k + \overline{H}_q \overline{d}_k + \overline{F}_q \tilde{f}_k \\ \underline{e}_{k+1} = (\underline{T}_q A_q - \underline{L}_q C)\underline{e}_k + \underline{H}_q \underline{d}_k + \underline{F}_q \tilde{f}_k, \end{cases} \tag{10}$$

where

$$\overline{H}_q = \begin{bmatrix} I_n \\ \overline{L}_q^T \\ \overline{N}_q^T \end{bmatrix}^T, \quad \underline{H}_q = \begin{bmatrix} I_n \\ \underline{L}_q^T \\ \underline{N}_q^T \end{bmatrix}^T, \quad \overline{F}_q = \begin{bmatrix} (\overline{L}_q F)^T \\ (\overline{N}_q F)^T \end{bmatrix}^T, \quad \underline{F}_q = \begin{bmatrix} -(\underline{L}_q F)^T \\ -(\underline{N}_q F)^T \end{bmatrix}^T.$$

Based on the error system (10), the nonnegativity, the stability and the robustness of the proposed interval observer are studied in the following theorems.

Assumption 3 The upper and lower bounds of the initial state, $\overline{x}_0$ and $\underline{x}_0$ are chosen such that $\underline{x}_0 \leq x_0 \leq \overline{x}_0$.

Lemma 2 *[13] Consider the system described by:*

$$x_{k+1} = Ax_k + u_k \ , \ u : \mathbb{Z}_+ \to \mathbb{R}^n_+, \ k \in \mathbb{Z}_+ \tag{11}$$

with $x \in \mathbb{R}^n$. The system (11) is said cooperative or non-negative if and only if $u_k \geq 0$ for all $k \geq 0$, $x_0 \geq 0$ and A is a non-negative matrix.

Theorem 1 *For system (1), $\overline{x}_k$ and $\underline{x}_k$ in (2) satisfy the following inequality*

$$\underline{x}_k \leq x_k \leq \overline{x}_k,$$

in the fault free case, if $\overline{T}_q A_q - \overline{L}_q C$ and $\underline{T}_q A_q - \underline{L}_q C$ are nonnegative for all $k \geq 0$ and $\overline{x}_0$, $\underline{x}_0$ are chosen such that $\underline{x}_0 \leq x_0 \leq \overline{x}_0$.

Proof In the fault free case ($f = 0$), according to Assumption 1, we have

$$\begin{aligned} \overline{\Delta} - \overline{T}_q D_q w_k + \overline{L}_q D_v v_k + \overline{N}_q D_v v_{k+1} &\geq 0 \\ -\underline{\Delta} + \underline{T}_q D_q w_k - \underline{L}_q D_v v_k - \underline{N}_q D_v v_{k+1} &\geq 0. \end{aligned}$$

In addition, let Assumption 3 be satisfied. Then, $\overline{e}_0 \geq 0$ and $\underline{e}_0 \geq 0$. Applying Lemma 3 to (9), the inclusion

$$\underline{x}_k \leq x_k \leq \overline{x}_k$$

holds for all $k \geq 0$ if $\overline{T}_q A_q - \overline{L}_q C$ and $\underline{T}_q A_q - \underline{L}_q C$ are nonnegative.

In order to study the stability of the proposed residual framers, we propose a new augmented state defined by $\mathcal{E}_k = [\overline{e}_k{}^T \ \ \underline{e}_k{}^T]^T$ and $\mathcal{R}_k = [\overline{r}_k{}^T \ \ \underline{r}_k{}^T]^T$. The corresponding augmented system can be deduced:

$$\begin{cases} \mathcal{E}_{k+1} = \mathcal{A}_q \mathcal{E}_k + \mathcal{H}_q d_k + \tilde{\mathcal{F}}_q \tilde{f}_k \\ \mathcal{R}_k = \mathcal{C}\mathcal{E}_k + \mathcal{V}\tilde{v}_k + \mathcal{F} f_k, \end{cases} \tag{12}$$

where

$$\mathcal{A}_q = \begin{bmatrix} \overline{T}_q A_q - \overline{L}_q C & 0 \\ 0 & \underline{T}_q A_q - \underline{L}_q C \end{bmatrix}, \ \mathcal{H}_q = \begin{bmatrix} \overline{H}_q & 0 \\ 0 & \underline{H}_q \end{bmatrix}, \ \tilde{\mathcal{F}}_q = \begin{bmatrix} \overline{F}_q \\ \underline{F}_q \end{bmatrix}, \ d_k = \begin{bmatrix} \overline{d}_k \\ \underline{d}_k \end{bmatrix},$$

$$\mathcal{F} = \begin{bmatrix} -F \\ -F \end{bmatrix}, \ \mathcal{C} = \begin{bmatrix} C^+ & C^- \\ -C^- & -C^+ \end{bmatrix}, \ \mathcal{V} = \begin{bmatrix} -D_v & D_v{}^+ & -D_v{}^- \\ -D_v & -D_v{}^- & D_v{}^+ \end{bmatrix}, \ \tilde{v}_k = \begin{bmatrix} v_k \\ \overline{v} \\ \underline{v} \end{bmatrix}.$$

The error dynamics in (12) can be split into two subsystems where (13) is decoupled from the effects of $f(k)$ and (14) is only affected by the sensor fault.

$$\begin{cases} \mathcal{E}^d_{k+1} = \mathcal{A}_q \mathcal{E}^d_k + \mathcal{H}_q d_k \\ \mathcal{R}^d_k = \mathcal{C}\mathcal{E}^d_k + \mathcal{V}\tilde{v}_k, \end{cases} \tag{13}$$

$$\begin{cases} \mathcal{E}^f_{k+1} = \mathcal{A}_q \mathcal{E}^f_k + \tilde{\mathcal{F}}_q \tilde{f}_k \\ \mathcal{R}^f_k = \mathcal{C}\mathcal{E}^f_k + \mathcal{F} f_k, \end{cases} \tag{14}$$

where $\mathcal{E}_k = \mathcal{E}^f_k + \mathcal{E}^d_k$.

The objective in the sequel is to design a FD observer (2) such that the error system in (13) is stable and the effect of disturbances is minimized.

Theorem 2 *Let Assumption 3 hold and suppose that there exists a piecewise Lyapunov function* $V_q(\mathcal{E}_k^d)$ *where* $V_q(\mathcal{E}_k^d) = \mathcal{E}_k^{d^T} P_q \mathcal{E}_k^d$. *Given scalars* $\gamma > 0$, $\gamma_1 > 0$, $\gamma_2 > 0$, $0 < \lambda < 1$ *and* $0 < \beta < 1$, *the error dynamics system in (13) are stable and* $\mathcal{R}^d$ *satisfies the* L_∞ *performance, if there exist a constant* $\mu > 0$, $a_2 > a_1 > 0$, *diagonal matrices* $P_q \in \mathbb{R}^{2n_x \times 2n_x}$ *such that* $P_q = \begin{bmatrix} P_{q1} & 0 \\ 0 & P_{q2} \end{bmatrix} > 0$, $P_q = P_q^T \succ 0$ *with diagonal matrices* P_{q1}, P_{q1} *and constant matrices* $M_l \in \mathbb{R}^{2n_x \times 2n_x}$, W_{q1}, $W_{q2} \in \mathbb{R}^{n_x \times n_y}$, Y_{q1}, $Y_{q2} \in \mathbb{R}^{n_x \times (n_x + n_y)}$, *for* $q = 1, 2, \ldots, N$ *such that:*

$$\min_{P_q, q \in I} \beta\rho + (1 - \beta)\mu, \tag{15}$$

$$a_1 I_{2n_x} \leq P_q \leq a_2 I_{2n_x}, \tag{16}$$

$$\begin{bmatrix} P_{q1}\Theta^{\dagger}\alpha_1 A_q + Y_{q1}\Psi\alpha_1 A_q - W_{q1}C & 0 \\ * & P_{q2}\Theta^{\dagger}\alpha_1 A_q + Y_{q2}\Psi\alpha_1 A_q - W_{q2}C \end{bmatrix} \geq 0, \tag{17}$$

$$\begin{bmatrix} \Upsilon_{q11} & 0 & \Upsilon_{q13} \\ * & -\mu I_{n_d} & \Upsilon_{q23} \\ * & * & \Upsilon_{q33} \end{bmatrix} \prec 0, \tag{18}$$

$$\begin{bmatrix} \lambda P_q & 0 & I_{2n_x} \\ * & (\gamma - \mu) I_{n_d} & 0 \\ * & * & \gamma I_{2n_x} \end{bmatrix} \succ 0, \tag{19}$$

$$\begin{bmatrix} C^T C - \gamma_1^2 I_{2n_x} & C^T \mathcal{V} \\ * & -\gamma_2^2 I_{2n_x} + \mathcal{V}^T \mathcal{V} \end{bmatrix} \prec 0, \tag{20}$$

$$\begin{bmatrix} M_l & P_q \\ P_q & P_q \end{bmatrix} \succeq 0, \tag{21}$$

hold for all $q, l \in I$, $q \neq l$ *where*

$$\Upsilon_{q11} = \begin{bmatrix} (\lambda - 1)P_{q1} & 0 \\ * & (\lambda - 1)P_{q2} \end{bmatrix}, \quad \Upsilon_{q33} = \begin{bmatrix} -P_{q1} & 0 \\ * & -P_{q2} \end{bmatrix},$$

$$\Upsilon_{q13} = \begin{bmatrix} P_{q1}\Theta^{\dagger}\alpha_1 A_q + Y_{q1}\Psi\alpha_1 A_q - W_{q1}C & 0 \\ * & P_{q2}\Theta^{\dagger}\alpha_1 A_q + Y_{q2}\Psi\alpha_1 A_q - W_{q2}C \end{bmatrix}^T,$$

$$\Upsilon_{q23} = \begin{bmatrix} P_{q1} & W_{q1} & P_{q1}\Theta^{\dagger}\alpha_2 + Y_{q1}\Psi\alpha_2 & 0 & 0 & 0 \\ 0 & 0 & 0 & P_{q2} & W_{q2} & P_{q2}\Theta^{\dagger}\alpha_2 + Y_{q2}\Psi\alpha_2 \end{bmatrix}^T$$

and

$$\alpha_1 = \begin{bmatrix} I_{n_x} \\ 0 \end{bmatrix}, \ \alpha_2 = \begin{bmatrix} 0 \\ I_{n_y} \end{bmatrix}, \ \Theta = \begin{bmatrix} I_{n_x} \\ C \end{bmatrix}, \ \Psi = I_{n_x+n_y} - \Theta\Theta^{\dagger}.$$

In addition, the estimation errors are stable under a switching signal with an ADT τ_a *satisfying:*

$$\tau_a > \tau_a^* = -\frac{ln(\rho)}{ln(1-\lambda)}, \tag{22}$$

where $\rho = \frac{a_2}{a_1}$ *and the observer gains* $\overline{L}_q$, $\underline{L}_q$, $\overline{T}_q$, $\underline{T}_q$, $\overline{N}_q$ *and* $\underline{N}_q$ *are given by:*

$$\begin{cases} \overline{L}_q = P_{q1}{}^{-1}W_{q1} \\ \underline{L}_q = P_{q2}{}^{-1}W_{q2} \\ \overline{T}_q = \Theta^{\dagger}\alpha_1 + P_{q1}{}^{-1}Y_{q1}\Psi\alpha_1 \\ \underline{T}_q = \Theta^{\dagger}\alpha_1 + P_{q2}{}^{-1}Y_{q2}\Psi\alpha_1 \\ \overline{N}_q = \Theta^{\dagger}\alpha_2 + P_{q1}{}^{-1}Y_{q1}\Psi\alpha_2 \\ \underline{N}_q = \Theta^{\dagger}\alpha_2 + P_{q2}{}^{-1}Y_{q2}\Psi\alpha_2. \end{cases} \tag{23}$$

Moreover, the interval error (13) satisfies:

$$\lim_{k\to\infty} \|\mathcal{E}_k^d\| < \frac{\mu}{a_1\lambda}\theta_d{}^2.$$

Proof The aim is to design an observer (2) satisfying the following conditions:

(i) The error system in (13) is stable.
(ii) The effect of disturbances is minimized using the L_∞ performance.

The following MQLF are chosen,

$$V_q(\mathcal{E}_k^d) = \mathcal{E}_k^{d\,T} P_q \mathcal{E}_k^d, \ P_q^T = P_q \succ 0, \ P_q \in \mathbb{R}^{2n_x \times 2n_x}.$$

It is worth noting that $P_q > 0$ since they are diagonal matrices. Consequently,

$$P_q \begin{bmatrix} \overline{T}_q A_q - \overline{L}_q C & 0 \\ 0 & \underline{T}_q A_q - \underline{L}_q C \end{bmatrix} \geq 0 \tag{24}$$

holds. Substituting (23) into (24) gives

$$\begin{bmatrix} P_{q1}\Theta^{\dagger}\alpha_1 A_q + Y_{q1}\Psi\alpha_1 A_q - W_{q1}C & 0 \\ * & P_{q2}\Theta^{\dagger}\alpha_1 A_q + Y_{q2}\Psi\alpha_1 A_q - W_{q2}C \end{bmatrix} \geq 0. \tag{25}$$

Therefore, the inequality (17) is satisfied.
The time difference of $V_q(\mathcal{E}_k^d)$ is given by

$$\begin{aligned} \Delta V_q(\mathcal{E}_k^d) &= V_q(\mathcal{E}_{k+1}^d) - V_q(\mathcal{E}_k^d) \\ &= \mathcal{E}_{k+1}^{d\ T} P_q \mathcal{E}_{k+1}^d - \mathcal{E}_k^{d^T} P_q \mathcal{E}_k^d \\ &= (\mathcal{A}_q \mathcal{E}_k^d + \mathcal{H}_q d_k)^T P_q (\mathcal{A}_q \mathcal{E}_k^d + \mathcal{H}_q d_k) - \mathcal{E}_k^{d^T} P_q \mathcal{E}_k^d \\ &= \mathcal{E}_k^{d^T} \mathcal{A}_q{}^T P_q \mathcal{A}_q \mathcal{E}_k^d + \mathcal{E}_k^{d^T} \mathcal{A}_q{}^T P_q \mathcal{H}_q d_k + d_k^T \mathcal{H}_q{}^T P_q \mathcal{A}_q \mathcal{E}_k^d \\ &\quad + d_k^T \mathcal{H}_q{}^T P_q \mathcal{H}_q d_k - \mathcal{E}_k^{d^T} P_q \mathcal{E}_k^d. \end{aligned} \tag{26}$$

Based on (26), the following equation can be obtained

$$\Delta V_q(\mathcal{E}_k^d) = \begin{bmatrix} \mathcal{E}_k^d \\ d_k \end{bmatrix}^T \begin{bmatrix} \mathcal{A}_q{}^T P_q \mathcal{A}_q - P_q & \mathcal{A}_q{}^T P_q \mathcal{H}_q \\ * & \mathcal{H}_q{}^T P_q \mathcal{H}_q \end{bmatrix} \begin{bmatrix} \mathcal{E}_k^d \\ d_k \end{bmatrix}.$$

If inequality (18) holds, then it can be rewritten as

$$\begin{bmatrix} (\lambda - 1)P_q & 0 & (P_q \mathcal{A}_q)^T \\ * & -\mu I_{n_d} & (P_q \mathcal{H}_q)^T \\ * & * & -P_q \end{bmatrix} \prec 0. \tag{27}$$

By pre- and post- multiplying (27) with $\begin{bmatrix} I_{2n_x} & 0 & \mathcal{A}_q^T \\ 0 & I_{n_d} & \mathcal{H}_q{}^T \end{bmatrix}$ and its transpose, respectively, the relation in (28) yields

$$\begin{bmatrix} \mathcal{A}_q{}^T P_q \mathcal{A}_q - P_q & \mathcal{A}_q{}^T P_q \mathcal{H}_q \\ * & \mathcal{H}_q{}^T P_q \mathcal{H}_q \end{bmatrix} + \begin{bmatrix} \lambda P_q & 0 \\ * & -\mu I_{n_d} \end{bmatrix} \prec 0. \tag{28}$$

Pre- and post-multiplying (28) with $[\mathcal{E}_k^{d^T} \quad d_k{}^T]$ and its transpose

$$\begin{bmatrix} \mathcal{E}_k^d \\ d_k \end{bmatrix}^T \begin{bmatrix} \mathcal{A}_q{}^T P_q \mathcal{A}_q - P_q & \mathcal{A}_q{}^T P_q \mathcal{H}_q \\ * & \mathcal{H}_q{}^T P_q \mathcal{H}_q \end{bmatrix} \begin{bmatrix} \mathcal{E}_k^d \\ d_k \end{bmatrix} + \begin{bmatrix} \mathcal{E}_k^d \\ d_k \end{bmatrix}^T \begin{bmatrix} \lambda P_q & 0 \\ * & -\mu I_{n_d} \end{bmatrix} \begin{bmatrix} \mathcal{E}_k^d \\ d_k \end{bmatrix} \prec 0,$$

it follows that

$$\begin{aligned} \Delta V_q(\mathcal{E}_k^d) + \lambda \mathcal{E}_k^{d^T} P_q \mathcal{E}_k^d - \mu d_k{}^T d_k &< 0 \\ \Delta V_q(\mathcal{E}_k^d) &< -\lambda V_q(\mathcal{E}_k^d) + \mu d_k{}^T d_k. \end{aligned} \tag{29}$$

When process disturbance w_k and measurement noise v_k are zero, $d_k = 0$ and the increment of Lyapunov function $V_q(\mathcal{E}_k^d)$ becomes

$$\Delta V_q(\mathcal{E}_k^d) = V_q(\mathcal{E}_{k+1}^d) - V_q(\mathcal{E}_k^d) < -\lambda V_q(\mathcal{E}_k^d) < 0.$$

Hence, the error system in (13) is stable.
Note that the following inequality can be derived from (29)

$$V_q(\mathcal{E}_{k+1}^d) < (1-\lambda)V_q(\mathcal{E}_k^d) + \mu\theta_d^2, \tag{30}$$

where θ_d is a known constant and represents the L_∞ norm of d. From (30), one can obtain

$$\begin{aligned} V_q(\mathcal{E}_k^d) &\leq (1-\lambda)^k V_q(\mathcal{E}_0^d) + \mu \textstyle\sum_{\tau=0}^{k-1}(1-\lambda)^\tau \theta_d^2 \\ &\leq (1-\lambda)^k V_q(\mathcal{E}_0^d) + \mu \frac{(1-\lambda^k)}{\lambda}\theta_d^2 \\ &\leq (1-\lambda)^k V_q(\mathcal{E}_0^d) + \frac{\mu\theta_d^2}{\lambda}. \end{aligned} \tag{31}$$

On the other hand, by using the Schur complement lemma, (19) is equivalent to

$$\begin{bmatrix} \lambda P_q & 0 \\ * & (\gamma-\mu)I_{n_d} \end{bmatrix} - \frac{1}{\gamma}\begin{bmatrix} I_{2n_x} \\ 0 \end{bmatrix}\begin{bmatrix} I_{2n_x} & 0 \end{bmatrix} \succ 0. \tag{32}$$

Then, pre-multiplying and post-multiplying (32) with $[\mathcal{E}_k^{d^T} \quad d_k{}^T]$ and its transpose, one can obtain

$$\mathcal{E}_k^{d^T}\mathcal{E}_k^d \leq \gamma\left(\lambda V_q(\mathcal{E}_k^d) + (\gamma-\mu)\theta_d^2\right). \tag{33}$$

Substituting (31) into (33) gives

$$\begin{aligned} \mathcal{E}_k^{d^T}\mathcal{E}_k^d &\leq \gamma\left(\lambda\left((1-\lambda)^k V_q(\mathcal{E}_0^d) + \frac{\mu\theta_d^2}{\lambda}\right) + (\gamma-\mu)\,\theta_d^2\right) \\ &\leq \gamma\left(\lambda\,(1-\lambda)^k\, V_q(\mathcal{E}_0^d) + \gamma\theta_d^2\right). \end{aligned}$$

In addition, the matrix inequality in (20) implies that

$$\begin{bmatrix} \mathcal{E}_k^d \\ \tilde{v}_k \end{bmatrix}^T \begin{bmatrix} C^TC - \gamma_1^2 I_{2n_x} & C^T\mathcal{V} \\ * & -\gamma_2^2 I_{2n_x} + \mathcal{V}^T\mathcal{V} \end{bmatrix}\begin{bmatrix} \mathcal{E}_k^d \\ \tilde{v}_k \end{bmatrix} < 0,$$

which follows

$$\begin{aligned} \mathcal{R}_k^{d^T}\mathcal{R}_k^d &\leq \gamma_1^2\mathcal{E}_k^{d^T}\mathcal{E}_k^d + \gamma_2^2\tilde{v}_k^T\tilde{v}_k \\ &\leq \gamma_1^2\gamma\left(\lambda(1-\lambda)^k V_q(\mathcal{E}_0^d) + \gamma\theta_d^2\right) + \gamma_2^2\theta_v^2. \end{aligned}$$

Therefore, the L_∞ criterion is satisfied.

Based on Lemma 1 in [14], the following inequality

$$a_1\|\mathcal{E}_k^d\| \leq V_q(\mathcal{E}_k^d), \tag{34}$$

holds. From (31), we have

$$V_q(\mathcal{E}_k^d) \leq (1-\lambda)^k V_q(\mathcal{E}_0^d) + \frac{\mu\theta_d^2}{\lambda}. \tag{35}$$

Thus, according to (34) and (35), it is easy to derive that

$$\|\mathcal{E}_k^d\| \leq \frac{1}{a_1}((1-\lambda)^k V_q(\mathcal{E}_0^d) + \frac{\mu\theta_d^2}{\lambda}).$$

Hence, when $k \to \infty$, $(1-\lambda)^k$ converge to zero, implies that:

$$\lim_{k\to\infty} \|\mathcal{E}_k^d\| < \frac{\mu}{a_1\lambda}{\theta_d}^2.$$

Note that with the proposed interval observer with the TNL structure, the boundedness of the resulting interval error is guaranteed. In addition, the design of a robust interval observer with a tight interval may be achieved optimally, if the bound $\frac{\mu}{a_1\lambda}{\theta_d}^2$ is minimized. Consequently, the problem of minimizing the interval width of the estimation error and thus the interval width of the residual signal is reduced to the minimization of the scalar μ for a given a_1 and λ. The second purpose consists in minimizing ρ to look for optimum dwell time. The resolution of such a problem leads to solving a problem of linear optimization which consists of seeking a minimization function. Then, the objective function can be added to the LMIs conditions and given by:

$$\beta\rho + (1-\beta)\mu,$$

where the weight β is in the range [0, 1].

Let us now focus on the stabilization of subsystems at the switching instants. It is straightforward to show that the third inequality in Lemma 1 in [14] becomes:

$$\rho P_l - P_q \succeq 0, \tag{36}$$

where $q, l \in \mathcal{I}, q \neq l$, q is the current mode. Then applying the Schur complement lemma, we obtain the following expression:

$$\begin{bmatrix} \rho P_l & I_{2n_x} \\ I_{2n_x} & P_q^{-1} \end{bmatrix} \succeq 0. \tag{37}$$

Multiplying the left and right by $\begin{bmatrix} I_{2n_x} & O_{2n_x} \\ O_{2n_x} & P_q \end{bmatrix}$ and defining $M_l = \rho P_l$, then (37) becomes:

$$\begin{bmatrix} M_l & P_q \\ P_q & P_q \end{bmatrix} \succeq 0.$$

Therefore, (21) is verified.

4 Residual Evaluation

Compared with the traditional methods of designing constant or time-varying thresholds, the present method provides a systematic way for residual evaluation based on a belonging test of the zero signal to the residual framers generated by the proposed FD observers. The corresponding FD decision scheme is based on determining whether the zero signal is excluded from the residual intervals or not such that:

$$\begin{cases} 0 \in [\underline{r}_k \;\; \overline{r}_k] & \text{Fault} - \text{free} \\ 0 \notin [\underline{r}_k \;\; \overline{r}_k] & \text{Faulty}. \end{cases} \tag{38}$$

The FD evaluation in (38) is deduced from the following relation

$$y_k \notin [\underline{y}_k \;\; \overline{y}_k]. \tag{39}$$

In fact, in the fault free case, the output signal is consistent with the estimation of the proposed interval observer, i.e. $y \in [\underline{y} \;\; \overline{y}]$. In contrary case, an inconsistency on the output signal is detected and it indicates the existence of a fault. Based on (39), the consistency test can be written as follow

$$0 \notin [\underline{y}_k \;\; \overline{y}_k] - y_k \Rightarrow 0 \notin [\underline{y}_k - y_k \;\; \overline{y}_k - y_k]. \tag{40}$$

If zero is contained in the estimated framers, the system is assumed fault free. Otherwise an alarm is triggered. In the next section, an illustrative example is introduced to show the efficiency of the developed results.

5 Numerical Example

The numerical example is considered for a discrete-time switched system (1) defined with three subsystems, $N = 3$, with:

$$A_1 = \begin{pmatrix} 0.6 & 0.5 & 1 \\ 0.2 & 0 & 1 \\ 0 & 0.2 & 0 \end{pmatrix} \; A_2 = \begin{pmatrix} 0.9 & -0.8 & 0 \\ 0.1 & 0 & 1 \\ 0 & 0.5 & 0.1 \end{pmatrix} \; A_3 = \begin{pmatrix} 0.5 & 0.5 & 0.1 \\ 0.4 & 0 & 0.5 \\ 0.1 & 0.2 & 0 \end{pmatrix}$$

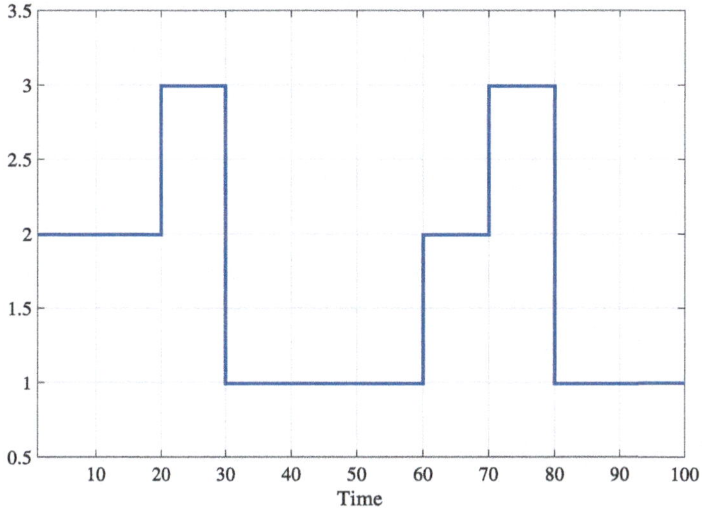

Fig. 1 Evolution of the switching signal

$$B_1 = \begin{pmatrix} 1 \\ 0.1 \\ 1.3 \end{pmatrix} B_2 = \begin{pmatrix} 0.1 \\ 1 \\ 1 \end{pmatrix} B_3 = \begin{pmatrix} 1.2 \\ 1 \\ 0.5 \end{pmatrix} D_1 = \begin{pmatrix} 0.05 \\ 0.1 \\ 0 \end{pmatrix} D_2 = \begin{pmatrix} 0.05 \\ 0.1 \\ 0 \end{pmatrix}$$

$$D_3 = \begin{pmatrix} 0 \\ 0.1 \\ 0.1 \end{pmatrix} D_v = \begin{pmatrix} 0.1 & 0 \\ 0 & 0.1 \end{pmatrix} F = \begin{pmatrix} 0.5 \\ -0.5 \end{pmatrix} C = \begin{pmatrix} 1.2 & 0.01 & 0 \\ 0.1 & 1.1 & 0.1 \end{pmatrix}$$

Remark 1 It is pointed out that the matrices C and F are considered to be constant and common for all modes. However, there is no theoretical difficulty with allowing them to be switched.

In this example, $w_k \in \mathbb{R}$ and $v_k \in \mathbb{R}^2$ are uniformly distributed bounded signals such that $|w_k| \leq 1$ and $|v_k| \leq [0.1 \quad 0.1]$. The state initial conditions are set as $x_0 = [0 \quad 0 \quad 0]^T$ $\underline{x}_0 = [-0.1 \quad -0.1 \quad -0.1]^T$ and $\overline{x}_0 = [0.1 \quad 0.1 \quad 0.1]^T$ such that $\underline{x}_0 \leq x_0 \leq \overline{x}_0$. Figure 1 shows the evolution of the switching signal. It indicates the active mode of the discrete-time switched system.

FD results are given in the sequel adopting MQLF with an ADT switching signal. The numerical simulation was carried out by using Matlab optimization tools (Yalmip/Sedumi). The conservatism of gain matrices is reduced by integrating weighted matrices $\overline{T}_q$, $\underline{T}_q$, $\overline{N}_q$ and $\underline{N}_q$. The existence of a solution for LMIs in Theorem 2 allows one to improve the accuracy of FD and to obtain an optimum dwell time. The resolution of the optimization problem in (15) can be then solved and the Lyapunov matrices are obtained:

$$P_{11} = \begin{pmatrix} 1.29 & 0 & 0 \\ 0 & 1.26 & 0 \\ 0 & 0 & 1.67 \end{pmatrix}, P_{12} = \begin{pmatrix} 1.5 & 0 & 0 \\ 0 & 1.5 & 0 \\ 0 & 0 & 1.92 \end{pmatrix}, P_{21} = \begin{pmatrix} 1.35 & 0 & 0 \\ 0 & 1.23 & 0 \\ 0 & 0 & 1.63 \end{pmatrix},$$

$$P_{22} = \begin{pmatrix} 1.5 & 0 & 0 \\ 0 & 1.49 & 0 \\ 0 & 0 & 1.84 \end{pmatrix}, P_{31} = \begin{pmatrix} 1.5 & 0 & 0 \\ 0 & 1.39 & 0 \\ 0 & 0 & 1.62 \end{pmatrix}, P_{32} = \begin{pmatrix} 1.49 & 0 & 0 \\ 0 & 1.58 & 0 \\ 0 & 0 & 1.77 \end{pmatrix}.$$

In the simulation study, $\mu = 2$ which leads to an ADT $\tau_a > 1$.
The set of observer gains $\overline{L}_q$, $\underline{L}_q$ and the weighted matrices $\overline{N}_q$, $\underline{N}_q$, $\overline{T}_q$ and $\underline{T}_q$ are computed according to (23):

$$\overline{L}_1 = \begin{pmatrix} 0.10 & 0.13 \\ -0.15 & -0.24 \\ -0.05 & 0.11 \end{pmatrix}, \underline{L}_1 = \begin{pmatrix} 0.13 & 0.18 \\ -0.11 & -0.20 \\ -0.04 & 0.11 \end{pmatrix}, \overline{L}_2 = \begin{pmatrix} 0.21 & -0.35 \\ -0.11 & -0.05 \\ 0.03 & 0.20 \end{pmatrix},$$

$$\underline{L}_2 = \begin{pmatrix} 0.18 & -0.36 \\ -0.05 & 0.01 \\ 0.08 & 0.25 \end{pmatrix}, \overline{L}_3 = \begin{pmatrix} 0.21 & 0.23 \\ 0.02 & -0.12 \\ 0.03 & 0.06 \end{pmatrix}, \underline{L}_3 = \begin{pmatrix} 0.22 & 0.23 \\ 0.05 & -0.10 \\ 0.04 & 0.07 \end{pmatrix},$$

$$\overline{N}_1 = \begin{pmatrix} 0.39 & 0.29 \\ 0.26 & 0.45 \\ -0.04 & -0.08 \end{pmatrix}, \underline{N}_1 = \begin{pmatrix} 0.42 & 0.29 \\ 0.29 & 0.45 \\ -0.03 & -0.08 \end{pmatrix}, \overline{N}_2 = \begin{pmatrix} 0.51 & -0.09 \\ 0 & 0.73 \\ -0.15 & -0.05 \end{pmatrix},$$

$$\underline{N}_2 = \begin{pmatrix} 0.48 & -0.06 \\ 0 & 0.71 \\ -0.14 & -0.08 \end{pmatrix}, \overline{N}_3 = \begin{pmatrix} 0.21 & -0.08 \\ 0 & 0.51 \\ 0.02 & -0.18 \end{pmatrix}, \underline{N}_3 = \begin{pmatrix} 0.21 & -0.08 \\ 0.02 & 0.5 \\ 0.03 & -0.16 \end{pmatrix},$$

$$\overline{T}_1 = \begin{pmatrix} 0.49 & -0.32 & -0.02 \\ -0.36 & 0.49 & -0.04 \\ 0.06 & 0.08 & 1 \end{pmatrix}, \underline{T}_1 = \begin{pmatrix} 0.45 & -0.32 & -0.02 \\ -0.39 & 0.49 & -0.04 \\ 0.054 & 0.09 & 1 \end{pmatrix},$$

$$\overline{T}_2 = \begin{pmatrix} 0.38 & 0.09 & 0 \\ -0.07 & 0.19 & -0.07 \\ 0.19 & 0.06 & 1 \end{pmatrix}, \underline{T}_2 = \begin{pmatrix} 0,42 & 0.06 & 0 \\ -0.07 & 0.21 & -0.07 \\ 0.18 & 0.09 & 1 \end{pmatrix},$$

$$\overline{T}_3 = \begin{pmatrix} 0.75 & 0.08 & 0 \\ -0.06 & 0.42 & -0.05 \\ -0.01 & 0.20 & 1.01 \end{pmatrix}, \underline{T}_3 = \begin{pmatrix} 0.75 & 0.09 & 0 \\ -0.08 & 0.44 & -0.05 \\ -0.01 & 0.18 & 1.01 \end{pmatrix}.$$

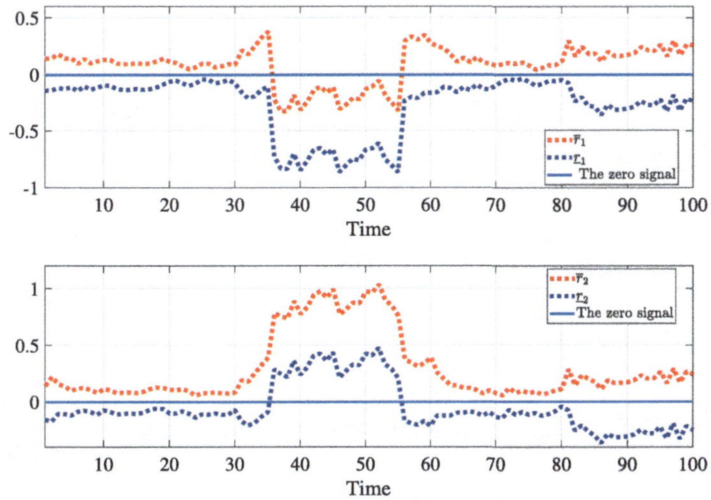

Fig. 2 Residual framers using fault detection interval observer

In the simulation, an abrupt sensor fault f_k is carried out and represented as follows:

$$f_k = \begin{cases} 1 & 35 \leq k \leq 55 \\ 0 & \text{otherwise} \end{cases}$$

Simulation results of the FD interval observer are depicted in Fig. 2 which illustrates the evolution of the residual signals. Under the proposed interval observer, the zero signal is excluded from the residual interval. Thus, the fault $f(k)$ can be detected.

The case of a small abrupt sensor fault is also considered and simulation results in Fig. 3 show that the small fault can be detected based on the TNL technique which is not the case when using the interval approach.

6 Conclusion

In this chapter, a new LMIs formulation has been presented to design a robust observer-based FD scheme for discrete-time switched systems with sensor faults. A new approach to construct residual framers has been considered. It is is based on a novel interval observer structure to relax the cooperativity constraints. The proposed technique offers more degrees of design freedom by integrating weighted matrices in the structure of the fault detection observer design. In addition, an L_∞ performance is introduced to improve the accuracy of fault detection. Cooperativity and stability conditions are expressed in terms of LMIs based on MQLF with an ADT control

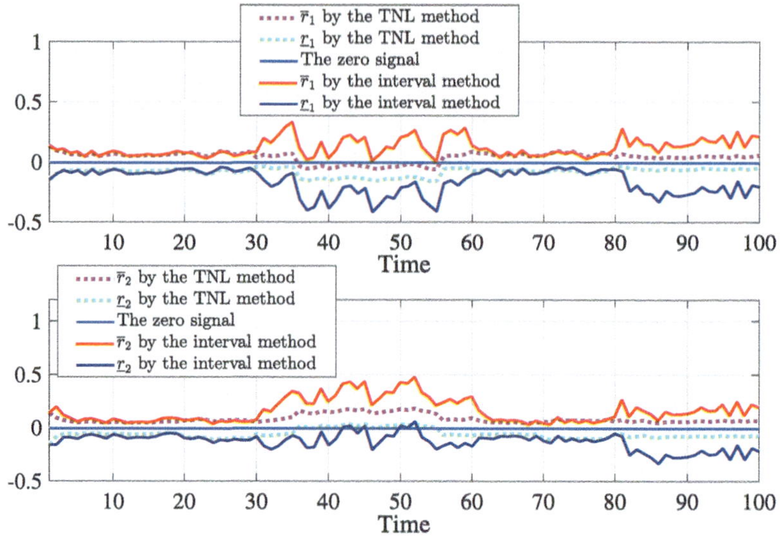

Fig. 3 Fault detection performance comparison between the TNL method and the interval approach (Small fault)

condition. Finally, the fault detection decision is based on determining whether the zero signal is excluded from the generated residual intervals when the faults occur. The proposed technique avoid the design of residual evaluation functions and threshold generators.

References

1. Zolghadri, A.: Advanced model-based fdir techniques for aerospace systems: today challenges and opportunities. Progress Aerosp. Sci. **53**, 18–29 (2012)
2. Chadli, M., Abdo, A., Ding, Steven X.: H_-/H_∞ fault detection filter design for discrete-time Takagi Sugeno fuzzy system. Automatica **49**, 1996–2005 (2013)
3. Zolghadri, A., Leberre, H., Goupil, P., et al.: Parametric approach to fault detection in aircraft control surfaces. J. Aircraft **53**, 846–855 (2016)
4. Tang, W., Wang, Z., Shen, Y.: Fault detection and isolation for discrete-time descriptor systems based on H_- / L_∞ observer and zonotopic residual evaluation. Int. J. Control **93**, 1867–1878 (2020)
5. Liberzon, D.: Switching in Systems and Control. Springer Science & Business Media (2003)
6. Zhang, Z., Yang, G.: Fault detection for discrete-time LPV systems using interval observers. Int. J. Syst. Sci. **14**, 2921–2935 (2017)
7. Su, Q., Fan, Z., Lu, T., et al.: Fault detection for switched systems with all modes unstable based on interval observer. Inf. Sci. **517**, 167–182 (2020)
8. Tian, Y., Zhang, K., Jiang, B., et al.: Interval observer and unknown input observer-based sensor fault estimation for high-speed railway traction motor. J. Franklin Inst. **357**, 1137–1154 (2020)

9. Chambon, E., Burlion, L., Apkarian, P.: Overview of linear time-invariant interval observer design: towards a non-smooth optimisation-based approach. IET Control Theory Appl. **10**, 1258–1268 (2016)
10. Belkhiat, D.E.C., Messai, N., Manamanni, N.: Design of a robust fault detection based observer for linear switched systems with external disturbances. Nonlinear Anal. Hybrid Syst. **5**, 206–219 (2011)
11. Zhai, D., Lu, A., Li, J., et al.: Simultaneous fault detection and control for switched linear systems with mode-dependent average dwell-time. Appl. Math. Comput. **273**, 767–792 (2016)
12. Wang, Z., Rodrigues, M., Theilliol, D., et al.: Fault estimation filter design for discrete-time descriptor systems. IET Control Theory Appl. **9**, 1587–1594 (2015)
13. Efimov, D., Raïssi, T.: Design of interval observers for uncertain dynamical systems. Automat. Remote Control **77**, 191–225 (2016)
14. Zhu, K., Song, Y., Ding, D., et al.: Robust MPC under event-triggered mechanism and Round-Robin protocol: an average dwell-time approach. Inf. Sci. **457**, 126–140 (2018)

A Regularized Inverse Problem Approach for Robust Condition Monitoring in Industrial Systems

Doniel Jiménez Sánchez, Marcos Quiñones-Grueiro, Antônio J. Silva Neto, and Orestes Llanes-Santiago

Abstract Condition monitoring is very important in modern industry in order to increase the safety of industrial plants and the economic benefits. Schemes based on model inversion or system inversion represent an important branch of the available solutions in model based condition monitoring. These techniques allow the development of detection, isolation and successful estimation of the fault magnitude. However, most of the proposed methods do not consider the noise present in industrial control systems which significantly affects the performance of the condition monitoring systems. They do not consider either the occurrence of multiple faults. In this paper, a proposal for robust condition monitoring, formulated as the solution of a regularized inverse problem in discrete linear time invariant systems is presented. Single and multiple faults are reconstructed by using the vector of residuals in the presence of noise. Tikhonov regularization is used to obtain a stable solution when noise in the measurements is considered. The proposed approach is applied to a case study with satisfactory results.

Keywords Robust condition monitoring · Inverse problem · Regularization · Fault estimation · Multiple faults

D. Jiménez Sánchez · O. Llanes-Santiago (✉)
Universidad Tecnológica de La Habana José Antonio Echeverría, CUJAE, Marianao, La Habana, Cuba
e-mail: orestes@tesla.cujae.edu.cu

M. Quiñones-Grueiro
Institute for Software Integrated Systems, Vanderbilt University, Nashville, Tennessee, USA
e-mail: marcos.quinones.grueiro@vanderbit.edu

A. J. Silva Neto
Instituto Politécnico-Universidade do Estado do Rio de Janeiro, Nova Friburgo, Rio de Janeiro, Brazil
e-mail: ajsneto@iprj.uerj.br

P. Shi et al. (eds.), *Complex Systems: Spanning Control and Computational Cybernetics: Applications*, Studies in Systems, Decision and Control 415,
https://doi.org/10.1007/978-3-031-00978-5_8

1 Introduction

A fault is a non permitted deviation of at least one characteristic, property or parameter of a system, from its normal or acceptable operating condition [29]. Faults can cause economic loss, as well as damage to human capital. equipment properties and the environment. For this reason, there is a great interest in searching new or improved methods for detection, isolation and estimation of faults in industrial systems [19].

In general, fault diagnosis methods are classified into two large groups: model based fault diagnosis and historical data or data driven based fault diagnosis [34–36].

Model-based methods, can be developed based on state observers or those based on parity spaces relationships [8, 16, 23, 27]. Most of these methods allow for the detection and isolation of single faults, and do not allow to estimate the fault magnitude [2, 6, 8]. Recently, in [39] multiple faults are estimated in presence of a disturbance but the proposed method is very complex. Parameter identification is another model-based technique which allows for estimation of the magnitude of the faults, but requires a prior knowledge or assumptions regarding the type of faults to be diagnosed [8, 20, 30].

An important group of strategies developed for fault diagnosis are schemes based on model inversion or system inversion [3–5, 10, 22, 32]. These techniques allow a successful detection, isolation and estimation of the fault magnitude without requiring any additional assumptions [17, 18, 33]. However, a very limited group among these strategies consider the noise present in most industrial control systems. Moreover, when noise is considered, the performance of the condition monitoring system decreases significantly. Such drawbacks indicate the need to develop new methods for fault detection and isolation (FDI) which are robust in the presence of noise and disturbances [2].

In this paper, a diagnosis scheme for additive faults in industrial control systems is proposed based on a vector of residues obtained from the difference between the process measured variables and the outputs calculated from a mathematical model of the process. The proposal is formulated as an inverse problem [5, 37]. The characteristics of the involved operator causes the appearance of an ill-posed problem which does not guarantee that for small variations in the data, which is usual in the presence of noise measurements, a satisfactory solution can be obtained [21]. A regularization technique is thus used to obtain a stable solution in the presence of noise in the measurements [25, 38]. This is the main contribution of the paper. The proposal allows the detection, location and estimation of multiple faults which is another contribution of the paper.

The structure of the paper is the following: in Sect. 2 the theoretical background of the formulation of the conditioning monitoring as an inverse problem with regularization is presented. In Sect. 3, the proposed strategy to design a fault diagnosis system is applied in a case study, and the results are discussed. Finally, the conclusions are presented.

2 Mathematical Foundation of the Robust Condition Monitoring

In this section, the conditioning monitoring strategy by using an inverse problem with regularization approach is presented.

2.1 Theoretical Formulation

Let Eq. (1) be the state space model for a discrete-time linear, time-invariant system

$$\begin{aligned} x(k+1) &= Ax(k) + Bu(k); \quad x(t_0) = x_0 \\ y(k) &= Cx(k) + Du(k) \end{aligned} \tag{1}$$

where $x(k) \in \mathbb{R}^n$ represents the state vector, $u(k) \in \mathbb{R}^p$ is the input vector, $y(k) \in \mathbb{R}^m$ is the measured output vector and x_0 represents the initial condition of the system, i.e. at time $t = t_0$. The matrices $A \in \mathbb{R}^{n\times n}$, $B \in \mathbb{R}^{n\times p}$, $C \in \mathbb{R}^{m\times n}$ and $D \in \mathbb{R}^{m\times p}$ guarantee the consistency of the equation system (1).

Considering that faults can occur in the actuators and in the sensor devices (additive faults), they represent additional inputs to the state and output equations. Then, the model of the system can be expressed as Eq. (2) [8, 19, 30]

$$\begin{aligned} x_1(k+1) &= Ax_1(k) + B[u_1(k) + f_a(k)]; \quad x_1(t_0) = x_{10} \\ y_1(k) &= Cx_1(k) + D[u_1(k) + f_a(k)] + f_s(k) \end{aligned} \tag{2}$$

where $f_a(k) \in \mathbb{R}^p$ is the vector representing the faults in the actuators, and $f_s(k) \in \mathbb{R}^m$ is the vector representing the faults which affect directly the measurements of the process variables, i.e. in the sensors.

If the fault vector is defined as:

$$f(k) = \begin{bmatrix} f_a(k) \\ f_s(k) \end{bmatrix} \in \mathbb{R}^{p+m} \tag{3}$$

and matrices E and F are constructed such that $E = \begin{bmatrix} B & \mathbf{0} \end{bmatrix}$ and $F = \begin{bmatrix} D & I \end{bmatrix}$ where $\mathbf{0} \in \mathbb{R}^{n\times m}$ is the null matrix and $I \in \mathbb{R}^{m\times m}$ is the identity matrix, then, the equation system (2) can be expressed as:

$$\begin{aligned} x_1(k+1) &= Ax_1(k) + Bu_1(k) + Ef(k); \quad x_1(t_0) = x_{10} \\ y_1(k) &= Cx_1(k) + Du_1(k) + Ff(k) \end{aligned} \tag{4}$$

Lets consider that the input vector u is the same for the systems (1) and (4), i.e. $u(k) = u_1(k)$, and both systems have the same initial condition x_0, i.e. $x_0 = X_{10}$. Then, subtracting (1) from (4), the following system (5) is obtained,

$$\begin{aligned} \Delta x(k+1) &= A\Delta x(k) + Ef(k); \quad \Delta x(t_0) = 0 \\ \Delta y(k) &= C\Delta x(k) + Ff(k) \end{aligned} \tag{5}$$

where $\Delta x(k+1) = x_1(k+1) - x(k+1)$ and $\Delta y(k) = y_1(k) - y(k)$.

The equation system (5) defines the dynamics of a system where the input is the fault vector $f(k)$ and the output is the vector of residues $\Delta y(k)$. Therefore, the process to estimate the faults from the vector of residues is an inverse problem. In order to obtain an explicit formulation for this inverse problem, it is necessary to find the operator that transforms the fault space into the space of residues. With this objective, the following assumptions about the system (5) are established:

(a) The system is observable, which guarantees that the outputs are sensitive to changes in the state.
(b) The system is input observable. This implies the possibility of reconstructing the fault vector from the residues [15].
(c) The system does not have invariant zeros. The existence of invariant zeros in Eq. (5) results in non-null inputs or states that produce null output. This indicates that operator injectivity defined as the transfer matrix of the system (5) is not guaranteed.
(d) $rank(C) = m \leq n$.
(e) $s \leq m$ faults can occur simultaneously. Then, $rank(E) = s \leq p \leq m$ if s faults occur in the actuators or $rank(F) = s$ if s faults occur in the sensor devices.
(f) There are $l \geq n$ available observations of the outputs, prior to the current instant.
(g) The eigenvalues of matrix A are found inside or on the unit circle of the complex plane.

Proposition 1 *Let a dynamical system be defined by Eq. (5). The vector equation*

$$\Delta y_l(k) = H_{o,l}\Delta x(k-l) + H_{f,l} f_l(k) \tag{6}$$

where $H_{o,l}$ is the extended observability matrix of the system (5), describes the relationship between the input $f_l(k)$ and the output $\Delta y_l(k)$ depending on the state $\Delta x(k-l)$.

Proof Let $l > 0$ previous sampling instants to the instant k. From Eq. (5), it is obtained to l:

$$\Delta y(k-l) = C\Delta x(k-l) + Ff(k-l) \tag{7}$$

for the instant $l-1$

$$\Delta y(k-l+1) = C\Delta x(k-l+1) + Ff(k-l+1) \tag{8}$$

Introducing the first equation of (5) in Eq. (8):

$$\Delta y(k-l+1) = CA\Delta x(k-l) + CEf(k-t) + Ff(k-l+1) \tag{9}$$

Repeating this process l times, it is obtained:

$$\Delta y(k) = CA^l \Delta x(k-l) + CA^{l-1}Ef(k-l) + \cdots + CEf(k+1) + Ff(k) \tag{10}$$

The following notations will be introduced:

$$\Delta y_l(k) = \begin{bmatrix} \Delta y(k-l) \\ \Delta y(k-l+1) \\ \vdots \\ \Delta y(k) \end{bmatrix}, \quad f_l(k) = \begin{bmatrix} f(k-l) \\ f(k-l+1) \\ \vdots \\ f(k) \end{bmatrix},$$

$$H_{o,l} = \begin{bmatrix} C \\ CA \\ \vdots \\ CA^l \end{bmatrix} \text{ and } H_{f,l} = \begin{bmatrix} F & 0 & \cdots & 0 \\ CE & F & \cdots & 0 \\ \vdots & \vdots & \ddots & \vdots \\ CA^{l-1}E & CE & \cdots & F \end{bmatrix}$$

where $H_{o,l}$ is the extended observability matrix of the system and $H_{f,l}$ is the matrix for the inversion of the system (5).

Taking into account the above notation, the $l+1$ last equations can be written in compact form as:

$$\Delta y_l(k) = H_{o,l}\Delta x(k-l) + H_{f,l} f_l(k)$$

□

Proposition 2 *The number of rows of the matrix $H_{o,l} > n$ and $rank(H_{o,l}) = n$*

Proof Since the system (5) is observable, thus $rank\mathcal{O} = n$ where $\mathcal{O}$ is its observability matrix defined as

$$\mathcal{O} = \begin{bmatrix} C \\ CA \\ \vdots \\ CA^{n-1} \end{bmatrix}$$

From the proof of the Proposition 1

$$H_{o,l} = \begin{bmatrix} C \\ CA \\ \vdots \\ CA^l \end{bmatrix}$$

If $l \geq n$, it is easy to conclude that the number of rows of $H_{o,l}$ is greater than n and $rank H_{o,l} = n$. □

Proposition 3 *Let $V_l \in \mathbb{R}^{(l+1)m \times (l+1)m}$ the orthogonal projection matrix of the space $\mathbb{R}^{(l+1)m}$ in $Ker(H_{o,l}^T)$, the following equality is satisfied*

$$V_l \cdot \Delta y_l(k) = V_l \cdot H_{f,l} f_l(k). \tag{11}$$

Proof The Proposition 2 guarantees that there exist at least one vector $v \in \mathbb{R}^{(l+1)m}$ such that $v^T \cdot H_{o,l} = H_{o,l}^T \cdot v = 0$. The set of vectors that satisfy this condition form the kernel subspace of $H_{o,l}^T$ which is defined as

$$Ker(H_{o,l}^T) = \{v \in \mathbb{R}^{(l+1)m} : H_{o,l}^T \cdot v = 0\}$$

Since $l \geq n$, it is guaranteed that the orthogonal projection matrix of the space $\mathbb{R}^{(l+1)m}$ in $Ker(H_{o,l}^T)$ exist, and it will be identified as $V_l \in \mathbb{R}^{(l+1)m \times (l+1)m}$.

Let M a matrix whose columns form a basis for a given subspace S. It is known from the linear algebra that the projection matrix P in subspace S is obtained as $P = M \cdot M^+$ where M^+ represents the pseudoinverse of M [31].

From the above, to obtain the orthogonal projection matrix V_l, it is necessary to obtain a matrix M whose columns form a basis for $Ker(H_{o,l}^T)$. Matrix M can be obtained by applying the Singular Value Decomposition Algorithm [11].

By multiplying the Eq. (6) by V_l, it is obtained:

$$V_l \cdot \Delta y_l(k) = \underbrace{V_l \cdot H_{o,l}}_{0} \Delta x(k-l) + V_l \cdot H_{f,l} f_l(k)$$

then

$$V_l \cdot \Delta y_l(k) = V_l \cdot H_{f,l} f_l(k)$$

□

Since the $H_{o,l}$ matrix represents the extended observability matrix of the system (5) and the rows of V_l are orthogonal to $Im(H_{o,l}) = \{v \in \mathbb{R}^{(l+1)m} : v = H_{o,l} \cdot u, u \in \mathbb{R}^n\}$, Eq. (11) explains that the orthogonal projection component of the output residuals $\Delta y_l(k)$ in $Ker(H_{o,l}^T)$ matches with the orthogonal projection component of the fault image for $H_{f,l}$, in $Ker(H_{o,l}^T)$.

Equation (11) is an explicit formulation of the inverse problem that consists in determining $f_l(k), k = 1, 2, \ldots; l \geq n$ from $\Delta y_l(k), k = 1, 2, \ldots; l \geq n$. The $V_l H_{f,l}$ matrix is the operator that transforms $f_l(k), k = 1, 2, \ldots; l \geq n$ in the orthogonal projection of $\Delta y_l(k), k = 1, 2, \ldots; l \geq n$ in $Ker(H_{o,l}^T)$.

2.2 Fault Reconstruction in Actuators

Assume that system (5) is only affected by the fault occurrences in the process actuators such that $rank(E) = s \leq p \leq m$. Under these conditions:

$$f(k) = \begin{bmatrix} f_a(k) \\ 0_{m\times 1} \end{bmatrix},$$

$$E = \begin{bmatrix} \bar{B} \; 0_{n\times m} \end{bmatrix}, \; F = \begin{bmatrix} \bar{D} \; 0_{m\times m} \end{bmatrix},$$

$$H_{f,l} = H_{fa,l} = \begin{bmatrix} \bar{D} & 0 & \cdots & 0 \\ C\bar{B} & \bar{D} & \cdots & 0 \\ \vdots & \vdots & \ddots & \vdots \\ CA^{l-1}\bar{B} & C\bar{B} & \cdots & \bar{D} \end{bmatrix}.$$

$\bar{B}$ and $\bar{D}$ are formed by those columns of B and D, respectively, which correspond to the actuators that fail.

From Eq.(11), it is obtained:

$$V_l \Delta y_l(k) = V_l H_{fa,l} f_{a,l}(k) \tag{12}$$

Since V_l is the orthogonal projection matrix of $\mathbb{R}^{(l+1)m}$ in $Ker(H_{o,l}^T)$, then:

$$\begin{bmatrix} f_a(k-l) \\ f_a(k-l+1) \\ \vdots \\ f_a(k) \end{bmatrix} = (V_l H_{fa,l})^+ V_l \begin{bmatrix} \Delta y(k-l) \\ \Delta y(k-l+1) \\ \vdots \\ \Delta y(k) \end{bmatrix} \tag{13}$$

where $(V_l H_{fa,l})^+$ is the pseudoinverse matrix of $V_l H_{fa,l}$.

If $(l + 1)$ solutions of Eq. (13) for f_a are calculated in the time interval from $k = l$ to $k = l_1$, then the vectorial succession (14) (for $k \geq l$):

$$\left\{ \begin{matrix} f_a(k-l) \\ f_a(k-l+1) \\ \vdots \\ f_a(k) \end{matrix} \right\}_{k=l}^{k=l_1} \tag{14}$$

constitutes, for each one of the $(l + 1)$ rows, an approximate solution of the inverse problem (12).

Remark In systems where one fault occurs in the actuator and there are invariant zeros, a linear combination of these rows may be suitable to compute the best solution.

2.3 Fault Reconstruction in Sensors

Assume that system (5) is only affected by the occurrence of faults in the sensors. In these conditions:

$$f(k) = \begin{bmatrix} 0_{p\times 1} \\ f_s(k) \end{bmatrix}, \ E = \begin{bmatrix} 0_{n\times(p+m)} \end{bmatrix}, \ F = \begin{bmatrix} 0_{m\times p} \ \bar{I} \end{bmatrix},$$

$$H_{f,l} = H_{fs,l} = \begin{bmatrix} I\bar{I} & 0 & \cdots & 0 \\ 0 & I^2\bar{I} & \cdots & 0 \\ \vdots & \vdots & \ddots & \vdots \\ 0 & 0 & \cdots & I^{l+1}\bar{I} \end{bmatrix}.$$

$\bar{I}$ contains the columns of I which correspond to the sensors that fail. Then:

$$V_l \Delta y_l(k) = V_l H_{fs,l} f_{s,l}(k) = \bar{V}_l f_{s,l}(k) \tag{15}$$

where $\bar{V}_l$ is formed from V_l taking the columns that correspond to the faulty sensors.

The above implies $f_s(k) = \Delta y(k)$. In this way, the solution of Eq. (15) is the projection of $\Delta y_l(k)$ in $Im(H_{o,l})$, according to the direction of the columns of $H^+_{fs,l}$:

$$\begin{bmatrix} f_s(k-l) \\ f_s(k-l+1) \\ \vdots \\ f_s(k) \end{bmatrix} = (I - V_l) H^+_{fs,l} \begin{bmatrix} \Delta y(k-l) \\ \Delta y(k-l+1) \\ \vdots \\ \Delta y(k) \end{bmatrix} \tag{16}$$

2.4 Reconstruction of Simultaneous Faults in the Actuator and Sensor

Assume that s faults in actuators and sensors occur simultaneously, so that $rank\left(\begin{bmatrix} E \\ F \end{bmatrix}\right) = s$. Then, in the Eq. (11), the matrix $H_{f,l}$ contains information about the faults in the actuator and the sensor. The solution of the Eq. (11) when there are not invariant zeros in the system is:

$$\begin{bmatrix} f(k-l) \\ f(k-l+1) \\ \vdots \\ f(k) \end{bmatrix} = (V_l H_{f,l})^+ V_l \begin{bmatrix} \Delta y(k-l) \\ \Delta y(k-l+1) \\ \vdots \\ \Delta y(k) \end{bmatrix} \tag{17}$$

If $(l+1)$ solutions of (17) are calculated for $f(k) = \begin{bmatrix} f_a(k) \\ f_s(k) \end{bmatrix}$ in the time interval from $k = l$ to $k = l_1$, then the vectorial succession:

$$\left\{ \begin{bmatrix} f(k-l) \\ f(k-l+1) \\ \vdots \\ f(k) \end{bmatrix} \right\}_{k=l}^{k=l_1}$$

is an approximate solution of the inverse problem (11) for each one of the $(l+1)$ rows.

2.5 *Tikhonov Regularization and Discrepancy Principle*

In practice, measurements of $\Delta y_l(k)$ are imprecise, contaminated with noise and influenced by disturbances that affect the process. Since the linear inverse problem (11) involves a compact operator, the problem is ill-posed [14]. The generalized inverse matrix can amplify the norm of the solution even for noise with small magnitude, thus causing an unstable solution of the inverse problem [1]. The Tikhonov regularization method is used to solve this difficulty. The solution obtained through this method is determined by maintaining the quadratic norm of the difference between the calculated solution and the measurements as small as possible, being stabilized by a penalty term. By applying this method, stable solutions are obtained for essentially ill-posed problems since the norm of the calculated solution and the norm of the difference between the solution and the measurements are controlled simultaneously [9, 21].

Let $\Delta y_l(k)^\delta$ be the measurement vector, where the superscript δ denotes the presence of inaccuracies. If $V_l \Delta y(k) \in Im(V_l H_{f,l})$, and the noise magnitude is bounded such that $\|\Delta y_l(k)^\delta - \Delta y(k)\| \leq \delta$, then, for some $\alpha > 0$, according to the Tikhonov's method, the solution of the problem (11) is rewritten as:

$$f_l(k) = \left[\alpha I + (V_l H_{f,l})^T V_l H_{f,l}\right]^{-1} (V_l H_{f,l})^T V_l \Delta y_l(k)^\delta \tag{18}$$

The discrepancy principle [24, 26] is an appropriate tool to select α in the Tikhonov's method. It establishes that the calculated solution cannot generate an error smaller than the noise present in the measurements, because otherwise the solution would be adjusted to the noise. With this principle α is obtained as the solution of the following equation:

$$\|V_l H_{f,l} f_l(k) - V_l \Delta y_l(k)^{\delta}\| = \delta \tag{19}$$

If $\|V_l \Delta y_l(k)^{\delta}\| \geq \delta$, then, Eq. (19) has unique solution with respect to α [28].

3 Experiments and Results

3.1 A Case Study: CD Motor AMIRA DR300

The AMIRA DR300 speed control system is a benchmark usually employed to test control and fault diagnosis strategies because of its similarity with industrial speed control systems.

The system consists of a permanent magnet coupled to a direct current generator. The main function of this generator is to simulate the effect of a fault, that results when a load torque is applied to the motor shaft. The speed measurement is acquired with a tachogenerator whose signal feeds a Proportional Integral (PI) controller for speed control. As usual in applications where speed control accuracy is required, the AMIRA DR300 system includes an internal armature current control loop. A complete description of this benchmark system can be found in [8].

In the present study, the internal current control loop, the CD motor and the tachogenerator are considered as a single block which constitutes the process to be controlled. In Fig. 1, it can be observed that the resulting closed-loop block diagram is composed by the process and the speed PI controller. It is also included in the diagram, the effect of applying a load torque to the motor shaft. The dynamic of the open loop control system is described by the Eq. (20) in the frequency domain.

$$U_T(s) = G_{yu}(s)U_C(s) + G_d(s)M_L(s) \tag{20}$$

$$G_{yu}(s) = \frac{8.75}{(1 + 1.225s)(1 + 0.03s)(1 + 0.005s)} \tag{21}$$

$$G_d(s) = \frac{-31.07}{s(1 + 0.005s)} \tag{22}$$

where U_T is the controlled variable, U_C is the control signal, M_L is the load torque applied to the motor shaft and G_d its respective transfer function.

The transfer function of the speed PI controller is:

$$G_C(s) = \frac{U_C(s)}{U_{ref} - U_T(s)} = 1.96 + \frac{1.6}{s} \tag{23}$$

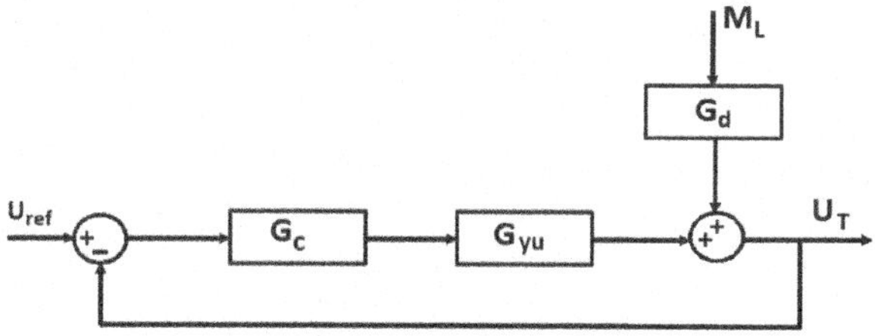

Fig. 1 Block diagram of the DC motor closed loop

The following is a discrete model that represents the open loop system, with a sampling period of 0.1 s, in which there are no load disturbances acting on the process ($M_L(s) = 0$):

$$
\begin{aligned}
x(k+1) &= \begin{bmatrix} -0.007853 & -1.418 & -0.5382 \\ 0.0001076 & 0.01529 & -0.7399 \\ 0.000148 & 0.03192 & 0.9476 \end{bmatrix} x(k) \\
&+ \begin{bmatrix} 0.0001076 \\ 0.000148 \\ 1.049 \cdot 10^{-5} \end{bmatrix} U_C(k) \\
U_T(k) &= \begin{bmatrix} 0 & 0 & 4.375 \cdot 10^4 \end{bmatrix} x(k).
\end{aligned}
\tag{24}
$$

3.2 *Fault Reconstruction in the Actuator Without Noise*

A fault in the actuator assumes that the system input (24) is $[U_c(k) + f_a(k)]$, where $f_a(k)$ is the fault in the actuator. The fault function described in [8] is a step type, and it is shown in Fig. 2. This function represents a deviation of 0.05 V in the motor armature voltage, which is the control signal.

The order of the system (24) is three, then, it is enough to take $l = 3$. The necessary matrices for the fault estimation in the actuator will be:

$$
H_{o,l} = \begin{bmatrix} C \\ CA \\ CA^2 \\ CA^3 \end{bmatrix} = \begin{bmatrix} 0 & 0 & 43750 \\ 6 & 1396 & 41457 \\ 6 & 1335 & 38248 \\ 6 & 1232 & 35253 \end{bmatrix} \tag{25}
$$

$$
V_l = \begin{bmatrix} 0.0003 & -0.0084 & 0.0134 & -0.0050 \\ -0.0084 & 0.2573 & -0.4096 & 0.1524 \\ 0.0134 & -0.4096 & 0.6522 & -3.2426 \\ -0.0050 & 0.1524 & -0.2426 & 0.0902 \end{bmatrix} \tag{26}
$$

$$V_l H_{fa,l} = \begin{bmatrix} 0 & 0 & 0 & 0 \\ 0 & 0.0033 & 0.0074 & 0 \\ -0.0001 & -0.1018 & -0.2291 & 0 \\ 0.0001 & 0.1066 & 0.2399 & 0 \end{bmatrix} \tag{27}$$

$$(V_l H_{fa,l})^+ = \begin{bmatrix} 7.75 \cdot 10^{-7} & 2 \cdot 10^4 & -2.4 \cdot 10^5 & -2.3 \cdot 10^5 \\ 4 \cdot 10^{-7} & 9 \cdot 10^4 & -6.2 \cdot 10^4 & -6.2 \cdot 10^4 \\ 1.8 \cdot 10^{-7} & -4 \cdot 10^4 & 2.7 \cdot 10^4 & 2.7 \cdot 10^4 \\ 0 & 0 & 0 & 0 \end{bmatrix} \tag{28}$$

The fault estimation in the actuator using Eq. (13), and acquired data without noise is shown in Fig. 3.

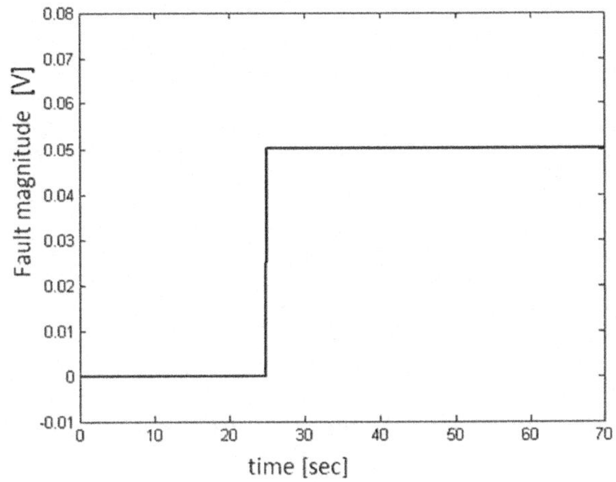

Fig. 2 Fault in the actuator

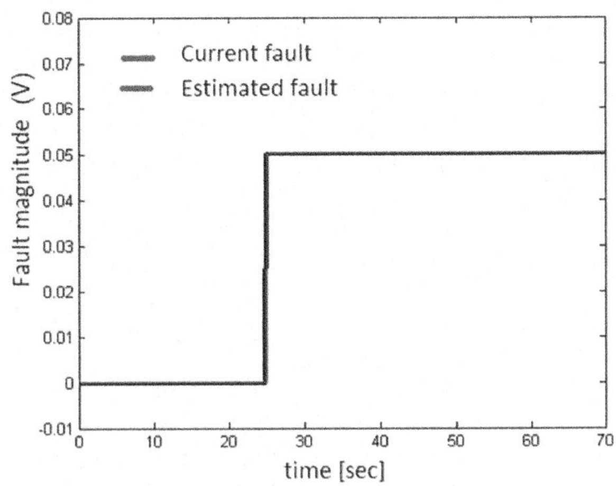

Fig. 3 Fault estimation in the actuator without noise

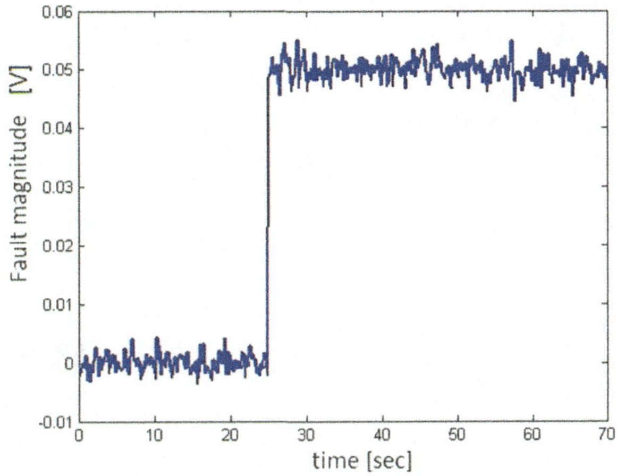

Fig. 4 Fault in the actuator in presence of noise

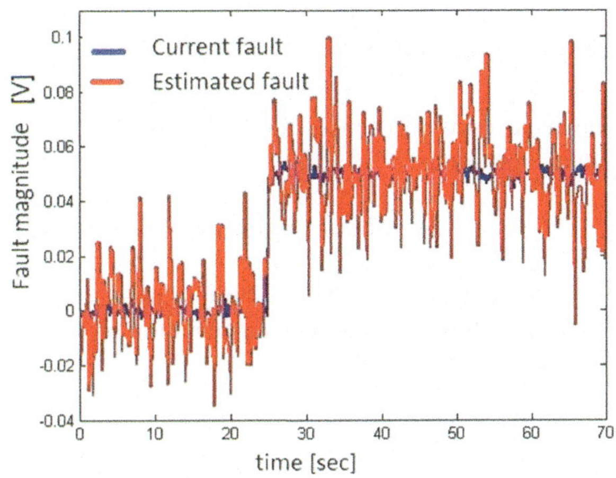

Fig. 5 Fault estimation in the actuator in presence of noise

Figure 4 shows the measured signal after adding a white noise with zero mean ($\mu = 0$) and standard deviation $\sigma = 0.04$ to represent a noise that affects the measurement.

The fault estimation using equation (13), and acquired data with noise is shown in Fig. 5.

The bad estimation of the fault in the actuator in the presence of noise is due to the ill-conditioning of the $V_l H_{fa,l}$ matrix (its condition number is very large).

Applying the Tikhonov regularization algorithm with $\alpha = 0.00059$ in Eq. (18), calculated considering the discrepancy principle given in Eq. (19), the fault estimation in the actuator is shown in Fig.6. This confirm the effectiveness of the diagnosis scheme proposed in this paper.

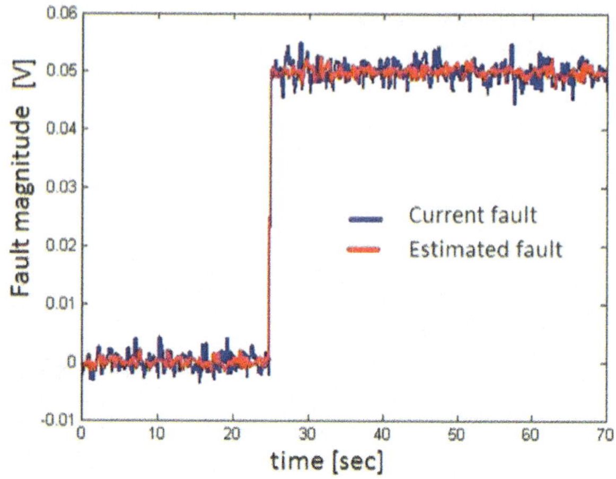

Fig. 6 Current and estimated fault in the actuator by using noisy data and the Tikhonov regularization algorithm

3.3 Fault Reconstruction in the Sensor

The occurrence of a sensor fault will now be considered. In this case, the system output equation is:

$$U_T(k) = \begin{bmatrix} 0 & 0 & 4.375 \cdot 10^4 \end{bmatrix} x(k) + f_s(k)$$

The fault type is similar to the one used in [8]. It is a step function shown in Fig. 7. This function represents a deviation of $-0.25\ V$ in the output voltage of the tachogenerator in the CD motor.

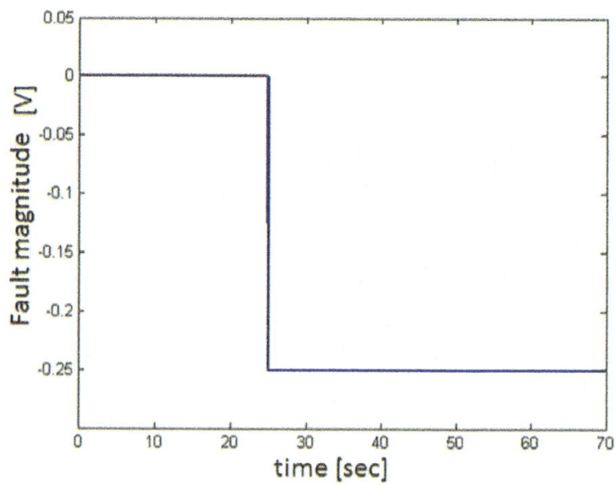

Fig. 7 Fault in the sensor without noise

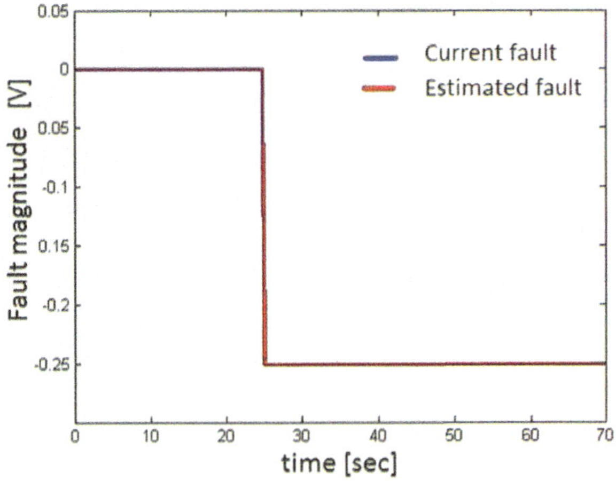

Fig. 8 Fault estimation in the sensor without noise

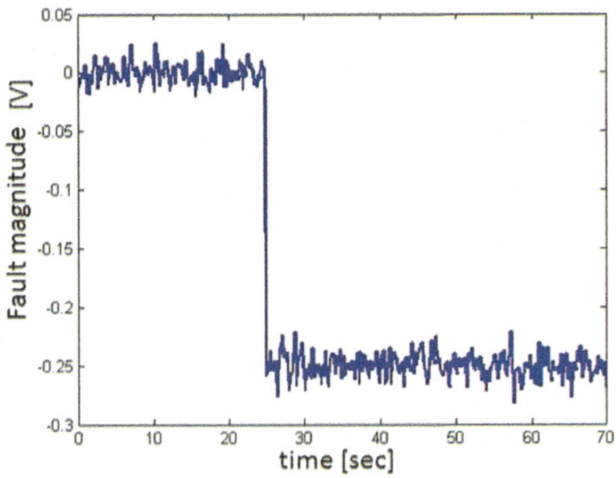

Fig. 9 Fault in the sensor in presence of noise

Taking $l = 3$, the matrix necessary to solve the inverse problem to estimate the fault that occurs in the sensor is

$$I - V_l = \begin{bmatrix} 1 & 0 & 0 & 0 \\ 0 & 0.9995 & 0.0154 & -0.0161 \\ 0 & 0.0154 & 0.5232 & 0.4992 \\ 0 & -0.0161 & 0.4992 & 0.4773 \end{bmatrix}$$

The actual fault function in the sensor and its estimate are shown in Fig. 8, being able to observe that are very similar.

Figure 9 shows the measured signal after adding a white noise with zero mean ($\mu = 0$) and standard deviation ($\sigma = 0.04$) to represent a noise that affects the measurements.

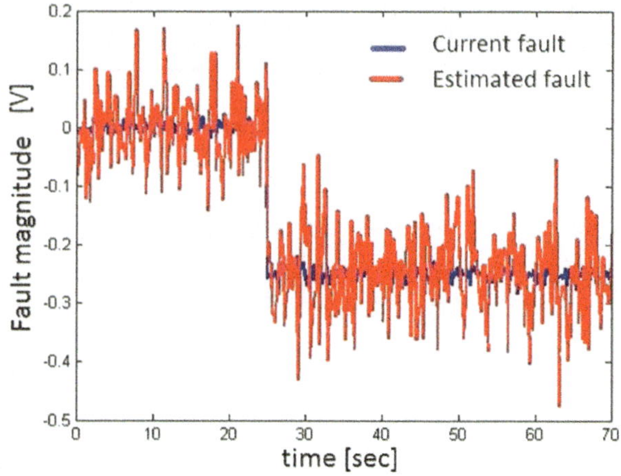

Fig. 10 Fault estimation in the sensor in presence of noise

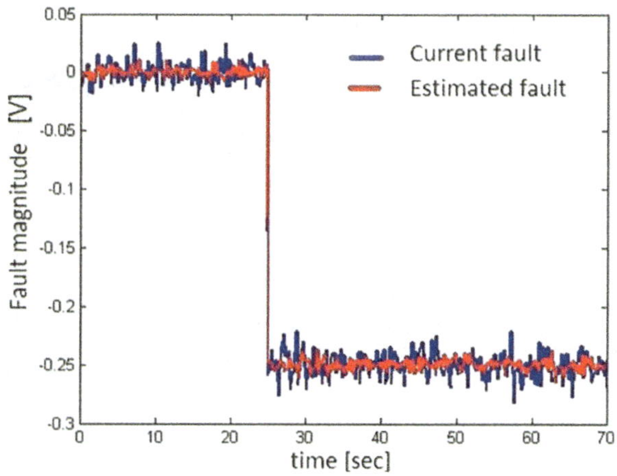

Fig. 11 Current and estimated fault in the sensor by using noisy data the Tikhonov regularization algorithm

The obtained result in the sensor fault estimation in the presence of noise is shown in Fig. 10.

The bad estimation of the sensor fault in the presence of noise is due to the ill-conditioning of the $I - V_l$ matrix (its condition number is large)

The application of Tikhonov regularization algorithm with $\alpha = 0.0098$ calculated by using the discrepancy principle, improves the sensor fault estimation as is shown in Fig. 11.

3.4 Fault Reconstruction in the Sensor and the Actuator

Now, the occurrence of a fault in the actuator and the sensor simultaneously is considered. These faults are represented by the same functions used previously and shown in Figs. 2 and 7. To guarantee the isolation of the two faults and that the system does not have invariant zeros, all the state variables need to be known [8]. Using the output of the system (the measurement of the angular velocity of the motor shaft), a Kalman filter (KF) is implemented to estimate the other two state variables of the system (angular position and armature current).

The estimation of the faults in the actuator and the sensor when the measurements are not affected by noise are shown in Fig. 12.

The fault estimation in the actuator and the sensor in the presence of white noise with mean ($\mu = 0$) and standard deviation ($\sigma = 0.04$) are shown in Fig. 13. The non-satisfactory fault estimation in the presence of noise is a result of the ill-conditioning of the matrix $V_l H_{f,l}$ (it condition number is very large).

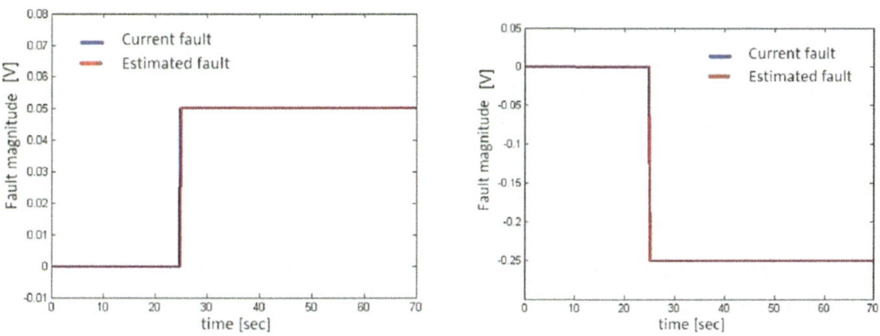

(a) Actuator (b) Sensor

Fig. 12 Fault estimation in the actuator **a** and in the sensor **b** without the presence of noise

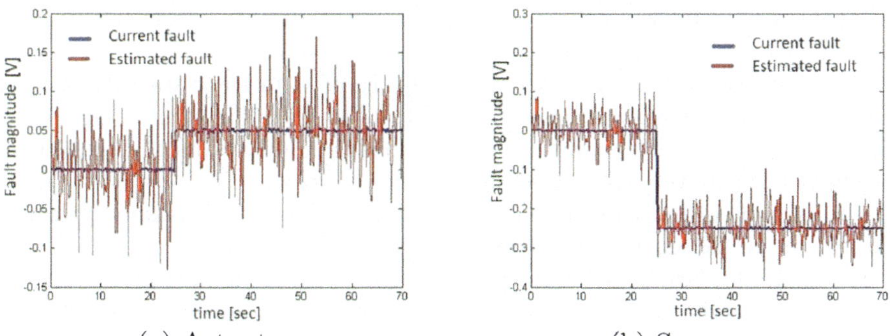

(a) Actuator (b) Sensor

Fig. 13 Fault estimation in the actuator **a** and the sensor **b** in presence of noise

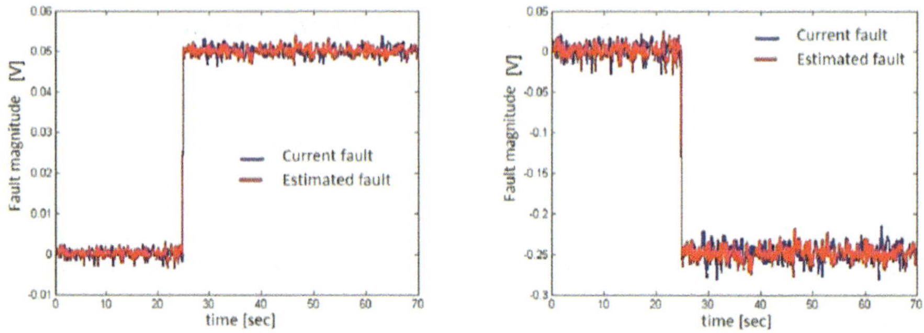

(a) Actuator (b) Sensor

Fig. 14 Fault estimation in the actuator **a** and the sensor **b** in presence of noise and by using the regularization algorithm

The regularization algorithm of Tikhonov is applied with $\alpha = 0.0001937$ obtained by the discrepancy principle. The result is shown in Fig. 14.

As it shown in Fig. 14, the explicit scheme proposed using Tikhonov regularization, allows for the successful solving of the inverse problem corresponding to the estimation of the simultaneous faults that occur in the actuator and in the sensor simultaneously in presence of noise, which constitutes one of the contributions of this paper.

3.5 Comparison with Other Known Fault Diagnosis Approaches

The parity equation method for fault detection and isolation was presented in [13]. It was one of the first techniques used in fault detection and isolation. This method is the direct implementation of the concept of analytical redundancy [7]. To achieve reasonable results in the presence of noise, the use of filters has been analyzed, but disadvantages are recognized due to computational complexities [12].

The objective of the state observer-based scheme is to reconstruct the state of the system from the measurements of the output [19, 29]. The disadvantage of the methods based on state observers is that they do not allow for estimation of the fault magnitude, and are very sensitive to the presence of noise.

The example of the actuator fault analyzed in Subsect. 3.2 is used to appreciate the behavior of the schemes based on parity spaces and state observers.

In Fig. 15, the results of the detection and isolation of the fault in the actuator are presented for the parity space and state observer schemes, respectively, when the measurements are not affected by noise.

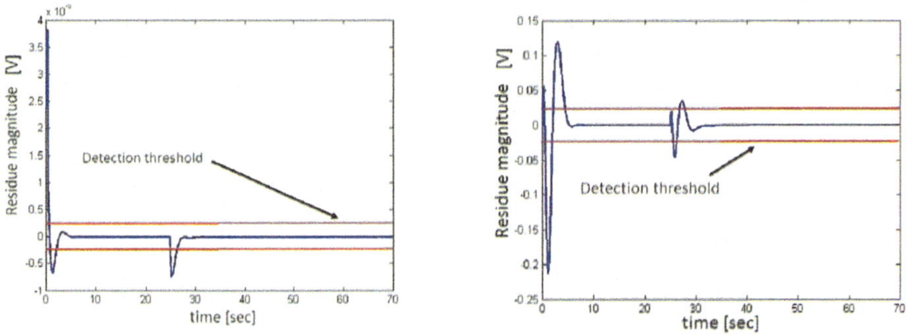

(a) Parity Space (b) State Observer

Fig. 15 Fault detection and isolation in the actuator based on parity space and state observer schemes without presence of noise

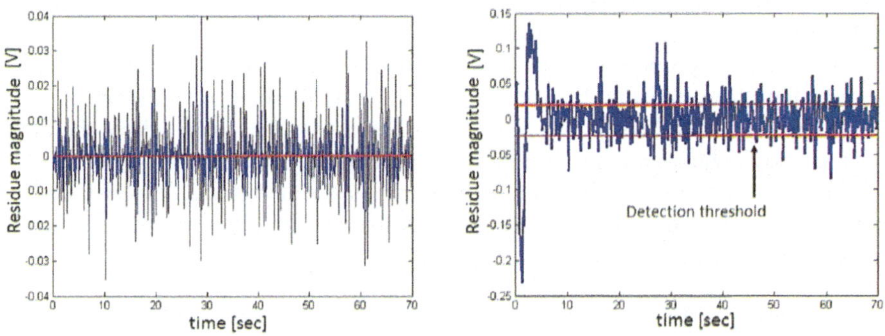

(a) Parity space (b) State Observer

Fig. 16 Fault detection and isolation in the actuator based on parity space and state observer schemes in the presence of noise

Figure 15 shows that the faults are detectable and isolated, but it is not possible to estimate their magnitude. The decision threshold to detect the fault using the residue signal can be small in both cases without implying that false alarms occur.

Figure 16 shows the results of the detection and isolation of the actuator fault using also the parity space and the state observer strategies, but with the presence of noise in the measurements.

In both schemes, it is not possible to detect or isolate the fault in the presence of noise by using the decision threshold selected in the previous case because the diagnostic system would always be giving a false alarm. If the decision threshold range is extended, then faults would not be detected without noise presence in the measurements.

An advantage of parity and observer space schemes is that they are easy to design, while the proposal in this paper needs additional tools for calculating the α regularization parameter. Nonetheless, the previous strategies fail for the same benchmark test problem even for relatively low level of noise in the data.

4 Conclusions

In this paper, a scheme based on model inversion for the diagnosis of additive faults in discrete linear systems has been presented. The main contribution of the proposal is the possibility of estimating the magnitude of the faults in addition to detecting and isolating them in the presence of noise which is an advantage over other based model methods present in the scientific literature. The diagnosis task is solved as an inverse problem with regularization, and this allow to reconstruct single and multiple faults, which constitute another contribution of the paper. The mathematical foundation of the proposed robust condition monitoring system constitutes itself a methodology for its design. In order to generate a stable solution in the presence of noise measurements, thus to detect, isolate and estimate the magnitude of the fault, the Tikhonov regularization technique was successfully applied.

The effectiveness and robustness of the proposed scheme was demonstrated by using a case study. Its better performance compared with other model-based techniques was also demonstrated.

Acknowledgements The authors acknowledge the decisive support provided by CAPES- Foundation for the Coordination and Improvement of Higher Level Education Personnel, through the project "Computational Modelling for Applications in Engineering and Environment", Program for Institutional Internationalization CAPES PrInt 41/2017, Processo No. 88887.311757/2018-00. Our gratitude is also due to Ministry of Science, Technology and Environment of Cuba through the Project No. 27 of National Program of Research and Innovation ARIA of CITMA, to CNPq—National Council for Scientific and Technological Development , to FAPERJ —Foundation Carlos Chagas Filho for Research Support of the State of Rio de Janeiro, as well as to the Cuban Ministry of Higher Education (MES) and Universidad Tecnológica de La Habana José Antonio Echeverría, CUJAE.

References

1. Aster, R., Borchers, B., Thurber, C.: Parameters Estimation and Inverse Problems. Academic Press, USA (2013)
2. Blanke, M., Kinnaret, M., Lunze, J., Staroswiecki, M.: Diagnosis and Fault-Tolerant Control, 2nd edn. Springer, Berlin (2006)
3. Camps-Echevarría, L., Llanes-Santiago, O., Silva Neto, A.: An approach for fault diagnosis based on bio-inspired strategies. In: IEEE Congress on Evolutionary Computation. IEEE (2010). https://doi.org/10.1109/CEC.2010.5586357

4. Camps-Echeverría, L., Campos Velho, H., Silva Neto, A., Llanes-Santiago, O.: The fault diagnosis inverse problem with ant colony optimization and ant colony optimization with dispersion. Appl. Math. Comput. **227**, 687–700 (2014)
5. Camps Echeverría, L., Llanes-Santiago, O., de Campos Velho, H., Silva Neto, A.: Fault Diagnosis Inverse Problems: Solution with Metaheuristics, Studies in Computational Intelligence, vol. 763. Springer International Springer International Publishing (2019). https://doi.org/10.1007/978-3-319-89978-7
6. Che Mid, E., Dua, V.: Model-based parameter estimation for fault detection using multiparametric programming. Indus. Eng. Chem. Res. **56**(28), 8000–8015 (2017). https://doi.org/10.1021/acs.iecr.7b00722
7. Chow, E., Willsky, A.: Analytical redundancy and the design of robust failure detection systems. IEEE Trans. Autom. Control **29**, 603–614 (1984)
8. Ding, S.: Model-Based Fault Diagnosis Techniques, Algorithm. Design Schemes and Tools, Springer, London (2008)
9. Doicu, A., Trautmann, T., Schreier, F.: Numerical Regularization for Atmospheric Inverse Problems. Springer, Germany (2010)
10. Edelmayer, A., Bokor, J., Szabó, Z.: Inversion-based residual generation for robust detection and isolation of faults by means of estimation of the inverse dynamics in linear dynamical systems. Int.J. Control **82**(8), 1526–1538 (2009)
11. Ford, W.: Numerical Linear Algebra with Applications Using MATLAB, 1st edn. Academic Press (2014)
12. Gertler, J.: Fault detection and isolation using parity relations. Control Eng. Pract. **5**(5), 653–661 (1997)
13. Gertler, J., Singer, D.: A new structural framework for parity equation based failure detection and isolation. Automatica **26**(381–388) (1990)
14. Hansen, C.: Rank Deficient and Discrete Ill-posed Problems. Numerical Aspects of Linear Inversion, SIAM, USA (1998)
15. Hou, M., Patton, R.: Input observability and input reconstruction. Automatica **34**(6), 789–794 (1998)
16. Hwang, W., Huh, K.: Fault detection and estimation for electromechanical brake systems using parity space approach. J. Dyn. Syst. Measur. Control **137**(1), 014504 (2015). https://doi.org/10.1115/1.4028184
17. Ireland, M., Mackenzie, R., Flessa, T., Worrall, K., Thomson, D., McGookin, E.: Inverse simulation as a tool for fault detection & isolation in planetary rovers. In: 10th International ESA Conference on Guidance, Navigation & Control Systems. Austria, May 2017
18. Ireland, M., Worrall, K., Mackenzie, R., Flessa, T., McGookin, E., Thomson, D.: A comparison of inverse simulation-based fault detection in a simple robotic rover with a traditional model-based method. Int. J. Mech. Mech. Eng. **11**(3), 607–615 (2017)
19. Isermann, R.: Model based fault detection and diagnosis. Status and applications. Ann. Rev. Control **29**(1), 71– 85 (2005)
20. Isermann, R., Freyermuth, B.: Process fault diagnosis based on process model knowledge: Part I principles for fault diagnosis with parameter estimation. J. Dyn. Syst. Measur. Control **113**(4), 620–626 (1991). https://doi.org/10.1115/1.2896466
21. Kirsh, A.: An Introduction to the Mathematical Theory of Inverse Problems, 2nd edn. Springer, New York (2011)
22. Kulcsar, B., Verhaegen, M.: Robust inversion based fault estimation for discrete-time lPV systems. IEEE Trans. Autom. Control **57**(6), 1581–1586 (2012)
23. Madrigal, G., Astorga, C., Vásquez, M., Osorio, G., Adam, M.: Fault diagnosis in sensors of boiler following control of a thermal power plant. IEEE Latin America Trans. **16**(6) (2018)
24. Morozov, V.: On the solution of functional equation by the method of regularization. Soviet Math Dokl **7**, 414–417 (1966)
25. Moura-Neto, F., Silva-Neto, A.: An Introduction to Inverse Problems with Applications. Springer, New York (2013)

26. Otero, F., Eliçbe, G., Frontini, G.: Comparación de técnicas para el cálculo del parámetro de regularización aplicado al problema inverso de dispersión de luz usando un modelo aproximado. Mecánica Computacional **XXXIII**, 1995–2008 (2014)
27. Patan, K.: Artificial Neural Networks for the Modelling and Fault Diagnosis of Technical Processes. Springer, Berlin (2008)
28. Petrov, Y., Sizikov, V.: Well-posed, ill-posed and Intermediate Problems with Applications. Koninklijke Brill NV, Netherlands (2005)
29. Simani, S., Fantuzzi, C., Patton, R.: Model-Based Fault Diagnosis in Dynamics Systems using Identification Techniques. Springer, London (2002)
30. Simani, S., Patton, R.: Fault diagnosis of an industrial gas turbine prototype using a system identification approach. Control Eng. Pract. **16**(7), 769–786 (2008)
31. Strang, G.: Linear Algebra and Its Applications, 3rd edn. Brace, Jovanovich, Publishers, Harcourt (1988)
32. Szabó, Z., Edelmayer, A., Bokor, J.: Inversion based FDI for sampled LPV systems. In: Conference on Control and Fault-Tolerant Systems (SysTol), October 2010, pp. 82–89. Nice, France (2010)
33. Tang, D., Patton, R., Wang, X.: A relaxed solution to unknown input observers for state and fault estimation. IFAC Symp. Fault Detect. Supervis. Saf. Tech. Processes **48**(21), 1048–1053 (2015)
34. Venkatasubramanian, V., Rengaswamy, R., Kavuri, S.N.: A review of process fault detection and diagnosis, part 1: quantitative model-based methods. Comput. Chem. Eng. **27**, 293–311 (2003)
35. Venkatasubramanian, V., Rengaswamy, R., Kavuri, S.N.: A review of process fault detection and diagnosis, part 2: qualitative models and search strategies. Comput. Chem. Eng. **27**, 313–326 (2003)
36. Venkatasubramanian, V., Rengaswamy, R., Kavuri, S.N.: A review of process fault detection and diagnosis, part 3: process history based methods. Comput. Chem. Eng. **27**, 327–346 (2003)
37. Vera, C., Edelmayer, A., Bokor, J., Szabó, Z., Szigeti, F.: Input reconstruction by means of system inversion: a geometric approach to fault detection and isilation in nonlinear systems. Int. J. Appl. Math. Comput. Sci. **14**(2), 189–199 (2004)
38. Wang, Y., Yagola, A., Yang, C.: Optimization and Regularization for Computational Inverse Problems and Applications. Higher Education Press, Beijing (2010)
39. Zhirabok, A.N., Shumsky, A.E., Zuev, A.V.: Fault diagnosis in linear systems via sliding mode observers. Int. J. Control **0**(0), 1–9 (2019). https://doi.org/10.1080/00207179.2019.1590738

On Robustification Based on Continuous Integral Sliding Modes

Juan-Eduardo Velázquez-Velázquez, Rosalba Galván-Guerra, Leonid Fridman, and Rafael Iriarte

Abstract The fundamental principles for designing robustification schemes for conventional and switched linear time-invariant systems based on continuous integral sliding modes are presented. State and output approaches that guarantee the exact compensation of matched uncertainties/perturbations are explained in detail, and a constructive methodology is given to facilitate the practical design of the controller and the observer involved in the robustification schemes.

1 Introduction

Many control algorithms are available in the literature, and every day more are published. All these algorithms are developed for specific tasks and require the fulfillment of some assumptions. Nevertheless, what can be done if a control algorithm in a real-world application is used and the uncertainties/perturbations affect the desired behavior of the system?. This problem can be solved by changing the used control algorithm to a robust one, minimizing uncertainties/perturbations under some conditions. However, what if the system already has an expensive implemented controller?.

J.-E. Velázquez-Velázquez (✉) · R. Galván-Guerra
Instituto Politécnico Nacional (IPN), Unidad Profesional Interdisciplinaria de Ingeniería campus Hidalgo, San Agustín Tlaxiaca 42162, Hidalgo, Mexico
e-mail: jvelazquezv@ipn.mx

R. Galván-Guerra
e-mail: rgalvang@ipn.mx

L. Fridman · R. Iriarte
Universidad Nacional Autónoma de México (UNAM), Facultad de Ingeniería, 04510, Mexico City, Mexico
e-mail: lfridman@unam.mx

R. Iriarte
e-mail: ririarte@unam.mx

P. Shi et al. (eds.), *Complex Systems: Spanning Control and Computational Cybernetics: Applications*, Studies in Systems, Decision and Control 415,
https://doi.org/10.1007/978-3-031-00978-5_9

In this case, it will be preferable to add an extra control loop that eliminates the effects of the unwanted uncertainties/perturbations, guaranteeing the desired behavior of the system.

Unlike Lyapunov's redesign (see [2] and [19, Sect. 14.2]) which aims to robustify the system maintaining its stability in the presence of uncertainties/perturbations. The purpose of this chapter is to robustify an existing nominal trajectory.

It is well known that the sliding mode theory can make the system insensitive to matched uncertainties/perturbations. In specific for uncertain linear time-invariant systems (ULTIS), the integral sliding modes (ISM) [31] make the system insensitive to matched uncertainties/perturbations right after the initial time while preserves the nominal behavior of the system. However, it uses first-order sliding mode control laws that are discontinuous and generate a high chattering level in systems with fast actuators [24]. The intrinsic characteristics of the ISM have been used to formulate robustification schemes [14, 26]. Using a stabilizing controller that guarantees a desired behavior of the system in the absence of uncertainties/perturbations, the ISM control law eliminates the effects of the matched uncertainties/perturbations while guarantees right after the initial time the system's behavior.

The high level of chattering makes the ISM unsuitable to be applied to systems where high-frequency signals may harm the actuators. The ISM approach has been improved to overcome this disadvantage, with the use of the super-twisting algorithm (STA) generating the continuous ISM (CISM) [8, 25]. This algorithm also preserves the nominal behavior of the system when it is affected by Lipschitz uncertainties/perturbations and diminishes the chattering effect [23]. Nevertheless, only finite-time convergence to the origin can be established, unless the system is not affected by the uncertainties/perturbations at the initial time or the initial conditions of the uncertainties/perturbations are known. However, it is possible to design robustification schemes that generate continuous control signals and achieved the system's desired behavior in finite time. It is worth mentioning that the STA has been combined with the Lyapunov redesign strategy [11] to compensate Lipschitz uncertainties/perturbations, guaranteeing the robustification of the system.

Both mentioned strategies (ISM and CISM) guarantee the exact compensation of the matched uncertainties/perturbations but require complete knowledge of the state vector. Output-based ISM and CISM strategies have been developed (OISM and COISM respectively) [5, 13, 15, 16] to eliminate this requirement. These strategies use an observer to reconstruct the unknown states even in the presence of uncertainties/perturbations. The OISM strategy reconstructs the state theoretically exactly right after the initial time, and compensates the matched uncertainties/perturbations in the same manner by using only output information. However, it requires filters to reconstruct the state, and in practice, the reconstruction error depends on the sample step. With the use of the COISM algorithm, the states are reconstructed theoretically exactly in finite time and eliminate the use of the filters, and the uncertainties/perturbations are also compensated theoretically exactly in finite time. Once more, the use of a continuous sliding mode strategy affects the converge properties of the COISM approach. Nevertheless, the high level of chattering is diminished, and the states are reconstructed without using filters (see [16] for more details).

The switched ISM (SISM) [15, 22] allows compensating the matched uncertainties/perturbations right after every switching for switched uncertain linear time-invariant systems (SULTIS). However, as its non-switched counterpart, it requires complete knowledge of the state vector and generates high-level chattering. Therefore, a continuous SISM (SCISM) has been proposed in [13] to generate continuous control signals and diminish the chattering, where the controller gains are designed to guarantee the convergence before every switching, and a switched gain strategy is used to diminish the chattering. Furthermore, under continuity conditions, it is possible to guarantee convergence in finite time. The output versions of those algorithms, namely the Switched OISM (SOISM) and the Switched COISM (SCOISM), are given in [13, 16]. In these algorithms, the states are reconstructed by using switched hierarchical observers. In the SOISM, the used observer is the same as in the OISM, allowing the reconstruction of the states theoretically exactly right after every switching, but uses filters. In the SCOISM, as in the COISM, the states are reconstructed theoretically exactly in finite time and eliminate the filters. However, the observer must converge before the controller can be turned on. Hence, the observer and the controller's gains must be designed to achieve the desired convergence time, and a switched gain strategy is needed to diminish the chattering.

This chapter provides an uncomplicated design procedure of robustification strategies for ULTIS and SULTIS that allows straightforward implementation of the continuous integral sliding mode approaches.

- For ULTIS two robustification scenarios are considered

  Scenario 1 *State based.* In this scenario, a CISM algorithm is used, the continuous control signals use the complete state vector, and the matched uncertainties/perturbations are compensated in finite time.

  Scenario 2 *Output based.* The COISM algorithm is employed, the states are reconstructed in finite time by using a continuous hierarchical observer. The matched uncertainties/perturbations are compensated in finite time.

- Also for SULTIS consider two scenarios

  Scenario 1 *State based.* In this scenario, the control signals depend on the state vector and are designed by the SCISM algorithm. The matched uncertainties/perturbations are compensated before every switching.

  Scenario 2 *Output based.* Following the SCOISM algorithm, the states are reconstructed using a hierarchical observer that converges before the controller is turned on. The matched uncertainties/perturbations are compensated before every switching. Both the observer and the controller re-converge after every switching unless the uncertainties/perturbations are continuous at the switching time.

Notation

Along this chapter the following notation is used.

- I_n denotes the identity matrix in $\mathbb{R}^{n\times n}$.
- X^+ is the Moore-Penrose pseudoinverse of the matrix X.
- $\lfloor r \rceil^q = |r|^q \text{sign}(r)$, $r \in \mathbb{R}$.
- $\lfloor w \rceil_n^q = \begin{bmatrix} \lfloor w_1 \rceil^q \\ \vdots \\ \lfloor w_n \rceil^q \end{bmatrix}$, with w_i the i-th element of the vector $w \in \mathbb{R}^n$.
- $X^\perp$ is the orthogonal complement of X, such that $X^T X^\perp = 0$.

2 Uncertain Linear Time Invariant Systems

Consider an application to be robustified represented by an ULTIS of the form

$$\begin{aligned} \dot{x}(t) &= Ax(t) + B\left(u(t) + \varphi(t)\right), \quad x(0) = x_0, \\ y(t) &= Cx(t), \end{aligned} \tag{1}$$

where $x(t) \in \mathbb{R}^n$ and $u(t) \in \mathbb{R}^m$ represent the state and control input vectors, respectively. The matched uncertainties/perturbations vector $\varphi(t) \in \mathbb{R}^m$ is formed by exogenous disturbances and unmodeled dynamics.

Now, consider a nominal system, i.e. the ULTIS system that is not affected by any uncertainty/perturbation. Assume that there is a nominal controller $u_n(t)$ such that it guarantees a desired nominal behavior of the system. Then the nominal linear time-invariant system (NLTIS) has the form

$$\begin{aligned} \dot{x}_n(t) &= Ax_n(t) + Bu_n(t), \quad x_n(0) = x_0, \\ y_n(t) &= Cx_n(t). \end{aligned} \tag{2}$$

Observe that the nominal controller can represent the existent controller of the application under consideration and can be designed by any chosen technique from the classical controller as PID to intelligent approaches like machine learning or robust designs as H_∞. Note also, that in general the trajectories of (1) and (2) are different. In particular, the trajectories of (1) deviate from the ones of (2) due to the effects of the uncertainties/perturbations $\varphi(t)$ affecting the desired behavior of the system.

Motivational example
To illustrate the effects that the matched uncertainties/perturbations can produce in a system, consider an NLTIS of the form (2) with

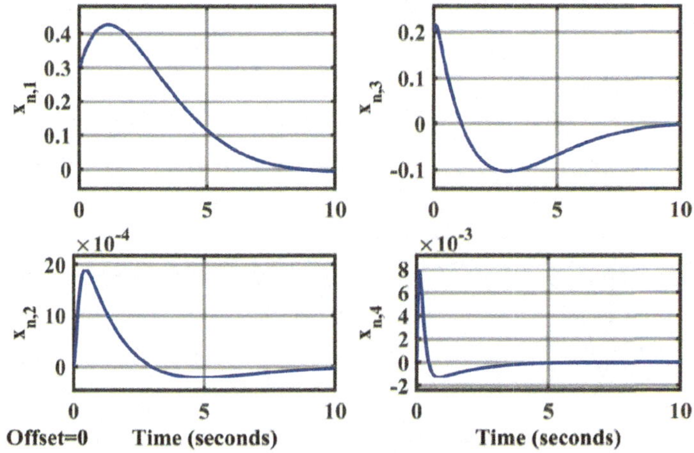

Fig. 1 Nominal behavior of the motivational example

$$A = \begin{bmatrix} 0 & 0 & 1 & 0 \\ 0 & 0 & 0 & 1 \\ 0 & 76.2884 & -0.5689 & 0 \\ 0 & 82.2655 & -0.2399 & 0 \end{bmatrix}, \quad B = \begin{bmatrix} 0 \\ 0 \\ 17.0842 \\ 7.2054 \end{bmatrix}, \quad C = I_n.$$

With a stabilizing state feedback nominal controller $u_n(t) = Kx_n(t)$, where

$$K = \begin{bmatrix} -0.0316 & 25.3471 & -0.1309 & 2.8824 \end{bmatrix}.$$

The nominal behavior of the system is depicted in Fig. 1. All the simulations of this chapter are done in Simulink, using the Euler method with a sample-step $\Delta t = 1e - 5$. Note that the regulation objective is achieved with the nominal controller in the absence of uncertainties/perturbations.

Now assume an ULTIS with the same dynamics and affected by unknown matched uncertainties/perturbations. For simulation purposes, assume

$$\varphi(t) = \frac{300}{\pi^2} \cos\left(\frac{\pi}{100} \cos\left(\frac{10\pi t}{3}\right) + 1.1\right) - 35.$$

The behavior of this system is given in Fig. 2. Under this conditions the nominal controller cannot guarantee the control design objective, making necessary to add a robustification scheme.

Control objective
The nominal controller cannot guarantee the desired behavior of the system in the presence of matched uncertainties/perturbations. The proposed robustification schemes give a control design procedure that guarantees the identity $x(t) = x_n(t)$

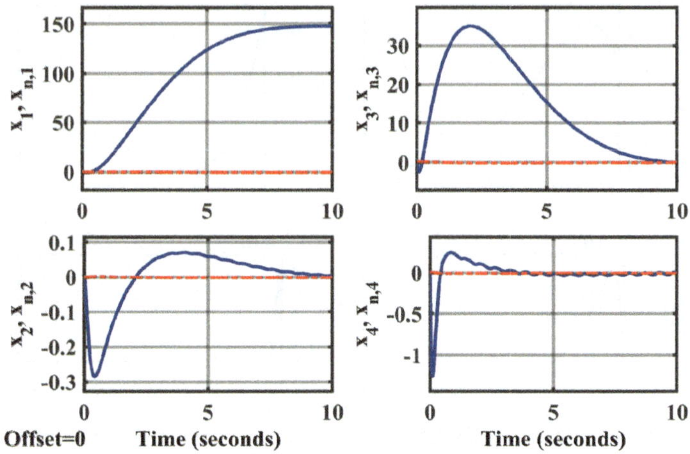

Fig. 2 Behavior of the ULTIS with only the nominal controller: NLTIS - red dashed line, ULTIS - blue continuous line

for all $t \geq t_r$; where t_r is the reaching time, that is the maximum time in which the sliding variable arrives and remains at the origin. Along this section, the control law $u(t)$ is composed of two control loops $u(t) = u_n(t) + u_{SM}(t)$ where u_{SM} is the continuous sliding mode control signal that would be designed depending on the specific scenario. Note that in general the control objective is guaranteed if $u_{SM}(t) = -\varphi(t)$ for all $t \geq t_r$.

2.1 State Based Robustification

Let start with a state based robustification scenario. For the design of the robustification scheme some general assumptions are needed:

A.2.1 Complete state information is available, i.e. $C = I_n$.
A.2.2 The ULTIS is controllable [27].
A.2.3 $\text{rank}B = m$.
A.2.4 The uncertainties/perturbations φ are Lipschitz unknown signals with a known derivative bound given by

$$\|\dot{\varphi}(t)\| \leq \Phi, \quad \forall t \geq 0.$$

The bound Φ can be found experimentally by a trial-error procedure. The last assumption is needed for the design of the CISM controller. The proposed robustification scheme is shown in Fig. 3.

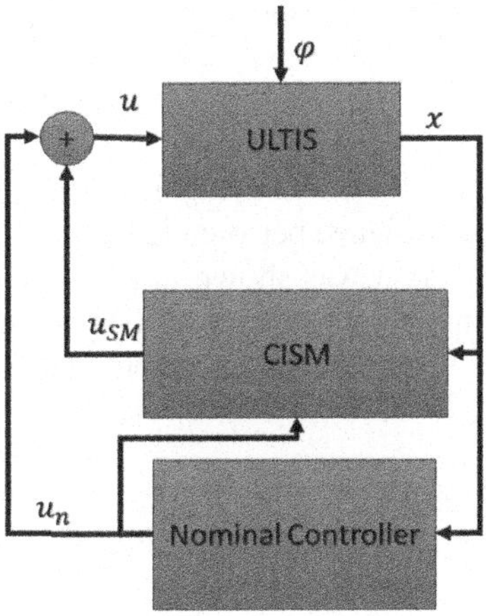

Fig. 3 Robustification scheme based on CISM

Consider that $u(t) = u_n(t) + u_{SM}(t)$, where u_{SM} is designed following a CISM algorithm [8, 16]. Hence, (1) takes the form

$$\dot{x}(t) = Ax(t) + B\left(u_n(t) + u_{SM}(t) + \varphi(t)\right), \quad x(0) = x_0. \tag{3}$$

To start designing the robutifying control loop, first it is necessary to define a virtual sliding variable of the form

$$s(t) = G(x(t) - x_0) - G\int_0^t \left(Ax(\tau) + Bu_n(\tau)\right) \mathrm{d}\tau, \tag{4}$$

where G is a design matrix. Note that this virtual variable contains the desired behavior (the dynamics of the nominal system) in such a way that if the sliding mode is guaranteed ($s = \dot{s} = \ddot{s} = 0$) the trajectory of the system will follows the nominal trajectory. Moreover, at the initial time the sliding variable fullfils $s(0) = 0$. Now, following the equivalent control method [32], the derivative of (4) is

$$\dot{s}(t) = GB\left(u_{SM}(t) + \varphi(t)\right), \quad s(0) = 0.$$

Assume the system is on the sliding mode, and G is designed such that $\det(GB) \neq 0$, the equivalent control takes the form

$$u_{SM_{eq}}(t) = -\varphi(t).$$

Please observe that the equivalent control cannot be constructed since it is the negative of the unknown uncertainties/perturbations φ. However, once the sliding variable reaches the origin, the sliding mode control behaves as the equivalent control [32].

Once the sliding mode is achieved, the dynamics of the ULTIS (1) are

$$\dot{x}(t) = Ax(t) + Bu_n(t),$$

and the system dynamics of (1) are equivalent to the nominal ones (2). With this, the equivalence between (2) and (1) on the sliding mode has been guaranteed.

As it was shown in [7] if $G = B^+$, all the possible non-considered unmatched uncertainties/perturbations are not increased. Hence, let choose for simplicity $G = B^+$. Under this condition, $GB = I_m$ and the derivative of s is simplified to

$$\dot{s}(t) = u_{SM}(t) + \varphi(t).$$

Let $u_{SM}(t)$ be a STA controller of the form

$$\begin{aligned} u_{SM}(t) &= -k_1 \lfloor s(t) \rceil_m^{\frac{1}{2}} + \omega(t), \\ \dot{\omega}(t) &= -k_2 \lfloor s(t) \rceil_m^0, \quad \omega(0) = 0. \end{aligned} \tag{5}$$

Then,

$$\begin{aligned} \dot{s}(t) &= -k_1 \lfloor s(t) \rceil_m^{\frac{1}{2}} + \Omega(t), \\ \dot{\Omega}(t) &= -k_2 \lfloor s(t) \rceil_m^0 + \dot{\varphi}(t); \end{aligned} \tag{6}$$

where $\Omega(t) = \omega(t) + \varphi(t)$. The gain design conditions are given in the following Theorem.

Theorem 1 ([21, 29]) *System* (6) *is finite-time stable if its parameters satisfy*

$$k_2 > \Phi, \quad k_1 > \sqrt{k_2 + \Phi}.$$

Then if the gains of the controller u_{SM} satisfies Theorem 1 the sliding variable s converge to the origin in finite time with a reaching time t_r, and the ULTIS (1) behaves as the NLTIS for all $t \geq t_r$. An upper bound of the reaching time can be calculated following the next Lemma.

Lemma 1 ([30]) *For any* $k_2 > \Phi$, *if* $s(0) = 0$ *and* $k_1 = \sqrt{8k_2}$. *System* (6) *reaches the origin at a reaching time*

$$t_r \leq \frac{\Omega_{max}}{k_2 - \Phi};$$

where Ω_{max} *is the upper-bound of the initial condition of* Ω, *i.e.* $|\Omega(0)| \leq \Omega_{max}$.

2.2 Output Based Robustification

A more realistic scenario is when only output measurements are available. Let us consider an ULTIS (1) and recall assumption **A.2.2**. In this scenario some extra assumptions are needed:

A.2.5 The initial conditions x_0 are unknown but with a known bound, $\|x_0\| \leq \mu$.
A.2.6 $\text{rank}B = \text{rank}(CB) = m$.
A.2.7 The uncertainties/perturbations φ conform a bounded Lipschitz function:

$$\|\varphi(t)\| \leq \phi, \; \|\dot{\varphi}(t)\| \leq \Phi;$$

where ϕ , $\Phi \in \mathbb{R}_+$ are given. These bounds can be obtained experimentally as in the state based scenario by a trial and error procedure.
A.2.8 The ULTIS has more outputs than inputs, i.e. $p > m$.
A.2.9 The ULTIS is strongly observable [18] with observability index l.

Assumptions **A.2.8** and **A.2.9** are needed to reconstruct the state in finite time in the presence of unknown inputs [14, 18, 20].

The proposed robustifying output-based scheme generates a continuous control signal that compensates theoretically exactly the matched uncertainties/perturbations in finite time by using only output information. The proposed scheme is based on a COISM controller, and is depicted in Fig. 4. The COISM controller generates a continuous control law that guarantees finite time convergence to the origin (see [16])

Fig. 4 Robustification scheme based on COISM

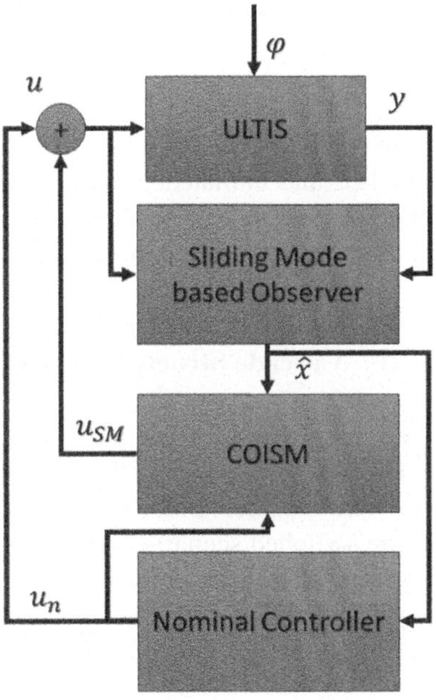

by using an observer capable of reconstructing theoretically exactly in finite time the states of the system. There are many observers in the literature capable of accomplishing this (see [3, 4, 10, 12, 28]). In this chapter, an STA-based cascade observer that reconstructs the states theoretically exactly in finite time without using filters is presented. Let us present the design methodology of the proposed robustifying scheme.

The ULTIS (1) is affected by matched uncertainties/perturbations, and they influence its response. That is why it is necessary to transform the system into a decoupled input-output form to reconstruct the states without perturbed dynamics. Let $\bar{x}(t) = T_x x(t)$ and an output transformation $\bar{y}(t) = T_y y(t)$, where

$$T_x = \begin{bmatrix} B^{\perp^+} \\ C_t B^{\perp^+} + B^+ \end{bmatrix},$$

with $C_t = (CB)^+ C(I_n - BB^+)B^\perp$, and

$$T_y = \begin{bmatrix} (CB)^{\perp^+} \\ (CB)^+ \end{bmatrix}.$$

Then the ULTIS (1) is transformed to

$$\begin{aligned} \dot{\bar{x}}(t) &= \underbrace{\begin{bmatrix} A_{11} & A_{12} \\ A_{21} & A_{22} \end{bmatrix}}_{\bar{A}} \underbrace{\begin{bmatrix} x_1(t) \\ x_2(t) \end{bmatrix}}_{\bar{x}(t)} + \underbrace{\begin{bmatrix} 0 \\ I_m \end{bmatrix}}_{\bar{B}} (u(t) + \varphi(t)), \\ \begin{bmatrix} y_1(t) \\ y_2(t) \end{bmatrix} &= \begin{bmatrix} C_1 \\ C_2 \end{bmatrix} \bar{x}(t) = \underbrace{\begin{bmatrix} C_{11} & 0 \\ 0 & I_m \end{bmatrix}}_{\bar{C}} \bar{x}(t). \end{aligned} \tag{7}$$

Note that with this transformation the outputs of the system have been separated into matched and unmatched parts. Due to Assumptions **A.2.8–A.2.9**, all the matched states are measured. Hence a reduced-order observer that reconstructs the unknown states x_1 is designed in the sequel.

2.2.1 Cascade Structure Observer

This observer is based on a step-by-step reconstruction and is based on the one given in [13, 16]. Every part of the observer depends on the convergence properties of the previous ones. It is possible to design the gains to reach a specific convergence time. In the switched scenarios part of this chapter, a switched gain design methodology is given. That methodology can be applied straight to this scenario.

The STA-based cascade observer is formed by two parts:

- A *Luenberger observer*

$$\dot{\tilde{x}}_1(t) = A_{11}\tilde{x}_1(t) + A_{12}y_2(t) + K\left(y_1(t) - C_{11}\tilde{x}_1(t)\right), \tag{8}$$

  that stabilizes the error $e_{LO} = x_1(t) - \tilde{x}_1(t)$ in a ball around the origin. The gain matrix K can be designed by using a typical pole placement approach. Hence the observation error is stable, such that $\|e_{LO}\| < \gamma$.
- A *hierarchical observer* composed by:

  - A family of STA-based observers

$$\dot{x}_{ak} = A_{11}\tilde{x}_1(t) + A_{12}y_2(t) - L_k(t)\left(C_{11}A_{11}^{k-1}L_k\right)^{-1}v_k(t), \tag{9}$$

    $k = 1, \ldots, l-1$, that reconstruct theoretically exactly, step by step, the output error uncertainties/perturbations free part and its derivatives. $L_k(t)$ is a design matrix and $v_k(t)$ is an output injection signal based on the STA. The convergence properties of this family of observers depends on the convergence of its previous elements, i.e. the $k+1$-th observer will converge after the k observer has converged.
  - And an algebraic part

$$\hat{x}(t) = T_x^{-1}\begin{bmatrix}\hat{x}_1(t)\\ y_2(t)\end{bmatrix}; \tag{10}$$

    where $\hat{x}_1(t) = \tilde{x}_1(t) - \mathcal{O}_l^+ v(t)$ with $\mathcal{O}_l$ the observability matrix of the pair (C_{11}, A_{11}) and

$$v(t) = \begin{bmatrix} C_{11}\tilde{x}_1(t) - y_1(t) \\ v_1(t) \\ v_2(t) \\ \vdots \\ v_{l-1}(t) \end{bmatrix}.$$

    The algebraic part (10) reconstructs theoretically exactly the states of the system in finite time.

The design of the family of STA observer is given in the following theorem

Theorem 2 *Assume*

(a) The auxiliary state vectors x_{ak}, for all $k = 1, \ldots, l-1$, is designed as in (9), *where $L_k(t) \in \mathbb{R}^{n-m\times p-m}$ is a design matrix such that det $\left(C_{11}A_{11}^{k-1}L_k\right) \neq 0$.*
(b) Let τ_k, $k = 2, \ldots, l-1$ be the convergence time of the $(k-1)$-th element of the hierarchical observer. At $t = \tau_k$ the k-th variable x_{ak} satisfies

$$C_{11}x_{a1}(0) = y_1(0),$$
$$C_{11}A_{11}^{k-1}x_{ak}(\tau_k) = C_{11}A_{11}^{k-1}\tilde{x}_1(\tau_k) - v_{k-1}(\tau_k).$$

(c) The sliding variables s_k are designed as

$$s_k(y_1(t), x_{ak}(t)) = \begin{cases} y_1(t) - C_{11}x_{a1}(t), & k = 1, \\ C_{11}A_{11}^{k-1}\tilde{x}_1(t) - v_{k-1}(t) - C_{11}A_{11}^{k-1}x_{ak}(t), & k = 2, \ldots, l-1. \end{cases} \tag{11}$$

(d) The output injection v_k is designed as an STA of the form

$$\begin{aligned} v_k(t) &= -\kappa_{k,1}\lceil s_k(y_1(t), t)\rfloor_{p-m}^{\frac{1}{2}} + \varpi_k(t), \\ \dot{\varpi}_k &= -\kappa_{k,2}\lceil s_k(y_1(t), t)\rfloor_{p-m}^{0}, \\ \varpi_k(0) &= 0. \end{aligned} \tag{12}$$

(e) And $(\kappa_{k,1}, \kappa_{k,2})$ are designed such that

$$\kappa_{k,2} > M_k, \quad \kappa_{k,1} > \sqrt{\kappa_{k,2} + M_k},$$

with

$$M_k \geq \left\| C_{11}A_{11}^{k} \right\| \left(\|A - KC\|\gamma + \|B\|\phi \right).$$

Then,

$$v_k(t) = -C_{11}A_{11}^{k}\left(x_1(t) - \tilde{x}_1(t)\right),$$

and it is possible to reconstruct theoretically exactly all the vector functions $C_{11}A_{11}^{k-1}x_1(t)$ in finite time.

Proof Recall that the observers are turning on sequentially whenever the observers that reconstruct the lowers derivatives have converged. The proof is constructive and can be obtained iteratively.

Let us start by recovering the first vector $C_{11}A_{11}x_1(t)$. Let k=1, since the conditions (a)-(c) are fulfilled; it is clear that $s_1(y_1(0)), x_{a1}(0) = 0$. Moreover, the time derivative of the sliding variable along the trajectories of (7) and (9) has the form

$$\dot{s}_1(y_1(t), x_{a1}(t)) = C_{11}A_{11}\left(x_1(t) - \tilde{x}_1(t)\right) + v_1(t), \tag{13}$$

and once it has converged, the equivalent control that maintains the trajectory on the origin has the form

$$v_{1_{eq}}(t) = -C_{11}A_{11}\left(x_1(t) - \tilde{x}_1(t)\right).$$

Now, the output injection (d) is designed assuring the sliding variable s_1 and its derivatives reach the origin in finite time. Substituting the STA output injection (12) on the auxiliary state vector x_{a1}, then (13) can be restated as

$$\dot{s}_1(y_1(t), x_{a1}(t)) = -\kappa_{1,1}\lceil s_1(y_1(t), t)\rfloor_{p-m}^{\frac{1}{2}} + \Lambda_1(t),$$
$$\dot{\Lambda}_1 = -\kappa_{1,2}\lceil s_1(y_1(t), t)\rfloor_{p-m}^{0} + C_{11}A_{11}\left(\dot{x}_1(t) - \dot{\tilde{x}}_1(t)\right),$$
$$\Lambda_1(\tau_1) = C_{11}A_{11}\left(x_1(\tau_k) - \tilde{x}_1(\tau_1)\right);$$

where $\Lambda_1(t) = C_{11}A_{11}\left(x_1(t) - \tilde{x}_1(t)\right) + \varpi_1(t)$. Observe that this dynamical system has the form (6).

Since $(\kappa_{1,1}, \kappa_{1,2})$ satisfies Theorem 1, s_1 converges to the origin in finite time. Then, exact convergence of s_1 and its derivatives to the origin in finite time is assured.

Assume the result is true for $k = \mathbf{k}$, and let us prove the result for $k = \mathbf{k} + 1$. Once again our aim is to recover the k-vector $C_{11}A_{11}^k\bar{x}(t)$. Conditions (a)-(c) are satisfied. Taking the time derivative of the sliding variable along the trajectories of (7) and (9),

$$\dot{s}_k(y_1(t), x_{a_k}(t)) = C_{11}A_{11}^k\left(x_1(t) - \tilde{x}_1(t)\right) + v_k(t). \tag{14}$$

Moreover, once the k-th sliding variable and its derivatives have converged,

$$C_{11}A_{11}^{\mathbf{k}}x_{a_k}(\tau_k) = C_{11}A_{11}^{\mathbf{k}}x_1(\tau_k),$$

and the equivalent control is

$$v_{\mathbf{k}_{eq}}(t) = -C_{11}A_{11}^k\left(\tilde{x}_1(t) - x_1(t)\right).$$

Designing the output injection (12), the sliding variable dynamics (14) can be rewritten as

$$\dot{s}_k(y_1(t), x_{a_k}(t)) = -\kappa_{k,1}\lceil s_k(y_1(t), t)\rfloor_{p-m}^{\frac{1}{2}} + \Lambda_k(t),$$
$$\dot{\Lambda}_k = -\kappa_{k,2}\lceil s_k(y_1(t), t)\rfloor_{p-m}^{0} + C_{11}A_{11}^k\left(\dot{x}_1(t) - \dot{\tilde{x}}_1(t)\right),$$
$$\Lambda_k(\tau_k) = C_{11}A_{11}^k x_1\left(\tau_k) - \tilde{x}_1(\tau_k)\right),$$

where $\Lambda_k(t) = C_{11}A_{11}^k\left(x_1(t) - \tilde{x}_1(t)\right) + \varpi_k(t)$.

Since condition (e) is satisfied and $s_k(y_1(\tau_{\mathbf{k}}), x_{ak}(\tau_{\mathbf{k}})) = 0$, once again from Theorem 1, s_k converges to the origin in finite time. The exact convergence of s_k and its derivatives to the origin in finite-time is guaranteed.

Note that the exact reconstruction of the output error and its $(l-1)$ time derivatives in finite time has been achieved. □

Remark 1 For simplicity the design matrices L_k can be chosen as $L_k = (C_{11}A_{11}^{k-1})^+$ or any other matrix such that $C_{11}A_{11}^{k-1}L_k = I_{p-m}$

The family of STA observers reconstructs the output y_1 and its $l-1$ time-derivatives in finite time. Using this information let us construct the vector

$$\mathcal{O}_l x_1(t) = \mathcal{O}_l\tilde{x}_1(t) - v(t).$$

Then the states can be reconstructed by

$$x_1(t) = \tilde{x}_1(t) - \mathcal{O}_l^+ v(t). \tag{15}$$

And the algebraic observer is suggested as

$$\hat{x}_1(t) = \tilde{x}_1(t) - \mathcal{O}_l^+ v(t); \tag{16}$$

Note that the proposed observer reconstructs theoretically exactly the states in finite time by using (10).

Remark 2 The proposed observer is not the only one that can be used in the proposed methodology. The only restriction that needs to be imposed on the chosen observer is reconstructing the state vector theoretically exactly in finite time.

2.2.2 Continuous Output Integral Sliding Mode

The proposed observer is capable to reconstruct theoretically exactly the states in finite time, i.e. $\hat{x}(t) = x(t)$ for all $t > \tau_l$. The controller is activated since the initial time. However, its reaching time is affected by the one of the observers. Let us design the integral part of the control law.

Consider the ULTIS (1) and define the output based integral sliding dynamics

$$s(y, t) = G\left(y(t) - y(0)\right) - \int_0^t GC(A\hat{x}(\tau) + Bu_n(\tau))d\tau; \tag{17}$$

where $G \in \mathcal{G}$ is a design matrix such that

$$\mathcal{G} = \{G \in \mathbb{R}^{m \times n} : \det D \neq 0, D = GCB\}.$$

Observe that $s\left(y\left(0\right), 0\right) = 0$. Assume the observer has converged. Then taking the first derivative of the sliding variable along the trajectory of (1)

$$\dot{s}(y, t) = GCA\left(x(t) - \hat{x}(t)\right) + D\left(u_{SM}(t) + \varphi(t)\right). \tag{18}$$

Since $x(t) = \hat{x}(t)$, then

$$\dot{s}(y, t) = D\left(u_{SM}(t) + \varphi(t)\right); \tag{19}$$

the equivalent control [32] that maintains the trajectory on the sliding mode is

$$u_{Ieq} = -D^{-1}\varphi(t), \tag{20}$$

and the sliding mode dynamics of the SULTIS takes the form

$$\begin{aligned} \dot{x}(t) &= Ax(t) + Bu_n(t), \\ y(t) &= Cx(t), \quad x(0) = x_0. \end{aligned} \tag{21}$$

For simplicity assume the projection matrix is designed such that $D = I_m$. A suitable selection is $D = (CB)^+$. As it was shown in [16], since the sliding variable depends on the non-Lipschitz variable $\hat{x}_1$, the STA controller needs to be modified to eliminate the non-Lipschitz term. Hence, the following STA controller is proposed

$$\begin{aligned} u_{SM} &= -GCAT_x^{-1} \begin{bmatrix} \mathcal{O}_l^+ v(t) \\ 0 \end{bmatrix} - \kappa_1 \lceil s(y(t), t) \rfloor_m^{\frac{1}{2}} + \omega(t), \\ \dot{\omega} &= -\kappa_2 \lceil s(y(t), t) \rfloor_m^0, \quad \omega(0) = 0; \end{aligned} \tag{22}$$

that compensates the non-Lipschitz term $\mathcal{O}_l^+ v(t)$, then the sliding dynamics has the typical STA form

$$\begin{aligned} \dot{s}(y,t) &= -\kappa_1 \lceil s(y(t), t) \rfloor_m^{\frac{1}{2}} + \Omega(t), \\ \dot{\Omega}(t) &= -\kappa_2 \lceil s(y(t), t) \rfloor_m^0 + GCAT_x^{-1} \begin{bmatrix} \left(\dot{x}_1(t) - \dot{\hat{x}}_1(t) \right) \\ 0 \end{bmatrix} + \dot{\varphi}(t), \\ s(y, 0) &= 0. \end{aligned} \tag{23}$$

Due to the integral structure of $s(y, t)$, the non-Lipschitz dynamics of the observer that affect the controller only depends on v, the continuous part of the observer dynamics. Hence, the continuity of u_{SM} is preserved. The following Lemma gives the design of the STA gains (κ_1, κ_2) for the controller.

Lemma 2 *Suppose assumptions of this section are satisfied and*

$$\kappa_2 > \mathcal{L}, \quad \kappa_1 > \sqrt{\kappa_2 + \mathcal{L}},$$

with

$$\mathcal{L} \geq \Phi + \|GCAT_x^{-1}\| \left(\|A - KC\| \gamma + \|B\| \phi \right).$$

Then, the sliding mode dynamics (23) *converges to the origin in finite time.*

Proof This result comes directly from Theorem 1. □

With this, the robustification of the nominal trajectory in finite time using only output information is achieved.

2.3 ULTIS Robustification Example

The COISM robustification scheme is not illustrated for brevity. Its design is very similar to the one presented in the switched section. To show the applicability of the CISM-based robustification scheme. Consider the system used in the Motivational Example. Assume that all the state vector is measured and that the system has a robustifying control loop designed following the CISM approach. Note that the considered perturbation is Lipschitz with a bound $\Phi = 5$. Then the CISM controller gains are chosen as $k_1 = 3.3541$ and $k_2 = 5.5$. Note that these gains satisfy Theorem 1. Moreover, $\Omega(0) = \varphi(0) \leq \Phi$, then the reaching time can be computed following Lemma 1 getting $t_r = 10$ s. However, it is well known that this is only an upper bound of the reaching time and that the sliding variable converges faster to the origin. A practical approach to estimate the reaching time (see [17]) is to define a practical convergence set $\mathcal{R}$ and to monitor the sliding variable to detect when the sliding variable enters and remains in the set. For the present simulation, the practical convergence set is defined as

$$\mathcal{R} = \{s : s \leq 1 \times 10^{-8}\},$$

and the estimated reaching time is $t_r = 1.15$ s.

The behavior of the robustified system is given in Fig. 5. Note that the system behaves as the nominal system for all $t \geq t_r$. In Fig. 6, the behavior of the sliding

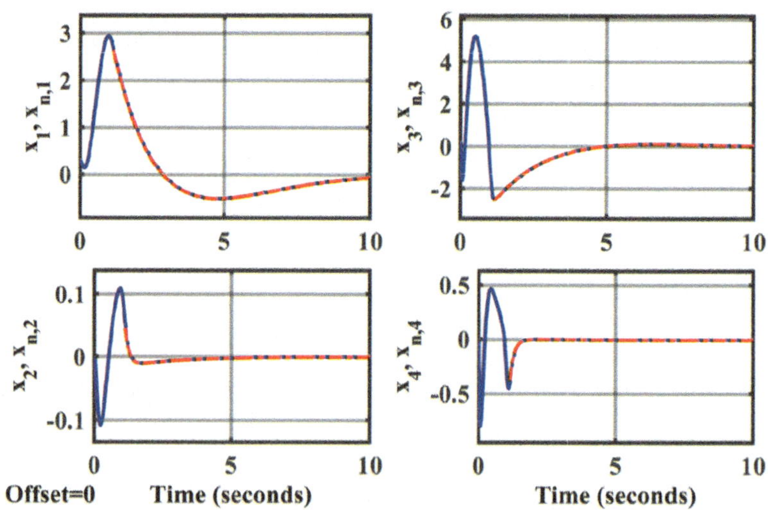

Fig. 5 Behavior of the ULTIS with the state based robustification scheme: NLTIS—red dashed line, ULTIS—blue continuous line

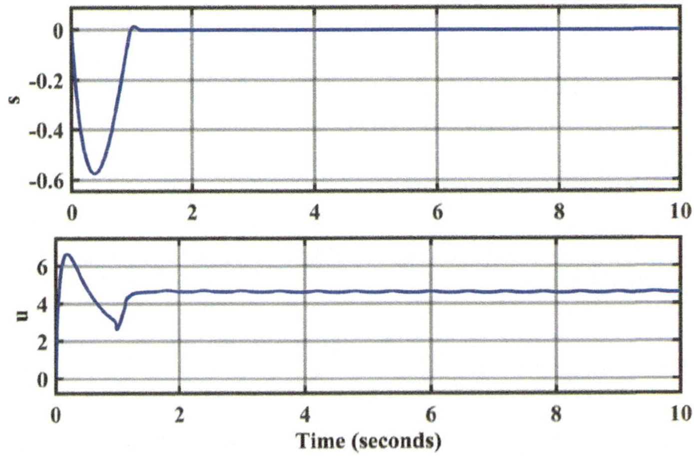

Fig. 6 Sliding variable and applied CISM control signal

variable and the applied control signal u are depicted. Observe that the control signal is continuous, and the sliding variable converges in finite time to the origin and before the computed reaching time t_r.

3 Switched Uncertain Linear Time Invariant Systems

This section deals with systems whose structure changes when a switching rule is fulfilled [6, 33]. In particular, SULTIS with state-dependent location transitions composed by r different locations are considered,

$$\begin{aligned} \dot{x}(t) &= A_i x(t) + B_i \left(u(t) + \varphi(t)\right), \\ y(t) &= C_i x(t), \qquad x(0) = x_0, \end{aligned} \tag{24}$$

where $i = \sigma(x, t) : \mathbb{R}^n \times \mathbb{R}^+ \to \mathcal{I} = \{1, 2, \ldots, r\}$ denotes the active location, $A_i \in \mathbb{R}^{n\times n}$, $B_i \in \mathbb{R}^{n\times m}$, and $C_i \in \mathbb{R}^{p\times n}$ are known matrices, $\varphi : \mathbb{R}_+ \to \mathbb{R}^m$ are matched uncertainties/perturbations and $u(t)$ is the control.

A transition between each location is ruled by a set of given switching manifolds $m_{i,i'} : \mathbb{R}^n \to \mathbb{R}$, with $i' \in \mathcal{I}$ and $i \neq i'$, such that

$$\bigcap_{\substack{i'\in\mathcal{I}\\ i'\neq i}} \{x : m_{i,i'}(x) = 0\} = \emptyset.$$

When the state x is in location i and $m_{i,i'}(x) = 0$, it is said that a transition to location i' occurs. It is considered that these switching manifolds are known.

Under the assumption that $\varphi(t) = 0$ a nominal switched system is defined as

$$\begin{aligned} \dot{x}_n(t) &= A_i x_n(t) + B_i u_n(t), \\ y_n(t) &= C_i x_n(t), \quad x_n(0) = x_0, \end{aligned} \tag{25}$$

where $u_n(t)$ is a given nominal control.

The switching moments $t_j,\ j = 1, 2, \ldots$ are not defined a priory but, by using the given switching manifolds and the nominal trajectory x_n, it is possible to detect these transitions such that

$$t_j = \min_{\substack{i' \in \mathcal{I} \\ i' \neq i}} \{t > t_j : m_{i,i'}(x_n(t)) = 0\}. \tag{26}$$

Suppose that the switching moments for (25) conform an ordered sequence

$$0 = t_0 < t_1 < t_2 < \cdots < t_{j-1} < t_j < \ldots,$$

such that $t_j - t_{j-1} > \delta$, i.e. the nominal switched system (25) satisfies the dwell time condition. To assure the nominal trajectory does not present Zeno behavior, we assume

$$\lim_{t \to t_j^-} \frac{\mathrm{d}}{\mathrm{d}t} m_{i,i'}(x_n(t)) > \rho \text{ and } \lim_{t \to t_j^+} \frac{\mathrm{d}}{\mathrm{d}t} m_{i,i'}(x_n(t)) > \rho$$

or

$$\lim_{t \to t_j^-} \frac{\mathrm{d}}{\mathrm{d}t} m_{i,i'}(x_n(t)) < -\rho \text{ and } \lim_{t \to t_j^+} \frac{\mathrm{d}}{\mathrm{d}t} m_{i,i'}(x_n(t)) < -\rho$$

with $\rho > 0$. Without loss of generality, the location trajectory of the nominal system (25) can be characterized by the following switching sequence

$$\Sigma = \{x_0; (i_0, t_0), \ldots, (i_j, t_j), \ldots | i_j \in \mathcal{I}, j = 0, 1, \ldots\}.$$

Control objective

For (24), in general, the switching sequence and the state trajectory are affected by the matched uncertainties/perturbations $\varphi(t)$. Moreover, in the presence of these uncertainties/perturbations, the nominal controller cannot guarantee the desired behavior of the system. Hence, this section aims to robustify a given nominal switched trajectory by compensating the matched uncertainties/perturbations $\varphi(t)$. Similar to Sect. 2, two scenarios are considered: complete state vector, and only output information.

3.1 State Based Robustification

For the case where complete state vector information is available, a synthesis of a sliding mode robustification control scheme for the SULTIS in (24) is considered. In this scenario, the initial conditions of the nominal system (25) are known and the switching moments can be precomputed off-line. The conditions to guarantee the nominal trajectory does not present Zeno behavior could be verified in the same manner.

Let $\sigma(x_n(0)) = i_0$. Then, the location trajectory σ is ruled by

$$\sigma(x_n(t)) := \begin{cases} i \text{ if } \sigma(x_n(t^-)) = i \text{ and } m_{i,i'}(x_n(t)) \neq 0, \\ i' \text{ if } \sigma(x_n(t^-)) = i \text{ and } m_{i,i'}(x_n(t)) = 0. \end{cases}$$

To compensate theoretically exactly the matched uncertainties/perturbations, let us assume the following.

A.3.1 The initial location and the initial condition are known.
A.3.2 $\text{rank} B_i = m$ for all $i \in \mathcal{I}$.
A.3.3 $C_i = I_n$ for all $i \in \mathcal{I}$.
A.3.4 The uncertainties/perturbations φ are bounded Lipschitz functions

$$\|\varphi(t)\| \leq \phi,$$
$$\|\dot{\varphi}(t)\| \leq \Phi,$$

where ϕ, and $\Phi \in \mathbb{R}_+$.
A.3.5 The switched system is controllable in every location [27], i.e. the pair (A_i, B_i) is controllable for $i \in \mathcal{I}$.

3.1.1 SCISM Control

Consider (24) and let $u(t) = u_n(t) + u_{SM}(t)$, here u_{SM} refers to the SCISM control part, which guarantees the compensation of the matched uncertainties/perturbations $\varphi(t)$, in the time interval $t \in (t_j, t_{j+1}]$.

Since, our objective is to robustify a given nominal trajectory, we define the following virtual sliding-mode variable

$$s(x,t) = G_i(x(t) - x(t_j)) - G_i \int_{t_j}^{t} (A_i x(\tau) + B_i u_n(\tau)) d\tau, \tag{27}$$

where $G_i \in \mathcal{G}_i$, $i \in \mathcal{I}$ is a projection matrix and

$$\mathcal{G}_i = \{G_i \in \mathbb{R}^{m \times n} : \det D_i \neq 0\},$$

with $D_i = G_i B_i$. Observe that at the switching times $s(x(t_j^+), t_j^+) = 0$, i.e. the system is in the sliding mode at every switching time and the sliding variable does not present reaching phase. Moreover, the virtual sliding variable is designed as the difference between the states of the SULTIS and the ones of the nominal system and it guarantees their equivalence when $s(x(t), t) = \dot{s}(x(t), t) = 0$. Following the equivalent control method [32] and selecting $G_i = B_i^+$, the first time derivative of s along the trajectories of (24) is

$$\dot{s}(x, t) = u_{SM} + \varphi(t). \tag{28}$$

The obtained equivalent control has the form $u_{SM_{eq}}(t) = -\varphi(t)$. Hence, the sliding mode dynamics of system (24) takes the form

$$\dot{x}(t) = A_i x(t) + B_i u_n,$$

which is equivalent to the nominal system (25) after $t = t_r$, where t_r is the reaching time of the SCISM controller with $x_n(t_r) = x(t_r)$ as initial condition.

Let $u_{SM}(t)$ be a STA based controller of the form

$$\begin{aligned} u_{SM} &= -k_1 \lfloor s(x, t) \rceil_m^{1/2} + \omega(t), \\ \dot{\omega}(t) &= -k_2 \lfloor s(x, t) \rceil_m^0, \quad \omega(t_j^+) = \omega(t_j^-); \end{aligned} \tag{29}$$

where gains $k_i \in \mathbb{R}_+, i = 1, 2$, must be designed guaranteeing a reaching time $t_r \leq \delta$. Substituting the control (29) in (28), the sliding dynamics takes the form

$$\begin{aligned} \dot{s}(x, t) &= -k_1 \lfloor s(x, t) \rceil_m^{1/2} + \Lambda(t), \\ \dot{\Lambda}(t) &= -k_2 \lfloor s(x, t) \rceil_m^0 + \dot{\varphi}(t), \end{aligned} \tag{30}$$

where $\Lambda(t) = \omega(t) + \varphi(t)$.

The next result gives sufficient conditions to assure (28) converge in finite time to the origin before the first switching.

Lemma 3 ([34])

Suppose Assumptions ***A.3.1–A.3.5*** *are satisfied and*

$$k_2 = \begin{cases} \frac{\phi}{\delta} + \Phi, \; t \leq \delta, \\ 1.1\Phi, \quad t > \delta, \end{cases}$$

and

$$k_1 = \begin{cases} \sqrt{8k_2}, \quad t \leq \delta, \\ 1.5\sqrt{\Phi}, \; t > \delta. \end{cases}$$

Then, the sliding mode dynamics (30) *converges to the origin in finite time with a reaching time* $t_r \leq \delta$.

With this, the robustification of the nominal trajectory before the dwell time is guaranteed.

Remark 3 The proposed switched gain strategy diminishes the chattering effect by switching to gains that consume less energy [23] once the sliding mode dynamics has converged to the origin before the dwell time.

Remark 4 The convergence of the sliding variable can be assured after the first switching time since the perturbations are continuous at the switching moments t_j and the switching conditions of the STA ensure the continuity of ω. Observe that, if uncertainties/perturbations were bounded in each location, a switched STA gains design can be used as in [13, 16].

3.2 Output Based Robustification

In order to robustify the SULTIS (24) with only output information, a state and an output transformations are needed. These transformations reveal the availability of states that are directly affected by the uncertainties/perturbations. Similar to Sect. 2.2, we consider $\bar{x}(t) = T_{x,i}x(t)$ and $\bar{y}(t) = T_{y,i}y(t)$, where

$$T_{x,i} = \begin{bmatrix} B_i^{\perp^+} \\ C_{t_i} B_i^{\perp^+} + B_i^+ \end{bmatrix},$$

with $C_{t_i} = (C_i B_i)^+ C_i (I_n - B_i B_i^+) B_i^{\perp}$, and

$$T_{y,i} = \begin{bmatrix} (C_i B_i)^{\perp^+} \\ (C_i B_i)^+ \end{bmatrix}.$$

Then the SULTIS (24) is transformed to

$$\begin{aligned} \dot{\bar{x}}(t) &= \underbrace{\begin{bmatrix} A_{11,i} & A_{12,i} \\ A_{21,i} & A_{22,i} \end{bmatrix}}_{\bar{A}_i} \bar{x}(t) + \underbrace{\begin{bmatrix} 0 \\ I_m \end{bmatrix}}_{\bar{B}_i} (u(t) + \varphi(t)), \\ \begin{bmatrix} y_1(t) \\ y_2(t) \end{bmatrix} &= \begin{bmatrix} C_{1,i} \\ C_{2,i} \end{bmatrix} \bar{x}(t) = \underbrace{\begin{bmatrix} C_{11,i} & 0 \\ 0 & I_m \end{bmatrix}}_{\bar{C}_i} \bar{x}(t), \qquad \bar{x}(0) = \bar{x}_0, \end{aligned} \tag{31}$$

with $i = \sigma(T_{x,i}^{-1}\bar{x}, t)$ and $\bar{x} = \left[x_1^T(t), x_2^T(t)\right]^T$.

The transition between locations is ruled by a set of given switching manifolds depending on $\bar{x}$. When the state $\bar{x}$ is in location i and $m_{i,i'}(\bar{x}) = 0$ it is said that a transition to location i' occurs. Considering that $\varphi(t) = 0$ a nominal switched system is defined as

$$\begin{aligned} \dot{\bar{x}}_n(t) &= \bar{A}_i \bar{x}_n(t) + \bar{B}_i u_n(t), \\ \bar{y}_n(t) &= \bar{C}_i \bar{x}_n(t), \quad \bar{x}_n(0) = \bar{x}_0, \end{aligned} \tag{32}$$

where $u_n(t)$ is a given nominal control.

The switching moments $t_j,\ j = 1, 2, \ldots$ can be detected by using the known switching manifolds as in (26) with the nominal trajectory, $\bar{x}_n$ of (32). We assume that the nominal trajectory (32) does not present Zeno behavior. Let $\sigma(T_{x,i_0}^{-1}\bar{x}_n(0), 0) = i_0 \in \mathcal{I}$. Then, the location trajectory $\sigma(T_{x,i}^{-1}\bar{x}_n(t), t)$ can be restated as

$$\sigma(T_{x,i}^{-1}\bar{x}_n(t), t) := \begin{cases} i \ \text{ if } \sigma(T_{x,i}^{-1}\bar{x}_n(t^-), t^-) = i \text{ and } m_{i,i'}(T_{x,i}^{-1}\bar{x}_n(t)) \neq 0, \\ i' \ \text{ if } \sigma(T_{x,i}^{-1}\bar{x}_n(t^-), t^-) = i \text{ and } m_{i,i'}(T_{x,i}^{-1}\bar{x}_n(t)) = 0. \end{cases}$$

Here is necessary to consider the following assumptions together with **A.3.4** and **A.3.5**.

A.3.6 The initial location is known and the initial condition is unknown but bounded, i.e. there exists $\mu \in \mathbb{R}_+$, such that $\|x(0)\| \leq \mu$.
A.3.7 The SULTIS without any controller satisfies the dwell time condition.
A.3.8 $\text{rank} B_i = \text{rank} C_i B_i = m$ for all $i \in \mathcal{I}$.
A.3.9 The switched system (24) has more outputs than inputs, i.e. $p > m$.
A.3.10 The switched system is strongly observable in every location [18] with observability index l, i.e. the triple (A_i, B_i, C_i) is strongly observable for $i \in \mathcal{I}$.

The problems considered are: suppression of the dependence on the state, chattering attenuation, and theoretically exact compensation of the matched uncertainties/perturbations before the dwell time. This methodology uses an STA-based cascade observer that reconstructs the state vector theoretically exactly before half of the dwell time. After the observer has converged, the SCOISM controller is turned on, assuring theoretically exact compensation of the matched uncertainties/perturbations before the dwell time. Assumption **A.3.7** ensures the absence of switching before the dwell time, even when the controller is turned off.

The main structure of the SCOISM robustifying methodology for the SULTIS is similar to the one presented in Sect. 2.2 for the ULTIS.

3.3 Cascade Structure Observer

Recall Assumptions **A.3.4, A.3.9, A.3.10**. Since (A_i, B_i, C_i) is strongly observable, the pair $(A_{11,i}, C_{11,i})$ is observable (see [5]). The design of the reduced-order observer is an extension of the one presented in Sect. 2.2 and is inspired in [13, 16]. The STA based cascade observer is formed by two parts:

- A *Luenberger observer*

$$\dot{\tilde{x}}_1(t) = A_{11,i}\tilde{x}_1(t) + A_{12,i}y_2(t) + K_i\left(y_1(t) - C_{11,i}\tilde{x}_1(t)\right). \tag{33}$$

The matrix K_i is designed such that the observation error dynamics

$$\dot{e}_1(t) = \left(A_{11,i} - K_i C_{11,i}\right) e_1(t) = \hat{A}_i e_1(t), \quad e_1(0) = x_1(0) - \tilde{x}_1(0),$$

is exponentially stable. The trajectory of the observer is assumed continuous, then at the switching moments

$$\tilde{x}_1(t_j^-) = \tilde{x}_1(t_j^+).$$

Since the dynamical error $e_1(t)$ is stable there exist positive constants γ_i, η_i such that

$$\|e_1(t)\| \leq \gamma_i e^{-\eta_i (t-t_j)} \|e_1(0)\| \leq \gamma_i e^{-\eta_i (t-t_j)} \left(\mu + \|\tilde{x}(0)\|\right) \leq \bar{\gamma}_i.$$

This Luenberger observer only assure that observation error $e_1(t) = x_1(t) - \tilde{x}_1(t)$ is bounded.

- A *hierarchical observer* composed by:

  - A family of STA based observers

$$\dot{x}_{ak} = A_{11,i}\tilde{x}_1(t) + A_{12,i} y_2(t) - L_{k,i}(t) \left(C_{11,i} A_{11,i}^{k-1} L_{k,i}\right)^{-1} v_k(t), \tag{34}$$

  $k = 1, \ldots, l-1,$, that reconstruct theoretically exactly, step by step, the output error $y_1(t) - C_{11,i}\tilde{x}_1(t)$ uncertainties/perturbations free part and its $l-1$ derivatives. Matrices $L_{k,i}(t)$ are design parameters and $v_k(t)$ is an output injection signal based on the STA.
  - And an algebraic part

$$\hat{x}(t) = T_{x,i}^{-1} \begin{bmatrix} \hat{x}_1(t) \\ y_2(t) \end{bmatrix}, \tag{35}$$

  where $\hat{x}_1(t) = \tilde{x}_1(t) - \mathcal{O}_{1,i,l}^{+} v(t)$ with $\mathcal{O}_{1,i,l}$ the observability matrix of the pair $(C_{11,i}, A_{11,i})$ and

$$v(t) = \begin{bmatrix} C_{11,i}\tilde{x}_1(t) - y_1(t) \\ v_1(t) \\ v_2(t) \\ \vdots \\ v_{l-1}(t) \end{bmatrix} \in \mathbb{R}^{(p-m)l}.$$

  The algebraic part (35) reconstructs theoretically exactly the states in finite time.

To reconstruct the states $x(t)$ theoretically exactly before half of the dwell time, it is necessary to recover the vectors $C_{11,i} A_{11,i}^{k-1} x_1(t)$ by using a family of STA based observers with a convergence time $t_r < \frac{\delta}{2(l-1)}$. Each observer is turn on once the lower derivative observers have converged. Assuring theoretically exact reconstruction of the output error and its $l-1$-derivatives for $t_{j-1}+\frac{\delta}{2}\leq t\leq t_j$. The design of the STA-based hierarchical observed (34) is given in the following theorem.

Theorem 3 *Assume*

(a) The auxiliary state vectors x_{ak}, for all $k=1,\ldots,l-1$, $i\in\mathcal{I}$ and $\tau_k=t_{j-1}+\frac{\delta(k-1)}{2(l-1)}\leq t\leq t_j$, are designed as in (34)*, where $L_{k,i}(t)\in\mathbb{R}^{n-m\times p-m}$ is a design matrix such that* det $\left(C_{11,i}A_{11,i}^{k-1}L_{k,i}\right)\neq 0$.

(b) Let τ_k, $k=2,\ldots,l-1$ be the convergence time of the $k-1$-th element of the hierarchical observer. At $t=\tau_k$ the k-th variable x_{ak} satisfies

$$
\begin{aligned}
C_{11,i}x_{a1}(t_{j-1}^{+}) &= y_1(t_{j-1}^{+}),\\
C_{11,i}A_{11,i}^{k-1}x_{ak}(\tau_k) &= C_{11,i}A_{11,i}^{k-1}\tilde{x}_1(\tau_k)-v_{k-1}(\tau_k);
\end{aligned}
$$

$i=\mathcal{I}$, and $j=1,2,\ldots$.

(c) The sliding variables s_k are designed as

$$
s_k(y_1(t),x_{ak}(t))=\begin{cases} y_1(t)-C_{11,i}x_{a1}(t), & k=1,\\ C_{11,i}A_{11,i}^{k-1}\tilde{x}_1(t)-v_{k-1}(t) & \\ -C_{11,i}A_{11,i}^{k-1}x_{ak}(t), & k{=}2,\ldots,l-1.\end{cases} \tag{36}
$$

(d) The output injection v_k is designed as an STA of the form

$$
\begin{aligned}
v_k(t) &= -\kappa_{k,i_1}\lceil s_k(y_1(t),t)\rfloor_{p-m}^{\frac{1}{2}}+\varpi_k(t),\\
\dot{\varpi}_k &= -\kappa_{k,i_2}\lceil s_k(y_1(t),t)\rfloor_{p-m}^{0},\\
\varpi_k(t_0)=0,\ \ &\varpi_k(\tau_k^{+})=\varpi_k(\tau_k^{-})+\bar{\varpi}_{i,k};
\end{aligned} \tag{37}
$$

where $\bar{\varpi}_{i,k}\in\mathbb{R}^{p-m}$ can be designed to guarantee that the observer does not lose the convergence at the switching times, for more details about this issue see discussion section of [16].

(e) And $(\kappa_{k,i_1},\kappa_{k,i_2})$ are designed such that

$$
\kappa_{k,i_2}=\begin{cases}\frac{2(l-1)M_{2,i,k}}{\delta}+M_{1,i,k}, & \tau_k\leq t\leq\tau_{k+1},\\ 1.1M_{1,i,k}, & t\geq\tau_{k+1},\end{cases}
$$

and

$$
\kappa_{k,i_1}=\begin{cases}\sqrt{8\kappa_{k,i_2}}, & \tau_k\leq t\leq\tau_{k+1},\\ 1.5\sqrt{M_{1,i,k}}, & t\geq\tau_{k+1}.\end{cases}
$$

with

$$
\begin{aligned}
M_{1,i,k} &\geq \left\|C_{11,i}A_{11,i}^{k}\right\|\left(\|A_i-K_iC_i\|\bar{\gamma}_i+\|B_i\|\phi\right);\\
M_{2,i,k} &\geq \left\|C_{11,i}A_{11,i}^{k}\right\|\bar{\gamma}_i.
\end{aligned}
$$

Then,

$$v_k(t) = -C_{11,i}A_{11,i}^k\left(x_1(t) - \tilde{x}_1(t)\right),$$

for $t_{j-1} + \frac{\delta}{2} \le t \le t_j$ *and it is possible to reconstruct theoretically exactly all the vector functions* $C_{11,i}A_{11,i}^{k-1}x_1(t)$ *in the same time interval.*

Proof This result is constructive and can be obtained straightforward following the proof of Theorem 2. □

Using the family of STA observers the output y_1 and its $l-1$ time-derivatives at every location have been reconstructed theoretically exactly in the time interval $[t_{j-1} + \frac{\delta}{2}, t_j]$. Using this information we can construct the vector

$$\mathcal{O}_{1,i,l}x_1(t) = \mathcal{O}_{1,i,l}\tilde{x}_1(t) - v(t).$$

Since the pair $(A_{11,i}, C_{11,i})$ is observable, the pseudo-inverse of $\mathcal{O}_{1,i,l}$ is well defined and the states can be recovered by means of the equation

$$x_1(t) = \tilde{x}_1(t) - \mathcal{O}_{1,i,l}^{+}v(t). \tag{38}$$

Then, the algebraic observer is suggested as

$$\hat{x}_1(t) = \tilde{x}_1(t) - \mathcal{O}_{1,i,l}^{+}v(t); \tag{39}$$

and we are able to reconstruct theoretically exactly the states for $t_{j-1} + \frac{\delta}{2} \le t \le t_j$ by using (35).

3.3.1 SCOISM Control

Once the states have been reconstructed, at $t = t_{j-1} + \frac{\delta}{2}$ the controllers are turn on. The design of the SCOISM is similar to the one presented in Sect. 2.2.2. Consider the SULTIS (24) and let $u(t) = u_n(t) + u_{SM}(t)$, here u_{SM} refers for SCOISM control part. Define the output based virtual sliding-mode variable

$$s(y,t) = G_i\left(y(t) - y\left(t_{j-1} + \frac{\delta}{2}\right)\right) - \int_{t_{j-1}+\frac{\delta}{2}}^{t} G_iC_i(A_i\hat{x}(\tau) + B_iu_n(\tau))\mathrm{d}\tau, \tag{40}$$

$i \in \mathcal{I}$; where $G_i \in \mathcal{G}_i$ is a design projection matrix, chosen, without loss of generality, such that $G_iC_iB_i = I$. Observe that $s\left(y\left(t_{j-1} + \frac{\delta}{2}\right), t_{j-1} + \frac{\delta}{2}\right) = 0$. Taking the first derivative of the sliding variable along the trajectory of (24) we obtain

$$\dot{s}(y,t) = G_iC_iA_i\left(x(t) - \hat{x}(t)\right) + u_{SM}(t) + \varphi(t). \tag{41}$$

Since $x(t) = \hat{x}(t)$, then

$$\dot{s}(y,t) = u_{SM}(t) + \varphi(t); \tag{42}$$

the equivalent control [32] that maintains the trajectory on the sliding mode is

$$u_{SMeq} = -\phi(t), \tag{43}$$

and the sliding mode dynamics of the SULTIS takes the form

$$\begin{aligned} \dot{x}(t) &= A_i x(t) + B_i u_n(t) \\ y(t) &= C_i x(t), \quad x(0) = x_0. \end{aligned} \tag{44}$$

The control law u_{SM} depends on the observed state $\hat{x}$. So it is necessary to design this controller in such a way that the observer STA dynamics does not affect the properties of the controller [9, 13, 16]. Hence, similar to Sect. 2.2.2 the following STA based controller is proposed

$$\begin{aligned} u_{SM} &= -G_i C_i A_i T_{x,i}^{-1} \begin{bmatrix} \mathcal{O}_{1,i,l}^{+} v(t) \\ 0 \end{bmatrix} - \kappa_{i_1} \lceil s(y(t),t) \rfloor_m^{\frac{1}{2}} + \omega(t), \\ \dot{\omega} &= -\kappa_{i_2} \lceil s(y(t),t) \rfloor_m^0, \quad \omega(0) = 0, \quad \omega(t_j^+) = \omega(t_j^-) + \bar{\omega}_i; \end{aligned} \tag{45}$$

we compensate the term $G_i C_i A_i T_{x,i}^{-1} \mathcal{O}_{i,l}^{+} v(t)$, the sliding dynamics have the typical STA form

$$\begin{aligned} \dot{s}(y,t) &= -\kappa_{i_1} \lceil s(y(t),t) \rfloor_m^{\frac{1}{2}} + \Omega(t), \\ \dot{\Omega}(t) &= -\kappa_{i_2} \lceil s(y(t),t) \rfloor_m^0 + G_i C_i A_i T_{x,i}^{-1} \begin{bmatrix} \left(\dot{x}_1(t) - \dot{\hat{x}}_1(t) \right) \\ 0 \end{bmatrix} + \dot{\varphi}(t), \\ s(y,t_{j-1}) &= 0; \end{aligned} \tag{46}$$

Thanks to the integral structure of $s(y,t)$ the non-Lipschitz dynamics of the observer that affect the controller, only depends on v, the continuous part of the observer dynamics. Hence, the continuity of u_{SM} is preserved.

The following Lemma gives the design of the STA gains $(\kappa_{i_1}, \kappa_{i_2})$ for the controller.

Lemma 4 *Suppose assumptions **A.3.4–A.3.9** are satisfied and*

$$\kappa_{i_2} = \begin{cases} \frac{2L_{2,i}}{\delta} + L_{1,i}, & t \leq \delta, \\ 1.1 L_{1,i}, & t > \delta, \end{cases}$$

and

$$\kappa_{i_1} = \begin{cases} \sqrt{8\kappa_{i_2}}, & t \leq \delta, \\ 1.5\sqrt{L_{1,i}}, & t > \delta. \end{cases}$$

with

$$L_{1,i} \geq \Phi + \|G_i C_i A_i T_{x,i}^{-1}\| \left(\|A_i - K_i C_i\|\gamma_i + \|B_i\|\phi\right);$$
$$L_{2,i} \geq \|G_i C_i A_i\|\gamma_i + \Phi.$$

Then, the sliding mode dynamics (46) *converges to the origin with a reaching time* $t_{jr} < t_{j-1} + \delta < t_j$.

Proof The proof of this results comes directly from Lemma 3. □

With this we assure the robustification of the nominal trajectory before the dwell time.

3.4 SULTIS Robustification Example

Now, lets recapitulate the design scheme presented in this section. For brevity, let us consider only the SCOISM-based robustification scheme for the transformed system. The SCISM design is similar.

Consider a system of the form (24) with two locations, where

$$\bar{A}_1 = \begin{bmatrix} 0\,1\,0\,0 \\ 0\,0\,1\,0 \\ 0\,0\,0\,1 \\ 1\,2\,3\,4 \end{bmatrix}, \quad \bar{A}_2 = \begin{bmatrix} 0 & 0 & -1 & 0 \\ -50 & -200 & -365 & -250 \\ 0 & 0 & 0 & -1 \\ 0 & -1 & 0 & 0 \end{bmatrix},$$
$$\bar{B}_1 = \begin{bmatrix} 0 \\ 0 \\ 0 \\ 1 \end{bmatrix}, \quad \bar{B}_2 = \begin{bmatrix} 0 \\ 2 \\ 0 \\ 0 \end{bmatrix}, \quad \bar{C}_1 = \begin{bmatrix} 0\,0\,0\,1 \\ 1\,0\,0\,0 \end{bmatrix}, \quad \bar{C}_2 = \begin{bmatrix} 1\,0\,0\,0 \\ 0\,1\,0\,0 \end{bmatrix},$$

and

$$\sigma(\bar{x}, t) = \begin{cases} 2 & \sigma(\bar{x}(t^-), t^-) = 1 \text{ and } |\bar{x}_4| \leq 0.5, \\ 1 & \sigma(\bar{x}(t^-), t^-) = 2 \text{ and } |\bar{x}_4| > 1, \\ \sigma(\bar{x}(t^-), t^-) & \text{otherwise.} \end{cases}$$

The system initiates on location $i_0 = 1$. The initial conditions $x_0 = [0.1, -0.1, 0.2, -1.5]^T$ are unknown with a known bound $\mu = 2$ and the uncertainties/perturbations used in the simulations are

$$\varphi(t) = 5\sin(\pi\cos(3\pi t)) + 10,$$

with known bounds given by $\phi = 15$ and $\Phi = 15\pi^2$. Note that the system is unstable.

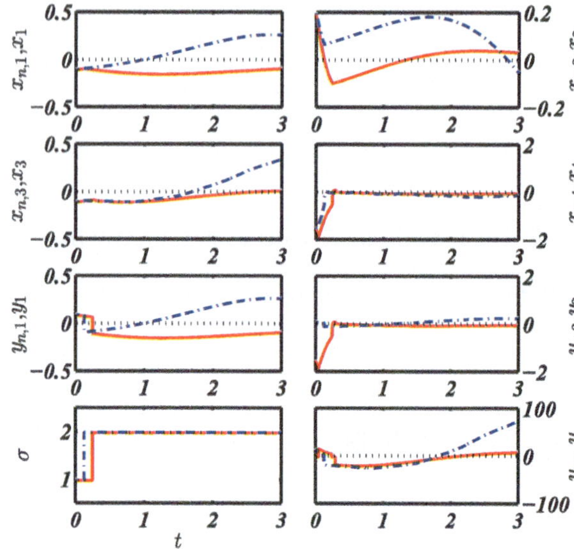

Fig. 7 Nominal behavior of (32) versus SULTIS (31) with the nominal controller only: Nominal behavior (continuous line), SULTIS (dashed line)

The system is transformed into the form (31), with

$$T_{x,1} = \begin{bmatrix} 0 & 1 & 0 & 0 \\ 0 & 0 & 1 & 0 \\ -1 & 0 & 0 & 0 \\ 0 & 0 & 0 & 1 \end{bmatrix}, \quad T_{x,2} = \begin{bmatrix} -1 & 0 & 0 & 0 \\ 0 & 0 & 1 & 0 \\ 0 & 0 & 0 & 1 \\ 0 & 0.5 & 0 & 0 \end{bmatrix},$$

$$T_{y,1} = \begin{bmatrix} 0 & 1 \\ 1 & 0 \end{bmatrix}, \quad \text{and} \quad T_{y,2} = \begin{bmatrix} -1 & 0 \\ 0 & 0.5 \end{bmatrix}.$$

The first step in the robustification scheme is the design of the nominal controller. Consider the transformed nominal system (32). An LQR nominal controller is designed in each location:

$$\bar{K}_1 = \begin{bmatrix} 2.4142 & 6.0210 & 9.0699 & 9.9279 \end{bmatrix},$$

and

$$\bar{K}_2 = \begin{bmatrix} -50.0200 & 1.2563 & 0.2688 & -251.8423 \end{bmatrix}.$$

The design of this nominal controller is done following the LMI strategy proposed in [1]. This nominal controller guarantees the stability of the nominal system (see Fig. 7) and satisfies the dwell time condition with $\delta = 0.1$. When the nominal controller is applied, and there are uncertainties/perturbations in the system, the nominal controller alone is incapable of taking the states to zero, see Fig. 7. However, any switched/hybrid controller design approach can be applied.

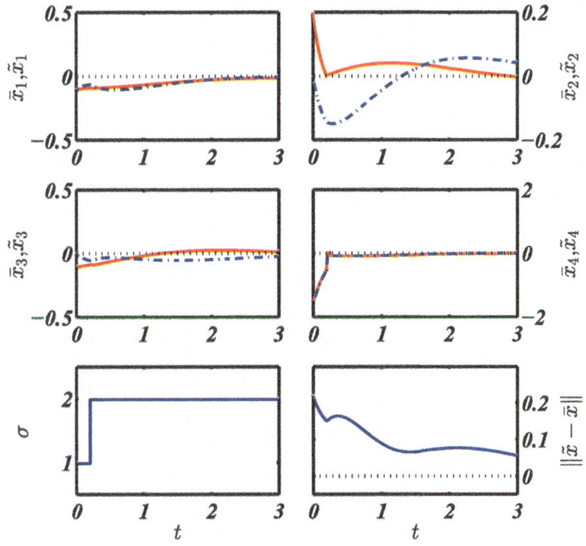

Fig. 8 Luenberger observer behavior: real state (continuous line), observed state $\tilde{x}$ (dashed line)

Now, let us continue with the design of the cascade observer. Recall that this observer is composed of two parts: The Luenberger-type error stabilizer and the hierarchical observer. The behavior of the stabilizer is shown in Fig. 8, and it is designed for the uncertainties/perturbations free part of the SULTIS, assuring the existence of a bound for the observation error e_1 and its first derivative.

Now, let design the second part of the cascade observer. Since the considered system has an observability index $l = 2$, a hierarchical observer composed of two sliding modes observers is necessary. The observers are designed such that they guarantee a convergence time lower than 0.025.

At every switching time, the controller is turned off, and the first observer starts to evolve. At $t = t_j + 0.025$, the second observer is turn on. The reaching phase of both observers is illustrated in Fig. 9. Notice that the first observer converges before $t = t_j + 0.025$. Moreover, the full observer converges before $t = t_j + 0.05$. A switched STA gains strategy [16] is applied to both observers to attenuate the chattering. This attenuation of the chattering is depicted in the second column of the figure. The complete behavior of the cascade observer is presented in Fig. 10. Hence, the system states have been reconstructed without filters, theoretically exactly before half of the dwell time.

Once the observer has converged, the SCOISM controller is turned on. Observe that if the controller is turned on since the switching time, the given reaching time results cannot guarantee the convergence time. The sliding variables of the controller are given in Fig. 11. The sliding variables converge before the dwell time, assuring exact compensation of the uncertainties/perturbations before the dwell time and the robustification of the nominal controller. This strategy helps to accomplish a convergence time $t_d = 0.05$, and after the dwell time, the gains are reduced to attenuate the chattering. Moreover, the generated control signal is continuous (see Fig. 12).

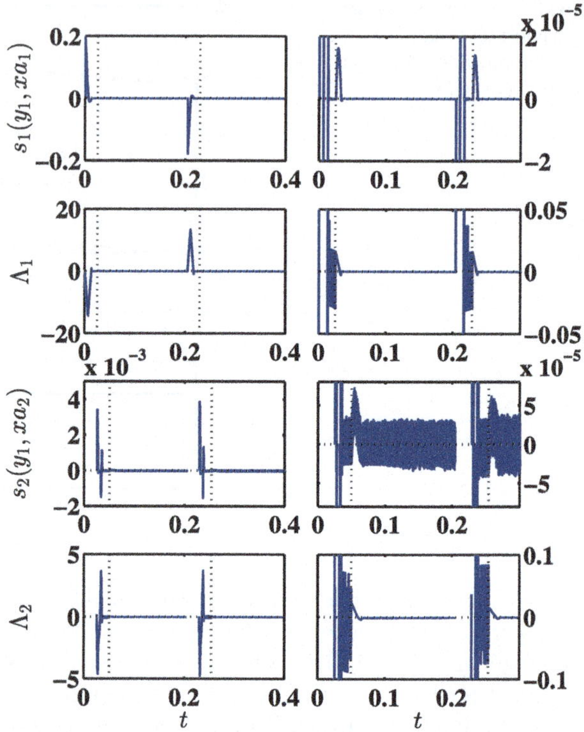

Fig. 9 Sliding variables of the hierarchical observer

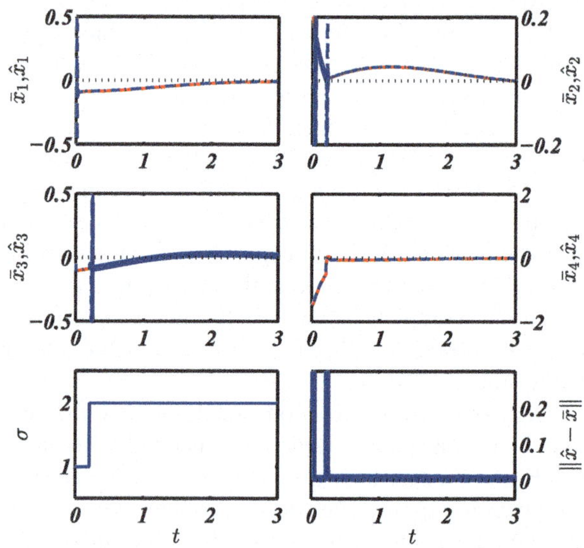

Fig. 10 Cascade observer: real states (continuous line) vs observed states $\hat{x}$ (dashed line)

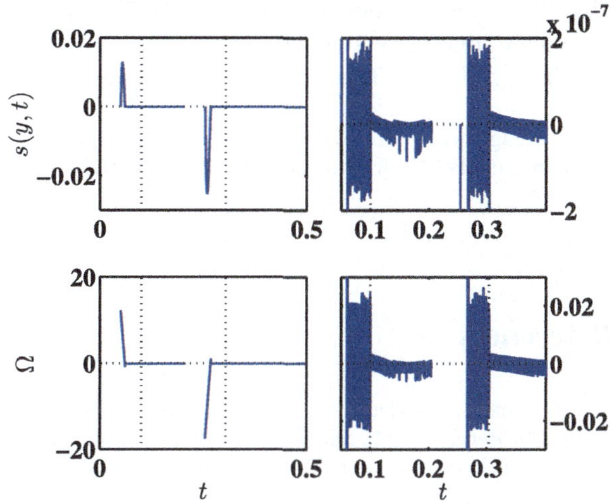

Fig. 11 Controller sliding variables

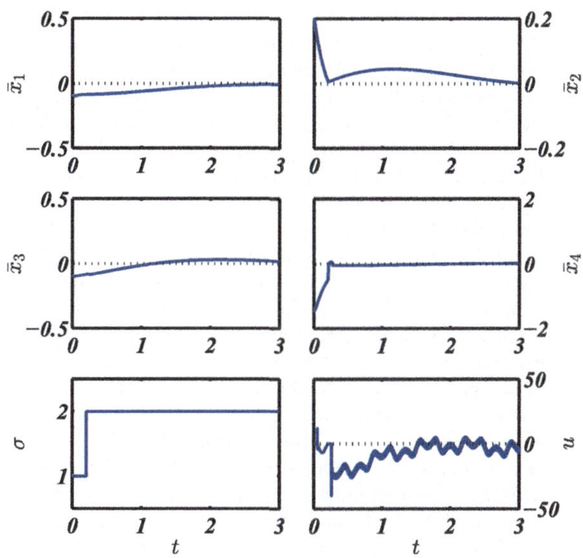

Fig. 12 SULTIS with the SCOISM controller

4 Conclusions

In this chapter, a robustification methodology applied to different scenarios has been explained. The proposed robustification schemes are based on CISM and COISM approaches and are applied to ULTIS and SULTIS. The full methodology is detailed, and the design steps are delineated. Four scenarios are considered. The first two scenarios are devoted to the design schemes for ULTIS and considering complete state information and only output measurements. The last two are dedicated to the

SULTIS with the same variants as the ULTIS, namely, complete state information and only output measurements. The design steps of the methodology are described, and the specific details are mentioned in the examples.

Acknowledgements This work was supported in part by the Secretaría de Investigación y Posgrado of the Instituto Politécnico Nacional under Grants 20220143 and 20221915. And PAPIIT-UNAM IN102621

References

1. Alessandri, A., Coletta, P.: Design of luenberger observers for a class of hybrid linear systems. In: Di Benedetto, M., Sangiovanni-Vincentelli, A. (eds.) Hybrid Systems: Computation and Control. Lecture Notes in Computer Science, vol. 2034, pp. 7–18. Springer, Berlin Heidelberg (2001)
2. Barmish, B.R., Corless, M., Leitmann, G.: A new class of stabilizing controllers for uncertain dynamical systems. SIAM J. Control. Optim. **21**(2), 246–255 (1983)
3. Bejarano, F., Pisano, A., Usai, E.: Finite-time converging jump observer for switched linear systems with unknown inputs. Nonlinear Anal. Hybrid Syst. **5**(2), 174–188 (2011). (Special Issue related to IFAC Conference on Analysis and Design of Hybrid Systems (ADHS09))
4. Bejarano, F.J., Fridman, L.: State exact reconstruction for switched linear systems via a super-twisting algorithm. Int. J. Syst. Sci. **42**(5), 717–724 (2011)
5. Bejarano, F.J., Fridman, L., Poznyak, A.S.: Output integral sliding mode control based on algebraic hierarchical observer. Int. J. Control **80**, 443–453 (2007)
6. Branicky, M.S.: Introduction to hybrid systems. In: Handbook of Networked and Embedded Control Systems, pp. 91–116. Birkhäuser Boston (2005)
7. Castaños, F., Fridman, L.: Analysis and design of integral sliding manifolds for systems with unmatched perturbations. IEEE Trans. Autom. Control **51**, 853–858 (2006)
8. Chalanga, A., Kamal, S., Bandyopadhyay, B.: Continuous integral sliding mode control: A chattering free approach. In: ISIE, Taiwan (2013)
9. Chalanga, A., Kamal, S., Fridman, L.M., Bandyopadhyay, B., Moreno, J.A.: Implementation of super-twisting control: Super-twisting and higher order sliding-mode observer-based approaches. IEEE Trans. Ind. Electron. **63**(6), 3677–3685 (2016)
10. Dávila, J., Ríos, H., Fridman, L.: State observation for nonlinear switched systems using non-homogeneous high-order sliding mode observers. Asian J. Control **14**(4), 911–923 (2012)
11. Estrada, M.A., Fridman, L., Moreno, J.A.: Control of fully actuated mechanical systems via super-twisting based lyapunov redesign. IFAC-PapersOnLine **53**(2), 5117–5121 (2020). (21th IFAC World Congress)
12. Floquet, T., Barbot, J.P.: Super twisting algorithm-based step-by-step sliding mode observers for nonlinear systems with unknown inputs. Int. J. Syst. Sci. **38**(10), 803–815 (2007)
13. Fridman, L., Galván-Guerra, R., Velázquez-Velázquez, J.t.E., Iriarte, R.: New Perspectives and Applications of Modern Control Theory, chap. Sliding Modes for Switched Uncertain Linear Time Invariant Systems: An Output Integral Sliding Mode Approach, pp. 153–185. Springer International Publishing, Cham (2018)
14. Fridman, L., Poznyak, A.: Bejarano Rodríguez. Robust Output LQ Optimal Control via Integral Sliding Modes. Birkhäuser, F.J. (2014)
15. Galván-Guerra, R., Fridman, L., Iriarte, R., Velázquez-Velázquez, J.E., Steinberger, M.: Integral sliding-mode observation and control for switched uncertain linear time invariant systems: a robustifying strategy. Asian J Control **20**(4), 1551–1565 (2018)

16. Galván-Guerra, R., Fridman, L., Velázquez-Velázquez, J.E., Kamal, S., Bandyopadhyay, B.: Continuous output integral sliding mode control for switched linear systems. Nonlinear Anal. Hybrid Syst. **22**, 284–305 (2016)
17. Gonzalez, T., Moreno, J., Fridman, L.: Variable gain super–twisting sliding mode control. IEEE Trans. Autom. Control **57**(8), 2100–2105 (2012)
18. Hautus, M.: Strong detectability and observers. Linear Algebra Appl. **50**, 353–368 (1983)
19. Khalil, H.K.: Nonlinear Systems, 3rd edn. Prentice Hall, Upper Saddle River, New Jersey, pp. 07458 (2002)
20. Kratz, W.: Characterization of strong observability and construction of an observer. Linear Algebra Appl. **221**, 31–40 (1995)
21. Levant, A.: Robust exact differentiation via sliding mode technique. Automatica **34**(3), 379–384 (1998)
22. Lian, J., Zhao, J., Dimirovski, G.M.: Integral sliding mode control for a class of uncertain switched nonlinear systems. Eur. J. Control. **16**(1), 16–22 (2010)
23. Pérez-Ventura, U., Fridman, L.: Design of super-twisting control gains: a describing function based methodology. Automatica **99**, 175–180 (2019)
24. Pérez-Ventura, U., Fridman, L.: When is it reasonable to implement the discontinuous sliding-mode controllers instead of the continuous ones? frequency domain criteria. Int. J. Robust Nonlinear Control **29**(3), 810–828 (2019)
25. Ríos, H., Kamal, S., Fridman, L.M., Zolghadri, A.: Fault tolerant control allocation via continuous integral sliding-modes: A hosm-observer approach. Automatica **51**, 318–325 (2015)
26. Rubagotti, M., Estrada, A., Castaños, F., Ferrara, A., Fridman, L.: Integral sliding mode control for nonlinear systems with matched and unmatched perturbations. IEEE Trans. Autom. Control **56**(11), 2699–2704 (2011)
27. Rugh, W.J.: Linear System Theory. Prentice Hall (1993)
28. Saadaoui, H., Manamanni, N., Djemaï, M., Barbot, J., Floquet, T.: Exact differentiation and sliding mode observers for switched Lagrangian systems. Nonlinear Anal. Theory Methods Appl. **65**(5), 1050–1069 (2006). (Hybrid Systems and Applications (4))
29. Seeber, R., Horn, M.: Stability proof for a well-established super-twisting parameter setting. Automatica **84**, 241–243 (2017)
30. Seeber, R., Horn, M., Fridman, L.: A novel method to estimate the reaching time of the super-twisting algorithm. IEEE Trans. Autom. Control, 1–8 (2018)
31. Utkin, V., Shi, J.: Integral sliding mode in systems operating under uncertainty conditions. In: Proceedings of the 35th IEEE Conference on Decision and Control, 1996, vol. 4, pp. 4591–4596 (1996)
32. Utkin, V.I.: Sliding Modes in Control and Optimization. Springer (1992)
33. Van der Schaft, A., Schumacher, H.: An Introduction to Hybrid Dynamical Systems, 1st edn. Lecture Notes in Control and Information Sciences, vol. 251. Springer, London (2000)
34. Velázquez-Velázquez, J.E., Galván-Guerra, R., Fridman, L., Iriarte, R.: Two relay control robustification by continuous switched integral sliding modes. IET Control Theory Appl. **13**(9), 1374–1382 (2019)

Novel Applications in Human Communities and Economic Systems

Model Predictive Tumour Volume Control Using Nonlinear Optimization

György Eigner, Máté Siket, Bence Czakó, Dániel András Drexler, Imre Rudas, Ákos Zarándy, and Levente Kovács

Abstract In this study, we developed a Nonlinear Model Predictive Control algorithm for tumour growth regulation. It is unique from the perspective that the subjects of the optimization were the feedback gains in the state feedback kind closed control loop. During the controller design, we utilized the results of our previous qualitative analysis of the model to be controlled. In order to realize the state feedback we implemented a discrete Extended Kalman Filter for state estimation purposes with slightly different model parameters compared to the virtual patient model to be controlled. The results show that the algorithm performed well within the predefined circumstances and it is able to satisfy the strict constraints.

Keywords Nonlinear model predictive control · Tumor growth control · Extended kalman filter · Nonlinear optimization · Cancer treatment

The paper was supported by the Eötvös Loránd Research Network Secretariat under grant agreement no. ELKH KÖ-40/2020 ('Development of cyber-medical systems based on AI and hybrid cloud methods'). This project received funding from the European Research Council (ERC) under the European Union's Horizon 2020 research and innovation programme (grant agreement No 679681). Project no. 2019-1.3.1-KK-2019-00007. has been implemented with the support provided from the National Research, Development and Innovation Fund of Hungary, financed under the 2019-1.3.1-KK funding scheme. M. Siket and B. Czakó were supported by the úNKP-20-3 New National Excellence Program of the Ministry for Innovation and Technology.

G. Eigner · M. Siket · B. Czakó · D. A. Drexler (✉) · I. Rudas · L. Kovács
Physiological Controls Research Center Research and Innovation Center, Óbuda University, 1096, Bécsi street 96/B., Budapest, Hungary
e-mail: drexler.daniel@nik.uni-obuda.hu

M. Siket · Á. Zarándy
Institute for Computer Science and Control (SZTAKI), Eötvös Lóránd Research Network (ELKH), Kende u. 13-17., 1111 Budapest, Hungary

P. Shi et al. (eds.), *Complex Systems: Spanning Control and Computational Cybernetics: Applications*, Studies in Systems, Decision and Control 415,
https://doi.org/10.1007/978-3-031-00978-5_10

1 Introduction

The control of physiological systems is often a challenging task due to inherent parametric differences among the subjects and limited measurement accessibility. Controlling the evolution of cancerous diseases form no exception, as scientist have been trying to tackle this issue for more than half a century now [1]. The goal is usually to optimize and individualize treatment strategies by developing a controller, that is able to determine an optimal amount of drug for each patient. In practice this involves developing a model that can accurately capture the main dynamical components of tumour growth which is also relatively simple to be utilized for control purposes. This model is then used in an optimizer, which is able to optimize the set of time instants where treatment should take place or the amount of drug that must be given to the patients. A summary of optimization models and control theoretical approaches were presented in [2] and [3]. A collection of more recent modelling techniques can also be found in the excellent review of [4]. In our work, we will focus on the optimization of chemotherapeutic treatment protocols using nonlinear model predictive control (NMPC) with impulsive input definition. Our approach is similar to what can be seen in [5, 6]. In previous works, a simple model was developed which is able to capture the main dynamics of tumour growth in [7]. The model describes the dynamics of the living and dead volume of the tumour with rates created with mass-action kinetics and Michaelis-Menten kinetics, augmented with an extra equation of the clearance of the drug. The model parameters were fitted to data from mice experiments [8], using the Stochastic Approximation of Expectation Maximization (SAEM) algorithm. Using this model, an NMPC was developed in [9], which could in silico solve the control problem for each virtual patient. The NMPC used an approximate impulse signal definition in conjunction with Direct Multiple Shooting (DMS) implementation, which is described in detail in [10]. In order to be a feasible design, the controller must be augmented with a state estimator as well.

2 Materials and Methods

2.1 Applied Tumor Growth Model

In this article we applied a minimal tumour growth model firstly appeared in [11] with two state variables describing the counter-effect and interaction between a specific drug agent and living tumour volume. Later on, the model has been extended with and additional state describing the necrotic tumour volume and further pharmacodynamical and pharmacokinematical interactions has been introduced in [12]. In this study, we applied the third order minimal model of tumour growth under treatment from [12–14].

The differential equations of the third order nonlinear model are the following [13, 14]:

$$\dot{x}_1(t) = (a - n)x_1(t) - b\frac{x_1(t)x_3(t)}{ED_{50} + x_3(t)}, \tag{1}$$

$$\dot{x}_2(t) = nx_1(t) + b\frac{x_1(t)x_3(t)}{ED_{50} + x_3(t)} - wx_2(t), \tag{2}$$

$$\dot{x}_3(t) = -c\frac{x_3(t)}{K_B + x_3(t)} - b_k\frac{x_1(t)x_3(t)}{ED_{50} + x_3(t)} + u(t), \tag{3}$$

where x_1 [mm^3] and x_2 [mm^3] are the time functions of the living and dead tumour volumes, respectively. The x_3 [mg/kg] describes the time function of the drug concentration in the subject. The u [mg/(kg · day)] is the input of the model (drug intake). The parameters and their explanations can be found in Table 1.

The measured output y mm^3 of the system is the sum of the proliferating (x_1) and necrotic (x_2) tumour volumes, namely

$$y = x_1 + x_2. \tag{4}$$

In this model, drug resistance of the system against chemotherapeutic drug is not modelled explicitly. However, the effect of resistance appears as a special combination of the parameters. When $a - b - n > 0$ then, the drug is not able to decrease the tumour volume due to the proliferation rate (a) is higher than the drug efficiency and necrotic rates (b and n). In this specific case the tumour volume increases without limits, however, the drug decreases the growth rate (slower growing dynamics). The qualitative analysis of the model has been previously done in [13]. The values of the parameters for the mice were identified based on a mixed-effect model [15], where the experimental data are from mice experiments published in [8]. The parameters are given in Table 2.

Table 1 The parameters of the third order tumour growth model

Parameter notation	Parameter name	Parameter dimension
a	Proliferation rate	$\frac{1}{\text{day}}$
b	Drug efficiency rate	$\frac{1}{\text{day}}$
c	Drug clearance	$\frac{1}{\text{day}}$
n	Necrosis rate	$\frac{1}{\text{day}}$
b_κ	Modified drug efficiency rate	$\frac{\text{mg}}{\text{kg}\cdot\text{day}\cdot\text{mm}^3}$
K_B	Michaelis–Menten constant of the drug	$\frac{\text{mg}}{\text{kg}}$
ED_{50}	Median effective dose of the drug	$\frac{\text{mg}}{\text{kg}}$

Table 2 Applied nominal parameter set and standard deviation as the results of the identification from [15]

Parameter	Nominal
a [1/day]	0.3098
b [1/day]	0.1799
c [1/day]	0.2717
n [1/day]	0.1732
$b_k \left[\frac{\text{mg}}{\text{kg}\cdot\text{day}\cdot\text{mm}^3}\right]$	$6.1 \cdot 10^{-7}$
K_B [mg/kg]	0.2296
ED_{50} [10^{-4}mg/kg]	1.3301
w [1/day]	0.3413

2.2 Controller Design

Basis of the optimization In this work, we exploit the results of the qualitative analysis of Drexler et al. introduced in [13] as the basis for the controller design and optimization. They proved that for (1)–(3) a stable equilibrium can be reached using state feedback stabilization in the case of specific circumstances.

The application of a linear state feedback controller is possible without using the volume of the dead tumour cell volume x_2 due to the fact that (1) and (3) do not depend on (2). Thus, we are able to omit the latter equation during the state feedback design. The control signal in this case can be the following, in accordance with [13]:

$$u = k_1 x_1 - k_3 x_3, \tag{5}$$

where k_1 and k_3 are the feedback gains. The values of k_1 and k_3 are positive, albeit, we use negative feedback (since increasing u results in a decrease in x_1).

By utilizing the aforementioned simplification during the controller design, from the model point of view (i.e., omitting x_2 from the feedback model) the closed-loop system dynamics can be written as

$$\dot{x}_1 = (a - n)\, x_1 - b \frac{x_1 x_3}{ED_{50} + x_3} \tag{6}$$

$$\dot{x}_3 = -c \frac{x_3}{K_B + x_3} - b_K \frac{x_1 x_3}{ED_{50} + x_3} + k_1 x_1 - k_3 x_3. \tag{7}$$

During the qualitative analysis, Drexler et al. transformed the model equilibria equations to the following form

$$0 = (a - n)\,x_1(ED_{50} + x_3) - bx_1x_3 \tag{8}$$

$$\begin{aligned} 0 = &-c\frac{x_3(ED_{50} + x_3)}{K_B + x_3} - b_\kappa x_1x_3 \\ &+(k_1x_1 - k_3x_3)(ED_{50} + x_3). \end{aligned} \tag{9}$$

which resulted in four equilibria from which one model equilibrium is exploitable for control purposes. The exact steps of the qualitative investigations can be found in [13].

In this model, the drug serum level equilibrium point can be calculated by

$$x_3^* = -ED_{50}\frac{a - n}{a - b - n}. \tag{10}$$

A physiological constraint is that if the tumour is not self-healing, namely, the $a > n$ inequality holds, then x_3^* can be positive if and only if

$$a - n - b < 0. \tag{11}$$

That means that the tumour volume can be decreased by using the given drug [16]. In the further work, we suppose that (11) is true for the parameters a, b, n.

The equilibrium of the tumour volume can be calculated by

$$x_1^* = \frac{x_{1,n}}{x_{1,d}} \tag{12}$$

where

$$\begin{aligned} x_{1,n} = &\; bED_{50}(a - n)\,(b(c + k_3K_B) \\ &-a(c + k_3(-ED_{50} + K_B)) \\ &+(c - ED_{50}k_3 + k_3K_B)n) \end{aligned} \tag{13}$$

$$\begin{aligned} x_{1,d} = &\; (a - b - n)(ab_\kappa - bk_1 - b_\kappa n)\,\cdot \\ &\cdot(a(ED_{50} - K_B) + bK_B \\ &+(-ED_{50} + K_B)n). \end{aligned} \tag{14}$$

Taking the aforementioned considerations, this equilibrium is positive if either

$$k_1 > b_\kappa\frac{a - n}{b} \tag{15}$$

$$k_3 > c\frac{a - b - n}{aED_{50} - aK_B + bK_B - ED_{50}n + K_Bn} \tag{16}$$

or

$$k_1 < b_\kappa \frac{a-n}{b} \tag{17}$$

$$k_3 < c\frac{a-b-n}{aED_{50} - aK_B + bK_B - ED_{50}n + K_Bn}. \tag{18}$$

The qualitative analysis of [13] showed that an other additional constraint should be considered with respect to k_3 to be sure that the $x_{1,3}*$ are positive:

$$k_3 > \frac{\varphi}{\omega} \tag{19}$$

with

$$\begin{aligned}\varphi = {} & c(-a+b+n)^2(a^2b_\kappa(ED_{50} - K_B) \\ & -b^2k_1K_B + 2ab_\kappa(-ED_{50} + K_B)n \\ & +b_\kappa(ED_{50} - K_B)n^2)\end{aligned} \tag{20}$$

$$\begin{aligned}\omega = {} & (a(ED_{50} - K_B) + bK_B \\ & +(-ED_{50} + K_B)n)^2(a^2b_\kappa \\ & +b^2k_1 - 2ab_\kappa n + b_\kappa n^2).\end{aligned} \tag{21}$$

In this case, the positive equilibrium is stable, otherwise it is unstable.

Taking into account the previously described conditions, the following theorem can be introduced regarding the closed-loop [13].

Theorem 1 ([13]) *Suppose that* $a - n > 0$ *and* $a - n - b < 0$ *are held, and the model parameters are positive. If the* k_1 *and* k_3 *feedback gains are chosen to satisfy (15)–(16) or (17)–(18) then there exists a positive equilibrium of the system described by (6)–(7). Furthermore, if* k_1 *satisfies (15) and* k_3 *satisfies (19), then this equilibrium is stable.*

In our study, the conditions of Theorem 1 are applied and embedded in the optimization process to find the best suitable gains which satisfy the predefined requirements.

Nonlinear Model Predictive Control Nonlinear Model Predictive Control (NMPC) is a widely used optimization based controller by which a given cost function can be minimized over a finite or infinite prediction horizon using the model of the controlled phenomena.

In this specific case, the NMPC takes the solution of the estimated states $\hat{\mathbf{x}}(t)$ of system (1)–(3) and predicts the evolution of the system under given external excitations on the given intervals, $(t_i, t_{i+1}], i \in \{0, 1, \ldots, N\}$ with N being the number of prediction intervals. In this study, we applied the following cost function

$$\mathcal{J}(y_{final}, u_i) = \alpha\, y_{final}^2 + \beta \int_{t_i}^{t_{i+1}} (u_i)^2 \,\mathrm{d}t, \tag{22}$$

where α and β are scalar control parameters, the y_{final} is the final output of the model at the end of the prediction horizon, u_i is a constant piecewise continuous control input defined by (5) on the interval $(t_i, t_{i+1}]$. In practice, the model is solved numerically while the integral is approximated by the trapezoidal rule.

With respect to the given control aspect described by (1)–(3) using (5) based on Theorem 1 the numerical optimization problem for the NMPC at t_r can be specified as follows.

$$\begin{aligned}
&\min k_{1,3} J_N(\mathbf{u}; t_i) = \sum_{r=0}^{N-1} \mathcal{J}(y_{i+r}, u_{i+r}) \\
&\text{s.t}\mathbf{u} \quad \in [\mathbf{0}, \bar{\mathbf{u}}] \\
&\mathbf{u^c} \quad \in [\mathbf{0}, \bar{\mathbf{u^c}}] \\
&\mathbf{k}_{1,3} \quad \in [\underline{\mathbf{k}_{1,3}}, \overline{\mathbf{k}_{1,3}}] \\
&(13), (14) \quad \text{are hold} \\
&x_{1,endmax} - \frac{x_{1,n}}{x_{1,d}} = 0 \\
&(15), (16), (19), (20), (21) \text{arehold}
\end{aligned} \tag{23}$$

where $\mathbf{u} = (u_i, u_{i+1}, \ldots u_{i+N-1})$ is the optimal input sequence where each element is constrained to lie in the interval $0 \le u_{i+r} \le \bar{u}$, which entails $\bar{\mathbf{u}} = \bar{u}\mathbf{1}$ (where $\mathbf{1}$ is an N dimensional column vector with all of its entries being equal to 1), and the second group of constraints $0 \le u_{i+r}^c \le \bar{u}^c$ denotes the cumulative dosage associated with each u_{i+r}, from which $\mathbf{u^c} = \bar{u}^c\mathbf{1}$, the $\overline{\mathbf{k}_{1,3}}$ is the upper limit of the feedback gains, the $\underline{\mathbf{k}_{1,3}}$ is the lower limit of the feedback gains, the $x_{1,endmax}$ is the maximum theoretical living tumour volume to be reached by the end of the therapy. The cumulative dosage u_i^c is defined here as the sum of doses in the past 10 days,

$$u_i^c = \sum_J u_j, \; J := \{\, j \mid t_i - 10 \le t_j \le t_i \,\}. \tag{24}$$

According to [8] the maximal tolerable dose (MTD) of PLD in mice is 8 mg/kg which could be repeated in every 10 days without triggering an irreversible weight loss. To define a safe cumulated dose threshold in our method, the maximum given PLD of 16 [mg/kg] was lowered to $\bar{u}_i^c = 14$ [mg/kg] in 10 days to minimize the possibility of severe systemic toxicity. The upper bound for each dose was set to $\bar{u}_i = 4$ [mg/kg] and is smaller than the 8 [mg/kg] which is the MTD that is given to a subject during a conventional course of therapy. Another important aspect is the days of injections (days of actuations). During a real therapy it is really hard to solve to administer the drug every day, since many logistical issues arise (e.g. capacity and availability of human resource,

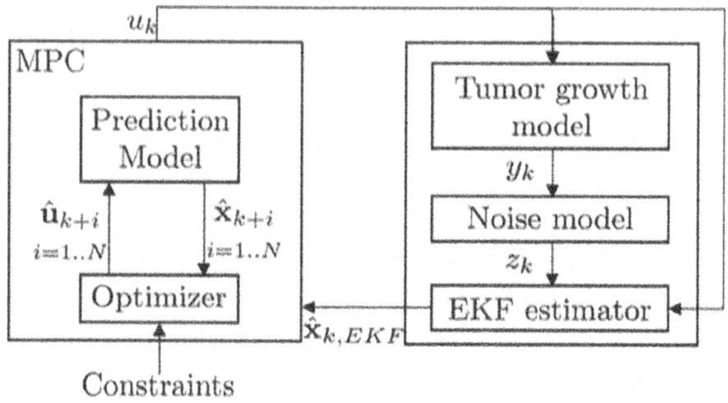

Fig. 1 Closed control-loop

limitation in the amount of drug). Thus, we embed another constraint in the algorithm, the "days of injections" (DOI). The DOI regulates that what days of the week can a given drug administration be executed. During the optimization that means only that we do the optimization every day, however, if the day is not equal with the assigned day of administration the drug injection is not executed. The algorithm and the optimization are not sensitive to such a hard saturation since the optimization is done on a daily basis where the initial conditions are changed in each iteration cycle.

Figure 1 shows the structure of the closed control-loop.

2.3 Extended Kalman Filter

An Extended Kalman Filter is implemented in order to estimate the unmeasured state variables. The method assumes that the state variables and the output have an additive Gaussian noise. Thus, a general, nonlinear system can be given as:

$$\dot{\mathbf{x}}(t) = f(\mathbf{x}(t), \mathbf{u}(t)) + \mathbf{w}(t), \tag{25}$$

$$\mathbf{y}(t) = h(\mathbf{x}(t), \mathbf{u}(t)) + \mathbf{v}(t), \tag{26}$$

where f describes the nonlinear system dynamics, h is the output equation, $\mathbf{w}$ is the additive process noise or disturbance and $\mathbf{v}$ is the measurement noise.

During the prediction step, the model and the covariances were solved numerically by the first order Euler method according to the sampling time of the measurements, resulting in a discrete EKF implementation. The prediction step is denoted as $r|r-1$ the previous step as $r-1$ and the current step as r:

$$\hat{\mathbf{x}}[r|r-1] = f(\hat{\mathbf{x}}[r-1], \mathbf{u}[r]), \tag{27}$$

$$y[r] = h(\hat{\mathbf{x}}[r|r-1], \mathbf{u}[r]), \tag{28}$$

$$\mathbf{P}[r|r-1] = \mathbf{F}[r]\mathbf{P}[r-1]\mathbf{F}^{\top}[r] + \mathbf{Q}[r], \tag{29}$$

$$\mathbf{F}[r] = \left.\frac{\partial f}{\partial \mathbf{x}}\right|_{\hat{\mathbf{x}}[r],\mathbf{u}[r]} \tag{30}$$

Table 3 Model parameters applied in the EKF

Parameter	EKF
a [1/day]	$0.306 + \mathcal{N}(0, 1) \cdot 10^{-2}$
b [1/day]	$0.166 + \mathcal{N}(0, 1) \cdot 10^{-2}$
c [1/day]	$0.257 + \mathcal{N}(0, 1) \cdot 10^{-2}$
n [1/day]	$0.144 + \mathcal{N}(0, 1) \cdot 10^{-2}$
$b_k \left[\frac{\text{mg}}{\text{kg}\cdot\text{day}\cdot\text{mm}^3}\right]$	$6.12 \cdot 10^{-7} + \mathcal{N}(0, 1) \cdot 10^{-8}$
K_B [mg/kg]	$0.36 + \mathcal{N}(0, 1) \cdot 10^{-2}$
ED_{50} [10^{-5}mg/kg]	$9.71 + \mathcal{N}(0, 1)$
w [1/day]	$0.34 + \mathcal{N}(0, 1) \cdot 10^{-2}$

where $\mathbf{P}$ is the error covariance matrix, $\mathbf{F}$ is the Jacobian matrix of the system and $\mathbf{Q}$ is the process noise covariance matrix. The update step in the r-th discrete step is defined by

$$\mathbf{K}[r] = \mathbf{P}[r|r-1]\mathbf{H}^\top[r]\big(\mathbf{H}[r]\mathbf{P}[r|r-1]\mathbf{H}^\top[r] + \mathbf{R}[r]\big)^{-1}, \tag{31}$$

$$\hat{\mathbf{x}}[r] = \hat{\mathbf{x}}[r|r-1] + \mathbf{K}[r]\big(\mathbf{z}[r] - h(\hat{\mathbf{x}}[r|r-1])\big), \tag{32}$$

$$\mathbf{P}[r] = \big(\mathbf{I} - \mathbf{K}[r]\mathbf{H}[r]\big)\mathbf{P}[r|r-1], \tag{33}$$

$$\mathbf{H}[r] = \left.\frac{\partial h}{\partial \mathbf{x}}\right|_{\hat{\mathbf{x}}[r|r-1]} \tag{34}$$

where $\mathbf{K}$ is the Kalman gain, $\mathbf{H}$ is the Jacobian of the h output equation, $\mathbf{z}$ is the measurement, $\mathbf{R}$ is the measurement noise covariance matrix.

The EKF structure has been implemented using the model (1)–(3). We applied the following process and noise covariance matrices based on our previous investigations [14]:

$$\mathbf{Q} = \begin{bmatrix} 10 & 0 & 0 \\ 0 & 10 & 0 \\ 0 & 0 & 0 \end{bmatrix}, \ \mathbf{R} = \begin{bmatrix} 10 \end{bmatrix} \tag{35}$$

The applied model parameter set in the EKF is given in Table 3. We have considered a slightly randomized parameter set in the EKF in order to make the investigated case more realistic.

3 Results

The NMPC environment is implemented in MATLAB R2020b. The numerical optimization in the NMPC design is made using $fmincon$ library. We applied first order forward Euler solver to solve the differential equations.

The developed simulation environment can be seen in Fig. 2. The parameters belonging to the constraints—CONINIT—are handled as global parameters which are not changed in each iteration cycle. The parameters belong to the model, EKF and the total simulated horizon can be changed in each iteration cycle. In this study, we considered permanent model and EKF parameters determined at the beginning of the simulation. Naturally, the initial conditions of the state variables ($\mathbf{x}(0)$,$\hat{\mathbf{x}}(0)_{EKF}$) are updated with the results of the last cycle in the model and the EKF, respectively. Furthermore, the N simulated horizon is continuously decreasing until 0 starting from N_{init} which is the total length of the therapy. That means that the simulated horizon is decreasing in each iteration cycle. In the simulations the DOI were limited to the first and fourth days of the week, thus the execution of the injection takes place only on these days. The summary of the constraints, initial conditions and parameters of optimization can be found in Table 4. The last important aspect is the applied noise model. We have selected the following noise model that is comparable with our observations during our previous animal experiments, where the "sensor" was a caliper which was applied to measure the dimension of the tumour injected in the subcutaneous tissue (the volume was estimated based on the two measurable dimensions). We also applied a saturation to guarantee the positivity of z_k.

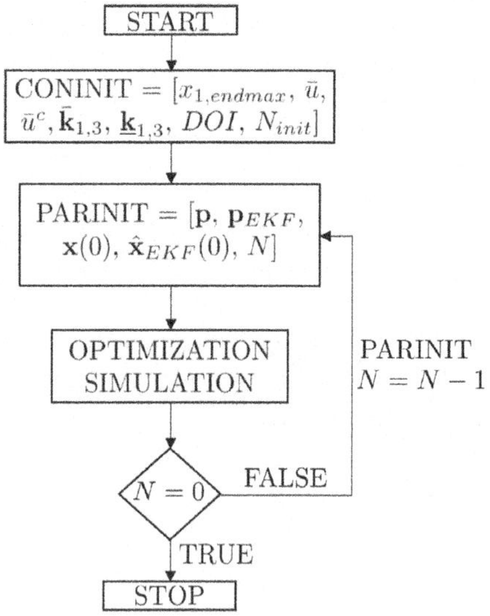

Fig. 2 Schematic process of the developed environment

Table 4 Initial conditions, constraints and parameters of optimization

Parameters	Numerical values
$x_{1,endmax}$ [mm^3]	0.1
$\mathbf{x}(0)$	$[1000,\ 50,\ 0]^\top$
$\hat{\mathbf{x}}(0)_{EKF}$	$[1100,\ 0,\ 0]^\top$
Sampling time [day]	1
N_{init} [day]	150
$\bar{u}$ [mg/kg]	4
$\bar{u}^c$ [mg/kg]	14
$\mathbf{k}_{1,3,init}$	$[10,\ 10]^\top$
$\underline{\mathbf{k}}_{1,3}$	$[100,\ 100]^\top$
$\underline{\mathbf{k}}_{1,3}$	$[0.01,\ 0.01]^\top$
DOI	$[1,\ 4]$

$$\begin{aligned} &z_k = y_k + 10 \cdot \mathcal{N}(0, 1), \\ &\text{if } z_k < 0 \text{ then } \ z_k = 0.01, \end{aligned} \tag{36}$$

where the 0.01 mm^3 means limitation of the sensor (i.e., the caliper).

Figure 3 introduces the results of the control action (applied therapy). The first row of diagrams shows the living tumour volumes of the model to be controlled ("subject") and the EKF ("estimator"). Due to the different initial conditions ($\mathbf{x}(0)$ and $\hat{\mathbf{x}}(0)_{EKF}$) the difference between the x_1 and $\hat{x}_1$ is bigger which is decreasing over time. Approximately at day 70 only the sensor noise is reflected in the deviation between x_1 and $\hat{x}_1$. Similar behaviour can be observed in the second row that belong to the x_2 and $\hat{x}_2$. The third row shows the drug concentration. At the start of the therapy, these values are equally zero, later on a slight estimation error is observable that disappears over time.

Figure 4 shows the drug administration realized by the NMPC. Due to the bigger state error, the algorithm administers the maximum authorized amount of drug at the beginning of the therapy. After that, a resting period can be seen where the result of the optimization was to administer no drug to reach the goals of control. The algorithm injected 0.05 mg/kg of drug at day 21 from which the control action continues till the end of the therapy administering low amount of the drug.

Figure 5 introduces the output of the model (blue line) and the disturbed, noisy measurements. Since the output of the model is $y = x_1 + x_2$, namely, the sum of the living and dead tumour cells, y increases at the beginning. The volume of the living tumour cells x_1 slightly increases, then it drastically diminishes due to the therapy while the volume of the dead tumour cells is approximately 0 at the beginning, then it drastically increases due to therapy. Therefore, the output increases at the beginning then owing to the dropping number of living and dead tumour cells it decreases quickly.

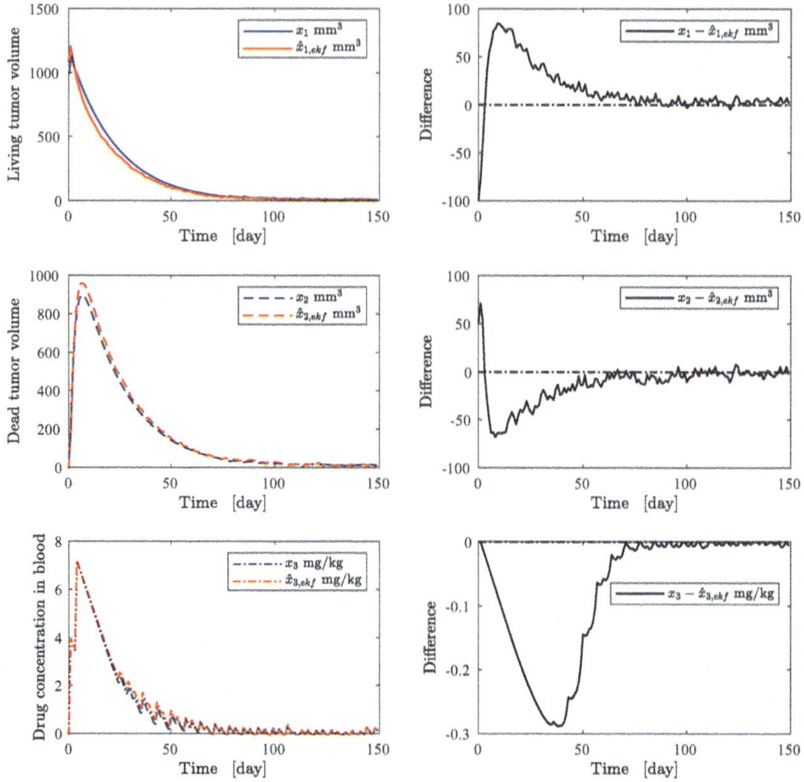

Fig. 3 State trajectories and estimated state variables during therapy. [Blue lines belong to the model (virtual subject), red lines belong to the EKF (state estimator)]

4 Discussion

The paper introduces a possible optimization based controller design approach using the NMPC scheme in a deeply nested structure, with several constraints coming from the qualitative analysis of the model and physiological limitations. An important aspect to be investigated is the computational environment and time to be sure the algorithm can be applied within everyday situation. We executed the simulation using a Dell Precision 7510 laptop computer (Intel Core i7 6820HQ @ 2.70GHz, 4095MB NVIDIA Quadro M2000M, 32GB RAM, 1.5TB SSD). Even though we did not implement object-oriented programming and parallelization, moreover, we did not use the GPU cores and executed the nested algorithm sequentially the simulated 150 days of therapy run under 11.7142 s which is an acceptable speed, especially, if we compare it to the timescale of the model.

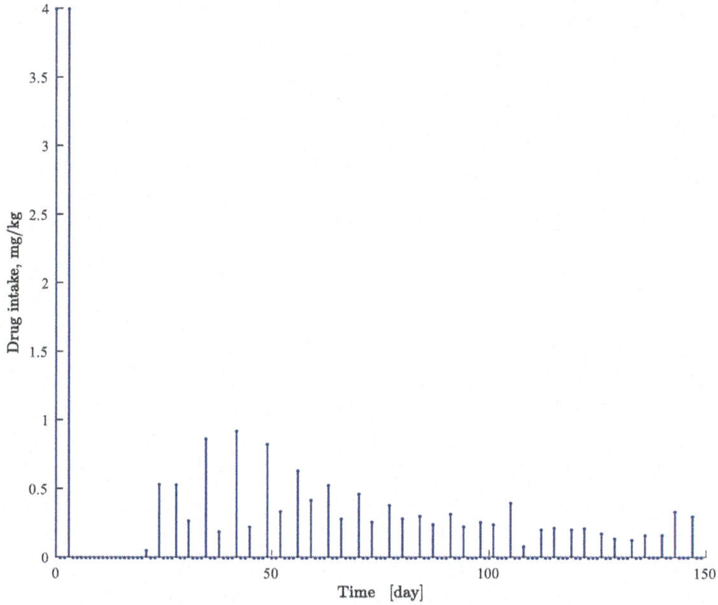

Fig. 4 Drug administration generated by the NMPC controller

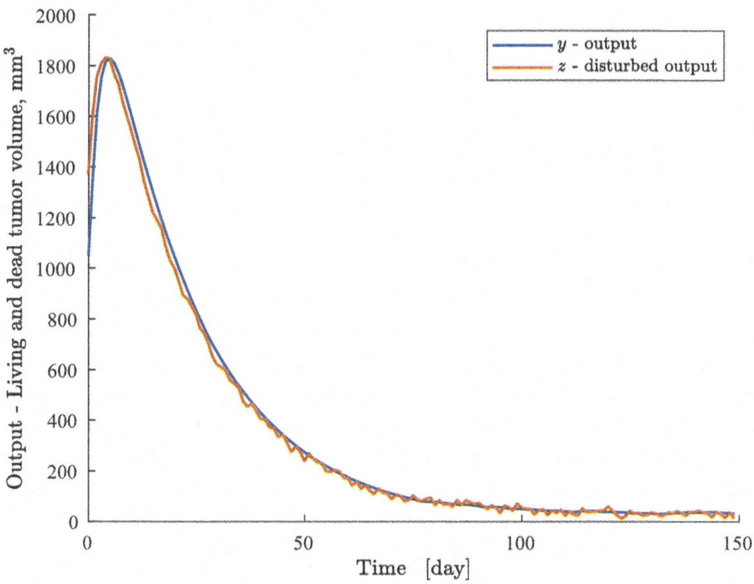

Fig. 5 Output of the model and the noisy measurements

Table 5 Final values of the simulation

Parameters	Numerical values
$\mathbf{x}(end)$	$[16.6746,\ 18.3124,\ 0.1441]^\top$
$\hat{\mathbf{x}}_{EKF}(end)$	$[14.9966,\ 20.2312,\ 0.1498]^\top$
$y(end)$ [mm^3]	34.9871
$z(end)$ [mm^3]	18.8763
Cumulative injected drug during the therapy [mg/kg]	20.3368

A main issue of the control problem is that none of the states are directly measurable, only the sum of the living and dead tumour cells can be measured. This fact includes high uncertainty in the system that requires well-performing models. This property should be considered during the application of the algorithm.

Another important aspect of the solution is the question of $x_{1,final}$ parameter. It is important in the optimization to be able to satisfy the (12)—this is a strict equality where x_1^* means the required equilibrium. By determining that final value to be reached by x_1 at the end of the therapy as the required equilibrium, i.e., $x_1^* = x_{1,final}$ then the following equality condition from (23) can be realized: $x_{1,final} - x_{1,n}/x_{1,d} = 0$. This constraint guarantees the requirements of Theorem 1 and besides drive the controller to reach this $x_{1,final}$ by the model. Naturally, we have information only from $\hat{x}_{1,EKF}$ during the operation that is loaded by noises (e.g., measurement noise). In practice, the estimated $\hat{x}_{1,EKF}$ will not be equal to $x_{1,final}$ because of the mentioned noises.

The final values of the state variables of the model and the EKF and the cumulative dosage injected by the NMPC during the therapy can be found in Table 5. The values of the table reflects that the algorithm performs well since the initial living tumour volume was significantly decreased. Albeit, it can be observed that the total length of the therapy (150 days) was not enough to fully eliminate the tumour volume and within the given circumstances longer therapy duration is needed.

In this study, we determined as constraint that the u^c cumulative drug dosage cannot be higher than 14 mg/kg within 10 days. The algorithm outperforms this requirement, since the amount of totally injected drug was 20.3368 mg/kg for 150 days. This fact is supported by Fig. 4 which shows that after day 21 small dosages only were administered. It is also visible that the controller satisfies the administration regimen by taking into account the DOI values, i.e., injection can be performed only on the first and fourth days of the week. Due to the high initial state error, the amount of the drug injection has been saturated in the first two occasions. Later on, alongside the diminishing state error, the calculated drug dosages became much lower. This property can be improved by embedding a trajectory generator path-tracking controller that can be the part of the future work.

5 Conclusion

An effective NMPC based closed-loop control algorithm was proposed in this study. The developed algorithm—despite the fact that several hard optimization constraints were utilized—was proven to provide stable control action during operation. There are non-covered aspects as well that will be the part of our further investigations. The robustness of the proposed solution is one corner point which is planned to be investigated using stability analysis against the inter-patient variability. Another issue is the applied noise model. We applied here a simple noise model, which should be improved in the future. This is also important because the controller is not able to drive the virtual patient's living tumour volume to the defined volume due to the noise effect coming from the estimator. This also causes a continuous excitation generating control intervention, since the state errors cannot be fully relaxed. Nevertheless, the proposed control architecture does have several benefits, and it is able to reach the main metrics expected from the stability and control action points of view.

References

1. Zietz, S., Nicolni, C.: Mathematical approaches to optimization of cancer chemotherapy. Bull. Math. Biol. **41**(3), 305–324 (1979)
2. Shi, J., Alagoz, O., Erenay, F.S., Su, Q.: A survey of optimization models on cancer chemotherapy treatment planning. Ann. Oper. Res. **221**(1), 331–356 (2011)
3. Sbeity, H., Younes, R.: Review of optimization methods for cancer chemotherapy treatment planning. J. Comput. Sci. Syst. Biol. **8**(2) (2015)
4. Altrock, P.M., Liu, L.L., Michor, F.: The mathematics of cancer: integrating quantitative models. Nat. Rev. Cancer **15**(12), 730–745 (2015)
5. João, P., Belfo, J.M.: Optimal Impulsive Control for Cancer Therapy. Springer International Publishing, Lemos (2021)
6. Chen, T., Kirkby, N.F., Jena, R.: Optimal dosing of cancer chemotherapy using model predictive control and moving horizon state/parameter estimation. Comput. Methods Progr. Biomed. **108**(3), 973–983 (2012)
7. Drexler, D.A., Ferenci, T., Lovrics, A., Kovács, L.: Modeling of tumor growth incorporating the effect of pegylated liposomal doxorubicin. In: Proceedings of the 2019 IEEE 23nd International Conference on Intelligent Engineering Systems, pp. 369–373. IEEE (2019)
8. Füredi, A., Szebényi, K., Tóth, S., Cserepes, M., Hámori, L., Nagy, V., Karai, E., Vajdovich, P., Imre, T., Szabó, P., Szüts, D., Tóvári, J., Szakács, G.: Pegylated liposomal formulation of doxorubicin overcomes drug resistance in a genetically engineered mouse model of breast cancer. J. Controlled Releas. **261**, 287–296 (2017)
9. Czakó, B.G., Drexler, D.A., Kovács, L.: Impulsive control of tumor growth via nonlinear model predictive control using direct multiple shooting. In: 2020 European Control Conference (ECC). IEEE (2020)
10. Diehl, M., Bock, H.G., Diedam, H., Wieber, P.-B.: Fast direct multiple shooting algorithms for optimal robot control. In: Lecture Notes in Control and Information Sciences, pp. 65–93. Springer, Berlin
11. Drexler, D., Sápi, J., Kovács, L.: A minimal model of tumor growth with angiogenic inhibition using bevacizumab. In: 15th IEEE International Symposium on Applied Machine Intelligence and Informatics (SAMI 2017), Herl'any, Slovakia, pp. 185–190

12. Drexler, D.A., Sápi, J., Kovács, L.: Modeling of tumor growth incorporating the effects of necrosis and the effect of bevacizumab. Complexity **2017** (2017)
13. Drexler, D.A., Nagy, I., Romanovski, V., Tóth, J., Kovács, L.: Qualitative analysis of a closed-loop model of tumor growth control. In: 2018 IEEE 18th International Symposium on Computational Intelligence and Informatics (CINTI), pp. 000329–000334. IEEE (2018)
14. Siket, M., Eigner, G., Drexler, D.A., Rudas, I., Kovács, L.: State and parameter estimation of the mathematical carcinoma model under chemotherapeutic treatment. Appl. Sci. **10**(24), 9046 (2020)
15. Drexler, D.A., Ferenci, T., Lovrics, A., Kovács, L.: Tumor dynamics modeling based on formal reaction kinetics. Acta Polytech. Hungarica **16**(10), 31–44 (2019)
16. Drexler, D.A., Sápi, J., Kovács, L.: Modeling of tumor growth incorporating the effects of necrosis and the effect of bevacizumab. Complexity 1–11 (2017)

Detection of Epilepsy Using Adaptive Neuro-Fuzzy Inference System and Comparative Analysis

Marjan Stoimchev and Vesna Ojleska Latkoska

Abstract This study presents the use of Adaptive Neuro-Fuzzy Inference System (ANFIS) for classification of the EEG signals. The data consists of two types of EEG signals, i.e. epileptic patients during epilepsy and healthy patients when their eyes are open. We propose two algorithms for the detection of epileptic patients. In the first algorithm we use Discrete Wavelet Transform (DWT) and statistical analysis for feature extraction, whereas Principal Component Analysis (PCA) is used in order to reduce the number of features in the second algorithm. ANFIS model learns how to classify the EEG signal, through the standard hybrid learning algorithm, whereas we use special form of ANFIS model, which depending on the number of inputs, splits the model into appropriate number of substructures (sub-ANFIS models). The algorithms were evaluated in terms of training performance and classification accuracies. From the simulation results it was concluded that the both algorithms have good potentials in classifying the EEG signals. Further, a comparative analysis for the influence of the tuning parameters was made, i.e. the influence of the different data splitting methods, the influence of the different input space partitioning methods, the usage of the different wavelet functions in the WT, the effects of normalization, as well as the effects of using different membership functions. From the analysis it was concluded that different combinations of input parameters differently classify the EEG signals. Lastly, a comparison of the both algorithms was made, in terms of training performance and classification accuracies, whereas it was concluded that the algorithm that uses PCA for feature extraction, in some cases, performs better than the algorithm that uses DWT, even though the number of features is significantly reduced (from 20 to 7).

Dedicated to Prof. Georgi M. Dimirovski on his anniversary.

M. Stoimchev · V. O. Latkoska (✉)
Faculty of Electrical Engineering and Information Technologies, "Ss. Cyril and Methodius" University in Skopje, 1000 Skopje, Republic of North Macedonia
e-mail: vojleska@feit.ukim.edu.mk

P. Shi et al. (eds.), *Complex Systems: Spanning Control and Computational Cybernetics: Applications*, Studies in Systems, Decision and Control 415,
https://doi.org/10.1007/978-3-031-00978-5_11

1 Introduction

Epilepsy is chronic brain disorder, characterized by seizures, which can affect any person at any age. It is characterized by recurrent convulsions over a time-period. Clinical diagnosis of epilepsy requires detailed history and also neurological examinations [1]. There are many techniques to investigate the recurrent epileptic convulsions (namely, Computer Tomography-CT, Magnetic Resonance Imaging-MRI and Electroencephalogram-EEG). As it is stated in [1], the most common effective diagnostic method for the detection of epilepsy is the analysis of EEG signals, which can be based on different types of approaches [2–4].

Although It is possible for experienced neurophysiologist to detect the epilepsy by visually scanning of the EEG signals for pre-ictal, inter-ictal and ictal activities [1], for a more objective analysis and reproducible results, it is always advantageous to detect these activities from the EEG signals through some computer methods, by extracting relevant features from the signals. Adeli et al. [2, 3] launched the field of automated EEG-based diagnosis by analyzing and characterizing epileptiform discharges using wavelet transform [1]. Some of their studies focus on detecting epilepsy by classifying only the normal and ictal stages (two-class problem), as proposed in these studies [4–7], and other studies present methods for classifying all three stages, namely, normal, interictal, and ictal (three-class problem) [8].

In order to solve the above-mentioned problems, i.e. to make an automated system for epilepsy detection, there are many proposed methodologies. In general, all of the techniques consist of several steps, i.e. from data preprocessing as the first step, through feature extraction as a second step, to classification as a third step.

EEG signals may be corrupted with noises, called artefacts which come from patient's body or instruments (examples include: the eyes, the heart, or the muscles movement, or the power line noise, etc.) [9]. Therefore, removal of these artefacts is a primary task and it forms a fundamental step for EEG signal preprocessing [9]. This task can be solved using conventional filtering methods, or filtering through wavelet analysis [10]. In this study for the purpose of data preprocessing we use conventional band-pass FIR filtering technique that uses Hamming window [11, 12].

For the second step, feature extraction takes place. There are many different methods that can be used for this task, whereas those techniques include frequency domain analysis, or time domain analysis, or both. Among the techniques that use time–frequency analysis is the wavelet transform (WT) [2, 3]. The results of the studies in the literature have demonstrated that the WT is the most promising method to extract relevant features from the EEG signals [2, 3, 7]. In this respect, in this study the WT was used for feature extraction from the EEG signals [2, 3, 7].

The final step is classifying the EEG signals. There are also different ways for classifying the EEG signals as proposed in [2]. Some studies use feature extraction method using genetic algorithm-based frequency domain (GAFD) feature search [13], the wavelet-based support vector machine (SVM) classifier [14], wavelet-based feed forward artificial neural network (FFANN) [15, 16], k-Nearest Neighbors classifier [17], fuzzy rule-based detection [18], Adaptive Neuro-Fuzzy Inference System (ANFIS) [4, 5, 7], and many others.

Artificial neural networks (ANNs) have been used as computational tools for pattern classification including diagnosis of diseases because of their great predictive power [19, 20]. On the other hand, fuzzy set theory plays an important role in dealing with uncertainty when making decisions in medical applications [21, 22]. Neuro-fuzzy systems are fuzzy systems, which use ANNs theory in order to determine their properties (fuzzy sets and fuzzy rules) by processing data samples [21]. A specific approach in neuro-fuzzy development is the ANFIS [21], which has shown significant results in modeling nonlinear functions. When input data is given, ANFIS learns features and adjust the parameters of the system according to a given error criterion [21]. There are many successful implementations of ANFIS in the biomedical field of engineering, i.e. classification of data [4, 7, 23], and data analysis [24].

This study is largely based on the work given in [25–27]. We firstly propose an algorithm for classification of EEG signals, that combines FIR filtering for artefact removal [5, 11, 12], discrete wavelet transform (DWT) for feature extraction [2–4, 6, 7], and ANFIS for classification [4, 6, 7]. The data consist of two types of EEG signals, i.e. epileptic patients during epilepsy and healthy patients when their eyes are open. We use two sets of data, described in [28], whereas S is the data set that contains epileptic patients and Z is the data set that contains healthy patients. We also perform normalization of the data, before it is used for the process of training in the ANFIS. This method is also called feature scaling method which will enable a better convergence when adapting the weighted factors in the ANFIS model during the training process. ANFIS model learns how to classify the EEG signal, through the standard hybrid learning algorithm. We use a special form of ANFIS model, which depending on the number of inputs, splits the model into appropriate number of substructures (sub-ANFIS models). ANFIS model was evaluated in terms of training performance and classification accuracies. From the simulation results it was concluded that the proposed algorithm has good potentials in classifying the EEG signals.

Authors in [4, 5] also use wavelet transform for feature extraction and ANFIS for classification of EEG signals, but our approach differs from theirs as we use conventional filtering method (FIR), as well as normalization of the data after feature extraction. Our approach gives similar, or somewhere even better results, which is summarized in Sect. 2.

We continue our analysis by making various modification to the parameters of the proposed algorithm, in order to make a comparative analysis for the influence of the tuning parameters on the overall accuracy. Firstly, we made an initial simulation analysis over the grid partitioning method [21] by dividing the dataset into various ways (using 70–30% and 50–50% ratio of the training and testing dataset, and using K-Fold cross validation technique) [29]. The second comparative analysis (again by using the grid partitioning method [21]) was based on the use of different types of wavelets (Daubechiesof order 1, 2 and 6–db1, db2, and db6; and Coifletsof order 4–coif4) [3, 30]. The next comparison was made by analyzing the ANFIS model through different input space partitioning methods (grid partitioning versus fuzzy c-means clustering, versus subtractive clustering) [21, 31, 32]. The fourth comparative analysis is based on the influence of normalization [33]. Lastly, we explored how the different membership functions (MFs) in the ANFIS model affect the accuracies.

As a next step to this study an introduction of the second (upgraded) algorithm was made, whereas we used PCA in order to reduce the number of features in the first algorithm. The researchers in [34–36] have also applied PCA to classify the epileptic EEG signals, using this time domain method to reduce the large number of data and select the most important components as feature vectors. The main difference of this papers with our study is that researchers in [34] used SVM based classification on EEG signals, whereas the researchers in [35] used EEG signal decomposition with the Wavelet Packet Decomposition (WPD) method, and the classification was obtained with the Gaussian Mixture Model (GMM) classifier and lastly, researchers in 0 have tested several classifiers, namely the K-Nearest Neighbour (KNN), SVM, the naive bayes (NB) classifier and the LDA method.

We continue our study with a comparison analysis of the first and the upgraded algorithm, i.e. evaluating the both algorithms in terms of training performance and classification accuracies. The comparative analysis of the both algorithmswas made using different data splitting methods, and different input space partitioning methods. Firstly, an initial simulation comparison analysis of the both algorithms over the grid partitioning method [21] was made, by dividing the dataset into various ways (using 70–30% and 50–50% ratio of the training and testing dataset, and using K-Fold cross validation technique) [29]. The second comparative analysis was made by analysing the first and the upgraded algorithm, through different input space partitioning methods (grid partitioning versus fuzzy c-means clustering, versus subtractive clustering) [21, 31, 32]. It was concluded that the upgraded algorithm has satisfactory performance, and in some cases performs even better than the old algorithm. Nevertheless, due to the reduced number of features (from 20 to 7), used for training the ANFIS network, the upgraded algorithm is much more resistant to overfitting.

This paper is organized as follows. In Sect. 2 we propose the overall algorithm for EEG signals classification, using ANFIS classifier. Section 3 presents an analyzes of the influence of the different splitting methods for the used dataset, as well as presenting how the different wavelet families, different types of input space partitioning, normalization, as well as the different MFs, affect the algorithm accuracies. In Sect. 4 the upgraded algorithm for epilepsy detection using PCA is presented. In Sect. 5 a comparative analysis for the influence of the training and testing data was made, for the both algorithms, as well as a comparative analysis of the both algorithms using different types of input space partitioning methods was presented. At the end, the study is summarized giving the necessary conclusions.

2 Algorithm for Epilepsy Detection Using ANFIS Classifier

Bellow we present an algorithm for detection of epilepsy with fuzzy-neural networks. The algorithm for classification of EEG signals consists of three main steps (Fig. 1):

(1) Filtering of the EEG signals with FIR filter
(2) Feature extraction and dimensionality reduction with discrete wavelet transform (DWT)
(3) Classification using ANFIS.

The general steps of the proposed algorithm are given in Fig. 1, with more details presented in Fig. 2.

Below we present a brief introduction for the used methodologies in the algorithm.

2.1 *Input Data and De-Noising of the EEG Signals*

The EEG data in this study was taken from the database of the university hospital in Bonn, Germany [28]. It consists of EEG signals that are recorded from three different events, namely, healthy subjects, epileptic subjects during seizure-free intervals (known as interictal states) and epileptic subjects during a seizure (ictal states).

The overall data consists of five subsets namely, O, Z, F, N and S. Each subset contains 100 segments of EEG signals, each with duration of 23.6 s. The sampling frequency of these signals is 173.61 Hz, so each segment contains 4097 samples. Set O and Z were obtained from healthy subjects with eyes open and closed respectively; sets F and N were obtained during interictal states in different zones of the brain and set S was taken from a subject during ictal state. In order to make comparison with

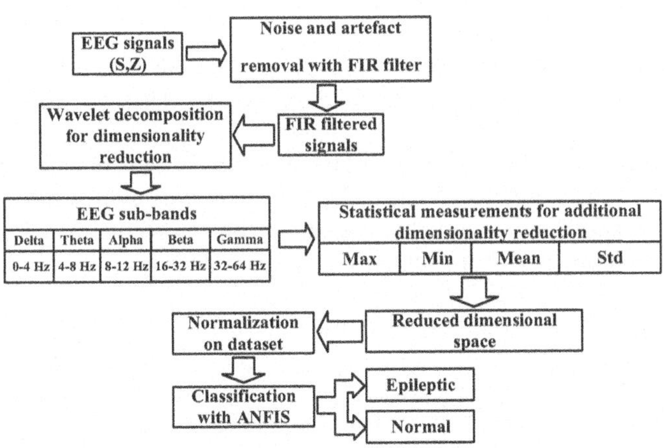

Fig. 1 Data flow for the proposed algorithm

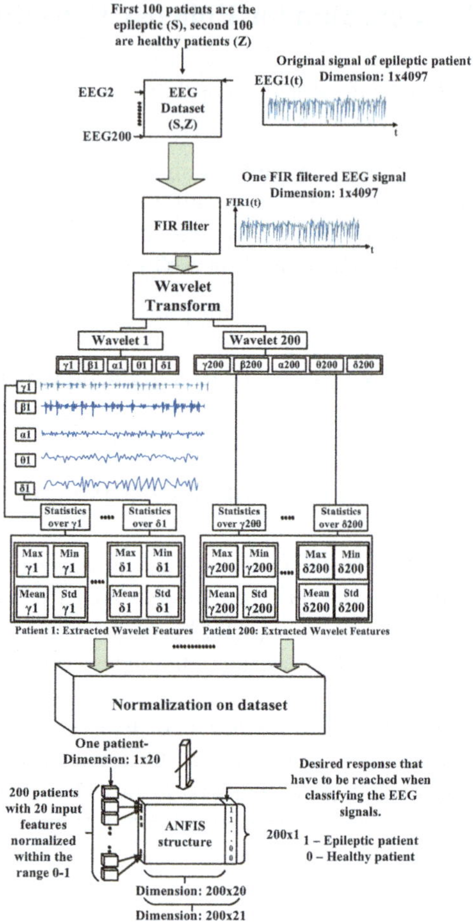

Fig. 2 Detailed analysis for the overall algorithm from Fig. 1

some of the works described in Sect. 1, sets Z and S were used only for the results reported here [4, 16]. As a result, the dimension of our dataset is 200 segments by 4097 samples [28].

In order to visualize the differences of healthy subject and epileptic subject, ratio between EEG signals of healthy and epileptic patient is presented in Fig. 3. As we can see, the signal from the epileptic patient has bigger oscillations, i.e. they have sudden transitions along with higher amplitude peaks than the signal from the healthy patient.

The next step of the overall algorithm is filtering of EEG signals, i.e. defining preprocessing technique which will allow reduction of artefacts, one of the major difficulties in analysis of EEG signals [5, 10]. This disturbance represents serious obstructing factor that prohibits further processing to identify useful diagnostic features [5].

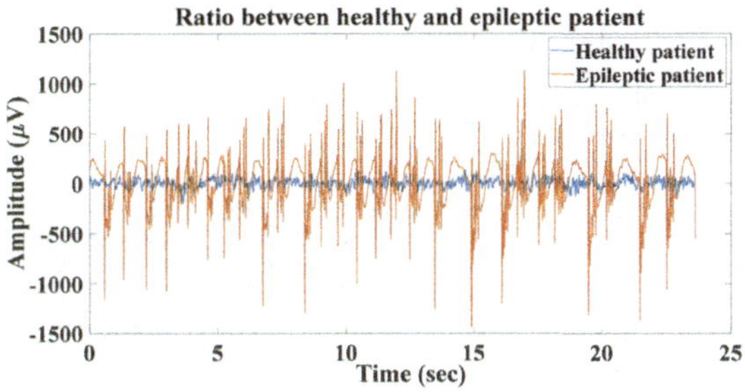

Fig. 3 Ratio between EEG signals of healthy and epileptic patient

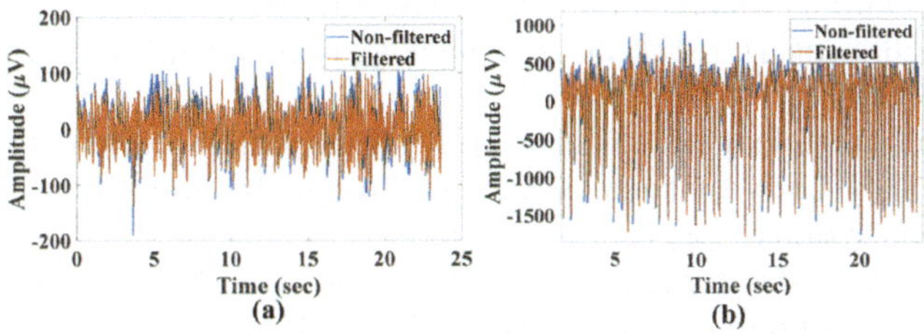

Fig. 4 Ratio between filtered and non-filtered EEG signals for **a** healthy patient and **b** epileptic patient

In our case, as a first step in the algorithm, presented in [25], we used the band-pass Finite Impulse Response (FIR) filter with the Hamming windowing method [11, 12]. The FIR filter is defined by two cutoff frequencies (in case of band-pass filtering), stopband attenuations and passband attenuation. The overall band of frequencies is defined by the Nyquist frequency, i.e. Fs/2 [11]. In our case we use 1 Hz and 60 Hz, respectively, and this is in order to eliminate the artefacts that have corrupted the EEG signals (below 1 Hz are the artefacts that are coming from the human body, and above 60 Hz is the power line noise) [9]. On Fig. 4a and b ratio between filtered and nonfiltered signals of healthy patient and epileptic patient is shown respectively.

2.2 *The Use of Discrete Wavelet Transform for Feature Extraction*

As a second step in the algorithm, we used the Discrete Wavelet Transform (DWT), which analyses the signal at different frequency bands, with different resolutions, in terms of approximation and detail coefficients [3, 6]. The number of decomposition levels is chosen to be 4. The levels are chosen such that those parts of the signal that correlate well with the frequencies required for classification of the signal are

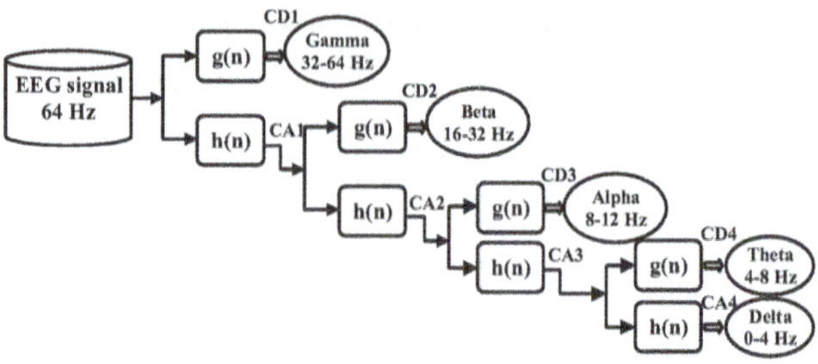

Fig. 5 Wavelet procedure for 4 levels of decomposition

retained in the wavelet coefficients [4, 6, 7]. After the DWT, the signal is decomposed in 4 detail coefficients (D1–D4) and one final approximation coefficient (A4), as it is shown in Fig. 5.

The wavelet coefficients are calculated using Daubechies wavelets of order 2 (db2) in MATLAB [30]. For visual analysis, the multistage decomposition on one signal from an epileptic patient are given on Fig. 6.

Table 1 represents the frequencies that correspond to the different levels of decompositions for db2 with sampling frequency of 173.61 Hz. We can see that the approximation coefficients-CA4 are correctly placed within the range of δ (0–4 Hz) brain waves, CD4 are placed in θ (4–8 Hz), CD3 in α (8–12 Hz), CD2 in β (16–32 Hz), and lastly, CD1 are placed within the range of γ (32–64 Hz) brain waves [4, 6].

For further dimensionality reduction, statistics over the extracted wavelet coefficients were made, namely maximum, minimum, mean value and standard deviation of the wavelet coefficients [4]. We represented the initial dataset into more compact representation, i.e. dataset with dimension of 200 × 20 (4 statistical measuremets × number of extracted coefficients = 20 features for each EEG segment). Now, those feature vectors were used as an inputs to the ANFIS model [4, 7].

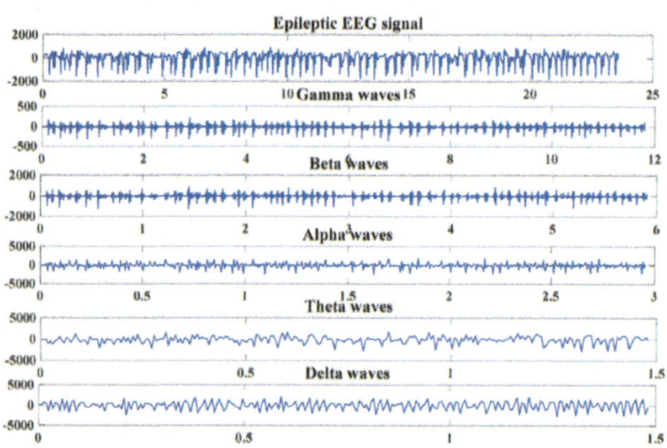

Fig. 6 Wavelet decomposition with Daubechies of order 2 on epileptic patient

Table 1 Frequencies corresponding to different levels of decomposition for Daubechies 2 with sampling frequency of 173.61 Hz

Decomposed signal	Frequency range (Hz)	Decomposition level
CD1	43.40–86.80	1 (gamma)
CD2	21.7–43.40	2 (beta)
CD3	10.85–21.7	3 (alpha)
CD4	5.425–10.85	4 (theta)
CA4	2.7125–5.425	4 (delta)

2.3 *ANFIS for Classification of EEG Signals*

ANFIS is an adaptive neural network that is based on a fusion of ideas from fuzzy control and neural networks and possesses the advantages of both [21]. ANFIS is used as a third step in this algorithm, in order to make the final classification of the EEG patients.

Before the process of training and testing on the ANFIS classifier, all the columns of our dataset, i.e. the features, are normalized within the range from 0 to 1, in order to achieve stable convergence on the weighted factors of the neural network during the training process. This is also called a feature scaling method which represents our next step of the overall algorithm. The min–max normalization technique is used for normalizing the input features [33].

The ANFIS classifier is trained with the hybrid learning algorithm [4, 6, 7, 21], whereas the 20 features were used as input patterns which represented the EEG signals, and output vector as the 21st column (epileptic patients are labeled with ones, and the healthy patients are labeled with zeros) which represented the desired response. Firstly, we performed the simulation analysis by dividing the dataset into ratio of 70–30% for training and testing dataset, respectively.

For our first results, the ANFIS model used the grid partitioning method [21] for input space partitioning. The ANFIS structure consists of membership functions divided in three regions, namely, small, medium and big [4] that are assigned to every input feature, i.e. generating $3^{20} = 3486784401$ if-than rules. According to this, we face the problem called "curse of dimensionality" [21]. In order to avoid this problem, we define a different way of dividing the ANFIS structure into many substructures [30]. The ANFIS structure is divided according to the input features, in our case it is divided into 7 substructures. The substructures from 1 to 6 receive 3 input features that will lead to $3^3 = 27$ if-than rules, and the last one will have $3^2 = 9$ if-than rules. With this, we surpassed this major obstacle (Fig. 7).

For the input membership functions we use generalized bell-shaped membership functions (gbellmf). From illustrative character, the initial membership functions for the second and seventh substructures are shown on Fig. 8a and b, respectively. In the second substructure the input parameters are: maximum of θ, maximum of δ and minimum of γ wavelet coefficients, and in the seventh substructure we have only two features as input parameters, namely: θ standard deviation and δ standard deviation of the wavelet coefficients.

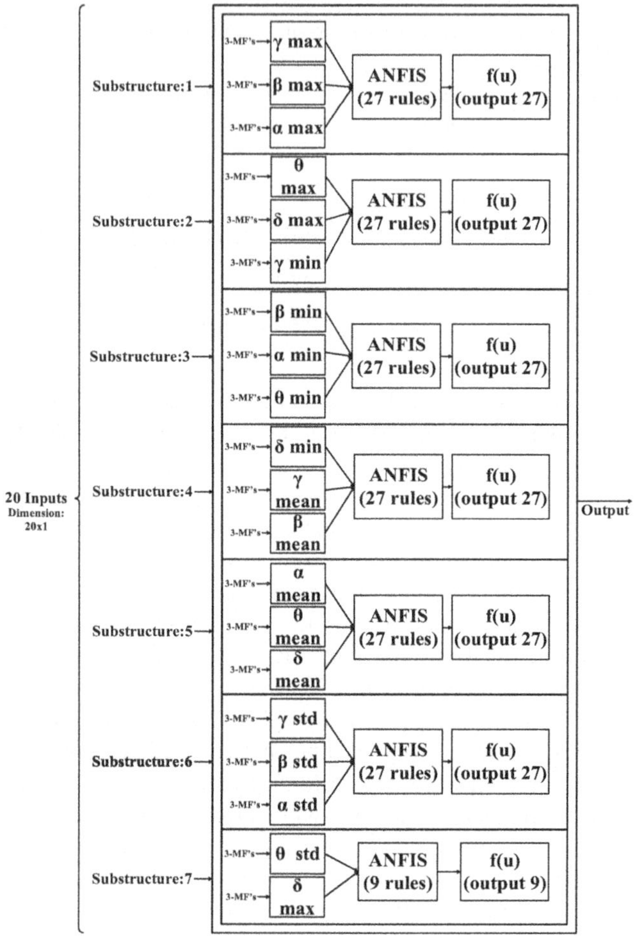

Fig. 7 Division of ANFIS onto 7 substructures

The model is trained with different number of epochs and the accuracy is defined according to the error measure-R for the overall ANFIS [4]. After the training, the parameters of the ANFISnetwork are changed, as shown in Fig. 9a for the second substructure and in Fig. 9b for the seventh substructure.

On Fig. 10a the ratio between the accuracies of the training and testing set are shown, while on Fig. 10b the ratio between root mean square errors (RMSE) on training and testing sets during different number of epoch is shown.

From Fig. 10 we see that in three moments we get 100% accuracy and zero RMSE on testing. This suggests us that if we over train the ANFIS model, it will result in overfitting [37], which will reduce the predictive power of the ANFIS or any neural network. According to this, for small datasets like in our case, it is enough to train

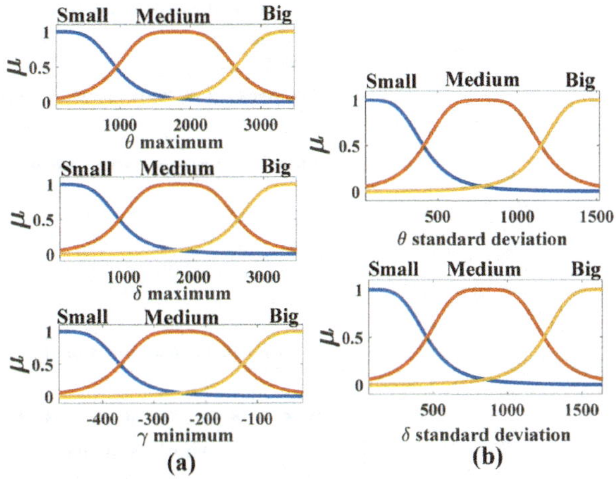

Fig. 8 Initial generalized bell shaped membership functions for substructure 2 (**a**) and substructure 7 (**b**)

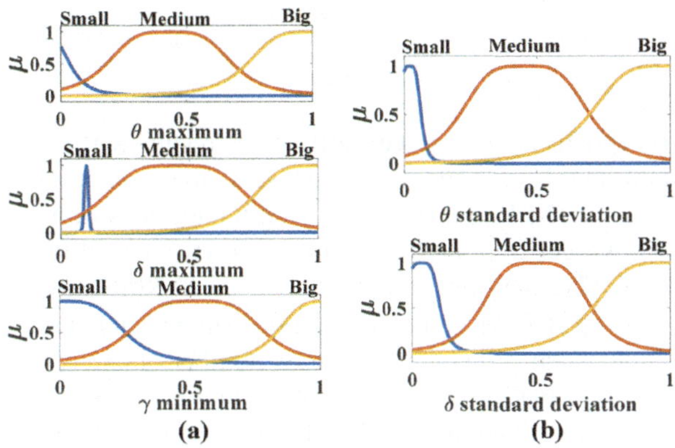

Fig. 9 Final generalized bell shaped membership functions after training for substructure 2 (**a**) and substructure 7 (**b**)

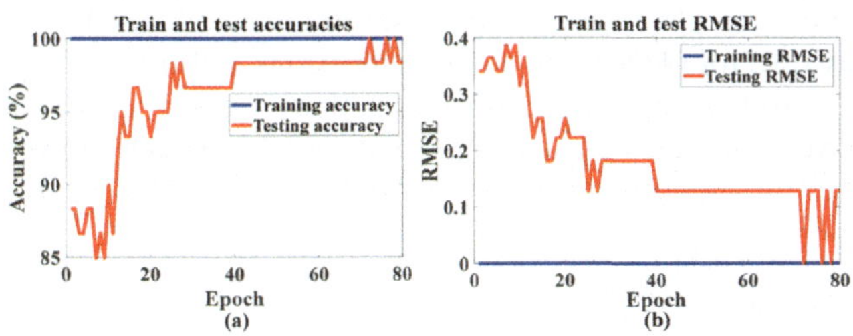

Fig. 10 Training and testing accuracies **a** Training and testing RMSEs **b** generated for different number of epochs

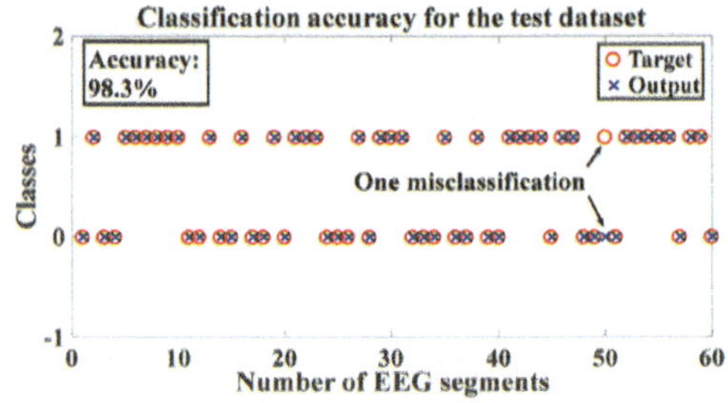

Fig. 11 Test set accuracy for 60 epochs

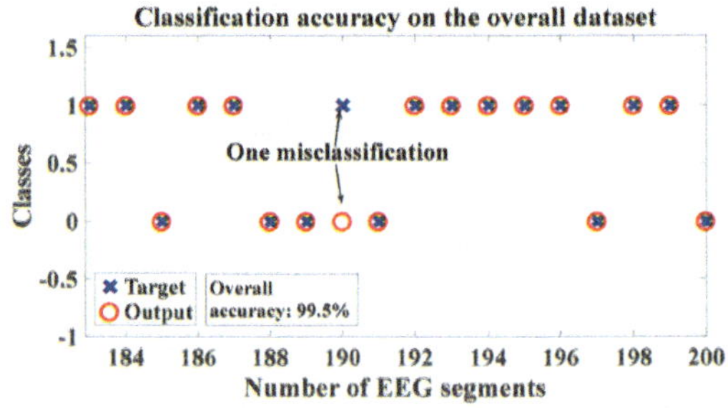

Fig. 12 Overall set accuracy for 60 epochs

the model with 60 epochs in order to get satisfactory results with testing accuracy of 98.3% and optimal RMSE of 0.191. Fig. 11 presents the accuracy for 60 epochs.

Also, we tested the ANFIS model on the overall dataset as shown in the Fig. 12.

As we can see, we get even better results on classifying all the EEG segments of 99.5% accuracy. The classification results for the ANFIS model are shown in Table 2 for the testing dataset (1 is an epileptic patient, and 0 denotes a healthy patient). In this table, each cell contains the raw number of exemplars classified for the corresponding combination of targeted and actual network outputs. As we can see, we make 1 misclassification, i.e. classifying an epileptic patient as a healthy patient.

We briefly describe some work that report results using Bonn database [28], which is used in our research.

Juarez-Guerra et al. [15] presented the results of a model based wavelet analysis and neural networks for identification of seizure events and epilepsy. They've tested several filters, wavelets and wavelet transformations, namely, Haar, Db2 and Db4. Six features have been used to train the Feed-Forward Artificial Neural Network (FF-ANN): mean, absolute median and variance of delta and alpha sub-bands. When using

Table 2 Confusion matrix for 60 epochs on the test dataset

Output	Set Z	Set S
Target		
Set Z	26	1
Set S	33	0

Table 3 Comparison of the accuracy results between our and other studies

Accuracy (%)	Test set accuracy	Overall set accuracy
This study	98.33	99.5
E. Juarez-FFANN [15] (WT and NN, using six features. Several filters and wavelets were used, namely, Haar, Db2 and Db4, getting 93.23% as the highest accuracy)	99.26	93.23
Omerhodzic-Wavelet + Neural Network [16] (Wavelet and NN. DWT withMultiresolution analysis (MRA), based on db4 was used)	/	94

the whole segments for training, 93.23% of accuracy has been achieved. Whereas when using sub-segments for training, 99.26% of accuracy has been achieved. Thus, the accuracy rates of the ANFIS model presented for this application were found to be higher when sub-segments for training are used (in our case we only use sub-segments for training) in Juarez-Guerra et al. [15].

Omerhodzic et al. [16] presented algorithm for classification of EEG signals based on Wavelet-Neural Network classifier. DWT with Multiresolution analysis (MRA) based on Daubechies of order 4 (db4) has been applied to decompose the EEG signals at resolution levels of the components of the EEG signal. They used percentage distribution of the energy as features of the EEG signals at different resolution levels. The classifier has been used to classify those extracted features to identify the EEGs type according to the percentage distribution of energy features. They achieved 94% overall accuracy, so that our ANFIS model showed higher accuracy than the study proposed by Omerhodzic [16] (Table 3).

We have to note that the authors in [4], also use WT and ANFIS, but our approach differs from theirs as we use conventional filtering method (FIR), as well as normalization of the data after feature extraction. When we compare only the numbers, our approach gives similar, or somewhere even better results from theirs. They get 98.63% test accuracy of the test set Z (healthy), and 98.25% test accuracy on the test set S (epileptic), whereas we get 98.33% accuracy on the test set containing both healthy and epileptic patients. Nevertheless, we have to strictly note that they make 5 class classification, which differs from our 2 class classification, for we did not summarize their results in Table 3.

3 Comparative Analysis for the Influence of the Different Tuning Parameters in the Algorithm for Epilepsy Detection Using ANFIS Classifier

In this section we present a comparative analysis for the influence of the tuning parameters in the algorithm for epilepsy detection using ANFIS classifier (presented in Sect. 2), i.e. the influence of the different data splitting methods, the influence of the different input space partitioning methods, result of the different wavelet functions in the WT, the effects of normalization, as well as the effects of using different membership functions.

3.1 *Influence of the Training and Testing Data*

In this section we present different approaches on dividing the dataset, i.e. how the size of the training and testing data influence the accuracy. In the initial simulation analysis, given in Sect. 2, we used 70–30% split ratio between the training and testing data, respectively. Here we will compare that splitting method with the splitting method that uses dataset divided into 50–50% ratio, as well as using the splitting method based on cross validation [29]. This initial comparative analysis is obtained using grid partitioning method.

When using the conventional splitting methods (70–30%, or 50–50% ratio of training and testing data, respectively) we simply divide the data into 2 appropriate sets. On the other hand, the K-Fold cross validation uses different approach.

With K-Fold cross validation, the available data is partitioned intoK separate sets of approximately equal size [29]. The procedure involves K learning iterations, where for every iteration K-1 subsets are used for training, and the remaining set is used as the testing data. Every iteration leaves out a different subset, which means that each subset is used as test subset only once. In the end, all accuracies obtained from each iteration (testing fold) are averaged in order to obtain a reliable estimate of the model performance [29]. In our case we use 3-Fold cross validation.

Figure 13 presents the test set Root Mean Square Errors (RMSEs) when using the grid partitioning method during 100 epoch period, by applying the three different data split methods (70–30%; 50–50%; 3-Fold).

As we can see from Fig. 13 the lowest RMSEs are obtained in different number of epochs during the three partitioning methods. In order to give an appropriate comparison between the methodologies used in the further analysis, we will train the ANFIS model with 40 epochs for all the data split methods. In Fig. 14 the test set accuracies are given, where the black bars represent the highest possible accuracy, and the red bars represent the accuracies obtained during 40 epoch of trainingfor each data split method. It can be concluded that the 3-Fold cross validation splitting method gives the most promising results.

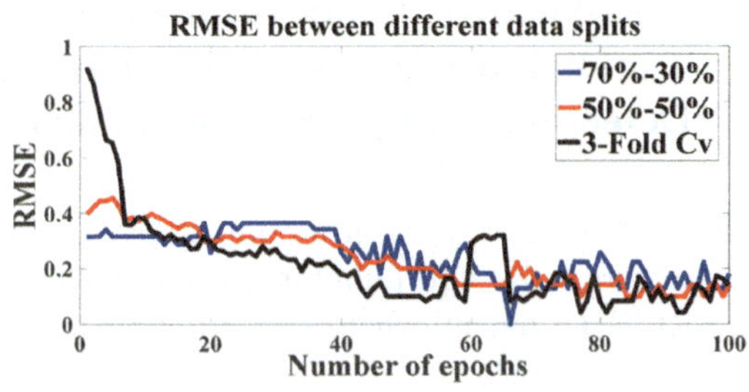

Fig. 13 Comparison of the test set RMSEs over the three data split methods, using grid partitioning

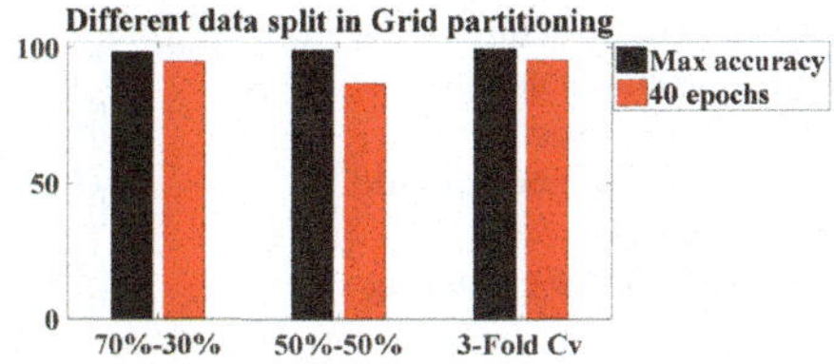

Fig. 14 Test set accuracies for the different data split methods using grid partitioning

Table 4 Accuracies for the different types of wavelet functions

Grid partitioning				
Filter	Wavelet	70–30%	50–50%	3-Fold
FIR	db1	95	90	98
FIR	db2	**95**	**87**	**95.51**
FIR	coif4	93.33	85	96.67
FIR	db6	91.67	88	95.97

3.2 Influence of the Different Wavelet Families

As a second comparative analysis (again by using the grid partitioning method [21]) we examine the influence of different types of wavelet functions, used for feature extraction, namely: Daubechies of order 1 (db1), Daubechies of order 6 (db6) and Coiflets of order 4 (coif4) in MATLAB [30]. Table 4 presents the test set accuracies for different data splits when trained with 40 epochs. As we can see, with 3-Fold cross validation method for all the wavelet families we reached highest accuracies. This was another comparison to prove that the cross validation method gives more satisfying results.

3.3 Different Types of Input Space Partitioning

This section presents different methods of input space partitioning in the ANFIS model and the overall overview is given in Fig. 15. In the algorithm presented in Sect. 2, we used only the grid partitioning method, as shown in Fig. 15a (i.e. Fig. 7), where in order to reduce the number of rules, sub-ANFIS models were formed. As it was noted in Sect. 2, and can be seen from Fig. 15a (i.e. Fig. 7), sub-ANFIS models from 1 to 6 accept 3 inputs and the last sub-ANFIS model accepts 2 inputs. In this section we present two new approaches for input space partitioning, namely, Fuzzy c-means (FCM) clustering and subtractive clustering [21]. The resulting ANFIS structure in this case is shown on Fig. 5b, where all the 20 inputs are passed at once, i.e. the number of rules is equal to the number of clusters, thus we do not face the problem called "curse of dimensionality" as presented in Sect. 2.

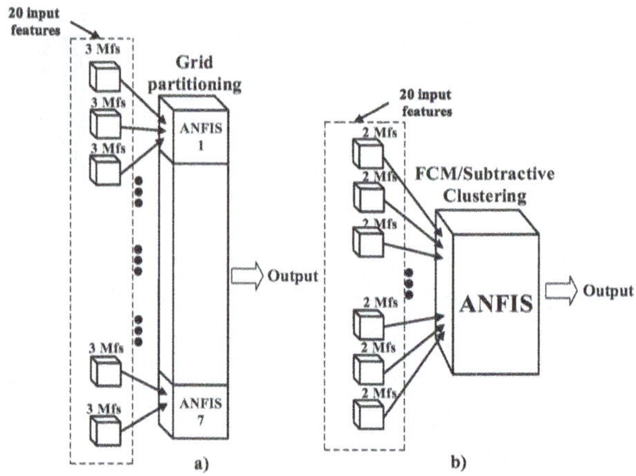

Fig. 15 The structure of the model generated using different input space partitioning methods: **a** Grid partitioning; **b** FCM/Subtractive clustering

Clustering is the process of grouping a set of objects in such a way that objects in the same group are more similar in some particular manner to each other than to those in the other groups [21, 32].

FCM is a data clustering algorithm in which each data point belongs to a cluster to a degree specified by a membership grade (i.e. given data point can belong to several groups with the degree of membership between 0 and 1). The cluster centers are manually specified, where the performance depends on the initial cluster centers [21].

Subtractive clustering, on the other hand, considers each data point as a potential cluster center, where the measure of potential is based on the distance of the data point from other data points (a data point located in a mound of different data points has a greater chance of being a cluster center) [21, 31].

Adequately to Fig. 13, Figs. 16 and 17 present the test set RMSEs for FCM clustering and Subtractive clustering, respectively, using different data split methods. The initial number of clusters for FCM clustering is 2, and the radius of coverage (influence) for the subtractive clustering is 0.8 (for more information of the parameters in these clustering algorithms see [21, 30]). In both cases we use Gauss MFs [30]. As we can see form Figs. 16 and 17, same as for the grid partitioning method (Fig. 13), we get satisfying results when trained with 40 epochs approximately (as for higher number of epochs, some of the methods over fit, we chose 40 epochs for all further calculations).

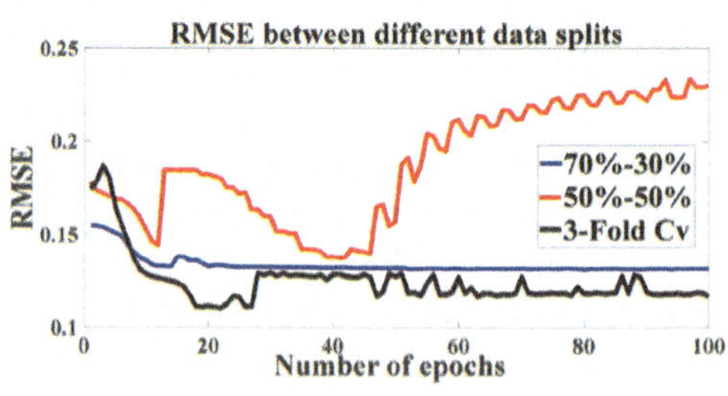

Fig. 16 RMSE values for FCM clustering for different data split methods

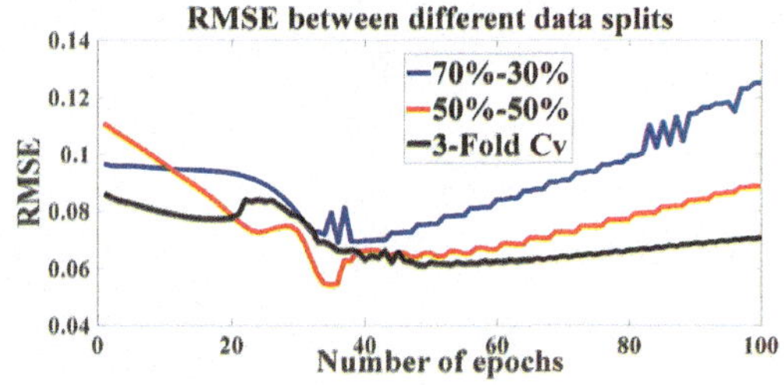

Fig. 17 RMSE values for Subtractive clustering for different data split methods

In the case of FCM clustering (Fig. 16) we get the best results for the threefold cross validation method. For the 50–50% method we tend to over fit the model when trained with large number of epochs [37].

In the case of subtractive clustering (Fig. 17), we also get the best results for the 3-Fold cross validation method, but for all the three cases of data splitting, the model tends to over fit when trained with large number of epochs [37], i.e. there is a slight growth in the RMSE values, after nearly 40 epochs (smallest in case of threefold cross validation).

3.4 *The Effect of Normalization*

In this section we present how the normalization affect the accuracies during the testing process over 200 epoch period. We use min–max normalization, i.e. we are normalizing the feature column vectors in the range from 0 to 1 [33, 38]. This method is also called "feature scaling", and represents a preprocessing technique [38]. On Figs. 18, 19 and 20 the RMSE values before and after normalization are shown, when grid partition, FCM clustering and Subtractive clustering are used, respectively.

We can notice that in all three methods, the RMSEs after the normalization have changed considerably, i.e. we get better results when using normalized data set for classifying the EEG segments. For the best of our knowledge, in all the relevant papers on this topic [4–6, 10, 13–16, 18], we did not find results where normalization technique was used, which further emphasizes the significance of this result.

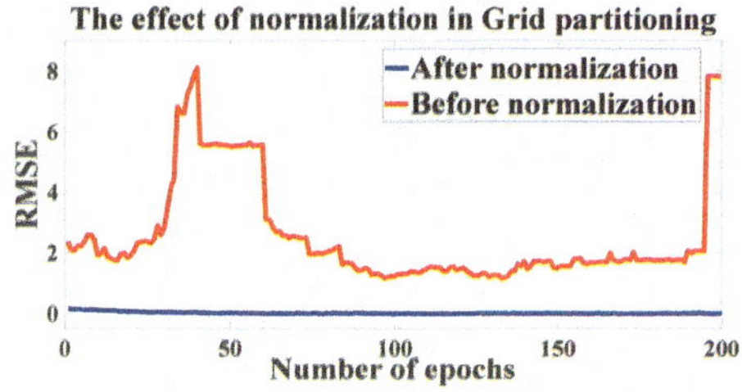

Fig. 18 RMSE values for grid partitioning before and after normalization

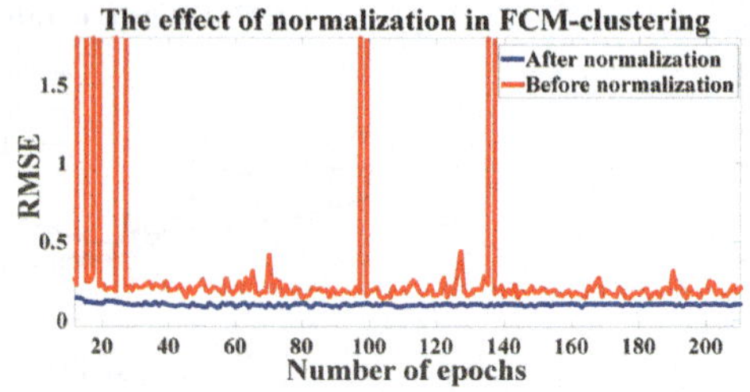

Fig. 19 RMSE values for FCM clustering before and after normalization

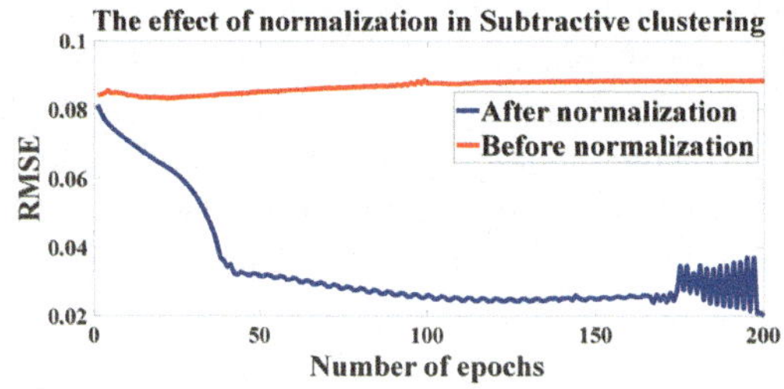

Fig. 20 RMSE values for Subtractive clustering before and after normalization

3.5 *The Effect of Using Different Number of Clusters, Different Radius of Coverage, and Different Types of Membership Functions*

Our next goal is to see how the number of clusters in FCM clusteringand the radius of coverage in Subtractive clustering affect the accuracies of the model (Fig. 21). As shown on Fig. 21a we get the highest accuracy possible for 3 clusters in FCM clustering, compared to our initial guess of 2 clusters (Sect. 3.3). Figure 21b presents the different radius sizes used in Subtractive clustering. We can see that 0.6 radius size gives the best results, compared to our initial guess of 0.8, used in Sect. 3.3.

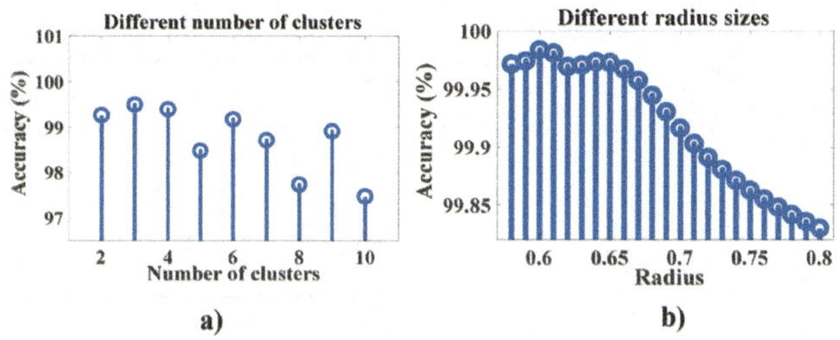

Fig. 21 Influence of: **a** Different number of clusters in FCM clustering; **b** Different radius sizes in Subtractive clustering

Table 5 Accuracies obtained for FCM and Subtractive clustering

Fuzzy c-means clustering		70–30%	50–50%	3-Fold
Filter	Wavelet			
FIR	db1	96.73	94.87	98.29
FIR	db2	98.25	95.41	96.39
FIR	coif4	98.32	97.12	99.03
FIR	db6	97.59	98.91	98.11
Subtractive clustering		70–30%	50–50%	3-Fold
Filter	Wavelet			
FIR	db1	98.80	99.45	99.40
FIR	**db2**	98.07	99.45	**99.59**
FIR	coif4	98.43	99.18	99.01
FIR	db6	98.52	99.50	99.02

Before we make the last comparison analysis (influence of the membership functions), we will conclude which input space partitioning method gives the best results, whereas we will use that method for the latest comparison. As we have already presented the influence of using different wavelet functions and different data split methods [29] for the grid partitioning method (Table 4), Table 5 shows the generated accuracies for different wavelets and data split methods, when FCM and Subtractive clustering are used. Compared to the initial results (Table 4), we get a maximum possible accuracy of 99.59% in Subtractive clustering when using the db2 wavelets and the 3-Fold cross validation method.

The last comparison is based on the different types of MFs used for the Subtractive clustering method. As all the results in Sects. 2 and 3 are based using Gauss MFs, here we present how the model accuracy is influenced by using other types of MFs, such as: psigmf, zmf, sigmf, gbellmf and dsigmf as defined in MATLAB [30]. Figure 22 presents the test accuracies when different types of MFs are used.

By this we conclude that our initial guess was correct, i.e. we get highest possible accuracy for the Gauss MFs.

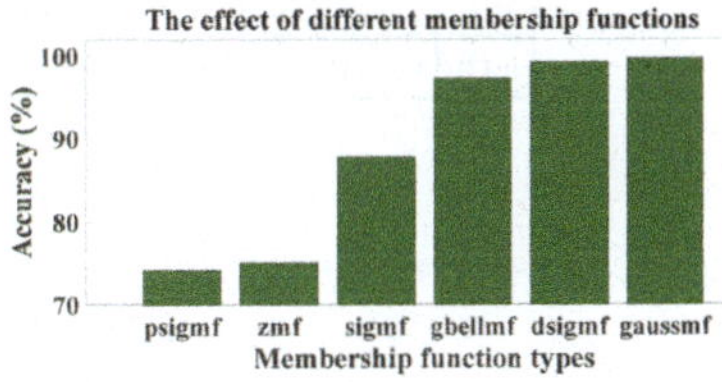

Fig. 22 Influence of the different types of MFs, when Subtractive clustering is used

4 Algorithm for Epilepsy Detection Using ANFIS Classifier, Extended with PCA

This section presents an upgrade to the algorithm, given in Sect. 2, using Principal Component Analysis (PCA) in order to reduce the number of features, used for training the ANFIS network.

The algorithm for epilepsy detection with fuzzy-neural networks for classification of EEG signals, presented in Sect. 2, consists of three main steps (Figs. 1 and 2):

(1) Filtering of the EEG signals with FIR filter
(2) Feature extraction and dimensionality reduction with discrete wavelet transform (DWT)
(3) Classification using ANFIS.

whereas the upgraded algorithm with PCA consists of one additional step, which implies the four following steps (Figs. 23 and 24):

(1) Filtering of the EEG signals with FIR filter
(2) Feature extraction and dimensionality reduction with DWT
(3) Using PCA for additional dimensionality reduction
(4) Classification using ANFIS.

Used methodologies in the upgraded algorithm for the Step 1, Step 2, and Step 4 are the same as the used methodologies in the algorithm, given in Sect. 2, for the Step 1, Step 2 and Step 3, respectively.

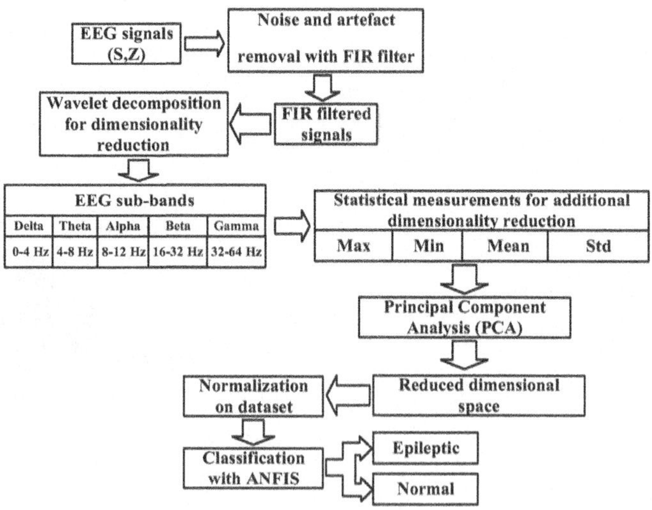

Fig. 23 Data flow for the upgraded algorithm, using PCA

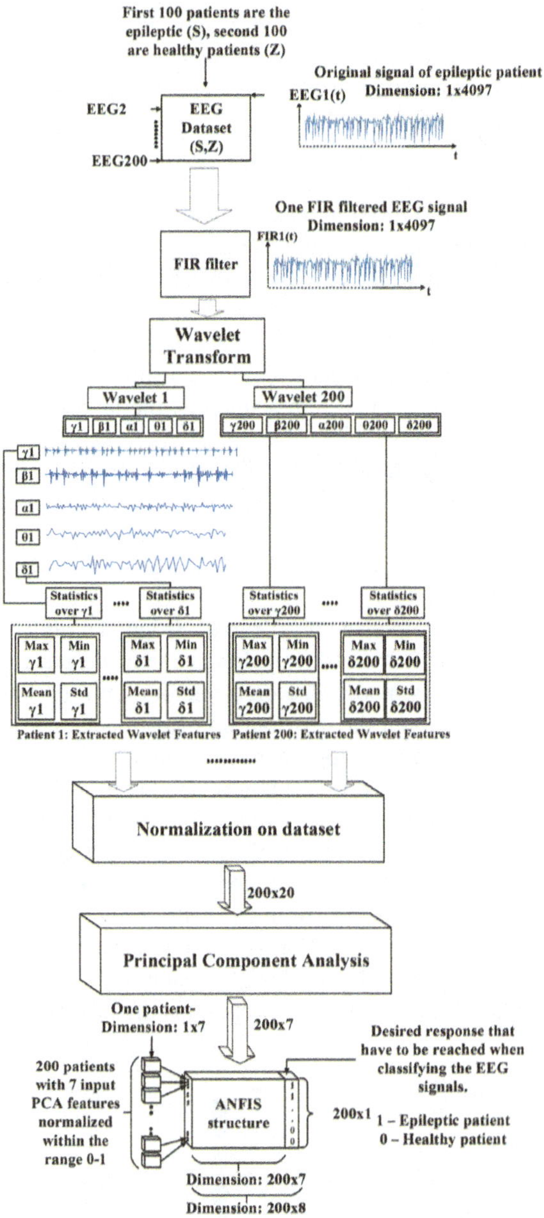

Fig. 24 Detailed analysis for the overall upgraded algorithm from Fig. 23

In Step 3, in the upgraded algorithm, PCA is used, which out of the given set of features selects the most important ones, in what way, further dimensionality reduction is performed. Application of PCA for classificationof epileptic EEG signals can also be found in [34–36].

In general, PCA is a time domain method which is used to reduce the large number of data and select the most important components as feature vectors. Since patterns in data can be hard to find in data of high dimensions, where the luxury of graphical representation is not available, PCA is a powerful tool for analyzing data. The objective is to represent data in a space that best expresses the variation in a sum-squared error sense.

First, the d-dimensional mean vector μ and $d \times d$ covariance matrix are computed for the full data set. Next, the eigenvectors and eigenvalues are computed, and sorted according to decreasing eigenvalue. Sub-sequently, the largest k such eigenvectors are chosen [34].

In this extended version of the algorithm, the PCA method comes after the dimensionally reduced space from the DWT. The PCA replaces original features with new features, called principal components, which are orthogonal and have eigenvectors with adequate eigenvalues. Eigenvalues for the given 20 features are shown in decreasing order on Fig. 25.

In order to decide which eigenvector(s) can be dropped without losing too much information from the construction of lower-dimensional subspace, inspection of the corresponding eigenvalues is made. The eigenvectors with the lowest eigenvalues bear the least information about the distribution of the data, and those are the ones that can be dropped. The common approach is to rank the eigenvalues from highest to lowest in order to choose the top k eigenvectors (as it is shown on Fig. 25). In order to choose adequate number of principal components, a useful measure, a so-called explained variance, is used. The total variance is the sum of variances of all individual principal components, so the fraction of variance explained by a principal component is the ratio between the variance of the principal component and the total variance, as shown on Fig. 26. The explained variance is a measure of how much information (variance) can be attributed to each of the principal components. It is clear that the first principal component explains the largest amount of variance

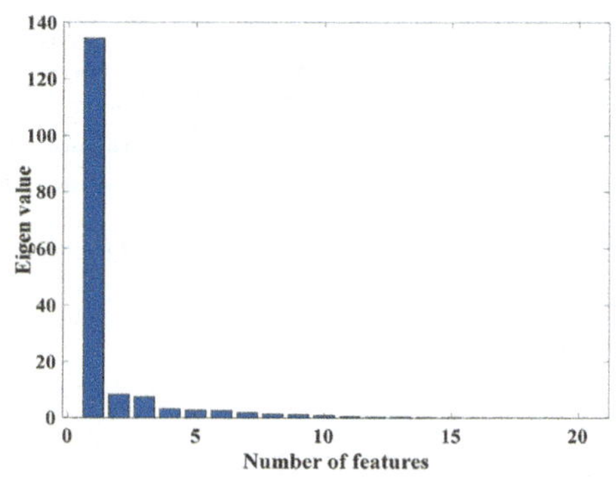

Fig. 25 Eigen values for the corresponding features sorted in descending order

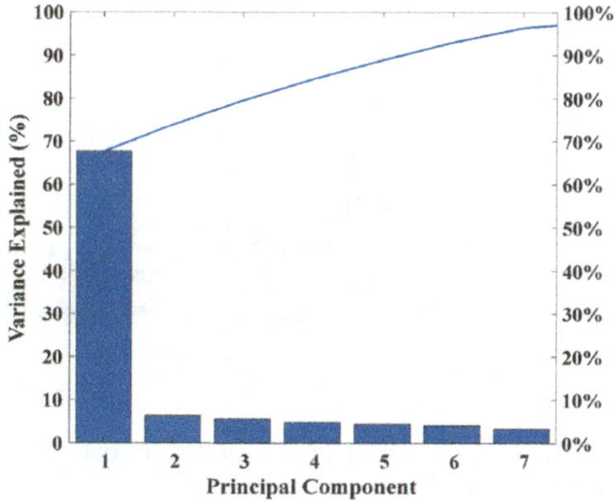

Fig. 26 Variance explained for each principal component

(67.66%) compared to other principal components. Together, the first seven principal components contain 96.40% of the information. According to this, the first 7 principal components as features are used, implying reduced dataset of 200 instances by 7 features. Subsequently, those 7 feature vectors are used as an inputs to the ANFIS model [4, 7].

It has to be emphasized that before the PCA, all the columns of the dataset have to be normalized within the range from 0 to 1 (Fig. 24), in order to achieve stable convergence on the weighted factors of the lately used neural network training [33]. Min–max normalization is used, i.e. we are normalizing the feature column vectors in the range from 0 to 1 [33, 38]. This method is also called "feature scaling", and represents a preprocessing technique [38].

We have to note that in the algorithm presented in Sect. 2, the ANFIS model used the grid partitioning method [21] for input space partitioning, where we presented a way of manipulating 20 inputs (with 3 MFs each) by partitioning the ANFIS model on sub-ANFIS models (as shown in Fig. 7, Fig. 15a), Fig. 28a, surpassing the major obstacle of "curse of dimensionality" [21]. In the upgraded algorithm with PCA simulations with grid partitioning method were also performed, but now only 7 inputs have to be manipulated (Fig. 28c).

5 Comparative Analysis for the Influence of the Different Data Splitting Methods, and Different Input Space Partitioning Methods, for the Both Algorithms

In order to make a comparison of the algorithm presented in Sect. 2, and the extended algorithm, presented in this section, i.e. evaluating the both algorithms in terms of training performance and classification accuracies, comparative analysis of the both algorithms was made, when using different data splitting methods, and different input space partitioning methods.

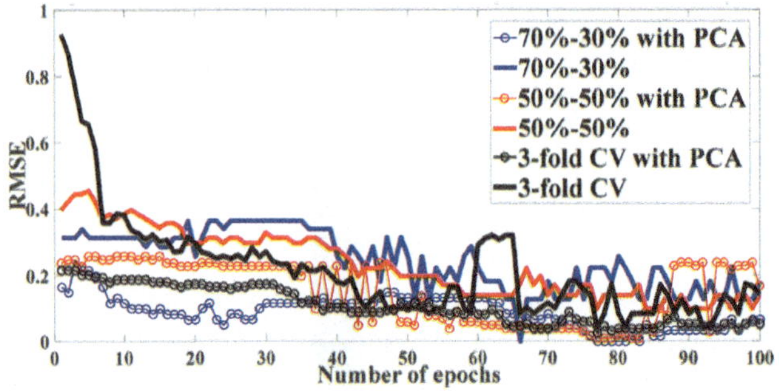

Fig. 27 Comparison of the test set RMSEs over the three data split methods, using grid partitioning, for the first and the upgraded algorithm

5.1 Comparative Analysis for the Influence of the Training and Testing Data, for the Both Algorithms

In Sect. 3 different approaches on dividing the dataset were presented, i.e. how the size of the training and testing data influence the accuracy of the algorithm, given in Sect. 2. In order to make a comparison on the performance of the Sect. 2 algorithm (that uses 20 features) and the upgraded algorithm with PCA (that uses 7 features) bellow 3 splitting methods were compared (divided dataset into 70–30% and 50–50% ratio, as well as using the splitting method based on cross validation [29]). This initial comparative analysis is obtained using grid partitioning method, and use of Gauss MFs [30], as this type of MFs gave best results in the analysis presented in Sect. 3.

Figure 27 presents the test set Root Mean Square Errors (RMSEs) when using the grid partitioning method during 100 epoch period, by applying the three different data split methods (70–30%; 50–50%; 3-Fold), for the first and the upgraded algorithm. As can be seen from Fig. 27, the lowest RMSEs are obtained in different number of epochs during the three partitioning methods. Nevertheless, it is evident that for the all three splitting methods, when the upgraded algorithm with PCA is used, RMSE values are lower than the RMSE values in the algorithm that does not use PCA. This implies that the use of PCA has a positive influence on the overall algorithm, when grid partitioning is used.

5.2 Influence of the Different Types of Input Space Partitioning

This section presents different methods of input space partitioning in the ANFIS model and the overall overview is given in Fig. 28. In Sect. 3 only the grid partitioning method as shown in Fig. 28a is used, where in order to reduce the number of rules,

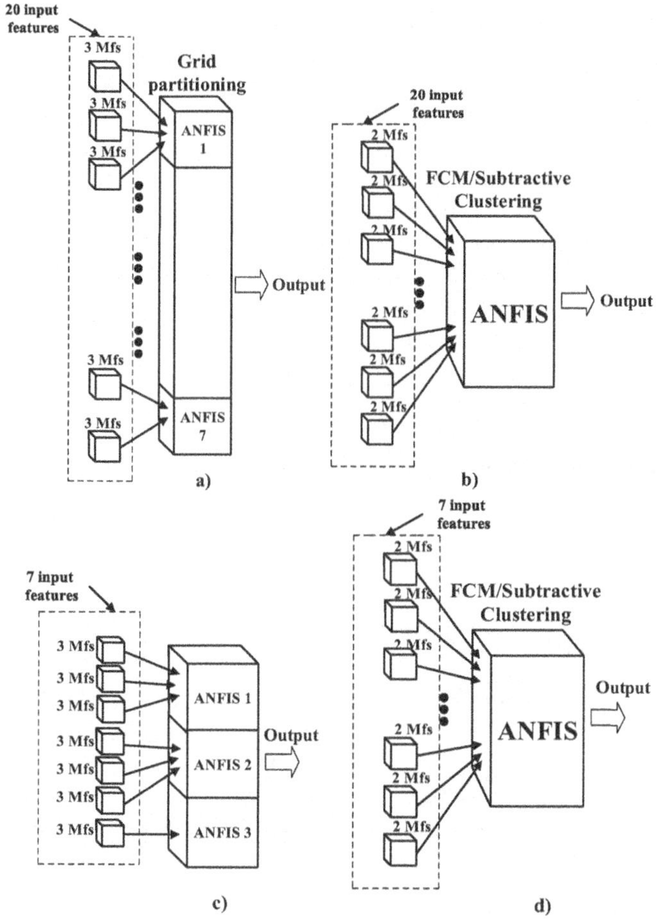

Fig. 28 The structure of the model generated using different input space partitioning methods: **a** Grid partitioning when the first algorithm with 20 input features is used (Fig. 7 in Sect. 2 and Fig. 15a in Sect. 3); **b** FCM/Subtractive clustering when the first algorithm with 20 input features is used (Fig. 15b in Sect. 3); **c** Grid partitioning when the upgraded algorithm with PCA and 7 input features is used; **d** FCM/Subtractive clustering when the upgraded algorithm with PCA and 7 input features is used

sub-ANFIS models are formed. As it is clear from Fig. 28a, sub-ANFIS models from 1 to 6 accept 3 inputs and the last sub-ANFIS model accepts 2 inputs. In the analysis in Sect. 3 two new approaches for input space partitioning, namely, fuzzy c-means (FCM) clustering and subtractive clustering [21] were used. The resulting ANFIS structure in that case are shown on Fig. 28b, where all the 20 inputs are passed at once, i.e. the number of rules is equal to the number of clusters, thus we do not face the problem called "curse of dimensionality". In this section, further reduction of the number of features (from 20 down to 7 features using PCA) was made, whereas the ANFIS input/output structure, when grid partitioning and FCM/Subtractive clustering are used is given on Fig. 28c and d, respectively. Short details for FCM and subtractive clustering can be found in Sect. 3.

Adequately to Fig. 27, the test set RMSEs for FCM clustering and Subtractive clustering, are given in Figs. 29 and 30, respectively, using different data split methods, when both algorithms are used. Two clusters for FCM clustering are used, and 0.8 radius of coverage (influence) for the subtractive clustering (for more information of the parameters in these clustering algorithms see [21, 30]). In both cases Gauss MFs [30] are used, and db2 wavelets, as these parameters gave best results in the analysis in Sect. 3. As we can see form Figs. 29 and 30, same as for the grid partitioning method (Fig. 27), satisfying results are obtained when trained between 40 and 60 epochs approximately (first algorithm performs better at about 40 epochs, but the upgraded one needs more epochs for training, which is expected as in the upgraded algorithm we use less data compared to the old algorithm).

In the case of FCM clustering (Fig. 29) different results are obtained, when the first and the upgraded algorithm are used, for the three splitting methods. At the beginning, for all the three splitting methods the upgraded algorithm shows worse

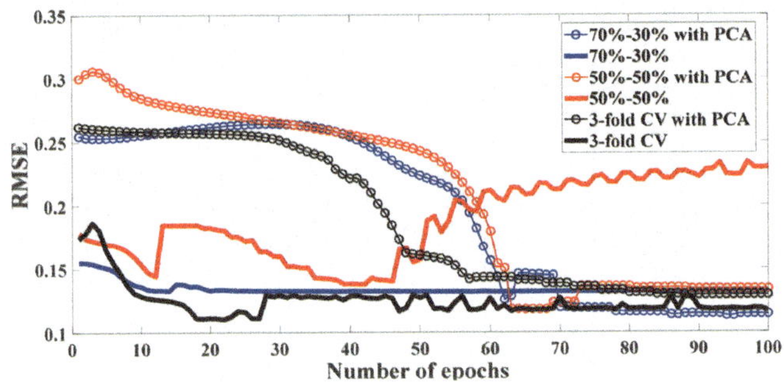

Fig. 29 RMSE values for FCM clustering for different data split methods (for the first and the upgraded with PCA algorithm)

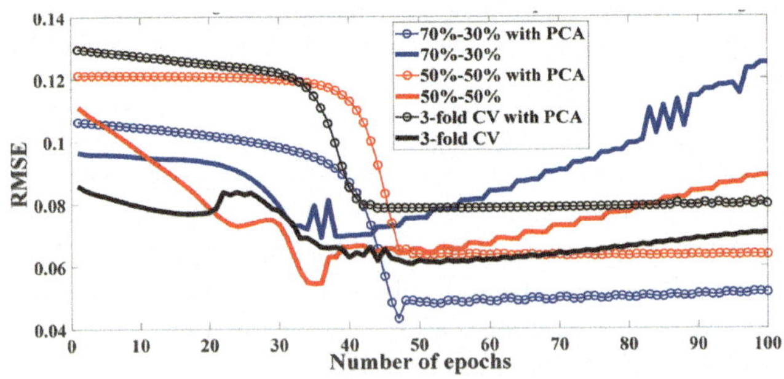

Fig. 30 RMSE values for Subtractive clustering for different data split methods (for the first and the upgraded with PCA algorithm)

behavior, but as the number of epochs increases this algorithm seams to stabilize after 60 epochs and does not show overfitting afterwards [37], as is the case with the first algorithm (for 50–50% splitting method—see Sect. 3).

In the case of subtractive clustering (Fig. 30), we also get different results when the first and the upgraded algorithm are used, for the three splitting methods. Nevertheless, it is evident that for this type of input space partitioning, the upgraded algorithm with PCA shows better results, as after about 45 epochs it is stable for all the three types of splitting, and it does not overfit, as is the case with the first algorithm.

In order to conclude the comparisons, in Tables 6 and 7 test set accuracies of the first and the upgraded algorithm are presented for the three different data splitting methods, using the three different input space partitioning methods.

Firstly, we choose to make comparative analysis using 40 training epochs (Table 6), as for some of the cases, the first algorithm overfits after 40 epochs (Figs. 27, 29, and 30). On the other hand, as the upgraded algorithm in some of the cases needs more than 40 epochs for training, we choose to make comparative analysis of the both algorithms using 60 epochs for training (Table 7).

Form Table 6 it is evident that the upgraded algorithm in most of the cases have worse performance than the first algorithm, but this is expected, as according to Figs. 27, 29, and 30, the upgraded algorithm needs more than 40 epochs for training. On the other hand, the test set accuracies for the both algorithms, trained with 60 epochs, show better performance in most of the cases for the upgraded algorithm (Table 7).

It general, it is evident that the upgraded algorithm has satisfactory performance, and in some cases performs even better than the old algorithm, although the number of features is significantly reduced (from 20 to 7), which also plays a crucial role in making the new algorithm more resistant to overfitting.

Table 6 Accuracies Obtained for Grid Partitioning, FCM and Subtractive clustering, for the first and the upgraded algorithm, when 40 epochs for training are used

Grid partitioning				70–30%	50–50%	3-Fold
Filter	Wavelet	MFs type	First versus upgraded algorithm			
FIR	db2	Gauss MFs	First algorithm from Sect. 2	95	87	95.51
			Upgraded algorithm with PCA	97.33	89	96.03
Fuzzy c-means clustering (number of clusters = 2)				70–30%	50–50%	3-Fold
Filter	Wavelet	MFs type	First versus upgraded algorithm			
FIR	db2	Gauss MFs	First algorithm from Sect. 2	98.25	95.41	96.39
			Upgraded algorithm with PCA	83.43	82.21	85.33
Subtractive clustering (radius of coverage—influence = 0.8)				70–30%	50–50%	3-Fold
Filter	Wavelet	MFs type	First versus upgraded algorithm			
FIR	db2	Gauss MFs	First algorithm from Sect. 2	98.07	99.45	99.59
			Upgraded algorithm with PCA	97.21	93.33	98.22

Table 7 Accuracies Obtained for Grid Partitioning, FCM and Subtractive clustering, for the first and the upgraded algorithm, when 60 epochs for training are used

Grid partitioning				70–30%	50–50%	3-Fold
Filter	Wavelet	MFs type	First versus upgraded algorithm			
FIR	db2	Gauss MFs	First algorithm from Sect. 2	98.33	97	92.96
			Upgraded algorithm with PCA	96.67	98	96.53
Fuzzy c-means clustering (number of clusters = 2)				70–30%	50–50%	3-Fold
Filter	Wavelet	MFs type	First versus upgraded algorithm			
FIR	db2	Gauss MFs	First algorithm from Sect. 2	99.74	90	98.98
			Upgraded algorithm with PCA	98.85	97.93	95.33
Subtractive clustering (radius of coverage—influence = 0.8)				70–30%	50–50%	3-Fold
Filter	Wavelet	MFs type	First versus upgraded algorithm			
FIR	db2	Gauss MFs	First algorithm from Sect. 2	92.21	93.33	97.43
			Upgraded algorithm with PCA	99.5	98.63	97.94

6 Conclusion

This studyfirstly presented the use of ANFIS for classification of the two classes EEG signals. The input parameters in the ANFIS model were the extracted features of the wavelet coefficients. The proposed ANFIS combined the adaptive capability of the neural networks and the qualitative approach in the fuzzy logics. The ANFIS classifier which was trained with 60 epochs reached 98.3% accuracy on the test set, and 99.5% classification accuracy on the overall dataset. We also made comparison with the results presented in other related works, whereas we can conclude that the proposed algorithm can be successfully used in classification of EEG signals. Secondly we made a comparative analysis through modification in the ANFIS parameters. Several comparisons were made: the usage of different wavelets, different data split methods, different MFs, the effect of normalization as well as the different types of input space partitioning methods (grid partitioning, versus FCM clustering, versus subtractive clustering). We concluded that the combination of db2 wavelets, the Gauss MFs and the 3-Fold cross validation method, using subtractive clustering, gives the best results, with accuracy of 99.59 when trained with 40 epochs. We also concluded that the effect of normalization made the biggest difference in our performance. Thirdly, weintroduced an extension of the previously presented algorithm, reducing the number of the used features, using PCA. In order to show that the upgraded algorithm acts similar, or even better than the firstly presented algorithm, several comparative analyses were made, comparing both algorithms for different data splitting methods (70–30%, 50–50%, and 3-Fold cross validation), and the different types of input space partitioningmethods (grid partitioning, versus FCM clustering, versus subtractive clustering). The comparisons were made using only db2 wavelets and Gauss MFs. The comparisons of the accuracies of the both algorithms for the different data split methods and the different input space partitioning were made both for 40 and 60 epochs, as the first algorithm in some of the cases overfits after 40 epochs, and 60 epochs was a reasonable sublimate according the RMSE simulation results. It is well known that the size reduction methods are preferred in applications where the number of input parameters in the data set is too large [39]. In this sense, the PCA in the upgraded algorithm plays a key role, i.e. although it sufficiently reduces the data dimension (number of features from 20 to 7), according to the results, the essence of the original data is preserved, but the algorithm has a potential of reducing computational costs.

References

1. Acharya, U.R., Sree, S.V., Swapna, G., Martis, R.J., Suri, J.S.: Automated EEG analysis of epilepsy: a review. Knowl. Based Syst. **45**, 147–165 (2013)
2. Adeli, H., Zhou, Z., Dadmehr, N.: Analysis of EEG records in an epileptic patient using wavelet transform. J. Neurosci. Methods **123**(1), 69–87 (2003)
3. Adeli, H., Dastidar, S.G.: Automated EEG-Based Diagnosis of Neurological Disorders: Inventing the Future of Neurology. Taylor and Francis Group (2010)
4. Guller, I., Ubeyli, E.D.: Adaptive neuro-fuzzy inference system for classification of EEG signals using wavelet coefficients. J. Neurosci. Methods **148**(2), 113–121 (2005)
5. Najumnissa, D., Rangaswamy, T.R.: Detection and classification of epilepsy seizures using wavelet feature extraction and adaptive neuro- fuzzy inverence system. Int. J. Comput. Eng. Res. **2**, 755–761 (2013)
6. Subasi, A.: Application of adaptive neuro-fuzzy inference system for epileptic seizure detection using wavelet feature extraction. Comput. Biol. Med. **37**(2), 227–244 (2005)
7. Guller, I., Ubeyli, E.D.: Application of adaptive neuro-fuzzy inference system for detection of electrocardiographic changes in patients with partial epilepsy using feature extraction. Expert Syst. Appl. **27**(3), 323–330 (2004)
8. Gajic, D., Djurovic, Z., Di Gennaro S., Gustafsson, F.: Classification of EEG signals based on wavelets and statistical pattern recognition. Biomed. Eng. Appl., Basis Commun. **26**(2), 1450021
9. Nakate, A., Bahirgonde, P.D.: Feature extraction of EEG signals using wavelet transform. Int. J. Comput. Appl. **124**(2), (2015)
10. Wang, L., Xue, W., Luo, Y.L.M., Huang, L., Cui, W., Huang, C.: Automatic epileptic seizure detection in EEG signals using multi-domain feature extraction and nonlinear analysis. Entropy **19**(6), (2017)
11. Sinha, P.: Speech Processing in Embedded Systems, pp. 25–32. Springer Science+Business Media, LLC (2010)
12. Mneney, S.H.: An Introduction to Digital Signal Processing: A Focus on Implementation, pp. 153–158. River Publishers (2008)
13. Wen, T., Zhang, Z.: Effective and Extensible Feature Extraction Method Using Genetic Algorithm-Based Frequency-Domain Feature Search for Epileptic EEG Multi-Classification. https://arxiv.org/abs/1701.06120v1 (2017)
14. Bhatia, P.K., Sharma, A.: Epilepsy seizure detection using wavelet support vector machine classifier. Int. J. Bio-Sci. Bio-Technol. **8**(2), 11–22 (2016)
15. Guerra, E.J., Aquino, V.A., Gil, P.G.: Epilepsy seizure detection in EEG signals using wavelet transforms and neural networks. Comput. Inf. Syst. Sci. Eng. (CISSE), 12–14 (2013)
16. Omerhodzic, I., Avdakovic, S., Nuhanovic, A., Dizdarevic, K.: Energy Distrubution of EEG Signals: EEG Signal Wavelet-Neural Network Classifier. https://arxiv.org/abs/1307.7897v1 (2013)

17. Kumar, A., Saini, L.M.: Detection of epileptic seizure using discrete wavelet transform of EEG signal. Int. J. Soft Comput. Artif. Intell. ISSN: 2321-404X (2015)
18. Rabbi, A.F., Rezai, R.F.: A fuzzy logic system for seizure onset detection in intracranial EEG. Comput. Intell. Neurosci. **2012**, 4 (2011)
19. Baxt, W.G.: Use of an artificial neural network for data analysis in clinical decision making: the diagnosis of acute coronary occlusion. Neural Comput. **2**, 480–489 (1990)
20. Miller, A.S., Blott, B.H., Hames, T.K.: Rewiev of neural network applications in medical imaging and signal processing. Med. Biol. Eng. Comput. **30**, 449–464 (1992)
21. Jang, J.S.R., Sun, C.T., Mizutani, E.: Neuro-Fuzzy and Soft Computing-A Computational Approach to Learning and Machine Intelligence. Prentice Hall Upper Saddle River (1997)
22. Kuncheva, L.I., Steimann, F.: Fuzzy diagnosis. Artif. Intell. Med. **16**, 121–128 (1992)
23. Belal, S.Y., Taktak, A.F.G., Nevill, A.J., Spencer, S.A., Roden, D., Bevan, S.: Automatic detection of distorted plethysmogram pulses in neonates and pediatric patients using an adaptive-network-based fuzzy inference system. Artif. Intell. Med. **24**, 149–165 (2002)
24. Virant-Klun, I., Virant, J.: Fuzzy logic alternative for analysis in the biomedical sciences, Comput. Biomed. Res.;**32**, 305–21 (1999)
25. Stoimchev, M., Ojleska Latkoska, V.: Detection of epilepsy using adaptive neuro-fuzzy inference system. J. Electr. Eng. Inf. Technol. **3**(1–2), 41–51 (2018)
26. Stoimchev, M., Ojleska Latkoska, V.: Comparative analysis for the influence of the tuning parameters in the algorithm for detection of epilepsy based on fuzzy neural networks. In: Proceedings of the 14th International Conference-ETAI 2018, Struga, R. Macedonia, September 20–22, (2018)
27. Stoimchev, M., Ojleska Latkoska, V.: Feature space reduction using PCA in the algorithm for epilepsy detection using adaptive neuro-fuzzy inference system and comparative analysis. Acta Polytech. Hung., Spec. Issue APH-ETAI, **17**(10), (2020)
28. Andrzejak, R.G., Lehnertz, K., Rieke, C., Mormann, F., David, P., Elger, C.E.: Indications of Nonlinear Deterministic and Finite Dimensional Structures in Time Series of Brain Electrical Activity. http://epileptologie-bonn.de/cms/front_content.php?idcat=193&lang=3
29. Omary, Z., Mtenzi, F.: Machine learning approach to identifying the dataset treshold for the performance estimators in supervised learning. Int. J. Inform. (IJI) **3**(3), (2010)
30. https://www.mathworks.com/products/matlab.html
31. Pandey, N., Tiwari, N.: Predictive accuracy of modified subtractive clustering algorithm on large dataset. Int. J. Res. Dev. Appl. Sci. Eng. (IJRDASE) **8**(2), (2015)
32. Ghuman, S.S.: Clustering techniques-a review. Int. J. Comput. Sci. Mob. Comput. **5**(5), (2016)
33. Mustaffa, Z., Yusof, Y.: A comparison of normalization techniques in predictiong dengue outbreak. Int. Conf. Bus. Econ. Res. **1**, (2011)
34. Subasi, A., Gursoy, M.I.: EEG signal classification using PCA, ICA, LDA and support vector machines. Expert Syst. Appl. **37**(12), 8659–8666 (2010)
35. Acharya, U.R., Sree, S.V., Alvin, A.P.C., et al.: Use of principal component analysis for automatic classification of epileptic EEG activities in wavelet framework. Expert Syst. Appl. **39**(10), 9072–9078 (2012)
36. Siuly, S., Li, Y.: Designing a robust feature extraction method based on optimum allocation and principal component analysis for epileptic EEG signal classification. Comput. Methods Programs Biomed. **119**(1), 29–42 (2015)
37. Domingos, P.: A few useful things to know about machine learning. Commun. ACM **55**(10), 78–87 (2012)
38. Singh, B.K., Verma, K., Thoke, A.S.: Investigations of impact of feature normalization techniques on classifier's performance in breast tumor classification. Int. J. Comput. Appl. **116**(19), (2015)
39. Yüksek, A., Arslan, H., Kaynar, O., Delibaş, E.: Comparison of the effects of different dimensional reduction algorithms on the training performance of anfis (adaptive neuro-fuzzy inference system) model. Cumhur. Sci. J. **38**(4), 716–730 (2017)

Improving Accuracy and Reducing Financial Risk When Forecasting Time Series of SIU0 Future Contracts Employing Neural Network with Word2vec Vector News

Nikolay Lomakin, Anastasia Kulachinskaya, Maxim Maramygin, and Elena Chernaya

Abstract The Bot-advisor system has been developed to forecast the closing price of the SIU0 futures contract on the Moscow stock exchange. It includes the Scraper programs, the Word2vec neural network, the Perceptron neural network on the Deductor platform, and the QUIK trading terminal with an integrated Lua-socket. The study is based on the theoretical background of forecasting time series of financial instruments. This article dwells on the experience of using artificial intelligence systems for collecting and processing BigData to forecast the time series of the SIU0 future contracts. The authors have put forward and proved the hypothesis that the use of a neural network makes it possible to forecast the closing price of the SIU0 futures contract on a 15-min timeframe. The proposed AI system improves accuracy and reduces financial risk when forecasting the time series of the SIU0 futures contract for exchange trading. To forecast the price of the SiU0 futures contract, the system uses the parameters of Japanese candlesticks and volume as well as «news fluctuations» from web sites. The Perceptron neural network, designed on the Deductor platform, was trained on two types of data: (1) cost-(Pclose) and (2) logarithmic-(ln). The proposed solution is of great practical value, since the developed AI system provides high forecast accuracy. Thus, the average size of the neural network error in the first case (Pclose) was 0.000927425, and in the second case (Pln) the average error size was 0.051026481. The variance of error values as a percentage of the closing price in the first case was 0.304107913, while in the second case it was 0.343654316, or 4 hundredths better. The first option proved to be more effective in

N. Lomakin
Volgograd State Technical University, Volgograd, Russian Federation

A. Kulachinskaya (✉)
Peter the Great St. Petersburg Polytechnic University, St. Petersburg, Russian Federation
e-mail: tel9033176642@yahoo.com; a.kulachinskaya@yandex.ru

M. Maramygin
Ural State University of Economics, Yekaterinburg, Russian Federation

E. Chernaya
Volga Institute of Economics, Pedagogy and Law, Volzhskiy, Russian Federation

P. Shi et al. (eds.), *Complex Systems: Spanning Control and Computational Cybernetics: Applications*, Studies in Systems, Decision and Control 415,
https://doi.org/10.1007/978-3-031-00978-5_12

terms of the amount of profit received (variation margin). In the first case, the profit amounted to 748.0 rubles, while in the second case the profit was only 548.0 rubles. The maximum drawdown in both variants was 169 rubles.

Keywords AI system · Big data · Financial risk · Forecast · Futures contract · Neural network · Quantile · SIU0 · VaR model

1 Introduction

1.1 Relevance and Practical Value

The article raises the problem of using artificial intelligence systems to forecast the parameters of the time series of an exchange-traded instrument. The solution to this problem is of great practical value since improving forecasting accuracy allows us to increase the efficiency of Stock exchange trading by minimizing financial risk. However, system dynamics in forecasting problems of large time-series data (let alone the contemporary Big-Data Sets) are notoriously known to be rather complex nonlinear and time-varying systems hence extremely difficult to carry out analysis by any methodology. In fact, study problem of this class have given rise to the discipline of time-series analysis, forecasting and control.

It has been long established understanding that the complexity dynamics of econimic and financial processes involve a variety of nonlinear [5] and time-varying phenomena [16, 22] regardless whether assumed stochastic, as they naturally are [27], or approximated deterministic. It is therefore such processes ought to be envisaged (Fig. 1) as general nonlinear time-varying multi–input–multi–output dynamical systems [1, 11, 13].

The authors propose an AI-system that uses the parameters of Japanese candlesticks and volume, as well as «news fluctuations» from websites, to forecast the price of the SiU0 futures contract. The Perceptron neural network designed on the Deductor platform was trained on two types of data (1) cost (Pclose) and (2) logarithm (ln).

The hypothesis was put forward and proved that the neural network makes it possible to forecast the closing price of the SIU0 futures contract on a 15-min time frame.

The proposed solution is of great practical value, since the developed AI system provides high forecast accuracy. Thus, the average size of the neural network error in the first case (Pclose) was 0.000927425, and in the second case the average error size was 0.051026481. The variance of error values as a percentage of the closing price in the first case was 0.304107913, while in the second case it was 0.343654316, or 4 hundredths better.

The sigma of the neural network error values in the first case (Pclose) was 0.551459802, while in the second case (Pln) it was 0.586220365. This suggests that in the first case there is lesser variance of values relative to the mean.

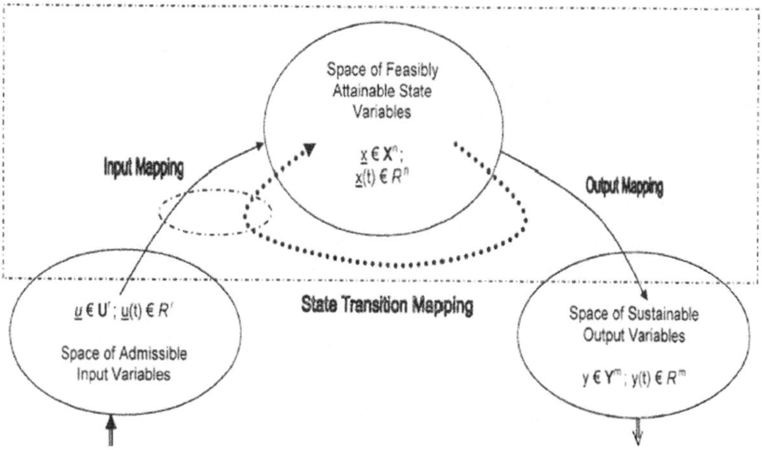

Fig. 1 An illustration of controlled general nonlinear systems in accordance to the fundamental laws of physics in Nature and Society [13, 14]; process input, state and output spaces in terms of involved classes of functions at any fixed instant of time all the vector-valued variables become real-valued vectors, (defined [11]), thus amenable to estimation and forecasting [1]

The first option turned out to be more efficient in terms of the amount of profit received (variation margin). In the first case, the profit amounted to 748.0 rubles, while in the second case the profit was only 548.0 rubles. The maximum drawdown in both variants was the same and amounted to 169 rubles.

1.2 Artificial Intelligence System Bot-Advisor Based on Big Data Processing

The advent and development of data science in terms of theoretical models, algorithms, experiments, applications and specific systems are described in the book Data Analytics. Springer Briefs is a concise study of the information and knowledge processing and, therefore, deserves the attention of researchers in this field.

The Hadoop platform is an open-source framework that allows you to make applications split into several fragments. Each of the fragments is processed on any node in the cluster of the developed computing system.

Today, there is a wide variety of programs for collecting, processing and analyzing big data. These programs include the following: frameworks (Hadoop, Spark, Storm), databases (Hive, Impala, Presto, Drill), analytical platforms (RapidMiner, IBMSPSS Modeler, KNIME, Qlik. Analytics Platform, STATISTICA DataMiner, WorldProgrammed an Intelligent platform, Deductor, SASEnterprise-Miner), and others.

It is possible to improve the accuracy of forecasting the price of the SIU0 futures contract, which is traded on the Moscow stock exchange using a Bot-advisor, the flow chart of which is shown in Fig. 1.

The Bot-adviser interacts with the exchange trading terminal QUIK via an integrated Lua socket. In the process of operation, the data of the time series of the SIU0 futures contract on a 15-min timeframe is imported online and transmitted for connection with a 300-dimensional vector from Word2vec to form a training sample for the Perceptron neural network. In this process, the following parameters are used: opening price (Po), closing price (Pc), maximum price (Ph), minimum price (Pl), as well as trading volume (V) in two formats: (1) cost (Pclose) and (2) logarithm (ln).

In the course of the study, the hypothesis was put forward and proved that the use of a neural network makes it possible to forecast the closing price of the SIU0 futures contract on a 15-min timeframe.

1.3 The Skraper Program for Collecting and Counting Words from News Websites for BI Processing

Words from news websites were collected using the Skraper program, the architecture of which is shown below (Fig. 2). BI-data collected by the Skraper program were transferred for further processing by the Word2vec program for their subsequent vector-valued ('vectorization') extension (Fig. 3).

Word2vec is a generic name for a set of artificial neural network-based models designed to obtain vector representations of words in a natural language. Word2vec is used to analyze the semantics of natural languages. Vector representation is a generic

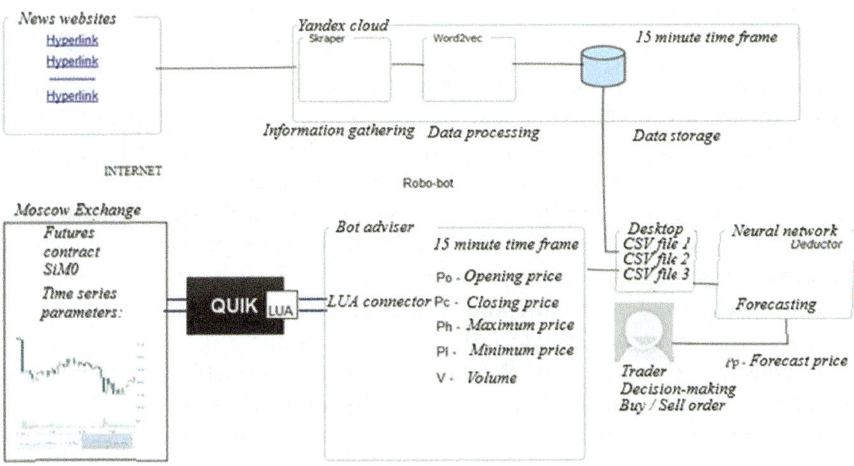

Fig. 2 Architecture of Bot-advisor platform

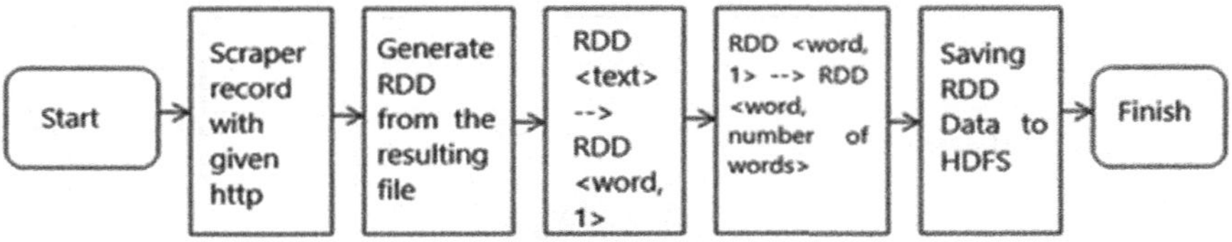

Fig. 3 Flow chart of the Skraper program

name for various approaches to language modeling and representation training in natural language processing, aimed at matching words (and possibly phrases) from the dictionary of vectors with a significantly smaller number of words from the language dictionary [36].

The theoretical basis for vector representations is distributive semantics [22].

Two data items are obtained from one context window (there are two related data items for one target word). The window size is usually user-defined. The larger the size of the context window, the more advanced is the model, but this affects the execution time of the algorithm. This model is known as the «skip-gram» algorithm, one of the word2vec algorithms. Another algorithm is known as the continuous bag-of-words model (CBOW) (see Fig. 4).

1.4 Perceptron Neural Network for Forecasting the Closing Price of the SIU0 Futures Contract for the Next 15 min Based on BI from News Websites and Time Series Data

The Perceptron program for forecasting the closing price of the SIU0 futures contract is based on the Deductor-Deductor platform is an analytical platform developed by Base. Group Labs. The most popular analytical algorithms (decision trees, neural networks, self-organizing maps, etc.) are built into Deductor. It provides dozens of visualization methods and integration with data sources/receivers [3].

The second type—«Logarithm-(ln)» was a 300-dimensional vector based on Word2vec and parameters with the natural logarithm of the price ratio P/Pt−1: openings (Po), closing price (Pc), maximum price (Ph), minimum price (Pl), trading volume (V). The forecast value was potentiated, i.e. that is to say Pprogn = e^Pprogn. Then the forecast value was compared with the actual closing price, and if the forecast value turned out to be higher, a long position was opened and vice versa. The first version of Perceptron was trained on a dataset that included 305-dimensional vectors and forecast values of the closing price Pt+1, the dataset contained data from April 14, 2020 to July 04, 2020 and consisted of 1930 rows (see Table 1).

The Perceptron neural network, designed on the Deductor platform, was trained on two types of data (1) cost (Pclose) and (2) logarithm (ln).

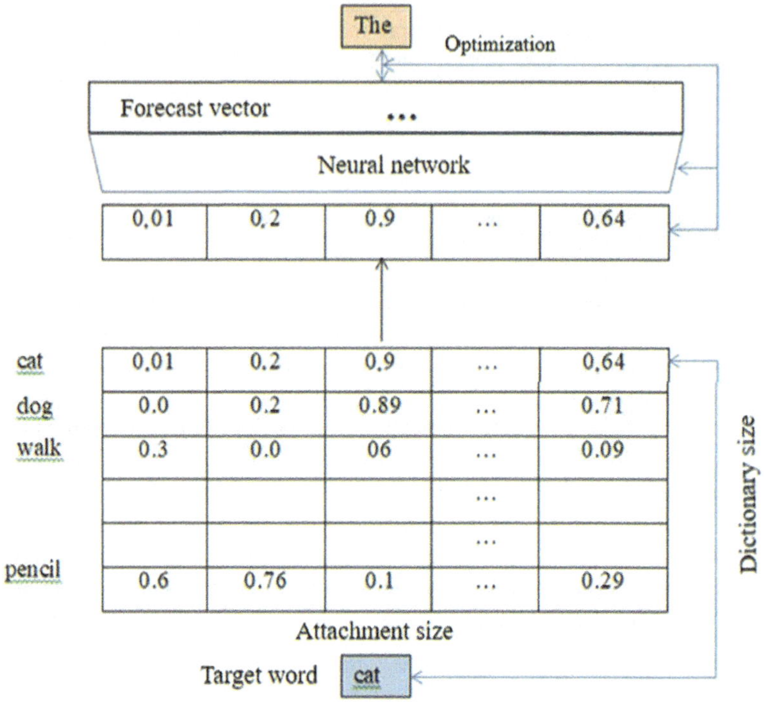

Fig. 4 Neural network background of the Word2vec

Table 1 Fragment of the training sample of the first variant of Perceptron

Date, time	X1	X2	…	X300
14.04.2020 22.00.00	−0.062179142	0.020744415	…	0.033938743
14.04.2020 21.45.00	0.059611453	−0.062379819	…	0.034278696
…	…	…	…	…
04.06.2020 16.15.00	0.058385131	−0.059321518	…	0.037604395

…	P_o	P_c	P_h	P_l	Volume	$P_{rognosis}$(Pt+1)
…	71,065	71,107	71,114	71,062	6721	71,140
…	71,065	71,067	71,080	71,035	12,248	71,107
…	…	…	…	…	…	…
…	74,849	74,717	74,861	74,715	28,285	74,878

The first type «Cost (Pclose)»is a 300-dimensional vector based on Word2vec and parameters: opening price (Po), closing price (Pc), maximum price (Ph), minimum price (Pl), and trading volume (V). The forecast value was compared with the actual closing price, and if the forecast value turned out to be higher, a long position was opened and vice versa.

The Deductor platform allows you to retrieve data from heterogeneous sources, consolidate it into a single storage and display it in the form of reports and OLAP cubes, and perform other functions.

The developed Perceptron contains 305 parameters on the input layer, two hidden layers of 100 and 10 nodes, respectively, and an output layer with one parameter—the forecast price. The Perceptron was designed and used on the Deductor platform. The neural network graph is shown in Fig. 5.

The characteristics of the statistical parameters of the dataset of the neural network employed are shown in Fig. 6.

The second type of dataset—«Logarithm-(ln)» is shown in Fig. 7.

The new generated dataset consists of the parameters of a 300-dimensional vector from the words retrieved by the Word2vec program and the logarithmic time series elements.

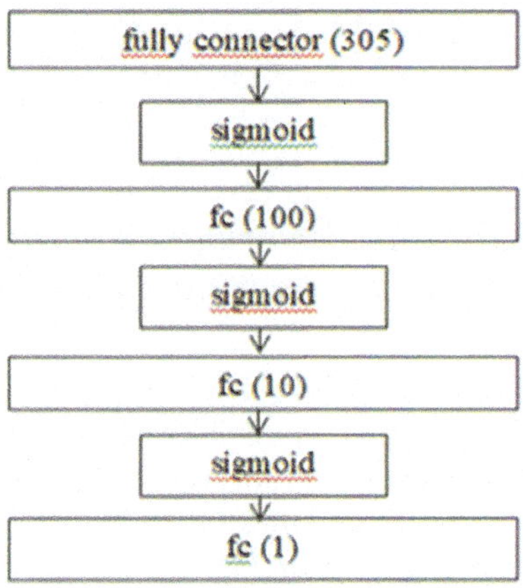

Fig. 5 Perceptron neural network flow-graph diagram

Fig. 6 Statistical parameters of the dataset of the neural networkemployed

9.0 Close

Range	Frequency
68348 : 69004.25	188 (9.7%)
69004.25 : 69660.5	228 (11.8%)
69660.5 : 70316.75	208 (10.8%)
70316.75 : 70973	164 (8.5%)
70973 : 71629.25	226 (11.7%)
71629.25 : 72285.5	169 (8.8%)
72285.5 : 72941.75	66 (3.4%)
72941.75 : 73598	102 (5.3%)
73598 : 74254.25	266 (13.8%)
74254.25 : 74910.5	218 (11.3%)
74910.5 : 75566.75	49 (2.5%)
75566.75 : 76223	46 (2.4%)

KJ	KK	KL	KM	KN	KO	KP	KQ	KR	KS	KT
X296	X297	X298	X299	X300	Ko	Kc	Kh	Kl	Kv	Kprogn
0,017904	-0.00692	0,01027	-0,00774	0,033938743	0	0,000562691	0,00047822	0,00038	-0,60012571	0,000463982
0,018052	-0,00578	0,010167	-0,00781	0,034278696	-0,00061896	1,40713E-05	-0,00059071	-0,0001548	0,31942375	0,000562691
0,016504	-0,00771	0,009248	-0,00609	0,034699797	0,000168769	-0,000590825	-0,00032334	-0,0003378	-0,36288106	1,40713E-05
0,018288	-0,00713	0,010215	-0,00822	0,033852436	0,000337624	-0,000731015	-0,00046373	0,0002111	-1,76618506	-0,000590825
0,017671	-0,0051	0,010171	-0,00827	0,035751723	0,001309371	0,001237415	0,000941745	0,0013097	0,32428176	-0,000731015
0,018663	-0,00503	0,01041	-0,00807	0,035947136	-0,00088718	0,001267124	0,000703373	9.865E-05	0,42771888	0,001237415
0,017874	-0,00573	0,010465	-0,00919	0,034497658	0,001464975	-0,000830851	9,85118E-05	0,000846	-0,36605271	0,001267124
0,017722	-0,00448	0,010538	-0,00808	0,035158633	0,000535815	0,001436822	0,001337906	0,0004938	0,24964734	-0,000830851
0,01809	-0,00609	0,009761	-0,00857	0,035959262	-0,000465329	0,000521711	0,000211412	-4,234E-05	0,44745977	0,001436822
0,018288	-0,00476	0,010401	-0,0082	0,035610339	-0,000817327	-0,00047942	-0,00135227	-0,0008604	-0,84799135	0,000521711
0,017547	-0,00611	0,010378	-0,00781	0,034824909	0,001000613	-0,000775046	0,000211171	0,0004654	0,14147116	-0,00047942
0,01829	-0,00468	0,010217	-0,00816	0,035190343	0,00094516	0,001000627	0,001141094	0,0010868	-0,1113302	-0,000775046
0,018127	-0,00552	0,009969	-0,00885	0,036060313	-0,000324558	0,000931059	0,000930744	0,0003107	0,15987984	0,001000627
0,018217	-0,00405	0,010539	-0,00809	0,034868728	-0.0018184	-0.000352776	-0,00183246	-0,0001412	-0,95146736	0.000931059

Fig. 7 Fragment of the second variant of the dataset with logarithmic elements of time series

This study of ours has shown that the developed AI-model for exchange trading of the SiU0 futures makes it possible to obtain forecast values and ensures better results in exchange trading.

1.5 The Performance of the Perceptron Neural Network in Forecasting the SIU0 Closing Price

The accuracy of the forecast and the economic performance of the neural network were studied on the data that were not included in either the trained sample or the test sample from July 03, 2020 from 10.00 to 15.30.

The graphs reflecting on the forecast accuracy for both options are shown in Fig. 8. The Graphs reflecting on the level of forecast error for both options are shown further below (Fig. 9).

The forecast accuracy and trading performance are presented in Table 2, which is given above.

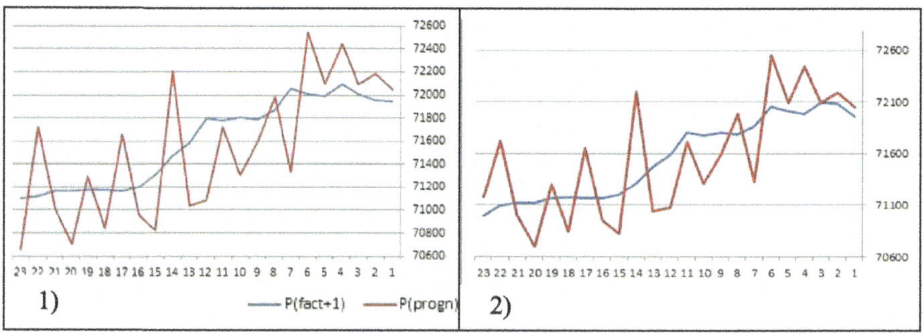

Fig. 8 «Fact-forecast» curves for two options

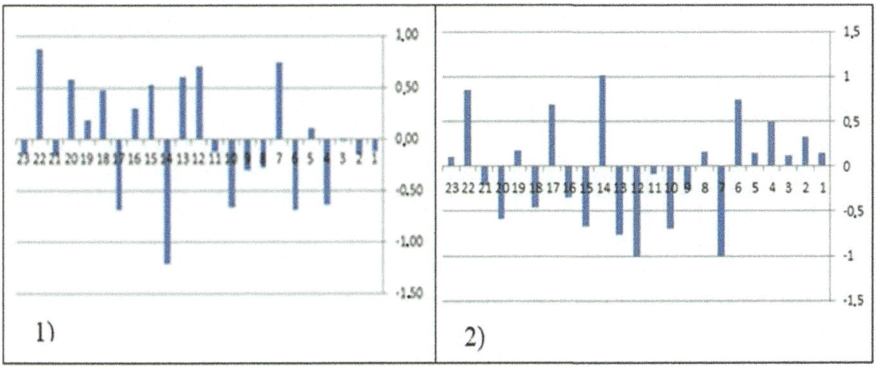

Fig. 9 The level of forecast errors for both options [%]

Table 2 Dynamics of forecast accuracy (fragment)

Name	Option		Deviation, %
	(1) P_{close}	(2) $P_{(ln)}$	
Number of 15 min time frames	24	24	0
Average size of the forecast error,%	0.000927425	−0.051026481	−55.019523
Maximum size of forecast error,%	0.881883305	1.02139141	1.1581934
Dispersion	0.304107913	0.343654316	1.1300407
Sigma	0.551459802	0.586220365	1.0630337
Profit—total, rub	748	548.00	0.73
The number of profitable trades, pcs	17	15	0.8823529
Profit on profitable trades, rub	1177	1077	0.91
Number of losing trades, pcs	8	8	1
Loss amount, rub	−429	−529	1.23
Maximum drawdown, rub	−169	−169	1
Maximum drawdown,% of GO(4810.13)	−3.51	−3.51	1
Profit per hour, rub	136	99.64	0.73
Profit-rate for 5.5 h,%	15.55	11.39	0.73
Profit-rate for 1 h,%	6.43	5.67	0.88

The deposit growth curves as a result of the Perceptron operation for one day of operation on July 3, 2020 are shown in Fig. 10.

The analysis of the experimental results showed that all the considered parameters in the first option turned out to be better in comparison with the second option. Thus, the average error size of the neural network in the first case (P_{close}) was 0.000927425, while in the second case (P_{ln}) the average error size was 0.051026481. The variance of error values as a percentage of the closing price in the first case was 0.304107913, while in the second case it was 0.343654316, or 4 hundredths better.

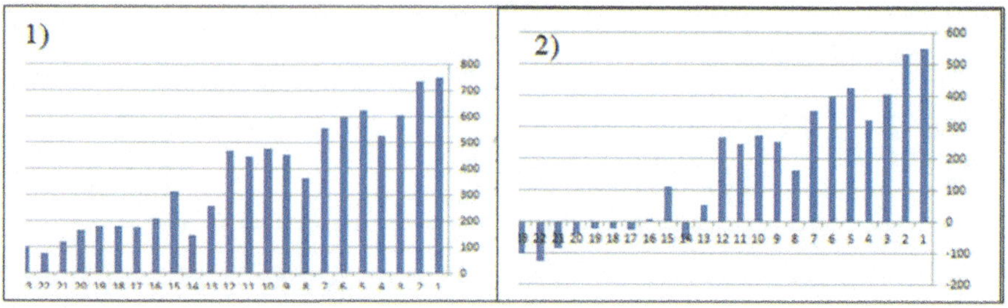

Fig. 10 Deposit growth curves for both options [%]

The sigma of the neural network error values in the first case (P_{close}) was 0.551459802, while in the second case (P_{ln}) it was 0.586220365. This suggests that in the first case, the variance of values relative to the mean was lesser.

The first option turned out to be more efficient in terms of the amount of profit received (variation margin). In the first case, the profit amounted to 748.0 rubles, while in the second case the profit was only 548.0 rubles. The maximum drawdown in both options was the same and amounted to 169 rubles.

The profitability for 1 h of work was 6.43% in the first case, and 5.67% in the second. Thus, the first version of the artificial intelligence system turned out to be more accurate and efficient.

2 Methods

2.1 Neural Network Modeling

The research methods used were the following: monographic, analytical, computational and constructive, as well as artificial intelligence. The authors used the facilities available on the Deductor platform and applied the methods of generalization and modeling.

3 Results and Discussion

3.1 Neural Network Model for Time Series Forecasting

It is assumed that a time series is a series of successive values that characterize the change in a certain indicator over time. Time series analysis is a set of mathematical and statistical methods of analysis aimed at identifying the structure of time series and forecasting them.

Time series mathematical models can take different forms and represent different stochastic processes. There are three broad classes of models in which subsequent data linearly depend on the previous ones: autoregressive models, integral models, and moving average models. Models of autoregressive moving average (Autoregressive Moving Average, ARMA) and model of autoregressive and integrated moving average (Autoregressive Integrated Moving Average, ARIMA) are built on their equivalent system-theoretic basis.

Nonlinear time series models include the following: GARCH, TARCH, EGARCH, FIGARCH, CGARCH, etc.

Let us consider the time series St, each member of which represents the value of some asset at the t-th moment in time.

Profit for a unit period of time (one-period simple return, linear return, or, in other words, the relative increment in value) is calculated by the formula:

$$R_n = (S_t - S_{t-1})/S_{t-1} = \frac{St}{St-1} - 1, \tag{1}$$

Although usually expressed as a percentage, in this case the result must be multiplied by 100. The profit for any period of time (k-period simple return) is calculated by the formula:

$$R_t = (S_t - S_{t-k})/S_{t-k} = \frac{St}{St-k} - 1, \tag{2}$$

$$R_t(k) = \prod_{j=0}^{k-1}\left(1 + R_{t-j}\right) - 1 \approx R_t + R_{t-1} + \cdots + R_{t-k+1}, \tag{3}$$

Logarithmic return over a unit time period (log-return, usually using the natural logarithm):

$$r_t = log(S_t/S_{t-1}) = \log(S_t) - \log(S_{t-1}), \tag{4}$$

$$r_t = \log(1 + R_t) \approx R_t, \tag{5}$$

Logarithmic return for any period of time (k-periodlog-return) is the following:

$$r_t(k) = r_t + r_{t-1} + \cdots + r_{t-1+1}, \tag{6}$$

The following inequality is always true:

$$r_t = \log(1 + R_t) \approx R_t, \tag{7}$$

It is important for stock trading to improve forecast accuracy and reduce the level of forecast error for a time series parameter. The level of error is due to risk and uncertainty. Therefore, it is advisable to use the generally accepted methodology for assessing financial risk. It is assumed that risk is variation (variance), which can be represented as a standard deviation. The root-mean-square deviation is calculated by using formula:

$$\sigma = \sqrt{\frac{\sum_{i=1}^{n} (r_i - \overline{r})^2}{n}} \tag{8}$$

where σ is the risk and other quantities are:
r, ri is the profitability of a financial instrument in the base and i-period;

n is the number of periods.

To assess the financial risk, the following models are widely used: VaR, SaR, GARSH, and others too. A number of researchers propose a wide range of financial mathematics tools to assess and minimize financial risk.

In the conducted experiment, the neural network model functioned as follows. The Skraper periodically collected words from news Web sites at 15 min intervals. Then the data was transferred to the Word2vec program, then, in the form of 300 numerical parameters, it entered the input layer of the Perceptron model. Before being sent to the input layer of the Perceptron, 5 parameters of the Japanese candlestick time series were added to the 300-dimensional vector. In the first variant of «Pclose», the neural network worked with five-digit numbers characterizing the SIU0 price in rubles, and the volume of transactions was represented in their actual value. In the second variant, the SIU0 price parameters were taken the logarithm using formula 4.

It should be noted that there were tools for creating vector-semantic models, but word2vec has become the first popular model implementation, primarily due to its ease of use, open source code and high performance. During the experiment, word2vec was hosted in a cloud service virtual machine.

3.2 *Risk Assessment When Forming a Time Series Forecast by the AI System*

Artificial intelligence systems are being increasingly used in solving practical problems at large enterprises, banks and IT companies in Russia. It should be noted that there is rather a low level of digital innovations in Russia, in comparison with the international companies in developed countries. To improve the situation, relevant regulatory documents were issued in May 2017 to stimulate organizations to digitalize all functional spheres of their activities. However, the reduction of financial riskstill remains an important issue.

In this respect, Ruppert's research study of statistics [28] and data analysis for financial engineering [29] and Jensen, Fischer and Myron's capital asset pricing model [22] are of great scientific interest.

Fama and MacBeth found it necessary to consider risk as a category that reflects return and equilibrium [16]. Frazinni and Pedersen took into account the role of beta portfolio of financial instruments [18].

Certain aspects of financial risk management under conditions of market uncertainty were considered by Shokhnekh et al. [32]. Considerable contribution to the study of financial risk management problems was made by Vasiliev, Pilchikov and Lyalin who proposed real options as a tool for assessing and hedging the risks of enterprises in the real economy [34].

Financial risk management is one of the most important aspects of any financial activity. Traditionally, the following stages of risk management are distinguished: identification, assessment, selection of a management technique, implementation of the selected technique, and assessment of results. The techniques of financial risk management include the following: (1) risk avoidance, (2) damage prevention, (3) risk acceptance, and (4) risk transfer. Three patterns of transferring financial risk are known, i.e. hedging, insurance, and diversification.

It should be noted that the Russian experience is important for world science, since the large-scale use of digital technologies will provide the possibility of using network interaction between the participants in the innovation process, which is especially important in the face of growing market uncertainty and all types of risk.

To minimize financial risk, some authors proposed a wide range of financial mathematics tools. Felmer and Shid proposed quantile hedging, hedging deficit with minimal risk, and optimal quadratic hedging [17].

Assets, portfolios, and arbitrage opportunities characterize the financial market model. We considered a financial market model with a $d + 1$ asset. Assets are stocks, bonds, commodities, and currencies. In a simple one-step model, these assets have different prices at the initial moment of $t = 0$ and at the final moment of $t = 1$. Assuming that at the time of $t = 0$, the price of the i-th asset is π ^ i ≥ 0, theprice range

$$\overline{\pi} = (\pi^0, \pi^1, \ldots, \pi^d) \in \mathbb{R}_+^{d+1} \tag{10}$$

At the time of $t = 1$, prices are usually not known, but they are known at the time of $t = 0$. To model this uncertainty, we fixed the probability space (Ω, F, P) and defined the prices of assets at the time of $t = 1$ as non-negative measurable functions

$$S^0, S^1, \ldots, S^d \tag{11}$$

on (Ω, F) with values in $[0, \infty]$. Each outcome $\omega \in \Omega$ corresponded to a certain scenario of market evolution, and $S^i(\omega)$ was the price of the *i*-th asset in the outcome (scenario) ω.

Thus, the level of financial risk under uncertainty can be evaluated using a probability space. It seems appropriate to use the artificial intelligence system to evaluate financial risk.

Algorithms based on data mining studied by P. Kumar, N. V. Kumar [23], Durg and Chauhan are of great interest for the researchers in this field [24]. Liu, Gibson and Osadchy noted that deep learning on large labeled datasets demonstrates very good performance [26]. Udomsak applied computational models and compared the Bayesian classifier and the support vector machine with respect to their ability to forecast the Thai stock exchange [33]. Shiralkar, Flammini, Menczer and Ciampaglia noted that it is necessary to identify flows in knowledge graphs [30] to support fact-checking [31].

The most important area in the development of artificial learning systems applied in financial risk management is machine learning [19]. Breiman identified obstacles to the use of machine learning [8]. Baltas studied the problem of stock selection using machine learning [4] and low-risk investment issues [6].

The experience has shown that the use of artificial intelligence systems can solve a wide range of problems. For example, it can help to find optimal solutions to optimize the management of innovation processes in a manufacturing enterprise [10].

Artificial intelligence is widely used today due to the introduction of Industry 4.0 technologies. The research of many scientists is focused on technological processes due to the introduction of the digital economy. So, Grishunin S., Suloeva S. and Nekrasova T. developed a planning mechanism taking into account risks and R&D budgeting in the field of telecommunications [20], Badenko V., Fedotov A. and Vinogradov K. developed algorithms for processing laser scanner data for the reconstruction of the earth's surface [3].

The results obtained by Apatova N.V., Boychenko O.V., Nekrasova T.P., Malkov S.V., who studied virtual telecommunications enterprises and assessed the risks, are also of great practical value [2].

The research of Georgiy Markovich Dimirovski dedicated to the complexity [35] and complex networks and systems in computational cybernetics and conducted in the form of the survey [11] and the work of Yang et al. [7] on the new delay-dependent stability criteria for recurrent neural networks with time-varying delays are of special importance. For, these pointed out to the superiority of recurrent neural networks and importance of involving time delays, which are to be used in the future research. Initially, via complexity [12] have shown that the same essentiality [15] is also underlying Kolmogorov Neural Networks and Process Characteristic Input–Output Modes, thus pointing out its importance for neural networks in general.

The studies conducted by V. Leventsov, A. Radaev and N. Nikolaevsky dealt with the design of information and communication systems for industrial enterprises of the new generation [25]. The studies of S. Grishunin and S. Suloev related to the development of the mechanism for assessing the credit risk of investment projects in the field of telecommunications [21], as well as the work of S. A. Chernogorskiy and Shvetsova K.V. regarding a game-theoretic model for investments in the telecommunications industry are important for further research in this field [9].

The areas of further research could include the attempt to combine the capabilities of Spark (Hadoop) and BERT to digitize financial news from Web sites, which can improve the processing of the collected data. Secondly, the use of TCN deep learning convolution networks designed for time series analysis can improve the forecast accuracy. Thirdly, it is necessary to «listen» to the financial market parsing stock exchanges, while getting the statistics. This will allow us to evaluate the financial trends on world exchanges and improve the accuracy of model forecast.

In order to hedge the risk, market-neutral portfolios of futures and options can be used, which will allows us to minimize financial risks and assess them using the VaR model and the Cox-Ross-Rubinstein model.

4 Conclusion

Based on the study presented above, the Authors believe apparently the following conclusions can be drawn straightforward.

4.1 Neural Network Models Have Been Developed

The Bot-advisor system has been designed to forecast the closing price of the SIU0 futures contract on the Moscow stock exchange. The system includes the Scraper programs, the Word2vec neural network, the Perceptron neural network on the Deductor platform, and the QUIK trading terminal with an integrated Lua-socket.

The Skraper program for counting words from news websites for BI processing has been developed.

The Perceptron neural network for forecasting the closing price of the SIU0 futures contract for the next 15 min based on BI from news websites and time series data has been designed.

4.2 The Theoretical Background Has Been Studied, the Calculation of the Logarithmic Return for a Unit Time Period Has Been Carried Out

The theoretical background of the artificial intelligence systems in the processing of Big Data and for forecasting the parameters of the time series of a financial instrument has been studied. A wide range of tools for data analysis and machine learning has been analyzed. A comparative analysis of the capabilities of the Hadoop framework and the Deductor analytical platform has been carried out.

The methodology of forecasting the parameters of the time series has been studied. Mathematical models of the time series, which can have various forms, have been discussed.

An approach to calculating the logarithmic return for a unit time period (log-return) has been tested using the natural logarithm.

4.3 The Financial Risk When Forecasting SIU0 Time Series Data Has Been Assessed

The approaches to assessing the financial risk when forecasting the time series by the AI-system have been discussed.

The study has shown that models such as VaR, SaR, GARSH and others are widely used to assess financial risk. In order to assess and minimize financial risk, the researchers propose a wide range of financial tools, including the following: quantile hedging, hedging with a minimal risk of deficit, quadratic optimal hedging.

4.4 The Proposed Hypothesis Has Been Proved

The following hypothesis has been proved: the developed AI-system makes it possible to forecast the SiU0 closing price and the price of the SiU0 futures contract using the parameters of Japanese candlesticks and volume, as well as «news fluctuations» from websites. The Perceptron neural network designed on the Deductor platform was trained on two types of data: (1) cost (Pclose) and (2) logarithm (ln).

The first type «Cost (Pclose)» was a 300-dimensional vector based on Word2vec and parameters: open price (Po), closing price (Pc), maximum price (Ph), minimum price (Pl) and trading volume (V). The forecast value was compared with the actual closing price, and if the forecast value turned out to be higher, a long position was opened and vice versa. The second type—«Logarithmic-(ln)» was a 300-dimensional vector based on Word2vec and parameters: natural logarithm of the price ratio P/Pt−1: opening price (Po), closing price (Pc), maximum price (Ph), minimum price (Pl), trading volume (V). The forecast value was potentiated, i.e., Pprogn = e^Pprogn. Then the forecast value was compared with the actual closing price, and if the forecast value turned out to be higher, a long position was opened and vice versa.

The study has shown that in both options the profit was received only from the transactions made against the prompts of the neural network. The profitability per hour of work was 6.43% in the first case, and 5.67% in the second. Thus, the first version of the artificial intelligence system turned out to be more accurate.

References

1. Andriyevsky, B.R., Matveev, A.S., Fradkov, A.L.: Control and estimation under information constraints: Toward a unified theory of control, computation and communications (in Russian). Avtomatika i Telemechanika **71**(4), 34–99 (2010)
2. Apatova, N.V., Boychenko, O.V., Nekrasova, T.P., Malkov, S.V.: Virtual telecommunication enterprises and their risk assessment. Lecture Notes in Computer Science (including subseries Lecture Notes in Artificial Intelligence and Lecture Notes in Bioinformatics), pp. 326–336. Springer (2017)
3. Badenko, V., Fedotov, A., Vinogradov, K.: Algorithms of laser scanner data processing for ground surface reconstruction. Lecture Notes in Computer Science (including subseries Lecture Notes in Artificial Intelligence and Lecture Notes in Bioinformatics), pp. 397–411. Springer (2018)
4. Baltas, N., Jessop, D., Jones, C., Lancetti, S., Winter, P., Holcroft, J.: Low-risk investing: perhaps not everywhere. Quantitative Monographs. UBS GlobalResearch (2015)
5. Baltas, N., Jessop, D., Jones, C., Lancetti, S., Winter, P., Holcroft, J., Gerken, J., Ivanova, J., Wu, S., Antrobus O., Stoltz, P.: Combining smart beta factors. Academic Research Monitor. UBS Global Research (2016)
6. Baltas, N., Jessop, D., Jones, S., Winter, P., Wu, S., Antrobus, O., Stoltz, P.: Quantitative monographs: stock selection using machine learning. Quantitative Monographs. UBS Global Research, (2015)
7. Breiman, L.: Bagging predictors. Machine learning, pp. 123–140 (1996)
8. Chernogorskiy, S.A., Shvetsov, K.V.: A game-theoretic model for investments in the telecommunications industry. Lecture Notes in Computer Science (including subseries Lecture Notes in Artificial Intelligence and Lecture Notes in Bioinformatics), pp. 351–364. Springer (2017)
9. Demidenko, D.S.: Optimization of the innovation process management at a manufacturing enterprise. In: Demidenko, D.S., Malevskaia-Malevich, E.D., Dubolazova, Y.A., Victorova, N.G. (eds.) Proceedings of the 31st International Business Information Management Association Conference, pp. 996–1003 (2018)
10. Dimirovski, G.M., Gough, N.E., Barnett, S.: Categories in systems and control theory. Int. J. Syst. Sci. **8**(9), 1081–1090 (1977)
11. Dimirovski, G.M., Yuan-Wei, J.: Kolmogorov networks and process characteristic input-output modes decomposition. In: Samad, T., Sgurev, V., Hadjiski, M. (eds.) Proceedings of the 1st IEEE international symposium on intelligent systems (Varna, BG, 10–12 September 2002). The IEEE, Piscataway, NJ, USA, Bulgarian Academy of Sciences and SAI Union, Sofia, BG, vol. 1, pp. 59–66 (2002)
12. Dimirovski, G.M. (ed.): Complex systems: relationships between control, communication and computing, Volume 55 in J. Kacprzyk Series Studies in Systems, Decision and Control, , pp. vii–xxvii. Springer International Publishing AG Switzerland, Cham (2016)
13. Dimirovski, G.M.: An overview of fascinating ideas on complexity and complex networks and systems in computational cybernetics (Invited Lecture). In: Proceedings of the 17th IEEE International Conference on Smart Technologies, EUROCON 2017 (6–8 July 2017, Ohrid, R. Macedonia). IEEE Republic of Macedonia Section and the IEEE, Piscataway, New Jersey, pp. 650–664 (2017)
14. Epstein, J.: Nonlinear Dynamics, Mathematical Biology, and Social Science. Santa Fe Institute Studies in Sciences of Complexity. Addison Wesley, Reading, Massachusetts (1997)
15. Fama, E.F., MacBeth, J.D.: Risk, return and equilibrium: empirical tests. J. Polit. Econ. **81**(3) (1973)
16. Felmer, G., Shid, A.: Introduction to Stochastic Finance: Discrete time. Munich, DE, ICMNO, p. 496 (2009)
17. Frazzini, A., Pedersen, L.H.: Betting Against Beta. NBER Working Paper (2010)
18. Gary, S.: Calculus of risk. Scientific American, pp. 92–97 (1998)

19. Grishunin, S., Suloeva, S., Nekrasova, T.: Development of the mechanism of risk-adjusted scheduling and cost budgeting of R & Damp D projects in telecommunications. Lecture Notes in Computer Science (including subseries Lecture Notes in Artificial Intelligence and Lecture Notes in Bioinformatics), pp. 456–470. Springer (2018)
20. Grishunin, S., Suloeva, S.: Development of the credit risk assessment mechanism of investment projects in telecommunications. Lecture Notes in Computer Science (including subseries Lecture Notes in Artificial Intelligence and Lecture Notes in Bioinformatics), pp. 300–314. Springer (2017)
21. Jensen, M., Fischer, V., Myron, V.: The Capital Asset Pricing Model: Some Empirical Tests. Praeger Publishers Inc. (1972)
22. Knight, F.: Risk, uncertainly and profit. Special Publication with the ISBN 978-0-9840614-2-6 (1921)
23. Kumar, P., Kumar, N.V., Durg, S., Chauhan, S.: A Benchmark to Select Data Mining Based Classification Algorithms For Business Intelligence And Decision Support Systems. DB. cs. LG (2012)
24. Leventsov, V., Radaev, A., Nikolaevskiy, N.: Design issues of information and communication systems for new generation industrial enterprises. Lecture Notes in Computer Science (including subseries Lecture Notes in Artificial Intelligence and Lecture Notes in Bioinformatics), pp. 142–150. Springer (2017)
25. Liu, S., Gibson, J., Osadchy M.: Learning to Support: Exploiting Structure Information in Support Sets for One-Shot Learning, L. cs.AI, stat.ML (2018)
26. LNCS Homepage, http://www.springer.com/lncs. Accessed 2016/11/21
27. Markowitz, H.: Portfolio selection. J. Financ. **253** (1952)
28. Ruppert, D.: Statistics and Data Analysis for Financial Engineering. Springer Ltd., London (2019)
29. Sharpe, W.F.: A simplified model for portfolio analysis. In: Portfolio Theory and Capital Markets. McGraw-Hill Co., New York, NY (1970)
30. Shiralkar, P., Flammini, A., Menczer, F., Ciampaglia, G.L.: Finding Streams in Knowledge Graphs to Support Fact Checking, cs.AI, cs.SI (2017)
31. Shokhnekh, A., Lomakin, N., Glushchenko, A., Kovalenko, O., Kosobokova, E., Sazonov, S.: Digital neural network for managing financial risk in business due to real options in the financial and economic system. In: Proceedings of the International Scientific-Practical Conference Business Cooperation as a Resource of Sustainable Economic Development and Investment Attraction, ISPCBC 2019. Pskov, Russia, Pskov State University, pp. 571–575. Publisher Atlantis Press (2019). https://www.atlantis-press.com/proceedings/ispcbc-19
32. Udomsak, N.: How do the Naive Bayes classifier and the Support Vector Machine Compare in their Ability to Forecast the Stock Exchange of Thailand? cs.LG, (2015)
33. Vasiliev, V.A., Pilchikov, A.F., Lyalin, V.E.: Mathematical models of risk assessment and management of business entities. Audit Financ. Anal. **4**, 200–237 (2005)
34. Weawer, W.: Science and complexity. Am. Sci. **36**, 538 (1948)
35. Word2Vec: How to work with vector representations of words [Electronic resource]. https://neurohive.io/ru/osnovy-data-science/word2vec-vektorrnyetnie-predstavlenija-slovdlja-mas hinnogo-obuchenija/ (Date of access May 2, 2020)
36. Yang, B., Wang, R., Shi, P., Dimirovski, G.M.: New delay-dependent stability criteria for recurrent neural networks with time-varying delays. Neurocomputing **151**, 1414–1422 (2015)

A Fuzzy Multistage Control Model for Stable Sustainable Agricultural Regional Development

Janusz Kacprzyk, Yuriy P. Kondratenko, José M. Merigó, Jorge Hernandez Hormazabal, Gia Sirbiladze, Alexander Bozhenyuk, Eulalia Szmidt, Sławomir Zadrożny, and Jan W. Owsiński

Abstract We propose a further extension of a multistage fuzzy control model of stable sustainable regional agriculture development that involves an additional capacity to reflect a stability requirement which addresses a clear preference of the stakeholders for a limited variability of crucial development indicators and parameters. We presented the use of fuzzy dynamic programming for solving the problem in which many crucial aspects, in particular life quality indicators, are subject to objective, by the authorities, and subjective, by the inhabitants, evaluations which are closely related to human perception and cognitive abilities. This model is then augmented with a requirement of a limited variability of crucial development indicators, parameters, etc. For illustration, we have shown a simple example in which the problem is to determine the best (optimal) investment policy under different development scenarios, and subject to objective and subjective evaluations.

J. Kacprzyk · E. Szmidt · S. Zadrożny · J. W. Owsiński
Systems Research Institute, Polish Academy of Sciences, ul. Newelska 6, 01-447 Warsaw, Poland
e-mail: szmidt@ibspan.waw.pl

S. Zadrożny
e-mail: zadrozny@ibspan.waw.pl

J. W. Owsiński
e-mail: owsinski@ibspan.waw.pl

J. Kacprzyk (✉)
WIT – Warsaw School of Information Technology, ul. Newelska 6, 01-447 Warsaw, Poland
e-mail: kacprzyk@ibspan.waw.pl

Y. P. Kondratenko
Intelligent Information Systems Dept., Petro Mohyla Black Sea National University 10, 68th Desantnykiv Str., Mykolaiv 54003, Ukraine
e-mail: yuriy.kondratenko@chmnu.edu.ua

J. M. Merigó
Department of Management Control and Information Systems, University of Chile, Av. Diagonal Paraguay 257, 8330015 Santiago, Chile
e-mail: jmerigo@fen.uchile.cl; Jose.Merigo@uts.edu.au

Faculty of Engineering and Information Technology, School of Information, Systems, and Modelling, University of Technology, 15 Broadway Ultimo, Sydney, NSW 2007, Australia

P. Shi et al. (eds.), *Complex Systems: Spanning Control and Computational Cybernetics: Applications*, Studies in Systems, Decision and Control 415,
https://doi.org/10.1007/978-3-031-00978-5_13

1 Introduction

The purpose of this paper is, first, to present an open loop multistage control (decision making) model for the planning of sustainable agricultural regional development. The essence of this model was proposed by Kacprzyk and Straszak [26] (cf. also Kacprzyk [15] as a result of regional agricultural project at the International Institute for Applied Systems Analysis (IIASA) in Laxenburg, Austria (www.iiasa.ac.at), and then applied over the next years for the planning of development of many agricultural regions all over the world. The model is based on a multistage optimal control (decision making) model (cf. Kacprzyk [15, 21]) in which the development is meant in terms of values of life quality indicators attained as results of some investments (expenditures, outlays). Second, this paper is an extension of our previous work Kacprzyk et al. [23] in which the use of this model within the project "RUC-APS: Enhancing and implementing Knowledge based ICT solutions within high Risk and Uncertain Conditions for Agriculture Production Systems" (www.ruc-aps.eu), funded by the EU under H2020-MSCA-RISE-2015 is dealt with, notably in the sense of handling the broadly perceived risk in the agricultural value chain. Third, while extending this model we propose here to take into account the so-called *stability* of the regional development (cf. Kacprzyk and Straszak [24–26], Kacprzyk [15]).

For our purposes, the *stability* of regional development is meant to concern the variability of (crucial) development indicators, characteristics, conditions, etc. Very often, a steady growth is looked for, and a high variability is not welcome by the humans. The problem of stability of (socio-) economic development has a long history and the limited variability is one of key elements. As some examples of a rich literature on this topic one can cite Asan et al. [4], Čajkóva and Čajka [6], Gu, Zhou and Ye [9], and Kubo [29].

Due to a lack of space, we will not discuss these foundational issues related to the stability in more detail and only assume an operational definition of the stability of sustainable regional agricultural development considered in this paper as some limitation of variability of some crucial indicators and other elements conditioning the development.

J. H. Hormazabal
Management School, University of Liverpool, Chatham Street, L69 7ZH Liverpool, UK
e-mail: J.E.Hernandez@liverpool.ac.uk

G. Sirbiladze
Department of Computer Sciences, Iv. Javakhishvili Tbilisi State University, University St. 13, Tbilisi 0186, Georgia
e-mail: gia.sirbiladze@tsu.ge

A. Bozhenyuk
Southern Federal University, Engineering and Technological Academy Taganrog, Rostov, Russia
e-mail: avb002@yandex.ru

The stability in the above sense is meant to involve:

- the *stability of development trajectory* that concerns the variability of development outcomes, i.e. the levels of the life quality indicators attained and their resulting social satisfactions, and
- the *stability of development "policy"* that concerns the variability of some development prerequisites, i.e. the imposed fuzzy constraints, fuzzy goals, and investment partitioning rules.

It is quite clear that both the above stability types are "soft" concepts, and their measures can be devised using tools and techniques of fuzzy logic.

It is interesting to note that the stability focused analysis to be shown in this paper is related to some extent to another extension of the sustainable regional agricultural development model proposed in our former papers (cf. Kacprzyk et al. [23], Kacprzyk [21]). In this paper it is proposed to explicitly take into account some human specific characteristic features, namely the so-called *status quo bias* the essence of which is that the humans usually and on the average prefer traditional, well established procedures, courses of action and solutions, no (or little) change in general.

Our focus is here a control (decision making) problem in the traditional Bellman and Zadeh's [5] sense in which there are some imprecisely specified constraints, the so-called fuzzy constraints, on the set of possible options (variants, choices, alternatives, ...) and some imprecisely specified goals, the so-called fuzzy goals, on the resulting consequences, i.e. outcomes implied by the selection of a particular option. A relation (mapping) from the set of options to the set of consequences is known, maybe also in an imprecise (fuzzy) manner. The goodness (quality) of applying an option, i.e making a decision, is evaluated by an aggregation of the degrees to which the fuzzy constraints and fuzzy goals are satisfied, and this is called a fuzzy decision. In our case, we assume a dynamic setting in which the fuzzy constraints and fuzzy goals are assumed at subsequent planning stages over some planning horizon, so that we have a multistage decision making (control) problem under fuzziness which can be conveniently formulated and solved in terms of dynamic programming, more specifically fuzzy dynamic programming originally introduced in the source paper by Bellman and Zadeh [5], and then considerably extended in Kacprzyk's books [13, 15]. The fuzzy dynamic programming will be presented as a powerful tool for the solution of multistage decision making and control problems under imprecise (fuzzy) information, i.e. under fuzzy goals and constraints. Of course, dynamic programming is plagued by the so called curse of dimensionality so that the solution of very large problems may be computationally difficult but this does not concern our model of sustainable agricultural regional development which is basically dealt with in terms of a few options or scenarios, as it usually happens in socioeconomic planning.

Many fuzzy dynamic programming models have found applications in diverse areas but for our purposes the use of fuzzy dynamic programming for solving sustainable agricultural regional development planning problems is primarily relevant. These works have been initiated by Kacprzyk and Straszak [24–26] at the

International Institute for Applied Systems Analysis (IIASA) in Laxenburg, Austria (www.iiasa.at) in the framework of development of an effective model based tool for designing optimum policies for an agricultural region (see, e.g. Albegov et al. [1–3], Kacprzyk et al. [12], Straszak, Kacprzyk and Owsiński [33].

Then the model has been employed in numerous agricultural regional planning projects in various countries, exemplified by the Upper Noteć Region in Poland, Tisza Region in Hungary, Kinki Region in Japan, and many other ones. These works have resulted in many publications as mentioned above and one should mention in our context Kacprzyk et al. [22] in which some models of both objective and subjective aspects of evaluations have been proposed. These models proposed and application results have been mentioned as one of the most successful examples of fuzzy systems modeling in a Special Volume on the Fiftieth Anniversary of the British Operational Research Society published in 1987 by Pergamon Press—cf. Thomas [34].

Recently, the above models have been extended along many lines, notably by using elements of Wang's [35, 36] *cognitive informatics* to better represent the role of human cognition for the modeling of sustainable agricultural development (cf. Kacprzyk [21]). The inclusion of some behavioral characteristics exemplified by the cognitive biases, notably the status quo bias, is important (cf. Kacprzyk et al. [23]).

A brief account of various approaches to a more general problem of the mathematical modeling, with a short summary of historical developments, in agriculture is given, for instance, in Jones et al. [10].

In the first part of this paper, we start with a short survey of main concepts, notations and properties of fuzzy sets theory, and then proceed to the fuzzy dynamic programming followed by a short description of the multistage control model for sustainable agricultural regional development planning considered in terms of expenditures, subsidies, life qualities, etc.

Then, we present in some detail the model of sustainable regional agricultural development in which the development proceeds in terms of changes of life quality indicators implied by some investments or expenditures (outlays). The model involves both an "objective" aspect in the sense of how some development goals are fulfilled in relation to the investments which must satisfy some constraints, and a "subjective" aspect in which a social satisfaction resulting from the results of development is accounted for. We illustrate the model with a simple yet illustrative and intuitively appealing example.

Then, we show how the concept of *stability* of sustainable regional agricultural development (cf. Kacprzyk and Straszak [26], Kacprzyk [15]) can be used to further extend the model by accounting for a natural human preference of a limited variability of crucial development indicators and parameters.

Finally, we provide short conclusions on both the fuzzy multistage control model for sustainable regional agricultural planning, and its human centric extensions that explicitly or implicitly take into account some human specific characteristics.

2 Brief Introduction to Fuzzy Sets Theory

The fuzzy set, introduced by Zadeh [39], may be viewed as a class of objects with unsharp boundaries, i.e. in which the transition from the belongingness to non-belongingness is gradual rather than abrupt, from the full belongingness to the full non-belongingness through all intermediate values. Formally, the fuzzy set A in a universe X can be modeled by a *membership function* defined as

$$\mu_A : X \longrightarrow [0, 1] \tag{1}$$

such that $\mu_A(x) \in [0, 1]$ is the degree to which an element $x \in X$ belongs to A: from $\mu_A(x) = 0$ for the full non-belongingness to $\mu_A(x) = 1$ for the full belongingness, through all intermediate $(0 < \mu_A(x) < 1)$ values.

The membership function is in practice, also here, usually assumed to be piecewise linear. Obviously, in many instances the actual shape of the membership function is subjective and models an individual's perception of a given gradual concept represented by a fuzzy set.

A *fuzzy set* A in a universe of discourse $X = \{x_i\}_{i \in I}$, A in X, can be conveniently defined as a set of pairs

$$A = \{(\mu_A(x_i), x_i)\}_{i \in I} \tag{2}$$

where $\mu_A : X \longrightarrow [0, 1]$ is the *membership function* of A and $\mu_A(x) \in [0, 1]$ is the *grade of membership* (or a *membership grade*) of an element $x \in X$ in A.

In practice, X is usually finite as, e.g., $X = \{x_1, \ldots, x_n\}$, and then A in X is written as

$$A = \{(\mu_A(x), x)\} = \{\mu_A(x)/x\} =$$
$$= \mu_A(x_1)/x_1 + \cdots + \mu_A(x_n)/x_n = \sum_{i=1}^{n} \mu_A(x_i)/x_i \tag{3}$$

where "+" and "$\sum$" are meant, slightly informally, in the set-theoretic sense, and by convention, the pairs "$\mu_A(x)/x$" with $\mu_A(x) = 0$ are omitted.

The basic definitions and properties related to fuzzy sets can be summarized as:

- A fuzzy set A is *empty*, $A = \emptyset$, if and only if $\mu_A(x) = 0, \forall x \in X$;
- Two fuzzy sets A and B in X are *equal*, $A = B$, if and only if $\mu_A(x) = \mu_B(x)$, $\forall x \in X$;
- A fuzzy set A in X is *contained in*, or is a *subset of*, a fuzzy set B in X, $A \subseteq B$, if and only if $\mu_A(x) \leq \mu_B(x), \forall x \in X$;
- A fuzzy set A in X is *normal* if and only if $\max_{x \in X} \mu_A(x) = 1$.

There are also some important non-fuzzy sets associated with a fuzzy set, notably the *α-cut*, or the *α-level set*, of A in X, A_α, defined as the following (non-fuzzy) set

$$A_\alpha = \{x \in X : \mu_A(x) \geq \alpha\}, \qquad \text{for } \alpha \in (0, 1] \tag{4}$$

Notice that the α-cuts (or the α-level sets), make it possible to uniquely replace a fuzzy set by a sequence of nonfuzzy sets (cf. Kacprzyk [15]) which is relevant both for the theory and applications.

An important concept is the *cardinality* of a fuzzy set which, in the simplest case, is the *nonfuzzy cardinality* of $A = \mu_A(x_1)/x_1 + \cdots + \mu_A(x_n)/x_n$, the so-called *sigma-count*, denoted $\sum\text{Count}(A)$, defined as $\sum\text{Count}(A) = \sum_{i=1}^{n} \mu_A(x_i)$.

The distance between two fuzzy sets, A and B, both defined in $X = \{x_1, \ldots, x_n\}$, is very relevant for our purposes and the two basic (normalized) distances are:

- the *normalized linear* (Hamming) *distance* between A and B in X defined as

$$l(A, B) = \frac{1}{n} \sum_{i=1}^{n} \mid \mu_A(x_i) - \mu_B(x_i) \mid \tag{5}$$

- the *normalized quadratic* (Euclidean) *distance* between A and B in X defined as

$$q(A, B) = \sqrt{\frac{1}{n} \sum_{i=1}^{n} [\mu_A(x_i) - \mu_B(x_i)]^2} \tag{6}$$

The basic operations on fuzzy sets are:

- The *complement* of a fuzzy set A in X, $\neg A$, is

$$\mu_{\neg A}(x) = 1 - \mu_A(x), \qquad \text{for each } x \in X \tag{7}$$

  and the complement corresponds to the connective of negation "not."
- The *intersection* of two fuzzy sets A and B in X, $A \cap B$, is

$$\mu_{A \cap B}(x) = \mu_A(x) \wedge \mu_B(x), \qquad \text{for each } x \in X \tag{8}$$

  where "$\wedge$" is the minimum, i.e. $a \wedge b = \min(a, b)$; the intersection of two fuzzy sets corresponds to the connective of conjunction "and."
- The *union* of two fuzzy sets A and B in X, $A \cup B$, is

$$\mu_{A \cup B}(x) = \mu_A(x) \vee \mu_B(x), \qquad \text{for each } x \in X \tag{9}$$

  where "$\vee$" is the maximum, i.e. $a \vee b = \max(a, b)$; the union of two fuzzy sets corresponds to the connective of disjunction "or."

The above definitions can be generalized to notably using the t-norms and s-norms (t-conorms), for the intersection and union, respectively, defined as follows.

A *t-norm* is defined as:

$$t : [0, 1] \times [0, 1] \longrightarrow [0, 1] \tag{10}$$

such that, for each $a, b, c \in [0, 1]$:

1. it has 1 as the unit element, i.e. $t(a, 1) = a$,
2. it is monotone, i.e. $a \leq b \Longrightarrow t(a, c) \leq t(b, c)$,
3. it is commutative, i.e. $t(a, b) = t(b, a)$, and
4. it is associative, i.e. $t[a, t(b, c)] = t[t(a, b), c]$.

Some more important t-norms are:

- the minimum (which is the most widely used)

$$t(a, b) = a \wedge b = \min(a, b) \tag{11}$$

- the algebraic product

$$t(a, b) = a \cdot b \tag{12}$$

- the Łukasiewicz t-norm

$$t(a, b) = \max(0, a + b - 1) \tag{13}$$

An *s-norm* (or a *t-conorm*) is defined as

$$s : [0, 1,] \times [0, 1] \longrightarrow [0, 1] \tag{14}$$

such that, for each $a, b, c \in [0, 1]$:

1. it has 0 as the unit element, i.e. $s(a, 0) = a$,
2. it is monotone, i.e. $a \leq b \Longrightarrow s(a, c) \leq s(b, c)$,
3. it is commutative, i.e. $s(a, b) = s(b, a)$, and
4. it is associative, i.e. $s[a, s(b, c)] = s[s(a, b), c]$.

Some more important s-norms are:

- the maximum (which is the most widely used, also here)

$$s(a, b) = a \vee b = \max(a, b) \tag{15}$$

- the probabilistic product

$$s(a, b) = a + b - ab \tag{16}$$

- the Łukasiewicz s-norm

$$s(a, b) = \min(a + b, 1) \tag{17}$$

Notice that a t-norm is *dual* to an s-norm (t-conorm) in that $s(a, b) = 1 - t(1 - a, 1 - b)$.

These are basic definitions and operations to be used in the paper, and for more details we refer the reader, for instance, to Kacprzyk [15].

3 Multistage Decision Making and Control via Fuzzy Dynamic Programming

The point of departure is here the well known and widely employed Bellman and Zadeh's [5] general model of decision making under fuzziness the essence of which can be stated as follows: we start with $X = \{x\}$ which is a set of possible *options* (alternatives, variants, choices, decisions, ...), and:

- the *fuzzy goal* to be attained is defined as a fuzzy set G in X, characterized by its membership function $\mu_G : X \longrightarrow [0, 1]$ such that $\mu_G(x) \in [0, 1]$ specifies the grade of membership (attainment) of a particular option $x \in X$ in the fuzzy goal G, and
- the *fuzzy constraint* is similarly defined as a fuzzy set C in the set of options X, characterized by $\mu_C : X \longrightarrow [0, 1]$ such that $\mu_C(x) \in [0, 1]$ specifies the grade of membership (satisfaction) of a particular option $x \in X$ in the fuzzy constraint C.

The problem is formulated in a natural way as:

$$\text{“Attain } G \ \underline{and} \text{ satisfy C”} \tag{18}$$

which is formalized by a *fuzzy decision* D which plays the role of a performance function

$$\mu_D(x) = \mu_G(x) \wedge \mu_C(x), \qquad \text{for each } x \in X \tag{19}$$

where “$\wedge$” stands for the minimum, which is traditionally used for the intersection of two fuzzy sets, but may be replaced, for instance, by another t-norm.

The *maximizing decision*, or *optimal decision*, which is the solution of the problem (18), is defined as an $x^* \in X$ such that

$$\mu_D(x^*) = \max_{x \in X} \mu_D(x) \tag{20}$$

This classic and basic Bellman and Zadeh's [5] framework can clearly be extended to cover more complex, realistic cases which are relevant for the problem considered by introducing:

- an *objective fuzzy goal* $\mu_{G_o}(x)$,
- a *subjective fuzzy goal* $\mu_{G_s}(x)$,
- an *objective fuzzy constraint* $\mu_{C_o}(x)$, and
- a *subjective fuzzy constraint* $\mu_{C_s}(x)$.

The problem—in its extended, “objective-and-subjective" form—is now

$$\text{“Attain } [G_o \ and \ G_s] \ \underline{and} \text{ satisfy } [C_o \ and \ C_s]\text{”}$$

which implies the following fuzzy decision D

$$\mu_D(x) = [\mu_{G_o}(x) \wedge \mu_{G_s}(x)] \wedge [\mu_{C_o}(x) \wedge \mu_{C_s}(x)], \qquad \text{for each } x \in X \quad (21)$$

and the *maximizing*, or *optimal*, decision is defined as in (20).

This conceptual framework can be extended to deal with multiple fuzzy constraints and fuzzy goals, and also fuzzy constraints and fuzzy goals defined in different spaces, cf. Kacprzyk [15, 21], as: if we have: $n_o > 1$ objective fuzzy goals – $G_o^1, \ldots, G_o^{n_o}$ defined in Y, $n_s > 1$ subjective fuzzy goals – $G_s^1, \ldots, G_s^{n_s}$ defined in Y, $m_o > 1$ objective fuzzy constraints – $C_o^1, \ldots, C_o^{m_o}$ defined in X, $m_s > 1$ subjective fuzzy constraints – $C_s^1, \ldots, C_s^{m_s}$ defined in X, and a function $f : X \longrightarrow Y$, $y = f(x)$, then

$$\begin{aligned}
&\mu_D(x) = \\
&= (\mu_{G_o^1}[f(x)] \wedge \cdots \wedge \mu_{G_o^{n_o}}[f(x)]) \wedge (\mu_{G_s^1}[f(x)] \wedge \cdots \wedge \mu_{G_s^{n_s}}[f(x)]) \wedge \\
&\quad \wedge([\mu_{C_o^1}(x) \wedge \cdots \wedge \mu_{C_o^{m_o}}(x)]) \wedge ([\mu_{C_s^1}(x) \wedge \cdots \wedge \mu_{C_s^{m_s}}(x)]), \qquad \text{for each } x \in X
\end{aligned}$$

and the *maximizing decision* is defined as (20), i.e. $\mu_D(x^*) = \max_{x \in X} \mu_D(x)$.

Further generalizations of the problem (18) with multiple goals/constraints are considered where particular goals/constraints are assigned priorities, explicitly or implicitly (cf. Kacprzyk [15]). In this context, various schemes of prioritized goals/constraints aggregation are possible. Among them we can mention a novel scheme based on the "and possibly" connective which provides a tool for requiring some, in a sense of a secondary importance, attainment/satisfaction of goals/constraints only if this is *possible* (cf. e.g., Zadrożny and Kacprzyk [40, 41]).

In a more specific context of a *control process* the decision (control) space is $U = \{c_i\}_{i=1,\ldots,m} = \{c_1, \ldots, c_m\}$, the state (output) space is $X = \{s_j\}_{j=1,\ldots,n} = \{s_1, \ldots, s_n\}$, and both are assumed here to be finite. The process starts from an initial state (t=0) $x_0 \in X$, a decision (control) $u_0 \in U$ is applied which is subjected to a fuzzy constraint $\mu_{C^0}(u_0)$, and a state $x_1 \in X$ is attained via a known state transition equation of the system under control S, and a fuzzy goal $\mu_{G^1}(x_1)$ is imposed on x_1. This is repeated for $t = 1, 2, \ldots$ until some termination time $t = N$.

The simplest case, assumed here, is the deterministic system under control, the dynamics of which is described by a *state transition equation*

$$x_{t+1} = f(x_t, u_t), \qquad t = 0, 1, \ldots \quad (22)$$

where $x_t, x_{t+1} \in X = \{s_1, \ldots, s_n\}$ are the states at time t and $t+1$, respectively, and $u_t \in U = \{c_1, \ldots, c_m\}$ is the decision (control) at t.

At $t, t = 0, 1, \ldots, u_t \in U$ is subjected to a *fuzzy constraint* $\mu_{C^t}(u_t)$, and on $x_{t+1} \in X$ a *fuzzy goal* is imposed, $\mu_{G^{t+1}}(x_{t+1})$. The *initial state* $x_0 \in X$ is fixed and specified in advance, and the *termination time* (planning horizon), $N \in \{1, 2, \ldots\}$, is finite, and fixed and specified in advance.

In the model considered the *performance* of the particular decision making (control) stage t, $t = 0, 1, \ldots, N-1$, is given as

$$v_t = \mu_{C^t}(u_t) \wedge \mu_{G^{t+1}}(x_{t+1}) = \mu_{C^t}(u_t) \wedge \mu_{G^{t+1}}[f(x_t, u_t)] \tag{23}$$

and the *performance* of the whole multistage decision making (control) process over the whole planning horizon is here assumed to be

$$\begin{aligned} \mu_D(u_0, \ldots, u_{N-1} \mid x_0) &= v_0 \wedge v_1 \wedge \ldots \wedge v_{N-1} = \\ &= [\mu_{C^0}(u_0) \wedge \mu_{G^1}(x_1)] \wedge \ldots \wedge [\mu_{C^{N-1}}(u_{N-1}) \wedge \mu_{G^N}(x_N)] \end{aligned} \tag{24}$$

which reflects a safety first attitude which is often proper for such problems involving socio-economic aspects.

The problem is to find an optimal sequence of decisions (controls) $u_0^*, \ldots, u_{N-1}^*$ such that

$$\mu_D(u_0^*, \ldots, u_{N-1}^* \mid x_0) = \max_{u_0, \ldots, u_{N-1} \in U} \mu_D(u_0, \ldots, u_{N-1} \mid x_0) \tag{25}$$

Kacprzyk's [15] book provides a comprehensive coverage of various aspects and extensions to this basic formulation, and solution techniques.

As already mentioned, in our inherently human centric context the objective and subjective fuzzy constraints and fuzzy goals are assumed, so that we have, at each $t = 0, 1, \ldots, N-1$: an objective fuzzy constraint $\mu_{C_t^1}(u_t)$ and a subjective fuzzy constraint $\mu_{C_s^t}(u_t)$, and an objective fuzzy goal $\mu_{G_o^{t+1}}(x_{t+1})$ and a subjective fuzzy goal $\mu_{G_s^{t+1}}(x_{t+1})$. Then, the (extended) performance of the particular stage t, $t = 0, 1, \ldots, N-1$, is

$$\overline{v}_t = [\mu_{C_o^t}(u_t) \wedge \mu_{C_s^t}(u_t)] \wedge [\mu_{G_o^t}(x_t) \wedge \mu_{G_s^t}(x_t)] \tag{26}$$

which can be schematically shown as in Fig. 1.

The (extended) performance of the whole multistage decision making (control) process is then given as

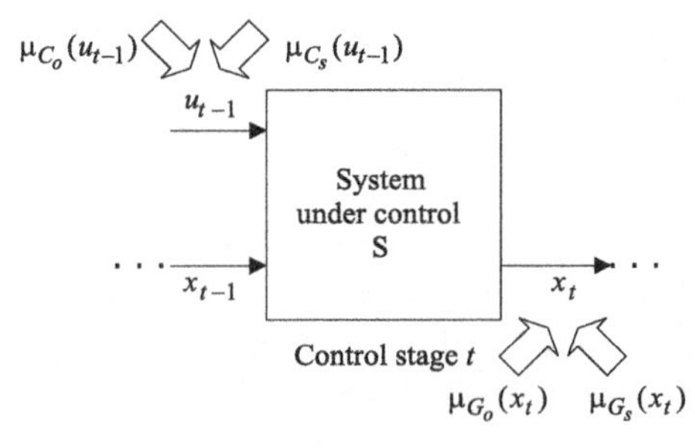

Fig. 1 Evaluation of (extended) performance of stage t

$$\begin{aligned}\mu_{\overline{D}}(u_0, \ldots, u_{N-1} \mid x_0) &= \overline{v}_0 \wedge \overline{v}_1 \wedge \ldots \wedge \overline{v}_{N-1} = \\ &= \{[\mu_{C_o^0}(u_0) \wedge \mu_{C_s^0}(u_0)] \wedge [\mu_{G_o^1}(x_1) \wedge \mu_{G_s^1}(x_1)]\} \wedge \ldots \\ &= \wedge\{[\mu_{C_o^{N-1}}(u_{N-1}) \wedge \mu_{C_s^{N-1}}(u_{N-1})] \wedge [\mu_{G_o^N}(x_N) \wedge \mu_{G_s^N}(x_N)]\}\end{aligned} \tag{27}$$

and we seek again a sequence of controls $u_0^*, \ldots, u_{N-1}^*$ such that

$$\mu_{\overline{D}}(u_0^*, \ldots, u_{N-1}^* \mid x_0) = \max_{u_0, \ldots, u_{N-1} \in U} \mu_{\overline{D}}(u_0, \ldots, u_{N-1} \mid x_0) \tag{28}$$

The (full) trajectory of the multistage decision making (control) process from $t = 0$ to a current stage $t = k > 0$ is

$$H_k = (x_0, u_0, C_o^0, C_s^0, x_1, G_o^1, G_s^1, \ldots, u_{k-1}, C_o^{k-1}, C_s^{k-1}, x_k, G_o^k, G_s^k) \tag{29}$$

but in most cases it is enough to consider, for simplicity, the *reduced trajectory*

$$h_k = (x_{k-2}, u_{k-2}, C_o^{k-2}, C_s^{k-2}, x_{k-1}, G_o^{k-1}, G_s^{k-1}, u_{k-1}, C_o^{k-1}, C_s^{k-1}, x_k, G_o^k, G_s^k) \tag{30}$$

which only takes into account – in the sense of outcomes – the current, $t = k$, and previous stage, $t = k - 1$ which is analogous to the Markov decision processes.

Since the problems considered in our context of socioeconomic development are very complex, we can further simplify them by associating with a trajectory, or reduced trajectory, an evaluation function, $E : S(H_k) \longrightarrow [0, 1]$ or $e : S(h_k) \longrightarrow [0, 1]$, where $S(H_k)$ and $S(h_k)$ are the sets (spaces) of all possible trajectories and reduced trajectories, respectively, such that $E(H_k) \in [0, 1]$ and $e(h_k) \in [0, 1]$ stands for a satisfaction of the past development, from 1 for full satisfaction to 0 for full dissatisfaction, through all intermediate values.

The subjective fuzzy constraints and fuzzy goals are now:

- when the (reduced) trajectory is accounted for

$$\begin{cases} \mu_{C_o^k}(u_k \mid h_k) & \text{and } \mu_{C_s^k}(u_k \mid h_k) \\ \mu_{G_o^{k+1}}(x_{k+1} \mid h_k) & \text{and } \mu_{G_s^{k+1}}(x_{k+1} \mid h_k) \end{cases} \tag{31}$$

- when the evaluation of the (reduced) trajectory is accounted for

$$\begin{cases} \mu_{C_o^k}[u_k \mid E(h_k)] & \text{and } \mu_{C_s^k}[u_k \mid E(h_k)] \\ \mu_{G_o^{k+1}}[x_{k+1} \mid E(h_k)] & \text{and } \mu_{G_s^{k+1}}[x_{k+1} \mid E(h_k)] \end{cases} \tag{32}$$

where "$(. \mid h_k(or\, E(h_k))$" means that the values of expression is conditioned on h_k in (31)), or on $E(h_k)$ in (32).

Problem (28) can be solved using the following two basic techniques: dynamic programming (cf. Bellman and Zadeh [5], Kacprzyk [13, 15]), and branch-and-bound (cf. Kacprzyk [15]), and also using the two new ones: a neural network (cf. Francelin et al. [7, 8]), and a genetic algorithm (cf. Kacprzyk [15]). In this paper we will only show the use of dynamic programming; other methods are described in Kacprzyk's [15] book or the papers cited above.

We start with rewriting (28) as to find $u_0^*, \ldots, u_{N-1}^*$ such that

$$\begin{aligned} &\mu_D(u_0^*, \ldots, u_{N-1}^* \mid x_0) = \\ &= \max_{u_0,\ldots,u_{N-1}} [\mu_{C^0}(u_0) \wedge \mu_{G^1}(f(x_0, u_0)) \wedge \ldots \\ &\ldots \wedge \mu_{C^{N-1}}(u_{N-1}) \wedge \mu_{G^N}(f(x_{N-1}, u_{N-1}))] \end{aligned} \tag{33}$$

and then, since

$$\mu_{C^{N-1}}(u_{N-1}) \wedge \mu_{G^N}(f(x_{N-1}, u_{N-1}))$$

depends only on u_{N-1}, then the maximization with respect to $u_0, \ldots, u_{N-1}$ in (33) can be split into:

- the maximization with respect to $u_0, \ldots, u_{N-2}$, and
- the maximization with respect to u_{N-1},

which can be written as

$$\begin{aligned} &\mu_D(u_0^*, \ldots, u_{N-1}^* \mid x_0) = \\ &= \max_{u_0,\ldots,u_{N-2}} \{\mu_{C^0}(u_0) \wedge \mu_{G^1}(f(x_0, u_0)) \wedge \ldots \\ &\ldots \wedge \mu_{C^{N-2}}(u_{N-2}) \wedge \mu_{G^{N-1}}(f(x_{N-2}, u_{N-2})) \wedge \\ &\wedge \max_{u_{N-1}}[\mu_{C^{N-1}}(u_{N-1}) \wedge \mu_{G^N}(f(x_{N-1}, u_{N-1}))]\} \end{aligned} \tag{34}$$

and this can be continued for u_{N-2}, u_{N-3}, etc.

This backward iteration leads to the following set of fuzzy dynamic programming recurrence equations:

$$\begin{cases} \mu_{\overline{G}^{N-i}}(x_{N-i}) = \\ \qquad = \max_{u_{N-i}}[\mu_{C^{N-i}}(u_{N-i}) \wedge \mu_{G^{N-i}}(x_{N-i}) \wedge \mu_{\overline{G}^{N-i+1}}(x_{N-i+1})] \\ x_{N-i+1} = f(x_{N-i}, u_{N-i}); \qquad i = 0, 1, \ldots, N \end{cases} \tag{35}$$

where $\mu_{\overline{G}^{N-i}}(x_{N-i})$ is viewed as a fuzzy goal at control stage $t = N - i$ induced by the fuzzy goal at $t = N - i + 1$, $i = 0, 1, \ldots, N$; $\mu_{\overline{G}^N}(x_N) = \mu_{G^N}(x_N)$.

The $u_0, \ldots, u_{N-1}$ sought is given by the successive maximizing values of u_{N-i}, $i = 1, \ldots, N$ in (35) which are obtained as functions of x_{N-i}, i.e. as an *optimal policy*, $a_{N-i} : X \longrightarrow U$, such that $u_{N-i} = a_{N-i}(x_{N-i})$.

Notice that the very idea of dynamic programming, i.e. the use of backward iteration via (35), prohibits the use of the subjective fuzzy constraints and subjective fuzzy goals defined as functions of the trajectory, or any evaluation of the trajectory since we proceed via backward iteration; the use of a genetic algorithm (cf. Kacprzyk [15]) or a neural network based approach by Francelin et al. [7] or Francelin et al. [8] can be employed.

Therefore, using the objective and subjective fuzzy constraints and fuzzy goals, of course depending on the current control and state, not on the history, we arrive at the following set of (extended) dynamic programming recurrent equations:

$$\begin{cases} \mu_{\overline{G}^{N-i}}(x_{N-i}) = \\ \qquad = \max_{u_{N-i}}\{[\mu_{C_o^{N-i}}(u_{N-i}) \wedge \mu_{C_s^{N-i}}(u_{N-i})] \wedge \\ \qquad\qquad [\mu_{G_o^{N-i}}(x_{N-i}) \wedge \mu_{G_s^{N-i}}(x_{N-i}) \wedge \mu_{\overline{G}^{N-i+1}}(x_{N-i+1})]\} \\ x_{N-i+1} = f(x_{N-i}, u_{N-i}); \qquad i = 0, 1, \ldots, N \end{cases} \tag{36}$$

Now, we will present the use of the above models for the planning of sustainable and stable socioeconomic regional agricultural development.

4 Sustainable Socioeconomic Regional Development Planning Under Fuzziness

The problem of regional development planning is, in spite of its crucial importance, not easy to state in more precise terms as it involves various aspects (political, economic, social, environmental, technological, etc.), different parties and agents (inhabitants, authorities of different levels, formal and informal groups, NGOs, etc.), just to mention a few. This is clearly even more so in the case of sustainable regional development planning (cf. Jovovic et al. [11], Roberts [31], Roberts and Colwell [32], Kubo [29]). The use of a fuzzy multistage decision making (control) model was proposed by Kacprzyk and Straszak [25, 26], and then extended by Kacprzyk [15], Kacprzyk et al. [22], Kacprzyk [21], etc. In this paper we will further develop this general model, more specifically in the context of agricultural regional planning, and in the next chapter we will add a stability related analysis.

We consider a (rural) region in which problems are implied by a relatively poor *life quality* perceived, and its improvement calls for some (mostly external) funds (investments) which are to be determined by the model.

The region is represented by a socioeconomic dynamic system under control in which the (multidimensional) state at the development (planning) stage $t-1$, represented by a vector X_{t-1}, is characterized by a set of relevant socioeconomic life quality indicators. Then, the decision (investment), at $t-1$, u_{t-1}, changes X_{t-1} to X_t; $t = 1, \ldots, N$; N is a finite, fixed and specified planning horizon.

In the evaluation of stage t, $t = 1, \ldots, N$, both the u_{t-1} applied (i.e. costs), and the X_t attained (i.e. benefits) are accounted for by how well some constraints are satisfied, and throw how well some goals are attained. For clarity and simplicity, only for the attainment of fuzzy goals the subjective evaluation will be dealt with.

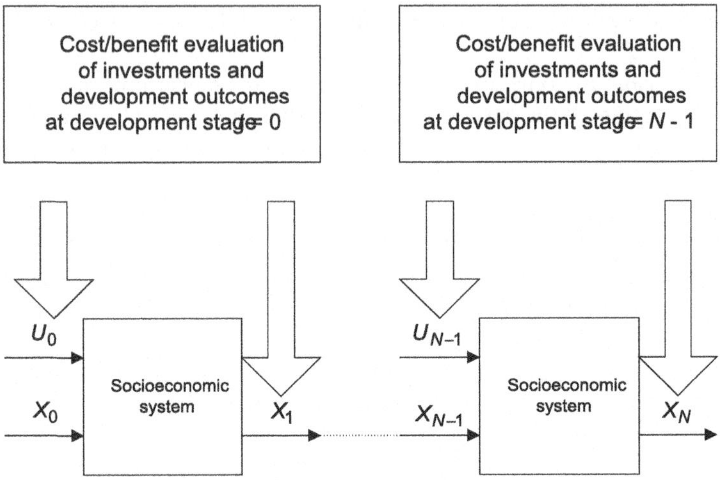

Fig. 2 Basic elements of the socioeconomic system under control

First, in the socioeconomic system shown in Fig. 2 its state (output) X_t is equated with a *life quality index* that includes the following seven *life quality indicators* (i.e. $X_t = [x_t^1, \ldots, x_t^7]$):

- x_t^1 – economic quality (e.g., wages, salaries, income, …),
- x_t^2 – environmental quality,
- x_t^3 – housing quality,
- x_t^4 – health service quality,
- x_t^5 – infrastructure quality,
- x_t^6 – work opportunity,
- x_t^7 – leisure time opportunity,

The decision at $t-1$, u_{t-1}, is investment, and we impose on u_{t-1} a fuzzy constraint $\mu_{C^{t-1}}(u_{t-1})$ in a piecewise linear form as shown in Fig. 3 to be read as follows. The u_{t-1} may be fully utilized up to u_{t-1}^p, the expected highest possible level of investment, so that $\mu_{C^{t-1}}(u_{t-1}) = 1$ for $0 < u_{t-1} < u_{t-1}^p$. This limit may be exceeded and some additional contingency investment, maximally up to u_{t-1}^c (the more the

Fig. 3 Fuzzy constraints on investment u_{t-1}

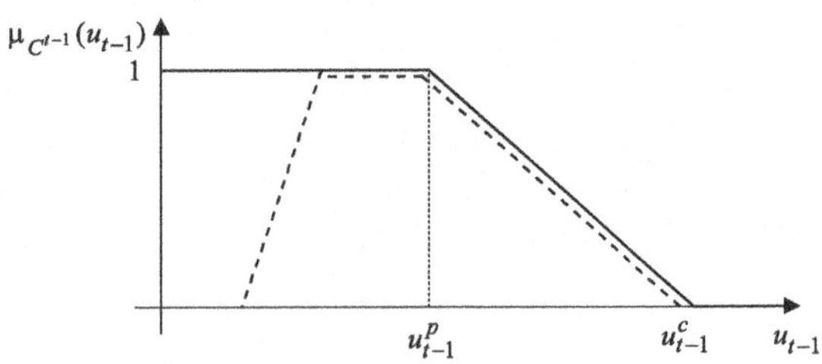

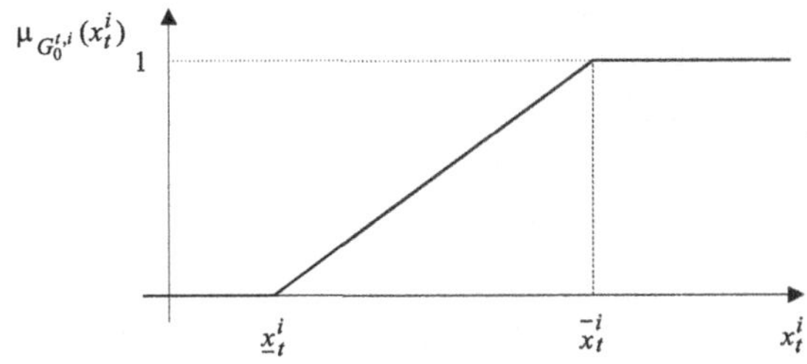

Fig. 4 Objective fuzzy subgoal

worse, of course), can be obtained. The fuzzy constraints are often as shown in the dotted line in Fig. 3 in that too low a use of available investments should also be avoided as in all public funding related cases. The u_{t-1} is partitioned, using some rule, into $u_{t-1}^1, \ldots, u_{t-1}^7$, devoted to an improvement of the respective life quality indicators.

The temporal evolution of the particular life quality indicators is assumed to be governed, for simplicity, by the state transition equation:

$$x_t^i = f_{t-1}^i(x_{t-1}^i, u_{t-1}^i), \qquad i = 1, \ldots, 7; t = 1, \ldots, N \tag{37}$$

which may be derived by, e.g., using experts' opinions, past experience, mathematical models, etc.

Now, we assume that the evaluation of development concerns first how well some predetermined goals are fulfilled, i.e. *effectiveness*, which are then related to the investment spent, i.e. *efficiency*—cf. Kacprzyk [15].

We start with the effectiveness of regional development that involves: the effectiveness of a particular development stage, and the effectiveness of the whole development trajectory. The effectiveness of a particular development stage has both an objective and subjective aspect.

For each life quality indicator at $t = 1, \ldots, N$, x_t^i, we define an *objective fuzzy subgoal* $G_o^{t,i}$ characterized by $\mu_{G_o^{t,i}}(x_t^i)$ as shown in Fig. 4 to be meant as: $G_o^{t,i}$ is fully satisfied for $x_t^i \geq \overline{x}_t^i$, where $\overline{x}_t^i$ is some *aspiration level* for x_t^i; therefore, $\mu_{G_o^{t,i}}(x_t^i) = 1$, for $x_t^i \geq \overline{x}_t^i$. Less preferable are $\underline{x}_t^i < x_t^i < \overline{x}_t^i$ for which $0 < \mu_{G_o^{t,i}}(x_t^i) < 1$, and $x_t^i \leq \underline{x}_t^i$ are impossible, so that $\mu_{G_o^{t,i}}(x_t^i) = 0$. Clearly, for the determination of an objective fuzzy (sub)goal two values, $\underline{x}_t^i$ and $\overline{x}_t^i$, are only needed which is very important when experts are to be employed.

To obtain the objective evaluation of the life quality index at t, $X_t = [x_t^1, \ldots, x_t^7]$, we use an aggregation of partial assessments of the particular life quality indicators, i.e.

$$\mu_{G_o^t}(X_t) = \mu_{G_o^{t,1}}(x_t^1) \wedge \ldots \wedge \mu_{G_o^{t,7}}(x_t^7) \tag{38}$$

and "$\wedge$", the minimum, may be replaced here and later on by another suitable operation as, e.g., another t-norm [cf. Kacprzyk (1997)] or an OWA operator but this will not be considered here. For more information, cf. Kacprzyk et al. [28]. Essen-

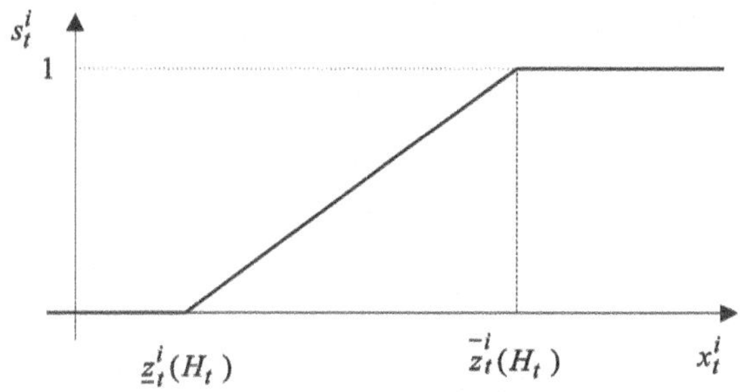

Fig. 5 Partial social satisfaction

tially, the use of "∧" (minimum) reflects a pessimistic, safety-first attitude, and a lack of substitutability (i.e. that a low value of one life quality indicator cannot be compensated by a higher value of another), which is often adequate.

Notice also that the objective evaluation concerns more the authorities than the inhabitants by just comparing the values of life quality indicators attained against some desired predetermined levels. The inhabitants' assessment of the development concerns in fact the (perception of) *social satisfaction* resulting from the life quality index attained which is clearly subjective. The attained value of a particular life quality indicator at t, x_t^i, implies its corresponding partial social satisfaction s_t^i depicted as in Fig. 5, and meant similarly as for the objective evaluation shown in Fig. 4.

Both $\underline{z}_t^i$ and $\overline{z}_t^i$ can generally be functions of the trajectory (history) of development, i.e.

$$H_t = [(X_1, S_1, \mu_{G_o^1}(X_1), \mu_{G_s^1}(S_1), \ldots, X_t, S_t, \mu_{G_o^t}(S_t), \mu_{G_s^t}(S_t))$$

where $S_k = [s_k^1, \ldots, s_k^7]$, $k = 1, \ldots, t$, is the social satisfaction from X_k. Basically, if H_t is encouraging, then the inhabitants may become more demanding, and $\underline{z}_t^i(H_t)$ and $\overline{z}_t^i(H_t)$ may move up while if H_t is discouraging, then $\underline{z}_t^i(H_t)$ and $\overline{z}_t^i(H_t)$ may move down (cf. Kacprzyk [13, 15]); and similarly for the reduced trajectory [cf. 30)].

The social satisfaction at t is now

$$\mu_{G_s^t}(S_t) = s_t = s_t^1 \wedge \ldots \wedge s_t^7 \tag{39}$$

where "∧" again reflects a pessimistic, safety-first attitude, and a lack of substitutability.

The social satisfaction s_t is subjected to a subjective fuzzy goal $\mu_{G_s^t}(s_t)$ which is meant similarly as its objective counterpart shown in Fig. 4.

The effectiveness of stage t is meant as a relation of what is attained (the values of life quality indices and their respective social satisfactions) to what is spent (the respective investments), i.e. is a *benefit–cost relationship*. Formally, the (fuzzy) effectiveness of stage t is

$$\mu_{E^t}(u_{t-1}, X_t, s_t) = \mu_{C^{t-1}}(u_{t-1}) \wedge \mu_{G_o^t}(X_t) \wedge \mu_{G_s^t}(s_t) \tag{40}$$

and the aggregation reflects the nature of a compromise between the interests of the authorities (for whom the fuzzy constraints and the objective fuzzy goal matter), and those of the inhabitants (for whom the subjective fuzzy goal, and to some extent the objective fuzzy goal, matter); the "$\wedge$" (minimum) reflects a safety-first attitude, hence a "more just" compromise.

Then, the effectiveness measures of the particular $t = 1, \ldots, N$, $\mu_{E^t}(u_{t-1}, X_t, s_t)$ given by (40), are aggregated to yield the fuzzy effectiveness measure for the whole development trajectory:

$$\mu_E(H_N) = \mu_{E^1}(u_0, X_1, s_1) \wedge \ldots \wedge \mu_{E^N}(u_{N-1}, X_N, s_n) \tag{41}$$

The fuzzy decision, for the whole development trajectory, is

$$\begin{aligned} &\mu_D(u_0, \ldots, u_{N-1} \mid X_0, B_N) = \\ &= [\mu_{C^0}(u_0) \wedge \mu_{G_o^1}(X_1) \wedge \mu_{G_s^1}(s_1)] \wedge \ldots \\ &\ldots \wedge [\mu_{C^{N-1}}(u_{N-1}) \wedge \mu_{G_o^N}(X_N) \wedge \mu_{G_s^N}(s_N)] \end{aligned} \tag{42}$$

and it expresses some crucial compromises between, e.g.:

- the fuzzy constraints and (objective and subjective) fuzzy goals,
- the interests of the authorities and inhabitants, etc.

We seek an optimal sequence of controls (investments) $u_0^*, \ldots, u_{N-1}^*$ under under a given development policy B_N; the optimization of development policy is a separate problem which will not be considered here, cf. Kacprzyk [15]):

$$\begin{aligned} &\mu_D(u_0^*, \ldots, u_{N-1}^* \mid X_0, B_N) = \\ &= \max_{u_0, \ldots, u_{N-1}} \{[\mu_{C^0}(u_0) \wedge \mu_{G_o^1}(X_1) \wedge \mu_{G_s^1}(s_1)] \wedge \ldots \\ &\ldots \wedge [\mu_{C^{N-1}}(u_{N-1}) \wedge \mu_{G_o^N}(X_N) \wedge \mu_{G_s^N}(s_N)]\} \end{aligned} \tag{43}$$

We can use fuzzy dynamic programming for the solution of this problem, though without accounting for evaluations of the history, but since in practice in virtually all such socioeconomic development problems a limited number of scenarios with respect to investments and other elements of the problem formulation is considered, then the solution boils down to an exhaustive search of alternatives and the main problem is not the solution technique itself, exemplified by dynamic programming, but an adequate assessment and evaluation of all relations between variables. This can be very difficult and costly as the expertise of many experts may be needed.

For illustration we will show a simple example the essence of which is as shown in Kacprzyk [15] but with more current data.

Example: We consider an agricultural region with ca. 120,000 inhabitants, and arable land of ca. 450,000 acres. For illustration, the region's development is dealt with over the next 3 development stages (years). The life quality index consists of the four life quality indicators:

- x_t^{I} – average subsidies in US$ per acre (per year),
- x_t^{II} – sanitation expenditures (water and sewage) in US$ per capital (per year),
- x_t^{III} – health care expenditures in US$ per capita (per year), and
- x_t^{IV} – expenditures for paved roads (new roads and maintenance of the existing ones) in US$ (per year).

The investments, to improve the life quality indicators, are partitioned due to the fixed partitioning rule $A_{t-1}(u_{t-1}, i)$: 5% for subsidies, 25% for sanitation, 45% for health care, and 25% for infrastructure.

At $t = 0$, the initial values of the life quality indicators are:

$$x_0^{\mathrm{I}} = 0.5 \quad x_0^{\mathrm{II}} = 15 \quad x_0^{\mathrm{III}} = 27 \quad x_0^{\mathrm{IV}} = 1,700,000$$

For simplicity, we consider the following two *scenarios* which are the investment in the 3 consecutive periods:

- Scenario 1: $u_0 = \$16,000,000$ $u_1 = \$16,000,000$ $u_2 = \$16,000,000$
- Scenario 2: $u_0 = \$15,000,000$ $u_1 = \$16,000,000$ $u_2 = \$17,000,000$

Under Scenario 1 and Scenario 2 we obtain:

Scenario 1:

Year(t)	u_t	x_t^{I}	x_t^{II}	x_t^{III}	x_t^{IV}
0	\$16,000,000				
1	\$16,000,000	0.88	16.7	30	\$4,000,000
2	\$16,000,000	0.88	16.7	30	\$4,000,000
3		0.88	16.7	30	\$4,000,000

Scanario 2:

Year(t)	u_t	x_t^{I}	x_t^{II}	x_t^{III}	x_t^{IV}
0	\$15,000,000				
1	\$16,000,000	0.83	15.6	28.1	\$3,500,000
2	\$17,000,000	0.88	16.7	30	\$8,000,000
3		0.94	17.7	31.9	\$2,250,000

For the evaluation of the above two development trajectories, for simplicity only the *effectiveness* of development, and the objective evaluation only, is assumed. The consecutive fuzzy constraints and objective fuzzy subgoals are assumed piecewise linear (cf. Figs. 3, and 4): the aspiration level (i.e. the fully acceptable value) and the lowest (or highest) possible (still acceptable) value) are:

$$
\begin{array}{llll}
t & & & \\
0 & C^0: u_0^p = \$15,000,000 & & \\
 & u_0^c = \$17,000,000 & & \\
1 & C^1: u_1^p = \$16,500,000 & & \\
 & u_1^c = \$18,000,000 & G_o^{1,\mathrm{I}}: \underline{x}_1^{\mathrm{I}} = 0.6 & \overline{x}_1^{\mathrm{I}} = 0.85 \\
 & & G_o^{1,\mathrm{II}}: \underline{x}_1^{\mathrm{II}} = 14 & \overline{x}_1^{\mathrm{II}} = 16 \\
 & & G_o^{1,\mathrm{III}}: \underline{x}_1^{\mathrm{III}} = 27 & \overline{x}_1^{\mathrm{III}} = 29 \\
 & & G_o^{1,\mathrm{IV}}: \underline{x}_1^{\mathrm{IV}} = \$3,600,000 & \overline{x}_1^{\mathrm{IV}} = \$3,800,000 \\
2 & C^2: u_2^p = \$16,000,000 & & \\
 & u_1^c = \$20,000,000 & G_o^{2,\mathrm{I}}: \underline{x}_2^{\mathrm{I}} = 0.7 & \overline{x}_1^{\mathrm{I}} = 0.9 \\
 & & G_o^{2,\mathrm{II}}: \underline{x}_2^{\mathrm{II}} = 15 & \overline{x}_1^{\mathrm{II}} = 17 \\
 & & G_o^{2,\mathrm{III}}: \underline{x}_2^{\mathrm{III}} = 28 & \overline{x}_1^{\mathrm{III}} = 30 \\
 & & G_o^{2,\mathrm{IV}}: \underline{x}_2^{\mathrm{IV}} = \$3,800,000 & \overline{x}_2^{\mathrm{IV}} = \$4,000,000 \\
3 & & G_o^{3,\mathrm{I}}: \underline{x}_3^{\mathrm{I}} = 0.75 & \overline{x}_3^{\mathrm{I}} = 1 \\
 & & G_o^{3,\mathrm{II}}: \underline{x}_3^{\mathrm{II}} = 16 & \overline{x}_1^{\mathrm{II}} = 18.5 \\
 & & G_o^{3,\mathrm{III}}: \underline{x}_3^{\mathrm{III}} = 29 & \overline{x}_1^{\mathrm{III}} = 31 \\
 & & G_o^{3,\mathrm{IV}}: \underline{x}_3^{\mathrm{IV}} = \$3,800,000 & \overline{x}_3^{\mathrm{IV}} = \$4,200,000
\end{array}
$$

Using the "$\wedge$" (minimum) to reflect a safety-first attitude, which is clearly preferable in the situation considered (a rural region plagued by aging of the society, out-migration to neighboring urban areas, etc.), the evaluation of the two investment scenarios is:

- Scenario 1

$$
\begin{aligned}
&\mu_D(\$16,000,000; \$16,000,000; \$16,000,000 \mid .) = \\
&= \mu_{C^0}(\$16,000,000) \wedge (\mu_{G_o^{1,\mathrm{I}}}(0.88) \wedge \\
&\quad \wedge \mu_{G_o^{1,\mathrm{II}}}(16.7) \wedge \mu_{G_o^{1,\mathrm{III}}}(30) \wedge \mu_{G_o^{1,\mathrm{IV}}}(\$4,000,000)) \wedge \\
&\quad \wedge \mu_{C^1}(\$16,000,000) \wedge (\mu_{G_o^{2,\mathrm{I}}}(0.88) \wedge \\
&\quad \wedge \mu_{G_o^{2,\mathrm{II}}}(16.7) \wedge \mu_{G_o^{2,\mathrm{III}}}(30) \wedge \mu_{G_o^{2,\mathrm{IV}}}(\$4,000,000)) \wedge \\
&\quad \wedge \mu_{C^2}(\$16,000,000) \wedge (\mu_{G_o^{3,\mathrm{I}}}(0.88) \wedge \\
&\quad \wedge \mu_{G_o^{3,\mathrm{II}}}(16.7) \wedge \mu_{G_o^{3,\mathrm{III}}}(30) \wedge \mu_{G_o^{3,\mathrm{IV}}}(\$4,000,000)) = \\
&= 0.5 \wedge (1 \wedge 1 \wedge 1 \wedge 1) \wedge 0.8 \wedge \\
&\quad \wedge (0.9 \wedge 0.85 \wedge 1 \wedge 1) \wedge 1 \wedge (0.52 \wedge 0.28 \wedge 0.5 \wedge 0.33) = \\
&= 0.5 \wedge 0.8 \wedge 0.28 = 0.28
\end{aligned}
$$

- Scenario 2

$$
\begin{aligned}
&\mu_D(\$15,000,000; \$16,000,000; \$15,500,000 \mid .) = \\
&= \mu_{C^0}(\$15,000,000) \wedge (\mu_{G_o^{1,\mathrm{I}}}(0.83) \wedge \\
&\quad \wedge \mu_{G_o^{1,\mathrm{II}}}(15.6) \wedge \mu_{G_o^{1,\mathrm{III}}}(28.1) \wedge \mu_{G_o^{1,\mathrm{IV}}}(\$3,750,000)) \wedge \\
&\quad \wedge \mu_{C^1}(\$16,000,000) \wedge (\mu_{G_o^{2,\mathrm{I}}}(0.88) \wedge \\
&\quad \wedge \mu_{G_o^{2,\mathrm{II}}}(16.7) \wedge \mu_{G_o^{2,\mathrm{III}}}(30) \wedge \mu_{G_o^{2,\mathrm{IV}}}(\$4,000,000)) \wedge \\
&\quad \wedge \mu_{C^2}(\$17,000,000) \wedge (\mu_{G_o^{3,\mathrm{I}}}(0.94) \wedge \\
&\quad \wedge \mu_{G_o^{3,\mathrm{II}}}(17.7) \wedge \mu_{G_o^{3,\mathrm{III}}}(31.9) \wedge \mu_{G_o^{3,\mathrm{IV}}}(\$4,250,000)) = \\
&= 1 \wedge (0.92 \wedge 0.8 \wedge 0.55 \wedge 0.75) \wedge 0.8 \wedge \\
&\quad \wedge (0.9 \wedge 0.85 \wedge 1 \wedge 1) \wedge 0.75 \wedge (0.76 \wedge 0.68 \wedge 1 \wedge 1) = \\
&= 0.55 \wedge 0.8 \wedge 0.68 = 0.55
\end{aligned}
$$

The second scenario is therefore better.

5 Stability of Development

Now we will briefly consider another aspect of the development, the so-called *stability*. The *stability* of regional development is here meant to concern the variability of (crucial) development indicators, characteristics, conditions, etc. Namely, it is known from experience, and this is confirmed by psychological investigations, that a limited variability usually implies a higher acceptance of development, by both the inhabitants and authorities, than a high variability.

The stability in the above sense is here meant to involve:

- the *stability of development trajectory* that concerns the variability of development outcomes, i.e. the life quality indicators attained and the resulting social satisfactions, and
- the *stability of development scenario (policy)* that concerns the variability of some development prerequisites, i.e. the imposed fuzzy constraints, fuzzy goals, and investment partitioning rules.

It is quite clear that both the above stability types are "soft" concepts, and their measures will now be developed by using fuzzy tools.

The stability of development trajectory is equivalent to the requirement that the variability of development outcomes (life quality indicators and their resulting social satisfactions) should be possibly low.

The variability of development trajectory now has the following four aspects:

- the variability of life quality indicators over time (development stages),
- the variability of social satisfactions over time,

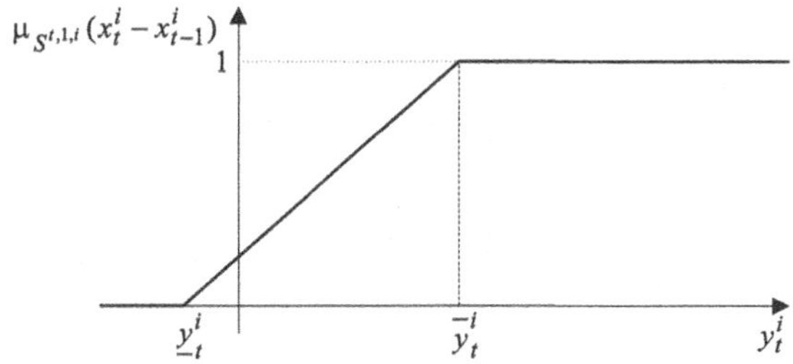

Fig. 6 Fuzzy limitation on the variability of x_t^i

- the variability "across" the life quality indices, and
- the variability "across" the social satisfactions.

The variability of a particular life quality indicator x_t^i—assuming for simplicity, but what is sufficient in virtually all applications, that the variability is meant only as a difference between the current and the previous value—is $x_t^i - x_{t-1}^i$, and is subjected to a *fuzzy limitation*

$$\mu_{S^{t,1,i}}(x_t^i - x_{t-1}^i)$$

of the form shown in Fig. 6.

This should be read as that we are fully satisfied with the variability (change) $x_t^i - x_{t-1}^i \geq \overline{y}_t^i$, partially satisfied with $\underline{y}_t^i \leq x_t^i - x_{t-1}^i < \overline{y}_t^i$, and fully dissatisfied with the values $x_t^i - x_{t-1}^i < \underline{y}_t^i$, i.e. such values are unacceptable. What is important, and what is consistent with the human perception, is that $\underline{y}_t^i < 0$ may occur, i.e. x_t^i may fall slightly below x_{t-1}^i, but not too much, that is, some interruption in the growth may happen but to a limited extent. Moreover, this may also be viewed to reflects also a natural human dissatisfaction of too high changes.

This measure of variability can also be viewed as a *growth requirement*, and the higher the planned or expected growth, the higher $\underline{y}_t^i$; $\underline{y}_t^i$ can even be positive if a constant growth is only to happen.

The variability limitation for the life quality index $X_t = (x_t^1, \ldots, x_t^7)$ is then

$$\mu_{S^{t,1}}(X_t - X_{t-1}) = \text{AGG}[\mu_{S^{t,1,1}}(x_t^1 - x_{t-1}^1), \ldots, \mu_{S^{t,1,7}}(x_t^7 - x_{t-1}^7)] \tag{44}$$

where AGG : $[0, 1]^7 \longrightarrow [0, 1]$ is some aggregation operation of the partial scores, for instance the "min" (minimum) for the safety first, pessimistic attitude.

On the other hand, the aggregation operations also reflect the substitutability of life quality indicators, from "$\wedge$" standing for the lowest (lack of) substitutability, through the weighted sum and other t-norms for an intermediate one, to "$\vee$" ("max") for the highest (full) one. Moreover the aggregation operation may be used to reflect some preferred growth pattern, from a balanced growth on the one extreme (for "$\wedge$"), to an imbalanced one on the other extreme ("$\vee$"), through all intermediate cases.

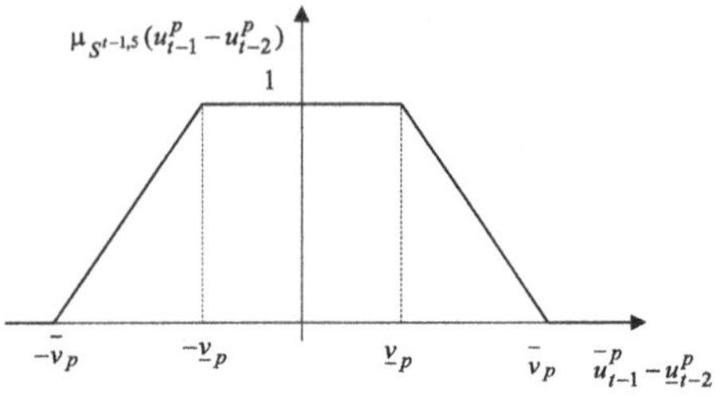

Fig. 7 Fuzzy limitation on the variability of the planned investment u_{t-1}^P, i.e. $u_{t-1}^P - u_{t-2}^P$

An analogous discussion remains valid for the variability of social satisfaction s_t over time which, for simplicity, involves the current and previous stage, i.e. t and $t-1$, only. The fuzzy limitation on this variability is in this case given by $\mu_{S^{t,2}}(s_t - s_{t-1})$ which is defined and elicited similarly as $\mu_{S^{t,1,i}}(x_t^i - x_{t-1}^i)$ shown in Fig. 7.

The second type of variability considered is what might be called "across." Basically, it concerns the variability of mutual "proportions (relations)" between the values of life quality indicators and their resulting partial social satisfactions which should also be limited. The derivation of these limitations is much more complicated and we refer the reader to Kacprzyk's [15] book for more detail.

Basically, the *variability "across" the life quality index* is evaluated by the the fuzzy limitation

$$\mu_{S^{t,3}}(X_t, X_{t-1}) = 1 - \text{AGG}[\mu_{Z_x^{t,1}}(x_t^1 - x_{t-1}^1), \ldots, \mu_{Z_x^{t,7}}(x_t^7 - x_{t-1}^7)]$$

and the *variability "across" the resulting social satisfaction* is evaluated by the fuzzy limitation

$$\mu_{S^{t,4}}(s_t, s_{t-1}) = 1 - \text{AGG}[\mu_{Z_s^{t,1}}(s_t^1 - s_{t-1}^1), \ldots, \mu_{Z_s^{t,7}}(s_t^7 - s_{t-1}^7)]$$

Here, again, the aggregation operations also reflect mainly the substitutability of life quality indicators, and a preferred growth pattern.

The total fuzzy limitation on the stability of development stage t is therefore

$$\begin{aligned} &\mu_{S^{t,d}}(X_t, s_t) = \\ &= \text{AGG}[\mu_{S^{t,1}}(X_t - X_{t-1}), \mu_{S^{t,2}}(s_t - s_{t-1}), \\ &\quad \mu_{S^{t,3}}(X_t, X_{t-1}), \mu_{S^{t,4}}(s_t, s_{t-1})] \end{aligned} \tag{45}$$

where the essence and choice of the AGG : $[0, 1]^4 \longrightarrow [0, 1]$ is analogous to the case of the variabilities across the life quality indicators and their resulting satisfactions given just before, and one may use the minimum, another t-norm, weighted sum, maximum, s-norm operation, or some OWA operator (cf. Kacprzyk et al. [28]).

And, finally, the fuzzy limitation on the stability of the whole development trajectory, i.e. for $t = 1, \ldots, N$, is evaluated by the following fuzzy limitation:

$$\mu_{S_d}(H_t) = \text{AGG}[\mu_{S^{1,d}}(X_1, s_1), \ldots, \mu_{S^{N,d}}(X_N, s_N)] \quad (46)$$

where the essence and choice of the AGG : $[0, 1]^N \longrightarrow [0, 1]$ is analogous to the case of the variabilities across the life quality index and the resulting social satisfactions, respectively.

We have therefore derived a measure for evaluating the stability of development strategy. Notice again that this stability is more relevant for the inhabitants than for the authorities. For the latter, the type of stability to be discussed below is more important.

5.1 Stability of Development Policy

This important aspect of development stability concerns some development prerequisites, or elements determining and conditioning the (pattern of) development. They are called here the *development policy* or *policy*, for short, and are meant to be the following sequence (from the beginning, i.e. $t = 0$, up to the development stage t, $t = 1, \ldots, N$):

$$B_t = [(C^0, A_0(u_0, i), G_o^1, G_s^1), \ldots, (C^{t-1}, A_{t-1}(u_{t-1}, i), G_o^t, G_s^t)] \quad (47)$$

where "$A_k(u_k, i)$" refers to the investment partitioning rule and specifies part of the investment at $t = k$ that is allocated to the i-th life quality indicator.

Therefore, the policy is here basically meant as how the successive investments are to be limited and partitioned (into parts devoted to the improvement of the particular life quality indicators), and how the successive life quality indices and their resulting partial social satisfactions should appear.

As previously, for simplicity we will prefer to use the *reduced development policy*:

$$b_t = [(C^{t-2}, A_{t-2}(u_{t-2}, i), G_o^{t-1}, G_s^{t-1}), (C^{t-1}, A_{t-1}(u_{t-1}, i), G_o^t, G_s^t)] \quad (48)$$

i.e. involving the last two development stages only. This is, fortunately enough, quite sufficient in virtually all applications.

The *stability of policy* concerns the variability of the following elements:

- the fuzzy constraints,
- the investment partitioning rules,
- the objective fuzzy goals, and
- the subjective fuzzy goals.

The first aspect is the stability of a fuzzy constraint. If we look at the form of a fuzzy constraint (cf. Fig. 3), we can recognize the following two elements: u_{t-1}^p and u_{t-1}^c. The planned investment, u_{t-1}^p, is much more crucial here. Namely, it is quite clear that high fluctuations of the planned investment, which may result from a varying attitude of higher level authorities with respect to the particular region, make the intra-regional planning, and then management, more difficult, and are therefore undesirable. The *variability of the planned investment* u^p_{t-1}, i.e. $u^p_{t-1} - u^p_{t-2}$, is hence subjected to a fuzzy limitation $\mu_{S^{t-1,5}}(u^p_{t-1} - u^p_{t-2})$ of the form shown in Fig. 7.

This limitation may be explained as follows: some variability, up to $\underline{v}_p$ in both directions (i.e. between $-\underline{v}_p$ and $\underline{v}_p$), is fully allowable and acceptable (with the grade of membership equal to 1). A higher variability, i.e. between $\underline{v}_p$ and $\overline{v}_p$ (and between $\underline{v}_p$ and $\overline{v}_p$) is still possible though less desirable (with the grade of membership between 0 and 1). Finally, the variability above $\overline{v}_p$ and below $-\overline{v}_p$ is unacceptable (with the grade of membership equal to 0). Notice that we define here again the variability with respect to the current and the last development stage only, i.e. $t-1$ and $t-2$, which is fully sufficient.

The second aspect of the stability of a fuzzy constraint is the variability of its "softness" (fuzziness). First, we define a *degree of softness* of a fuzzy constraint at stage $t-1$ as

$$ds_{t-1} = \frac{u^c_{t-1} - u^p_{t-1}}{u^p_{t-1}} \tag{49}$$

that is, the higher the relative difference between the maximum contingency investment and the planned investment, i.e. the "flatter" the membership function of the fuzzy constraint, the higher its softness.

The *variability of softness* of a fuzzy constraint may now be expressed as

$$vf_{t-1} = | ds_{t-1} - ds_{t-2} | \tag{50}$$

which is subjected to a fuzzy limitation $\mu_{S^{t-1,6}}(vf_{t-1})$ defined analogously as in Fig. 7. Some variability of the softness is therefore also allowable but not too high.

The stability limitation on the variability of a fuzzy constraint at stage $t-1$ is now

$$\mu_{S^{t-1,C}}(b_t) = \text{AGG}[\mu_{S^{t-1,5}}(u^p_{t-1} - u^p_{t-2}), \mu_{S^{t-1,6}}(vf_{t-1})] \tag{51}$$

where AGG : $[0, 1]^2 \longrightarrow [0, 1]$ is an aggregation operation exemplified by:

- the minimum operation

$$\mu_{S^{t-1,C}}(b_t) = \mu_{S^{t-1,5}}(u^p_{t-1} - u^p_{t-2}) \wedge \mu_{S^{t-1,6}}(vf_{t-1}) \tag{52}$$

- a t-norm, in general,

$$\mu_{S^{t-1,C}}(b_t) = \mu_{S^{t-1,5}}(u^p_{t-1} - u^p_{t-2})\, t\, \mu_{S^{t-1,6}}(vf_{t-1}) \tag{53}$$

- the weighted sum

$$\mu_{S^{t-1,C}}(b_t) = w_1 \cdot \mu_{S^{t-1,5}}(u^p_{t-1} - u^p_{t-2}) + w_2 \cdot \mu_{S^{t-1,6}}(vf_{t-1}) \tag{54}$$

where $w_1, w_2 \in [0, 1]$, and $w_1 + w_2 = 1$,

- the maximum operation

$$\mu_{S^{t-1},C}(b_t) = \mu_{S^{t-1},5}(u^p_{t-1} - u^p_{t-2}) \vee \mu_{S^{t-1},6}(vf_{t-1}) \tag{55}$$

- an s-norm, in general,

$$\mu_{S^{t-1},C}(b_t) = \mu_{S^{t-1},5}(u^p_{t-1} - u^p_{t-2})\, s\, \mu_{S^{t-1},6}(vf_{t-1}) \tag{56}$$

and the aggregation operations range from the most conservative, pessimistic, and of a safety-first type, i.e. "$\wedge$," on the one extreme, to the most optimistic one, i.e. "$\vee$," on the other extreme. Evidently, the linguistic quantifier based and OWA based aggregation makes here no sense as there are only two terms to be aggregated.

The fuzzy limitation (51) expresses therefore a desire of the authorities to operate under possibly low varying available investments and possibly high allowable flexibility.

An analogous argument can be applied for the objective and subjective fuzzy goals:

- the stability limitation on the objective fuzzy goal at stage t is

$$\begin{aligned}\mu_{S^t,G_O}(b_t) &= \\ &= \mathrm{AGG}\{\mathrm{AGG}_1[\mu_{S^t,G_O,1}(\overline{x}^1_t - \overline{x}^1_{t-1}), \ldots, \mu_{S^t,G_O,7}(\overline{x}^7_t - \overline{x}^7_{t-1})], \\ &\quad \mathrm{AGG}_2[\mu_{S^t,G_O,f_1}(gf^1_t), \ldots, \mu_{S^t,G_O,f_7}(gf^7_t)]\}\end{aligned} \tag{57}$$

  where

$$gf^i_t = \left| \frac{(\overline{x}^i_t - \underline{x}^i_t)}{\overline{x}^i_t} - \frac{(\overline{x}^i_{t-1} - \underline{x}^i_{t-1})}{\overline{x}^i_{t-1}} \right|, \qquad i = 1, \ldots, 7 \tag{58}$$

  and both the aggregation operations $\mathrm{AGG} : [0,1]^2 \longrightarrow [0,1]$ and $\mathrm{AGG}_1, \mathrm{AGG}_2 : [0,1,]^7 \longrightarrow [0,1]$ can be represented by, e.g., the minimum, another t-norm, weighted sum, maximum, another s-norm etc.;
- the stability limitation on the subjective fuzzy goal at stage t is

$$\mu_{S^t,G_S}(b_t) = \mathrm{AGG}[\mu_{S^t,G_S,s}(\overline{s}_t - \overline{s}_{t-1}), \mu_{S^t,G_S,f}(sf_t)] \tag{59}$$

  where

$$sf_t = \left| \frac{(\overline{s}_t - \underline{s}_t)}{\overline{s}_t} - \frac{(\overline{s}_{t-1} - \underline{s}_{t-1})}{\overline{s}_{t-1}} \right| \tag{60}$$

  and $\mathrm{AGG} : [0,1]^2 \longrightarrow [0,1]$ is an aggregation operation which can be exemplified by the minimum, another t-norm, weighted sum, maximum, another s-norm etc.

Therefore

$$\mu_{S^t,G}(b_t) = \mathrm{AGG}[\mu_{S^t,G_O}(b_t), \mu_{S^t,G_S}(b_t)] \tag{61}$$

and the aggregation operations $\text{AGG} : [0, 1]^2 \longrightarrow [0, 1]$ and $\text{AGG}_1, \text{AGG}_2 : [0, 1]^7 \longrightarrow [0, 1]$ are meant as before, and may be represented by, e.g., the minimum, another t-norm, weighted sum, maximum, another s-norm, etc.

The expression (61) represents a desire that the development (objective and subjective) goals be possibly low varying with respect to their aspiration levels and softness.

Finally, let us proceed to the last, very important aspect of the stability of policy, namely the *stability of the investment partitioning rules.*

First, we define a *degree of variability of an investment partitioning rule* at stage $t-1$ as

$$vp_{t-1} = \sum_{i=1}^{7} \mid A_{t-1}(u_{t-1}, i) - A_{t-2}(u_{t-2}, i) \mid \tag{62}$$

where the i-th partitioning rule $A_{t-1}(u_{t-1}, i)$, $i = 1, \ldots, 7$, is meant as a part (percentage) of the total investment at stage $t-1$ allotted to the improvement of the life quality indicator x_t^i.

The stability limitation on the investment partitioning rule $A_{t-1}(.)$ is now $\mu_{S^{t-1,A}}(b_{t-1}) = \mu_{S^{t-1,A}}(vp_{t-1})$ whose form is analogous to that shown in Fig. 7, i.e. that some variability is allowed but should be basically kept as low as possible.

Hence, the *stability of development policy* at development stage t is

$$\mu_{S^{t,P}}(b_t) = \text{AGG}[\mu_{S^{t-1,C}}(b_{t-1}), \mu_{S^{t-1,A}}(b_{t-1}), \mu_{S^{t,G_o}}(b_t), \mu_{S^{t,G_s}}(b_t)] \tag{63}$$

where the aggregation operation $\text{AGG} : [0, 1]^4 \longrightarrow [0, 1]$ is meant as previously, and also the fuzzy linguistic quantifier based and OWA based aggregations may be employed.

The *stability of development policy* over the whole development planning horizon, i.e. for $t = 1, \ldots, N$, is therefore

$$\mu_{S_p}(B_N) = \text{AGG}[\mu_{S^{1,P}}(b_1), \ldots, \mu_{S^{N,P}}(b_N)] \tag{64}$$

where the aggregation operation $\text{AGG} : [0, 1]^N \longrightarrow [0, 1]$ has at least a twofold character. First, it may be viewed as related to how the partial (from the particular stages) scores are aggregated as in the fuzzy decision. One can also employ aggregation based on a fuzzy linguistic quantifier or an OWA operator as before. Second, the aggregation operation may also be intended to capture the discounting aspect in that the terms concerning earlier development stages should have more impact on the value of $\mu_{S_p}(.)$.

We have therefore developed a measure for the evaluation of the stability of policy meant as some basic conditions under which the development is to proceed, and this type of stability is clearly more relevant for the authorities and other bodies responsible for the development than for the inhabitants.

The *total stability limitation of the development* is now defined as

$$\mu_S(H_N, B_N) = \text{AGG}[\mu_{S_d}(N_N), \mu_{S_p}(B_N)] \tag{65}$$

where the aggregation operation AGG : $[0, 1]^2 \longrightarrow [0, 1]$, as previously, is intended to properly reflect a compromise between the interests of authorities and inhabitants, ranging from a more "just" and balanced one using the minimum, to an "unjust," unbalanced one using the maximum, through a whole array of intermediate cases.

We are now in a position to formulate the multistage control problem for the stable sustainable regional agricultural development planning considered.

The fuzzy decision which is used for the evaluation of both the effectiveness and stability of the regional development is as follows:

$$\begin{aligned}&\mu_D(u_0, \ldots, u_{N-1} \mid X_0, B_N) = \\&= \text{AGG}_1[\mu_E(H_N), \mu_S(H_N, B_N)] = \\&= \text{AGG}_2[\mu_{C^0}(u_0), \mu_{G_o^1}(X_1), \mu_{G_s^1}(s_1), \mu_{S^{1,d}}(X_1, s_1), \mu_{S^{1,p}}(b_1), \ldots \\&\quad \ldots, \mu_{C^{N-1}}(u_{N-1}), \mu_{G_o^N}(X_N), \mu_{G_s^N}(s_N), \mu_{S^{N,d}}(X_N, s_N), \mu_{S^{N,p}}(b_N)]\end{aligned} \tag{66}$$

where $\text{AGG}_1 : [0, 1]^2 \longrightarrow [0, 1]$ and $\text{AGG}_2 : [0, 1]^N \longrightarrow [0, 1]$ are some aggregation operations meant analogously as mentioned before.

The fuzzy decision expresses therefore some crucial compromises between, e.g.:

- the fuzzy constraints, fuzzy goals, and fuzzy stability limitations,
- the effectiveness and stability of development,
- the interests of the authorities and inhabitants, etc.

An adequate choice of the aggregation operation(s) is very important not only to properly reflect the intended types of the above compromises, but also to reflect many other factors as, e.g., general attitudes as to various regional development factors, preferred growth patterns, a possibility of compensating false decisions (by future decisions), etc. Rules for choosing an appropriate aggregation operation follow the lines of reasoning for choosing these operations.

The problem is now to find an optimal sequence of controls (investments) $u_0^*, \ldots, u_{N-1}^*$ such that (under a given policy B_N)

$$\begin{aligned}&\mu_D(u_0^*, \ldots, u_{N-1}^* \mid X_0, B_N) = \\&= \max_{u_0,\ldots,u_{N-1}} \mu_D(u_0, \ldots, u_{N-1} \mid X_0, B_N)\end{aligned} \tag{67}$$

or, in the case when additionally the policy optimization is involved, to find an optimal sequence of controls (investments) $u_0^*, \ldots, u_{N-1}^*$ and an optimal policy B_N^* such that

$$\mu_D(u_0^*, \ldots, u_{N-1}^* \mid X_0, B_N^*) =$$
$$= \max_{u_0, \ldots, u_{N-1}, B_N} \mu_D(u_0, \ldots, u_{N-1} \mid X_0, B_N) \tag{68}$$

It is obvious that the policy optimization is in virtually all practical cases limited to the consideration of a very short list of policy scenarios.

The problem (67) [and eventually also (68)] may be solved, at least for some types of the aggregation operations. For instance, in the case of employing the minimum operation the problem (67) becomes that of finding an optimal sequence of controls $u_0^*, \ldots, u_{N-1}^*$ such that

$$\begin{aligned} \mu_D(u_0^*, \ldots, u_{N-1}^* \mid X_0, B_N) &= \max_{u_0, \ldots, u_{N-1}} [\mu_E(H_N) \wedge \mu_S(H_N, B_N)] = \\ &= \max_{u_0, \ldots, u_{N-1}} [\mu_{C^0}(u_0) \wedge \mu_{G_o^1}(X_1) \wedge \mu_{G_s^1}(s_1) \wedge \\ &\wedge \mu_{S^{1,d}}(X_1, s_1) \wedge \mu_{S^{1,p}}(b_1) \wedge \ldots \\ &\ldots \wedge \mu_{C^{N-1}}(u_{N-1}) \wedge \mu_{G_o^N}(X_N) \wedge \mu_{G_s^N}(s_N) \wedge \\ &\wedge \mu_{S^{N,d}}(X_N, s_N) \wedge \mu_{S^{N,p}}(b_N)] \end{aligned} \tag{69}$$

and similarly for the case with the additional policy optimization.

6 Concluding Remarks

We have shown a further extension of a multistage fuzzy control model of stable sustainable regional agriculture development that involves an additional capacity to reflect a stability requirement which expresses a clear preference for a limited variability of crucial development indicators and parameters. We have presented the use of fuzzy dynamic programming for solving the stable sustainable regional agriculture development problem in which many crucial aspects, in particular life quality indicators, are subject to objective, by the authorities, and subjective, by the inhabitants, evaluations which are closely related to human perception and cognitive abilities. For illustration, we have shown a simple example in which the problem is to determine the best (optimal) investment policy under different development scenarios with a consideration of objective and subjective evaluations.

Acknowledgements The contribution of the Project 691249,RUC-APS: Enhancing and implementing Knowledge based ICT solutions within high Risk and Uncertain Conditions for Agriculture Production Systems (www.rucaps.eu), funded by the European Union under their funding scheme H2020-MSCARISE-2015 is acknowledged by Jorge Hernandez Hormazabal, Janusz Kacprzyk and Sławomir Zadrożny.

References

1. Albegow, M., Kacprzyk, J., Owsinski, J.W., Straszak, A.: Regional agricultural policy making on the basis of a detailed economic and agrotechnical model. WP-80-18. IIASA, Laxenburg (1980)
2. Albegow, M.M., Kacprzyk, J., Owsinski, J.W., Straszak, A.: Regional agricultural policy design on the basis of a detailed linear economic and agrotechnical model. In: Janssena, J.M.L., Pau, L.F., Straszak, A. (eds.) Dynamic Modelling and Control of National Economies, pp. 221–229. Pergamon Press, Oxford (1981)
3. Albegow, M., Kacprzyk, J., Orchard-Hays, W.M., Owsinski, J.W., Straszak, A.: A General Regional Agricultural Model (GRAM) Applied to a Region in Poland. RR-82-16. IIASA, Laxenburg (1982)
4. Asan, D., Batyrova, N., Aubakirova, A., Abishov, N., Kuralbayev, A.: Implementation of a comparative evaluation method of stable socio-economic development. Int. J. Econ. Financ. Issues **6**(S2), 9–13 (2016). ISSN: 2146-4138
5. Bellman, R.E., Zadeh, L.A.: Decision making in a fuzzy environment. Manage. Sci. **17**, 141–164 (1970)
6. Čajková, A., Čajka, P.: Challenges and sustainability of china's socio-economic stability in the context of its demographic development. Societies **11**, 22 (2021). https://doi.org/10.3390/soc11010022
7. Francelin, R.A., Gomide, F.A.C., Kacprzyk, J.: A biologically inspired neural network for dynamic programming. Int. J. Neural Syst. **11**, 561–572 (2001)
8. Francelin, R.A., Kacprzyk, J., Gomide, F.A.C.: Neural network based algorithm for dynamic system optimization. Asian J. Control **3**(2), 131–142 (2001)
9. Gu, J., Zhou, S., Ye, X.: Uneven Regional Development Under Balanced Development Strategies: Space-Time Paths of Regional Development in Guangdong, China. Tijdschrift voor economische en sociale geografie (J. Econ. Soc. Geography) **107**, 596–610 (2016). https://doi.org/10.1111/tesg.12200
10. Jones, J.W., et al.: Brief history of agricultural systems modeling. Agricult. Syst. **155**, 240–254 (2017)
11. Jovovic, R., Draskovic, M., Delibasic, M., Jovovic, M.: The concept of sustainable regional development - institutional aspects, policies and prospects. J. Int. Stud. **10**(1), 255–266 (2017). https://doi.org/10.14254/2071-8330.2017/10-1/18
12. Kacprzyk, J., Owsinski, J.W., Straszak, A.: Agricultural policy making for integrated regional development in a mixed economy through a large-scale LP model. In: Titli, A., Singh, M.G., (eds.), Proceedings of the 2nd IFAC LSSTA Symposium, Toulouse. Pergamon Press (1980)
13. Kacprzyk, J.: Multistage Decision Making under Fuzziness. Verlag TÜV Rheinland, Cologne (1983)
14. Kacprzyk, J.: Multistage control under fuzziness using genetic algorithms. Control Cybern. **25**, 1181–1215 (1996)
15. Kacprzyk, J.: Multistage Fuzzy Control. Wiley, Chichester (1997)
16. Kacprzyk, J.: A genetic algorithm for the multistage control of a fuzzy system in a fuzzy environment. Mathw. Soft Comput. **IV**, 219–232 (1997)
17. Kacprzyk, J.: Multistage control of a stochastic system in a fuzzy environment using a genetic algorithm. Int. J. Intell. Syst. **13**, 1011–1023 (1998)
18. Kacprzyk, J.: Including socio-economic aspects in a fuzzy multistage decision making model of regional development planning. In: Reznik, L., Dimitrov, V., Kacprzyk, J. (eds.) Fuzzy Systems Design, pp. 86–102. Physica-Verlag (Springer), Heidelberg and New York (1998)
19. Kacprzyk, J.: Towards perception-based fuzzy modelling: an extended multistage fuzzy control model and its use in sustainable regional development planning. In: Sinčak, P., Vaščak, J., Hirota, K. (eds.) Machine Intelligence - Quo Vadis?, pp. 301–337. World Scientific, Singapore (2004)
20. Kacprzyk, J.: Fuzzy dynamic programming: interpolative reasoning for an efficient derivation of optimal control policies. Control Cybern. **42**(1), 63–84 (2013)

21. Kacprzyk, J.: Fuzzy dynamic programming for the modeling of sustainable regional development. Appl. Comput. Math. **14**(2), 107–124 (2015)
22. Kacprzyk, J., Francelin, R.A., Gomide, F.A.C.: Involving objective and subjective aspects in multistage decision making and control under fuzziness: dynamic programming and neural networks. Int. J. Intell. Syst. **14**, 79–104 (1999)
23. Kacprzyk, J., Kondratenko, Y.P., Merigó, J.M., Hormazabal, J.H., Sirbiladze, G., Gil-Lafuente, A.M.: A status quo biased multistage decision model for regional agricultural socioeconomic planning under fuzzy information. In: Kondratenko, Y., Chikrii, A., Gubarev, V., Kacprzyk, J. (eds.) Advanced Control Techniques in Complex Engineering Systems: Theory and Applications. Studies in Systems, Decision and Control, vol. 203, pp. 201–2225. Springer, Cham (2019)
24. Kacprzyk, J., Straszak, A.: Application of fuzzy decision making models for determining optimal policies in 'stable' integrated regional development. In: Wang, P.P., Chang, S.K. (eds.) Fuzzy Sets Theory and Applications to Policy Analysis and Information Systems, pp. 321–328. Plenum, New York (1980)
25. Kacprzyk, J., Straszak, A.: A fuzzy approach to the stability of integrated regional development. In: Lasker, G.E. (ed.) Applied Systems and Cybernetics, vol. 6, pp. 2997–3004. Pergamon Press, New York (1982)
26. Kacprzyk, J., Straszak, A. (1984) Determination of stable trajectories for integrated regional development using fuzzy decision models. IEEE Trans. Syst. Man Cybern. **SMC-14**, 310–313
27. Kacprzyk, J., Sugianto, L.F.: Multistage fuzzy control involving objective and subjective aspects. In: Proceedings of the 2nd International Conference on Knowledge-Based Intelligent Electronic Systems KES'98. (Adelaide, Australia), pp. 564–573 (1998)
28. Kacprzyk, J., Yager, R.R., Merigó, J.M.: Towards human-centric aggregation via ordered weighted aggregation operators and linguistic data summaries: a new perspective on Zadeh's inspirations. IEEE Comput. Intell. Mag. **14**(1), 16–30 (2019)
29. Kubo, Y.: Scale economies, regional externalities and the possibility of uneven regional development. J. Region. Sci. **35**, 29–42 (1995). https://doi.org/10.1111/j.1467-9787.1995.tb01398.x
30. Lezoche, M., Hernandez, J.E.H., Alemany, M.E., Diaz, P.H., Kacprzyk, J.: Agri-food 4.0: a survey of the supply chains and technologies for the future agriculture. Comput. Indust. **117**, 103187 (2020)
31. Roberts, P.: Sustainable regional planning. Region. Stud. **28**(8), 781–787 (1994). https://doi.org/10.1080/00343409412331348666
32. Roberts, P., Colwell, A.: Moving the environment to centre stage: a new approach to planning and development at European and regional levels. Local Environ. **6**(4), 421–437 (2001)
33. Straszak, A., Kacprzyk, J., Owsinski, J.W.: Multicriterial policy assessment for regional agriculture through detailed optimization models. In: Applications of Systems Theory to Economics, Management and Technology. Proceedings of the 5th Polish-Italian Symposium. PWN, Warszawa, 1980, pp. 563–580
34. Thomas, L.C. (ed.): Golden Developments in Operational Research. Pergamon Press, New York (1987)
35. Wang, Y.: On cognitive informatics. Brain and mind. Transdiscip. J. Neurosci. Neurophilos. **4**(2), 151–167 (2003)
36. Wang, Y.: The theoretical framework of cognitive informatics. Int. J. Cognit. Inf. Nat. Intell. **1**(1), 1–27 (2007)
37. Wang, Y., Karray, F., Kwong, S., Plataniotis, K.N., Leung, H., Hou, M., Tunstel, E.W., Rudas, I.J., Trajkovic, L., Kaynak, O., Kacprzyk, J., Zhou, M., Smith, M.H., Chen, Ph., Patel, S.: On the Philosophical, Cognitive and Mathematical Foundations of Symbiotic Autonomous Systems (SAS). Philos. Trans. Ten R. Soc. A, Math. Eng. Sci. **379**, 20200362 (2021)
38. Wang, Y., Ruhe, G.: The cognitive process of decision making. Int. J. Cognit. Inf. Nat. Intell. **1**(2), 73–85 (2007)
39. Zadeh, L.A.: Fuzzy sets. Inf. Control **8**, 338–353 (1965)

40. Zadrożny, S., Kacprzyk, J.: Bipolar queries: an approach and its various interpretations. In: IFSA/EUSFLAT Conference, vol. 2009, pp. 1288–1293 (2009)
41. Zadrożny, S., Kacprzyk, J.: Bipolar queries: an aggregation operator focused perspective. Fuzzy Sets Syst. **196**, 69–81 (2012)

Novel Applications in Infrastructure and Manufacturing Industry

Multicriteria Optimal Control of Industrial Thermal Processes with Distributed Parameters Under Variable Operational Conditions

Vassil Sgurev, Mincho Hadjiski, and Nencho Deliiski

Abstract An intelligent system for control of the thermal treatment process (TTP) of wood materials addressed toward manufacturing with necessity of often rescheduling is proposed via combination of model based and data-driven approaches. Using First-principle mathematical model of TTP presented by Partial Differential Equations in 2D space with suboptimal model-based control algorithm and Case-Based Reasoning (CBR) approach an explicit suboptimal control system is investigated in different operational conditions. A set of virtual subspaces for feasible operational situations for variety of objective criteria of value assessment is created using traditional "problem-decision" representation. As the search spaces are well structured, the search procedure based on traditional K–NN algorithm is strongly simplified. In this way the complicated computer simulation of the TTP at each time step due to the plant's parameter distribution, nonlinearity and operational or environmental disturbances are fulfilled off-line. On-line is accomplished relatively small part of the calculations connected with the traditional R^4–operations in CBR, objective functions estimation, some data-based and rule-based control parameter corrections and possible adaptation from charge to charge. Some results of the simulation experiments are presented and analyzed.

V. Sgurev (✉) · M. Hadjiski
Institute of Information and Communication Technologies, BAS, Acad. G. Bonchev Str., Bl. 2, 1113 Sofia, Bulgaria
e-mail: vsgurev@gmail.com

M. Hadjiski
e-mail: zdravkah@abv.bg

M. Hadjiski
Department of Industrial Automation, University of Chemical Technology and Metallurgy, St. Kliment Ohridski blvd. 8, 1796 Sofia, Bulgaria

N. Deliiski
Faculty of Forest Industry, University of Forestry, St. Kliment Ohridski blvd. 10, 1796 Sofia, Bulgaria
e-mail: deliiski@netbg.com

P. Shi et al. (eds.), *Complex Systems: Spanning Control and Computational Cybernetics: Applications*, Studies in Systems, Decision and Control 415,
https://doi.org/10.1007/978-3-031-00978-5_14

Keywords Case-Base Reasoning (CBR) · Mathematical modeling · Operational conditions · Scheduling · Suboptimal control · Thermal Treatment Process (TTP)

1 Introduction

Modern industrial production is facing increasingly acute challenges—competitiveness, dealing with recurring crises of various kinds (financial, economic, political, epidemiological, etc.), the need to take into account increasingly persistent eco-oriented factors (harmful emissions, heat pollution, green industry), protection against cyber-attacks.

This applies to both innovative sectors and traditional industries. The main approach to answer these challenges is to attract modern methods of digitalization mainly through the application of elements of artificial intelligence (AI).

The main focus for the application of AI-based technologies is the optimization of the operational management [1, 2, 5], which has a direct impact on the Key Performance Indicators (KPI). A representative survey [3] shows that "60% of respondents believe that improving operational efficiency is one of the most important results they can get from their investment in AI".

The risk of inefficient implementation of certain AI-based algorithms is in many cases significant due to the still existing insurmountable shortcomings of machine learning [4]. This is especially characteristic of traditional technological processes, in the operational optimization of which, both at the level of individual site and production system, there is significant experience [2, 5].

The search for adequate solutions for this type of objects is undoubtedly a topical issue, because a number of business, technical and social aspects remain poorly studied and unresolved [2, 6, 7].

As for these objects more than 95% of the basic regulation is carried out with proven PID algorithms, solutions for the integration of AI-based technologies in estab-lished management structures are vigorously sought [6–9].

The present paper considers the possibility of combining three approaches for implementation of suboptimal control of the thermal treatment process (TTP) of timber, which are based on a mathematical model, data flow and expertise with direct application of appropriate algorithms, based on AI technologies.

2 Problem Statement

A. *Object of Research*

The object of the research is a system of parallel autoclaves for thermal treatment of wood materials (Fig. 1) with a common steam supply line (Fig. 2).

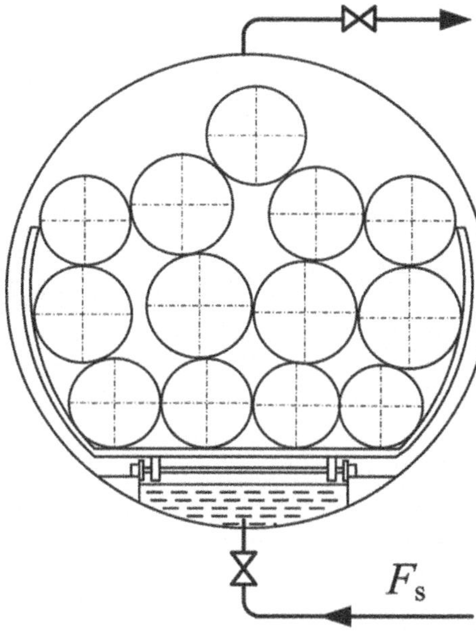

Fig. 1 Cross section of an autoclave for steaming of wood materials

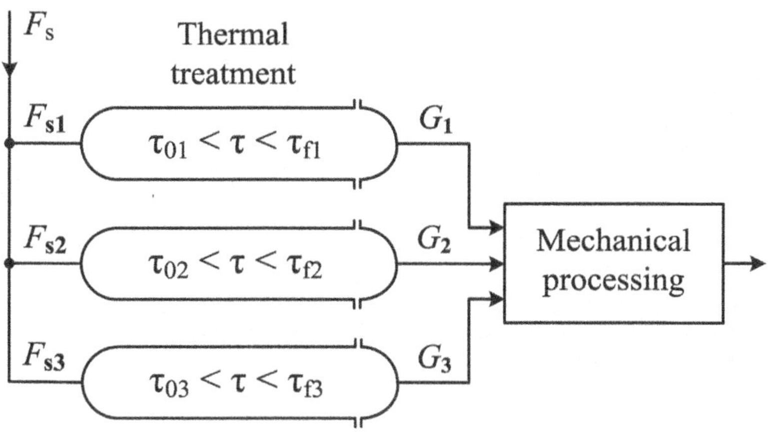

Fig. 2 Scheme with 3 parallel autoclaves

Each autoclave (Fig. 2) functions autonomously and its work is synchronized with the other autoclaves based on the current planning with a horizon of 1 ÷ 2 weeks and an operational schedule with a horizon of 2 ÷ 5 days (Fig. 3).

The thermal treatment is a periodic process that includes 5 consecutive stages (Fig. 4): intensive heating ($0 - \tau_1$); heating at constant temperature ($\tau_1 - \tau_2$) and three stages of exponential temperature degradation ($\tau_2 - \tau_3$), ($\tau_3 - \tau_4$) and ($\tau_4 - \tau_f$) with switching of the heat exchange conditions.

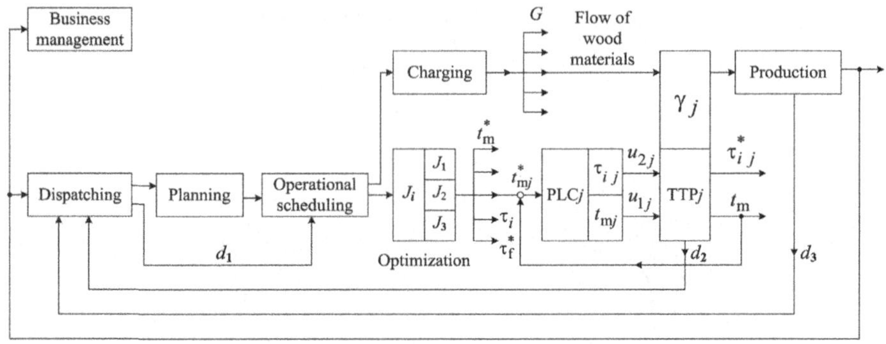

Fig. 3 Scheme of operational management in TTP of wood materials in an autoclave with switching parameterization

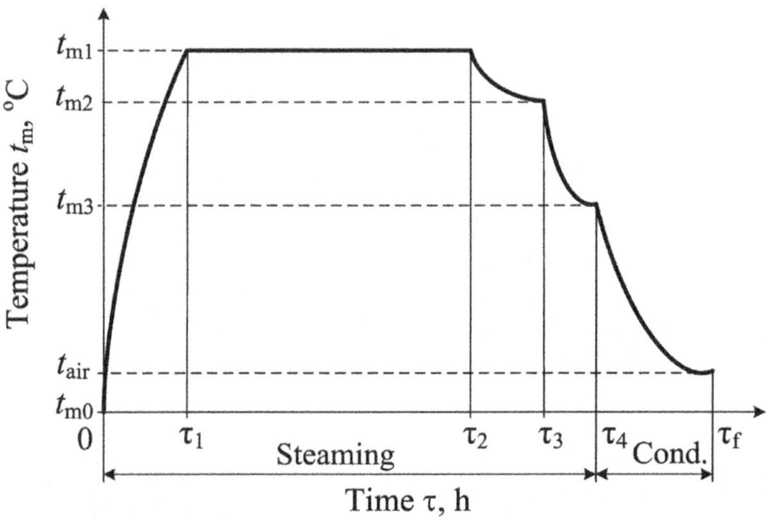

Fig. 4 Typical time profile for change of the processing medium temperature during steaming of wood materials in an autoclave

A detailed description of the TTP of wood and its energy characteristics is made in [7, 8, 10–12].

B. Characteristics of the Autoclave as an Object of Control

1. The process of thermal treatment of round and prismatic timber is described by parabolic nonlinear partial differential equations (PDEs) of the type

$$c_{\mathrm{w}} \cdot \rho_{\mathrm{w}} \frac{\partial t_{\mathrm{w}}(\xi, \tau)}{\partial \tau} = \mathrm{div}\ (\lambda_{\mathrm{w}} \mathrm{grad}\ t) \tag{1}$$

under the following conditions of unambiguity:

- Initial condition

$$t_{\mathrm{w}}(\xi, 0) = t_{\mathrm{w}}^{0}, \tag{2}$$

- Conductive boundary conditions in the time interval $0 - \tau_4$ (Fig. 4)

$$t_{\mathrm{w}}(\zeta_s, \tau) = t_{\mathrm{m}}, \tag{3}$$

- Convective boundary conditions in the time interval $\tau_4 - \tau_{\mathrm{f}}$

$$\frac{\partial t_{\mathrm{w}}(\zeta_{\mathrm{s}}, \tau)}{\partial \zeta} = -\frac{\alpha_{\mathrm{ws}}(\zeta_{\mathrm{s}}, \tau)}{\lambda_{\mathrm{ws}}(\zeta_{\mathrm{s}}, \tau)}[t(\zeta_{\mathrm{s}}, \tau) - t_{\mathrm{air}}(\tau)], \tag{4}$$

- Terminal conditions

$$\tau_{\mathrm{f}} = \tau_{\mathrm{f}}^{0}, \tag{5}$$

$$t_{\mathrm{w}}(\xi_{\mathrm{s}}, \tau_{\mathrm{f}}) < t_{\mathrm{w}}^{\max 0}(\tau_{\mathrm{f}}) = t_{\mathrm{wf}}^{\max}, \tag{6}$$

$$t_{\mathrm{w}}(\xi_{\mathrm{c}}, \tau_{\mathrm{f}}) > t_{\mathrm{w}}^{\min 0}(\tau_{\mathrm{f}}) = t_{\mathrm{wf}}^{\min}, \tag{7}$$

where c_{w} is the specific heat capacity of the wood; λ_{w}—thermal conductivity of the wood in separate anatomical directions; λ_{ws}—thermal conductivity of the wood on its surface; α_{ws}—convective heat transfer coefficient of the wood surface; ρ_{w}—density of the wood; $\xi = (x, y, z)$—space coordinate; ξ_{s}—coordinate of the surface layer of the f(22 wood; ξ_{c}—coordinate of the central layer of the wood; τ and τ_{f}—current and terminal time; t_{w}^{0}—average mass temperature of the wood material in the beginning of TTP; $t_{\mathrm{w}}(\xi, \tau)$—temperature in the wood material at point ξ at time τ; t_{m}—temperature of the processing medium; t_{air}—temperature of the surrounding air during conditioning of the heated wood materials out of the autoclave before their subsequent mechanical processing; $t_{\mathrm{w}}^{\max 0}(\xi_{\mathrm{s}}, \tau_{\mathrm{f}})$ and $t_{\mathrm{w}}^{\min 0}(\xi_{\mathrm{c}}, \tau_{\mathrm{f}})$– terminal restrictions.

According to Eq. (3), the temperature at the prisms surfaces being in contact with the processing medium is equal to the temperature t_{m} due to the extremely high values of the heat transfer coefficient between the condensed water steam and the wooden surfaces.

Mathematical descriptions of t_{m} and thermo-physical properties of wood, which participate in (1), (3), and (4) have been suggested in [16–18].

The non-stationary heat balance of the autoclave for each moment $n \cdot \Delta\tau$ of the steaming process can be described by the following mathematical model:

$$Q_{\text{ha}}^{n} = Q_{\text{hw}}^{n} + Q_{\text{hf}}^{n} + Q_{\text{hil}}^{n} + Q_{\text{he}}^{n} + Q_{\text{hfv}}^{n} + Q_{\text{hcw}}^{n}, \tag{8}$$

where Q_{ha} is the specific (for 1 m^3 wood) heat energy, which is supplied into the autoclave by the introduced in it water steam; Q_{hw}—energy used for heating of the subjected to steaming wood materials; Q_{hf}—energy used for heating of the body of the autoclave and of the situated in it metal trolleys for positioning of the wood materials; Q_{hil}—energy used for heating of the insulating layer of the autoclave; Q_{he}—energy used for covering of the heat emission from the autoclave in the surrounding air; Q_{hfv}—energy used for filling in with steam the free (unoccupied by wood materials) part of the working volume of the autoclave; Q_{hcv}—energy, which is accumulated in the gathered in the lower part of the autoclave condense water; n—current number of the step along the time coordinate, $\Delta\tau$, with the help of which the solving of the model is carried out: $n = 0, 1, 2, 3, \ldots$

The whole mathematical model (8) with descriptions of all components in it, depending on the influencing factors, has been given in [11, 17].

2. Favorable features of TTP:

   - TTP is a slow process;
   - The boundaries (6) and (7) are relatively wide and represent "soft" constraints;
   - TTP is a periodic process and allows the accumulation of archival data for various similar conditions and requirements;
   - The process of non-stationary heat transfer is well studied and its analytical modeling has been developed;
   - TTP is a process without degradation of wood;
   - The constants involved in Eq. (1) (density ρ, specific heat capacity c and thermal conductivity coefficients λ) can be represented as deterministic functions of the basic initial thermos-physical and operational characteristics of timber from different wood species.

3. Unfavorable features of TTP:

   - The current state $t_{\text{w}}(\xi, \tau)$ cannot be measured, therefore:
     - Introducing a condition observer or a Kalman filter greatly complicates the computational procedures;
     - Using model based soft-sensing on objects with distributed parameters requires a large volume of on-line calculations;
   - The heat exchange process is nonlinear and spatially distributed;
   - The initial conditions for each batch of materials loaded in the autoclave are different and they require special measurements and on-line procedures for their reconstruction;

- The process can be influenced by various disturbances—flow rate F_s and parameters P_s (pressure and temperature) of the water vapor used to heat the materials; change in ambient air temperature t_{air} dispatching influences; damage to the machines performing the subsequent mechanical processing of the heated wood materials, etc.

C. ***Operational Modes and Criteria for Optimal Operational Management***

The thermal treatment process takes place in the following three main modes with criteria J_1, J_2 and J_3:

1. J_1—Maximal productivity:

$$J_1 \rightarrow \tau_f = \tau_f^{min}. \tag{9}$$

2. J_2—Set productivity with minimization of the total heat consumption from the autoclave Q_a:

$$J_2 \rightarrow \tau_f = \tau_f^0 \text{ at } Q_a \rightarrow \min . \tag{10}$$

3. J_3—Interrupt mode of the initial operational schedule with set performance $\tau_f = \tau_f^0$ at time τ_t (refer to Fig. 6):

$$\tau_f = \tau_f^N \text{ at } Q_a \rightarrow \min . \tag{11}$$

D. ***Purpose of the Research***

A detailed study of the problems for optimization of TTP with criteria J_1 and J_2 is presented in [12, 13]. The object of the present work is to find a suitable solution for criterion J_3 with the occurrence of a change assignment, according to the operational condition (11). We search a TTP management that:

1. To use for the case of changed operational conditions of type (11) the same conceptual framework for combining the model-based approach and method of precedents applied in [12, 13] (Case-Based Reasoning—CBR).
2. To ensure a suboptimal continuation of the trajectory

$$t_w(\xi,\tau) \text{ for } \tau_t < \tau < \tau_f^N \tag{12}$$

so that the total heat consumption Q_a for both sections of the trajectory.

$$\begin{aligned} &\text{(a) } t_{w1}(\xi,\tau) \text{ at } 0 < \tau < \tau_t, \\ &\text{(b) } t_{w2}(\xi,\tau) \text{ at } \tau_t < \tau < \tau_f^N \end{aligned} \tag{13}$$

to be close to the minimal one.

In (12) and (13) t_w is the temperature of the wood, τ_t is the time (moment) of the operational intervention, and N is the index of the new operational situation.

3. The control impacts in the control system should be the following two types—u_1 and u_2:

    (a) Constantly $u_1 = t_m^{N*}(\tau)$ at $t_m^{N*} = \text{const}$ determines certain heat consumption. Here t_m denotes the temperature of the heating steam;
    (b) Structurally $u_2 = < \tau_2^N, \tau_3^N, \tau_4^N >$ changing the nature of the boundary conditions at times τ_2^N, τ_3^N and τ_4^N.

4. The system should be implemented by switching re-parameterization of the same PLC-controller, which is accepted in the control schemes for criteria J_1 and J_2 and in [12, 13] (see Fig. 3).

3 Preliminary Considerations

A. *Operational Management*

The operational management (OM) scheme shown in Fig. 3 includes two streams—material and informational. OM has the task to ensure efficient decision-making procedures in the functional blocks of dispatching, production planning, operational scheduling and basic management in both streams.

The goal is the transformation of natural wood through TTP into an intermediate product that meets the regulated requirements for plasticity, color, lack of cracks, etc. It is achieved by correcting the initial operational schedule (rescheduling) and optimizing according to some of the criteria J_1, J_2, J_3 in the optimization block.

The criteria for optimization of the operational management include:

1. Cover the entire optimization area with effective solutions for J_1, J_2, or J_3.
2. Rapid response to disturbances occurred:

    (a) On the part of the dispatcher—d_1;
    (b) In the separate parallel autoclaves (flow rate F_s and parameters P_s of the incoming steam)—d_2
    (c) In the process of subsequent machining (damage, accidents, forced interrupt-tion)—d_3.

3. Robustness to unforeseen disturbances (d_1, d_2, d_3).

In the present work, instead of using universal optimization units (Optimization Solvers—OS), which would unnecessarily complicate the whole scheme, simpler approaches for suboptimal operational management are used.

They operate with a limited number of scheduling parameters $\left(t_m^N, \tau_2^N, \tau_3^N, \tau_4^N\right)$ and the corresponding intermediate quasi-stationary states (t_{w3}, t_{w4}, t_w^*) but take into account all the main constraints in TTP (5)–(7).

B. *Suboptimal Control*

Following the approach adopted in our previous work [12, 13], in the present study the problem of suboptimal OM is solved by applying the case-based method (Case-Based Reasoning—CBR) for search and decision making in a limited oper-ational space, obtained on the basis of simulation of virtual states of the system "process model–control system". The result is an explicit form of suboptimal control.

The considerations for choosing suboptimal instead of strictly optimal control are the following:

1. The data show that the most of the real industrial automation systems are suboptimal.
2. The thermal treatment process depends on a large number of factors, which is a prerequisite for a large dimension of the operational reach space. The suboptimal approach allows its significant limitation and thus reduces the volume of on-line calculations in the control system.
3. The efficiency criteria J_1, J_2 and J_3 in (9)–(11) have little sensitivity to the different suboptimal control options.
4. Suboptimal control solves in an acceptable way a number of problems outside the task of achieving high functional accuracy, namely:
    - The system is designed with minimal complexity using a traditional PLC controller (Fig. 3);
    - The procedure for complex management of the process model is avoided (maintenance of constant accuracy);
    - The development is seamlessly integrated into an existing SCADA system;
    - The overall investment for implementation of the suboptimal management (hardware, software, staff training, maintenance, etc.) is acceptable.
5. The corporate risk of adopting suboptimal management is low.
6. The implementation of suboptimal management can be considered as a step in the right direction in view of the prospects for total digitalization of the industry according to IR4.0.
7. Many of the unsolved problems of a number of modern AI-based algorithms are avoided [4].
8. The suboptimality of the management system in the solution proposed in the pre-sent research is due to the following:
    (a) Although an accurate process model is used [10, 11, 14], the control algorithm in the virtual precedent database (CB) is suboptimal ($t_{\mathrm{m}}^{*} =$ const, $\tau_i = \tau_i^{*}$).
    (b) In the CBR approach, there is always a difference between the defining parameters of the real $\mathrm{P^E}$ process:

$$\mathrm{P^E} = (\pi^{\mathrm{E}}, a^{\mathrm{E}}, \gamma^{\mathrm{E}}, w^{\mathrm{E}}, t_{\mathrm{w}}^{0}) \tag{14}$$

and those of the P^v virtual process:

$$P^v = (\pi^v, a^v, \gamma^v, w^v, t_w^0) \tag{15}$$

which give rise to the suboptimal control

$$u = \begin{vmatrix} u_1 \\ u_2 \end{vmatrix} = \begin{vmatrix} t_m^0 \\ \tau_1^*, \tau_2^*, \tau_3^*, \tau_4^*, \tau_f^* \end{vmatrix}. \tag{16}$$

(c) Error due to all unaccounted and dynamically changing factors not included in the reduced vector of the real process (14), such as the degree of wood icing ψ, the specific micro-structural characteristics υ of the individual specimens of the same wood species π, etc.

Thus, suboptimal error of the suboptimal control can be represented as

$$\delta = \frac{\sqrt{|\delta_e|^2 + |\delta_r|^2 + |\delta_w|^2}}{|\delta_e| + |\delta_r| + |\delta_w|}. \tag{17}$$

In (14)–(17) π, a, γ, w, t_w denote the determining parameters of TTP, respectively wood species π, the representative size of materials a, humidity w and temperature t_w of the wood; with τ_i $(i = 1 \div 4)$—switching times (see Fig. 6); δ_e, δ_r, δ_w are the errors of the suboptimal control, mismatch of the real P^E and the virtual P^v parameters, and the influence of the unreported characteristics of the wood species respectively.

A detailed research [13] shows that the suboptimal error can be kept within acceptable limits because:

1. The error δ_e is small because the suboptimal and optimal trajectories differ slightly from each other [7, 8].
2. The parametric error δ_r can be reduced by appropriately dividing the search space of the defining parameters a, γ, w and t_w.
3. Much of the secondary regime parameters t_{air}, steam parameters (t_s, P_s) and others can be taken into account using neural networks.
4. The nonlinear effects of the phase transition of the ice in wood in the range from −1 to 0 °C can be taken into account using an analytical model.

C. ***Method of Precedents (Case-Based Reasoning—CBR)***

By precedent C is meant the couple

$$C = < Pr, S >, \tag{18}$$

where Pr is the problem and S is the solution to the problem.

Fundamental to CBR is the premise that with CB—a sufficiently complete base of known precedents

$$C_i^B \in CB \tag{19}$$

for each new precedent $C^N(Pr^N, S^N)$ with a known problem Pr^N and an unknown solution S^N in the precedent database, one with the closest problem Pr_N^B will be found so that

$$\left| \overset{N}{Pr} - Pr_N^B \right| \to \min \tag{20}$$

and then a similar problem $Pr^N \approx Pr_N^B$ is assumed to correspond to a similar solution

$$S^N \approx S_N^B. \tag{21}$$

The similarity (21) can be found with different types of search procedures [9, 14], among which the most popular is the one for N–nearest neighbors (K–NN), which is used in the case under consideration.

In our task we use an approach close to process-oriented CBR (PO-CBR) [9], where in the precedent C^N the component "problem" Pr^N contains the defining parameters of the current batch of heat-treated materials P^E (11), and the solution is suboptimal control $S^N = u$, according to Eq. (16).

The CBR for the timber TTP in question has the following advantages justifying its choice:

1. The CB precedent base can be generated based on an accurate analytical model of TTP [7, 10, 11, 14] with a randomly selected density of situations in the opera-tional space.
2. The part "Solution S^N" (21) has a simplified form (16).
3. The loss of the suboptimal solution is completely acceptable for the specific class of objects [13].
4. The adoption of CBR as the main approach for suboptimal management of TTP of timber allows the required amount of computer calculations to be performed once in off-line mode, and in real operation to perform only R^4 procedure, which is standard for CBR [9, 15].

D. *Existing Results*

The direct use of model predictive control (MPC) or CBR-based approach is impossible because one of the key characteristics of wood—its humidity w cannot be measured directly, and in addition it can vary significantly for individual elements of each batch of loaded materials.

In our previous works [7, 12, 13] a two-stage procedure was proposed, in which the moisture of the wood is first evaluated, and then the CBR-based suboptimal control for the criteria J_1 and J_2 is realized. A simplified diagram of the proposed procedure is shown in Fig. 5 and is performed in the following sequence:

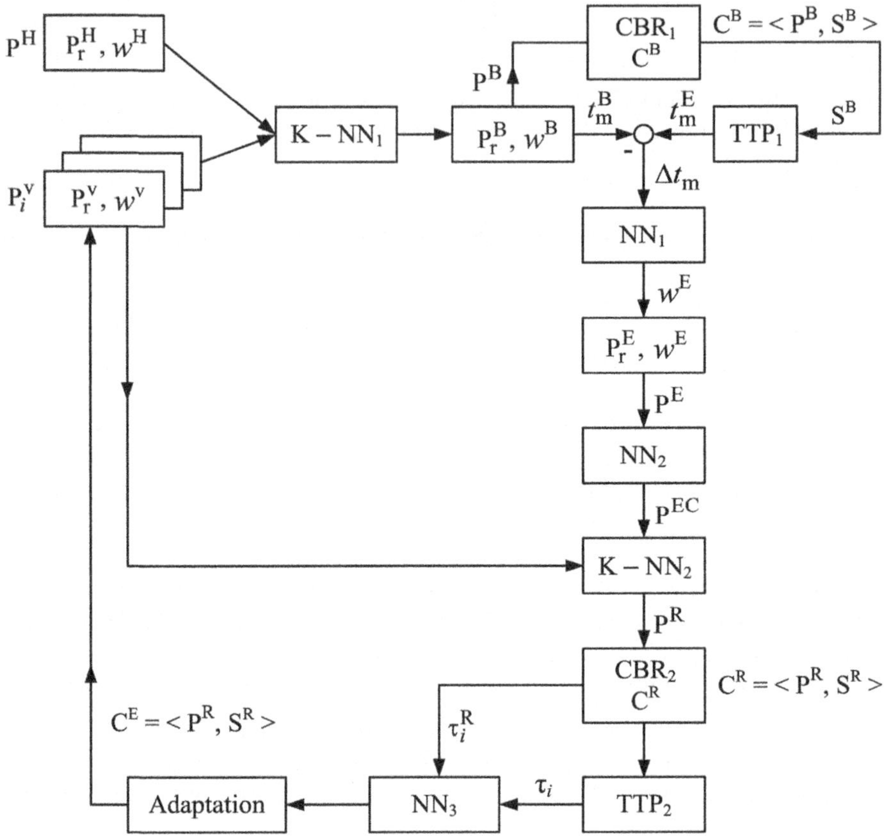

Fig. 5 Scheme of the procedure for suboptimal control of TTP in an autoclave according to criteria J_1 and J_2

1. Part of the characteristics of the materials loaded for thermal treatment is measured: $\pi^E, a^E, \gamma^E, t_w^{0E}$.
2. An expert assessment of the humidity w^H and the temperature of the heating medium in the autoclave t_m^H is added. The parametric vector is obtained

$$P^H = \left(\pi^H, a^H, \gamma^H, w^H, t_m^H, t_w^{OH}\right). \tag{22}$$

3. With the procedures (K–NN)₁ and CBR₁ is the closest situation in the virtual database CB^v:

$$P^B = \left(\pi^B, a^B, \gamma^B, w^B, t_m^B, t_w^{OB}\right). \tag{23}$$

4. The actual thermal treatment process (TTP) starts. The experimental values of the heating medium temperature $t_m^E(\tau)$ are measured.

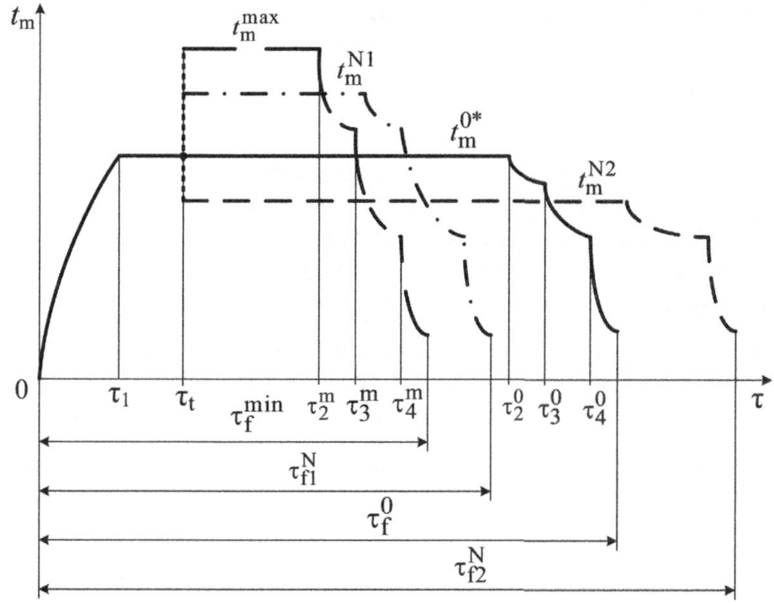

Fig. 6 Scheme of situations with reduction and increase of the autoclave productivity

5. Based on the differences $\Delta t_m(\tau) = t_m^B(\tau) - t_m^E(\tau)$ the average calculated humidity of the wood w^E is estimated using the NN_1 neural network. The vector with the real features for the *i*th charge is formed:

$$P^E = \left(\pi^E, a^E, \gamma^E, w^E, t_w^{0E}\right). \tag{24}$$

6. The P^E vector is corrected with the NN_2 neural network.
7. With the procedure $(K\text{–}NN)_2$ a new closest situation P^v is found in the CB^v precedent database.
8. Using CBR_2, the object-oriented part (τ_2, τ_3, τ_4, τ_f) of $C^R = (P^R, S^R)$ is determined, which is realized in the autoclave.
9. A new precedent is obtained, which is recorded in the CB precedent database.

The described procedure implements suboptimal control of the autoclave with respect to the optimality criteria J_1 and J_2, where $J_1 \to \tau_f^{min}$ (maximal productivity) and $J_2 \to \tau_f^R$ at $Q_a \to \min$ (minimal heat consumption at a given productivity τ_f^R).

If there are no dispatching (d_1) or other operational actions (d_2, d_3) (see Fig. 3) the TTP in the autoclave ends according to the controls S_1^R or S_2^R depending on the accepted criterion.

4 Features of the Proposed Method

A. *Operational Modes*

The aim of this research is to find the optimal continuation of the process of thermal treatment of wood materials in autoclaves, when due to a dispatching intervention d_1 to change the schedule of an autoclave, a damage in the autoclave d_2 or in the next line for machining d_3 (see Fig. 3) at the moment of interruption τ_t (Fig. 6) a change in the final duration of the process is required: $\tau_f^N \neq \tau_f^0$.

If the requirement $\tau_{f1}^N < \tau_f^{min}$ is set then it is necessary to do the following:

(a) To increase the steam temperature t_m^N in the autoclave so that $t_m^N > t_m^0$;
(b) To check that at the set value of τ_1 and the maximally allowable steam temperature in the autoclave t_m^{max} the desired process duration τ_{f1}^N is achievable.

Figure 7 shows the area of inaccessibility of TTP of wood materials in relative coordinates α, β, where

$$\alpha = \frac{\tau_f^N - \tau_f^{min}}{\tau_f^{min}}, \tag{25}$$

$$\beta = \frac{\tau_{BP} - \tau_1^0}{\tau_f^{min} - \tau_1^0}. \tag{26}$$

In (25) and (26) with τ_1^0 is indicated the time for reaching the set suboptimal trajectory of increase of t_m at the beginning of TTP in the autoclave; τ_f^{min} is the minimal possible time for TTP at the maximally allowable temperature of the heating

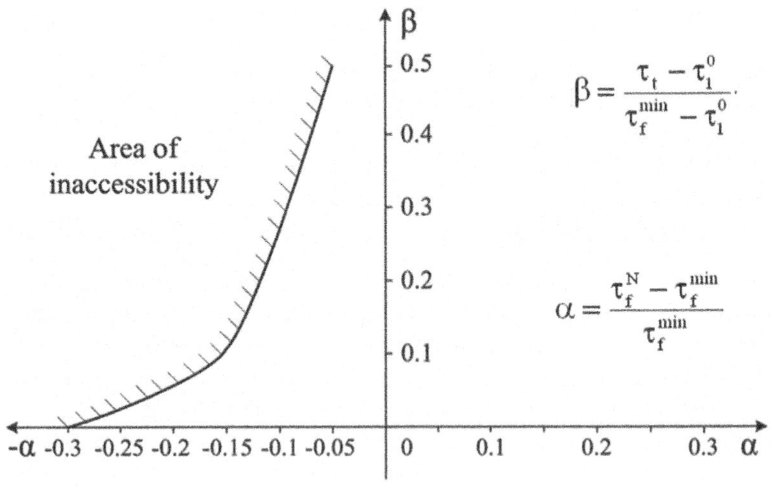

Fig. 7 Area of inaccessibility of TTP in an autoclave

medium in the autoclave $t_{\mathrm{m}}^{\max}$ and at parameters according to the nearest virtual situation P^{B} for which there is a maximal similarity with the current values of the parameters of the loaded batch P^{E}.

As can be seen in Figs. 6 and 7, when switching is requested in order to dramatically increase productivity ($\tau_{\mathrm{f1}}^{\mathrm{N}} < \tau_{\mathrm{f}}^{\min}$), this is far from always achievable. Therefore, the reachability limit must be determined by solving the equation

$$F[\alpha(\tau_{\mathrm{f}}^{\mathrm{N}}, \mathrm{P}^{\mathrm{B}}, t_{\mathrm{m}}^{\max}), \beta(\tau_{\mathrm{BP}}, \mathrm{P}^{\mathrm{B}}, t_{\mathrm{m}}^{\max})] = 0. \tag{27}$$

This limit should be used by dispatchers or the autonomous system for optimal operational management [7] to form achievable goals.

2. A situation in which $\tau_{\mathrm{f2}}^{\mathrm{N}} > \tau_{\mathrm{f}}^{\min}$, i.e. reducing productivity at the request of dispatchers. Such an increase in the duration of the TTP is always possible, as criterion J_3 includes the requirement for minimal heat consumption.

The general scheme of operating modes of the autoclave is shown in Fig. 8.

B. ***Management and Control***

As shown in Figs. 4 and 8 the thermal treatment process (TTP) of the wood materials is characterized by the following features:

1. It is periodic with duration $\tau_{\mathrm{f}}^{\mathrm{R}}$ and consists of successive batches, which have various loading conditions (wood type π, representative size a, degree of filling of the autoclave with materials γ, humidity w and initial temperature of the materials t_{w}^{0}).
2. Due to the lack of possibility to measure the temperature field $t_{\mathrm{w}}\ (x,y,z,\tau)$ in the wood materials, the process is controlled as an open system with a predetermined trajectory for each load.
3. The temperature of the heating medium in the autoclave $t_{\mathrm{m}}(\tau)$ is measured and used as feedback in a local stabilizing system containing a standard PLC controller.
4. Control impacts are of two types:

   (a) Continuous, providing change of the primary heat carrier (steam consumption F_{s}) for temperature regulation $t_{\mathrm{m}}(\tau)$ in the autoclave;
   (b) Discrete, providing switching of the steaming regime at the moments τ_2, τ_3, and τ_4 (Fig. 1).

5. Indirect data on the quality of TTP (electricity consumption for subsequent mechanical treatment of the heat-treated materials, lack of cracks in them, change in their natural color, etc.) are used to adjust the tasks of the open system in the mode "batch to batch".

TTP management is carried out in the following two (at J_1 and J_2) or three (at J_3) stages:

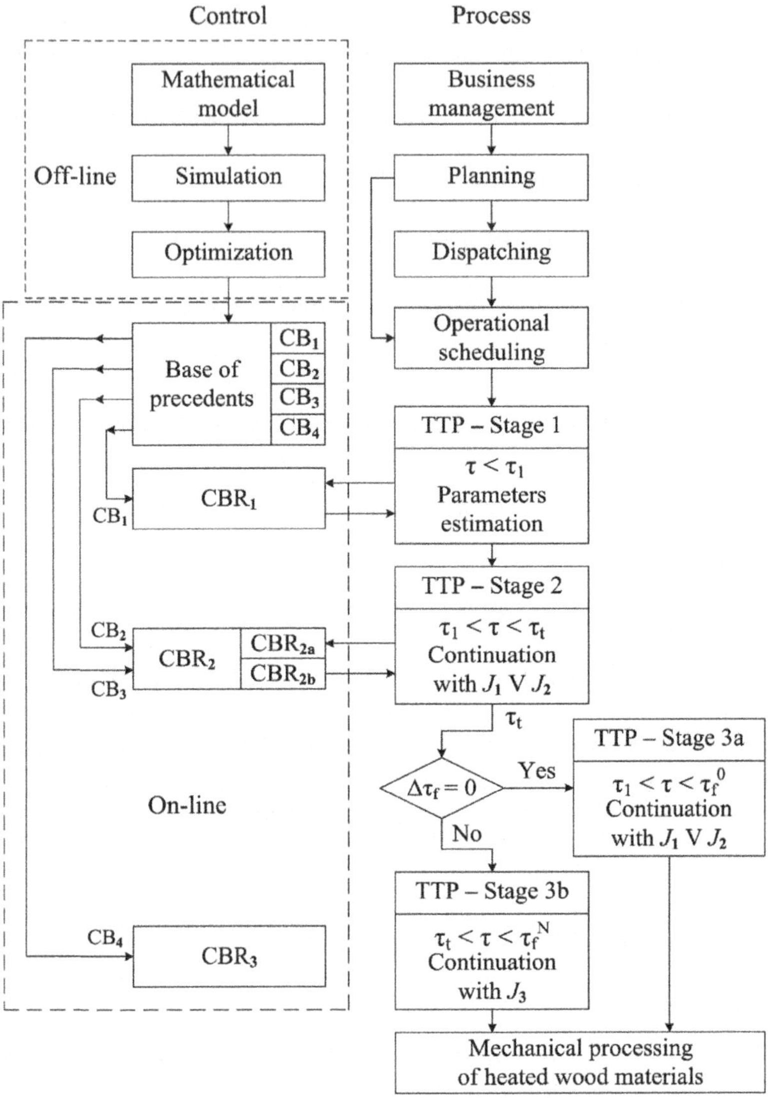

Fig. 8 Scheme of the operating modes of an autoclave for heat treatment of wood

1. The first stage consists of initial intensive heating and evaluation of the parameters of the mathematical model for the specific jth batch. It includes:

   - Measurement of the main characteristics of the loading—wood species π, representative size a, degree of filling of the autoclave γ, initial temperature of the loaded materials $t_w(0)$;

- Expert setting of the initial humidity w^{H} and the initial temperature of the processing medium t_{m}^{0};
- From the base of precedents CB_1 through the procedure CBR_1 [9, 15] the closest precedent C^{B0} is found;
- The differences between the real temperature of the heating medium in the autoclave $t_{\mathrm{m}}^{\mathrm{E}}(\tau)$ and the calculated operating temperature $t_{\mathrm{m}}^{\mathrm{B}}(\tau)$ of CB_1 determine the estimate of the actual humidity of the wood materials w^{E} and respectively the real vector of the loading parameters $\mathrm{P}^{\mathrm{E}} = \left(\pi^{\mathrm{E}}, a^{\mathrm{E}}, \gamma^{\mathrm{E}}, w^{\mathrm{E}}, t_{\mathrm{w}}^{\mathrm{E}}(0)\right)$.

2. The second stage is realized in two varieties depending on the criterion chosen by the dispatchers $J_1 \rightarrow (\tau_{\mathrm{f}}^{\min})$ or $J_2 \rightarrow \tau_{\mathrm{f}}^{\mathrm{R}}$ at $Q_{\mathrm{a}} \rightarrow \min$.
3. The third stage occurs in the case of an operational disturbance at time τ_{t} with a requirement to change the time to complete the TTP of a batch of loaded materials $\tau_{\mathrm{f}}^{\mathrm{N}}$ (Fig. 6).

The suboptimal control involves the preliminary preparation of four precedent bases for each of the situations listed above, namely:

(a) For the first stage, the dimension of the search space is $n_1 = 6$: $(\pi, a, \gamma, w, t_{\mathrm{w}}^{0}, t_{\mathrm{m}}^{0})$;
(b) The second stage is in two varieties:

  - suboptimal control at criterion maximal productivity $J_1 \rightarrow \tau_{\mathrm{f}}^{\min}$ at $t_{\mathrm{m}}^{\max}$ with dimension of search space $n_2 = 5$: $(\pi, a, \gamma, w, t_{\mathrm{w}}^{0})$;
  - suboptimal control for criterion J_2—set productivity ($\tau_{\mathrm{f}} = \tau_{\mathrm{f}}^{0}$) at minimal steam consumption ($Q_{\mathrm{a}} \rightarrow \min$). The dimension of the search space is the same $n_3 = 5$: $(\pi, a, \gamma, w, t_{\mathrm{w}}^{0})$;

(c) Third stage—the criterion is J_3, and the search space has a dimension $n_4 = 7$: $(\pi, a, \gamma, w, t_{\mathrm{w}}^{0}, \alpha, \beta)$.

C. ***Optimization***

Schematically, the procedure for searching for control effects for the third stage, on which the present research is focused, is shown in Fig. 9. It contains an address part (π, a, γ, w), the initial temperature of the heating medium t_{m}^{0} as part of the temperature profile corresponding (according to criteria J_1 or J_2) to the initial schedule and an optimization part in which, depending on the operating parameters α (25) and β (26), the new control impacts $t_{\mathrm{m}}^{\mathrm{N}}$, τ_2^{N}, τ_3^{N},τ_4^{N} are determined at a set value of $\tau_{\mathrm{f}}^{\mathrm{N}}$.

As can be seen from Fig. 8, the implementation of the suboptimal control system [1] of TTP of wood materials in an autoclave requires pre-generation of four different precedent bases $\mathrm{CB}_1 \div \mathrm{CB}_4$, each of which has a tree structure with 6, 5, 5 and 7 layers respectively.

The density of available precedents $\mathrm{C}_i(\mathrm{P})$ in the search spaces is different depending on the private derivatives:

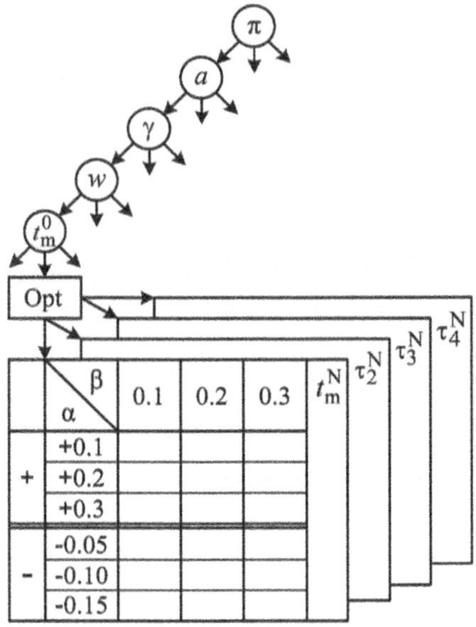

Fig. 9 Scheme of the procedure for searching management influences for the third stage

$$g_{ij} = \frac{\partial J_i}{\partial P_j}, \tag{28}$$

which for each wood species π_k are estimated in [12] in the following hierarchical order: t_m, a, γ, w, t_w^0, α, β. The CBR_1 procedure is identification—to determine the assessment of the actual moisture content of the wood w^E in the loaded batch of timber. Therefore, the solution S_1 contains only the address part of the nearest point $C^v = C^B (P^B)$.

The other procedures CBR_{2a}, CBR_{2b} и CBR_3 contain address S_c, temperature S_t S_t and temporal S_τ parts:

$$S = < S_c, S_t, S_\tau >, \tag{29}$$

where

$$S_c = \pi^v, a^v, \gamma^v, w^v, t_w^{0v},$$

$$S_t = t_m^*, \tag{30}$$

$$S_\tau = \tau_2^*, \tau_3^*, \tau_4^*, \tau_f^*.$$

The procedure for K–N-nearest neighbors (K–NN), which is set by the problematic part of the P^E precedent, applies only to the address part so that

$$\arg\min_{S_c}\left|P^E - S_c\right| = S_c^v. \tag{31}$$

Since in the solution of the virtual spaces proposed by CBR_i^v the temperature S_t and the temporal S_τ components of the solution are suboptimal, the overall closest proposal for the solution S^v is also suboptimal.

As noted above, to the error of the suboptimal virtual space of situations C^v is added the error of mismatch of the real and the nearest neighbor according to the (K–NN) procedure (31).

Figure 10 shows the generalized functional diagram of the proposed system for suboptimal control of TTP of timber in an autoclave.

D. ***Advantages of the Proposed System for Intelligent Suboptimal TTP Management and Control***

For the considered class of objects for thermal treatment of wood materials the application of a strictly optimal control system is complex, expensive, unreliable and irrational from a business point of view.

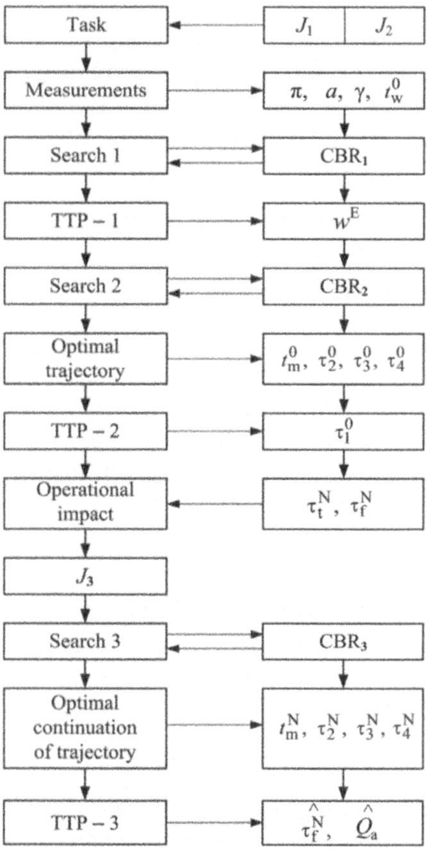

Fig. 10 Scheme of the system for suboptimal control of TTP of wood materials in an autoclave

The proposed suboptimal control system is an upgrade of the existing model-based system. It has the following advantages:

1. The system is adaptive and close to the optimal for each batch of heat-treated wood materials, due to the inclusion of the initial stage of evaluation of the loading parameters $(\pi, a, \gamma, w, t_{\mathrm{w}}^{0})$.
2. The system has a small degree of suboptimality due to the specific features of the object, mainly due to the exponential reduction of heat consumption [10–12, 16] after reaching the time τ_1 in TTP (see Fig. 4).
3. While maintaining the model-based approach, complex computer calculations for an object with distributed parameters, described by partial differential equations (PDEs), are exported in off-line mode, and on-line procedures are extremely simplified and reduced.
4. The development in off-line mode of virtual spaces of possible operational states is carried out once and for all for the materials of the covered type of wood species.
5. The system has simplified functionality and is cost-effective in terms of benefit/price, as it:

   - reduces energy costs while maintaining the quality of heat treatment compared to the requirements and restrictions;
   - has simplified additional hardware and software;
   - is easily installed as an upgrade of an existing system;
   - does not require special training of the operating personnel.

6. The system covers all operating modes and guarantees flexibility and robust-ness in automatic mode.

The required additional investments to the system are as follows:

- means for video-based measurement of the loading conditions of the wood materials (video camera, image processing software);
- simplified thermal imaging system for measuring the initial temperature of the wood materials;
- elaborating with an available mathematical model of simulation-based four bases with precedents.

5 Simulation Results

A small part of the obtained simulation results is shown below as an illustration of the possibilities of the proposed suboptimal control system in several aspects.

A. ***Modes Without Dispatch Interference During a Given Charge***

Figure 11 shows the temperature profiles of the heating medium t_{m}, as well as the temperatures at four representative points of a beech prism with a cross section of

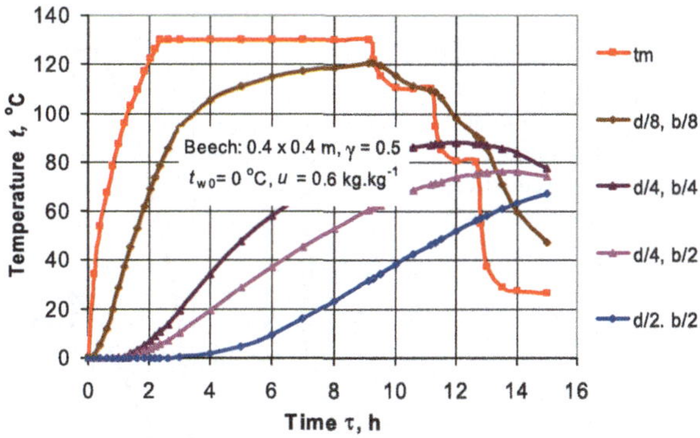

Fig. 11 Change of t_m and t at 4 points of beech prism with cross section 0.4 × 0.4 m during TTP in the autoclave

0.4 × 0.4 m, humidity $w = 0.6\ \text{kg} \cdot \text{kg}^{-1}$ and initial temperature $t_{\text{w}}^{0} = 0\ °\text{C}$ at a PLC heating temperature $t_{\text{m}}^{0} = 130\ °\text{C}$ during its steaming in an autoclave with an inner diameter $D = 2.4$ m, length $L = 9.0$ m and degree of filling with prisms $\gamma = 0.5$.

The analysis of Fig. 11 shows that the use of a simplified model with concentrated parameters of the TTP is impossible, and the application of model predictive control with calculation of the temperature $t_{\text{m}}(\tau)$ for each time step would take unacceptably long machine time without providing any significant advantage over the suboptimal control.

B. ***Dispatch Interference Regimes***

Figures 12 and 13 show the change in the temperature of the heating medium in the autoclave $t_{\text{m}}(\tau)$ during the TTP of the same beech prisms in the presence of dispatching disturbances with reduction of t_{m} from 130 to 110 °C (Fig. 12) and from 130 to 100 °C (Fig. 13) at times $\tau_{\text{t}} = 3$, 5 and 7 h from the beginning of the process. The same figures show the change in the specific (for 1 m^3 of wood) heat consumption of the autoclave Q_{a} in the different explored modes.

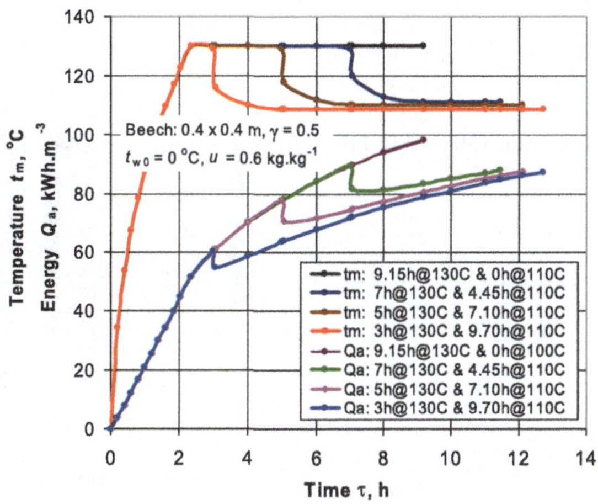

Fig. 12 Change of t_{m} and Q_{a} during TTP of beech prisms with cross section 0.4 × 0.4 m at 3 types of dispatching influences for reduction of t_{m} from 130 to 110 °C

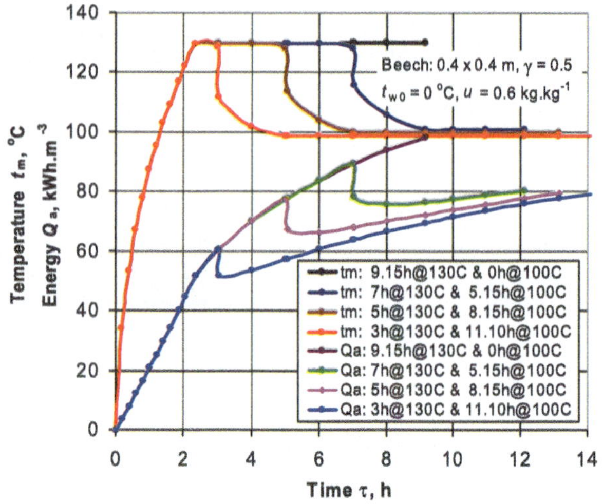

Fig. 13 Change of t_m and Q_a during TTP of beech prisms with cross section 0.4 × 0.4 m at 3 types of dispatching influences for reduction of t_m from 130 to 100 °C

Figures 12 and 13 show that lowering the t_m from $t_m^{max} = 130$ to 110 or 100 °C in the range from τ_1 to τ_2 (refer to Figs. 4 and 6) causes an increase in the time τ_2, during which steam is introduced into the autoclave, as well as a decrease in the specific energy consumption of the autoclave Q_a. When lowering t_m from 130 to 110 °C the time τ_2 increases in the range up to 39% compared to τ_2 of the steaming regime with $t_m = 130$ °C = const and when lowering t_m from 130 to 100 °C the increase of τ_2 is in the range up to 45%. The reduction Q_a in the first case is about 12%, and in the second case it is about 19%.

Figure 14 shows the change of Q_a (in kWh · m^{-3}) in the 9 types of dispatching disturbances, causing a decrease in t_m from 130 to 100, 110 and 120 °C, occurring at the moments $t_m = 3$ h, 5 h and 7 h from the beginning of the TTP.

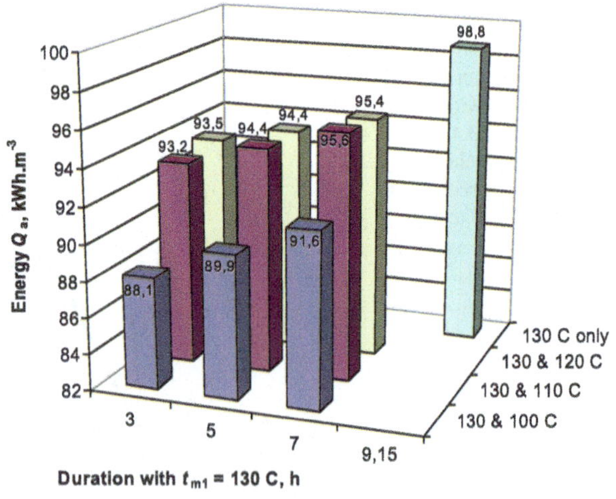

Fig. 14 Change of Q_a in TTP of beech prisms with cross section 0.4 × 0.4 m at 9 types of dispatching influences for reduction of t_m from 130 to 100, 110 and 120 °C.

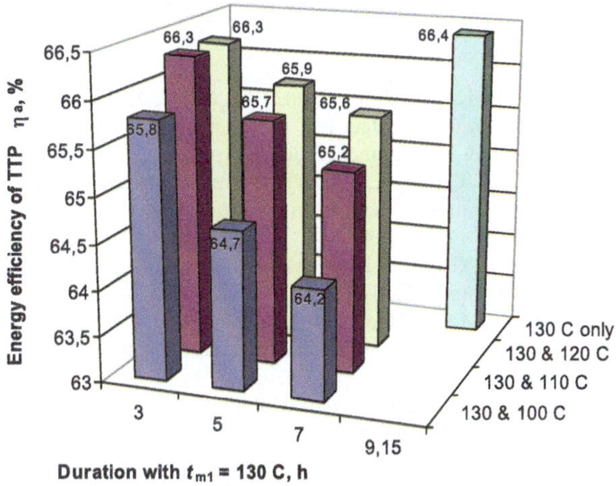

Fig. 15 Change of TTP efficiency of beech prisms with cross section 0.4 × 0.4 m at 9 types of dispatching influences for reduction of t_m from 130 to 100, 110 and 120 °C

Figure 15 shows the change in the energy efficiency η (in%) of TTP in operational disturbances of the type discussed above.

It can be seen that the dispatching effects can cause a reduction of the energy efficiency up to 2.2%, but their main purpose is to ensure timely supply of heat-treated materials for subsequent mechanical processing (e.g. veneer cutting), despite some deterioration of the energy efficiency of the process.

6 Conclusions

Solutions in which there is a rational compromise between acquired business advantages and additional costs are of interest to industry.

The proposed control system ensures a reasonable balance between complexity and simplification, optimality and losses from suboptimality, new methods and existing knowledge, portability and uniqueness, intelligent and traditional techniques, concrete solution and more general applicability.

The main result of the research is to transform the model-based management approach to modern procedures based on data flows using CBR. Developments of a number of techniques in the field of artificial intelligence are proposed—formation of virtual search spaces with sufficient density compared to expected operational situations, attracting effective search methods in the space of possible solutions, reducing complexity by extracting and reducing the main features determining the process.

Exportation of complex calculations from on-line into off-line regime has been achieved. The efficiency of using different forms of the existing knowledge about the object is shown—models, rules, restrictions, preferences.

References

1. Choi, T.-M., Gao, J., Lambert, J.H., Ng, C.-K., Wang, J. (eds.) Optimization and control for systems in the big-data era: theory and applications. In: International Series in Operations Research & Management Science. Springer (2017)
2. Fei, H., Li, Q., Sun, D.: A survey of recent research on optimization models and algorithms for operations management from the process view. Sci Program **2017**, Article ID 7219656, 1–19 (2017)
3. McComic, C.: Optimization technology and AI in the real world: operational efficiency stories. www.ibmbigdatahub.com
4. Hadjiski, M.: Problems of machine learning. J. Bulgarian Acad. Sci. (1) (2020) (in Bulgarian)
5. Rao, A.: A survey of numerical methods for optimal control. Adv. Astronaut. Sci. **135**, 497–528 (2009)
6. Sanders, D., Gegov, A.: Artificial intelligence for control engineering. Paper Based J. Control Eng. (2), 38–40 (2015)
7. Hadjiski, M., Deliiski, N.: Cost oriented suboptimal control of the thermal treatment of wood materials. In: Proceedings of the 16th IFAC conference on technology, culture, and international stability, Sozopol, Bugaria, PapersOnLine, pp. 48−24, 54−59 (2015). https://doi.org/10.1016/j.ifacol.2015.12.056
8. Hadjiski, M., Deliiski, N.: Advanced control of the wood thermal treatment processing. Cybern. Inform. Technol. **16**(2), 179–197 (2016)
9. Muller, G., Bergmann, R.: Generalization of workflows in process-oriented case-based reasoning. In: Association for the Advancement of Artificial Intelligence (AAAI), Proceedings of 28th International Florida Artificial Intelligence Research Society Conference, pp. 391–396 (2015)
10. Deliiski, N., Dzurenda, L.: Modeling of the thermal treatment in the technologies for wood processing. University of Forestry, Sofia, 299 p. (2010) (in Bulgarian)
11. Deliiski, N.,:Modeling of the energy needed for heating of capillary porous bodies in Frozen and Non-frozen states. Lambert Academic Publishing, Scholars' Press, Saarbrücken, 116 p., Germany (2013). http://www.scholars-ress.com//system/covergenerator/build/1060
12. Hadjiski, M., Deliiski, N., Tumbarkova, N.: Intelligent hybrid control of thermal treatment processes of wood. In: Proceedings of the International Conference on Intelligent Systems IS 2020, Varna, pp. 482–489 (2020)
13. Hadjiski, M., Deliiski, N.: Advanced process control of distributed parameter plants by integration first principle modeling and case-based reasoning. Part 2: Case-Based Reasoning Control of DPP. In: Proceedings of the International Conference of Automatics and Informatics "D. Atanassov", 1–3 October 2020, Varna, 6 p. (2020)
14. Deliiski, N.: Computation of the 2-dimensional temperature distribution and heat energy consumption of Frozen and Non-frozen logs. Wood Res. **54**(3), 67–78 (2009)
15. Aamod, A., Plaza, E.: Case-Based Reasoning: Fundamental Issues, Methodological Variations and System Approaches, vol. 1, pp. 39–59. AI Communications IOS Press (1994)
16. Deliiski, N., Dzurenda, L., Angelski, D., Tumbarkova, N.: Computing the energy for warming up of prisms for veneer production during autoclave steaming with a limited power of the heat generator. Acta Facultatis Xilologiae Zvolen **61**(1), 63–74 (2019)
17. Deliiski, N.: Modeling and technologies for steaming wood materials in autoclaves. Dissertation for D.Sc., University of Forestry, Sofia, 358 p. (2003) (in Bulgarian)
18. Deliiski, N.: Transient heat conduction in capillary porous bodies. In: Ahsan, A. (eds.) Convection and Conduction Heat Transfer, pp. 149–176. InTech Publishing House, Rieka (2011). https://doi.org/10.5772/21424

Deep Learning Based Multimodal Information Fusion for Near-Miss Event Detection in Intelligent Traffic Monitoring Systems

Nikolaj Apostolovski, Naum Trajanovski, Marko Chavdar, Tomislav Kartalov, Branislav Gerazov, and Zoran Ivanovski

Abstract Detection and marking of road accident "black spots" are of paramount importance for road safety. Their detection and localization are based on accident statistics for certain road segments. Following established safety standards today, this approach of "waiting for accidents to happen" in order to prevent future events is unacceptable. Behind any accident statistics there is an even larger near-miss statistics that could be exploited for black spots detection, before a significant accident statistics increase. Detecting near-miss events is difficult due to their vague definition and even more vague manifestation. Therefore, in this paper we propose the detection of near-miss events based on the simultaneous occurrence of rapid deceleration and skidding of vehicles. Rapid deceleration of vehicles could easily be detected in video; however, the occurrence of rapid deceleration alone does not always imply a near-miss event. Detection of skidding, on the other hand, is very challenging in video due to the occlusion of wheels. The occurrence of skidding has a distinctive audio signature; nevertheless, the spatial location of the audio source is very difficult to extract from an audio-only data stream. Thus a novel development of a multi-modal algorithm for near-miss detection based on audio and video information fusion is proposed here. The information extraction in both domains, audio and video, is performed using deep convolutional neural networks (CNN) combined with different pre- and post-processing techniques. Deceleration of vehicles is estimated using video from calibrated surveillance cameras. Vehicle positions are estimated

N. Apostolovski (✉) · N. Trajanovski · M. Chavdar
iTek Systems, Skopje 1000, North Macedonia
e-mail: apostolovski.nikolaj@outlook.com

T. Kartalov · B. Gerazov · Z. Ivanovski
Faculty of Electrical Engineering and Information Technology, University Ss Cyril and Methodius in Skopje, Skopje 1000, North Macedonia
e-mail: kartalov@feit.ukim.edu.mk

B. Gerazov
e-mail: gerazov@feit.ukim.edu.mk

Z. Ivanovski
e-mail: mars@feit.ukim.edu.mk

P. Shi et al. (eds.), *Complex Systems: Spanning Control and Computational Cybernetics: Applications*, Studies in Systems, Decision and Control 415,
https://doi.org/10.1007/978-3-031-00978-5_15

using a CNN-based estimator with the output corrected via content matching techniques and a predictor Kalman filter. Audio events are detected using CNNs applied on Mel-frequency cepstral coefficients. In the final information fusion step, a SVM classifier is used for decision on the occurrence of potentially dangerous near-miss event.

Keywords Vehicle detection · Vehicle tracking · Motion parameters estimation · Sound event detection · Mel-Frequency Cepstral coefficients · Convolutional neural networks · Kalman filtering · Support vector machines

1 Introduction

Vehicle detection in image and video has been in the focus of attention of the scientific community for a long time. Its application and usefulness have been recognized by the public and the industry, and the expectation from systems for vehicle detection and tracking are set high.

Many different approaches to vehicle detection problem has been developed in the last few decades. The real performance boost comes with the development of machine learning based algorithms, and particularly in the last decade, with the development of Deep Neural Networks (DNN). Usually, this class of algorithms are developed for general object recognition, and almost unavoidably, applied to vehicle detection. Finally, the attention turned to Convolutional Neural Networks (CNNs) as toll of choice for most of the signal processing problems. A large attention boost came with the ImageNet Large Scale Visual Recognition Challenge (ILSVRC) series, in which vehicles were part of the required 1000 recognition classes. The AlexNet, [1], is considered as milestone of the development of the CNN for object recognition. It definitely put the CNNs in the focus of interest by exceeding the previous state of the art algorithms with the error rate margin of over 10%. A real standard in CNNs architecture was set with the VGG, [2], achieving half the error rate of AlexNet. The authors showed the importance of the depth of the CNN and offered rules for structure building that are used in many contemporary designs. The authors of GoogLeNet [3] offered a more appropriate solution to the problem of objects with different scales and sizes, by building a structure consisting of stem based on convolutional layers and followed by inception modules. The architecture enables adaptation of the support of filter kernels to the size of the objects. In the attempt to lower the computational requirements of the architecture, the authors argue that Fully Connected Network (FCN) as a classifier is unnecessary complex and powerful for simple object recognition, and therefore, they replace it with averaging layer. The layered structure of the network is not very optimal for different levels of abstraction. Different concepts could be learned by networks with different depth, therefore the ResNet, [4], enables different depth by introducing skip connections. The layers are packed in residual modules and skip connections can be of any length. The ResNet achieved error rate of 3.6%, which is close to human performance.

All of the above pointed networks were designed for object recognition and applied on different classes of object, among which also vehicles. They only perform recognition based on image of an object. For efficient object localisation in global image, it is necessary to extract possible object positions and then decide which extracted block contains an object. This approach is used in the design of Fast R-CNN, [5], and Faster R-CNN, [6], networks. Fast-RCNN uses fully convolutional network that is applied to an input image and multiple regions of interest (RoIs). Each RoI is a fixed-size feature map that is converted into a feature vector using FCN. The feature vector is used to generate two output vectors for every RoI: class probabilities and per-class bounding-box offsets. This concept is further developed in Faster R-CNN by introducing Region Proposal Network (RPN) that shares convolutional layers with detection network, avoiding the costly region proposal computations. Applied on vehicle detection problem Faster R-CNN achieves processing rate at the level of 10 fps. We use similar concept, as explained below.

The other approach to fast object recognition and localization is avoiding the concept of region proposal and unifies all the components of object detection in a single network. Following this approach, YOLO, [7], achieves high speed of detection by slightly sacrificing the accuracy. Tested on vehicle detection problem, YOLO performs fast enough to be applied in real time, e.g., 45 fps, however, it is apparently less accurate than Faster R-CNN. The newer version of YOLO, YOLOv2, [8], offers a trade-off between speed and accuracy by removing the fully connected layers and uses an anchor box. This is the fastest architecture of all mentioned in this paper and when applied to vehicle detection problem processes roughly 65 fps. The real leap in the accuracy comes with YOLOv3, [9]. It uses residual network and combines multi-scale prediction network to achieve significantly higher accuracy, sacrificing speed in comparison with YOLOv2.

The SSD network, [10], further improves the structure allowing features from multiple feature layers to be used in the estimation of output scales, aspect ratios and confidence of the bounding boxes. This architecture performs relatively well both in speed and accuracy; however, when used for vehicle detection it's prone to errors with small vehicles and produces multiple scale bounding boxes for one vehicle. An extension of the SSD network is proposed in [11], denoted as DP-SSD. It proposes the use of different feature extractors for localization and classification tasks in a single network, and enhances the two feature extractors through deconvolution and pooling between layers in the feature pyramid. It shows comparable or better performance on UA-DETRAC, [14], dataset than aforementioned architectures, except for YOLOv3.

Another approach based on three networks called Deep Convolutional Network, Proposal Network, and Fine-Tuning Network, was proposed in [12]. The architecture outperforms Faster R-CNN, however, it is less accurate compared to some of the later architectures mentioned above.

After the initial vehicle detection, the next step is vehicle tracking. Multi-object tracking (MOT) is one of the most important and most essential tasks in video processing. Many studies and applications that improve and implement MOT in video processing using different algorithms have been done. For example, OpenCV has several types of MOT using different algorithms, such as MIL, BOOSTING,

TLD, KCF, etc., [17]. Choosing the right MOT algorithm can be crucial in dealing with the biggest problems with object tracking, such as occlusion between objects and object confusion.

Since the ITS systems saw a rise in popularity, there are a lot of studies proposing different approach in MOT, each of them bringing their own advantages and shortages. The authors in [18] proposed a histogram-based tracking algorithm by finding the minimum Chi-squared distance in the histogram domain among a group of candidates. The Mean-Shift algorithm to track every blob in the subsequent frame is used in [19]. In [20], the authors propose fusion of visual and semantic features for both single-camera tracking (SCT) and inter-camera tracking (ICT). They introduce a histogram-based adaptive appearance model to learn long-term history of visual features for each vehicle target and incorporate semantic features into a bottom-up clustering strategy for data association in each single camera view. Bayesian inference methods are proposed in [21]. Automatically generated 3D vehicle models are used in [22, 23].

However, the most popular tool in MOT has to be the Kalman Filter [24]. Kalman filter, as an optimal estimator, filters out the noise from noisy data and finds the best estimate. It is used for tracking multiple objects, identity resolution and to manage occlusion scenarios. Studies that show the usefulness of the Kalman filter are [25, 26]. The study in [27] use a combination of the Kalman Filter and the Hungarian algorithm [28] to resolve occlusions. In [29], the Kalman Filter is used with compressive tracking, thus turning the high dimensional signals into image features in low dimensional space for processing by random projection.

The next problem that needs to be solved after the tracking is the vehicle speed estimation. Camera calibration is required so that the speed obtained from the pixel domain can be converted to the real world. Different methods have been used, both automatic and manual. As examples of the automatic methods, Cathey and Dailey [30] used detections of the vanishing point in the direction of vehicle movement. To obtain this vanishing point, detected line markings are used and their intersection in the least squares manner. Dubska et al. [31] detected two vanishing points based on vehicle movement. Schoepflin et al. [32] use an activity map (by detecting the vehicles as the moving foreground) to obtain lane boundaries, thus obtaining the first vanishing point. The second vanishing point is detected from the intersection of lines formed by the bottom edges of vehicles.

The manual methods use known real-time metrics, such as Maduro et al. [33], who assume two known arbitrary angles on the ground plane to calibrate the camera and use lengths of line markings' stripes to obtain the camera scale for the given scene. Another example is Nurhadiyatna et al. [34] who use a calibrated pinhole camera with zero pan and known distances in the real world. In this paper, the Direct Linear Transformation (DLT), [35], is used, by assuming a stationary camera and knowing the precise distances between 4 or more points in the real-world coordinates.

The results obtained with our algorithm are comparable with some of the algorithms discussed in the aforementioned papers. A significant number of methods were also implemented during our work in order to find the most suitable sequence of methods for our needs. It is debatable that some of the methods discussed show greater accuracy and lower susceptibility to error. However, the most important goal set at the start of our work was the real-time implementation of our system, therefore certain trade-offs were made to gain lower computational complexity, and thus higher estimation speed.

A sound event is a segment of audio which can be characterized and identified by a textual label [36]. Sound events usually overlap in real-life recordings and show wide variations in their frequency content and temporal structure, all of which makes the task of recognizing each separate event difficult. The field of automatic sound event detection (SED) is focused on the recognition of sound events from continuous acoustic signals, both monophonic and polyphonic—the former being concerned with the most discernible sound event at a given time instance and the latter with recognizing multiple simultaneous sound events.

Applications of sound event detection include audio surveillance systems and context-based indexing and retrieval in multimedia databases [36]. SED can be posed as a scene-dependent or scene-independent problem. In the scene-dependent approach, the system is made aware of the acoustic scene at training and test time, and thus different models can be trained for different scenes. In the scene-independent approach, the system lacks any information about the acoustic scene [37].

Sound event detection is a developing research field that has traditionally benefited from advances in more mature research areas, such as automatic speech recognition (ASR) [38]. Deep neural networks have recently achieved remarkable success in speech recognition [39]; specifically, CNNs have outperformed dense neural networks in this task [40, 41]. In line with these results, we use CNNs for monophonic, scene-dependent sound event detection in simulated traffic acoustic signals. We base our neural network architecture on our previous work on traffic sound event detection [42].

The SVM classifiers were widely used in different classification problems ever since their introduction, [43]. They have beenappliedsuccessfully in vehicle detection problems as well, [44]. The riseof neural networks did not diminish their importance. Moreover, in vehicle detection systems the combination of neural networks and SVM proved to be considerably beneficial [46]. In this work we utilize SVM based classifier for near-miss event detection applied on combined information extracted from video and audio streams.

2 Video Analysis

The video analysis part of the system is designed to detect the traffic participants through an object detection procedure and to estimate their motion parameters. Part of the estimated parameters are used as input parameters in the skidding detection system presented here.

2.1 Vehicle Detection

The object detection procedure consists of machine learning-based techniques that are relatively well-understood and are often used in similar problems, appropriately adapted and/or redesigned to the problem at hand. The motivation behind this approach is in the required high performance of the system. Namely, the system should achieve very high accuracy in traffic participants classification and localization following from its importance in traffic safety, while operating in real time. The architecture of the object detection procedure is based on a convolutional neural network, consisting of convolutional layers for feature extraction and a fully connected neural network for object classification. This type of architecture has been utilized for different image classification problems and the achievable performance of well-trained models is high. Moreover, using a FCN for classification purposes only is considered an overshoot, [3]; nevertheless, the required high accuracy of the system and the availability of parallel processing hardware in the form of Graphic Processing Units does justify the chosen approach. The procedure detects objects of different sizes and performs their localization using the sliding window technique. Of course, direct application of the sliding window technique would result in large computational complexity due to information redundancy in the overlapping image blocks. In that direction, we developed an architecture that lowers the computational complexity similar to [6]. Namely, the feature extraction phase is performed on the complete image and the sliding window technique for dividing the image into overlapping blocks is applied on the feature map generated in the feature extraction phase, thereby avoiding redundant filtering of overlapping image blocks. The structure of the approach is depicted in Fig. 1.

The designed architecture of the multiple output feature extraction part of the network is presented in Fig. 2. As can be seen, the feature extraction is performed using convolutional layers followed by down-sampling layers implemented using the max pooling technique. The last down-sampling layer is different for different outputs. Each output generates feature maps with different resolutions. The feature maps are then divided into overlapping blocks of fixed size using the sliding window technique, and classified using a fully connected network classifier. Each output from the fully connected networks corresponds to a different size of the block in the original image.

In order to classify the information obtained in the feature extraction phase, a fully connected neural network is used. Each output from the feature extractor is connected to its own FCN. Although one FCN could be trained for all outputs (the size of the blocks is the same), the reason for using different FCNs is better adaptation of the classifier to the signal characteristics. The parallel implementation of the algorithm allows this. The structure of the FCN is shown in Fig. 3. The multiple output classifier classifies input blocks in one of the four classes: Cars, Trucks, Background and Non-vehicles. The first three classes are self-explanatory, and the last class is comprised of blocks containing partial and multiple overlapping objects. The introduction of

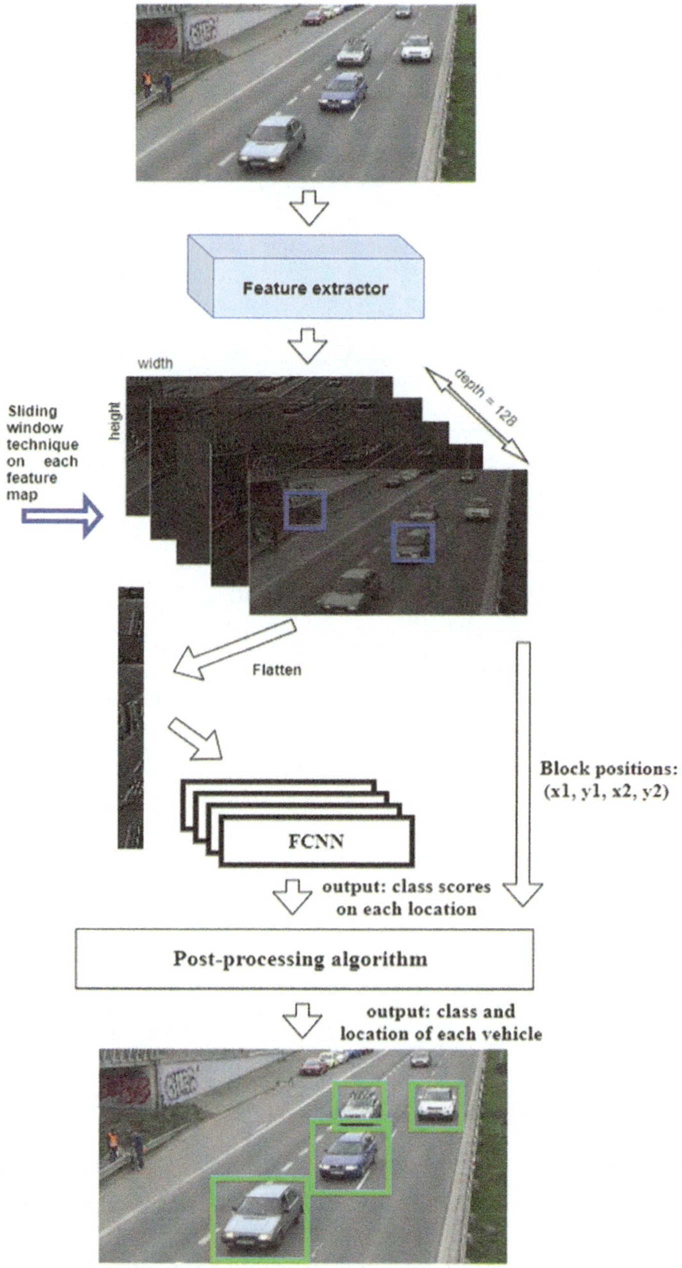

Fig. 1 The structure of the vehicle detection algorithm

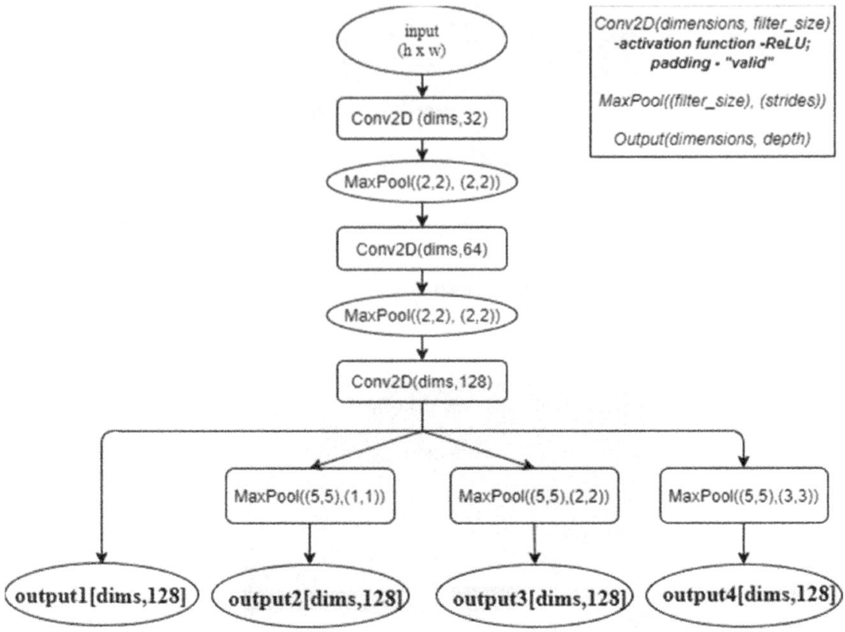

Fig. 2 Multiple output feature extractor

Fig. 3 The structure of the classifier

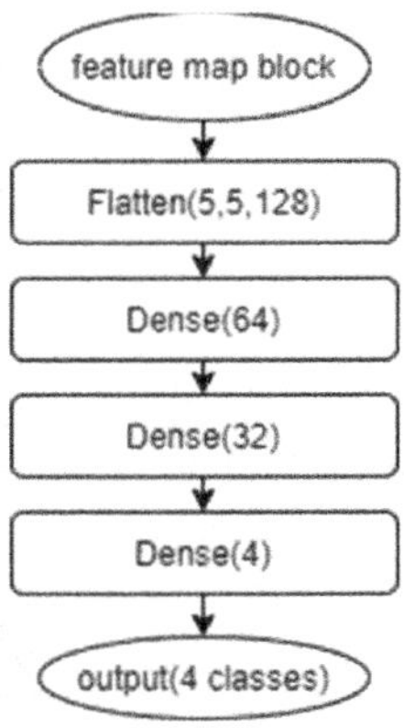

this class significantly increased the classifier accuracy. More details on this class are presented later.

The architecture of the object detection algorithm has another significant advantage: it can be easily extended to include additional classes of objects with different window sizes and ratios. For example, in order to recognize motorbikes, it is sufficient to add a new type of block in the sliding window phase applied on the feature maps (another output in the diagram in Fig. 2) and an appropriately trained FCN.

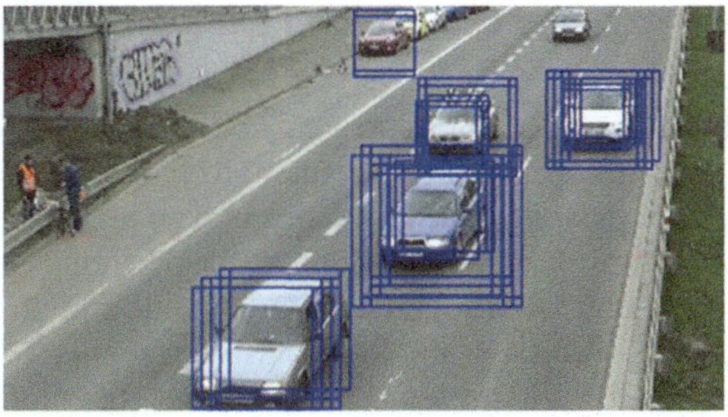

Fig. 4 Outcome of the classifier

Following from the appearance of the motorbikes the added block would be with a different size and aspect ratio. The rest of the architecture is unchanged and, thus, the performance of the rest of the system is not compromised. One could argue that the convolutional part of the network would not be optimal to the newly added classes since they were not used in its training, which of course is true; however, the difference in the signal characteristics for different classes is small enough to be overcome by the strong classifier like FCN.

The final decision of the vehicle detector is reached through postprocessing of the classifier output. As can be seen in Fig. 4, detected vehicles are covered, and thus localized with multiple blocks of various sizes. In order to generate unique block for each vehicle, blocks are clustered based on the mutual distance (threshold on overlapping area). For each group of blocks the class of the final block is calculated using majority vote. Finally, the location and size of the block is calculated using averaging over the blocks with two biggest sizes of the major class. The reason for using two biggest sizes of blocks only is in the self-similarity of the vehicles that can sometimes generate multiple small detections inside the vehicle. These detections can significantly lower the final size of the block, to the level that the estimated block can be much smaller than the vehicle.

The result of the postprocessing of the classifier output from Fig. 4 is shown in Fig. 5.

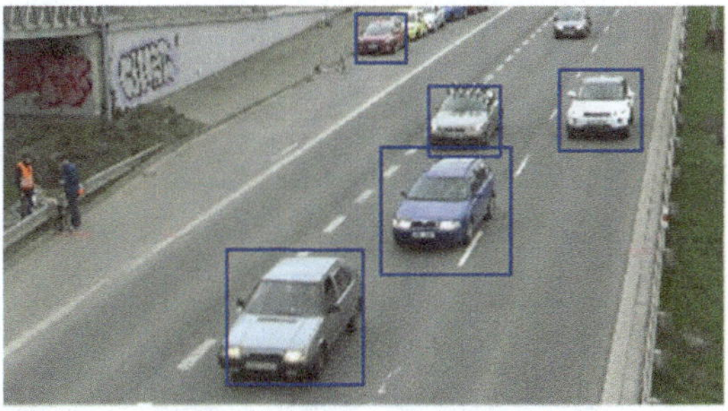

Fig. 5 Detected vehicles after postprocessing

2.2 Object Classes and Database Compilation

Four classes of objects are considered in the video detection algorithm: Car, Truck, None-vehicle and Background. The first two classes are important in statistical analysis of the traffic used for efficient road maintenance. The third class, "Non-vehicle" class, was introduced to improve the performance of the classifier in ambiguous situations of blocks containing partial and/or multiple overlapping objects of first two classes. Although the FCN classifier could be adequately trained to resolve these cases too with three classes only, the addition of extra class enabled improved accuracy with much smaller number of blocks in that class in comparison with the number of blocks that should be added to Background class to achieve the same goal. When the algorithm is used for prediction this class is considered as background.

The Car class consists of images of cars showing front/front-left/front-right view of the car. The images were obtained in two different ways. First, the two publicly available datasets, [13, 14], were used. The CompCars dataset, [13], consists of different types of cropped images of the front part of the car. The UA-detrac dataset, [14], consists of images of the motorway with annotated vehicles. The annotations (XML files) were used for extraction of the vehicle images.

The second method for obtaining vehicle images was the application of the trained algorithm (up to that point) for extraction of vehicle images from the images of the Miladonovci-Stip motorway in Republic of North Macedonia, recorded by our team. The same method was also applied to the publicly available dataset BrnoCompSpeed [15].

In total for training the algorithm 72,846 images of cars were used, 62,194 images from [13], 971 images from [14] and the 9,581 are obtained using the second method. Separate dataset was used for validation, consisting of 18,284 images, out of which 13,333 images were obtained from [13] and the rest were obtained using the second method. Examples of images are given in Fig. 6.

For the training dataset of the class Truck publicly available dataset IRVD, [16], was used, out of which 1003 images were selected. The images were additionally

Fig. 6 Examples of images in the car class

augmented. Rotation and low pass filtering were applied on some images. Thus, the initial class Trucks consists of 9503 images. Additionally, video sequences recorded on the motorway Miladinovci–Shtip and from the publicly available BrnoCompSpeed dataset [15] were processed using the algorithm trained on the initial training dataset and images of detected vehicles were added to the training database in order to increase its accuracy. Examples of images used in this class are given in Fig. 7.

The class Non-vehicle consists of images that contain partial objects of the previous classes, as well as multiple objects. This type of images is likely to occur as we slide the window through the feature map in the detection process. The images were generated using sliding window technique applied on the aforementioned datasets. In all 35,315 images are used for training and 4267 for validation. Examples of images of this class are shown in Fig. 8.

Finally, the class Background consists of images with the content that cannot be categorized as vehicle, yet it is inside the field of view of the camera, in different real-world scenarios (different weather and lighting conditions, etc.). In order to cover such a diverse content 186,699 images were extracted from the publicly available datasets for the training dataset and 40,380 for the validation dataset. Examples are given in Fig. 9.

Fig. 7 Examples of images in the truck class

Fig. 8 Examples of images in the non-vehicle class

Fig. 9 Examples of images in the background class

2.3 Detection of the Front Part of the Vehicle

Vehicle detection procedure can only localize vehicles to some predetermined accuracy due to the sliding window technique applied to the feature map. Two subsampling layers (max-pooling) in the stem of the structure that are lowering the size of the feature map by 4 are causing sliding of the window in the feature map with stride 1 to be equivalent to sliding of the block in the input image with stride 4. This localization accuracy is insufficient for accurate speed and acceleration estimation, for which pixel accuracy is necessary.

The required accuracy is achieved by additional localization of the lower front part of the vehicle based on output of the previous step. The image of the complete front part of the vehicle detected in the previous procedure is used as input in the CNN that detects the lower front part of the vehicle. The structure of the CNN is shown in Fig. 10. Again, the sliding window technique is used, however, this time it is applied on the input image and the stride is 1, assuring the required localization accuracy. The hight of the block for detecting the lower front part is set to 1/3 of the hight of the detected vehicle and the width is the same as the detected vehicle width. The search area is set to lower 2/3 of the hight of the detected vehicle, additionally widened 3 pixels in each direction. The size of the input image is small relative to the complete input image resulting in low computational burden to overall algorithm.

The postprocessing of the classifier output is performed using averaging over the sizes and locations of the detected positive blocks. The example of the outcome of the classifier and the postprocessing is shown in Fig. 11. The detected parts of the vehicle are depicted in yellow and the averaged value is depicted in red. The vehicle is marked with dark blue rectangle, and the light blue line marks the 2/3 of the hight of the vehicle below which the front part is searched.

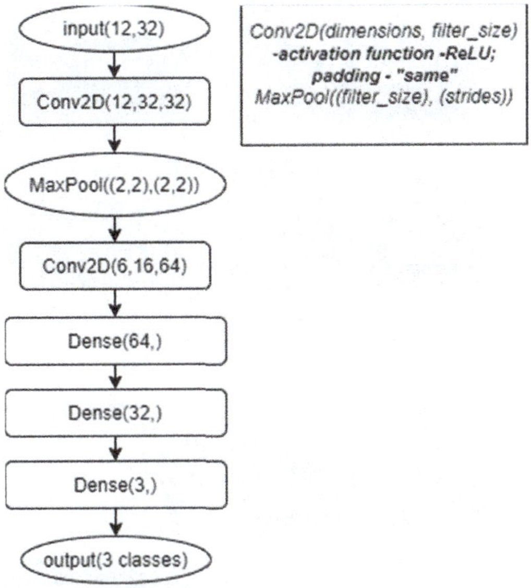

Fig. 10 The structure of the CNN for vehicles lower front part detection

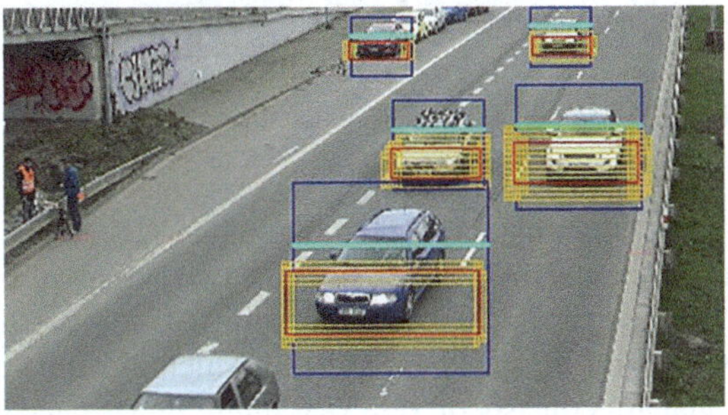

Fig. 11 Outcome of the lower front part detector (yellow) and the postprocessing (red)

2.4 Object Classes and Database Compilation

This part of the algorithm considers three classes: Lower front part of the vehicle, Upper front part of the vehicle, and Background. The motivation for introducing the second class, "Upper front part of the vehicle" class, was similar to the introduction of the "Non-vehicle" class in the vehicle detection algorithm: to improve the performance of the classifier in ambiguous situations, e.g., high variance upper front part of the vehicle due to windshield reflections, and to achieve that with relatively small number of blocks in that class. When the algorithm is used for prediction this class is considered as background. The background class is the one used in the previous algorithm.

Fig. 12 Examples of images in the class lower front part of the vehicle

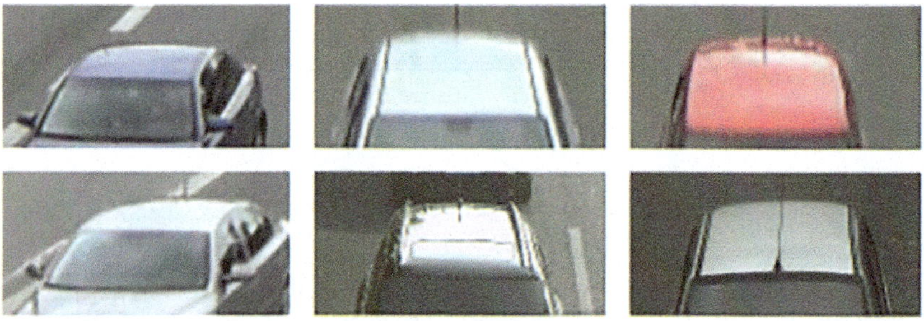

Fig. 13 Examples of images in the class upper front part of the vehicle

The training and validation images for the first and second class are compiled by cropping the images from the first two classes of the vehicle detection algorithm. The database for the Lower front part class consists of 7553 training images and 1970 validation images, examples of which are shown in Fig. 12. The Upper front part database consists of 12,837 training images and 3548 validation images. Examples are shown in Fig. 13.

2.5 Vehicle Trajectory Tracking and Parameters Estimation

The coordinates of the detected vehicles, as well as the coordinates of their lower front parts, calculated by the procedure explained above, are passed to the algorithm for vehicle tracking (VT), alongside the bounding boxes that determine the content of the frame covered by these parts, Fig. 14.

The algorithm itself will be explained in detail below, however, the accompanying elements necessary for its operation should be established beforehand.

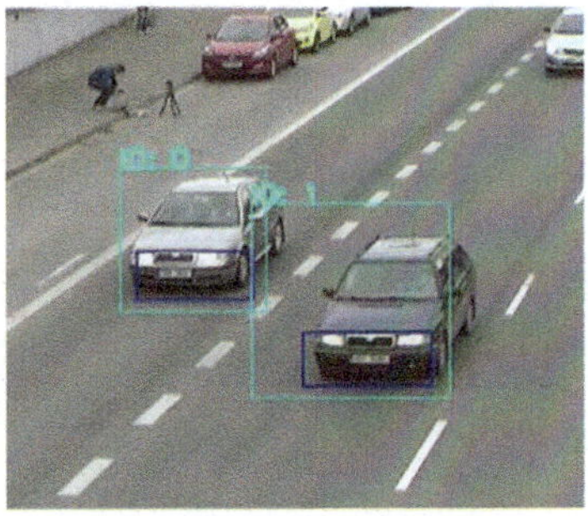

Fig. 14 Bounding boxes of the detected vehicles (cyan) and of the lower front parts (blue), passed to the VT algorithm

2.5.1 Kalman Filtering

The filtering procedure itself can be divided in two stages:

1. *Prediction Stage*: The Kalman filter model assumes that the state of a system x_t at a time t evolved from the prior state at time $t-1$ according to the Eq. (1).

$$x_t = F_t x_{t-1} \tag{1}$$

The state model devised assumes the motion of the object to be an accelerating one with constant acceleration and 6 state variables $s_x, s_y, v_x, v_y, a_x, a_y$ modeled as in (2).

$$x_t = \begin{pmatrix} s_x \\ s_y \\ v_x \\ v_y \\ a_x \\ a_y \end{pmatrix}, F_t = \begin{pmatrix} 1 & 0 & \Delta t & 0 & \frac{\Delta t^2}{2} & 0 \\ 0 & 1 & 0 & \Delta t & 0 & \frac{\Delta t^2}{2} \\ 0 & 0 & 1 & 0 & \Delta t & 0 \\ 0 & 0 & 0 & 1 & 0 & \Delta t \\ 0 & 0 & 0 & 0 & 1 & 0 \\ 0 & 0 & 0 & 0 & 0 & 1 \end{pmatrix} \tag{2}$$

The covariance matrix for the constant acceleration model is given in (3)

$$E_z = \begin{pmatrix} \frac{\Delta t^5}{20} & 0 & \frac{\Delta t^4}{8} & 0 & \frac{\Delta t^3}{6} & 0 \\ 0 & \frac{\Delta t^5}{20} & 0 & \frac{\Delta t^4}{8} & 0 & \frac{\Delta t^3}{6} \\ \frac{\Delta t^4}{8} & 0 & \frac{\Delta t^3}{6} & 0 & \frac{\Delta t^2}{2} & 0 \\ 0 & \frac{\Delta t^4}{8} & 0 & \frac{\Delta t^3}{6} & 0 & \frac{\Delta t^2}{2} \\ \frac{\Delta t^3}{6} & 0 & \frac{\Delta t^2}{2} & 0 & \Delta t & 0 \\ 0 & \frac{\Delta t^3}{6} & 0 & \frac{\Delta t^2}{2} & 0 & \Delta t \end{pmatrix} \tag{3}$$

2. *Measurement Update*: Only the position measurements in pixels are given to the Kalman filter for measurement update, being applied only on s_x and s_y. Since the Kalman filter prediction is used as a feedback loop for outlier detections,

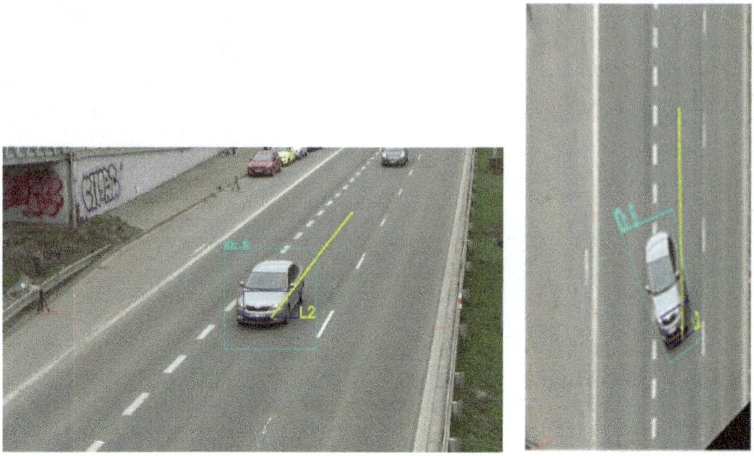

Fig. 15 Left: the scene as seen from the camera perspective. Right: top-view projection in which measurements in pixels correspond to real world distances

the covariance matrix values for the measurement noise are assumed to be two times higher than the ones for the process noise.

2.5.2 Homography

Planar homography is a transformation occurring between two planes, i.e., the projective mapping of the image from one plane to another. Given a set of points x_i in P_2 and a corresponding set of points $x_i^{'}$ in P_2, the projective transformation that maps each x_i to $x_i^{'}$ is computed. In a practical situation, the points x_i and $x_i^{'}$ are points in two images, each image being considered as a projective plane P_2.

In our case, the projection is done by carefully selecting 4 or more points in the image between which the real-world distances are known. The goal is to transform the perspective view in the image to a top-view projection, as shown in Fig. 15. A simple linear algorithm is used for determining H given a set of four 2D-to-2D point correspondences, $x_i \leftrightarrow x_i^{'}$. The transformation is described by the Eq. (4).

$$x_i^{'} = \left[x^{'} y^{'} 1\right] = Hx_i = [h_{11}h_{12}h_{13}h_{21}h_{22}h_{23}h_{31}h_{32}h_{33}]x_i \tag{4}$$

2.5.3 VT Algorithm: Motion Parameters Estimation

In the VT algorithm, the received coordinates are used to deduce the motion of the objects in the frame, and for the vehicles entering the frame, the coordinates are assigned to the initialization process—new objects are introduced. The motion of the vehicles is calculated using the bounding boxes for the lower front parts of the vehicles, applying a block matching algorithm between the frames. This approach yields very high precision of the tracked points.

In order to track the object movement in every frame, in addition to the comparison of the content of the bounding boxes, the Euclidian distances between the instances of the tracked objects are calculated and used as constraints to limit possible misdetections. A mutual similarity matrix is formed, in which all detected objects in the frame are compared one-vs-all, and their Euclidian distances are calculated. Then, indexes are generated, and used for matching of the detected objects in the current frame, to the objects that exist in the previous frame. After the connections (matches) between objects are made, the next part is to update the object positions and to verify their consistency. For every tracker object which position is being updated, the first check is for its initialization status.

If the object is already initialized in the previous frames, the Kalman filter prediction is used to check for possible outliers in the coordinates of the vehicle bounding box by generating a loose area where the resulting block match is expected. The bounding box is considered as an outlier if the intersection between the bounding box coordinates and the generated area is <70%. In this case, new coordinates are generated around the predicted point from the Kalman filter by keeping the ratio of the positions of the tracked point from the past frame with the coordinates of the bounding box. However, if the intersection between the bounding box coordinates and the generated area is <30%, it reveals some serious inconsistency in the vehicle movement, and in this case a misdetection is assumed.

Because the vehicle detection procedure works separately on every frame, it is possible for it to output different bounding box dimensions for the same vehicle in consecutive frames. Since the block content information is of utmost importance for the block matching and object tracking, these jumps in bounding box dimensions should be suppressed. This is achieved by passing the box dimensions to a separate Kalman filter that predicts the dimensions in the future frames, ensuring their consistency.

The block matching algorithm itself is implemented using the template Match function of the OpenCV library and taking the lowest central point of the resulting block as a point on which all movement calculations are based. Again, the Kalman filter correction step is also used here, in order to check for a possible outlier in the block matching result. A possible error in the measurement point will be most reflected on the object's acceleration, the second derivative of the position in the state vector. In this approach, the resulting point is passed as the correction measurement to the already made prediction, and empirically obtained thresholds are applied for the state vector's accelerations on both axes. If at least one of the acceleration values is higher than the threshold, it is considered as an outlier. These outliers are handled by completely discarding the resulting block and generating another one with the same dimensions around the Kalman filter prediction, Fig. 16. All further calculations are based on that prediction.

The VT algorithm also tackles the problem of possible missing detections in order to keep the continuity of the objects during their complete trajectories. Missing

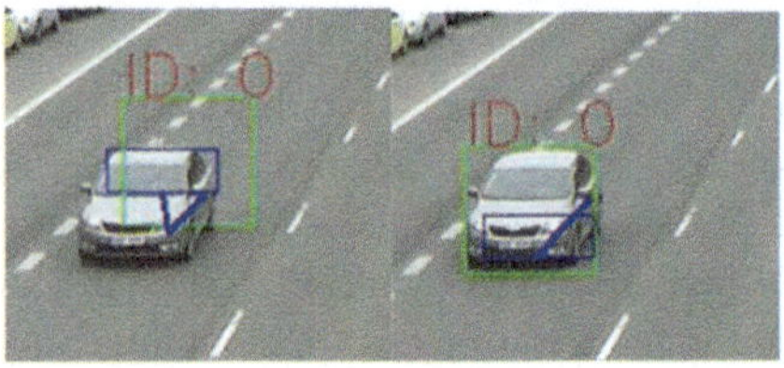

Fig. 16 Wrongly matched block by the matching algorithm (left) and the correction by the Kalman filter prediction (right)

locations in the trajectories are generated using the Kalman filter, in a similar way as the outliers in the bounding box coordinates are handled. The prediction from the Kalman filter is taken as the point around which the bounding box is generated. After obtaining the bounding box coordinates, the rest is the same as the regular detections, except of course checking for possible outliers.

The number of possible consecutive missed detections that can be generated using the Kalman filter is fixed at three, so that the VT algorithm will not rely on the state predictions for too long. If the object exceeds the number of consecutive missed detections, or the coordinates of the generated bounding box are out of the frame dimensions, it is regarded as an object that left the frame and it is removed from further tracking. Also, if an outlier is found in the full vehicle detection coordinates, the counter for the consecutive missed detections is incremented, in order to limit larger dependence on the Kalman filter prediction.

The use of the object tracker is expanded by adding an object buffer, another feature to help solve the potential problem with missing detections and possible re-identification of the tracked objects. The buffer's purpose is to keep the objects that surpassed the maximum number of generated bounding boxes separate from the objects currently in the object tracker. By keeping the objects separate, the positions of the new detections can be compared with the predictions of the positions of the objects already in the buffer, and it can be decided if the new detection could be the same object that was tracked (and lost) in the past frames.

The image projection is separated in 4 quadrants, Fig. 17, and each of the new detections is placed in one of those 4 quadrants, by projecting the central point of the coordinates of the bounding box. In the same manner, the tracker objects which are supposed to be removed from the tracker (misdetected objects that surpassed the maximum number of generated bounding boxes, or objects that left the frame dimensions), are placed somewhere in the 4 quadrants.

If the last known position of the object from the previous frame is in any of the first three quadrants, it is placed in the buffer and a countdown counter is initialized (for counting the remaining frames during which the object can stay in the buffer). The initialized number in the third quadrant is the lowest, since the object is expected to leave the scene sooner; and it is highest in the first quadrant, because the first quadrant is on the start of the trajectory.

The object buffer relies solely on the Kalman filter prediction. After an object is pushed to the buffer, the object's position is updated in each consequent frame by taking the prediction without the correction step, since the actual object positions are not available from the vehicle detection procedure. First, the object tracker is used to

Fig. 17 The separation of the most usable portion of the scene into four quadrants

generate the respective connections between tracked objects. If some objects remain without a connection with the currently tracked objects (possible new objects, or lost objects), the location quadrant of the new detection is checked, whether it coincides with the location quadrant of any of the objects in the buffer. If that is the case, the same principle of generating an area around the Kalman filter prediction is used, in order to calculate the intersection with the bounding box coordinates. This time, if the area threshold is met, the connection between the object and the new detection is established the same way as explained above, and the object is reinstated from the buffer to the main object tracker. Objects being in the buffer that have not been connected prior to depleting their countdown counters are deleted from consequent tracking. Unconnected new detections found in the first quadrant are not compared with the objects in the buffer and they are initialized immediately as new objects. Misdetected objects that reached the last quadrant and also met the maximum number of trajectory points replaced with Kalman filter predictions, are not compared with the new detections, but they are immediately deleted from further tracking. The entire set of possible decisions for object placementin the tracker and the buffer, can be visualized observing the flowchart shown in Fig. 18.

Before the object is deleted from the object tracker or the buffer tracker, the information from the object trajectory is extracted, and the average speed of the object is estimated, Fig. 19. In the figures, the averages are shown in coloured flat lines, the ground truth (measured with a radar) shown in yellow; the calculated average by the VT algorithm, without corrections, shown in green; and the calculated average by the VT algorithm with corrections shown in red. The speed is estimated by taking the point position difference between continuous frames and converting it into km/h.

Having the vehicle speeds and trajectories, several traffic events can be registered, like overspeed, underspeed, stopping, lane change, driving off the road, crashes and wrong direction driving. A registration of some of the events are depicted in Fig. 20.

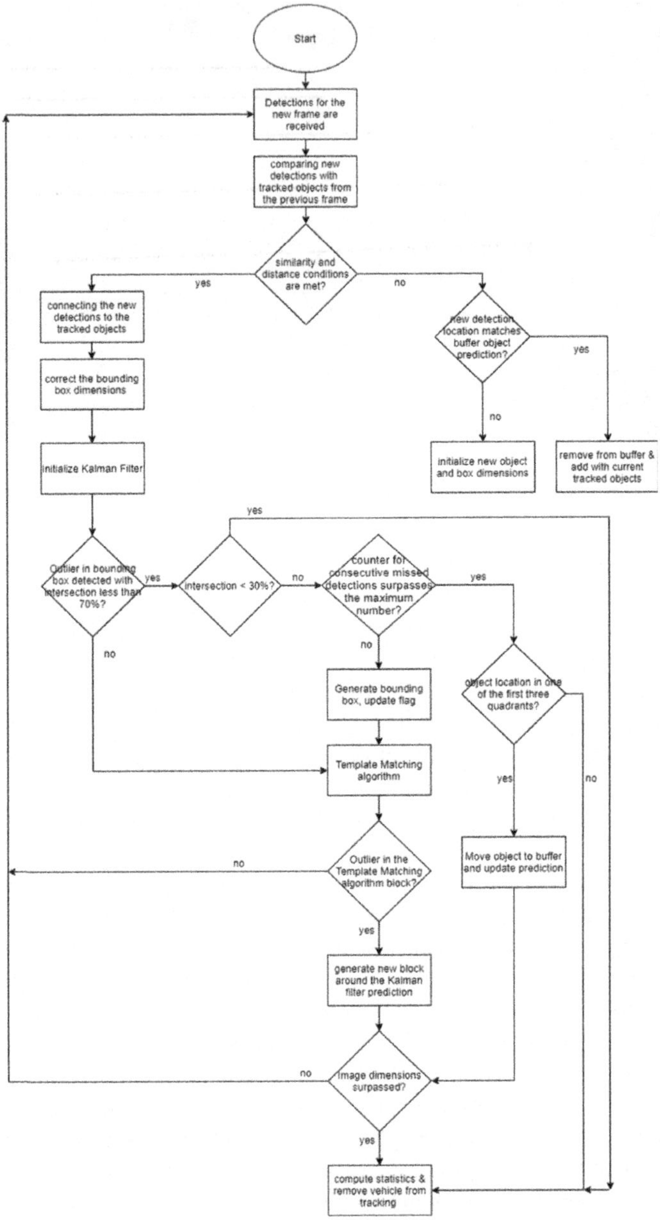

Fig. 18 The decision tree associated with the operation of the object tracker and the buffer

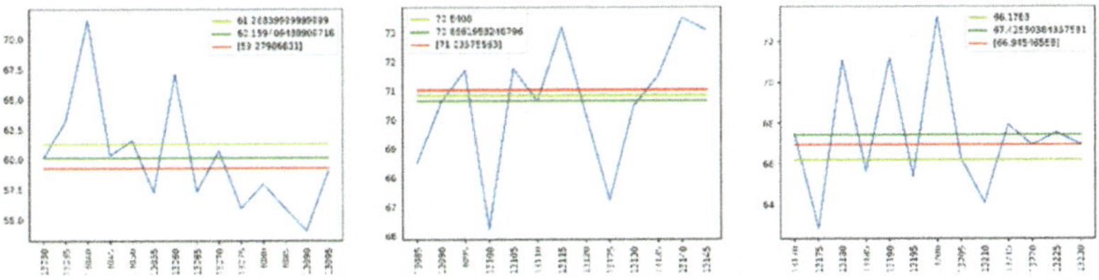

Fig. 19 Vehicle speed measurement, point to point, and average, for 3 different vehicles

Fig. 20 Examples of typical traffic events detected by the VT algorithm: lane change from L2 to L3 (left), overspeed 92 km/h (middle), underspeed 48.6 km/h (right)

In the moment of obtaining the vehicle trajectory, the registration of the events and possible signalization for them is immediate, allowing the whole system to operate in real-time.

3 Audio Analysis

The audio analysis part of the system is designed to detect alarming sound events in traffic, specifically we are interested in detecting sounds of a siren, car horn, skidding and car crash. We use a scene-dependent approach and train and test our audio recognition model on a highway soundscape, noting that we expect misclassifications when applied to a noisier city soundscape richer with sound events. Thus our system needs to solve a simplified monophonic soundscape problem, where we use only one label per data sample.

3.1 Dataset

A typical highway traffic acoustic signal can be thought of as a superposition of two acoustic signals: an ambient noise signal and critical sound event signals. The ambient noise is made up of the sound of different types of vehicles passing by, wildlife sounds (such as bird chirping), acoustic noise from different weather conditions (rain, wind,

Table 1 Dataset properties

Event	Average length (s)	Maximum length (s)	Minimum length (s)	Total length (s)	Number of events
Crash	0.36	0.74	0.093	36.54	100
Horn	1.58	4	0.096	158.32	100
Siren	3.65	4	0.27	365.51	100
Skidding	2.07	9.19	0.36	207.07	100

thunder). On the other hand, the sound of a siren, crash, horn and skidding are rare critical sound events and provide important information about any alarming events in the environment.

We created a custom dataset of alarm sound events for the development of our system. In order to simulate a natural environment we added recordings of the four target sound events to ambient traffic noise. As ambient traffic noise we used two audio tracks with a duration of one hour each—one recorded in wet road conditions and one in dry road conditions. We collected a total of 400 alarm events, 100 of each of the four classes (Table 1). The crash and skidding recordings were gathered from YouTube and the siren and car horn were obtained from the UrbanSound 8K dataset [43]. All the audio data is sampled at 22,050 Hz. The event and background classes were normalized to a maximum amplitude of −3 dB. This was done because the system's reliability upon deployment will be guaranteed only within a certain distance from the microphone. The normalized event sounds were added randomly to the two ambient tracks at 4 s apart. We use 60% of the data for training, 20% for validation and 20% for testing.

3.2 Pre-processing and Feature Extraction

In the pre-processing stage the input acoustic signal is normalized to −3 dB and divided into frames of equal length using a short-time Fourier transform (STFT) with a Hamming window 2048 samples (92.87 ms) long, using a hop size of 512 samples (23.21 ms). In order to extract the MFCCs features, we use the mel-scale, which models the human non-linear sound perception, to position and scale a triangular filterbank with 20 filters to the spectrogram. After summing the energy in each filter and taking the logarithm of the filterbank energies, we take the discrete cosine transform of the log-mel-spectrogram and obtain 20 Mel-Frequency Cepstral Coefficients.

The siren, horn and skidding sounds have a similar waveform which is characterized by the presence of harmonics. The spectrogram of the horn and skidding is characterized by a fundamental frequency and its harmonics, with the difference being that the skidding sound has an irregular fundamental frequency and fewer harmonics. Similar to the horn and skidding sounds, the spectrogram of the siren

crash horn siren skidding

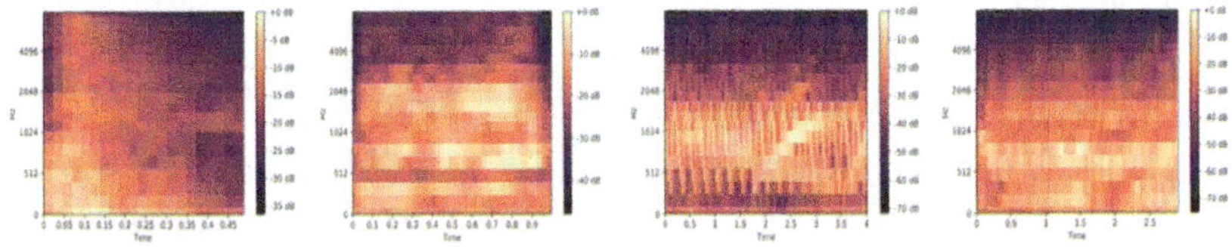

Background:

dry dry car passing rain rain car passing

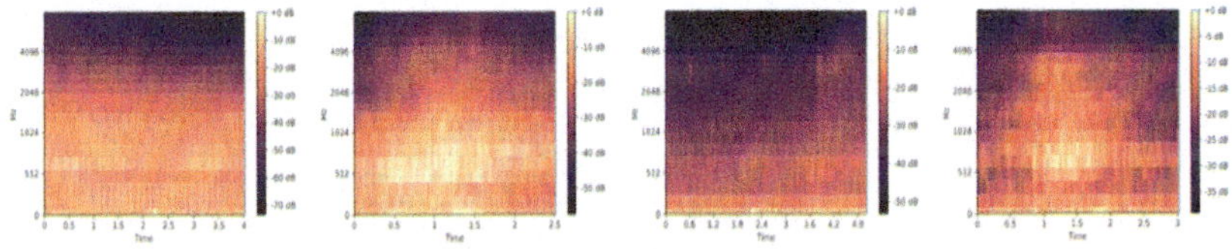

Fig. 21 Mel-spectrograms of background sounds including noise, rain, and a car passing, contrasted to the 4 alarm classes: siren, horn, skidding and car crash

has a combination of the horn pattern and a triangular pattern with harmonics. The crash is an instantaneous high-amplitude burst of energy with an approximately flat frequency response over the spectrum or impulse noise. However, short horn honks have a spectrogram which is similar to the spectrogram of the crash, as the harmonics are indiscernible and resemble impulse noise. This overlap between the spectral-temporal characteristics of the critical events makes their classification a hard problem to solve (Fig. 21).

3.3 Classification Model

We built a Convolutional Neural network sound event detection model. The architecture of the model is shown in Fig. 22. In contrast to our previous work [42], the present model has reduced complexity in terms of the number of filters in the convolutional layers, as well as the number of hidden units in the dense layers. These modifications reduced the overall number of parameters by 10x, specifically from 1,025,765 to 17,305. In this way, the new model has reduced resource utilization, optimizing the overall system's computing power requirements.

Additional design changes to the model include the omission of dropout layers and the use of the leaky rectified linear unit (ReLU) activation function The ReLU function is a piecewise linear activation function whose derivative is 0 in the negative part. This can lead to dead neurons in the process of training the neural network. The leaky ReLU circumvents this problem by substituting zeros with small non-zero values, thus precluding vanishing gradients in model training.

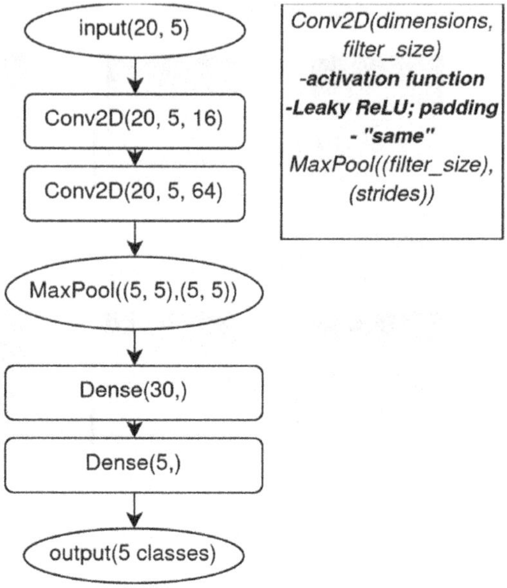

Fig. 22 Neural network architecture diagram

We train our model by minimizing the binary cross-entropy loss function, which is usually used for multi-label classification. In this way we make the computed loss for each CNN output to not be affected by other component values. We also obtained higher accuracy using this loss function in our experiments, in contrast to using categorical cross-entropy. We train the model using the Adam optimizer [48] with a learning rate of 0.0001.

4 Information Fusion

The information fusion is performed on the event level. The term "event" here refers to the passing of the vehicle in the field of view of the camera. Vehicle behaviour is analysed in video, as well as in audio domain and the results in both domains are averaged before information fusion and final decision. Following from the camera position and the speed of the vehicle, the time interval of the event has order of magnitude seconds, enabling significant changes in the speed and acceleration of the vehicle, as well as multiple occurrences of the skidding that could be detected in the audio analysis. This fact is rising suspicion to some extent in the approach based on averaging of the parameters; however, averaging is alleviating the effects of local errors and increasing the accuracy of the estimated parameters. It also enables avoiding multiple detection of the essentially single event. Namely, one vehicle could perform multiple skidding events in the single pass in the field of view of the camera, which from statistical point of view should be considered as single event.

The occurrence of multiple vehicles in the field view of the camera is also possible. Since the microphones that were used for audio recording have no significant directionality separating the vehicles in the audio domain is almost impossible. Therefore, it is assumed that only the vehicle with biggest deceleration is participating in the potentially dangerous event and its motion parameters are used for classification.

The feature vector for the event detection consists of 9 parameters: the speed and acceleration of the vehicle, both in x and y direction, and the probabilities of occurrence of five audio events (skidding, horn, siren, crash and background) estimated in the audio analysis.

The final decision of occurrence of dangerous event is reached using Support Vector Machine (SVM) as classifier.

4.1 Class Parameters and Database Compilation

The parameters, as stated earlier, are estimated using time frames of 100 ms. The final parameters of the event that are used for decision of occurrence of dangerous event are calculated by averaging frame parameters. Given the time interval of the event, the deceleration and skidding could occur only in part of the event time interval. Short occurrence of skidding is regarded as random instance and not considered as dangerous events.

Both training and testing of the approach is performed using simulated data. In video domain simulation of different acceleration is achieved by dropping frames from the original sequences. Simulated video sequences are then processed by video detection algorithm, motion parameters are estimated and averaged over the time interval of the event. During the simulation care has been taken to keep the acceleration in the range 0 to −13 m/s2, that corresponds with real occurrence of skidding. 30 original video sequences are used to generate 150 simulated video sequences. The acceleration in the Non-dangerous event class is in the interval −0.017 to −4.15 m/s2, with average of −1.26 m/s2 and variance of 0.49. Similarly, in the Dangerous event class the acceleration is in the interval −3.22 to −12.15 m/s2, with average of −6.32 m/s2 and variance of 1.78. The starting speed (before the beginning of the deceleration) of all vehicles in x direction is roughly in the interval between 60 and 100 km/h, and close to 0 km/h in y direction.

Audio sequences are generated by combining sounds with different lengths and intensities from all five audio classes. Each audio class is represented with 50 original sounds to choose from, excluding the Skidding class. This class is represented with 103 original sounds, out of which 8 are with duration <0.6 s, which is considered too short to represent danger. The combined audio sequences are than truncated to the time frame of the event and are processed with audio analysis algorithm. Estimated class probabilities are averaged over the time frame of the event.

In total 303 simulated audio-visual events are generated. Their distribution is shown in the Table 2.

Table 2 Distribution of samples in the database

	Non-dangerous event	Dangerous event	Total
Training	155	60	215
Validation	23	15	38
Test	26	20	46
Total	208	95	303

5 Results and Analysis

Vehicle detection algorithm was tested using images and videos from different sources recorded in different conditions. Here as illustration, we shortly present only part of the experimental results from application of the algorithm to videoframes, due to intended application of the vehicle detection algorithm in dangerous events detection. The detailed complete results and analysis are outside the scope of the paper. The test dataset for vehicle detection algorithm consists of 100 videoframes from different video sources. In total 206 vehicles present in the frames were annotated. Evaluation of the accuracy is performed using intersection over union (IoU) method and the threshold was set to 0.7. The number of true positive detections was 191. There were 8 false positive detections and 7 false negative detections, thusthe achieved *precision* is 0.9598 and the *recall* is 0.9646. It can be concluded that the achieved performance of the vehicle detector is high and its applicability in the foreseen scenario is confirmed.

The detector of the lower front part of the vehicle was tested in a similar way, using 879 annotated vehicles and the threshold for the IoU was set to 0.7. The achieved *precision* is 0.8434 and the *recall* is 0.9738. Again, it can be concluded that the achieved performance is sufficient for the intended usage of the detector. Th conclusion is additionally supported by the fact that the localization of the lower front part of the vehicle is further corrected in the following stage of the algorithm.

The performance of the vehicle tracking algorithm was tested using the BrnoCompSpeed dataset [15], which contains information about the speed of the recorded vehicles, obtained by radar. In the performed experiments, the radar measured speed was considered as ground truth. The vehicle speed calculated by the VT algorithm was compared to the ground truth and that comparison was used to determine the accuracy of the algorithm. A total of 100 randomly chosen vehicles was used for the testing. The mean absolute error, MAE_{VT}, and the overall accuracy of the VT algorithm, C_{VT}, are expressed by the formulas:

$$MAE_{VT} = \frac{\sum_{i=1}^{N} v_{VT} - v_{GT}}{N} = 1.59 \text{ km/h} \tag{5}$$

$$C_{VT} = 100\% - \frac{\sum_{i=1}^{N} \left| \frac{v_{VT} - v_{GT}}{v_{GT}} \right|}{N} = 97.6\% \tag{6}$$

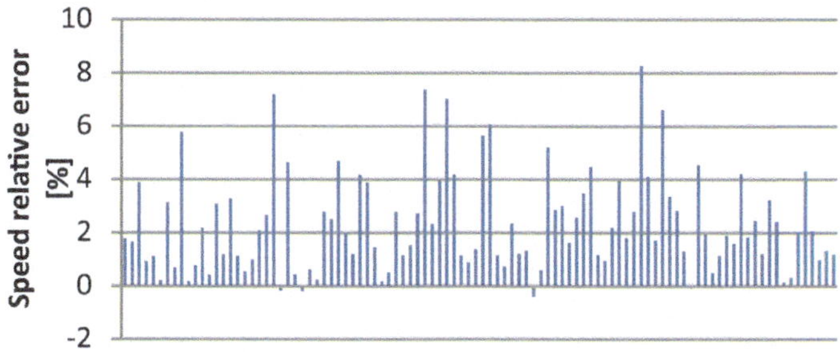

Fig. 23 The relative errors in the calculated speed for 100 tested vehicles

where N is the total number of tested vehicles (100), v_{VT} is the vehicle speed calculated by the VT algorithm, and v_{GT} is the ground truth speed.

These results show that the VT algorithm possesses very high accuracy, and it's error in the calculated speed is comparable to the typical errors produced by the radar speed guns that are used by the police force. This makes the VT algorithm highly usable for traffic control and intelligent transport systems.

However, the relative error of the speed calculated by the VT algorithm shows somewhat odd behaviour, as can be seen in Fig. 23, where the relative errors in [%] of the ground truth speed are shown for all 100 tested vehicles. It can be easily noticed that in the vast majority of the measurements the relative error is positive (the calculated speed is higher than the ground truth speed), which is an unusual offset, considering that the error in detected vehicle location could happen in any direction. This out of the ordinary conduct of the VT algorithm can be explained by the discrepancy between the real 3D scene and its homographic projection on a flat plane. Namely, when working with the Direct Linear Transform (DLT), it is assumed that the topology of the scene itself is in 3D, however, that concept is not translated for the objects in the scene. In the homography based speed estimation, the value of the speed is obtained from a point with a fixed location on the vehicle. In the proposed VT algorithm, the experiments showed that the ideal point for speed estimation would be the point on the plane between the front wheels of the vehicle. This makes the tracking to be performed on a point located above the ground (above the road) in the real world. However, no physical height for the vehicles is assumed in the process of the homohraphic flat plane projection, and all their points project as they would be on the road itself. The higher on the vehicle body is the tracking point, the more prominent is the resulting error in the projected flat plane position of that point, always pointing towards the camera. Thus, although minimal in terms of absolute value and percentage, a positive offset in the calculated speed is constantly present.

In the audio analysis, the proposed CNN model achieved 99.01% accuracy of sound event detection. This shows that the new model, even with the speed optimization, i.e. reduction of the number of CNN filters, still succeeds in maintaining a high level of accuracy, comparable to the 99.1% of our previous more complex model [42].

Table 3 Computed metrics for each target class

Class	Precision	Recall	F1-score
Background	0.9919	0.9980	0.9949
Crash	0.7690	0.8957	0.8275
Horn	0.9840	0.9291	0.9557
Siren	0.9886	0.9241	0.9553
Skidding	0.8786	0.9557	0.9155

The precision, recall and F1-score metrics of the classifier per class are shown in Table 3. It can be seen that the model makes classification errors within the alarm sound events due to the similarity of the horn, siren and skidding classes. Another source of errors could be the short duration of some of the audio samples. This is an area in which information fusion has the potential to improve sound event detection and the robustness of the overall system.

Four different kernels were used during the testing of the SVM classifier:linear, polynomial, radial basis function (RBF) and sigmoid. The highest accuracy of 0.9783 was achieved using linear and polynomial kernels. A closer look to particular errors reveals that there is one misclassification of the Dangerous event. The acceleration of the vehicle is -4.13 m/s^2, which is in the overlapping interval of the acceleration of the classes. The classification of the events in this interval is real challenge for the SVM classifier. The RBF kernel SVM achieved lower accuracy, 0.8913, and the lowest accuracy of 0.5217 was achieved using sigmoid kernel.

The achieved highest accuracy of 0.9783 is sufficient for practical usage of the system. This statement is justified by the fact that the system is designed to be used as off-line support in the process of detection and localization of "black spots", and thus it is not affecting the road safety directly in real time. However, it should also be pointed out that the performance estimation of the last component of the system, the information fusion and event classification, is based on simulated data only, therefore the overall system performance in real world scenario cannot be guaranteed. Practical usage of the system should be justified through thorough field testing in different environmental conditions, for which currently there is no enough available data. Special attention should be put to testing in winter conditions. These conditions are complex and difficult for simulation; therefore, real-world tests are necessary.

6 Conclusions

The paper presents a multi-modal algorithm for detection of potentially dangerous events based on audio and video information fusion. The information extraction in both audio and video domains is performed using deep CNN. The motion parameters of vehicles are estimated using video from calibrated surveillance cameras. In the first step vehicle positions are estimated using a CNN-based estimator. Appropriate design modifications were applied to well understood and proven classifier configurations in order to lower the computational complexity and enable simple extendibility of the system. The achieved performance of the vehicle detector, $precision = 0.9598$ and $recall = 0.9646$, confirm the usability of the detector as part of the dangerous events detection system. The recognition and localisation of the lower front part of the vehicle is also performed using a CNN-based estimator with relatively low computational requirements due to the small size of the input image. The achieved $precision = 0.8434$ and $recall = 0.9738$ enable its usage in the system.

In the second step, the vehicle tracking, the motion parameters of the objects in the scene are calculated, based on content matching techniques and a Kalman filtering. This procedure yields an excellent *accuracy* of 97.6%, and averages absolute error in the *estimated speeds*, below 2 km/h. With the trajectories and the speeds of the vehicles known, a set of traffic parameters can be easily determined, like, overspeed, underspeed, illegal stopping, overtaking, lane changes, wrong direction driving, etc.

Although highly accurate, the vehicle tracking procedure and the speed estimation can be made even more accurate. An improvement of the speed estimation accuracy could be achieved since the initial block being tracked is taken from the front vehicle coordinates, if some procedure for block location correction is introduced. Eventually, this would lower (or rise) the initially chosen block on the vehicle body, in order to minimize the error produced by the real- world physical distance of the tracked point from the road surface. This is one idea that remains to be addressed in our future work.

Audio events detections performed using CNN applied in Mel-frequency cepstral coefficients domain. We have shown that our lean model for sound event detection performs at a high level of accuracy, comparable to that of more complex CNN models. In this way, it does not seriously impact overall system resource utilization. It should be noted nonetheless, recently developed advanced compound CNN such as in [49] may well improve considerably the estimation of near-miss events.

The final stage of the system performs information fusion and classification on the event level. Each event is described using 9-element feature vector consisting of averaged motion parameters and class probabilities of the audio event classes. The feature vector is fed to the SVM based classifier to classify the event as Dangerous or Non-dangerous event. The achieved level of *accuracy* of 0.9783 confirms the usability of the system for assessment of occurrence of potentially dangerous events of skidding. The statistics of this near-miss events could enable reliable detection end localization of road accident "black spots".

The design of the systems considering rare events is difficult due to lack of information and recorded data. To our best knowledge the research described in the paper is first attempt to detect near-miss event based on audio-visual detection of skidding and there is no publicly available database containing necessary audio-visual information. Therefore, all experimental evaluation is performed based on simulated data. For practical usage of the system a thorough experimental evaluation in real-world conditions is necessary.

Acknowledgements This work is supported by the Fund for Innovation and Technology Development of Republic of North Macedonia. The authors would like to express special gratitude to Prof. Georgi M. Dimirovski for his constant encouragement, patience and interest for our work.

References

1. Krizhevsky, A., Sutskever, I., Hinton, G.: Imagenet classification with deep convolutional neural networks. In: NIPS Conference, pp. 1097–1105 (2012)
2. Simonyan, K., Zisserman, A.: Very deep convolutional networks for large-scale image recognition. In: Proceedings of 3rd International Conference on Learning Representations, (ICLR 2015), San Diego, CA, USA, May 7–9 (2015)
3. Szegedy, C., et al.: Going deeper with convolutions. In: 2015 IEEE Conference on Computer Vision and Pattern Recognition (CVPR), pp. 1–9 (2015). https://doi.org/10.1109/CVPR.2015.7298594
4. He, K., Zhang, X., Ren, S., Sun, J.: Deep residual learning for image recognition. In: Proceeding of IEEE Conference on Computer Vision and Pattern Recognition, pp. 770–778 (2016)
5. Girshick, R.: Fast R-CNN. In: Proceedings of the IEEE International Conference on Computer Vision (ICCV), pp. 1440–1448 (2015)
6. Ren, S., He, K., Girshick, R., Sun, J.: Faster R-CNN: towards real-time object detection with region proposal networks. In: Proceedings of the 28th International Conference on Neural Information Processing Systems 2015, pp. 91–99
7. Redmon, J., Divvala, S., Girshick, R., Farhadi, A.: You only look once: unified, real-time object detection. In: 2016 IEEE Conference on Computer Vision and Pattern Recognition (CVPR), pp. 779–788 (2016). https://doi.org/10.1109/CVPR.2016.91
8. Redmon, J., and Farhadi, A.: YOLO9000: better, faster, stronger. In: 2017 IEEE Conference on Computer Vision and Pattern Recognition (CVPR), pp. 6517–6525 (2017). https://doi.org/10.1109/CVPR.2017.690
9. Redmon, J., Farhadi, A.: YOLOv3: an incremental improvement (2018). ArXiv abs/1804.02767, n. pag
10. Liu, W., et al.: SSD: single shot MultiBox detector. In: Leibe, B., Matas, J., Sebe, N., Welling, M. (eds.) Computer Vision—ECCV 2016. ECCV 2016. Lecture Notes in Computer Science, vol. 9905. Springer, Cham (2016). https://doi.org/10.1007/978-3-319-46448-0_2
11. Zhang, F., Li, C., Yang, F.: Vehicle detection in urban traffic surveillance images based on convolutional neural networks with feature concatenation. Sensors **19**, 594 (2019). https://doi.org/10.3390/s19030594
12. Wang, L., Lu, Y., Wang, H., Zheng, Y., Ye, H., Xue, X.: Evolving boxes for fast vehicle detection. In: 2017 IEEE International Conference on Multimedia and Expo (ICME), pp. 1135–1140 (2017). https://doi.org/10.1109/ICME.2017.8019461
13. Yang, L., Luo, P., Loy, C.C., Tang, X.: CompCars: a large-scale car dataset for fine-grained categorization and verification. In: 2015 IEEE Conference on Computer Vision and Pattern Recognition (CVPR), pp. 3973–3981 (2015). https://doi.org/10.1109/CVPR.2015.7299023
14. Wen, L., Du, D., Cai, Z., Lei, Z., Chang, M.-C., Qi, H., Lim, J. Yang, M.-H., Lyu, S.: UA-DETRAC: a new benchmark and protocol for multi-object detection and tracking. Comput. Vis. Image Underst. 193 (2015). https://doi.org/10.1016/j.cviu.2020.102907
15. Sochor, J., et al.: Brno CompSpeed (Dataset): comprehensive data set for automatic single camera visual speed measurement. In: IEEE Transactions on Intelligent Transportation Systems, vol. 20, no. 5, pp. 1633–1643, May 2019 (2019). https://doi.org/10.1109/TITS.2018.2825609
16. Khosravi, H., Gholamalinejad, H.: IRVd: a large-scale dataset for classification of iranian vehicles in urban streets (2020). https://doi.org/10.22044/JADM.2020.8438.1982

17. Object Tracking using OpenCV [Online]. https://learnopencv.com/object-tracking-using-opencv-cpp-python/
18. Huang, T.: Traffic speed estimation from surveillance video data: for the 2nd NVIDIA AI city challenge track 1. In: IEEE International Conference on Computer Vision Workshops (2018)
19. Mao, H., Ye, C., Song, M., Bu, J., Li, N.: Viewpoint independent vehicle speed estimation from uncalibrated traffic surveillance cameras, 4920–4925 (2009). https://doi.org/10.1109/ICSMC.2009.5346288
20. Tang, Z., Wang, G., Xiao, H., Zheng, A., Hwang, J.-N.: Single-camera and inter-camera vehicle tracking and 3D speed estimation based on fusion of visual and semantic features. In: CVPR Workshop (CVPRW) on the AI City Challenge (2018)
21. Liu, T., Liu, Y., Tang,Z., Hwang, J.-N.: Adaptive ground plane estimation for moving camera-based 3D object tracking. In: IEEE International Workshop Multimedia Signal Processing (2017)
22. Sochor, J., ŠpaĖhel, J., Herout, A.: BoxCars: improving fine-grained recognition of vehicles using 3-d bounding boxes in traffic surveillance. IEEE Trans. Intell. Transp. Syst. (2018)
23. Lee, K.-H., Hwang, J.-N., Chen, S.-I.: Model-based vehicle localization based on three-dimensional constrained multiple-kernel tracking. IEEE Trans. Circ. Syst. Video Technol. **25**(1), 38–50 (2014)
24. Faragher, R.: Understanding the basis of the Kalman filter via a simple and intuitive derivation [Lecture Notes]. In: IEEE Sig. Process. Mag. **29**(5), 128–132, September 2012. https://doi.org/10.1109/MSP.2012.2203621
25. Chen, Z., Ellis, T., Velastin, S.: Vehicle detection, tracking and classification in urban traffic. In: 15th ITSC, pp. 951–956 (2012)
26. Nurhadiyatna, A., Hardjono, B., Wibisono, A., Sina, I., Jatmiko, W.,Ma'sum, M.A., Mursanto, P.: Improved vehicle speed estimation using Gaussian mixture model and hole filling algorithm. In: Proceedings of the 2013 International Conference on Advanced Computer Science and Information Systems (ICACSIS), Bali, Indonesia, 28–29 September 2013, pp. 451–456 (2013)
27. Anil Rao, Y.G., Kumar, N.S., Amaresh H.S., Chirag, H.V.: Real-time speed estimation of vehicles from uncalibrated view-independent traffic cameras. In: TENCON 2015—2015 IEEE Region 10 Conference, pp. 1–6 (2015). https://doi.org/10.1109/TENCON.2015.7373162
28. Kuhn, H.W.: The Hungarian method for the assignment problem. Naval Res Logist Quar. **2**(1–2), 83–97, March 1955
29. Li, H., Bai, P., Song, H.: Car tracking algorithm based on Kalman filter and compressive tracking. In: 7th International Congress on Image and Signal Processing, pp. 27–31 (2014). https://doi.org/10.1109/CISP.2014.7003744
30. Cathey, F.W., Dailey, D.: A novel technique to dynamically measure vehicle speed using uncalibrated roadway cameras. In: Intelligent Vehicles Symposium, pp. 777–782 (2005)
31. Dubska, M., Sochor, J., Herout, A.: Automatic camera calibration for traffic understanding. In: BMVC (2014)
32. Schoepflin, T., Dailey, D.: Dynamic camera calibration of roadside traffic management cameras for vehicle speed estimation. IEEE Trans. Intell. Transp. Syst. **4**(2), 90–98, June 2003
33. Maduro, C., Batista, K., Peixoto, P., Batista, J.: Estimation of vehicle velocity and traffic intensity using rectified images. In: IEEE International Conference on Image Processing, ICIP 2008, 15 October 2008, pp. 777–780 (2008)
34. Nurhadiyatna, A., Hardjono, B., Wibisono, A., Sina, I., Jatmiko, W., Ma'sum, M., Mursanto, P.: Improved vehicle speed estimation using Gaussian mixture model and hole filling algorithm. In: International Conference on Advanced Computer Science and Information Systems (ICACSIS), September 2013, pp. 451–456 (2013)
35. Hartley, R., Zisserman, A.: Multiple View Geometry in Computer Vision, 2nd edn. Cambridge University Press, New York, NY, USA (2003)
36. Heittola, T., Mesaros, A., Virtanen, T., Gabbouj, M.: Supervised model training for overlapping sound events based on unsupervised source separation. In: 2013 IEEE International Conference on Acoustics, Speech and Signal Processing, Vancouver, BC, Canada, May 2013, pp. 8677–8681 (2013). https://doi.org/10.1109/ICASSP.2013.6639360

37. Valenzise, G., Gerosa, L., Tagliasacchi, M., Antonacci, F., Sarti, A.: Scream and gunshot detection and localization for audio-surveillance systems. In: 2007 IEEE Conference on Advanced Video and Signal Based Surveillance, London, UK, September 2007, pp. 21–26 (2017). https://doi.org/10.1109/AVSS.2007.4425280
38. Cakir, E., Parascandolo, G., Heittola, T., Huttunen, H., Virtanen, T.: Convolutional recurrent neural networks for polyphonic sound event detection. IEEEACM Trans. Audio Speech Lang. Process. **25**(6), 1291–1303, June 2017. https://doi.org/10.1109/TASLP.2017.2690575
39. Zhang, H., McLoughlin, I., Song, Y.: Robust sound event recognition using convolutional neural networks. In: 2015 IEEE International Conference on Acoustics, Speech and Signal Processing (ICASSP), South Brisbane, Queensland, Australia, April 2015, pp. 559–563 (2015). https://doi.org/10.1109/ICASSP.2015.7178031
40. Deng, L., Hinton, G., Kingsbury, B.: New types of deep neural network learning for speech recognition and related applications: an overview. In: 2013 IEEE International Conference on Acoustics, Speech and Signal Processing, Vancouver, BC, Canada, May 2013, pp. 8599–8603 (2013). https://doi.org/10.1109/ICASSP.2013.6639344
41. Sainath, T.N., Mohamed, A., Kingsbury, B., Ramabhadran, B.: Deep convolutional neural networks for LVCSR. In: 2013 IEEE International Conference on Acoustics, Speech and Signal Processing, Vancouver, BC, Canada, May 2013, pp. 8614–8618 (2013). https://doi.org/10.1109/ICASSP.2013.6639347
42. Tóth, L.: Phone recognition with hierarchical convolutional deep maxout networks. EURASIP J. Audio Speech Music Process. **2015**(1), 1–13 (2015)
43. Chavdar, M., Gerazov, B., Ivanovski, Z., Kartalov, T.: Towards a system for automatic traffic sound event detection. In: 2020 28th Telecommunication Forum TELFOR, pp. 1–4 (2020). https://doi.org/10.1109/TELFOR51502.2020.9306592
44. Cortes, C., Vapnik, V.: Support-vector networks. Mach. Learn. **20**, 273–297 (1995). https://doi.org/10.1007/BF00994018
45. Velazquez-Pupo, R., Sierra-Romero, A., Torres-Roman, D., Shkvarko, Y.V., Santiago-Paz, J., Gómez-Gutiérrez, D., Robles-Valdez, D., Hermosillo-Reynoso, F., Romero-Delgado, M.: Vehicle detection with occlusion handling, tracking, and OC-SVM classification: a high performance vision-based system. Sensors **18**, 374 (2018). https://doi.org/10.3390/s18020374
46. Karungaru, S., Dongyang, L., Terada, K.: Vehicle detection and type classification based on CNN-SVM. Int. J. Mach. Learn. Comput. **11**(4), 304–310 (2021)
47. Salamon, J., Jacoby, C., Bello, J.P.: A dataset and taxonomy for urban sound research. In: Proceedings of the 22nd ACM international conference on Multimedia, Orlando Florida USA, November 2014, pp. 1041–1044 (2014). https://doi.org/10.1145/2647868.2655045
48. Kingma, D.P., Ba, J.: Adam: a method for stochastic optimization (2017)
49. Cheng, D.-D., Liu, L.-J., Yu, Z., CNN-based intelligent fault-tolerant control design for turbofan engines with actuator failures. IEEE Access (2021). https://doi.org/10.1109/ACCESS.2021.3058387. Accessed 10 Feb 2021

Modeling and Feedback Control for Development of Mobile Technologies in Virtual Education Environments

Nurassyl Kerimbayev, Vladimir Jotsov, Aliya Akramova, and Nurdaulet Nurym

Abstract Contemporary Information Technologies (IT) became one of the main communication means establishing the contact between the teacher and student. Modeling and feedback control are one of the main features to be explored in this direction aiming at high quality results. The communication in real time conditions is established using a broad variety of web based hardware like notebooks, smartphones, and other innovation gadgets improving the education process via modern visualization and virtualization tools. For better results the communication feedback should use Data Science methods to model and control the contemporary IT education process, especially in case of education in Control Systems. Virtual education environments are explored in this case. The usage of web based and mobile technologies allows to improve the quality of education process and to organize more efficient understanding and active learning conditions. On the other hand, this feedback control includes a quick and efficient estimation/scoring process. The proposed complex of mobile technologies and platforms visualize different processes of communication activities in the interactive learning environments. The functional value of the mobile technologies gives an opportunity to apply visual search, voice recognition, mirror display and so on. The elaborated feedback tools backed by mobile technologies in virtual learning environment may be broadly used in the contemporary education sphere. It improves the efficiency of the tutoring/learning process and to enhance the better upbringing process.

Keywords Virtual education environment · Virtual reality · Data science · Modeling · Constraint · Feedback control · Education · M-Learning · Data-driven education

---

N. Kerimbayev · V. Jotsov (✉) · A. Akramova · N. Nurym
Al-Farabi Kazakh National University, Almaty, Kazakhstan
e-mail: v.jotsov@unibit.bg

V. Jotsov
University of Library Studies and Information Technologies, Sofia, Bulgaria

P. Shi et al. (eds.), *Complex Systems: Spanning Control and Computational Cybernetics: Applications*, Studies in Systems, Decision and Control 415,
https://doi.org/10.1007/978-3-031-00978-5_16

1 Introduction

The issues and problems of mobile learning in virtual environments have gained widespread academic and commercial recognition in recent years. However, the issues of modeling and feedback management for mobile technologies in virtual learning environments are not well understood. This work makes an attempt to show the change in the mechanisms of functioning and implementation of the education system in the conditions of virtual learning environments. The proposed work is expected to make the virtual learning organization effective, practical and ergonomic. Modeling and feedback management, in turn, provide a learning process aimed at mastering the result. Mobile technologies are technical support for the learning virtual environment, under the influence of which significant changes occur in the process of assimilating knowledge, implementing mobile, highly effective feedback.

The problem field and purpose of this chapter is to explore the process of modeling and feedback management for the development of mobile technologies in virtual learning environments. The scientific and theoretical assumptions put forward in this study are translated into the field of practical application of the developed mobile technologies, their transfer to a virtual learning environment, while simulating feedback management processes.

This chapter outlines the prospects for further research in the development of technologies for augmented and virtual reality. The relevance of this issue is determined by the modern rapid development of information and communication technologies in all spheres of human activity. The tasks that were solved in the course of this work are aimed at identifying the current state of mobile technologies in the field of higher education in foreign and domestic practice, presenting the experience of scientific and practical activities in the studied area and the considered problem of modeling and managing mobile technologies in a virtual educational environment.

With the rise of mobile gadgets, social networking and rapid technology innovation, new possibilities for improvement of the quality of education arise. The contemporary education is widely using different intelligent tools in m-learning, e-learning, SCRUM, virtual education, and so on. One of the most advanced of them concern Data Science technologies. To construct and/or use them we need strong deep modeling tools used with other non-classical logic applications, Description Logics. In these conditions original deep modeling tools have been elaborated and considered, they are combinable with classical constraint satisfaction cases. Deep modeling methods are fit to describe and gradually sustain various thinking and reasoning processes. They are effective in conditions when the logical reasoning is naturally constrained and the machine learning using artificial Neural networks (ANNs) couldn't be widely applied. It is considered that both classical and non-classical logical operators may be applied in one scheme. On the other hand, the effectiveness of data selection and data processing in the Data Science cycle significantly increases. As a result, the education becomes more transparent, adaptive and effective. In Sect. 4, aiming to improve the quality of education on IT and a broad range of different fields, a special preprocessing stage of the data science cycle is elaborated and analyzed. It is based on deep knowledge modeling skills and concerns other corresponding synthetic/data-driven puzzle methods. It is shown that the skillful/correctly designed/successful combination of Data Science preprocessing methods doesn't infer higher demands to the IT/intelligent background of the students of any specialty and significantly diminishes the problems occurring from learning contemporary rapidly developing fields like Data Science as the center of contemporary education, cyber security, virtual reality and other innovative research areas. This is executed via higher qualification of the teacher and/or application of methods transferring formal representations to ontology-based graphical and/or virtual interpretations of the presented IT or other types of material.

Mobile technologies facilitate the processes of management, control, and feedback support. In recent decades, mobile phones have been used as an interface for providing feedback. For example, a study of the effectiveness of mobile devices in the transmission of electricity information is described in the work Weiss, Loock, Staake, Mattern and Fleisch, «Evaluating Mobile Phones as Energy Consumption Feedback Devices» [1].

Wechtitsch, Fassold, Thaler, Kozłowski, Bailer, in work «Quality Analysis on Mobile Devices for Real-Time Feedback» offer methods for visual quality analysis on mobile devices, in order to provide direct feedback to the contributing user about the quality of the captured content. The proposed developments can be used in real time at concerts, e.g., Music Festival [2].

The educational sector today makes extensive use of mobile technologies in virtual learning environments. Many studies are devoted to the development of mobile technologies and their application in the educational process of higher education, modeling and feedback management. These processes are considered taking into account objective factors and take into account the subjective preferences of the participants in the educational process. The development of mobile applications, as a rule, occurs based on the modern needs and interests of students, who can be attributed to the new digital generation.

The contradictions and problems of the use of mobile technologies in modern higher education are considered in a number of studies. Rius, Masip, & Clarisó, studying mobile learning, note: «Educational institutions are facing the challenge of providing students with tools for mobile learning (m-learning). However, the evolution of technology makes the development and continuous improvement of these tools rather expensive» [3].

Evans and Johri offer strategy for integrating mobile technology-based learning experiences in higher education. The authors researched how mobile technologies and social software can be used to (a) facilitate guided participation among undergraduate engineering students within classes and (b) teach graduate students in instructional technology to design for guided participation. Researchers Raise the Problem designing creative learning environments for self and other, sure, that mobile technologies provide a substantive, fertile, and invigorating area for teaching and research in higher education for the foreseeable future [4].

The proposed study summarizes the experience of developing mobile technologies in virtual learning environments through simulation and feedback management.

Therefore, the logic of constructing the study will correspond to the main problems disclosed in this work. In the following sections, feedback management in mobile technologies used in the field of higher education will be considered. The Modeling and Feedback Management section covers modeling and feedback management in virtual learning environments using mobile technologies.

The virtual educational environment allows for interactive communication using LMS Moodle [5]. When conducting lecture lessons, there is a necessity to search for motivational methods for students, to provide a fast feedback and assistance in identifying gaps the students demonstrate while acquiring a new material. The students use links to choose the given chapter of the material tested and participate in on-line questionnaire answering questions in their tablets, laptops or smart phones. The model of virtual learning offered serves as a basis for continuous monitoring and assessment, which provides an immediate feedback.

The developed mobile technologies are considered as a way of professional communication of participants in the educational process in a virtual learning environment represented as a "teacher–student" and various invariants of the elements of this system.

2 Mobile Technologies in Virtual Learning Environments: Overview and Analytics

In this section, an attempt is made to highlight the issues of mobile technologies in virtual learning environments, to reveal the features and ideas of scientific research in the field of pedagogical and information technology design, development and design of mobile applications. Today, mobile learning (m-learning–Mobile Learning) has become one of the options for modern learning and a source of fast knowledge. M-learning technologies related to e-learning and distance learning technologies are becoming the most widespread. Mobile technologies are becoming an integral part of modern educational technologies including in virtual learning environments [6].

Mobile learning is implemented using various mobile portable devices: gadgets, laptops, tablet personal computers, smartphones, e-books, portable audio and video guides, and modern game consoles. Mobile technologies implement a variety of support mechanisms for e-learning and have much more advantages over traditional training which required preparation of the workplace using a personal computer or computer technology.

Mobile learning is characterized by the following features:

- implementation of communication regardless of the geographical location of the participants in the educational process, the absence of time dependence for obtaining information (training can occur asynchronously);
- access of students to all necessary information/content of educational content;

- inclusion and activation of students' activities in the educational process, the ability to communicate online and offline, ask questions through chats, instant messengers which contributes to the expansion and growth of student cooperation;
- personal mobile devices act as an individual technical means for obtaining and accessing network information, providing feedback, using multimedia and other educational content;
- collaboration, continuity, situation-oriented solutions, individualization and personalization of the learning process consisting in the ability to learn at your own pace according to an individual plan and schedule.

Mobile devices can use the services of teachers in an information environment which can be a virtual educational environment. Interaction and collaboration in such an environment allows teaching various content, organizing project training, and performing interactive tasks using mobile devices. Today the main develop of mobile applications concerns the design of accessible training applications, the assessment and adjustment the used tools.

Android Accessibility Designer Toolkit and the Accessibility Advisor tool is presented by Gemou et al. [7]. Authors consider toolkits for developing third-generation Android accessible mobile applications. The research reveals that the developer tools have a big potential to help developers create easily accessible applications.

Topical questions of development, and evaluation of a mobile learning application are the following. A mobile learning application, MobileEdu, for computing education in the Nigerian higher education is context-developed by Oyelere et al. Mobileedu facilitates the learning of computer science courses on mobile devices [8]. The application supports ubiquitous, collaborative, and social aspects of learning among higher education students. The study offered suggestions for how to implement effectively a mobile learning-supported course in computing curriculum.

When working with mobile devices, it is necessary to raise awareness, motivation to use mobile devices in a learning environment, organization and development of mobile technologies to support collaborative learning [9]. In group communications, the user must adapt to the system, and the behaviour of the system, providing group feedback in accordance with the services.

Mobile technologies in virtual environments based on augmented reality technologies and on mobile learning are widely represented in educational practices. Development of a virtual butterfly ecological system based on augmented reality and mobile learning technologies in research Tarng et al. is a virtual process of breeding butterflies and observing their life cycle [10]. This example clearly illustrates that «the virtual butterfly ecological system can increase the learning motivation and interest of students through virtual breeding and observation activities, so it is a suitable assistant tool for science education».

Three-dimensional simulated systems in virtual environments are also becoming one of the main technologies for mobile learning. Virtual laboratories and excursions are training, in which not only the processes/events occurring in the environment can be recreated, but also various tools for joint activities can be provided. As noted by a number of researchers, 3D-modeled applications increases mobility, decreases set-up and breakdown time, and has less spatial requirements [11].

Virtual reality in the learning process is a complex universal technology, improved work and improvement of the learning process, skills and abilities of the student. The inclusion of students in the virtual educational environment has a positive effect on the motivational and cognitive sphere of personal development, and on the process of their socialization in general. Getting positive emotions while learning with the help of reality effectively affects the success of learning [12].

Article Nurym et al. "Virtual Reality and Using the Unity 3D Platform for Android Games" explores how computer technology and virtual 3D devices extensively applied in the industry of video games play roles of great importance in the development of virtual reality [13]. Training in virtual reality (VR) has shown that this type of training reduces the cost of attracting expensive equipment, specialists and teachers. Immersing a person into an immersive virtual world is possible using specialized devices (glasses, virtual reality helmets). It improves memorization and understanding of spatial and visual information; psychomotor skills: observation, emotional reactions. Augmented Reality (AR) technology, provides additional computer information, virtual examples, and introduces game elements to support educational materials. Virtual and augmented reality are not limited for use in a specific academic subject, age group, or educational level. The availability and portability of teaching materials in VR and AR makes them especially attractive for use in education (Fig. 1).

Virtual learning in a modern university is widespread both in the educational process and in the preparation of specialists for training in a virtual environment [14]. The development of the Virtual Learning Center including educational/cognitive content made it possible to conduct training in a virtual environment, and implement joint training with foreign partners. The collaborative environment reinforced educators to look for new approaches, ways, and methods of teaching. There was a need to conduct feedback in a virtual environment, which was successfully resolved by the development and management of mobile technologies in VS.

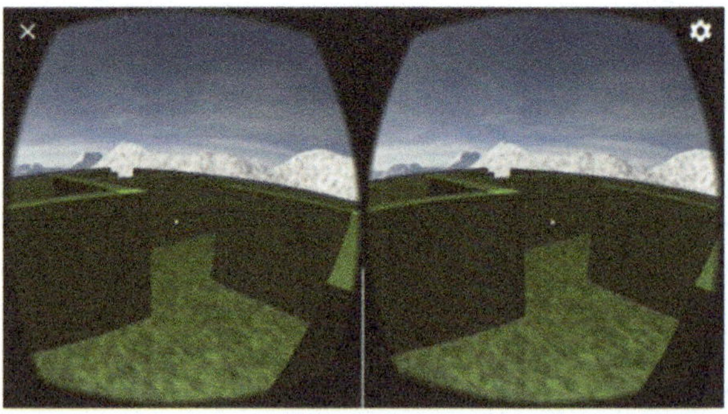

Fig. 1 Form of the game for virtual reality glasses

Thus, modern digital learning determines the nature of the educational process/educational schemes. Students today use different mobile devices to take courses. Therefore, mobile versions of courses and applications are becoming a necessary requirement.

Training during the Covid-19 pandemic transferred to the format of distance learning. The main form of training was web meetings, video lectures, webinars: "virtual" seminars organized by IT, distance learning in the form of video conferencing, and virtual classes. The wide range of mobile learning tools includes audio podcasts, recording, archiving, collaboration, software for demonstration, drawing and sketching, chatting, etc. Comparison of four digital learning technologies: e-lectures, classroom response system, classroom chat, and mobile virtual reality in terms of their technology acceptance had been conducted by Sprenger and Schwaninger, [15]. According to this study, students prefer to study in the format «classroom response system» closely followed by e-lectures, then was the CC and in the end mobile VR. Today mobile learning in the cloud is also receiving a widespread academic and commercial recognition [16].

3 Feedback: Modeling and Management in Virtual Learning Environments

As part of this study, we will demonstrate a developed mobile application with feedback elements. The mobile learning model using the developed application is shown in Fig. 2.

The mobile learning model consists of 4 components: pedagogical, informational, technical, and organizational. Considering the structure of a mobile application, the trajectories "Student" and "Teacher" can be distinguished. The podcasts of these tractors are presented in Figs. 3 and 4.

The "Student" mode assumes two modes of operation "Class selection (Classroom)" and "Self-study (Extracurricular)".

Class «Selection mode».

- Completing the topics of the computer science course for a specific class using various teaching methods;
- Presentation of new materials;
- Practical assignments;
- Control;
- The transition from one topic to another will be possible after the completion of the previous one;
- Assignments are graded and student progress is recorded;
- Feedback from the teacher.

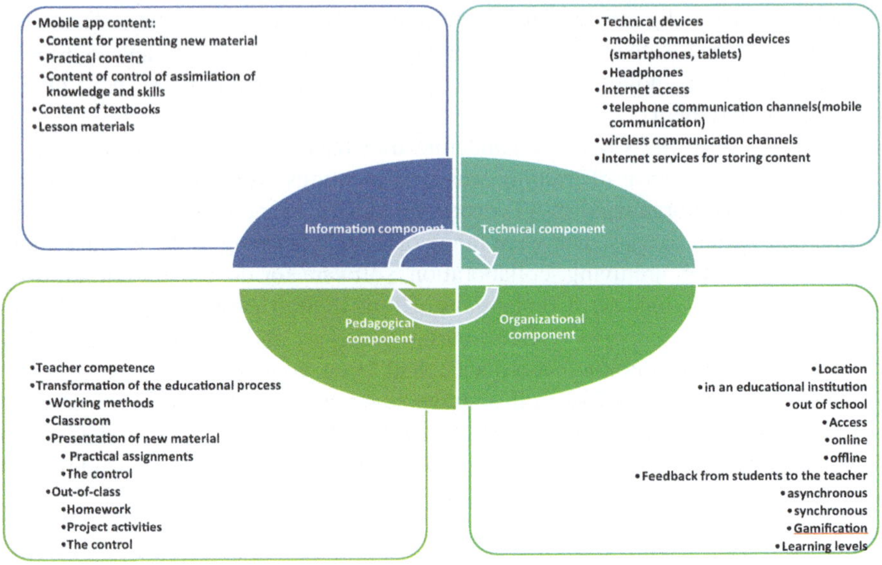

Fig. 2 Mobile learning model

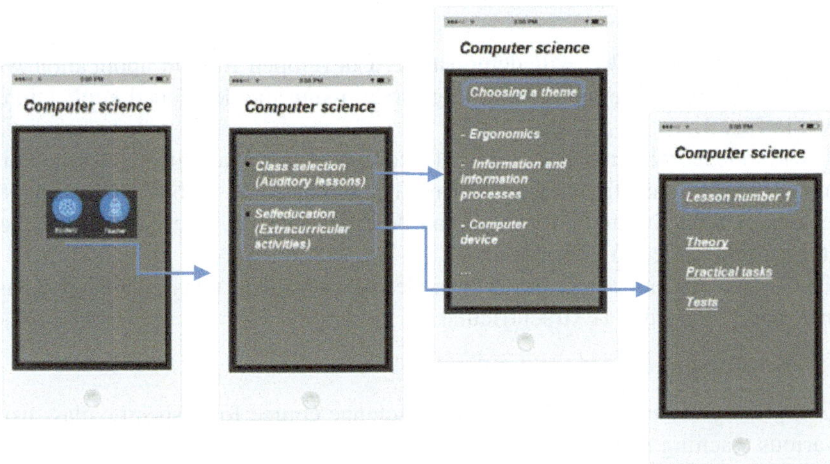

Fig. 3 The structure of the student's mobile application

"Selfeducation":

- Passage of training in an arbitrary mode, provides for the choice of any topic
- Progress does not sync
- Feedback from the teacher.

The "Teacher" mode assumes:

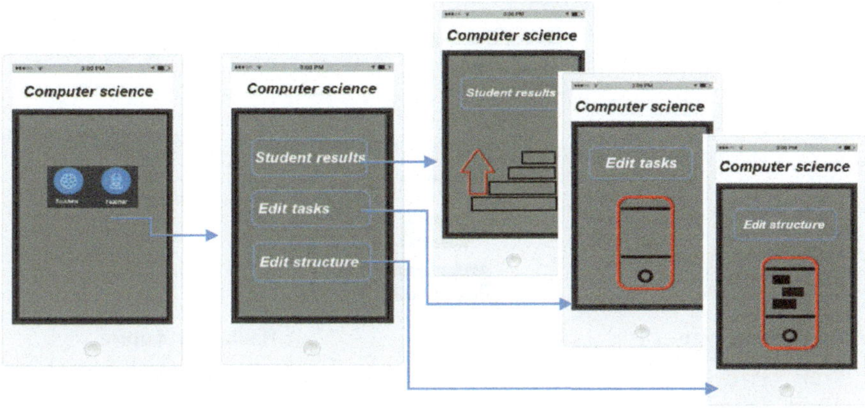

Fig. 4 Structure of the teacher's mobile application

- The ability to edit the learning path in the mobile application;
- The ability to edit and add tasks;
- View student progress;
- Implementation of feedback from students.

3.1 Feedback Control Technology

Currentlythe learning process is being activated with the help of information technology, and new opportunities have appeared in the e-learning system for the development of various digital tools to achieve the learning outcome of students. The teacher/student uses one of these tools, Voting, to support student motivation and the process of knowledgeself-regulation. The interface of the virtual environment platform developed by us for ease of the application consists of two parts: a web application for the teacher, and a mobile application purposed for students.

Teacher interface. The teacher in the platform creates his account or he is authorized in the system. The teacher has the ability to create, add and delete test assignments, determine the date of the test, edit, send a link to students, visualize the results (Figs. 5, 6 and 7).

Student interface. Currently, many existing platforms require student login upon entry. This takes time and prevents quick feedback from the teacher. In order to bypass this obstacle, we decided to leave the student part of the platform open. The teacher sends the students an access link to the assignments, and the students via received linkanswer the test assignments anonymously or by writing their first/last name. Theycanseetheiranswersonthesmartphonescreenaftercompletingthesurveyandcancomparethemwiththecorrectanswer (Fig. 8).

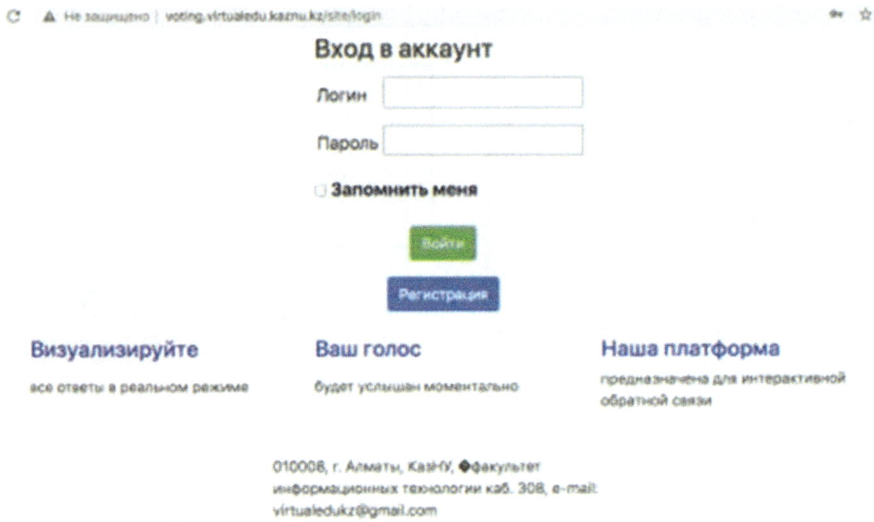

Fig. 5 Authorization or registration

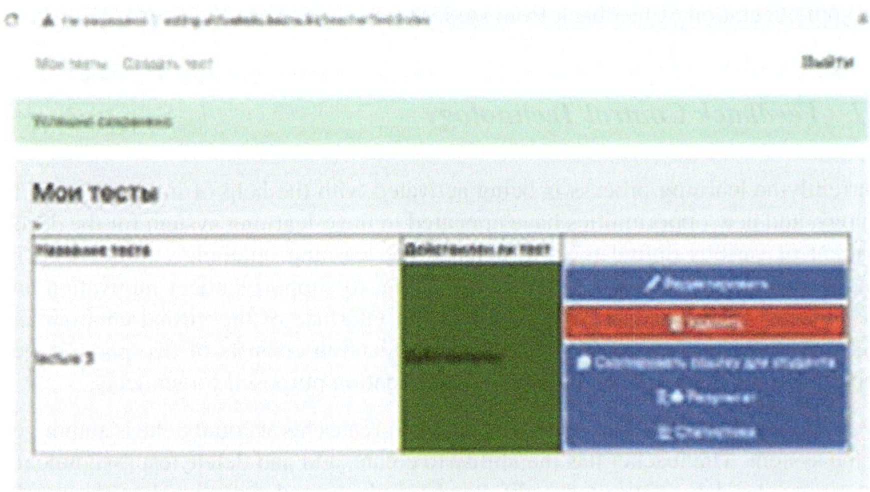

Fig. 6 Login to “My tests”

This allows students to instantly see their answer visually in a real time, without authorization, aiming to establish feedback with the teacher. The system allows passing tests only once. In the traditional teaching format, Voting of the teacher/student is used during lectures to establish feedback with students, as well as to check the intermediate learning outcome. A survey conducted among students and teachers using this feedback form revealed a positive attitude towards this platform.

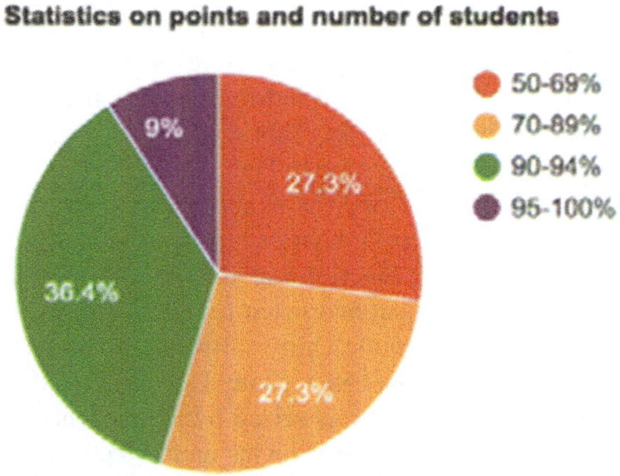

Fig. 7 Results visualization

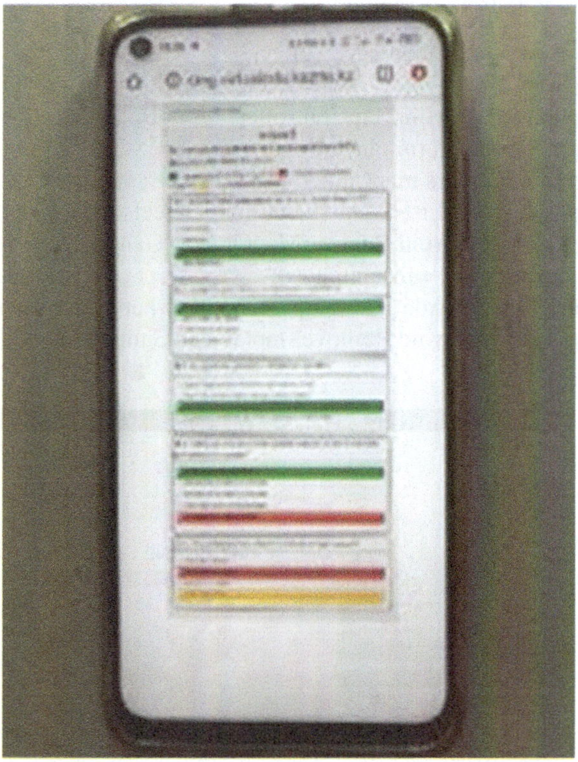

Fig. 8 Test result on a mobile device

Students themselves are involved in the development of mobile applications or virtual educational games. Here is an example of an application developed on the Apache Cordova platform "MyFhoneEnglish" for learning English using a mobile device (Fig. 9). This application acts as a program for study of speech communications in a foreign language.

The application runs in wrappers designed for each platform and relies on a standard API to access device sensors, data, and network status. The application has an audio recording function that allows the listener to pronounce words, phrases, and if necessary to repeat and listen to the recording. The software is developed in the Kazakh language in the Unity software environment. The interface of the software product has 4 buttons: New game, Download, Exit, Settings (Fig. 10).

Such games are characterized by the presence of a mini-map on which you can follow the navigation of vehicles and objects on the territory. Below is an excerpt from the code that performs the function of calculating the distance/approach of the vehicle from the target:

```
public Image icon {get; set;}
public GameObject owner {get; set;} }
public class MiniMapController: MonoBehaviour {
public Transform playerPos;
public Camera mapCamera;
public static List<MapObject>mapObjects=new List<MapObject>();
public static void RegisterMapObject (GameObject o, Image i){
Image image =Instantiate(i);
mapObjects.Add(new MapObject(){owner=o, icon=image});}
public static void RemoveMapObject(GameObject o){
```

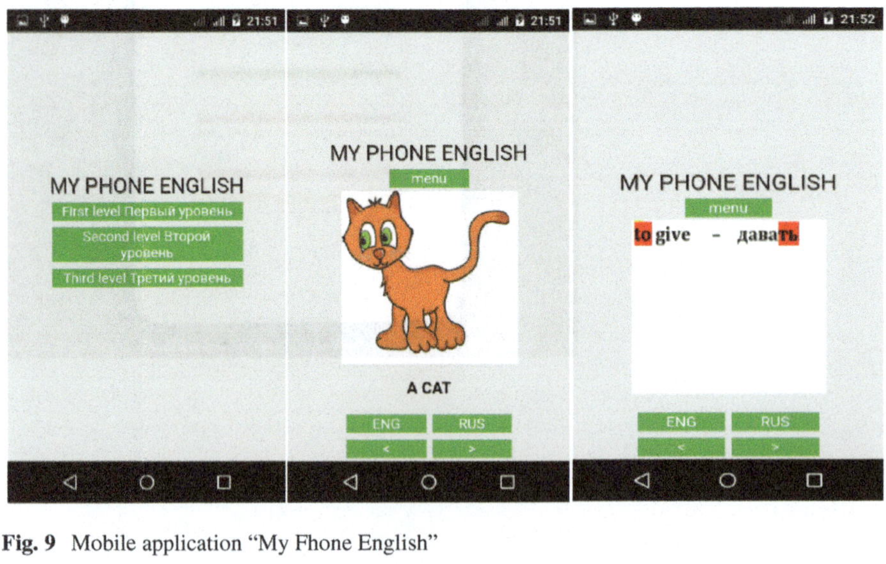

Fig. 9 Mobile application "My Fhone English"

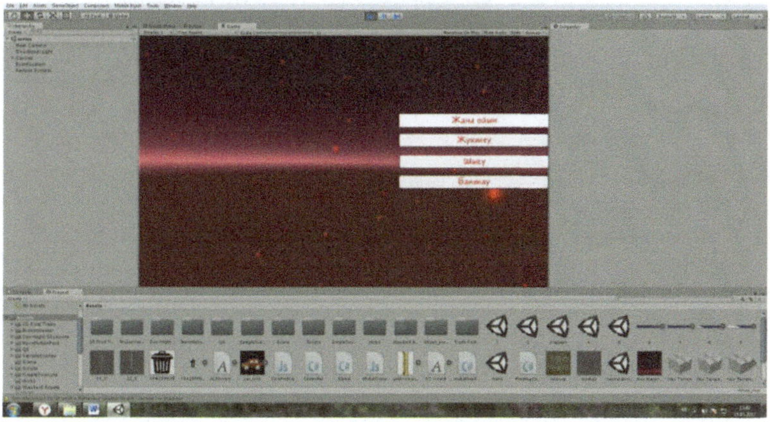

Fig. 10 Main menu window

```
List<MapObject>newList=new List<MapObject>();
for (int i=0; i<mapObjects.Count;i++){
if(mapObjects[i].owner==o){
Destroy(mapObjects[i].icon);
continue;}
else newList.Add(mapObjects[i]);}
mapObjects.RemoveRange(0, mapObjects.Count);
mapObjects.AddRange(newList);}
void DrawMapIcons(){
foreach (MapObject mo in mapObjects){
Vector3 screenPos=mapCamera.WorldToViewportPoint
(mo.owner.transform.position);
mo.icon.transform.SetParent(this.transform);
RectTransform rt=this.GetComponent<RectTransform>();
Vector3 [] corners=new Vector3[4];
rt.GetWorldCorners(corners);
screenPos.x=Mathf.Clamp(screenPos.x*rt.rect.width+corners[0].x,corners[0].x,
corners[2].x);
screenPos.y = Mathf.Clamp(screenPos.y*rt.rect.height+corners[0].y,corners[0].y,
corners[1].y);
screenPos.z=0;
mo.icon.transform.position=screenPos;
}}
void Update(){
DrawMapIcons();
}}
```

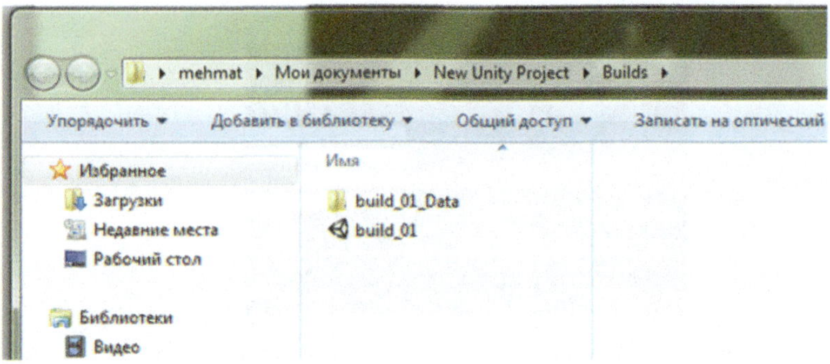

Fig. 11 The interface of one of the presented games

You can also create an exe file to package the entire project. To do this, use the commands File-> Build Settings. Then a dialog box will appear as in the picture below.

After clicking the Build and Run button, the program will ask you to save and name the project. After that, the exe file will appear (Fig. 11).

3.2 Robotic Mobile Technologies in Teaching

Nowadays the robots are becoming more common, there is an increasing amount of research aimed at simulation of the process of automatic movement of a robotic system, navigation and control of mobile robots. Due to the large amount of data required to train AI-based approaches, development of robotic mobile devices is an attractive choice for teaching robots.

Multimodal feedback is carried out in the development of mobile robots where the human operator receives the necessary and timely information about the object [17]. User-friendly interfaces with powerful capabilities integrate into the medical and healthcare industry. Mobile applications are being developed for health monitoring, collaboration between doctors and the public [18].

Smartphone-based assistive technologies promotes independence, ease of use and usability resulting in improved quality of life yet poses several challenging opportunities for blind people [19].

Robotics has positively established itself in the international educational space. Teaching robotics, cooperation between children/students from different countries can take place in a virtual environment, in the process of communication, interaction between students and teachers from different countries. The integration of robotics in the international educational space expands the horizons of the activity in this area enriching theory and practice [20]. The study proved that the experience of scientific pedagogical cooperation and the exchange of knowledge and achievements of children from different countries leads to an understanding of most phenomena and processes, teaches them friendship and tolerance in modern society and in the world.

Work experience has shown the effectiveness of involving students in the development of robotic mobile devices for various spheres of human activity.

The robotic mobile device developed by us is designed to assist in the movement of blind people. When developing a mobile device, we used a small microcontroller from Arduino Nano, an electronic board, a buzzer, a power supply, an ultrasonic sensor to determine the distance to an obstacle.

This is due to the fact that the ultrasonic sensor has high accuracy, stability of readings, does not depend on external influences, and also:

- this is the most harmless in relation to air pollution;
- the paint does not affect the body at a distance from it;
- it uses a wide temperature range;
- small size;
- no special experience is required to work with it;
- the build quality is good as there are no moving parts.

At the initial start-up, the device automatically calibrates, individually adjusts to the height of each user. During a full calibration, it determines the distance to the ground and fixes this distance as zero level. When this distance changes, the device gives more or less 10 cm from the zero level an alarm signal about the obstacle: when the distance decreases with one frequency, and when the distance increases with another frequency.

In order to calibrate sensors, it is important to have accurate physical standards to simulate the corresponding external influences. Since the sensor is included in measuring systems with feedback, we needed to give its phase characteristics (Figs. 12 and 13).

A surface with obstacles can be thought of as uneven by replacing the obstacles with a flat "bump". Obstacles are associated with different resource values e.g., asphalt, sieves, sand, grass, wetlands, etc.

The following approach is suggested here. There is an easy way: find the trajectory using variational methods.

Fig. 12 External view of the mobile device

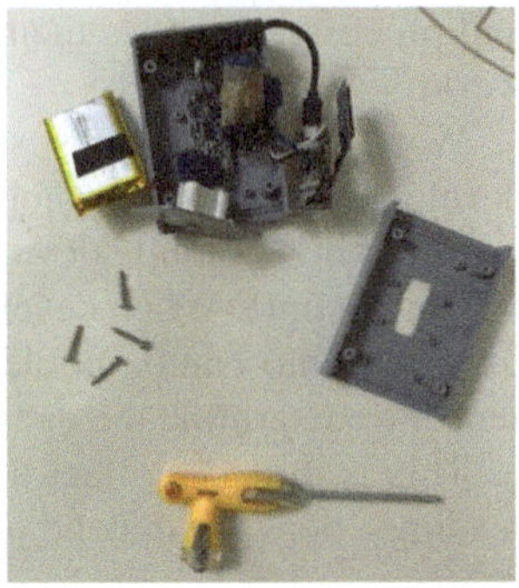

Fig. 13 Disassembled mobile device

Think of the work area as a function (not smooth).

$$z = f(x, y),$$

where: z is the "height" of the trajectory of this point.

x and y are the Cartesian coordinates of the projection of the point onto the plane.

Working area function (smooth) in a uniform plane:

$$\mathrm{y} = \mathrm{f}\,(\mathrm{x})$$

We obtained a numerical solution to this problem and used it when writing the program code. Robotic mobile device.

When creating real-time systems, it was necessary to bind intra-system events to moments in time, timely capture and release of system resources, synchronization of computing processes, buffering of data streams, etc.

In virtual reality used for education purpose and especially when intelligent/mobile devices have been applied, the classical modeling possibilities are insufficient to cover the demand for data = driven learning depending not by the plan, emotional control, transfer of information by sense, personalization through the help of software agents, and the control of accumulated big data from the cloud services and learning systems. Aiming to achieve these and other similar goals, three new large logic-based groups of constraints have been introduced and applied.

4 Deep Modeling Through Constraint Satisfaction in Virtual Learning Environments: Overview

Two groups of students have been analyzed in the considered research where modern technologies are used in the learning process. The first group consisted of students studying Data Science at B.S. or M.S. stage. The second group was from Education/Teacher Preparation scope. As a result of the below described deep modeling, the concentration of attention and the teamwork is increased in both groups.

Deep modeling and data preprocessing may significantly improve the versatility of various education applications, and diminish the preparation work much of which is hidden behind the perfect mathematically described apparatus. One of the problems in Data Science concerning deep modeling problems is related to the selection of data in order to process them more efficiently. How could we select the constantly changing data in the warehouse environment? In these conditions our reasoning will be mainly outdated or totally wrong. To pay the attention to the important details and to avoid over attention on little issues, the learner should fully know/understand the scope specifics. The proposed set of methods could be used for both data analysis and correction of weak/wrong results. These issues are directly related to the methods for deep modeling of data/knowledge without which the application cycle for logical and statistical data processing in order to extract hidden patterns/deep knowledge cannot be used. Different types of non-classical logics, especially data-driven approaches and/or machine learning methods have been used for this purpose [21–24]. Classical selection tools include both various statistical applications and applications of non-classical logics: Description Logics, Modal Logics, some Paraconsistent Logic elements. There are many studies for solving crossword puzzles [22–26], but in most of them the problems are resolved in a probabilistic way, by chance, by using random numbers and their combinations, which is not effective, and the computational complexity is inappropriately high. This chapter proposes to increase the efficiency by using logically-oriented modeling tools. To solve the problem, it is proposed to use three groups of non-classical constraints. They are considered in connection with the research of the puzzle methods using the system of classical and new introduced types of constraints [27–31]. The goal is to form a closed area aiming to focus attention to certain/interesting to the system data including two objects X and Y (Fig. 14).

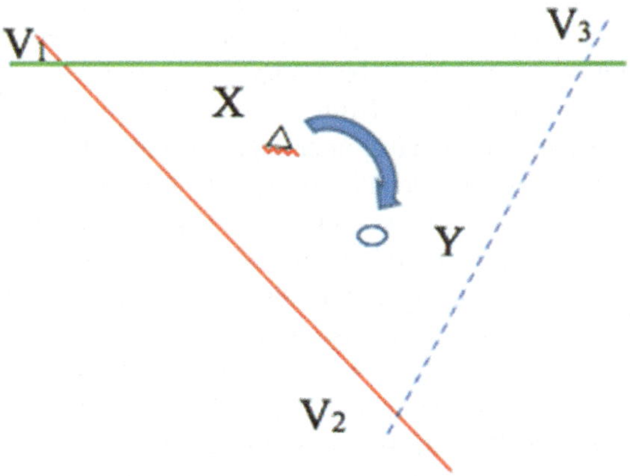

Fig. 14 A system of constraints including objects X and Y

The specified area contains the required data for objects denoted by X and Y. The closed area contains much less data than in the initial case but still it may be too large to explore, in some cases it could contain billions of data, examples, and cases. If the data selection process is enough effective, the focusing process/area closure helps to reveal a new information concerning data, metadata, knowledge and/or metaknowledge for objects X and Y, for example: 'it follows from X that Y is true' or 'there is a significant association/causal relation between X and Y' or, as will be explained later 'X points or is binded to Y'. As a result, a new knowledge is born as a result of the data selection and analysis process.

Some disadvantages of the classical approaches: the constraint satisfaction is a rather static process. The set of constraints cannot be modified during the resolution process. On the other hand, what is the constraint in the application sense? If this is a line/curve drawn on a surface, it should be enough to notice it to prevent crossing it. And what about its functionality? If the constraint cannot be crossed/intruded/destroyed, we have the classical case but when its crossing is harmful not just to the intruder agent but only to the other persons/objects/victims, in many cases the violation will be a fact. Many corresponding application problems also arise. Below are the proposed solutions to the quoted problems.

The constraint satisfaction solution should take in the account *why* does the constraint appear, *when* it could disappear, *how* does it apply, etc. Aiming at the data selection flexibility, new ontology-resembling metaknowledge forms have been elaborated and applied concerning logical-based control of the constraint application process.

First of all, one and the same constraint can be used/treated differently depending on the current situation/conditions. The result depends from the point of view of the resolver [agent] and from the situation.

Denote the object X is a city museum area rounded by ancient city walls with a visitor entrance, service entrance and a hidden tunnel to inner rooms beneath the walls. Let the population P of individuals A, B, C, D, Ir $\in$ P consists of individuals A and B with some official and many hidden secret intelligence functions, a museum guardian C, a set of ordinary visitors D and an intruder Ir disguised as a tourist. Every individual can reveal a place on the wall where one can jump over the walls/constraints but he cannot return back using the same way. Hence the constraint barrier could be broken only in one direction and this could be done only in an illegal/brute force attack manner. On the other hand, the individuals A, B, C could penetrate through the set of constraints legally through the service entrance, and all considered individuals could go through the constraints via the official entrance. The intruder appeared at the site X because of a gathered information that something peculiar happens inside the inner rooms. Could the drawing attention event be described using the standard set of objective or fitness functions? In general case this is possible. A generalized case is described below, it supports a more deep and effective modeling.

Let only A has special means to penetrate and use the hidden tunnel. Every such usage could attract the attention of Ir. Also Ir could conclude its existence analyzing the distribution of external guarding devices, and other collateral information. As a result Ir will reveal the most possible location of the tunnel entrance. This is modeled as a binding region with the maximum probability value in its centre. The usage of visual surveillance/drones/satellite information constantly or temporary extends the guarded region outside the walls. The described toy problem case aims to reveal the need of an introduction of new, logical-based types of constraints. Turning to education examples, we may find many examples when the constraint violation is almost not visible to the others to the contrary with the above example: carefully/completely prepare the home works, don't use mobile devices at the lecture, don't play games too much, etc.

Aiming at better deep modeling, it is proposed to use three non classical groups of constraints: binding, pointing and crossword constraints. Binding constraints model situations where the solution is/was not far or close to the center of the binding area. A case is best investigated wren the farther from the center the less likely it is to find the desired solution. There are several types of binding constraints in this group. It is explored that the binding process may be unconditional or conditional, it may invalidate the linearity/range/the region of the usage and/or other properties of other constraints intersecting the binding domain. In Fig. 15 one of the most explored groups of binding constraints is represented by the areas {B, D} intersecting the surface of the required/searched solution G_1, the searched goal to find the best teaching tool or something else.

In Fig. 16 another group of binding constraints is represented in the situation where C points to A which is hiddenly binded to B. In this case A has a hidden interrelation [a tunnel connection] to B. For example, let A-B denotes 'students with high competence in Data Science applications [A] will have a brilliant future [B]'. Here we have not a rule, and every student should find his own way to a success

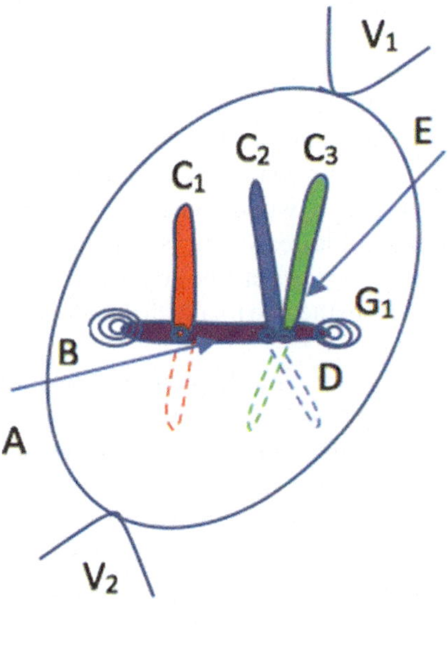

Fig. 15 A system of nonlinear constraints and the three groups of binding, pointing and crossword constraints

Fig. 16 Type 2 binding example

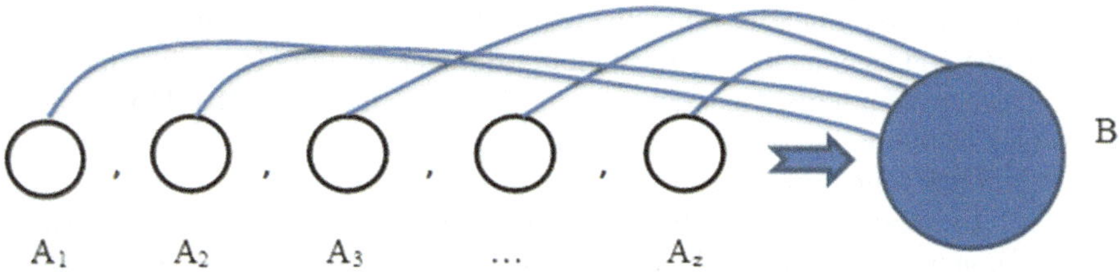

Fig. 17 Internal relationships in the rule A → B

to fulfill the sense of A-B binding. This type 2 binding could be easily interrupted, changed or modified. The range of the binding area may be undefined.

The investigation of different types of binding and pointing constraints still is an open question. For example, different types of binding constraints could be found in every rule as depicted in Fig. 17.

Every conjunct A_i from the antecedent is binded to the conclusion/consequent B. Some of the conjuncts are more important for B than the others: they are strongly binded to B while some of the other A_i are not binded to B at all. As a result, a defeasible reasoning is extended in this book chapter [31]. It is shown that the binding process can be controlled which can defeat one parts or whole rules and can transform them to totally different rules. This is important in contemporary education practice.

Together with the binding constraints from Fig. 15, it is convenient to use the pointing (indicating) constraints {A, E} in order to determine not only the area, but also the direction of the search. The pointing constraints could be in a form of any [higher order] curve.

The group of pointing constraints can be considered as a generalization of the classical systems of goal, target or fitness functions. To the contrary, pointing constraints can change the direction of the search depending on the accumulated data or the knowledge (circumstances), in other words, the logic of data and based on them events can be followed. For example, if there is suspect that there was an illegal copying in the class, the data on his coordinates are probably undefined or no longer valid. In this case, the direction has only prognostic character and the exact result is in doubt until the exact proof is found.

The last group of constraints is the crossword one depicted by $\{C_1, C_2, C_3\}$ from Fig. 15. Unlike $\{A, E\}$, they reveal not only the direction, but also parts of the desired solution G_1. The purpose of their application is to solve the problem G_1 when only some parts of the solution $\{C_1, C_2, C_3\}$ are known. For this purpose, combinations of other types of constraints are frequently used, the relationships between $\{C_1, C_2, C_3\}$ and other knowledge are studied. $\{C_1, C_2, C_3\}$ belongs to G_1 but is only its subset. Many algorithms may be investigated how to calculate the remainder, the unknown parts of the goal, and how to estimate their fitness to the solution of the problem. As a whole, the quoted problem is how to make links from the known set $\{C_1, C_2, C_3\}$ from the example to the unknown knowledge, how to use the pointing/binding and other constraints aiming to diminish the set of possible solutions. No matter what is the application algorithm, its goal is to produce a set of knowledge which is the most appropriate to $\{C_1, C_2, C_3\}$. The considered descriptions do not have a purpose to comprise all possible methods and applications, but to consider a synthetic view of the field as a whole.

When any combination of the proposed new constraint types is applied, many of the formal descriptions can be easily transformed into ontology-type interpretations.

A wide range of tools for modeling and presentation of material is used in this section. The modeling must be used not only for software agents but also for people: teachers, students. Respectively it is easier to introduce audio/multimedia/AR/VR ontologies, transmitting the meaning of things. For example, in Fig. 18 the answer to the question 'How the passive defense works' is given at a schematic level, in the form of a picture, not via formal descriptions. To present the meaning of things it is enough to show one or several key points on the subject altogether with corresponding explanations. Sometimes the result is better than an entire movie or lecture. The relevant ontology is not limited by the photo which has just a specifying role. More important are the descriptions to the picture made by a professional. The picture explains many details clear to people, but intelligent agents will learn nothing from this picture, especially if the agents are designed for cyber security.

Presenting information by meaning is a key element of modern higher and university education. The introduced new types of constraints help both teachers and students to process the information at a high speed, and to reach high quality

Fig. 18 Schematic (ontological) view of a passive defense

results. The proposed methods help the teachers to introduce game elements in the contemporary classrooms, the experiments will continue.

5 Conclusions

In this chapter the experience is represented on modeling and feedback management for mobile technology development in virtual learning environments. The research topic covers various aspects of the mobile learning problem which we tried to reveal when describing the sections. The use of mobile technologies in teaching requires modeling and feedback management which allows for a quick exchange of information, control of the assimilation of knowledge and an instant result. Mobile technologies in virtual environments are transforming the learning process, expanding spatial opportunities, changing the forms of learning and control. This transformation contributes to the development of new forms of cognition, thinking of students, the formation of a new view on the modern learning process. This is especially true for teachers who need to be mobilized in the assimilation and ability to use modern computer technology, to be in demand in the modern rapidly developing and changing world of digital technologies. The educational, professional and personal impact of the teacher on the student in mobile learning is seen as moderation of the process.

Mobile technologies in virtual learning environments provide wide access to the active use of interactive and simulation models, the possibility of tactile, and sensory interactions. The feedback in mobile technology development of virtual learning environments is dynamic and integrative. The set of data, domains, participants in the learning process, in which communication takes place, expands the cognitive capabilities of students, form certain personality traits, and generates various scenarios of learning and pedagogical design.

Thus, the broad technical and functional capabilities of mobile devices in modern educational environments, including virtual ones, provide a high level of adaptability and interactivity of students. Mobile technologies in virtual environments remove space–time restrictions, providing fast and high-quality access to information.

Application examples are considered of new constraint types from binding and/or pointing groups. Their usage doesn't change the considered education results but makes them more data-driven, effective and the modeling tools mode deep and universal.

Acknowledgements This work was carried out as part of project No. AP09259370 "Development of a technological platform for virtual learning based on artificial intelligence approaches" due to grant funding from the Ministry of Education and Science of the Republic of Kazakhstan and by Project "CyberTwin: Reinforcing the Scientific Excellence and Innovation Capacity in Cyber Security through Twinning", funded by Bulgarian National Research Program "European Scientific Networks".

References

1. Weiss, M., Loock, C.M., Staake, T., Mattern, F., Fleisch, E.: Evaluating mobile phones as energy consumption feedback devices. In: International Conference on Mobile and Ubiquitous Systems: Computing, Networking, and Services pp. 63–77. Springer, Berlin, Heidelberg. (2010)
2. Wechtitsch, S., Fassold, H., Thaler, M., Kozłowski, K., Bailer, W.: Quality analysis on mobile devices for real-time feedback. In: Tian, Q., Sebe, N., Qi, G.J., Huet, B., , R., Liu, X. (eds.) MultiMedia Modeling. MMM 2016. Lecture Notes in Computer Science, vol. 9516. Springer, Cham, (2016). https://doi.org/10.1007/978-3-319-27671-7_30
3. Rius, À., Masip, D., Clarisó, R.: Student projects empowering mobile learning in higher education. Int. J. Educ. Technol. High Educ. **11**, 192–207 (2014). https://doi.org/10.7238/rusc.v11i1.1901
4. Evans, M.A., Johri, A.: Facilitating guided participation through mobile technologies: designing creative learning environments for self and others. J. Comp. High Educ. **20**, 92–105 (2008). https://doi.org/10.1007/s12528-008-9004-1
5. Kerimbayev, N., Nurym, N., Akramova, A., et al.: Virtual educational environment: interactive communication using LMS Moodle. Educ. Inform. Technol. **25**, 1965–1982 (2020). https://doi.org/10.1007/s10639-019-10067-5
6. Kerimbayev, N.: Formats of virtual learning. In: Tatnall, A., (eds) Encyclopedia of Education and Information Technologies. Springer, Cham, 2020. https://doi.org/10.1007/978-3-319-60013-0_201-1
7. Gemou, M., Montalva Colomer, J.B., Cabrera-Umpierrez, M.F., et al.: Validation of toolkits for developing third-generation android accessible mobile applications. Univ. Access Inform. Soc. **15**, 101–127 (2016). https://doi.org/10.1007/s10209-014-0377-9
8. Oyelere, S.S., Suhonen, J., Wajiga, G.M., et al.: Design, development, and evaluation of a mobile learning application for computing education. Educ. Inform. Technol. **23**, 467–495 (2018). https://doi.org/10.1007/s10639-017-9613-2
9. Cheverst, K., Blair, G., Davies, N., et al.: The support of mobile-awareness in collaborative groupware. Pers. Technol. **3**, 33–42 (1999). https://doi.org/10.1007/BF01305318
10. Cole, H., Stanton, D.: Designing mobile technologies to support co-present collaboration. Pers. Ubiquit. Comp. **7**, 365–371 (2003). https://doi.org/10.1007/s00779-003-0249-4
11. Tarng, W., Ou, K.L., Yu, C.S., et al.: Development of a virtual butterfly ecological system based on augmented reality and mobile learning technologies. Virt. Real. **19**, 253–266 (2015). https://doi.org/10.1007/s10055-015-0265-5
12. Ritter, K.A., Chambers, T.L.: Three-dimensional modeled environments versus 360 degree panoramas for mobile virtual reality training. Virt. Real. (2021). https://doi.org/10.1007/s10055-021-00502-9
13. Nurym, N., Sambetova, R., Azybaev, M., Kerimbayev, N.: Virtual Reality and Using the Unity 3D Platform for Android Games. In: 2020 IEEE 10th International Conference on Intelligent Systems (IS), pp. 539–544. The IEEE. (2020). https://doi.org/10.1109/IS48319.2020.9199959
14. Kerimbayev, N.: Virtual learning: possibilities and realization. Educ. Inform. Technol. **21**, 1521–1533 (2016). https://doi.org/10.1007/s10639-015-9397-1
15. Harley, J.M., Poitras, E.G., Jarrell, A., et al.: Comparing virtual and location-based augmented reality mobile learning: emotions and learning outcomes. Educ. Technol. Res. Dev. **64**, 359–388 (2016). https://doi.org/10.1007/s11423-015-9420-7
16. Sprenger, D.A., Schwaninger, A.: Technology acceptance of four digital learning technologies (classroom response system, classroom chat, e-lectures, and mobile virtual reality) after three months' usage. Int. J. Educ. Technol. High Educ. **18**(8) (2021). https://doi.org/10.1186/s41239-021-00243-4
17. Mitra, S., Gupta, S.: Mobile learning under personal cloud with a virtualization framework for outcome based education. Educ. Inform. Technol. **25**, 2129–2156 (2020). https://doi.org/10.1007/s10639-019-10043-z

18. Mata, P., Chamney, A., Viner, G., et al.: A development framework for mobile healthcare monitoring apps. Pers Ubiquit Comput. **19**, 623–633 (2015). https://doi.org/10.1007/s00779-015-0849-9
19. Caballero, P., Ortiz, G., Garcia-de-Prado, A., et al.: Paving the way to collaborative context-aware mobile applications: a case study on preventing worsening of allergy symptoms. Multimed. Tools Appl. **80**, 21101–21133 (2021). https://doi.org/10.1007/s11042-021-10759-6
20. Kerimbayev, N., Beisov, N., Kovtun, A., et al.: Robotics in the international educational space: integration and the experience. Educ. Inform. Technol. **25**, 5835–5851 (2020). https://doi.org/10.1007/s10639-020-10257-6
21. Makridis, M., Papamarkos, N.: A new technique for solving puzzles. IEEE Trans. Syst. Man Cybernet. Pt B Cybernet. Publ. IEEE Syst. Man Cybernet. Soc. **39**(1), 1–10 (2009)
22. Kochan, O., et al.: Methods of reducing the effect of the acquired thermoelectric inhomogeneity of thermocouples on temperature measurement error. J. Meas. Techn. **58**, 327–331 (2015)
23. Levitin, A.: Algorithmic puzzles: history , taxonomies, and applications in human problem solving. J. Prbl. Solv. **10**, 1–15 (2017)
24. Jotsov, V.: Emotion-aware education and research systems. J. Issues Inform. Sci. Inform. Technol. USA **6**, 779–794 (2009)
25. Jotsov, V.: Semantic conflict resolution using ontologies. In: Proceedings of the 2nd International Conference on System Analysis and Information Technologies, SAIT 2007, pp. 83–88. RAS, Obninsk. vol. 1, (2007)
26. Alajlan, N.: Solving square jigsaw puzzles using dynamic programming and the "Hungarian Procedure." Am. J. Appl. Sci. **6**(11), 1941–1947 (2009)
27. Jotsov, V., Sgurev, V.: Applications in intelligent systems of knowledge discovery methods based on human-machine interaction. In: International Journal of Intelligent Systems (IJIS), vol 23, pp. 588–606. Wiley Press (USA), NJ, (2008)
28. Jotsov, V.: Machine self-learning applications in security systems. In Proceedings of the 6th IEEE International Conference on Intelligent Data Acquisition and Advanced Computing Systems, pp. 727–732. Prague, Czech Republic. (2011)
29. Afanasyev, I., et al.: Blockchain solutions for multi-agent robotic systems: related work and open questions In: Balandin S., Deart V., Tyutina, T. (Eds.), Proceedings of the 24th Conference of Open Innovations Association FRUCT, the FRUCT'24, Article No. 76. (2019)
30. Jotsov, V.: Evolutionary parallels. In: Proceedings of the 1st International Symp. On Intelligent Systems. Varna, Bulgaria, (**ISBN:**0–7803–7134–8). (2002)
31. Jotsov, V.: New proposals for knowledge driven and data driven applications in security systems, innovative issues in intelligent systems. In: Sgurev, V., Yager, R., Kacprzyk, J., Jotsov, V. (eds.) Studies in Computational Intelligence, vol. 623, pp. 231–294. Springer, Berlin Heidelberg (2016)

Control of Power Consumption with Integrated System of Technology, Regulation and Consumer Behavior Management

Igor Bimbiloski, Valentin Rakovic, and Aleksandar Risteski

Abstract The environmental and social issues that we are facing these days, like climate change and pandemic, are showing that the science, technology, social behaviour, and rules of law must act together and integrated to reach the final goal in any complex system. As the climate changes are mostly affected by irrational consumption of energy, the integrated system of technology, regulation, and consumer behaviour management could contribute to improve efficiency and reduce the CO_2 emission. Massive control ICT systems are currently deployed in Smart Energy Networks, combining the big data management, automation, artificial intelligence support and advanced IoT systems. However, the important component to make this system more efficient is to control the consumer behaviour and stimulate investment in green energy production. The dependency of the consumer of the smart devices and social networks can be used as potential platform to increase the awareness and change the consumption behaviour. Positive regulation of green energy investments on consumer side is also potential tool for improving, if it is considered an overall financial and environmental interests. The models of alignment of financial and environmental benefits are using the Game Theory approach to find the best scenarios and influencing consumer behaviour, with so called "transfer of benefits" policy. This policy is considering that the different players on the electricity market e.g., companies and consumers, should be ruled and one system with joint environmental benefit, not as separate players with oppose financial interest. The management of such complex systems is ruled by advanced ICT system.

I. Bimbiloski (✉) · V. Rakovic · A. Risteski
Faculty of Electrical Engineering and Information Technologies, SSs. Cyril and Methodius University, Ruger Boskovic 18, 1000 Skopje, North Macedonia
e-mail: igor.bimbiloski@gmail.com

V. Rakovic
e-mail: valentin@feit.ukim.edu.mk

A. Risteski
e-mail: acerist@feit.ukim.edu.mk

P. Shi et al. (eds.), *Complex Systems: Spanning Control and Computational Cybernetics: Applications*, Studies in Systems, Decision and Control 415,
https://doi.org/10.1007/978-3-031-00978-5_17

1 Introduction

Climate changes are becoming one of the biggest issues in the modern world, affecting all aspects of human living, and it's a target of many industries, economy and society at all. ICT industry, as one of the most progressive one, is getting more important role in the mission against the climate changes and changing of the energy sector business model.

The role of the ICT industry in that mission is best explained by the Global e-Sustainability Initiative (GeSI) [1]. Namely, ICT has the potential to reduce global greenhouse gas (GHG) emissions by 20% by 2030, by helping companies and consumers to use more intelligently and save energy. GeSI is optimistic about the industry's ability to be fully sustainable. The findings show that with ICT support, the world will be cleaner, healthier and more prosperous by 2030, offering greater opportunities for individuals everywhere. The emissions avoided by using ICT are already ten times higher than the emissions generated by their installation. The sector can help avoid producing about 12 gigatons CO_2 by 2030, and halt further emissions.

In this study, we consider the role of the ICT industry in supporting climate change, primarily as a tool that will improve the processes in industry and businesses, i.e. its indirect contribution to increased energy efficiency. At the same time, the importance of digitalization occupies a central place in that intention, both for all and for the energy industry. Several auspicious technologies, which are recognized as part of the 4th industrial revolution, namely IoT, artificial intelligence, large cloud databases and their processing are pinpointed as the core facilitators in the energy industry. The utilization of these technologies, results in energy networks to transform into smart grids, which are in fact a combination of energy and information-communication networks. However, technology is only one part of the topics covered in the study. As usual for all transformation processes, the key challenge is to change the habits of the users [2]. Therefore, we believe that the role of consumers in improving environmental protection measures is crucial. This study provides an overview of consumer behavior, their communication with energy providers, the importance of that communication, and possible channels and communication technologies. Additionally, the studyelaborates on the importance of smartphones and mobile applications in this whole process [3–5].

Game theory is most suitable for analysis in scenarios where there exists a need for synergy between the ICT technology and the consumers who are part of that ecosystem [6]. Namely, the study discusses the use cases where game theory can be exploited, and define its basic formulation which is later used in the design of the model and performance assessment. Also, we explain the model with its elements, the system architecture, and the way of connecting with the users. It provides an overview of the technologies that can be used, i.e. mobile networks and their role in the model, the potentials of using 5G and IoT technologies, and especially mobile phones and their role in the entire ecosystem. The model of information-telecommunication system (ITSM) is explained through the three states in which the user can be found, the role of smartphones as means of localization and communication and their impact on changing user behavior [7–11].

2 Basic Idea and Elements of the System ITSM

In this work, the consumer is placed in the core of the energy ecosystem, and the proposed model ITSM (IT Systems model) is user oriented. The assumption is that the consumer can be in three types of state (state of motion, state of work, and state of privacy, as shown in Fig. 1), according to which the appropriate model of energy efficiency is determined. If we consider that 80% of energy consumption is realized due to transport, working conditions and domestic needs, it is concluded that by defining these three conditions of the consumer, a segment of 80% of the total energy consumptionare targeted, in which this model can be applied.

The idea also includes the use of smartphones, as a tool to identify the situation of the consumer, and at the same time will serve as a means of communication through which to receive guidance (messages, information, etc.) on consumer behaviour.Considering the daily use of smartphones and mobile applications, as well as consumer dependence and usage time, it is expected that their use will strongly influence energy efficiency habits.

The C-consumer can be in one of three states, which can change over time (t -time). For example, when consumer C_1 is in state of work, he uses energy fromprovider P_1. The consumer is changing its status depending on the location and the state, and it can move from one to another provider. On the supplier's side P_1 there is a measuring device that measures energy consumption, where:

$$P_{1(C_1)} = f(\{C_1\}, t) \tag{1}$$

$$P_{2(C_2)} = f(\{C_2\}, t) \tag{2}$$

$$P_{3(C_3)} = f(\{C_3\}, t) \tag{3}$$

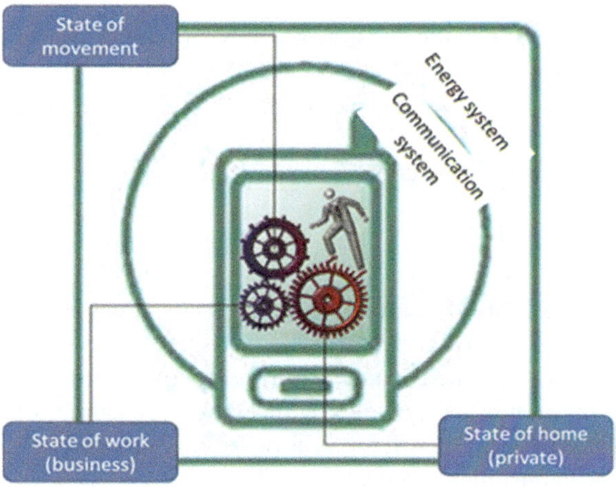

Fig. 1 Conditions of the energy consumer in a user-oriented model for energy efficiency

where C_1, C_2, C_3 denotes consumer in 3 different stages: private, business, movement.

At time t_0 there could be N consumers connected to the same meter, hence the total energy consumed by provider P_1 will be:

$$P_1 = \sum_{x=1}^{N} P_{1(C1_x)} \tag{4}$$

And the total energy consumption of all M providers will be:

$$P_{total} = \sum_{x=1}^{M} P_x \tag{5}$$

All parameters on the supplier side P_1 *to* P_M are measurable quantities in time.

On the left side, i.e., the consumer side, only the state in which the consumer is located is known and it is valid that the energy they use is equal to P_{total}:

$$P_{total} = \sum_{\substack{x=1 \\ y=1}}^{\substack{N \\ M}} P_{x(C_y)} \tag{6}$$

The main goal of ITSM is to make the predictions about:

- State of the consumer in dependence of time:

$$C_{x(t_a)} = C_1 \; or \; C_2 \; or \; C_3 \tag{7}$$

- Location of the consumer and the provider from which it is consuming the energy:

$$P_x = f(C_x, geolocation) \tag{8}$$

- Volume of the energy consumed:

$$P_{x(C_y)} = f\{C_y, time\} \tag{9}$$

Since the assumption is that each consumer is identified via a smartphone, it will be used to predict the three elements listed above.

If it is assumed that there is a register of all electrical meters and real-time metering, a register of all vehicles with average consumption and that all consumers have smartphones, then there is a possibility to make a forecast of the consumption of each consumer at a given time.

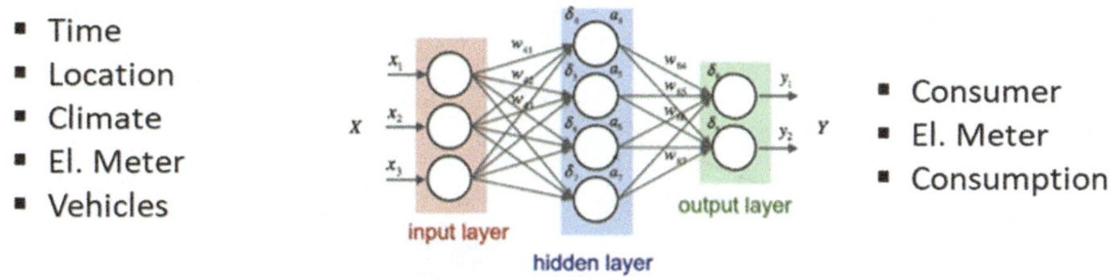

Fig. 2 Input and output parameters of the artificial neural network

In doing so, a model of artificial neural networks can be used to predict user behaviour. We suggest using 5 (five) input and 3 (three) output parameters in ITSM (Fig. 2).

The output parameters refer to the consumer, as they should be used to send information and instructions that should direct his behaviour towards energy efficiency. Hence, the output parameters of the artificial neural network are proposed to be:

- The condition of the consumer in which he is
- The location or supplier from which it uses energy
- The amount of energy it uses

$$\mathrm{P}_{\mathrm{x}(C_y)} = f\{C_{y,}time\} \tag{10}$$

There are several elements of the architecture for the proposed ITSM, shown in Fig. 3. Within the smart home environment, the network to be used is the existing Wi-Fi network and will collect and consolidate data from all smart devices. The second element is a smartphone, which is in constant communication at the IP level via Wi-Fi network. A smartphone is an element that defines the mode / phase in which the consumer is, in this case the home mode of the consumer. The third element is the back-end system, where data is stored and processed with an artificial neural network (artificial intelligence–AI). Following the processing rules, this server communicates via the consumer's 4G / 5G network (smartphone) or home devices on any IoT platform [12, 13].

3 Model Analysis with Competitive Game Theory

This subchapter provides a comprehensive analysis of the ITSM model in terms of the effects it would have on electricity providers and consumers, as well as its contribution to environmental goals. Namely, by applying the methodology of competitive game theory, we analyse ITSM from its reference version with two players (provider and consumer) to a version with multi providers and many users. By applying the classic elements of game theory, the model is scaled, both in terms of the number of users and in terms of technologies that can be used (Fig. 4). It starts with analysing

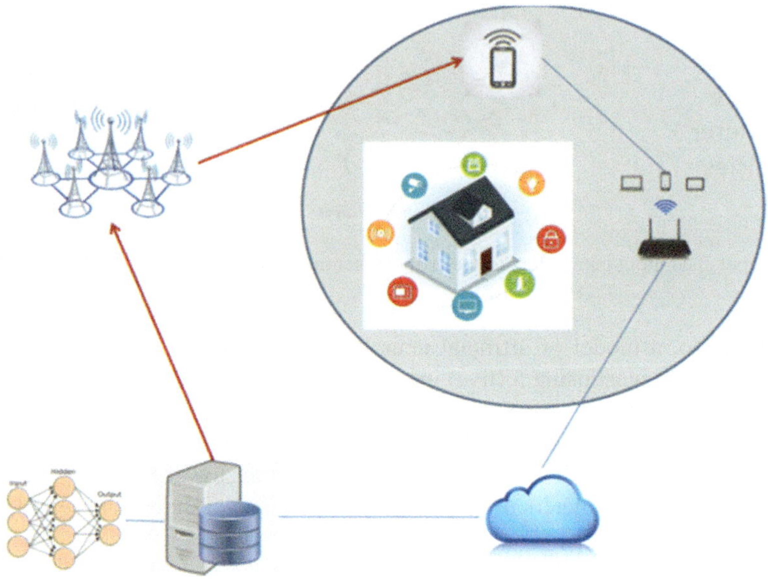

Fig. 3 ITSM architecture

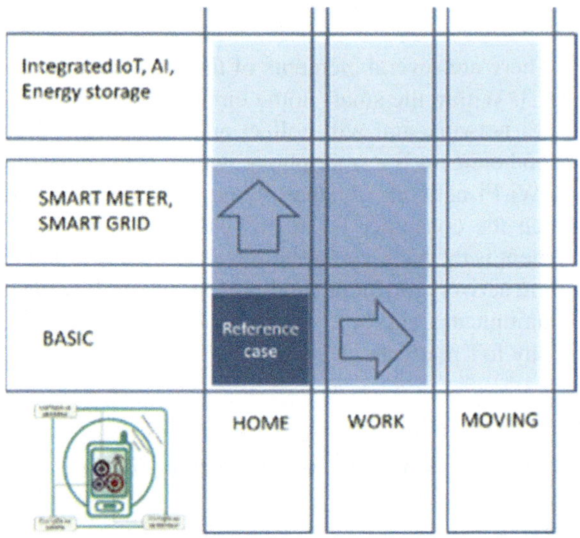

Fig. 4 Building blocks of ITSM and matrix of areas of applicability

a simple scenario of a basic market, and scales to a market with smart grids, green energy and distributed generation. Thereby, an analysis is also made of the market with single-tariff meters and scaled to the market with multi-tariff meters [14–19]. In short, the analysis covers realistic market scenarios, which seeks to show the financial effects of the use of ITSM on providers and consumers. This chapter also presents a mathematical model of ITSM which explains the conditions for its successful operation and the role of market elements.

We will start the analysis with the basic ITSM for potential energy savings, mainly for saving electricity in the home. The goal is to investigate whether this ITSM is useful to use by electricity providers to implement and use, especially for load balancing purposes, and to anticipate or avoid power outages.

Given that game theory is increasingly used as a tool for analysing and defining strategy in energy systems [20], in our case we will define the game according to the standard formulation. The players in the game theory used in the proposed ITSM are:

- Electricity provider, and
- Electricity consumer.

The motives and consequently the strategies of these elements are opposite to each other: the motive of the service provider is to maximize profitability with optimal consumption and appropriate distribution throughout its time, and user motives to reduce consumption, but in a reasonable way that will satisfy standard of living. One of the main elements in this case is the protection of the environment, which means reduction and efficiency in energy consumption.

The assumption is that the provider has real-time information on the total electricity consumption that is provided in the predefined area with a certain number of consumers. The time frame (resolution) for measuring electricity consumption is predetermined by the provider (minutes or hours). The main goal of the provider is to avoid overloading the electricity consumption, because the price it pays for electricity is higher than the regular price, but it is not compensated by the end consumers. We assume that end consumers have a flat price for electricity (same price all the time) and have no information on electricity consumption, neither for their own home, nor for total consumption.

The proposed ITSM strategy/solution/approach/system envisages the use of a mobile platform that consists of several elements (Fig. 5):

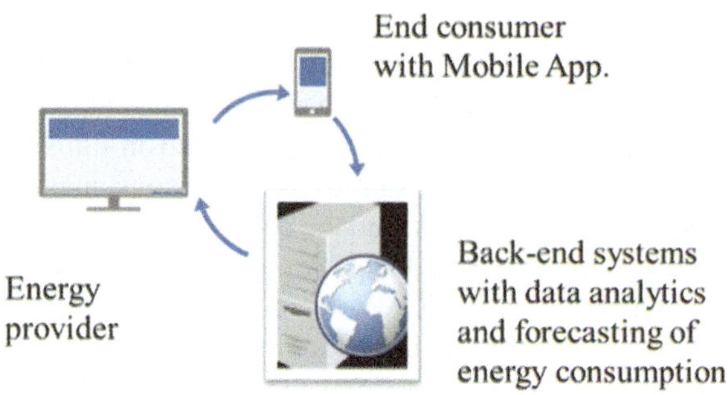

Fig. 5 ITSM architecture with mobile application

- Storage and processing unit, with modules for forecasting electricity consumption. This information is available to the operator.
- Consumer notification unit, operated by the provider.
- Mobile application, used by end users to obtain information on energy savings and efficiency.

There are two states of the environment in which the ecosystem (nature) operates: (i) the normal state when electricity consumption is normal, and no abnormal changes are expected; (ii) abnormal state of the ecosystem (nature), when extraordinary deviations in electricity consumption are predicted (state of high load), such as extreme temperature changes, events (concerts, sports, meetings…), natural disasters, etc.

Provider P can have two actions to choose from (send a request to the consumer or not), in two types of environments (normal load and high load):

(a) send information to the final consumer C and ask him to reduce or switch on the electricity consumption when it is in normal load mode.
(b) to send a request for reduction or change of consumption from high to normal load.
(c) the option not to send a request or information, it is considered that there is no action on the consumer side in any case.

Consumer C has three (three) options to choose from:

- accept the provider's request and take action to reduce (reduce consumption),
- to shift (transfer) consumption to another time, from high load mode to normal,
- not to take any action, whether the request was submitted in normal or high operating mode.

Based on the above, we can define the following sets of the game:

- *Set of provider's actions:* A_P= {send request in normal or high mode, do not send request}
- *Set of environment conditions:* A_E= {normal load mode, high load mode}
- *Set of consumer's actions:* A_C= {acts and reduces consumption, acts and migrates consumption, and does not act}.

As a note for the further course of the game in the work, the case of non-submission of a request from the provider to the consumer will be treated with equal values for benefit (cost) as if in the case a request was sent, and the consumer did not accept it.

Possible player pay-outs are measured in units of money, i.e., money saved or spent. We will assume that the cost to the consumer per unit of electricity (e.g. KWh) is equal to Y at all times, and the cost of electricity for the electricity service provider in normal mode is X and in peak mode (mode of overload) is V per unit of electricity. Assume that $Y = a * X$, where $a > 1$, i.e. that the provider has a profit $Y–X = (a-1) * X$ for each energy unit sold in normal mode, and assume that $V = b * X$, where $b > a$, i.e. that $Y–V= (ab) * X$ is the loss of the provider in load mode.

Table 1 Game theory for the provider-consumer relationship in the basic case of ITSM–P1K1T1

		Consumer		
		Act to reduce	Act to migrate	Not acting
Provider	Normal regime	Reduce units (−1, −1)	Migrate units (1,1)	
		Payoff (−1X, + 2X)	Payoff (0, 0)	Payoff (0, 0)
	Peak regime	Reduce units (−1, −1)	Migrate units (1,1)	
		Payoff (+2X, + 2X)	Payoff (+4X, 0)	Payoff (−2X, 0)

To clarify the representation of possible scenarios in ITSM game theory, we will consider a situation in which a = 2 or Y = 2 * X, and b = 4 or V = 4 * X. According to this example, it follows that in normal operating mode the provider makes a profit Y–X = (a–1) * X = 1 * X, i.e. that in high load mode it makes a loss of Y-V = (ab) * X = −2 * X. In Table 1, we will assume that the average volume of electricity that is subject to a reduction or change request is 1 (one) unit of electricity (for example, 1 kWh) over a period of time (for example, 1 h).

In normal mode, the provider's interest is to maintain a higher level of consumption and to earn maximum revenue from consumers within the planned limits of the total network load. If the provider sends a request for reduction of consumption and consumers accept it, the provider will have a lost opportunity of −1 * X profit, and the consumer will benefit from + 2 * X savings. So, the pay-off in this scenario will be

$$\Pi(P,\ C) = ((1 - a),\ a) = (-1 * X,\ +2 * X) \tag{11}$$

where P (P, C) shows the pay-off of the provider (P) and the consumer (C), respectively (Table 1). So, this is a mode that does not benefit the provider to take any measures to reduce consumption. But when providers ask the consumer to distribute consumption in another time interval, then the provider's pay-off is neutral P (P, C) = (0.0). Also, if the consumer does not take action, the effect on both sides is neutral P (P, C) = (0.0).

It is very important for the service provider the top network mode (under load). In this case, when the provider has an overload for an additional load (where V = b * X = 4 * X is the price for additional electricity from producers) it is useful to ask the consumer to reduce the energy. It is more convenient for the provider to switch the power consumption from load mode to normal mode in another time interval. The provider will generate losses of Y-V = (a–b) * X = −2 * X if the consumer does not take any action, ie

$$\Pi\ (P,\ C) = ((a - b) * X,\ 0) = (-2X,\ 0) \tag{12}$$

So, when the provider enters (or predicts) network overload, then it is useful to take action. If the consumer reduces consumption, pay-off is equal on both sides

$$\Pi(P, C) = ((b - a) * X, a * X) = (+2X, +2X) \quad (13)$$

and when the consumer switches consumption from high mode to normal mode, the provider has a maximum profitability of + 3 * X, i.e.

$$\Pi(P, C) = ((b - a) * X + (a - 1) * X, 0) = (+3X, 0) \quad (14)$$

For the consumer, each scenario is useful or neutral, so there is no doubt that consumers will benefit from the implementation of ITSM. The service provider will benefit if it manages the consumption load in an appropriate way using the proposed ITSM.

If for the needs of the game we take the probability of the scenarios according to the tree in Fig. 6, and if we follow the assumptions from above, we could calculate the profitability of the provider from the use of ITSM. The total payment to the service provider is.

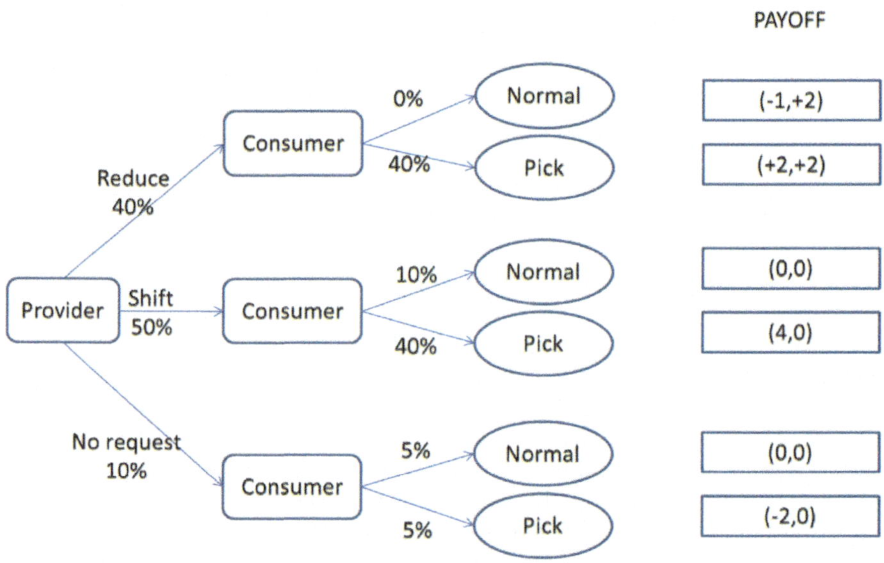

Fig. 6 Game theory tree for actions between provider and consumer

$$
\begin{aligned}
\Pi(provider) &= \sum_{n=1}^{N} \Pi(\mathrm{P})_n * probability(n) = \\
&= 2\mathrm{X} * 0.4 + 3\mathrm{X} * 0.4 + (-2\mathrm{X}) * 0.05 \\
&= +1,9 * \mathrm{X}\ \text{(money units)}
\end{aligned} \tag{15}
$$

$$
\begin{aligned}
\Pi(consumer) &= \sum_{n=1}^{N} \Pi(C)_n * probability(n) = \\
&= 2\mathrm{X} * 0.4 = 0,8\ \mathrm{X}\ \text{(money units)}
\end{aligned} \tag{16}
$$

where n = {1, …, N} is the set of all possible game scenarios, $\Pi(\mathrm{P})_n$ e is the pay-off of the provider in the n-th scenario, $\Pi(C)_n$ is the pay-off of the consumer in n-th scenario, and probability (n) is the probability of the n-th scenario occurring.

We see that by applying ITSM the pay-off of the provider for the level of profitability with the above assumptions is + 1.9 * X units of money, upon request for 1 unit of electricity at a given time. The consumer side is certainly profitable, so we show that ITSM is acceptable to all stakeholders in the electricity market.

Figure 7 shows that the discontinuation of the pay-off of the provider using ITSM with the assumptions of Fig. 6, is achieved at a price level of 1.5 * X in high operating mode, which is lower than the consumer price of Y = 2 * X. We could also see that the benefit of the service provider is higher than the benefit of the consumer, when the price in high mode is 2.7 times higher than the normal cost of production.

If we extend this logic to the market with N providers, then for each provider f_n we will have the following profitability:

$$
M_{cfn}(j,t_r) = \sum_{j=1}^{J}\sum_{r=1}^{R} \left\{ \begin{array}{c} \left. \left\{ \begin{array}{l} [p_{nr} - i_{fn}(j,t_r)] * l_c(j,t_r),\ when\ i_f(j,t_r) < i_c \\ [p_{1r} - i_c] * l_c(j,t_r),\ when\ i_f(j,t_r) > i_c \end{array} \right. \right\} when\ p_{nr} < p_{mr}\ for\ any\ m \neq n\ i\ m = \{1,2,..,N\} \\ 0 \ldots\ldots\ldots\ldots\ldots\, in\ any\ other\ case \end{array} \right\} \tag{17}
$$

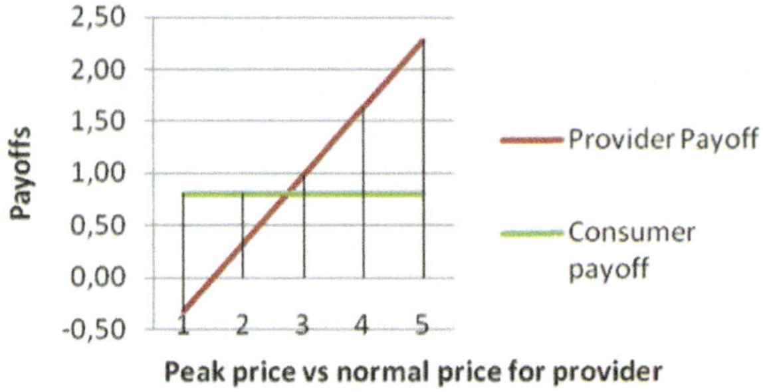

Fig. 7 Provider payoff as function of cost of electricity in peak load (consumer price is Y = 2X, normal provider cost for electricity production is 1X)

On the consumer side, if we take the equation from the previous calculations and adapt it to the new situation when the consumer has a choice of multiple providers, then consequently the equations for all 4 types of user scenarios would look like the following:

(A) energy reduction

$$B_{cf}(j,t_r)=\sum_{j=1}^{J}\sum_{r=1}^{R}\sum_{n=1}^{N}\left\{\begin{array}{c} p_{nr}*\Delta l_c(j,t_r);\ when\ p_{nr}<p_{mr}\ for\ any\ m\neq n im=\{1,2,..,N\} \\ 0.................in\ any\ other\ case \end{array}\right\} \tag{18}$$

(B) transfer of energy consumption from one to another interval, with possibility to change the provider

$$B_{cf}(j,t_{w\rightarrow q})=\sum_{w=1}^{R}\sum_{q=1}^{R}\sum_{m=1}^{N}\sum_{n=1}^{N}\left\{\left\{\begin{array}{c} (p_{wn}-p_{qm})*\Delta l_c(j,t_{w\rightarrow q})when\ p_{wn}>p_{qm} \\ 0,\ when\ p_{wn}\leq p_{qm} \end{array}\right\}\right\} \tag{19}$$

(C) energy consumption reduction and replacement with green energy consumption

$$B_{cf}(j,t_r)=\sum_{j=1}^{J}\sum_{r=1}^{R}\sum_{n=1}^{N}\left\{\begin{array}{c} p_{nr}*\Delta l_c(j,t_r);\ when\ p_{nr}<p_{mr}\ for\ any\ m\neq n\ im=\{1,2,..,N\} \\ 0;\ in\ any\ other\ case \end{array}\right\}-I_{p,s} \tag{20}$$

(D) when consumer is producing and selling green energy

$$\begin{aligned} B_{cf}(j,t_r)=\sum_{j=1}^{J}\sum_{r=1}^{R}\sum_{n=1}^{N}\{&i_{fn}*l_{sale}(j,t_r), \\ &when\left\{\begin{array}{c} i_{fn}(j,t_r)>i_c \\ l_{prod.}(j,t_r)>l_{sale}(j,t_r) \end{array}\right\}i_{fn}(j,t_r) \\ &>i_{fm}(j,t_r)m\neq n,m=\{1,2,\ldots,N\}\}-I_{sale} \end{aligned} \tag{21}$$

Accordingly, regarding the pay-off of the consumer in the market with N providers, multitariff meters and smart grid, we are reaching the following formula:

$$
\begin{aligned}
&B_{cf}(j,t_r)\\
&=\sum_{j=1}^{J}\sum_{r=1}^{R}\sum_{n=1}^{N}\left\{\begin{matrix} p_{nr}*\Delta l_c(j,t_r);\, when\ p_{nr}<p_{mr}\, for\, any\, m\neq n\, i\, m=\{1,2,..,N\}\\ 0;\ in\ any\ other\ case\end{matrix}\right\}\\
&+\sum_{w=1}^{R}\sum_{q=1}^{R}\sum_{m=1}^{N}\sum_{n=1}^{N}\left\{\left\{\begin{matrix}(p_{wn}-p_{qm})*\Delta l_c\big(j,t_{w\to q}\big) when\ p_{wn}>p_{qm}\\ 0,\ when\ p_{wn}\le p_{qm}\end{matrix}\right\}\right\}\\
&+\sum_{j=1}^{J}\sum_{r=1}^{R}\sum_{n=1}^{N}\left\{\begin{matrix} p_{nr}*\Delta l_c(j,t_r);\, when\ p_{nr}<p_{mr}\, for\, any\, m\neq n\, i\, m=\{1,2,..,N\}\\ 0;\ in\ any\ other\ case\end{matrix}\right\}\\
&-I_{p,s}\\
&+\sum_{j=1}^{J}\sum_{r=1}^{R}\sum_{n=1}^{N}\{i_{fn}*l_{sale}(j,t_r),\\
&when\left\{\begin{matrix} i_{fn}(j,t_r)>i_c\\ l_{prod.}(j,t_r)>l_{sale}(j,t_r)\end{matrix}\right\} and\, i_{fn}(j,t_r)>i_{fm}(j,t_r)\, for\ m\neq n,\\
&m=\{1,2,\ldots,N\}\}-I_{sale}
\end{aligned}
\tag{22}
$$

4 Benefits Transfer via Cooperative Game Theory

However, the analysis of ITSM with a competitive game theory would not be complete, if the climate and environmental goals are not added with the highest priority [21]. Therefore, in this part, based on the findings obtained in the previous chapter, an analysis with cooperative game theory is applied. This approach may not be common in game theory scenarios, but in our case the assessment is that these two opposing approaches, at first glance, give positive effects to the use of ITSM with proper application and management. Namely, applying the principles of "fair" distribution of benefits, we explain the mathematical model for the distribution of ITSM benefits among market players (Fig. 8).

In order to develop ITSM and test cooperative game theory for it, it is necessary to take a minimal case of players in cooperative play. As before, we will define a minimum case of cooperation game with a transferable benefit, with the following characteristics: The number of players in the market will be 3, of which we will have 1 provider (provider) and 2 energy consumers, i.e. N = {1, 2, 3}, where 1 is the supplier and 2 and 3 are players, i.e.N = {P, C_1, C_2}. The second element of the cooperative game, which is a characteristic functionc : $2^N \rightarrow \mathbb{R}$ denoting the profit of the players, will have 8 possibilities for coalition S as a result of the competing reference case:

According to the figures in the table above, it is obvious that the biggest environmental outburst is in the grand coalition {P, C_1, C_2},which means that this cooperative

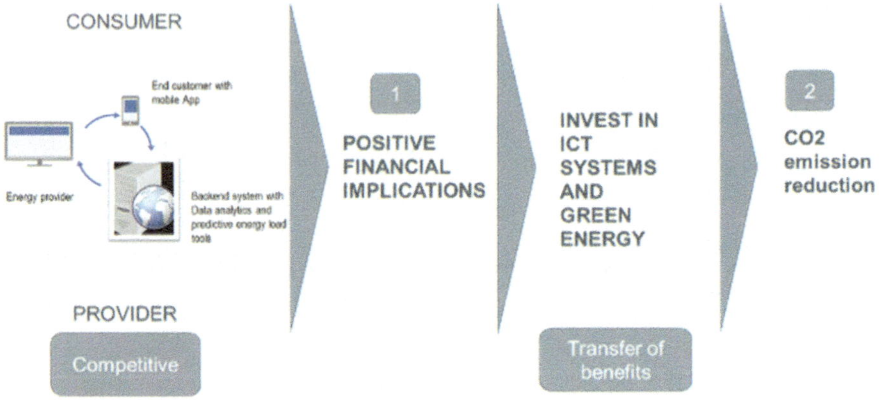

Fig. 8 Transfer of financial in environmental benefits

game is essential:

$$v(N) > \sum_{i \in N} v(\{i\}) \tag{23}$$

where V (N) is a payoff of the grand coalition.We can also find that this game is concave, that for each additional member of the coalition, the value of the characteristic function increases - which is a property of super-additivity.

$$v(S_1) + v(S_2) \leq v\left(S_1 \bigcup S_2\right) S_1, S_2 \subseteq N, S_1 \cap S_2 = 0 \tag{24}$$

in order to create the effect of "snowballs" - the motive for joining the coalition grows as the coalition grows

$$v(S_1) + v(S_2) \leq v(S_1 \cup S_2) + v(S_1 \cap S_2) S_1, S_2 \subseteq N \tag{25}$$

n such a cooperative game, the distribution of the benefit as a division vector $\pi = (\pi_1, \pi_2, \ldots, \pi_N)$ is of importance for the game. In this case, we should bear in mind that in the case of dividing benefits $(\pi_1, \pi_2, \ldots, \pi_N)$ it is a value that is not related to their financial benefits. We considered the game of financial benefits in the previous section as a strategic and non-cooperative game, and in this section, we consider the benefits of reduced radiation of harmful gases, which are benefits derived from and for the environment. The distribution of the benefit between the players, in order for the game to be stable, it is necessary to fulfill the following requirement.

$$\sum_{i \in N} \pi_i \leq v(N) \tag{26}$$

that is to say that the individual set of benefit to the players is less than or equal to the benefit within the grand coalition. Bearing in mind that the individual benefit according to the table above in a two-player coalition is 0.8, the maximum value each player can have is less than that value. So for example, if we have a coalition between {P, C_1}, and C_2 is an independent player, then we will have that $v(S = \{P, C_1\}) = 0.8$, that is max. $\pi_P = 0.8$, and that max. $\pi_C = 0.8$, and that v(C_2) = 0.

If this coalition joins C_2, and if we know that v(S = $\{P, C_1, C_2\}$) = 2.4 then it is obvious that

$$\pi_i \geq v(\{N\}) - max.\pi_P - max.\pi_C \geq 0.8, i \in N \tag{27}$$

that is, the individual benefit in the big coalition in a case is greater than the benefit of the smaller coalition, that is, individual rationality and stability of the game is ensured.

If we calculate the core of the cooperative game, we will found that the Core of the cooperative game for the imputation x = (x_1, x_2, x_3) and the figures from the Table 2, the results shows that the Core in in the area (Fig. 9)

$$0 \leq x_1 \leq 1.6; \quad 0 \leq x_2 \leq 1.6; \quad 0 \leq x_3 \leq 1.6 \tag{28}$$

This confirms that the x = (0.8; 0.8; 0.8) imputation belongs to the core of the cooperative game. It can also be noted that the core of the game is not empty, because for all imputation with individual values greater than 0.8, the core requirements are satisfied, that is, the core in this case is a three-dimensional space defined as a set of characteristic function

$$C(N = 3, v) = \left\{ \pi \in \mathbb{R}^N | \sum_{i \in S} \pi_i \geq v(S) S \subseteq N,\ S \neq 0 \right\} \tag{29}$$

which in three-dimensional form can be shown as in the Fig. 10.

Table 2 Coalition pay-off in ITSM

Coalition (S)	Payoff (v)
0	0
{P}	0
{C_1}	0
{C_2}	0
{P,C_1}	0.8
{P,C_2}	0.8
{C_1,C_2}	0.8
{P, C_1, C_2}	2.4

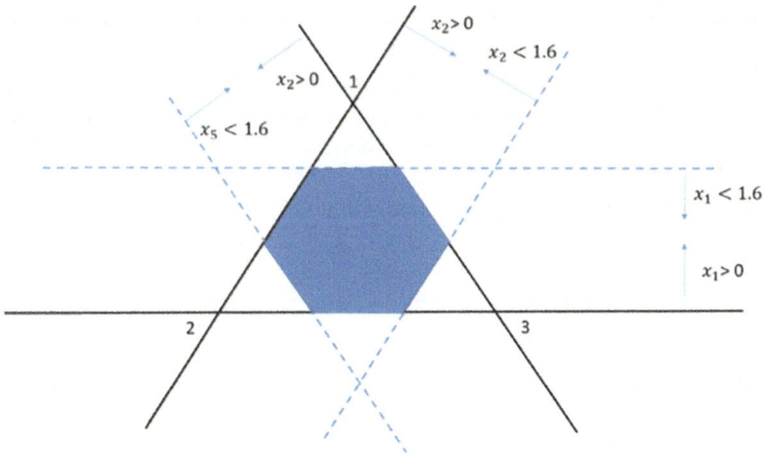

Fig. 9 Core plane of ITSM

Fig. 10 Core of ITSM in 3-dimensions

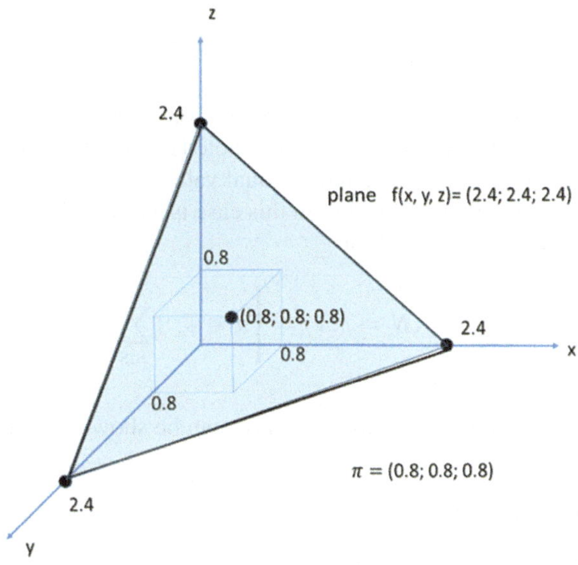

We have seen that the competitive game theory approach could be a tool for calculating the financial benefits of using ITSM, both for service providers and consumers, in competitive market conditions. We also saw that ITSM as a model can be used to position electricity prices (normal production price, selling price and maximum production price) in order to find interest in all stakeholders involved. The generalization of ITSM shows the conditions under which the financial benefits of the consumer side can be transferred to investments in production and storage of green energy and a distributed production system. The ultimate goal of ITSM is not financial gain, but its transformation into the reduced emissions. To that end, providers and consumers need to work together, and we have combined the ITSM approach to cooperative game theory. We proved that the cooperative approach provided a stable core as a solution for the game and that the payments from the competitive game are part of the core of the cooperative game. This proves that ITSM is the only stable model with combined effects: competitive (to bring financial benefits) and cooperation (to support environmental protection).

5 Conclusions

One of the greatest challenges of the modern world and of humanity is the protection of the environment and the maintenance of living conditions, i.e. the fight against climate change and the harmful effects of industrialization and overpopulation. We actually address this topic here, with the main emphasis on energy consumption and the possible use of new ICT technologies for its control and efficient use. Like many other researches on this topic, an attempt is made to define a system–ITSM that will make the use of energy more efficient, and thus will contribute to the preservation of the environment.

There are many scientific studies and models that deal with this topic, and usually, each of them achieves its goal with different success. For this ITSM to be successful, we believe that it should satisfy some prerequisites. Those preconditions are necessary for the successful implementation of the model, as well as its transformation from theoretical to practical environment.

First condition is that the model should be acceptable to all parties present in the energy market. We have achieved this in ITSM, because with the mathematical method of game theory we have shown that ITSM is acceptable for both end users and energy providers, and also contributes to reducing emissions into the atmosphere.

As we are dealing with business environment, model should be financially viable, i.e. it can be implemented with a small investment. And we have shown that it is valid for ITSM, because it uses the existing infrastructure and technology, and is a reward that connects the elements in an efficient way. The investment in new ICT elements of ITSM is insignificantly small, and it uses primarily existing investments such as smartphones and telecommunication networks.

Also, it should be easy to implement, which is also achieved with this ITSM, as it is based on cloud technology and internet platform, and is available to all consumers globally or in smaller market segments. ITSM can penetrate the market quickly, because it provides for the use of a mobile application, and at the same time it is simple to use.

The implementation of the solution always starts on a smaller scale and grows on a larger scale, which means that ITSM should be scalable or upgradeable. ITSM is scalable in two dimensions. The first dimension is technical scalability. It is applicable to technologies that are simple and even obsolete, such as power grids operating in a monopoly situation where there is only simple energy distribution technology, single-tariff pricing models, and outdated metrics. The same ITSM, we saw that it can be technologically developed to the most modern technologies, taking into account the latest communication technologies such as 5G and IoT, such as artificial intelligence, smart power grids, renewable energy sources, remote control systems, etc. The second dimension is commercial scalability, i.e. scalability on the part of the provider and the side of consumers. As we have seen from the dissertation, ITSM supports markets with an unlimited number of consumers and an unlimited number of providers, as well as numerous market segments. ITSM in the dissertation uses examples of electricity with residential consumers, but it can be extended to business consumers, as well as with more types of energy to be distributed. Given the latter, ITSM is also expandable for mobile consumers, who can use electricity or fossil fuels.

ITSM focuses on energy efficiency by coordinating the activities of providers and consumers, i.e. the consumer is placed in the center of ITSM and is a user-oriented system. ITSM achieves energy efficiency by changing user behaviour, as opposed to technological investments that are financially intensive in the energy sector.

ITSM is within the policies of the world and European regulatory processes, which means that it is ready to be institutionally supported. Namely, in the official announcement for energy efficiency of these bodies, support is already given to research and projects that focus on achieving energy efficiency goals by changing consumer habits by applying information and telecommunications technologies.

All the above conditions are met in the proposed system–ITSM, and most of them are confirmed by applying the game theory. Namely, two forms of game theory are applied in the study: competitive and cooperative game theory.

Competitive game theory is used to analyse the economic or financial benefits of using the proposed ITSM for energy efficiency. First, we confirmed that the choice of competitive game theory, with the characteristics of the Stokeberg game of leader–follower, is the right model for this game, and that in such an environment all players have the benefits of using ITSM. This game confirmed that ITSM can be used in a variety of markets, from the simplest to the most developed, from energy markets with a simple single-tariff meter to markets with smart grids, renewables and two-way energy transport. We also showed that in ITSM with more providers and consumers, the benefits of using ITSM exist.

In the next stage using the cooperative type of games, we showed that in addition to the financial effects of using ITSM, environmental benefits can also be obtained. This means that if the market entities manage to find a model of cooperation, that mutual cooperation increases the effect of energy saving, i.e. stimulates the use of green energy instead of energy from fossil fuels.

In this study we tried to treat a specific technological-economic-social problem with the parallel application of competitive and cooperative game theory. Namely, in the professional and scientific literature, there are many examples that treat game theory and its application, but models of game theory are rarely combined. With this dissertation, we have shown that there is not always a uniform approach that should be accepted to solve problems, but that it is necessary to look for the best models that correspond to the current situation and environmental conditions. As strange and contradictory as the use of these two models in game theory may seem, in this particular example and proposed ICT concept, this is what gives the best results in achieving the ultimate goals. We started the analysis without assuming and limiting the possible results in advance, so that the analysis itself, which started with a competitive model of two-player simple game theory, did not lead to a situation to combine it with a cooperative form of the same game, but with another goal, while retaining all the benefits of achieving the other goals. As a final confirmation, the game theory model showed that a combination of competitive and cooperative play is possible, to achieve different goals in the same environment at the same (parallel) time.

Another contribution of the paper, which can be transformed from theoretical to practical, is the part for transferring benefits from a competitive game in a cooperative form. The paper gives and confirms the idea that the proposed ITSM can provide financial benefits for both providers and players. But at the same time, there is a need for reduced emissions that can be addressed by changing consumer habits and investing in green energy. This paper, with the proposed ITSM, proposes that the financial benefits obtained from the use of ITSM be used for investments in green energy sources. This solves the economic dilemma for the cost-effectiveness of green energy that we face today. Namely, with this, we offer a solution that instead of allocating additional funds for investments in green energy, through a cooperative model to use the already existing budgets, while providers and consumers are not at a loss.

References

1. ICT Sector Helping to Tackle Climate Change. https://unfccc.int/news/ict-sector-helping-to-tackle-climate-change
2. NASA Global Climate Change, Vital Signs of the Planet. https://climate.nasa.gov/evidence/
3. Igor Bimbiloski, Aleksandar Risteski, "Model for improving of energy efficiency with smart phone usage and artificial intelligence", ETAI Macedonia (Sept. 2018)
4. Bimbiloski, I., Risteski, A.: Modeling of customer centric energy efficiency system with mobile technology and artificial intelligence. ENAR Turkey (Nov. 2018)
5. Bimbiloski, I., Risteski, A.: Model for energy efficiency improvement by using mobile technologies and IoT. J. Electric. Eng. Inf. Technol. **3**(1–2) (2018)
6. Zhao, Z., Neighbour, G., Han, J., McGuire, M., Deutz, P.: Using game theory to describe strategy election for environmental risk and carbon emissions reduction in the green supply chain. J. Loss Prev. Process Ind. **25**, 927e936 (2012)
7. Bimbiloski, I., Rakovic, V., Risteski, A.: Providers' and Consumers' mutual benefits in energy efficiency model with elements of cooperative game theory. In: FABULOUS 2019-4th EAI International Conference on Future Access Enablers of Ubiquitous and Intelligent Infrastructures, Sofia (2019)
8. Bimbiloski, I., Rakovic, V., Sefidanoski, A., Risteski, A.: Competitive game theory approach of energy efficiency ICT model in multiplayer market. In: IEEE EUROCON 2019, Novi Sad (2019)
9. Bimbiloski, I., Risteski, A.: Matching competitive and cooperative game theory in single ICT model for energy efficiency. In: Submitted, BalkanCom 2019, Third International Balkan Conference on Communications and Networking, Skopje, North Macedonia (June 2019)

10. Bimbiloski, I., Risteski, A.: Matching economic and enviromental aspects of energy efficiency in single ICT model. In: Submitted, 16th International Conference on Informatics and Information Technologies, Mavrovo, North Macedonia (2019)
11. Bimbiloski, I., Risteski, A.: ICT and game theory models for energy efficiency with IoT, AI and blockchain perspectives. In: Submitted, 2nd International Conference on Energy Research (ENRES-2019) Marmaris/Turkey (Apr. 2019)
12. Shroufab, F., Miragliotta, G.: Energy management based on Internet of Things: practices and framework for adoption in production management J. Clean. Prod. **100**, 235–246 (1 Aug. 2015)
13. Jhaa, S.K., Bilalovic, J., Jhab, A., Patelc, N., Zhangd, H.: Renewable energy: present research and future scope of Artificial Intelligence Renew. Sustain. Energy Rev. **77**, 297–317 (Sept. 2017)
14. Han, S., Lu, Y., Yang, S., Mu, X.: Game theory-based energy efficiency optimization in multiuser cognitive MIMO interference channel. School of Information Engineering, Zhengzhou University, Zhengzhou 450001, China, 2016. IEEE MTT-S International Wireless Symposium (IWS) (2016)
15. Wang, Y.: Behavioral Game Theory for Smart Grid Energy Management. Ph.D. dissertation, University of Miami (2016)
16. Zhao, R., Neighbour, G., Han, J., McGuire, M., Deutz, P.: Using game theory to describe strategy selection for environmental risk and carbon emissions reduction in the green supply chain. J. Loss Prev. Process Ind. **25**, 927–936 (2012)
17. Soner Aplak, H., Ziya Sogut, M.: Game theory approach in decisional process of energy management for industrial sector. Energy Convers. Manag. **74**, 70–80 (2013)
18. Ludkovski, M., Sircar, R.: Game theoretic models for energy production. Commodities, Energy and Environmental Finance (2015)
19. Bauso, D.: Game Theory: Models, Numerical Methods and Applications-Dipartimento di Ingegneria Chimica Gestionale, Informatica, Meccanica, Università di Palermo. Found. Trends R Syst. Control **1**(4), 379–522 (2014)
20. Saad, W., Han, Z., Vincent Poor, H., Basar, T.: Game theoretic methods for the smart grid, 2012 in IEEE Signal Processing Magazine
21. Grueneich, D.M.: The next level of energy efficiency: the five challenges ahead. Electric. J. **28**(7), 44–56 (Aug.–Sept. 2015)

Novel Applications in Computer Networks and Telecommunications

Double-Router TCP/AQM Network Systems: Backstepping Communication Control Design

Yuanwei Jing, Yan Zheng, Wenjuan Xu, Zanhua Li, and Kun Wang

Abstract The problem on congestion communication control for double-router TCP/AQM network systems is investigated. By employing the idea on the window model of single-router TCP/AQM network system, a double-router TCP/AQM network system model is constructed, which follows the "priority selecting, random transferring" principle of data packet transmission. By using a class of state transformation, the state space model of a nonlinear system with lower triangle form is obtained. An active queue management congestion control algorithm for double-router network system is proposed by using the backstepping design method. An innovated state feedback controller is designed to make the TCP/AQM network system asymptotically stable. The simulation results verify the feasibility and effectiveness of the proposed method.

Honoring Professor Georgi M. Dimirovski for his many Academic contributions and merits to our community

---

Y. Jing (✉) · Y. Zheng
College of Information Science and Engineering, Northeastern University, Shenyang 110004, China
e-mail: ywjjing@mail.neu.edu.cn

Y. Zheng
e-mail: zhengyan1@ise.neu.edu.cn

W. Xu
Sergeant School, PLA Army Academy of Artillery and Air Defense, Shenyang 110161, China
e-mail: q35341355@qq.com

Z. Li
College of Science, Shenyang Ligong University, Shenyang 110159, China
e-mail: lizanhuamd@163.com

K. Wang
School of Control Science and Engineering, Tiangong University, Tianjin 300387, China
e-mail: wkwfg@126.com

P. Shi et al. (eds.), *Complex Systems: Spanning Control and Computational Cybernetics: Applications*, Studies in Systems, Decision and Control 415,
https://doi.org/10.1007/978-3-031-00978-5_18

1 Introduction

Since the emergence of the Internet, how to avoid network congestion and improve network performance has always been the goal of people. The data transmission protocol widely used in the Internet is the end-to-end transmission control protocol (TCP), which can effectively avoid network crash [1]. A technology to ensure that the end-to-end data transmission queue is stable to the desired value is a class of algorithms called Active Queue Management (AQM), which appeared at the end of last century [3]. The TCP/AQM system combining the two technologies (TCP and AQM) can realize the reasonable resource utilization, the acceptable packet dropping, and the stable and reliable network operation [21]. Active queue management (AQM) mechanism is to send out congestion alerts before the total load of the routing buffer reaches full load, and then disposes the subsequent packets with a certain probability of discarding labels.

At present, many AQM schemes have been proposed. Based on the difference of congestion detection methods, AQM schemes can be divided into three kinds, such as the queue-based AQM, the rate-based AQM, and the queue-rate-based AQM. There are the time-driven AQM and the event-driven AQM based on different updating methods of indication probability. Based on the different research methods, there are mainly heuristic methods developed by computer scholars, methods based on mathematical tools such as game theory, and methods based on control theory, etc. [4].

To study the problem of congestion control in TCP/AQM network system, and design an efficient, stable and robust AQM algorithm, one of the most important things is to understand the dynamic characteristics of the system. That is to say, a mathematical model describing the dynamic characteristics of the TCP/AQM network system is needed. As early as the end of last century and the beginning of this century, scholars have proposed various models based on certain assumptions (see [7, 9, 11–14] etc.). These models have their own characteristics. Among them, Mathis et al. presented a simple model which is easy to design based on some simplified assumptions [13], but these assumptions also cause significant errors; Kelly et al. and Low proposed models [9, 11], respectively, which are closer to the actual system but do not have scalability; Marsan et al. presented a TCP model in the form of partial differential equation being better to draw a large-scale network [12], but it is too complex to analyse.

The fluid flow model [14] proposed by Misra et al. can well analyse the performance of TCP/AQM systems, which provides a theoretical basis for the numerical analysis and synthesis of the TCP/AQM systems. Later on, Hollot et al. gave an improved simplified fluid model to make it more convenient for application [7]. It is one of the most widely used models for congestion control of TCP/AQM network systems, based on which many new AQM schemes have been presented (see [5, 6, 8, 10, 15, 16, 19, 20, 22, 23] etc.). In many schemes, the control theory analysis and design methods such as the linearization technology or the nonlinear control technology are applied. Those controls include generalized predictive control, T-S fuzzy control, model predictive control, optimal control, H-infinity control, PI control, sliding mode control, data-driven control, and backstepping control etc.

These research results are almost all aimed at the end-to-end transmission structure, that is, the so-called single-router network topology, or single-class data flow situation. There is little information about multi-routing/multi-bottleneck/multi-TCP-flow network transmission. However, the network is composed of many routers which are coupled with each other, so it is more practical to study the situation of multi-router. Bauso et al. first studied the multi-bottleneck problem, pointing out that the results of single-bottleneck network cannot be directly applied to multi-bottleneck network, and we must find out the solutions to the stability of multi-bottleneck network, which is an important open problem that needs to be further explored [2]. Wang et al. studied the stability of networks with multi-bottleneck topology, and proposed a model of multi-link transmission of multi-inhomogeneous flows [18]. However, the proposed model is a parallel combination of multiple routes, ignoring the interaction between router queues.

Xu et al. [21] analyzed the four stages of data transmission, i.e. slow start, congestion avoidance, fast retransmit and fast recovery. In view of the two working conditions of TCP data transmission, two sets of mathematical models were established respectively, which is an improvement of the model presented by Masra et al. in [14]. Alaoui et al. extended the model presented in [21] to the multi-bottleneck networks, that is, multi-TCP flows are transmitted to the receiver through multi-router [1]. This multi-bottleneck model is composed of N single-bottleneck models in parallel, and constitutes a composite large-scale system, which establishes the relationship between subsystems through the probability of packet drop indication. In the process of data transmission, there will be a phenomenon, that is, when a packet arrives at a certain router which is with congestion, it will be labelled as discarded. At the same time, another (or several) routers may not reach the upper limit of the expected queue number, which results in idle resources.

In order to overcome the phenomenon of data discarding while routing redundancy, a model with the structure of parallel topology and cascade operation is proposed in this chapter. Based on Masra model, the models of two subsystems are combined in parallel to form a double-router network system. Then, a state feedback controller is designed by using backstepping control technology to make the nonlinear double-router TCP/AQM system stable. The simulation example shows the feasibility and validity of the proposed model and the designed controller.

2 State Space Dynamics of Double Router TCP/AQM Network Systems

Consider the dynamic model of single-router TCP network system described by the following set of two differential equations [14].

$$\dot{W}_1(t) = \frac{1}{R_1(t)} - \frac{W_1(t)W_1(t - R_1(t))}{2R_1(t)} p_1(t - R_1(t)),$$
$$\dot{q}_1(t) = \frac{N(t)}{R_1(t)} W_1(t) - C(t), \tag{1}$$

where, $W(t)$ is the size of TCP source window measured by the number of packets, $Q(t)$ is the instantaneous queue length of routers measured by the number of packets, $N(t)$ is the load factor of TCP network, $C(t)$ is the link bandwidth measured by the number of packets, and $R(t)$ is the round trip transmission time, which is given by the following formula.

$$R_1(t) = \frac{q_1(t)}{C(t)} + T_p.$$

T_p is the propagation delay, $p(t)$ is the probability of congestion indication, and the value is taken in the interval [0,1].

In (1), the window change equation describes the mechanism of additive increase and multiplicative decrease of window. The first item on the right side indicates the principle of additive increase in bandwidth detection phase, and the second item is multiplication reduction, which is a response to the probability of packet drop marking. It reflects the rule of Random Early Detection (RED) method, that is, once congestion is being detected, the window size will be halved.

Based on this window model, aiming at the problem of active queue management in network congestion control, modern control theory and methods are used to study the problem, and quite a number of research results have emerged. The core element of communication network system is the router. Many routers and sending/receiving nodes constitute a complex communication network system. Therefore, the research results based on single routing are very necessary to extend to multi-routing. However, the current research results on multi-routing are rare. This chapter intends to study the congestion control problem of TCP network system from the double-routing scenario. It is assumed that there are two routers with the same technical specifications between the sender and the receiver, which are represented by two single router models of (1) and (2), respectively.

$$\dot{W}_2(t) = \frac{1}{R_2(t)} - \frac{W_2(t)W_2(t - R(t))}{2R_2(t)} p_2(t - R_2(t)),$$
$$\dot{q}_2(t) = \frac{N(t)}{R_2(t)} W_2(t) - C(t), \tag{2}$$

where, each symbol expresses the same meaning as in (1).

For the parallel transmission links composed of these two routers, we give out a packet transmission rule as "priority selection, random transfer". That is, if a data packet is sent by the sender, Router 1 is selected first. If the Router 1 is close to

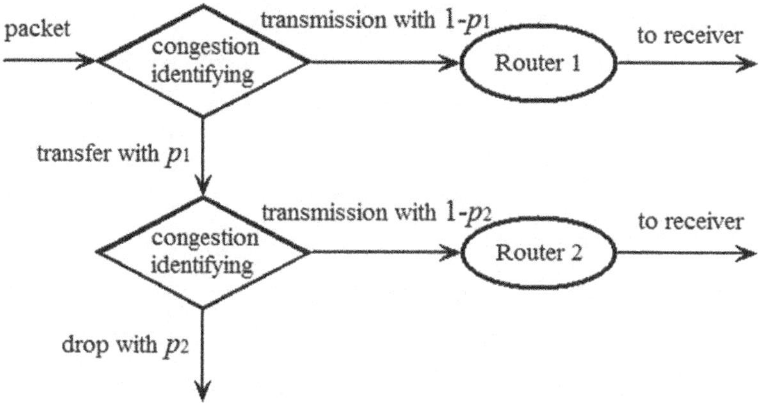

Fig. 1 Directions of data transfer via the double-router

congestion at the time, the packet is transferred to Router 2 with a certain "drop" probability, and Router 2 completes the transmission. If Router 2 is also close to congestion, the packet will be drop-marked with a certain indication probability. The flow directions of data transfer via double-router can be illustrated in Fig. 1.

The probability of "packet drop" transfer at Router 1 is related to the queue length of Router 2, so the dynamic model of TCP system for Router 1 can be rewritten as follows.

$$\begin{aligned}\dot{W}_1(t) &= \frac{1}{R_1(t)} - \frac{W_1(t)W_1(t-R_1(t))}{2R_1(t)}\left(1-\frac{q_2(t)}{q_{\text{d}}}\right),\\ \dot{q}_1(t) &= \frac{N(t)}{R_1(t)}W_1(t) - C(t),\end{aligned} \tag{3}$$

where, q_{d} is the expected queue length. In the rewritten model (3), the "packet drop" probability is characterized by the instantaneous queue length and bandwidth of Router 2.

A double-router TCP network system model can be obtained by connecting (3) and (2) in parallel form. In order to facilitate the use of control theory to study congestion control problems, it is necessary to transform the TCP network system model into a state space representation. The queue length and window size of the two routers are selected as the state variables, and the congestion drop marking probability of router 2 is selected as the control input, *i.e.*

$$x_1 = q_1(t) - q_{\text{d}},\ x_2 = W_1(t),\ x_3 = q_2(t) - q_{\text{d}},\ x_4 = W_2(t),\ u(t) = p_2(t).$$

Let $\mathbf{x}=(x_1, x_2, x_3, x_4)^{\text{T}}$. For convenience, ignoring the effect of delay, the following set of state equations can be obtained.

$$
\begin{aligned}
\dot{x}_1 &= f_1(t) + g_1(t, x_1)x_2 - C(t) \\
\dot{x}_2 &= f_2(t) + g_2(t, x_2)x_3 \\
\dot{x}_3 &= f_3(t) + g_3(t, x_3)x_4 - C(t) \\
\dot{x}_4 &= f_4(t) + g_4(t, x_4)u(t)
\end{aligned} \tag{4}
$$

where

$$f_1(t) = f_3(t) = 0,\ f_2(t) = \frac{1}{R_1(t)},\ f_4(t) = \frac{1}{R_2(t)},$$

$$g_1(t) = \frac{N(t)}{R_1(t)},\ g_2(t, x_2) = \frac{x_2^2}{2R_1(t)q_d},\ g_3(t) = \frac{N(t)}{R_2(t)},\ g_4(t, x_2) = -\frac{x_4^2}{2R_2(t)}.$$

The Eq. (4) is the state space representation of double-router TCP network systems. The controller design and performance analysis for (4) can be carried out by using the nonlinear system control method. Our goal is to design controllers to make the state variables tend to the expected values, that is, to make the TCP network system (4) or (3) and (2) stable at its equilibrium point.

3 Backstepping Design of the Control Law

In this section, we will give the controller design process of the system (4). The system (4) has a strict lower triangular structure, so the nonlinear control method, backstepping design method, can be used to control the system. According to the design idea of the backstepping method, firstly, the subsystem of the system (4) concerning the state variable $x_1$is considered, to which the controller is designed. Then the rest subsystems of (4) are designed in turn. Finally, the controller $u(t)$ is designed. For this purpose, the following state transformations are defined.

$$e_1 = x_1,\ e_2 = x_2 - x_2^*,\ e_3 = x_3 - x_3^*,\ e_4 = x_4 - x_4^* \tag{5}$$

It is therefore that we have

$$\dot{e}_1 = \dot{x}_1,\ \dot{e}_2 = \dot{x}_2 - \dot{x}_2^*,\ \dot{e}_3 = \dot{x}_3 - \dot{x}_3^*,\ \dot{e}_4 = \dot{x}_4 - \dot{x}_4^* \tag{6}$$

The process of stabilization controller design by making use of the backstepping method is organized into four steps.

3.1 Step 1

Consider the first subsystem of the system (6)

$$\dot{e}_1 = \dot{x}_1 = g_1(t, x_1)x_2 - C(t) = g_1(t, x_1)(e_2 + x_2^*) - C(t) \tag{7}$$

where, x_2^* can be regarded as the virtual control variable ofsubsystem (7), and e_2 as the error of system state x_2 and virtual control x_2^*. The purpose of this step is to design a virtual feedback control scheme x_2^* making $e_1 \to 0$. Select the following Lyapunov function

$$V_1 = \frac{1}{2}e_1^2, \tag{8}$$

Calculating the time derivative of V_1 along (7) yields

$$\dot{V}_1 = e_1\dot{e}_1 = e_1[g_1(t, x_1)(e_2 + x_2^*) - C(t)] = g_1(t, x_1)e_1e_2 + g_1(t, x_1)e_1x_2^* - e_1C(t). \tag{9}$$

Select a proper virtual control law

$$x_2^* = \frac{1}{g_1(t, x_1)}(C(t) - l_1e_1) \tag{10}$$

to make $\dot{V}_1 < 0$, where, l_1 is a positive design parameter. Substituting (10) into (9), we can get

$$\dot{V}_1 = -l_1e_1^2 + g_1(t, x_1)e_1e_2 \tag{11}$$

If $e_2 = 0$, then $\dot{V}_1 = -l_1e_1^2 < 0$, from which the error state is asymptotically stable. Generally, the state e_2 cannot be guaranteed to be zero. Therefore, we need to further consider the subsystem including the state variable x_2 and design the corresponding controller to make the error state variable e_2 possess the desired asymptotic stability.

3.2 Step 2

Consider the second subsystem of the system (6)

$$\dot{e}_2 = \dot{x}_2 - \dot{x}_2^* = f_2(t) + g_2(t, x_2)e_3 + g_2(t, x_2)x_3^* - \dot{x}_2^* \tag{12}$$

It can be further rewritten as

$$\dot{e}_2 = \dot{x}_2 - \dot{x}_2^* = f_2(t) + g_2(t, x_2)e_3 + g_2(t, x_2)x_3^* + \frac{\dot{g}_1(t, x_1)}{g_1^2(t, x_1)}(C(t) - l_1 e_1) - \frac{1}{g_1(t, x_1)}(\dot{C}(t) - l_1 \dot{e}_1) \tag{13}$$

Construct the following Lyapunov function

$$V_2 = V_1 + \frac{1}{2}e_2^2, \tag{14}$$

Then the time derivative of V_2 along (7) and (13) appears as follows:

$$\begin{aligned}
\dot{V}_2 &= \dot{V}_1 + e_2\dot{e}_2 \\
&= -l_1 e_1^2 + g_1(t, x_1)e_1 e_2 + e_2[f_2(t) + g_2(t, x_2)e_3 + g_2(t, x_2)x_3^* \\
&\quad + \frac{\dot{g}_1(t, x_1)}{g_1^2(t, x_1)}(C(t) - l_1 e_1) - \frac{1}{g_1(t, x_1)}(\dot{C}(t) - l_1 \dot{e}_1)] \\
&= -l_1 e_1^2 + e_2[g_1(t, x_1)e_1 + f_2(t) + g_2(t, x_2)e_3 + g_2(t, x_2)x_3^* \\
&\quad + \frac{\dot{g}_1(t, x_1)}{g_1^2(t, x_1)}(C(t) - l_1 e_1) - \frac{1}{g_1(t, x_1)}(\dot{C}(t) - l_1 \dot{e}_1)]
\end{aligned} \tag{15}$$

According to the structure of (15), we can get the following virtual feedback control law

$$\begin{aligned}
x_3^*(t) = &-\frac{1}{g_2(t, x_2)}[g_1(t, x_1)e_1 + f_2(t) + l_2 e_2 \\
&+ \frac{\dot{g}_1(t, x_1)}{g_1^2(t, x_1)}(C(t) - l_1 e_1) - \frac{1}{g_1(t, x_1)}(\dot{C}(t) - l_1 \dot{e}_1)]
\end{aligned} \tag{16}$$

to make $\dot{V}_2 < 0$, where, l_2 is a positive design parameter. Substituting (10) into (9), we can get

$$\dot{V}_2 = -l_1 e_1^2 - l_2 e_2^2 + g_2(t, x_2)e_2 e_3 \tag{17}$$

If e_3=0, then $\dot{V}_2 = -l_1 e_1^2 - l_2 e_2^2 < 0$. We can see that the error state e_1 and e_2 are asymptotically stable. Therefore, we need to further consider the next subsystem and design the controller to make the error state variable e_3 possess the desired asymptotic stability.

3.3 Step 3

Consider the third subsystem of the system (6)

$$\dot{e}_3 = \dot{x}_3 - \dot{x}_3^* = g_3(t, x_3)e_4 + g_3(t, x_3)x_4^* - C(t) - \dot{x}_3^* \tag{18}$$

Denote

$$\begin{aligned} Q(t) = {} & [g_1(t, x_1)e_1 + f_2(t) \\ & + \frac{\dot{g}_1(t, x_1)}{g_1^2(t, x_1)}(C(t) - l_1 e_1) - \frac{1}{g_1(t, x_1)}(\dot{C}(t) - l_1 \dot{e}_1)] \end{aligned} \tag{19}$$

Then, (16) can be rewritten as

$$x_3^*(t) = -\frac{1}{g_2(t, x_2)}[Q(t) + l_2 e_2] \tag{20}$$

and so the subsystem (18) can be further rewritten as

$$\begin{aligned} \dot{e}_3 = {} & \dot{x}_3 - \dot{x}_3^* = g_3(t, x_3)e_4 + g_3(t, x_3)x_4^* - C(t) \\ & - \frac{\dot{g}_2(t, x_2)}{g_2^2(t, x_2)}(Q(t) + l_2 e_2) + \frac{1}{g_2(t, x_2)}(\dot{Q}(t) + l_2 \dot{e}_2) \end{aligned} \tag{21}$$

Construct the following Lyapunov function

$$V_3 = V_2 + \frac{1}{2}e_3^2, \tag{22}$$

Calculate the time derivative of V_3 along (7), (13) and (21)

$$\begin{aligned} \dot{V}_3 = {} & \dot{V}_2 + e_3 \dot{e}_3 \\ = {} & - l_1 e_1^2 - l_2 e_2^2 + g_2(t, x_2)e_2 e_3 \\ & + e_3[g_3(t, x_3)e_4 + g_3(t, x_3)x_4^* - C(t) \\ & - \frac{\dot{g}_2(t, x_2)}{g_2^2(t, x_2)}(Q(t) + l_2 e_2) + \frac{1}{g_2(t, x_2)}(\dot{Q}(t) + l_2 \dot{e}_2)] \\ = {} & -l_1 e_1^2 - l_2 e_2^2 + e_3[g_2(t, x_2)e_2 \\ & + g_3(t, x_3)e_4 + g_3(t, x_3)x_4^* - C(t) \\ & - \frac{\dot{g}_2(t, x_2)}{g_2^2(t, x_2)}(Q(t) + l_2 e_2) + \frac{1}{g_2(t, x_2)}(\dot{Q}(t) + l_2 \dot{e}_2)] \end{aligned} \tag{23}$$

Design the following control law

$$\begin{aligned} x_4^*(t) = {} & -\frac{1}{g_3(t, x_3)}[g_2(t, x_2)e_2 - C(t) + l_3 e_3 \\ & - \frac{\dot{g}_2(t, x_2)}{g_2^2(t, x_2)}(Q(t) + l_2 e_2) + \frac{1}{g_2(t, x_2)}(\dot{Q}(t) + l_2 \dot{e}_2)] \end{aligned} \tag{24}$$

to make $\dot{V}_3 < 0$, where, l_3 is a positive design parameter. Substituting (24) into (23) yields

$$\dot{V}_3 = -l_1 e_1^2 - l_2 e_2^2 - l_3 e_3^2 + g_3(t, x_3) e_3 e_4 \tag{25}$$

If $e_4 = 0$, then $\dot{V}_2 = -l_1 e_1^2 - l_2 e_2^2 - l_3 e_3^2 < 0$, and the error states e_1, e_2 and e_3 are asymptotically stable. We need to consider the next subsystem and design the controller to make the error state variable e_4 possess the desired asymptotic stability.

3.4 Step 4

Consider the fourth subsystem of the system (6)

$$\dot{e}_4 = \dot{x}_4 - \dot{x}_4^* = f_4(t) + g_4(t, x_4) u(t) - \dot{x}_4^* \tag{26}$$

In the last step of controller design process, when the control input $u(t)$ appears in the state equation of this subsystem. Let denote

$$P(t) = [g_2(t, x_2) e_2 - C(t) - \frac{\dot{g}_2(t, x_2)}{g_2^2(t, x_2)}(Q(t) + l_2 e_2) + \frac{1}{g_2(t, x_2)}(\dot{Q}(t) + l_2 \dot{e}_2)] \tag{27}$$

Thus the Eq. (24) can be written as

$$x_4^*(t) = -\frac{1}{g_3(t, x_3)}[P(t) + l_3 e_3] \tag{28}$$

Therefore, the subsystem (26) can be further rewritten as

$$\begin{aligned} \dot{e}_4 &= \dot{x}_4 - \dot{x}_4^* \\ &= f_4(t) + g_4(t, x_4) u(t) - \frac{\dot{g}_3(t, x_3)}{g_3^2(t, x_3)}(P(t) + l_3 e_3) \\ &+ \frac{1}{g_3(t, x_3)}(\dot{P}(t) + l_3 \dot{e}_3) \end{aligned} \tag{29}$$

Then construct the following Lyapunov function

$$V = V_3 + \frac{1}{2} e_4^2, \tag{30}$$

Calculate the derivatives of V along (7), (13), (21) and (29)

$$
\begin{aligned}
\dot{V} &= \dot{V}_3 + e_4\dot{e}_4 \\
&= -l_1e_1^2 - l_2e_2^2 - l_3e_3^2 + g_3(t, x_3)e_3e_4 \\
&\quad + e_4[f_4(t) + g_4(t, x_4)u(t) \\
&\quad - \frac{\dot{g}_3(t, x_3)}{g_3^2(t, x_3)}(P(t) + l_3e_3) + \frac{1}{g_3(t, x_3)}(\dot{P}(t) + l_3\dot{e}_3)] \\
&= -l_1e_1^2 - l_2e_2^2 - l_3e_3^2 + e_4[g_3(t, x_3)e_3 \\
&\quad + f_4(t) + g_4(t, x_4)u(t) \\
&\quad - \frac{\dot{g}_3(t, x_3)}{g_3^2(t, x_3)}(P(t) + l_3e_3) + \frac{1}{g_3(t, x_3)}(\dot{P}(t) + l_3\dot{e}_3)]
\end{aligned} \tag{31}
$$

Design the following feedback control law

$$
\begin{aligned}
u(t) = &-\frac{1}{g_4(t, x_4)}[g_3(t, x_3)e_3 + f_4(t) + l_4e_4 \\
&- \frac{\dot{g}_3(t, x_3)}{g_3^2(t, x_3)}(P(t) + l_3e_3) + \frac{1}{g_3(t, x_3)}(\dot{P}(t) + l_3\dot{e}_3)]
\end{aligned} \tag{32}
$$

to make $\dot{V} < 0$, where, l_4 is a positive design parameter. Substituting the Eq. (32) into (31) yields

$$
\dot{V} = -l_1e_1^2 - l_2e_2^2 - l_3e_3^2 - l_4e_4^2 \tag{33}
$$

From (33), it is clearly seen that $\dot{V} < 0$. Then the error states e_1, e_2, e_3 and e_4 vanish asymptotically hence network system stability is guaranteed.

From the above deduction and analysis process, it can be seen that the system (4) is asymptotically stable under the action of the designed backstepping feedback controller. In other words, the system (3) and (2) achieves the active queue management congestion control, which means that we can get the desired queue length and appropriate window size.

4 Simulation Analysis

For the parallel transmission link system consisting of two routers, the corresponding simulation is given in this section. In the simulation, the link capacity $C(t)$ is taken as a constant, and the parameters of the system are as follows.

$$
N = 60,\ q_d = 375 \text{ packets},\ C = 1750 \text{ packets/s}
$$

The control parameters of the system are

$$l_1 = 0.2,\ l_2 = 0.5,\ l_3 = 1.2,\ l_4 = 0.8.$$

The simulation results are given as follows.

From the simulation result shown in Fig. 2, it can be seen that the tracking error of system state x_2 reaches zero in 6 s through the packet-loss-transfer mechanism of double-router network, which makes the state x_2 reach a stable steady-state equilibrium.

From Fig. 3, it can be seen that some data packets needed to be dropped in Router 1 are transferred to Router 2, and make the window of Router 2 fluctuate greatly, which conforms to the operation mechanism of double router networks. Even so, under the controller designed in this chapter, the system state x_4 is stable in 11 s. Furthermore, Fig. 4 shows this controller does enforce very small packet loss rate. All these results demonstrate the effectiveness of the proposed control design.

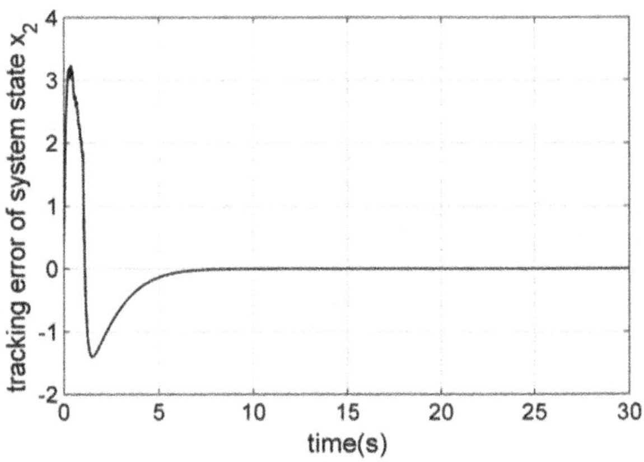

Fig. 2 Error tracking time-response for the states x_2

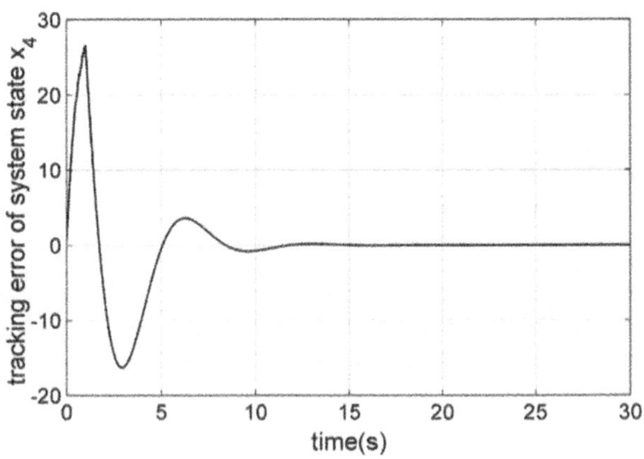

Fig. 3 Error tracking time-response for the state x_4

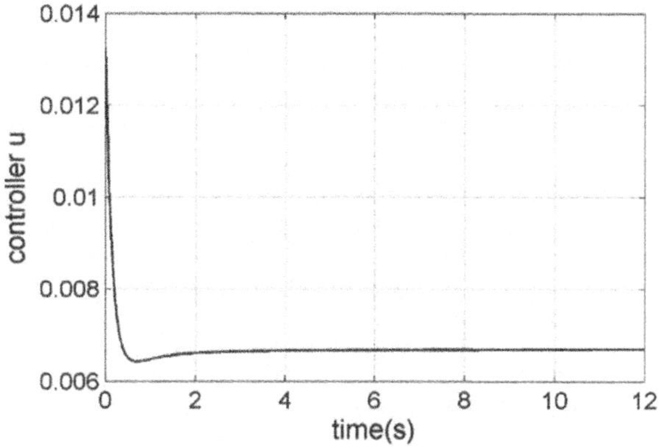

Fig. 4 Time-response of the synthesized control law (32)

5 Conclusions

Based on the analysis of various existing TCP/AQM network system models, this paper proposes a novel modelling idea of "parallel topology, cascade operation" for double-router TCP/AQM network systems. The data packet transmission rule of "priority selection, random transfer" can overcome the phenomenon that on the one hand the router is idle, while on the other, the data packet is still dropped. Furthermore, dynamic model of the double-router network system is transformed into a state space model of the nonlinear system with lower triangle form, which makes it suitable for the design of the feedback controller for the congestion control algorithm by using the nonlinear backstepping design method, and makes the nonlinear double-router TCP/AQM network system asymptotically stable. The simulation results verify the feasibility and effectiveness of the proposed method.

Due to space limitation, we only discussed the case of double routers and delay free. The research approach, the underlying idea and most results can be readily extended to the case of multi-routers. The case with delay and uncertainty too could be considered by using similar control methodology. At the same time, based on other existing models, the single router models can be combined in parallel interconnection to form a multi-router network system. Then, the controller can be designed with advanced control technology employing computational intelligence [17] in order to achieve the active flexible queue management scheme for TCP/AQM network systems.

Acknowledgements This work is supported by the National Natural Science Foundation of China (No. 61773108). Yuanwei Jing and Yan Zheng want to acknowledge their former doctoral and postgraduate students, respectively, for carrying out computer simulations. Inspiring issues emerged while discussing TCP/AQM network systems.

References

1. Alaoui, S.B., Tissira, E.H., Chaibi, N.: Active queue management based feedback control for TCP with successive delays in single and multiple bottleneck topology. Comput. Commun. **2018**(117), 58–70 (2018)
2. Bauso, D., Giarre, L., Neglia, G.: Active queue management stability in multiple bottleneck networks. In: Proceedings of IEEE ICC'04, pp. 2267–2271 (2004)
3. Braden, B., Clark, D., Crowcroft, J., Davie, B., Deering, S., Estrin, D., et al.: Recommendations on queue management and congestion avoidance in the internet. Project RFC 2309 Informational Report, (1998)
4. Domanska, J., Domanski, A., Nowak, S.: Performance modelling of selected AQM mechanisms in TCP/IP network. In: Internet: Technical Development and Applications, pp. 11–20. Springer, Berlin, Heidelberg (2009)
5. Firuzi, M., Haeri, M.: Adaptive generalized predictive control of active queue management in TCP networks. In: Proceedings of the International Conference on Computer as a Tool (EUROCON 2005), pp. 676–679 (2005)
6. Han, C.W., Li, M.Q., Chang, S.R., Liu, L.: Simulation of congestion control based on slide-mode observer. Comput. Simul. **34**(9), 265–269 (2017)
7. Hollot, C.V., Misra, V., Towsley, D., Gong, W.B.: A control theoretic analysis of RED. In: Proceedings of the 20th IEEE International Conference on INFOCOM, pp. 1510–1519. Anchorage, Alaska, USA (2001)
8. Jing, Y.W., Chen, B., Dimirovski, G.M., Sohraby, K.: On leader-follower model of traffic rate control for networks. Control Theory Appl. **18**(6), 817–822 (2001)
9. Kelly, F.P., Maulloo, A.K., Tan, D.K.H.: Rate control for communication networks: shadow prices, proportional fairness and stability. J. Op. Res. Soc. **49**(3), 237–252 (1998)
10. Liu, Y., Liu, X.P., Jing, Y.W., Zhou, S.W.: Adaptive backstepping H-infinity tracking control with prescribed performance for internet congestion. ISA Trans. **2018**(72), 92–99 (2018)
11. Low, S.: A duality model of TCP and queue management algorithms. IEEE/ACM Trans. Netw. **11**(4), 525–536 (2003)
12. Marsan, M.A., Garetto, M., Giaccone, P., Leonardi, E., Schiattarella, E., Tarello, A.: Using partial differential equations to model TCP mice and elephants in large IP networks. IEEE/ACM Trans. Netw. **13**(6), 1289–1301 (2005)
13. Mathis, M., Semke, J., Mahdavi, J., Ott, T.: The macroscopic behavior of the TCP congestion avoidance algorithm. SIGCOMM Comput. Commun. Rev. **27**(3), 67–82 (1997)
14. Misra, V., Gong, W.B., Towsley, D.: Fluid-based analysis of a network of AQM routers supporting TCP flows with an application to RED. In: Proceedings of the 19th IEEE International Conference on SIGCOMM, vol. 30(4), pp. 151–160. Stockholm, Sweden (2000)
15. Unal, H.U., Melchor-Aguilar, D., Ustebay, D., Niculescu, S., Ozbay, H.: Comparison of PI controllers designed for the delay model of TCP/AQM networks. Comput. Commun. **2013**(36), 1225–1234 (2013)
16. Wang, H.W., Yu, C., Jing, Y.W.: Network congestion control algorithm based on T-S fuzzy observer. J. Northeast. Univ. Nat. Sci. **31**(4), 461–464 (2010)
17. Wang, K., Jing, Y.W., Liu, Y., Liu, X., Dimirovski, G.M.: Adaptive finite-time congestion controller design of TCP/AQM systems based on neural network and funnel control. Neural Comput. Appl. **32**(13), 9471–9478 (2020)
18. Wang, L.J., Cai, L., Liu, X.Z., Shen, X.M., Zhang, J.S.: Stability analysis of multiple-bottleneck networks. Comput. Netw. **53**(3), 338–352 (2009)
19. Wang, P., Chen, H., Yang, X.P., Ma, Y.: Design and analysis of a model predictive controller for active queue management. ISA Trans. **51**(1), 120–131 (2011)
20. Wang, P., Zhu, D.J., Lu, X.H.: Active queue management algorithm based on data-driven predictive control. Telecommun. Syst. **2017**(64), 103–111 (2017)
21. Xu, Q., Li, F., Sun, J.S., Zukerman, M.: A new TCP/AQM system analysis. J. Netw. Comput. Appl. **2015**(57), 43–60 (2015)

22. Ye, C.Y., Jing, Y.W.: Finite-time congestion control based on terminal sliding mode control. J. Northeast. Univ. Nat. Sci. **35**(6), 761–765 (2014)
23. Yu, L., Ma, M.D., Hu, W.D., Shi, Z.B., Shu, T.T.: Design of parameter tunable robust controller for active queue management based on H-infinity control theory. J. Netw. Comput. Appl. **34**(2), 750–764 (2011)

Noise-Robust and Secure Communication Protocol for Industrial Networked Control Systems

Gorjan Nadzinski and Mile Stankovski

Abstract The increasing and rapid development of technology and complex systems and the coming of the fourth industrial revolution has, among many other areas, also influenced industrial automation. Industrial processes are no longer isolated entities; they now represent complex systems and subsystems which collect an abundance of data and are in constant communication with each other. Intelligent control, machine learning, BigData, and the constant improvement of measurement and control equipment all play a significant role in this development, but communication still remains one of the crucial aspects of these new generations of industrial systems. While the concept of industrial networked control systems brings along many benefits such as flexibility, speed, and modularity, the key role communication plays makes it a segment in these systems which is now vulnerable to both environmental and man-made interference. We are presenting a method of increasing the level of security of industrial communication protocols by developing an algorithm which uses the coupling functions between two dynamical systems for data encryption, and dynamical Bayesian inference for data decryption. The algorithm has been used to develop a communication protocol whose performance has been tested and verified in real-world experimental conditions, in the presence of both white Gaussian and coloured low frequency noise. This approach results in communication that is both cryptographically secure and noise robust, that is capable of functioning in industrial environment, and that potentially has reduced data transmission power and increased energy efficiency.

G. Nadzinski (✉) · M. Stankovski
Faculty of Electrical Engineering and Information Technologies, Ss. Cyril and Methodius University in Skopje, Ruger Boskovic 18, 1000 Skopje, North Macedonia
e-mail: gorjan@feit.ukim.edu.mk

M. Stankovski
e-mail: milestk@feit.ukim.edu.mk

P. Shi et al. (eds.), *Complex Systems: Spanning Control and Computational Cybernetics: Applications*, Studies in Systems, Decision and Control 415,
https://doi.org/10.1007/978-3-031-00978-5_19

1 Introduction

The rapid advancement of technology has ushered complex systems at the front lines of the fourth industrial revolution. It has brought about a situation where the vast and intricate theories and frameworks regarding complex systems that have been developed in the past decades, are finally more accessible and easier to implement on hardware and technology available today [1]. This has influenced many technical, commercial, and social areas, but none as much as industrial automation. As a result, industrial processes have ceased to be isolated objects and now represent complex systems and subsystems which constantly communicate with each other and manipulate an abundance of data. Along with concepts such as artificial intelligence and machine learning, BigData, or the enhanced measurement and control equipment, communication is still one of the crucial aspects of the new generation of industrial systems. While the concept of networking these industrial control systems has contributed to many benefits such as flexibility, efficiency, speed, and modularity, the key role communication plays makes the process of information exchange vulnerable to environmental, technical, and malicious man-made interference. All of these problems result in increase of noise and interference, reduction of communication quality, and distortion of often valuable data and information.

Many different communication protocols have been designed throughout the years, based on different concepts: mathematical, logical, signal processing, dynamical chaotic systems, quantum information approaches [2–8]. Even when considering the effects of noise on a typical industrial networked control system with a commonly used communication protocol (such as industrial Ethernet, for example), accents the need for both constant improvement of the existing protocols and also for the development of new communication protocols which would deliver better and/or more efficient noise robustness [9].

Our work has so far been done on developing a secure communications protocol based on coupling functions between dynamical systems [10]. The main subject of this paper is to attempt to unify and completely present the work on this protocol from the past several years, going over the theoretical background and especially focusing on the practical implementation of the protocol; it will also present new results regarding the influence of non-Gaussian noise on the information transfer, and the resulting benefits regarding energy efficiency.

The proposed protocol for secure and noise robust communication results in cryptographically safe communication, secured by using coupling functions for data encryption at the transmitter, and dynamical Bayesian inference for data decryption at the receiver. The use of the latter also contributes to an effective separation between the deterministic signals which carry the information and the dynamical perturbations in the communication channel, thus making the protocol highly noise robust and capable of functioning in interference and noise polluted industrial environments. The paper will also show the practical implementation of the protocol on commercially available hardware, and will present the results of the experiments examining its robustness to the effects of real world and simulated white Gaussian and low frequency noise. It will explore the possibilities of more energy-efficient communication as well, in the sense of using reduced transmission power caused by the communication protocol's high robustness in low signal-to-noise ratio conditions.

We will first discuss the theoretical and mathematical basics of the secure communications protocol based on coupling functions in Sect. 2, before presenting its practical realization using commercially available devices and circuitry in Sect. 3, outlining its potential for real-world implementation. Here we will also analyse the protocol's performance in the presence of both white Gaussian noise and low frequency Ornstein–Uhlenbeck noise, and discuss the results and implications regarding noise robustness and energy efficiency. We will give the concluding remarks in Sect. 4.

2 A Secure Communication Protocol Based on Coupling Functions

A coupling function gives a detailed description of the physical rules of the occurrence and manifestation of interaction between systems [11]. It is described in terms of the strength and form of the coupling, prescribing how the input influence from one of the coupled systems translates into the output from the other system. This way the coupling function determines the possibility of qualitative transitions between states of the systems. Furthermore, a description of the functional contributions from each subsystem within the coupling relationship can be obtained by decomposition of a coupling function. There have been a few different methods for coupling function reconstruction from data, with dynamical Bayesian inference [12] being used in this particular protocol.

Considering all this, coupling functions provide an effective way of encrypting information transfer between dynamical systems. This encryption would be extremely hard to break unless the exact coupling functions are known, because a set of linearly independent coupling functions between self-sustained dynamical systems can provide complex non-linear mixing of information. In addition, highly noise-robust communications are achieved, resulting from the use of dynamical Bayesian inference for stochastic processes which allows for effective separation between the deterministic information signals and the dynamical noise perturbations [13]. The starting inspiration for the protocol was an (at first sight) unrelated finding of the time-varying, decomposable, biological coupling functions of the human cardiorespiratory interaction [12, 14]. The communication protocol uses the same analysis methods developed for and used on the biological signals.

An experimental implementation of the protocol was also developed and robustness tests involving real analogue noise were performed. In the analogue electronic experiments the dynamical systems have truly continuous states and measurement noise and other imperfections are unavoidable, making the conditions very close to those of many real applications [15–17]. In this experiment, the transmitter and receiver were developed on two Raspberry PI 2 single-board computers, demonstrating the possibility for implementation of the protocol on low-cost devices commonly available in general use and comparable to smart-phones and devices in general [18]. Analogue electronic noise was added in order to simulate the reality of communications conditions and robustness was evaluated for different levels of perturbation noise.

2.1 The Concept of the Protocol

An overview of the experimental setup of the entire communications system is given in Fig. 1 [13]. A set of information carrying signals s_i come from different channels or devices and are to be transmitted simultaneously. Each of these signals acts as a scale parameter in the non-linear coupling functions between two self-sustained systems located in the transmitter. One signal from each of these systems is then transmitted through a public channel and is then used for enslaving and completely synchronizing two similar coupled systems at the receiving end. Finally, dynamical Bayesian inference is applied, so that the model parameters can be inferred and the information signals s_i can be decrypted.

An important thing to note here is that, even though the number of coupling functions is always finite and depends on the number of required information channels, the choice of forms for the linearly independent coupling functions gives an unbounded number of possible combinations.

The model that is to be inferred consists of two noisy M-dimensional interacting systems that can be described by the stochastic differential equation:

$$\dot{x_i} = f\left(x_i, x_j|c\right) + \sqrt{D}\xi_i = g(x_i|c_1) + q\left(x_i, x_j|c_2\right) + \sqrt{D}. \tag{1}$$

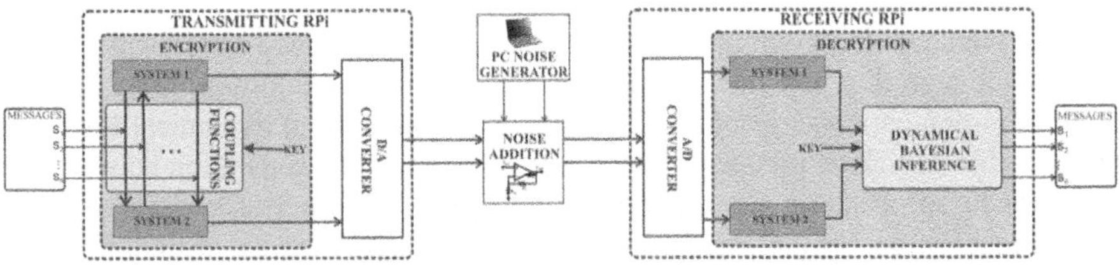

Fig. 1 A diagram of the practical implementation of the coupling function communication protocol. Information carrying signals s_i are used as scale parameters in the non-linear coupling functions between two self-sustained systems in the transmitter. The signals being sent are actually two signals (one from each system), which have generated noise added to them and then they enslave and completely synchronize two coupled systems at the receiver. Dynamical Bayesian inference then infers the model parameters, which include the signal s_i themselves [13]

Here, c is the parameter vector, $f(x_i, x_j|c)$ are the base functions which describe the autonomous dynamics $g(x_i)$ and the coupling functions $g(x_i, x_j)$, ξ_i is white Gaussian noise with autocorrelation $<\xi(t)\xi(t')> = D\delta(t - t')$, and D is the noise diffusion matrix for white Gaussian noise, and $i \neq j = 1, 2$.

The self-sustained dynamical systems don't need to be chaotic in general, but chaotic systems offer an additional level of complexity and therefore encryption security because they appear random-like and unpredictable, despite their underlying deterministic nature [19]. Additionally, the Bayesian inference framework used for the decryption of the signals favours the fact that attractors of chaotic coupled dynamical systems typically span a relatively large state space area.

In this setup, a system of two coupled chaotic Lorenz systems was used, and two binary signals $s_1(t)$ and $s_2(t)$ were to be transmitted. The first Lorenz system is given by:

$$\begin{aligned} \dot{x_1} &= 10x_2 + s_1(t)cos(y_1)x_2 + s_2(t)x_1y_2/y_3 \\ \dot{x_2} &= 28x_1 - x_1x_3 - x_2 \\ \dot{x_3} &= x_1x_2 - 2.67x_3, \end{aligned} \tag{2}$$

and the second one by:

$$\begin{aligned} \dot{y_1} &= 10y_2 - 10y_1 \\ \dot{y_2} &= 28y_1 - y_1y_3 - y_2 \\ \dot{y_3} &= y_1y_2 - 2.67y_3. \end{aligned} \tag{3}$$

In the expression for x_1 of the first oscillator, two coupling functions are comprised of variables from both the first and the second system (these are just examples of non-linear coupling functions and other choices of linearly independent functions can also be used). The behaviour of the systems is shown in Fig. 2 [13], where the relationships between two of the states in each system are shown by the Lissajous curves—Fig. 2a, b; it can be seen that the coupling has an effect on the first system in the shape of relatively minor disturbances of its attractor (Fig. 2b). The dependence between the mutually coupled states of the two systems during the process of data transmission is given in Fig. 2c. The convoluted inter-trajectories are the main property used for scrambling the information signals.

The signals x_1 and y_2 are the only ones which are then transmitted, with added noise during the transmission. On the receiving end both chaotic systems are completely synchronized [20]: x_1 drives the system u to become effectively identical to the system x:

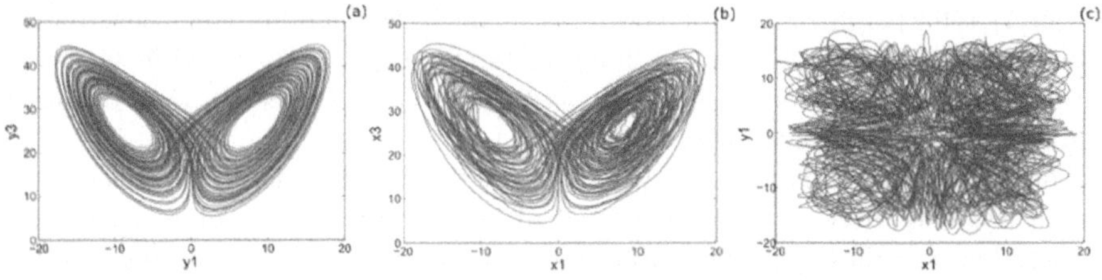

Fig. 2 The trajectories of the Lorenz systems at the transmitting side: **a** The trajectories of y_1 and y_3 from the second autonomous system (3). The trajectories are expectedly attracted to stable points. **b** The trajectories of x_1 and x_3 from the first autonomous system (2). The situation is similar to the previous case, but with a distinctive difference in the roughness of the trajectories, caused by the fact that x_1 contains the coupling elements. **c** The trajectories of x_1 and y_1 from both systems during data transmission. The coupling Lissajous curve is chaotic and the solution is not narrowed to a bounded and predictable trajectory space, but is instead well hidden in the chaotic and convoluted coupled trajectories and non-linear dependences between the systems [13]

$$u_1 = x_1$$

$$\dot{u_2} = 28x_1 - x_1u_3 - u_2 \tag{4}$$

$$\dot{u_3} = x_1u_2 - 2.67u_3,$$

and y_2 drives the system w to become effectively identical to system y:

$$\dot{w_1} = 10y_2 - 10w_1$$

$$w_2 = y_2 \tag{5}$$

$$\dot{w_3} = w_1y_2 - 2.67w_3.$$

Finally, the signals from the reconstructed dynamical systems u and w are the six inputs for the dynamical Bayesian inference framework.

2.2 Dynamical Bayesian Inference

The decryption of the signals s_i from the two reconstructed coupled systems u and w is done by state-space dynamical Bayesian inference, mostly outlined in detail in [10, 12, 21]. Here, the model that is supposed to be inferred is given by Eq. (1) and the coupling functions $q(x_i, x_j)$ actually represent the encryption key. When given a time-series $X = \{x_n \equiv x(t_n)\}$ $(t_n = nh)$ as input, the Bayesian dynamical inference's task is to infer the unknown model parameters and the noise diffusion matrix $M = \{c, D\}$. This is done by maximization of the posterior conditional probability $p_X(M|X)$ of observing the parameters M when given the data X [21]. The prior density $p_{prior}(M)$ represents the prior knowledge of the unknown parameters based on observations, and its relationship with the posterior conditional probability and to the likelihood function $l(X|M)$ (the conditional probability density to observe X given choice M), is given by the well known Bayes' theorem:

$$p_X(M|X) = \frac{l(X|M)p_{prior}(M)}{\int l(X|M)p_{prior}(M)dM} \tag{6}$$

With dense enough sampling h the problem is solved with the Euler midpoint $x_n{}^* = (x_{n+1} + x_n)/2$ discretization of Eq. (1):

$$x_{i,n+1} = x_{i,n} + hf\left(x_{i,n}{}^*, x_{j,n}{}^*|c\right) + h\sqrt{D}z_n. \tag{7}$$

Here $z_n \equiv \int_{t_n}^{t_{n+1}} z(t)dt$ is the stochastic integral of the noise term over time, which is statistically independent and the likelihood is given by a product over n of the probability at each moment of time of observing x_{n+1} [13]. The joint probability density of z_n is used to find the joint probability density of the process in respect of $x_{n+1} - x_n$. The negative log-likelihood function $S = -lnl(X|M)$ can be expressed as:

$$S = \frac{N}{2}ln|D| + \frac{h}{2}\sum_{n=0}^{N-1}\left(c_k\frac{\partial f_k(x_{\cdot,n})}{\partial x} + [\dot{x}_n - c_k f_k(x_{\cdot}{}^*{}_{,n})]^T\left(D^{-1}\right)[\dot{x}_n - c_k f_k(x_{\cdot}{}^*{}_{,n})]\right) \tag{8}$$

where $\dot{x}_n = (x_{n+1} - x_n)/h$, with implicit summation over the repeated index k.

Given a multivariate normal distribution for the prior probability of the parameters c, with mean $\overline{c}$, covariance matrix Σ_{prior}, and concentration matrix $\Xi_{prior} \equiv \Sigma^{-1}{}_{prior}$, the posterior multivariate probability density of each parameter of (1) $N_X(c|\overline{c}, \Xi)$ can be evaluated by applying the following four equations [21] to each sequential block of data X:

$$D = \frac{h}{N}\left[\dot{x}_n - c_k f_k\left(x_{\cdot}{}^*{}_{,n}\right)\right]^T\left[\dot{x}_n - c_k f_k\left(x_{\cdot}{}^*{}_{,n}\right)\right]$$

$$\Xi_{k\omega} = \left(\Xi_{prior}\right)_{k\omega} + hf_k\left(x_{\cdot}{}^*{}_{,n}\right)\left(D^{-1}\right)f_\omega\left(x_{\cdot}{}^*{}_{,n}\right) \tag{9}$$

$$r_\omega = \left(\Xi_{prior}\right)_{k\omega} + hf_k\left(x_{\cdot}{}^*{}_{,n}\right)\left(D^{-1}\right)\dot{x}_n - \frac{h}{2}\frac{\partial f_k\left(x_{\cdot,n}\right)}{\partial x}$$

$$c_k = \left(\Xi^{-1}\right)_{k\omega} r_\omega.$$

Here, summation over $n = 1, \ldots, N$ is assumed, the initial prior is set to be the non-informative flat normal distribution given by $\Xi_{prior} = 0$ and $\overline{c}_{prior} = 0$, and summation over repeated indices k and w is implicit.

The inference simultaneously follows the time evolution of c and separates its dynamical effects from the accompanying noise. For that purpose, the time-series are separated into sequential blocks and the evaluation of every next block of data uses the evaluation results of the previous block, so that the process of information propagation between the n posterior distribution and the $n + 1$ prior distribution allows for the time-variability of the parameters to be followed. A squared symmetric positive definite matrix Σ_{diff} shows how much each parameter diffuses normally, so the next prior probability of the parameters is the convolution of two current normal multivariate distributions, Σ_{post} and Σ_{diff}: ${\Sigma_{prior}}^{n+1} = {\Sigma_{post}}^{n} + {\Sigma_{diff}}^{n}$. The diffusion matrix is $\Sigma_{diff\,i,j} = \rho_{ij}\sigma_i\sigma_j$, where σ_i is the standard deviation of the diffusion of c_i in the current time window, and ρ_{ij} is the correlation between the changes in the parameters c_i and c_j [21].

The dynamical Bayesian inference is applied at the receiving unit to the six time series $u_1, u_2, u_3, w_1, w_2, w_3$ from both reconstructed coupled systems, which decrypts the information carrying binary signals $s_1(t)$ and $s_2(t)$. The functions on the right-hand sides of Eqs. (2) and (3) are the base functions for the inference of the model in the Bayesian framework.

Much more detailed information about the dynamic Bayesian inference, its development, application, and practical implementation can be found in [10, 12, 21, 22].

3 Implementation of the Communication Protocol

3.1 Analogue Electronic Implementation

The transmitter and receiver were implemented on two Raspberry PI 2 Model B single-board computers to test the feasibility and the efficiency of the approach on commercially available, broadly used hardware, with similar performance as other common low-cost devices such as smart phones. Figure 3 [13] gives a detailed schematic circuit diagram, with the two Raspberry PIs on the left and right sides. The signals were generated in the transmitting unit and reconstructed in the receiving unit using a fourth order Runge–Kutta scheme with sampling of $h = 0.01$.

The two random binary signals $s_1(t)$ and $s_2(t)$ were generated in the transmitter and then used as scale parameters in the non-linear coupling functions between the systems (2) and (3). A digital-to-analogue converter converted x_1 and y_2 into analogue signals, before they were amplified and transmitted by wires to the receiving unit. While in analogue form, independent white noise was added to both signals; both noise signals were generated in Matlab with a 100 kHz sampling frequency and the same amplitude. On the receiving side, both analogue signals were converted back to digital converter and then used to synchronize the chaotic systems in the receiver

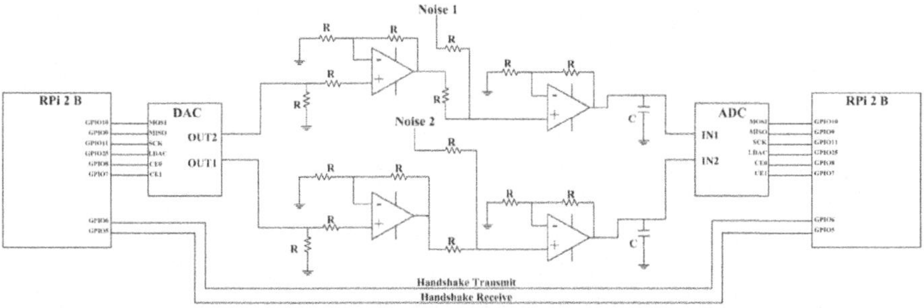

Fig. 3 A detailed diagram of the electronic implementation of the protocol. The signals are generated in the transmitting Rpi unit and are sent after being converted to analogue signals, amplified, and have noise added to them. At the receiving side they are converted back into digital form before synchronizing the local chaotic systems [13]

(4) and (5). The transmission also included a handshake setup with direct digital input/output connections between the transmitter and receiver. two Raspberry PIs (indicated by the two lines on the bottom of Fig. 4). The time window in which Bayesian inference was applied was 250 s, i.e. each bit {0 or 1} was transmitted within this window length, as communication speed was not the primary focus of the experiment.

Samples of the three signal time-series captured by an oscilloscope at an arbitrary time can be seen in Fig. 4 [13]. The top signal shows the effect of the added analogue white noise (middle signal) to the original information-carrying transmitted analogue signal $y_2(t)$ (bottom signal).

Fig. 4 Real-time capture of the transmitted signal $y_2(t)$ (bottom), the generated white noise (middle), and the signal $y_2(t)$ with the noise added (top) [13]

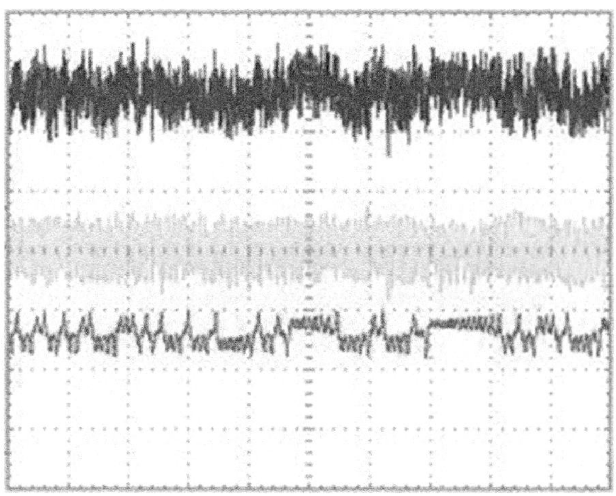

3.2 Signal Analysis

With all signals recorded for offline analysis, it was possible to analyse the effects of noise on the time-series of the transmitted signals. Signals $x_1(t)$ and $y_2(t)$ are shown before and after the noise was added on Fig. 5a, b [13] respectively, while the digitized time-series of the analogue white noise is shown in (c). The corresponding fast Fourier transforms of both signals and the noise are given in panels (d), (e), and (f) respectively. The spectra of the chaotic signals $x_1(t)$ and $y_2(t)$ are clearly broadened but without visible harmonics, while the noise spectrum contained all frequencies and is spread across the entire observed domain, as is expected with white noise processes.

Since the main goal of the experiment was testing the effectiveness and robustness of coupling-function based encryption for communication in noisy conditions, the strength of the noise was gradually increased (decreasing the SNR), while randomly generated binary data was transmitted. Figure 6 [13] shows the deviations of the decrypted binary signal $s_1(t)$ (the inferred parameter c_1) and signal $s_2(t)$ (the inferred parameter c_2) as functions of the SNR. For each examined value of SNR the parameters c_1 and c_2 are plotted as two boxplots showing the mean and the distribution of all the 1 and 0 values within that set. Expectedly, for small noise (high SNR),

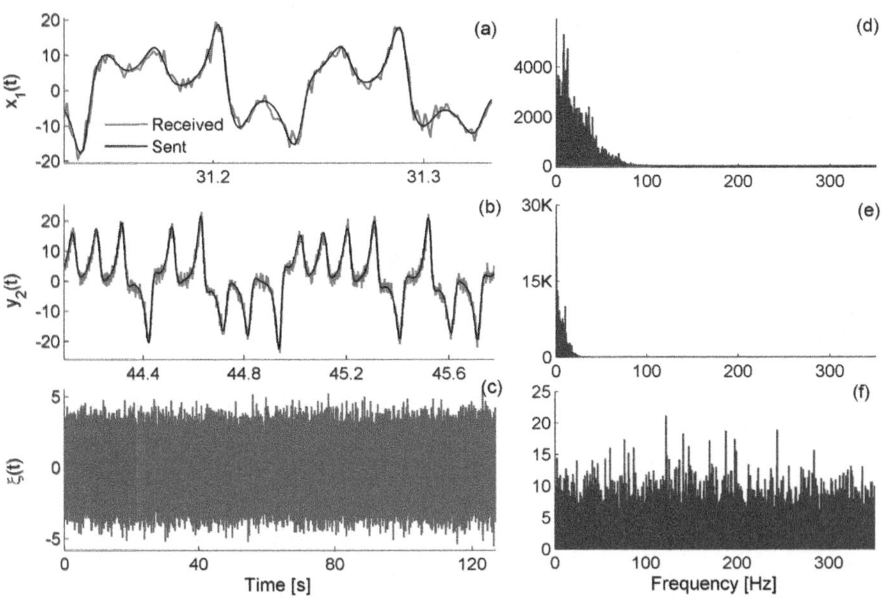

Fig. 5 Time-series and spectral analysis of the transmitted, received, and noise signal. **a** Time-series of the received signal $x_1(t)$ (blue) and its original transmitted version (black). **b** Time-series of the received signal $y_2(t)$ (blue) and its original transmitted version (black). **c** The discretized analogue white noise time-series. **d** The FFT frequency spectrum of $x_1(t)$. **e** The FFT frequency spectrum of $y_2(t)$. **f** The FFT frequency spectrum of the noise signal [13]

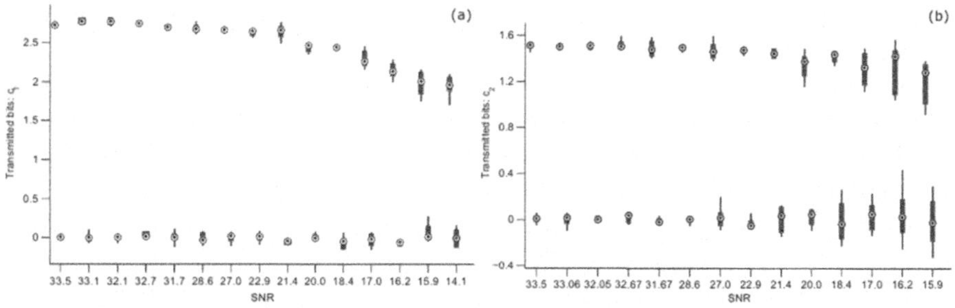

Fig. 6 The noise-caused deviations of the decrypted signal from its initial binary states, presented as compact box plots (median, quartiles, max, and min), as a function of signal-to-noise ratio (SNR). **a** For the first information signal $s_1(t)$ (inferred parameter c_1) the binary value for $\{1\}$ is represented by $c_1 = 2.7$. **b** For the second information signal $s_2(t)$ (inferred parameter c_2) the binary $\{1\}$ is represented by $c_2 = 1.5$. As SNR drops down to around14–15 dB, there is still no overlap in the distributions and the bit-error-rate BER is zero [13]

the boxplot distributions are compressed around the original values; for higher noise levels (low SNR), the boxplot distributions become much wider. The observer can not separate the binary 1 from the binary 0 when the two distributions for each SNR trial begin to overlap, and the bit-error-rate (BER) then becomes non-zero. Figure 6 shows that in all of the investigated scenarios the distributions are non-overlapping and BER is zero. This process was performed with increasing noise causing SNR to drop down to around 14–15 dB for the simultaneous parameters c_1 and c_2, after which the experiment became impossible due to experimental limitations. This is a relatively high level of noise for real communications, with the SNR threshold for finite BER being around 15 dB for wireless transmission and around 40 dB for wireline communication [23].

In order to determine the effectiveness of the coupling function protocol, it was compared with one of the most used protocols for secure communication with (chaotic) dynamical systems based on complete synchronization—the signal masking protocol [24]. This is a less complex protocol since it uses complete synchronization without dynamical Bayesian inference for decryption. With the same number of random bits transmitted within the same window length and by using the same hardware setup, it was found that a finite BER appeared at around SNR = 20 dB, which is a significantly higher threshold indicating lower robustness than the values obtained with the coupling function protocol.

3.3 Influence of Low-Frequency Non-Gaussian Noise on the Information Transfer

Interference in communications networks is often modelled as a Gaussian random process to which the central limit theorem is applicable, and this is appropriate when the noise is caused by different mutually independent, uncorrelated, and non-dominating signals (e.g. thermal noise in electronics). However, in many real-world situations dominant sources of interference occur and the probability density function of the interference features a heavier tail than the one predicted by the Gaussian model. In such cases, alternative models have been proposed, such as the Ornstein–Uhlenbeck process [25], 1/f noise [26], the spatial Poisson process [27], etc.

Therefore, an analysis with low-frequency Ornstein–Uhlenbeck noise $\eta(t)$ added to the sent signals x_1 and y_2 from the coupled systems (2) and (3) was performed, in order to examine the robustness of the coupling function protocol to noise more commonly found in real systems. During the simulated transmission we have

$$x_1 = x_1 + \eta_1(t)$$

$$y_2 = y_2 + \eta_2(t), \tag{10}$$

where $\eta_1(t)$ and $\eta_2(t)$ are Ornstein–Uhlenbeck noise signals that influence the respective channels. The Ornstein–Uhlenbeck noise can be defined as:

$$\dot{\eta}(t) = \xi(t) - \frac{1}{\gamma}\eta(t). \tag{11}$$

Here, ξ is white Gaussian noise and γ is the correlation time of the random Ornstein–Uhlenbeck process. This converges to white Gaussian noise for the limiting case $\gamma \to 0$, but in reality the noise has a non-neglectable, non-zero correlation time. Therefore, the Ornstein–Uhlenbeck process can be used to represent the noise that occurs in real-world communication systems [28]. It tends to drift towards a long-term mean over time, which is neglectable here because of the relatively brief time windows in which the communication takes place.

The noise signals generated with (11) were also subjected to both the Kolmogorov–Smirnov and the Anderson–Darling tests to examine the similarity of their distributions to the standard normal distribution. Both tests showed that, for the used time windows and for correlation times $\gamma \geq 0.09$, the generated noises indeed do not come from a Gaussian distribution.

The numerical simulations with Ornstein–Uhlenbeck noise enforced at the communication channel were run for times of 20,000 s, with 400 data bits sent and decrypted over 400 Bayesian windows of 50 s each, and sampling time of 0.01 s. The time-series of the transmitted and received signals for both $x_1(t)$ and $y_2(t)$ and the time-series and the corresponding FFT power spectra for the noise are given in Fig. 7 [13]. The noise signal $\eta_1(t)$ applied to $x_1(t)$ had a strength of $D_1 = 20$ and a correlation time of $\gamma_1 = 30$, while the noise signal $\eta_2(t)$ applied to $y_2(t)$ had the same strength of $D_2 = 20$ with a much shorter correlation time of $\gamma_2 = 0.09$ (the limit determined by the Kolmogorov–Smirnov and Anderson–Darling tests). As shown in Fig. 7a and c respectively, the longer correlation time contributed towards a more visible drift of the noise within the time domain, and towards a power spectrum

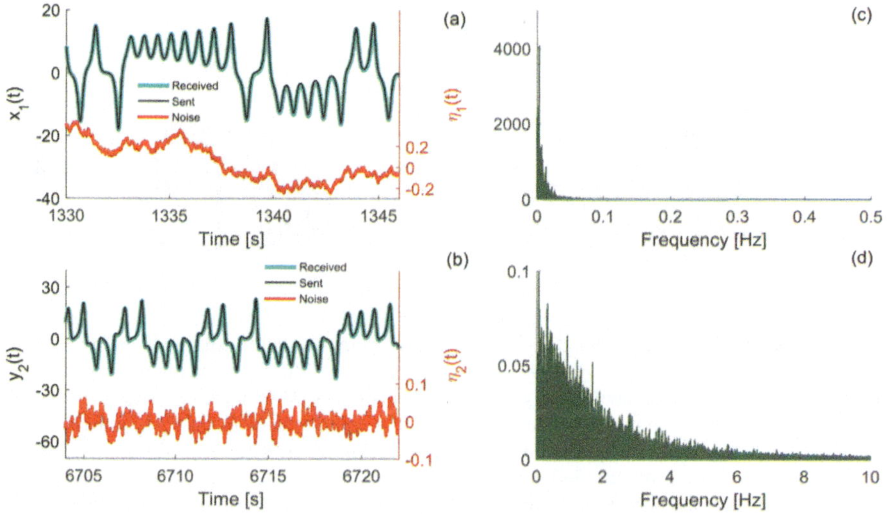

Fig. 7 Time-series and spectral analyses of the transmitted, received, and noise signals. **a** Time-series of the received signal $x_1(t)$ (blue), its original transmitted version (black), and the applied Ornstein–Uhlenbeck noise $\eta_1(t)$ with strength $D_1 = 20$ and correlation time $\gamma_1 = 30$ (red trace, and right-hand ordinate axis). **b** Time-series of the received signal $y_2(t)$ (blue), its original transmitted version (black), and the applied Ornstein–Uhlenbeck noise $\eta_2(t)$ with strength $D_2 = 20$ and correlation time $\gamma_2 = 0.09$ (red trace and right-hand ordinate axis). **c** The FFT frequency spectrum of $\eta_1(t)$. **d** The FFT frequency spectrum of $\eta_2(t)$. It can be noted that $\eta_2(t)$ has a shorter correlation time and therefore looks more like white noise than $\eta_1(t)$, which can seen in both the time evolutions and the frequency spectra [13]

restricted to much lower frequencies, confirming the Ornstein–Uhlenbeck process as an adequate model of a low-pass-filtered white noise. On the other hand, Fig. 7b and (d) show $\eta_2(t)$ to be more reminiscent of standard white Gaussian noise, both in the time domain and in the power spectrum respectively, due to the shorter correlation time.

Next, Fig. 8 [13] shows a few examples of the performance of the dynamical Bayesian inference when dealing with noise of this nature, using the mean-square error between the sent and the received values of the decrypted bits $c_1(t)$ and $c_2(t)$:

$$MSE(c) = \frac{1}{n}\sum_{i=1}^{n}\left(c_{j,decrypted} - c_{j,encrypted}\right)^2. \tag{12}$$

At first, simulations were run where $\eta_2(t)$ was kept in a constant state of $D_2 = \gamma_2 = 1$, while the parameters of $\eta_1(t)$ were being changed in the ranges of [0;22] for D_1 and [0;52] for γ_1. The mean-square error of the decrypted bit $c_2(t)$ was calculated for each simulation scenario and plotted as a function of D_1 and γ_1. Figure 8a shows that this error increases with the increase of either the noise strength or the correlation

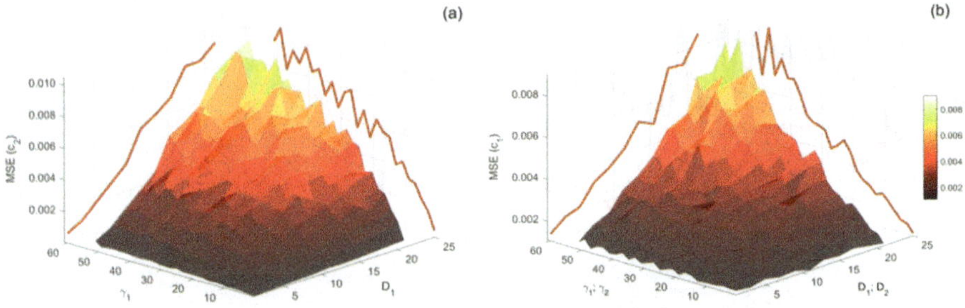

Fig. 8 The dependence of the mean-square error (MSE) of the transmitted bits when Ornstein–Uhlenbeck noise with strength D and correlation time γ is applied to the communication channels. **a** The MSE of $c_2(t)$ as a function of D_1 and γ_1 of the noise signal $\eta_1(t)$ applied to $x_1(t)$, while the parameters of the noise applied to y_2 remain constant. The MSE increases with the strength and correlation time of the noise. **b** The MSE of $c_1(t)$ as a function of D_1 and γ_1, and of D_2 and γ_2, of the noise signals $\eta_1(t)$ applied to $x_1(t)$ and $\eta_2(t)$ applied to $y_2(t)$, respectively. Again, the MSE increases with the strengths and the correlation times of the noise signals. Similar plots were obtained for the dependence of the MSE of the bit $c_1(t)$ on the change of the parameters of $\eta_1(t)$, and for the dependence of the MSE of both $c_1(t)$ and $c_2(t)$ on the change of the parameters of $\eta_2(t)$. These plots have been omitted in view of space considerations [13]

time. The projections of the three-dimensional surface on the side plains show the error rising more steadily and linearly with the increase of D_1, than with the increase of γ_1.

Also simulated was the simultaneous increase in strength and correlation time of both noise signals $\eta_1(t)$ and $\eta_2(t)$. As shown in Fig. 8b, the mean-square error of the decrypted bit $c_1(t)$ increases with the increase of these parameters. The change of the noise strengths again causes a steadier and more linear-like increase than the change of the correlation times.

In continuation, Figs. 9 and 10 show the deviations of the encrypted signals for bits c_1 and c_2 during communication in the presence of low frequency noise modelled by the Ornstein–Uhlenbeck random process. It can be seen that the overlapping of the deviations occurs for values of SNR between 20 and 25 dB, which is somewhat higher than the results obtained in the previous section, indicating and indeed confirming again that the communication protocol has a lower tolerance towards noise with higher correlation times (less similar to Gaussian).

Nevertheless, even these SNR levels of bit-error rate are satisfactory when compared with conventional communication protocols and their noise robustness [23].

In conclusion, it can be said that dynamical Bayesian inference is, with its stochastic nature, better capable of dealing with noise resembling Gaussian noise (i.e. with a shorter correlation time). However, we also find that it exhibits satisfactory results when more realistic forms of noise are applied.

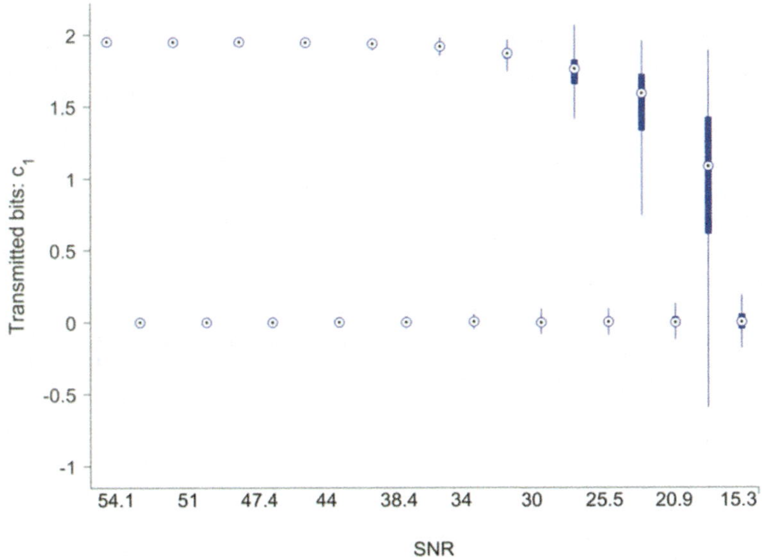

Fig. 9 Deviations of the decrypted signal $s_1(t)$ (inferred parameter c_1) from the initial binary states due to low frequency Ornstein–Uhlenbeck noise, presented as compact box plots (median, quartiles, max, and min), as a function of signal-to-noise ratio (SNR). The overlap starts to occur at SNR between 20 and 25 dB, which shows lower tolerance resistance compared to white noise, but still satisfactory when compared to conventional protocols and their nose robustness

3.4 Data Transmission Power and Energy Efficiency

Figures 11 and 12 give yet another interpretation of the results, this time showing the influence of the white Gaussian noise and the low frequency noise respectively via the SNR level, on the mean squared error of the decrypted data bits, calculated with Eq. (12). As expected, the error steadily drops down with the increase of the SNR level (as the noise intensity decreases) in both scenarios and for both signals.

One interesting aspect of all these results is the option of energy conservation when using the proposed protocol. Since SNR represents the ratio between the intensity of the signal and the noise, reduction of the data transmission power would result in a decrease of the SNR level and thus the mean squared error would rise. However, knowing the tolerance range of the protocol, if the ratio remains within a margin where the noise effects can be dealt with, then the communication process would still work properly but with less power used for data transmission.

The concept can also be seen from Fig. 13, which shows the mean squared error as a function of the transmission power ratio between the used and the nominal transmission power, under the influence of low frequency noise modelled with the Ornstein–Uhlenbeck process with strength $D = 100$ and correlation time $\gamma = 10$. For the given noise parameters, the SNR ratio in the system is almost 31 dB, close

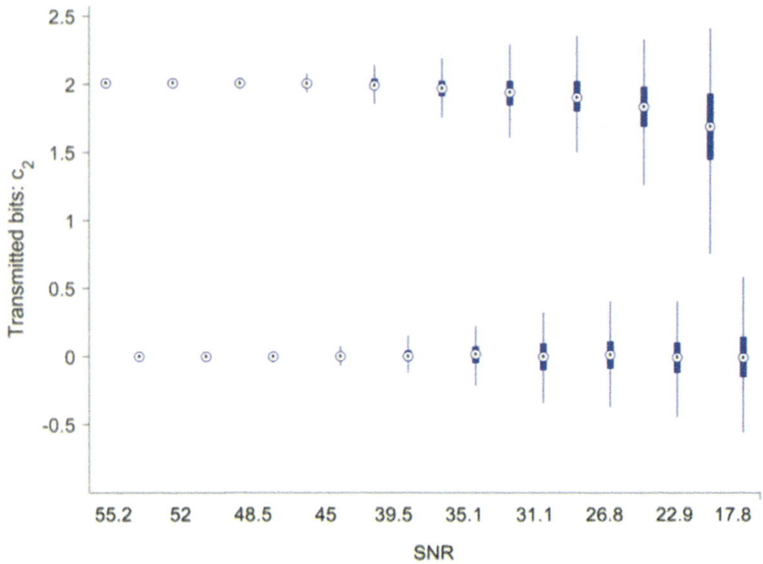

Fig. 10 Deviations of the decrypted signal $s_2(t)$ (inferred parameter c_2) from the initial binary states due to low frequency Ornstein–Uhlenbeck noise, presented as compact box plots (median, quartiles, max, and min), as a function of signal-to-noise ratio (SNR). The overlap starts to occur at SNR between 20 and 25 dB, which shows lower tolerance resistance compared to white noise, but still satisfactory when compared to conventional protocols and their nose robustness

to the margins of acceptable noise levels for the protocol (as can be seen on Fig. 10). Naturally, the reduction of this ratio results in a higher error and in less successful communication.

This presented opportunity can be used in the future to adapt the data transmission power as a function of the noise in the channel/communication medium, or as a function of the information increment of the data being decrypted [29]. The possibility of adaptive data transmission power can also counterweight the protocol's one significant drawback, which is the higher complexity and requirements of computational power, and can thus contribute to a very welcomed power savings and to a higher rate of energy efficiency for the system which utilizes the proposed communication protocol.

4 Conclusion

The proposed protocol was shown to increase the level of security and noise robustness of industrial communication by using coupling functions and dynamical Bayesian inference for data encryption and decryption. Safe and noise robust communication protocols in industry do exist, but the protocol uses the interaction

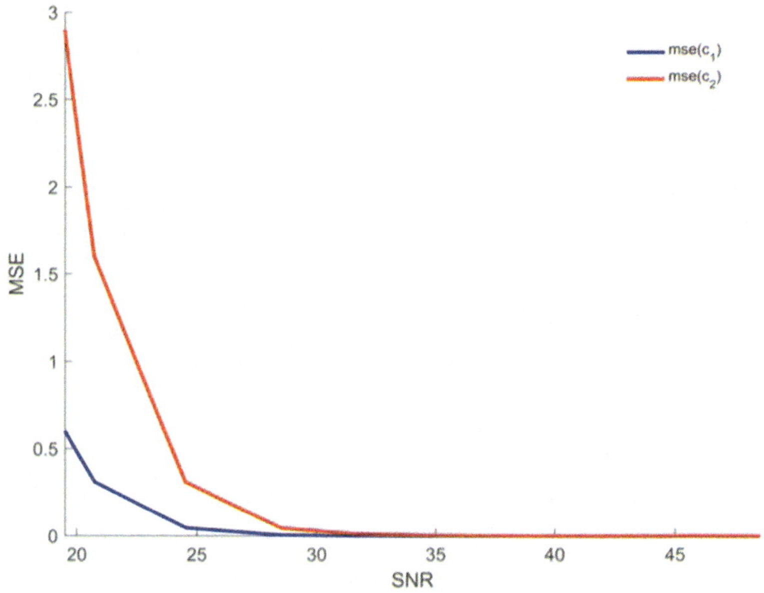

Fig. 11 The mean squared error of decrypted data c_1 (blue) and c_2 (red) as a function of the SNR level in a communication channel influenced by white Gaussian noise. Knowing the tolerance range of the protocol, if the ratio remains within a margin where the noise effects can be dealt with, then the communication process would still work properly but with less power used for data transmission

defined by coupling functions to provide non-linear mixing for encrypting the information. The encryption key is actually acquired from an unbounded set of possible forms and values, which makes the communication extremely secure. The decryption is done with dynamical Bayesian inference, itself stochastic by nature and thus able to deal with the effects of random noise and environmental inference.

A simple experimental proof of concept also tested the noise robustness of the protocol in a more realistic environment, giving results consistent with the theoretical findings. The setup was simple because the coupling function protocol was tested on a device with performance and features similar to those of widely used, low-cost smart devices. There were of course some inevitable and expected practical limitations regarding conversion rates, faster processing, or remote synchronization. All these can be overcome by using more expensive and more complex higher-performance hardware. The small amounts of encountered measurement noise also reduced the precision of the inference at the decrypting side, so its decomposition should also be taken into account in the future.

The coupling function protocol is effective and functioning in the presence of relatively high noise, and it was shown that it can be in fact used in current communications applications. It was also successful in dealing with low-frequency non-Gaussian noise, which is a common occurrence in many practical situations where there are dominant interference sources.

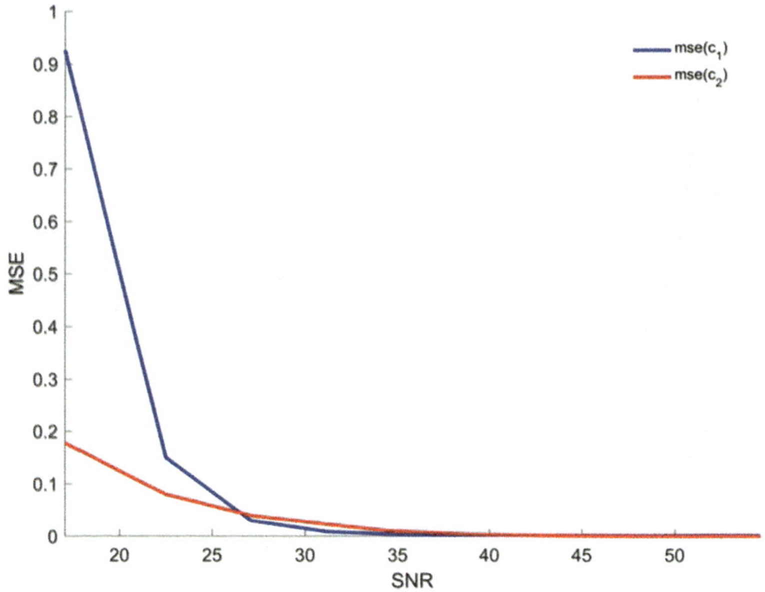

Fig. 12 The mean squared error of decrypted data c_1 (blue) and c_2 (red) as a function of the SNR level in a communication channel influenced by low frequency Ornstein–Uhlenbeck noise. Knowing the tolerance range of the protocol, if the ratio remains within a margin where the noise effects can be dealt with, then the communication process would still work properly but with less power used for data transmission

Finally, it was shown that the high noise robustness provides a possibility for implementing adaptive data transmission power, considering the noise in the communication medium or the information increment of the decrypted data. This could balance the relatively higher requirements of the protocol for computational power, and result in a communication method with better (or at least controllable) energy efficiency.

All these features make the proposed coupling functions secure communications protocol a promising solution in critical, information-sensitive, noise-heavy applications such as modern complex systems, especially networked industrial control systems.

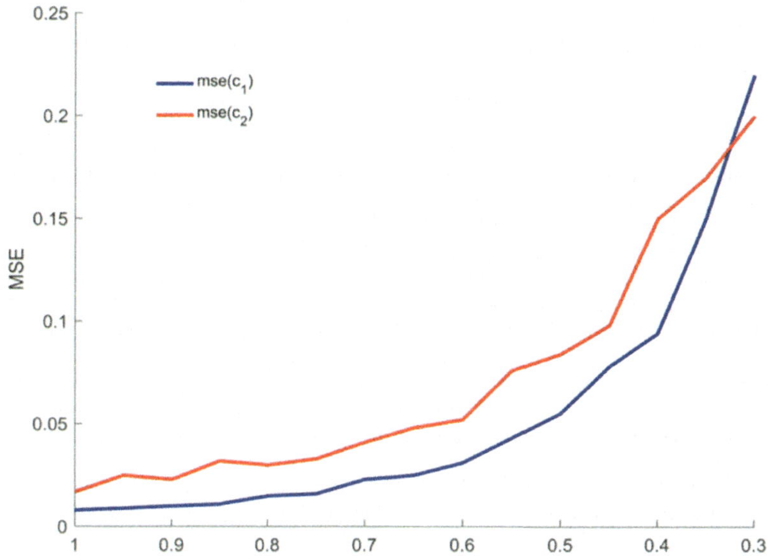

Fig. 13 The mean squared error of decrypted data c_1 (blue) and c_2 (red) as a function of the data transmission power ratio [0;1] under the influence of low frequency noise modelled with Ornstein–Uhlenbeck process with strength $D = 100$ and correlation time $\gamma = 10$ (which contributes to SNR of around 31 dB). Reducing the data transmission power ration to values of around 0.7 still doesn't result in a significant increase in MSE, which indicates possibilities for more energy-efficient data transfer

Acknowledgements The authors are grateful to Tomislav Stankovski, Peter V. E. McClintock, and Aneta Stefanovska. The work started and done by them has produced many of the results given in this paper. Along with their colleagues and research teams they have developed this protocol and have laid the groundwork for its future application.

References

1. Dimirovski, G.M.: Complex systems: relationships between control, communications and computing. Springer: Studies in Systems, Decision and Control (2016)
2. Kish, L.B.: Totally secure classical communication utilizing Johnson(-like) noise and Kirchoff's law. Phys. Lett. A **352**(3), 178–182 (2006)
3. Kaddoum, G.: Wireless chaos-based communication systems: a comprehensive survey. IEEE Access **4**, 2621–2648 (2016)
4. Li, N., Martinez-Ortega, J.-F., Diaz, V.H., Chaus, J.M.M.: A new high-efficiency multilevel frequency-modulation different chaos shift keying communication system. IEEE Syst. J. **12**, 3334–3345 (2018)
5. Liu, Y.-Z., Xie, Y.-Y., Ye, Y.-C., Zhang, J.-P., Wang, S.-J., Liu, Y., Pan, G.-F., Zhang, J.-L.: Exploiting optical chaos with time-delay signature suppression for long-distance secure communication. IEEE Photonics J. **9**(1), 1–12 (2017)
6. Bennett, C.H.: Quantum information and computation. Nature **404**(6775), 247–255 (2000)

7. Patel, K.A., Dynes, J.F., Choi, I., Sharpe, A.W., Dixon, A.R., Yuan, Z.L., Penty, R.V., Shields, A.J.: Coexistence of high-bit-rate quantum key distribution and data on optical fiber. Phys. Rev. X **2**, 041010 (2012)
8. Haroun, M.F., Gulliver, T.A.: Secret key generation using chaotic signals over frequency selective fading channels. IEEE Trans. Inf. Forensics Secur. **10**(8), 1764–1775 (2015)
9. Nadzinski, G., Stankovski, M., Ojleska Latkoska, V., Gochev, I.: Experimental test of the effects of electrostatic discharge on an industrial networked control system. In: 13th IEEE International Conference on Control & Automation (ICCA), pp. 82–87 (2017)
10. Stankovski, T., McClintock, P.V.E., Stefanovska, A.: Coupling functions enable secure communications. Phys. Rev. X **4**, 011026 (2014)
11. Stankovski, T., Pereira, T., McClintock, P.V.E., Stefanovska, A.: Coupling functions: universal insights into dynamical interaction mechanisms. Rev. Mod. Phys. **89**(33), 045001 (2017)
12. Stankovski, T., Duggento, A., McClintock, P.V.E., Stefanovska, A.: Inference of time-evolving coupled dynamical systems in the presence of noise. Phys. Rev. Lett. **109**, 024101 (2012)
13. Nadzinski, G., Dobrevski, M., Anderson, C., McClintock, P.V., Stefanovska, A., Stankovski, M., Stankovski, T.: Experimental realization of the coupling function secure communications protocol and analysis of its noise robustness. IEEE Trans. Inf. Forensics Secur. **13**(10), 2591–2601 (2018)
14. Iatsenko, D., Bernjak, A., Stankovski, T., Shiogai, Y., Owen-Lynch, P.J., Clarkson, P.B.M., McClintock, P.V.E., Stefanovska, A.: Evolution of cardio-respiratory interactions with age. Philos. Trans. R. Soc. Lond. A **371**(1997), 20110622 (2013)
15. Luchinsky, D.G., McClintock, P.V., Dykman, M.I.: Analogue studies of nonlinear systems. Rep. Prog. Phys. **61**(8), 889–997 (1998)
16. Millerioux, G., Amigo, J.M., Daafouz, J.: A connection between chaotic and conventional cryptography. IEEE Trans. Circuits Syst. I Regul. Pap. **55**(6), 1695–1703 (2008)
17. Cuomo, K.M., Oppenheim, A.V., Strogatz, S.H.: Synchronization of Lorenz-based chaotic circuits with applications to communications. IEEE Trans. Circuits Syst. II **40**(10), 626–633 (1993)
18. Irakiza, D., Karim, M.E., Phoha, V.V.: A Non-interactive dual channel continuous traffic authentication protocol. IEEE Trans. Inf. Forensics Secur. **9**(7), 1133–1140 (2014)
19. Crutchfield, J.P.: Between order and chaos. Nat. Phys. **8**(1), 17–24 (2012)
20. Pecora, L.M., Carroll, T.L.: Synchronization in chaotic systems. Phys. Rev. Lett. **64**(8), 821–824 (1990)
21. Stankovski, T., Duggento, A., McClintock, P.V., Stefanovska, A.: A tutorial on time-evolving dynamical Bayesian inference. Eur. Phys. J. Spec. Top. **223**(13), 2685–2703 (2014)
22. Smelyanskiy, V.N., Luchinsky, D.G., Stefanovska, A., McClintock, P.V.E.: Inference of a nonlinear stochastic model of the cardiorespiratory interaction. Phys. Rev. Lett. **94**(9), 098101 (2005)
23. Alvarez, G., Li, S.: Some basic cryptographic requirements for chaos-based cryptosystems. Int. J. Bifurc. Chaos **16**(8), 2129–2151 (2006)
24. Cuomo, K.M., Oppenheim, A.V.: Circuit implementation of synchronized chaos with application to communications. Phys. Rev. Lett. **71**, 65–68 (1993)
25. Hanggi, P., Mroczkowski, T.J., Moss, F., McClintock, P.V.: Bistability driven by colored noise: theory and experiment. Phys. Rev. A **32**, 695–698 (1985)
26. Bak, P., Tang, C., Wiesenfeld, K.: Self-organized criticality: an explanation of 1/f noise. Phys. Rev. Lett. **59**, 381–384 (1987)
27. Win, M.Z., Pinto, P.C., Shepp, L.A.: A mathematical theory of network interference and its applications. Proc. IEEE **97**(2), 205–230 (2009)
28. Lehle, B., Peinke, J.: Analyzing a stochastic process driven by Ornstein-Uhlenbeck noise. Phys. Rev. E **97**(1), 012113 (2018)
29. Wu, J., Li, Y., Quevedo, D.E., Lau, V., Shi, L.: Data-driven power control for state estimation. Automatica **54**, 332–339 (2015)

Distributed Adaptive NN Finite-Time Congestion Control for Multiple TCP/AWM Networks

Yang Liu, Hongyi Li, Yuanwei Jing, Xiaoping Liu, and Renquan Lu

Abstract A distributed adaptive NN congestion control problem is considered for TCP/AWM networks in this paper on the basis of practical finite-time criterion. First, inspired by multi-agent systems, TCP/AWM networks are modeled as a multiple network case. And then, under the framework of recursion algorithm, we extend the practical finite-time criterion to multiple TCP/AWM networks. Furthermore, by the cooperative control agreement among sub-networks, the queue length of all TCP/AWM networks reaches the consensus. In addition, a simple adaptive law is designed and singular issue is avoided. Finally, simulation results are used to demonstrate the effectiveness of the proposed scheme.

Keywords Multiple TCP/AWM networks · Practical finite-time control · Adaptive backstepping · Distributed control

---

Honoring Professor Georgi M. Dimirovski for his many Academic contributions and merits to our community.

Y. Liu (✉)
School of Automation Science and Technology, Qingdao University of Science and Technology, Qingdao 266100, Shandong, China
e-mail: liuyang0595@163.com

H. Li · R. Lu
School of Automation, Guangdong University of Technology, Guangzhou 510006, Guangdong, China
e-mail: lihongyi2009@gmail.com

R. Lu
e-mail: rqlu@gdut.edu.cn

Y. Jing
College of Information Science and Engineering, Northeastern University, Shenyang 110004, China
e-mail: ywjjing@mail.neu.edu.cn

X. Liu
Department of Electrical Engineering, Lakehead University, Thunder Bay P7B 5E1, Canada
e-mail: xliu2@lakeheadu.ca

P. Shi et al. (eds.), *Complex Systems: Spanning Control and Computational Cybernetics: Applications*, Studies in Systems, Decision and Control 415,
https://doi.org/10.1007/978-3-031-00978-5_20

1 Introduction

With the increasing information and data transmission amounts, the network congestion is getting worse. Compared with the passivity queue management (PQM), the active queue management (AQM) has certain advantages in terms of addressing congestion problems. But, the early algorithms exist some problems, for instance, RED and its modified methods are sensitive to parameters etc. Therefore, combining with the modern control theory, numerous active congestion control results are provided in [1–5], and a TCP/AQM model is established [6]. However, as pointed out in [7], AQM mechanism breaks down windows to make TCP react, which results in queue instability. Up to now, the explicit control protocol (XCP) and active window management (AWM) are two main mechanisms. Since XCP is difficult to configure, AWM is broadly employed. In [8, 9], authors propose AWM operation and design mechanism, and introduce the corresponding merits by comparing with XCP strategy. Furthermore, authors in [10] deeply investigate the impact of parameters in AWM approach on network performance. Authors of [11] consider a novel AWM algorithm to cut down the energy loss. In comparison with AQM method, the corresponding reports about the congestion control of TCP/AWM network are less. Therefore, how to further study this field to solve congestion problem is still an open area.

In the realm of finite-time control [12–14], it is a fact that it has many good performances such as fast convergence, strong robustness, high precision etc. Thus far, there have been two mainstream finite-time stability criteria. One is that when $\dot{V}(x) + cV^{\alpha}(x) \leq 0$ holds, we can say that the system is finite-time stable where $V(x)$ is a Lyapunov function, $c > 0$ and $0 < \alpha < 1$. However, it follows from [15, 16] that if the Lyapunov function is far away from the origin, the convergence rate may be not superior to the linear one in such a case. As a result, [16] proposes a fast finite-time control method, i.e., $\dot{V}(z) + cV^{h}(z) + bV(z) \leq 0$, which has been widely used in [17–19]. It is worth noting that system uncertainties and disturbances have an impact on the system performance in practice, which leads to that the above two criterions is difficult to realize. Therefore, some scholars begin to investigate the practical finite-time stability [20] depending on adaptive fuzzy control and neural networks. Afterwards, numerous achievements come to the fore [21–26]. A new criterion with respect to semi-global practical finite-time stability is introduced in [21] where an adaptive finite-time fuzzy control method is investigated. Based on the criterion of [21], an observer-based adaptive finite-time approach is provided in [22]. And then, it is extended to multi-input multi-output systems [23]. Further, a fast practical finite-time control scheme is presented in [24], and a novel finite-time error compensation strategy is designed to achieve a faster convergence rate and higher tracking precision. For a kind of nonlinear systems with unknown actuator faults, an adaptive fault tolerant technique is considered in [25], and the issue on explosion of complexity is solved well. Authors of [26] propose a new fast practical finite-time stability criterion to cope with the above problem.

Based on the aforesaid analysis, this work is to the address distributed adaptive NN finite-time congestion problem for multiple TCP/AWM networks. The main work is that:

(1) Inspired by the idea of multi-agent systems, a model of multiple TCP/AWM networks is constructed and the result of [26] is extended to this kind of more complex systems. Meanwhile, the congestion control design becomes more difficult than the single network [1–5].

(2) Under the framework of backstepping, the singularity issue has not been well addressed when the adaptive practical finite-time control is considered with the approximation strategy [25], which motivates us to try to solve it.

(3) Besides, we propose a simpler design method of adaptive laws without the power term, which avoids the singularity case arisen from the adaptive law term to some extent.

2 Preliminaries

Consider a multiple TCP/AWM network including N followers, and the dynamics of the ith network is described by

$$\begin{aligned} \dot{W}_{i,1}(t) &= u_i(t) \\ \dot{q}_{i,2}(t) &= \frac{N_i(t)}{R_i(t)} W_{i,1}(t) - C_{i,0}(t) \\ R_i(t) &= \frac{q_{i,2}(t)}{C_{i,0}(t)} + T_{p,i} \\ y_i &= q_{i,2}(t) \end{aligned} \tag{1}$$

where $i = 1, 2, \cdots, N$, $W_{i,1}(t)$ and $q_{i,2}(t)$ are the TCP window size and queue length, respectively, $R_i(t)$ is the round-trip time, $C_{i,0}(t)$ is available link, $N_i(t)$ is the number of TCP sessions, $T_{p,i}$ is the propagation delay, u_i is the control law. Without loss of generality, we assume that $N_i(t)$ and $C_{i,0}(t)$ are constants, which can be rewirtten as N_i and $C_{i,0}$.

In what follows, let $x_{i,1} = q_{i,2}(t)$ and $x_{i,2} = W_{i,1}(t)$, then the system (1) is changed to the following form

$$\begin{aligned} \dot{x}_{i,1} &= \phi_{i,1}\left(x_{i,1}\right) x_{i,2}(t) - C_{i,0} \\ \dot{x}_{i,2} &= u_i(t) \\ y_i &= x_{i,1} \end{aligned} \tag{2}$$

with $\phi_{i,1} = \frac{N_i C_{i,0}}{x_{i,1} + T_{p,i} C_{i,0}}$.

The control target is to develop an adaptive fast practical finite-time controller to make sure that all queue lengths $q_{i,2}(t)$ can track the desired value q_d. Therefore, some necessary assumption and lemmas are first introduced.

Assumption 1 The desired queue length q_d and its k-order derivative $q_d^{(k)}$ are bounded, that is, $|q_d| \leq Q_d$ and $\left|q_d^{(k)}\right| \leq Q_d$.

Lemma 1 ([27]) *For any $a_k \in \mathcal{R}$, $0 < \tau \leq 1$ with $k = 1, 2, \cdots, n$, we have*

$$\left(\sum_{k=1}^{n} |a_k|\right)^{\tau} \leq \sum_{k=1}^{n} |a_k|^{\tau} \leq n^{1-\tau} \left(\sum_{k=1}^{n} |a_k|\right)^{\tau} \tag{3}$$

Lemma 2 ([26]) *Consider a nonlinear system $\dot{x} = f(x)$. Let $V(x) > 0$ be a smooth function. If there are scalars $p_1 > 0$, $p_2 > 0$, $0 < \alpha < 1$ and $p_3 > 0$ such that*

$$\dot{V}(x) \leq -p_1 V^{\alpha}(x) - p_2 V(x) + p_3 \tag{4}$$

holds. Then, there is a constant $T > 0$ such that

$$V(x) \leq \frac{p_3}{p_2 - \varsigma} \tag{5}$$

for all $t \geq T$ and T is defined by

$$T = t_0 + \frac{1}{\varsigma(1-\alpha)} \ln\left[\frac{\frac{p_1}{\varsigma} + V(t_0)^{1-\alpha}}{\frac{p_1}{\varsigma} + \left(\frac{p_3}{p_2 - \varsigma}\right)^{1-\alpha}}\right] \tag{6}$$

with ς being $0 < \varsigma < p_2$ and t_0 being the initial time.

Remark 1 All assumption and lemmas are essential and can be found in the existing results [26, 28–30]. Thus, we omit their proof processes in this paper. It is noted that the merits of Lemma 2 can be observed from Remark 1 in [26]. But, the main difference between our method and [26] is that we extend its criterion to more complex systems, that is, multiple TCP/AWM networks. Moreover, the singular problem is tackled and the simpler adaptive law is established (see Remark 4 for details).

It is well known that the graph theory is a mature theory (see [31–33]). In this paper, $G = (V, E, A)$ is a direct graph containing N followers, $V = \{V_1, V_2, \cdots, V_N\}$, $E \subseteq V \times V$, $A = \left[a_{i,j}\right]_{N\times N}$ stand for sets of the node and edge as well as adjacency matrix, respectively. If the note i is able to acquire the note j information, then $\left(\mathcal{V}_j, \mathcal{V}_i\right) \in \mathcal{E}$ and $a_{i,j} > 0$, otherwise $a_{i,j} = 0$. It should be emphasized that $a_{i,i} = 0$. $N_i = \left\{j | \left(V_j, V_i\right) \in E\right\}$ denotes the set of neighbor node i. $L = D - A$ is the Laplacian matrix and $\mathcal{D}$ =diag$(d_1, d_2, \cdots, d_N)$ and $d_i = \sum_{j\in\mathcal{N}_i} a_{i,j}$.

3 Distributed Cooperative Control of Multiple TCP/AWM Networks

In this section, the control design process and stability analysis are given to reach the desired performance under the backstepping framework.

3.1 Distributed Cooperative Controller Design

Define the following change of coordinates

$$e_i = \sum\nolimits_{j \in \mathcal{N}_i} a_{i,j}\left(y_i - y_j\right) + a_{i,0}\left(y_i - q_d\right) \tag{7}$$

$$z_{i,1} = e_i \tag{8}$$

$$z_{i,2} = x_{i,2} - \eta_{i,1} \tag{9}$$

with $\eta_{i,1}$ being a virtual controller.

Step 1. A Lyapunov function is defined by

$$V_{i,1} = \frac{1}{2} z_{i,1}^2 + \frac{1}{2\upsilon_{i,1}} \tilde{\theta}_{i,1}^2 \tag{10}$$

where $\upsilon_{i,1} > 0$ is a constant, $\tilde{\theta}_{i,1} = \theta_{i,1} - \hat{\theta}_{i,1}$ with $\hat{\theta}_{i,1}$ being an estimation of the unknown constant $\theta_{i,1}$ to be specified later.

The time-derivative of $V_{i,1}$ with respect to time is calculated as

$$\begin{aligned} \dot{V}_{i,1} &= z_{i,1}\dot{z}_{i,1} - \frac{1}{\upsilon_{i,1}}\tilde{\theta}_{i,1}\dot{\hat{\theta}}_{i,1} \\ &= z_{i,1}\left[\sum\nolimits_{j \in \mathcal{N}_i} a_{i,j}\left(\dot{x}_{i,1} - \dot{x}_{j,1}\right) + a_{i,0}\left(\dot{x}_{i,1} - \dot{q}_d\right)\right] - \frac{1}{\upsilon_{i,1}}\tilde{\theta}_{i,1}\dot{\hat{\theta}}_{i,1} \\ &= z_{i,1}\left[\left(d_i + a_{i,0}\right)\phi_{i,1}\left(z_{i,2} + \eta_{i,1}\right) - \left(d_i + a_{i,0}\right)C_{i,0} - a_{i,0}\dot{q}_d + \psi_{i,1}\right] - \frac{1}{\upsilon_{i,1}}\tilde{\theta}_{i,1}\dot{\hat{\theta}}_{i,1} \end{aligned} \tag{11}$$

where $\psi_{i,1} = d_i C_{j,0} - d_i \phi_{j,1}\left(x_{j,1}\right) x_{j,2}(t)$ is required to estimate by the following RBF NN (see [31, 34, 35]).

$$\psi_{i,1} = W_{i,1}^{*\mathrm{T}} S_{i,1}\left(\vec{X}_{i,1}\right) + \delta_{i,1}\left(\vec{X}_{i,1}\right) \tag{12}$$

where $\vec{X}_{i,1} = \left[x_{j,1}, x_{j,2}\right]^{\mathrm{T}}$, $\left|\delta_{i,1}\left(\vec{X}_{i,1}\right)\right| \le \bar{\delta}_{i,1}$. Furthermore, the following inequality holds.

$$
\begin{aligned}
z_{i,1}\psi_{i,1} &= z_{i,1}W_{i,1}^{*\mathrm{T}}S_{i,1}\left(\vec{X}_{i,1}\right)+z_{i,1}\delta_{i,1}\left(\vec{X}_{i,1}\right)\\
&\leq \frac{z_{i,1}^{2}\left\|W_{i,1}^{*}\right\|^{2}S_{i,1}^{\mathrm{T}}\left(\vec{X}_{i,1}\right)S_{i,1}\left(\vec{X}_{i,1}\right)}{2\varrho_{i,1}^{2}}+\frac{1}{2}\varrho_{i,1}^{2}+\frac{1}{2}z_{i,1}^{2}+\frac{1}{2}\bar{\delta}_{i,1}^{2}\\
&\leq \frac{z_{i,1}^{2}\theta_{i,1}S_{i,1}^{\mathrm{T}}\left(\bar{X}_{i,1}\right)S_{i,1}\left(\bar{X}_{i,1}\right)}{2\varrho_{i,1}^{2}}+\frac{1}{2}\varrho_{i,1}^{2}+\frac{1}{2}z_{i,1}^{2}+\frac{1}{2}\bar{\delta}_{i,1}^{2}
\end{aligned}
\tag{13}
$$

with $\theta_{i,1}=\left\|W_{i,1}^{*}\right\|^{2}$, $\varrho_{i,1}>0$ being a design parameter and $\bar{X}_{i,1}=\left[x_{j,1}\right]^{\mathrm{T}}$. Then, substituting (13) into (11) produces

$$
\begin{aligned}
\dot{V}_{i,1} \leq & z_{i,1}\Bigg[\left(d_{i}+a_{i,0}\right)\phi_{i,1}\left(z_{i,2}+\eta_{i,1}\right)-\left(d_{i}+a_{i,0}\right)C_{i,0}-a_{i,0}\dot{q}_{d}+\frac{1}{2}z_{i,1}\\
&+\frac{z_{i,1}\theta_{i,1}S_{i,1}^{\mathrm{T}}\left(\bar{X}_{i,1}\right)S_{i,1}\left(\bar{X}_{i,1}\right)}{2\varrho_{i,1}^{2}}\Bigg]+\frac{1}{2}\varrho_{i,1}^{2}+\frac{1}{2}\bar{\delta}_{i,1}^{2}-\frac{1}{\upsilon_{i,1}}\tilde{\theta}_{i,1}\dot{\hat{\theta}}_{i,1}
\end{aligned}
\tag{14}
$$

Next, a virtual controller $\eta_{i,1}$ with $\kappa_{i,1}>0$ is designed as

$$
\begin{aligned}
\left(d_{i}+a_{i,0}\right)\phi_{i,1}\eta_{i,1}= & -\kappa_{i,1}\eta_{i,1}-0.5z_{i,1}+a_{i,0}\dot{q}_{d}-\frac{z_{i,1}\hat{\theta}_{i,1}S_{i,1}^{\mathrm{T}}\left(\bar{X}_{i,1}\right)S_{i,1}\left(\bar{X}_{i,1}\right)}{2\varrho_{i,1}^{2}}\\
&+\left(d_{i}+a_{i,0}\right)C_{i,0}
\end{aligned}
\tag{15}
$$

which is substituted into (14) produces

$$
\dot{V}_{i,1}\leq-\kappa_{i,1}z_{i,1}^{2}+\left(d_{i}+a_{i,0}\right)\phi_{i,1}z_{i,1}z_{i,2}+\frac{1}{2}\varrho_{i,1}^{2}+\frac{1}{2}\bar{\delta}_{i,1}^{2}+\frac{\lambda_{i,1}}{\upsilon_{i,1}}\tilde{\theta}_{i,1}\hat{\theta}_{i,1}
\tag{16}
$$

with the following adaptive law

$$
\dot{\hat{\theta}}_{i,1}=\frac{\upsilon_{i,1}z_{i,1}^{2}S_{i,1}^{\mathrm{T}}\left(\bar{X}_{i,1}\right)S_{i,1}\left(\bar{X}_{i,1}\right)}{2\varrho_{i,1}^{2}}-\lambda_{i,1}\hat{\theta}_{i,1}
\tag{17}
$$

To cope with the singularity problem, a processing way for (16) is given in this paper, that is,

$$
\dot{V}_{i,1}\leq-\left(\kappa_{i,1}-\kappa_{i,1,1}\right)z_{i,1}^{2}-\bar{\kappa}_{i,1}z_{i,1}^{2\alpha}+\left(d_{i}+a_{i,0}\right)\phi_{i,1}z_{i,1}z_{i,2}+\Theta_{i,1}+\frac{\lambda_{i,1}}{\upsilon_{i,1}}\tilde{\theta}_{i,1}\hat{\theta}_{i,1}
\tag{18}
$$

where $\bar{\kappa}_{i,1}>0$ is a design parameter, $\bar{\kappa}_{i,1}z_{i,1}^{2\alpha}\leq\kappa_{i,1,1}z_{i,1}^{2}+\kappa_{i,1,2}$, $\kappa_{i,1}>\kappa_{i,1,1}$, $\kappa_{i,1,1}>0$, $\kappa_{i,1,2}=(1-\alpha)\left(\frac{\alpha}{\kappa_{i,1,1}}\right)^{\frac{\alpha}{1-\alpha}}_{i,1}\bar{\kappa}^{\frac{1}{1-\alpha}}$ and $\Theta_{i,1}=\kappa_{i,1,2}+\frac{1}{2}\varrho_{i,1}^{2}+\frac{1}{2}\bar{\delta}_{i,1}^{2}$.

Remark 2 In the existing results [24–26], for completing finite-time control objective, the term $z_{i,1}^{2\gamma-1}$ is designed in virtual control signal, however, the singularity problem may arise due to $0 < \gamma < 1$, if the virtual controller is repeatedly differentiated. For avoiding this issue, the power γ is introduced by adding and subtracting the term $\bar{\kappa}_{i,1} z_{i,1}^{2\alpha}$, and then the power γ is further transformed as 2. Note that the similar process is made in the following steps.

Step 2. Design a Lyapunov function as

$$V_{i,2} = V_{i,1} + \frac{1}{2} z_{i,2}^2 + \frac{1}{2\upsilon_{i,2}} \tilde{\theta}_{i,2}^2 \tag{19}$$

where $\upsilon_{i,2} > 0$ is a design parameter, differentiating $\dot{V}_{i,n}$ yields

$$\begin{aligned} \dot{V}_{i,2} =& \dot{V}_{i,1} + z_{i,2}\dot{z}_{i,2} - \frac{1}{\upsilon_{i,2}} \tilde{\theta}_{i,2} \dot{\hat{\theta}}_{i,2} \\ \leq& -\left(\kappa_{i,1} - \kappa_{i,1,1}\right) z_{i,1}^2 - \bar{\kappa}_{i,1} z_{i,1}^{2\alpha} + \Theta_{i,1} + \frac{\lambda_{i,1}}{\upsilon_{i,1}} \tilde{\theta}_{i,1} \hat{\theta}_{i,1} - \frac{1}{\upsilon_{i,2}} \tilde{\theta}_{i,2} \dot{\hat{\theta}}_{i,2} \\ &+ z_{i,2} \left(\left(d_i + a_{i,0}\right) \phi_{i,1} z_{i,1} + u_i + \psi_{i,2} \right) \end{aligned} \tag{20}$$

where RBF NN is used to approximate $\psi_{i,2} = -\dot{\eta}_{i,1}$, i.e.,

$$\psi_{i,2} = W_{i,2}^{*\mathrm{T}} S_{i,2}\left(\bar{Z}_{i,2}\right) + \delta_{i,2}\left(\bar{X}_{i,2}\right) \tag{21}$$

with $\bar{X}_{i,2} = \left[\bar{z}_{i,2}^{\mathrm{T}}, \bar{\theta}_{i,2}^{\mathrm{T}}, \bar{y}_{d,2}^{\mathrm{T}}\right]^{\mathrm{T}}$, $\left|\delta_{i,2}\left(\bar{X}_{i,2}\right)\right| \leq \bar{\delta}_{i,2}$. It follows from Young's inequality that

$$z_{i,2}\psi_{i,2} \leq \frac{z_{i,2}^2 \theta_{i,2} S_{i,2}^{\mathrm{T}}\left(\bar{X}_{i,2}\right) S_{i,2}\left(\bar{X}_{i,2}\right)}{2\varrho_{i,2}^2} + \frac{1}{2}\varrho_{i,2}^2 + \frac{1}{2} z_{i,2}^2 + \frac{1}{2}\bar{\delta}_{i,2}^2 \tag{22}$$

Then, one has

$$\begin{aligned} \dot{V}_{i,2} \leq& -\left(\kappa_{i,1} - \kappa_{i,1,1}\right) z_{i,1}^2 - \bar{\kappa}_{i,1} z_{i,1}^{2\alpha} + \Theta_{i,1} + \frac{\lambda_{i,1}}{\upsilon_{i,1}} \tilde{\theta}_{i,1} \hat{\theta}_{i,1} - \frac{1}{\upsilon_{i,2}} \tilde{\theta}_{i,2} \dot{\hat{\theta}}_{i,2} \\ &+ z_{i,2} \left[\left(d_i + a_{i,0}\right) \phi_{i,1} z_{i,1} + u_i + \frac{1}{2} z_{i,2} + \frac{z_{i,2}\theta_{i,2} S_{i,2}^{\mathrm{T}}\left(\bar{X}_{i,2}\right) S_{i,2}\left(\bar{X}_{i,2}\right)}{2\varrho_{i,2}^2} \right] \\ &+ \frac{1}{2}\bar{\delta}_{i,2}^2 + \frac{1}{2}\varrho_{i,2}^2 \end{aligned} \tag{23}$$

In the following, the controller u_i is constructed by

$$u_i = -\kappa_{i,2} z_{i,2} - \left(d_i + a_{i,0}\right) \phi_{i,1} z_{i,1} - \frac{1}{2} z_{i,2} - \frac{z_{i,2}\hat{\theta}_{i,2} S_{i,2}^{\mathrm{T}}\left(\bar{X}_{i,2}\right) S_{i,2}\left(\bar{X}_{i,2}\right)}{2\varrho_{i,2}^2} \tag{24}$$

which is substituted into (23) gives

$$\dot{V}_{i,n} \leq -\left(\kappa_{i,1}-\kappa_{i,1,1}\right) z_{i,1}^{2}-\bar{\kappa}_{i,1} z_{i,1}^{2\alpha}+\frac{1}{2} \bar{\delta}_{i,2}^{2}+\frac{1}{2} \varrho_{i,2}^{2}+\Theta_{i,1}+\frac{\lambda_{i,1}}{\upsilon_{i,1}} \tilde{\theta}_{i,1} \hat{\theta}_{i,1} \\ -\kappa_{i,2} z_{i,2}^{2}-\frac{1}{\upsilon_{i,2}} \tilde{\theta}_{i,2}\left[\dot{\hat{\theta}}_{i,2}-\frac{\upsilon_{i,2} z_{i,2}^{2} S_{i,2}^{\mathrm{T}}\left(\bar{X}_{i,2}\right) S_{i,2}\left(\bar{X}_{i,2}\right)}{2 \varrho_{i,2}^{2}}\right] \tag{25}$$

with the following adaptive law

$$\dot{\hat{\theta}}_{i,2}=\frac{\upsilon_{i,2} z_{i,2}^{2} S_{i,2}^{\mathrm{T}}\left(\bar{X}_{i,2}\right) S_{i,2}\left(\bar{X}_{i,2}\right)}{2 \varrho_{i,2}^{2}}-\delta_{i,2} \hat{\theta}_{i,2} \tag{26}$$

As a result, (25) can be written as

$$\dot{V}_{i,2} \leq-\sum_{j=1}^{2}\left(\kappa_{i,j}-\kappa_{i,j,1}\right) z_{i,j}^{2}-\sum_{j=1}^{2} \bar{\kappa}_{i,j} z_{i,j}^{2\alpha}+\sum_{j=1}^{2} \Theta_{i,j}+\sum_{j=1}^{2} \frac{\delta_{i,j}}{\upsilon_{i,j}} \tilde{\theta}_{i,j} \hat{\theta}_{i,j} \tag{27}$$

where $\bar{\kappa}_{i,2}>0$ is a design parameter, $\bar{\kappa}_{i,2} z_{i,2}^{2\alpha} \leq \kappa_{i,2,1} \eta_{i,n}^{2}+\kappa_{i,2,2}$, $\kappa_{i,2}>\kappa_{i,2,1}$, $\kappa_{i,2,1}>0$, $\kappa_{i,2,2}=(1-\alpha)\left(\frac{\alpha}{\kappa_{i,2,1}}\right)^{\frac{\alpha}{1-\alpha}} \bar{\kappa}_{i,2}^{\frac{1}{1-\alpha}}$ and $\Theta_{i,2}=\kappa_{i,2,2}+\frac{1}{2} \bar{\delta}_{i,2}^{2}+\frac{1}{2} \varrho_{i,2}^{2}$.

3.2 Stability Analysis

So far, an adaptive NN finite-time control process has been proposed. The corresponding stability analysis will be made in this subsection, and the main result is summarized by Theorem 1.

Theorem 1 *Under Assumption 1, with the virtual controller (15), control input signal (24) and adaptive laws (17), (26), the plant (1) satisfies the desired performances.*

Proof From (27) and (19), we can know that

$$V_{i,2}=\sum_{j=1}^{2} \frac{1}{2} z_{i,j}^{2}+\sum_{j=1}^{2} \frac{1}{2 \upsilon_{i,j}} \tilde{\theta}_{i,j}^{2} \tag{28}$$

and

$$\dot{V}_{i,2} \leq-\sum_{j=1}^{2}\left(\kappa_{i,j}-\kappa_{i,j,1}\right) z_{i,j}^{2}-\sum_{j=1}^{2} \bar{\kappa}_{i,j} z_{i,j}^{2\alpha}+\sum_{j=1}^{2} \Theta_{i,j}+\sum_{j=1}^{2} \frac{\delta_{i,j}}{\upsilon_{i,j}} \tilde{\theta}_{i,j} \hat{\theta}_{i,j} \tag{29}$$

Let $\chi_{i,j} = \frac{\delta_{i,j}}{\upsilon_{i,j}} = \chi_{i,j,1} + \chi_{i,j,2}$ with $\chi_{i,j,1} > 0$, $\chi_{i,j,2} > 0$, then one has

$$\dot{V}_{i,2} \leq -\sum_{j=1}^{2}\left(\kappa_{i,j} - \kappa_{i,j,1}\right) z_{i,j}^2 - \sum_{j=1}^{2}\bar{\kappa}_{i,j} z_{i,j}^{2\alpha} + \sum_{j=1}^{2}\Theta_{i,j} + \sum_{j=1}^{2}\left[\chi_{i,j,1} + \chi_{i,j,2}\right]\tilde{\theta}_{i,j}\hat{\theta}_{i,j} \tag{30}$$

For the term $\tilde{\theta}_{i,j}\hat{\theta}_{i,j}$, it gives

$$\tilde{\theta}_{i,j}\hat{\theta}_{i,j} \leq -\frac{1}{2}\tilde{\theta}_{i,j}^2 + \frac{1}{2}\theta_{i,j}^2 \tag{31}$$

$$\left(\chi_{i,j,2}\tilde{\theta}_{i,j}^2\right)^{\alpha} \leq \bar{\chi}_{i,j,2}\tilde{\theta}_{i,j}^2 + \tilde{\chi}_{i,j,2} \tag{32}$$

with $\bar{\chi}_{i,j,2} > 0$, $\tilde{\chi}_{i,j,2} = (1-\alpha)\left(\frac{\alpha}{\bar{\chi}_{i,j,2}}\right)^{\frac{\alpha}{1-\alpha}} \chi_{i,j,2}^{\frac{\alpha}{1-\alpha}}$.

Therefore,

$$\sum_{j=1}^{2}\chi_{i,j,1}\tilde{\theta}_{i,j}\hat{\theta}_{i,j} \leq -\sum_{j=1}^{2}\frac{\chi_{i,j,1}}{2}\tilde{\theta}_{i,j}^2 + \sum_{j=1}^{2}\frac{\chi_{i,j,1}}{2}\theta_{i,j}^2 \tag{33}$$

$$\begin{aligned}\sum_{j=1}^{2}\chi_{i,j,2}\tilde{\theta}_{i,j}\hat{\theta}_{i,j} &\leq -\sum_{j=1}^{2}\frac{\chi_{i,j,2}}{2}\tilde{\theta}_{i,j}^2 + \sum_{j=1}^{2}\frac{\chi_{i,j,2}}{2}\theta_{i,j}^2 \\ &\leq -\sum_{j=1}^{2}\frac{1}{2}\left(\chi_{i,j,2}\tilde{\theta}_{i,j}^2\right)^{\alpha} + \sum_{j=1}^{2}\frac{1}{2}\tilde{\chi}_{i,j,2} + \sum_{j=1}^{2}\frac{\chi_{i,j,2}}{2}\theta_{i,j}^2\end{aligned} \tag{34}$$

So far, it follows from Lemma 1 that (30) can be written as

$$\begin{aligned}\dot{V}_{i,2} \leq &-\sum_{j=1}^{2}\left(\kappa_{i,j} - \kappa_{i,j,1}\right) z_{i,j}^2 - \sum_{j=1}^{2}\frac{\chi_{i,j,1}}{2}\tilde{\theta}_{i,j}^2 - \sum_{j=1}^{2}\bar{\kappa}_{i,j} z_{i,j}^{2\alpha} - \sum_{j=1}^{2}\frac{1}{2}\left(\chi_{i,j,2}\tilde{\theta}_{i,j}^2\right)^{\alpha} \\ \leq &-\rho_i\left[\sum_{j=1}^{2}\frac{1}{2}z_{i,j}^2 + \sum_{j=1}^{n}\frac{1}{2\upsilon_{i,j}}\tilde{\theta}_{i,j}^2\right] - \vartheta_i 2^{1-\alpha}\left[\sum_{j=1}^{2}\frac{1}{2}z_{i,j}^2 + \sum_{j=1}^{2}\frac{1}{2\varepsilon_{i,j}}\tilde{\theta}_{i,j}^2\right]^{\alpha} + \bar{\Theta}_i \\ = &-\rho_i V_{i,2} - \xi_i V_{i,2}^{\alpha} + \bar{\Theta}_i\end{aligned} \tag{35}$$

with $\kappa_{i,j} > \kappa_{i,j,1}$, $\bar{\Theta}_i = \sum_{j=1}^{2}\frac{\chi_{i,j,1}}{2}\theta_{i,j}^2 + \sum_{j=1}^{2}\frac{\chi_{i,j,2}}{2}\theta_{i,j}^2 + \sum_{j=1}^{2}\frac{1}{2}\tilde{\chi}_{i,j,2} + \sum_{j=1}^{2}\Theta_{i,j}$, $\rho_i = \min\left[2\left(\kappa_{i,j} - \kappa_{i,j,1}\right), \upsilon_{i,j}\chi_{i,j,1}\right]$, $\vartheta_i = \min\left(\bar{\kappa}_{i,j}2^{\alpha}, 2^{\alpha-1}\upsilon_{i,j}^{\alpha}\chi_{i,j,2}^{\alpha}\right)$, $\xi_i = \vartheta_i 2^{1-\alpha}$.

From $V = \sum_{i=1}^{N} V_{i,2}$ and (35), we have

$$\begin{aligned}\dot{V} =& \dot{V}_{1,2} + \cdots + \dot{V}_{N,2} \\ \leq& -\rho_1 V_{1,2} - \xi_1 V_{1,2}^{\alpha} + \bar{\Theta}_1 - \cdots - \rho_N V_{N,2} - \xi_N V_{N,2}^{\alpha} + \bar{\Theta}_N \\ \leq& -\rho V - \xi V^{\alpha} + \Theta\end{aligned}$$

where $\Theta = \sum_{i=1}^{N} \bar{\Theta}_i$, $\rho = \min(\rho_1, \cdots, \rho_N)$, $\xi = \min(\xi_1, \cdots, \xi_N) N^{1-\alpha}$.

This completes the proof based on Lemma 2. ■

Remark 3 In [26], the terms $\hat{\theta}_i$ and $\hat{\theta}_i^{2h-1}$ contain in the adaptive law simultaneously for obtaining the finite-time performance. Due to the existence of $\hat{\theta}_i^{2h-1}$, the singularity problem may also arise in the traditional backstepping method. Hence, just one term $\hat{\theta}_i$ is employed to establish adaptive laws, which not only simplify design process, but also avoid the singularity problem. The specific case can be seen in the controller design and stability analysis.

4 Simulation Results

For illustrating the presented algorithm, we choose five TCP/AWM networks including one leader and four followers. The communication topology is illustrated in Fig. 1.

The initial values are set as

$$\begin{aligned}x_{i,1}(0) =& [50, 50, 50, 50]^{\mathrm{T}}, x_{i,2}(0) = [0, 0, 0, 0]^{\mathrm{T}} \\ \hat{\theta}_{i,1}(0) =& [0.5, 0, 0.2, 0]^{\mathrm{T}}, \hat{\theta}_{i,2}(0) = [0.5, 0, 0.1, 0]^{\mathrm{T}}\end{aligned}$$

with $i = 1, 2, 3, 4$.

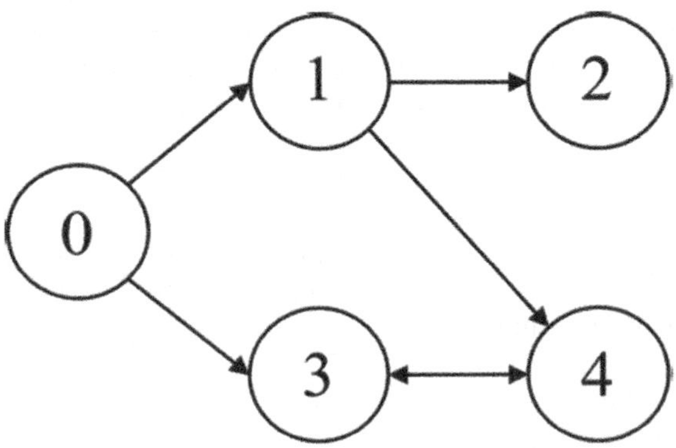

Fig. 1 Communication topology

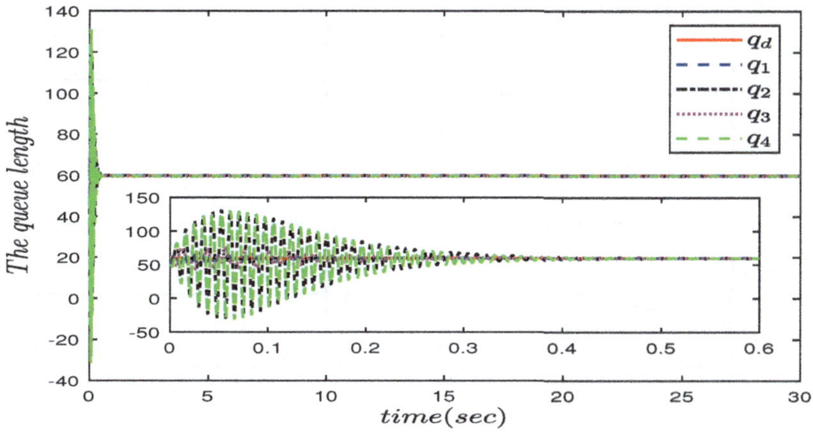

Fig. 2 The queue length

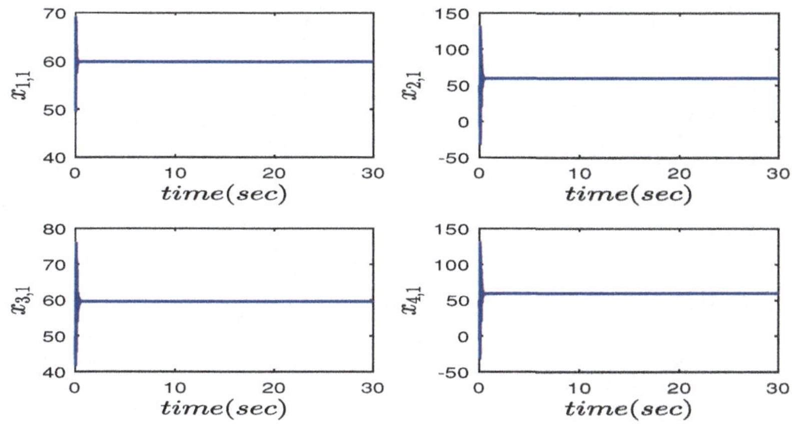

Fig. 3 States $x_{i,1}$ with $i = 1, 2, 3, 4$

The design parameters are selected as $C_{i,0} = 1750$, $N_i = 100$, $T_{p,i} = 0.1$, $q_d = 60$, and

$$\begin{array}{llll}\delta_{1,1} = \delta_{1,2} = 0.5, & \delta_{2,1} = \delta_{2,2} = 0.8, & \delta_{3,1} = \delta_{3,2} = 0.8, & \delta_{4,1} = \delta_{4,2} = 0.8 \\ \varrho_{1,1} = \varrho_{1,2} = 3, & \varrho_{2,1} = \varrho_{2,2} = 10, & \varrho_{3,1} = \varrho_{3,2} = 7, & \varrho_{4,1} = \varrho_{4,2} = 5 \\ \kappa_{1,1} = \kappa_{1,2} = 25, & \kappa_{2,1} = \kappa_{2,2} = 25, & \kappa_{3,1} = \kappa_{3,2} = 10, & \kappa_{4,1} = \kappa_{4,2} = 25 \\ \upsilon_{1,1} = \upsilon_{1,2} = 8, & \upsilon_{2,1} = \upsilon_{2,2} = 2, & \upsilon_{3,1} = \upsilon_{3,2} = 6, & \upsilon_{4,1} = \upsilon_{4,2} = 7\end{array}$$

The simulation results are reported in Figs. 2, 3, 4, 5, 6 and 7. From Fig. 2, we can see that four networks track well the desired queue length within finite-time. It

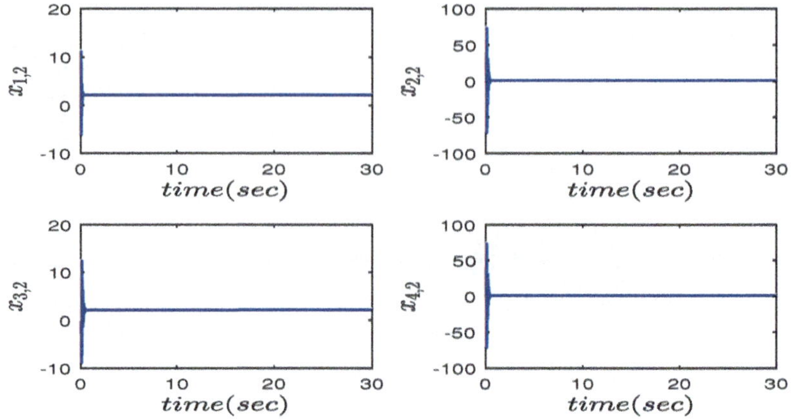

Fig. 4 States $x_{i,2}$ with $i = 1, 2, 3, 4$

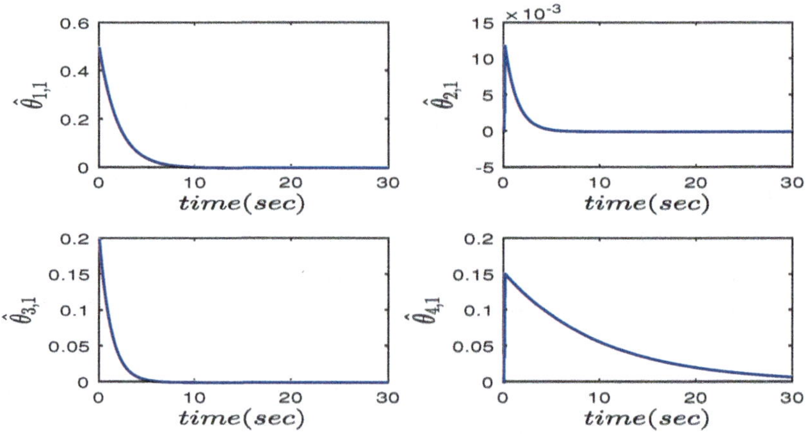

Fig. 5 $\hat{\theta}_{i,1}$ with $i = 1, 2, 3, 4$

can be observed from Figs. 3 and 4 that the states $x_{i,1}$ and $x_{i,2}$ with $i = 1, 2, 3, 4$ are bounded. Figures 5 and 6 show the adaptive law trajectories $\hat{\theta}_{i,1}$ and $\hat{\theta}_{i,2}$ with $i = 1, 2, 3, 4$. The control input signals u_i with $i = 1, 2, 3, 4$ of are described in Fig. 7.

5 Conclusion

The congestion control problem has been solved for multiple TCP/AWM networks in this paper. A fast finite-time criterion has been extended to networks to deal with consensus. Based on the adaptive backstepping, a practical finite-time control

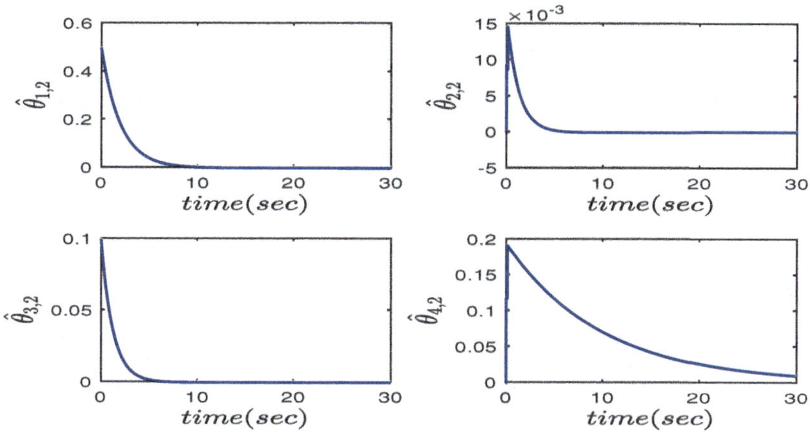

Fig. 6 $\hat{\theta}_{i,2}$ with $i = 1, 2, 3, 4$

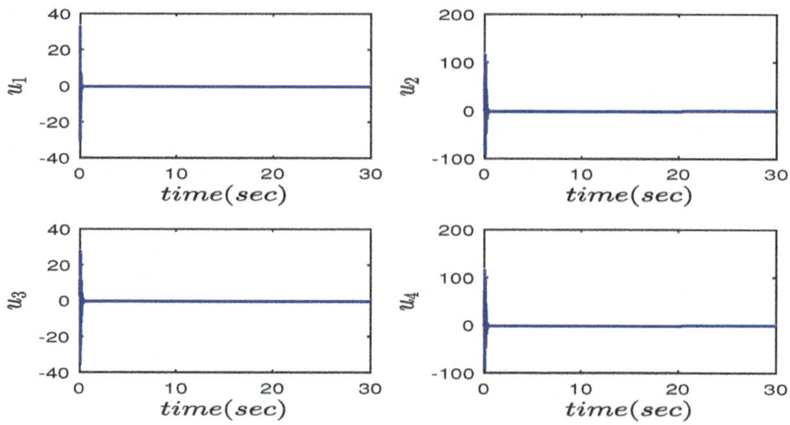

Fig. 7 Input signals u_i with $i = 1, 2, 3, 4$

method has been shown, by which the finite-time tracking performance has also been obtained. Moreover, the boundedness of closed-loop system has been achieved. Simulation experiment has been utilized to illustrate the effectiveness of the presented method.

Acknowledgements This work was supported in part by the National Natural Science Foundation of China under Grant 62033003, Grant 62003097, and Grant 61773108; in part by the Local Innovative and Research Teams Project of Guangdong Special Support Program under Grant 2019BT02X353; in part by the China Postdoctoral Science Foundation under Grant 2019M662812; and in part by the Joint Funds of Guangdong Basic and Applied Basic Research Foundation under Grant 2019A1515110505.

References

1. Liu, Y., Jing, Y.W., Chen, X.Y.: Adaptive neural practically finite-time congestion control for tcp/aqm network. Neurocomputing **351**, 26–32 (2019)
2. Li, F., Sun, J.S., Zukerman, M., Liu, Q., Xu, Z.F., Chan, S., Chen, G.R., Ko, K.T.: A comparative simulation study of tcp/aqm systems for evaluating the potential of neuro-based aqm scheme. J. Netw. Comput. Appl. **41**, 274–299 (2014)
3. Sadek, B.A., Houssaine, T.E., Noreddine, C.: Small-gain theorem and finite-frequency analysis of tcp/aqm system with time varying delay. IET Control Theory Appl. **13**, 1971-C1982 (2019)
4. Wang, F., Liu, Y., Liu, X.P., Jing, Y.W., Zhang, S.Y.: Adaptive fuzzy funnel congestion control for tcp/aqm network. ISA Trans. **95**, 11-C17 (2019)
5. Liu, Y., Liu, X.P., Jing, Y.W., Zhou, S.W.: Adaptive backstepping H_∞ tracking control with prescribed performance for internet congestion. ISA Trans. **72**, 92-C99 (2018)
6. Misra, V., Gong, W.B., Towsley, D.: Fluid-based analysis of a network of aqm routers supporting tcp flows with an application to red. In: Proceedings ACM/SIGCOM. IEEE 2000, p. 151–C160 (2000)
7. Li, Z.H., Chen, X.Y., Ding, S.H., Liu, Y., Qiu, J.L.: Tcp/awm network congestion algorithm with funnel control and arbitrary setting time. Appl. Math. Comput. **385**, 125410 (2020)
8. Barbera, M., Lombardo, A., Panarello, C., Schembra, G.: Active window management: an efficient gateway mechanism for tcp traffic control. In: Proceedings of the IEEE International Conference on Communications. IEEE 2007, pp. 6141–C6148 (2007)
9. Barbera, M., Lombardo, A., Panarello, C., Sanadidi, M., Schembra, G.: Active window management: performance assessment through an extensive comparison with xcp. In: Proceedings of the International Conference on Research in Networking, Springer, 2008, pp. 679–C690 (2007)
10. Barbera, M., Lombardo, A., Panarello, C., Schembra, G.: Queue stability analysis and performance evaluation of a tcp-compliant window management mechanism. IEEE/ACM Trans. Netw. **18**, 1275-C1288 (2010)
11. Bruschi, R., Lombardo, A., Panarello, C., Podda, F., Santagati, G., Schembra, G.: Active window management: reducing energy consumption of tcp congestion control. In: Proceedings of the IEEE International Conference on Communications, IEEE, 2013, pp. 4154–C4158 (2013)
12. Bhat, S.P., Bernstein, D.S.: Finite-time stability of continuous autonomous systems. SIAM J. Control. Optim. **38**(3), 751–766 (2000)
13. Huang, X.Q., Lin, W., Yang, B.: Global finite-time stabilization of a class of uncertain nonlinear systems. Automatica **41**(5), 881–888 (2005)
14. Sun, Z.Y., Xue, L.R., Zhang, K.M.: A new approach to finite-time adaptive stabilization of high-order uncertain nonlinear system. Automatica **58**, 60–66 (2015)
15. Shen, Y.J., Huang, Y.H.: Uniformly observable and globally lipschitzian nonlinear systems admit global finite-time observers. IEEE Trans. Autom. Control **54**(11), 2621–2625 (2009)
16. Yu, S., Yu, X., Shirinzadeh, B., Man, Z.: Continuous finite-time control for robotic manipulators with terminal silding mode. Automatica **41**(11), 1957–1964 (2005)
17. Xiao, Q.Y., Wu, Z.H., Peng, L.: Fast finite-time consensus tracking of first-order multi-agent systems with a virtual leader. Appl. Mech. Mater. **596**, 552–559 (2014)

18. Liu, Y., Liu, X.P., Jing, Y.W., Zhang, Z.Y.: Design of finite-time H_∞ controller for uncertain nonlinear systems and its application. Int. J. Control **92**(12), 2928–2938 (2019)
19. Liu, Y., Liu, X.P., Jing, Y.W., Zhang, Z.Y.: Semi-globally practical finite-time stability for uncertain nonlinear systems based on dynamic surface control. Int. J. Control **94**(2), 476–485 (2021)
20. Zhu, Z., Xia, Y., Fu, M.: Attitude stabilization of rigid spacecraft with finite-time convergence. Int. J. Robust Nonlinear Control **21**(6), 686–702 (2011)
21. Wang, F., Chen, B., Liu, X.P., Lin, C.: Finite-time adaptive fuzzy tracking control design for nonlinear systems. IEEE Trans. Fuzzy Syst. **26**(3), 1207–1216 (2018)
22. Yu, Z.X., Yang, Y.K., Li, S.G., Sun, J.T.: Observer-based adaptive finite-time quantized tracking control of nonstrict-feedback nonlinear systems with asymmetric actuator saturation. IEEE Trans. Syst. Man Cybern. Syst. **50**(11), 4545–4556 (2020)
23. Li, Y.M., Li, K.W., Tong, S.C.: Finite-time adaptive fuzzy output feedback dynamic surface control for mimo nonstrict feedback systems. IEEE Trans. Fuzzy Syst. **27**(1), 96–110 (2019)
24. Yu, J.P., Shi, P., Zhao, L.: Finite-time command filtered backstepping control for a class of nonlinear systems. Automatica **92**, 173–180 (2018)
25. Li, Y.X.: Finite time command filtered adaptive fault tolerant control for a class of uncertain nonlinear systems. Automatica **106**, 117–123 (2019)
26. Chen, B., Lin, C.: Finite-time stabilization-based adaptive fuzzy control design. IEEE Trans. Fuzzy Syst. https://doi.org/10.1109/TFUZZ.2020.2991153
27. Hua, C.C., Li, Y.F., Guan, X.P.: Finite/fixed-time stabilization for nonlinear interconnected systems with dead-zone input. IEEE Trans. Autom. Control **62**(5), 2554–2560 (2017)
28. Wang, Y.J., Song, Y.D., Wen, C.Y., Kristic, M.: Fault-tolerant finite time consensus for multiple uncertain nonlinear mechanical systems under single-way directed communication interactions and actuation failures. Automatica **63**, 374–383 (2016)
29. Littlewood, J.E., Hardy, G.H., Polya, G.: Inequality. Cambridge University Press, Cambridge, U.K. (1952)
30. Wang, F., Lai, G.Y.: Fixed-time control design for nonlinear uncertain systems via adaptive method. Syst. Control Lett. **140**, 104704 (2020)
31. Dong, G.W., Li, H.Y., Ma, H., Lu, R.Q.: Finite-time consensus tracking neural network ftc of multi-agent systems. IEEE Trans. Neural Netw. Learn. Syst. **32**(2), 653–662 (2021)
32. Li, H.Y., Wu, Y., Chen, M.: Adaptive fault-tolerant tracking control for discrete-time multi-agent systems via reinforcement learning algorithm. IEEE Trans. Cybern. **51**(3), 1163–1174 (2021)
33. Lin, G.H., Li, H.Y., Ma, H., Yao, D.Y., Lu, R.Q.: Human-in-the-loop consensus control for nonlinear multi-agent systems with actuator faults. IEEE/CAA J. Automatica Sinica. https://doi.org/10.1109/JAS.2020.1003596
34. Liu, Y., Liu, X.P., Jing, Y.W., Chen, X.Y., Qiu, J.L.: Direct adaptive preassigned finite-time control with time-delay and quantized input using neural network. IEEE Trans. Neural Netw. Learn. Syst. **31**(4), 1222–1231 (2020)
35. Liu, Y., Yao, D.Y., Li, H.Y., Lu, R.Q.: Distributed cooperative compound tracking control for a platoon of vehicles with adaptive NN. IEEE Trans. Cybern. https://doi.org/10.1109/JAS.2020.1003596

Network Traffic Classification Using Supervised Machine Learning Algorithms in Systems with NFV Architecture

Gjorgji Ilievski and Pero Latkoski

Abstract Network functions virtualization architecture concept is gaining more popularity and it is used in different systems. Together with the cloudification within public, private and mixed clouds it is becoming a base for the future development of the digital world. The concepts of containers, virtual network functions, application functions are cohered within the clouds and guided with the NFV systems. Another aspect which is developing rapidly are the access technologies, especially the 5G, which is the all expected enabler of the IoT. Within such circumstances, most of the network traffic is expected to flow in the east–west direction, never leaving the cloud. Our work if focused on preparation of experimental environment that will simulate such traffic. We are analysing the traffic by making classification of the network data flows, using a selected set of six supervised machine learning (ML) algorithms. The goal of our research is to find the algorithm with the best performance within the prepared environment. We define the performance as a combination of the ML algorithm's classification precision, and the time consumption of the algorithm, which bears a great significance, especially from a point of 5G, where any packet delay introduced within the system may compromise the 5G specification calls for latency. From the research we conclude that out of the 6 explored ML algorithms, the Decision Tree algorithms is the most suitable classifier that fits within the needed precision across all classes, but also within the time consumption needs. Our approach also considers the regulatory point of view for automated data analysis within systems, and we deal only with statistical features of the network flows, while the payload data, the source and destination information, as well as the network port, are excluded

---

Honoring Professor Georgi M. Dimirovski on his anniversary.

G. Ilievski
Makedonski Telekom AD Skopje, Kej 13-ti Noemvri 6, 1000 Skopje, North Macedonia
e-mail: gjorgji.ilievski@telekom.mk

P. Latkoski (✉)
Faculty of Electrical Engineering and Information Technologies, Ss Cyril and Methodius University, Rugjer Boshkovic 18, 1000 Skopje, North Macedonia
e-mail: pero@feit.ukim.edu.mk

P. Shi et al. (eds.), *Complex Systems: Spanning Control and Computational Cybernetics: Applications*, Studies in Systems, Decision and Control 415,
https://doi.org/10.1007/978-3-031-00978-5_21

as attributes used for classification, especially as we deal with VoIP and encrypted VoIP data that is used in 5G.

1 Introduction

Network architectures are changing continuously, especially now when Virtual Machines (VM), Software Defined Networking (SDN), private, public and mixed clouds are common in the IT world. The trend is now moving towards microservices, containers, application functions, network functions in Network Functions Virtualization (NFV) environments [1], which is adding more complexity to the network flows. In such scenario, majority of the network traffic is moving in the cloud, usually within the same datacentre, in the east–west direction. This traffic never leaves the virtual plane and is often managed by the SDN elements in the NFV environment, which obstructs the capture or any other operation over the traffic. This is important both for the cloud operators and for entities using the services provided by the public clouds. Operations which are common practice and are considered trivial, such as implementing Quality of Service (QoS), network security, optimization, application management and monitoring, are becoming a challenge.

In this paper, we are performing an experimental test to reveal the network traffic classification efficiency of several supervised machine learning (ML) algorithms. We have created a unique test environment that reassembles real life processes and simulates the east–west traffic in the virtual plane among virtual hosts where NFV is established. The efficiency of the ML algorithms is explored from a point of classification precision, but also from a point of calculation speed. This is very important when we take into consideration the penetration of 5G, because it is tightly integrated with the mentioned cloudification and the new technologies used in it. For example, 5G specification calls for user plane latency of just 1 ms for ultra-reliable low-latency communications (URLLC) [2]. This is why the speed of the ML algorithm is crucial and must be performed in a manner that will minimize the expected latency added by the classification.

The study we have conducted provides a novel scenario that is comparable to the emerging architectures where NFV and 5G are implemented. It involves 6 different supervised ML algorithms: Bayes Net, NaiveBayes, J48, K-Nearest Neighbours (K-NN), Decision Tree and AdaBoost because they are widely used in the traditional computer networks, are proven to be reliable while providing a valid classification, and are not expensive to be implemented in practice. We have used Weka [3] as a tool for classification.

There are a variety of works that are using ML algorithms to perform packet inspection [4–7]. What distinguishes or work is the novel experimental testbed, as well as the approach to classify the network data based only on statistical parameters of the packets and the packet flows, without the use of source and destination addresses (both MAC and IP addresses), without any examination of the payload, and without the communication ports.

The encrypted network traffic is in a rapid rise. Significant number of services and applications are using encryption as a primary method of securing information. But this has made traffic classification a challenge. The solution for traffic classification that we propose is applicable in practice without compromising privacy and data integrity, it provides an insight into the performance of supervised ML algorithms and determines which one is the most suitable for NFV based environment.

There are also many examples of ML Algorithms used for Deep Packet Inspection (DPI) in traditional networks [3, 8, 9]. Compared to these works, we focus on virtualization and the NFV environment. In such scenario, the network packets are mostly moving in the east–west direction and are often encrypted, so classical DPI is impossible to be conducted. In our approach, it is not important whether the payload is encrypted or not. Also, the legal aspect of performing DPI in a cloud environment (especially public one) is satisfied, as the data carried within the payload is not compromised. We are using only statistical features of the network packets and the network flows to create datasets that are later used for training and testing of the ML algorithms. During the testing phase, we are evaluating the efficiency of the algorithm from a point of precision, but also from a point of speed. Network traffic is sniffed inside an open vSwitch directly. We are not introducing additional probe or SDN element to capture the traffic. We consider all network traffic, between the virtual elements inside the environment, but also the traffic that is used for management of the environment (including the one form the controllers), as well as the traffic that is going in and out to internet. This is a realistic scenario with most cloud solutions in the practice.

Besides the precision, the ML algorithm speed in many cases is even more important. If the time consumed to classify the data is adding significant latency in the network traffic and is consuming resources (CPU cycles, memory) of the cloud, the classification precision loses its relevance.

In the reminder of the paper we will go through the related work on the subject, briefly explained in the next section. The experimental setup and the dataset creation are explained in Sect. 3, while the results are analysed in Sect. 4. Section 5 is reserved for the conclusion and our plans for future work.

2 Related Work

Many researches are focused on the DPI aspects in scenarios involving SDN elements [10–12]. Others are researching the security aspects when performing DPI [13, 14] by using SDN probes for network traffic sniffing and data processing. Our work distinguishes in terms of the NFV-based setup, while targeting a complete isolation of the packet payload. Some authors consider the classification of network traffic in traditional networks [15, 16] without tackling the specifics of the trendy virtualization, which on the other hand is an important aspect of our work.

Mohammad Reza Parsaei et al. [17] are using SDN to categorize traffic by application, by applying different variants of Neural Network estimator. They are using data mining techniques based on different ML algorithms and propose a controller that could dynamically allocate bandwidth on network flows and optimize resource allocation. They achieve classification accuracy of over 97%. Distinct to our work, they are using source and destination IPs, as well as the transport layer port for classification.

In Karakus and Durresi [18], QoS in an SDN based network is being researched with an accent on overcoming the limitations of traditional networking architectures. Different flow routing mechanisms are categorized. In our research, we are exploring classification as a basic concept from which QoS can benefit significantly.

Vergara-Reyes et al. [4] is a study where NFV environment is prepared to classify different types of TCP traffic using three supervised ML algorithms: NaiveBayes, Bayes Net and J48. Network packets are analysed individually resulting in three different datasets: traditional, virtual and combined, in order to compare the classification performance. Only statistical parameters of the packets are used. In our case, we use TCP and UDP based traffic and we analyse the statistical parameters of the packet flows within an NFV environment that closely reassembles to cloud platforms.

Le et al. [19] applied big data, ML algorithms, SDN, and NFV to build a practical and powerful framework for clustering, forecasting, and managing traffic behaviour for a huge number of base stations with different statistical traffic characteristics of different types of cells (GSM, 3G, 4G). The framework is intended to be used for developing future 5G Self-Organizing Network (SON) applications. Five ML algorithms are used to classify the traffic generated by the mobile applications, with QoS implemented to enable bandwidth guarantees. The conclusion is that from the selected algorithms, Decision Tree has the best overall performance. Our experiment is bound on the transport network layer with aim to classify the traffic that is mostly east–west based using ML algorithms, but also to evaluate the time needed for classification which is crucial for the future 5G environments.

Alshammari et al. [5] is focused on VoIP traffic within traditional networks. Data is extracted from existing network environment with a complex topology. The authors evaluate classification of both encrypted and unencrypted VoIP using three ML algorithms: C5.0, ADA Boost and GP Classifier, and using subset sampling technique. In the experiments, C5.0 had the best performance and the highest precision rate. In our case, cloud-based environment with NFV implemented is used.

In Zander and Armitage [20], Machine Learning classification of multi-service internet traffic is used to evaluate resources consumption (CPU time and usage of system memory). We are complementing this research, as we are evaluating the ML algorithms time needed to perform the classification.

Shu et al. [21] proposes network traffic classification based on deep learning network structure. The experimental dataset is created from ten types of data, each of which is abstracted from a complete TCP bidirectional stream containing 249 network flow attributes. Google's TensorFlow deep learning framework is used in the experimental environment. NaiveBayes and Decision Tree ML algorithms are used to compare the classification efficiency in respect to the deep learning network. Compared to this work, we are targeting six different supervised ML algorithms classification, having in mind that not only classification precision, but also the time needed to perform the classification.

The effect of NFV elements placement on the network traffic, especially on the increase or decrease of the volume of the processed traffic, is researched in [22]. The authors develop an algorithm that determines the flow path and then propose a Least-First-Greatest-Last routing.

Bonfiglio et al. [23] are researching the traffic specifics of Skype as an application that is based on encrypted VoIP for voice calls. The traffic is explored in real time, with two different approaches by using the statistical parameters of the traffic generated by Skype. The approaches are then assessed using flow correlation.

To summarize, our testing setup is similar to that introduced in Vergara-Reyes et al. [4], with additional elements added to the environment. Both TCP and UDP traffic is generated, with and without encryption. The classification groups and labels are chosen in a manner that various traffic is classified. Viber and Skype are used to generate VoIP traffic, whereas scripts are used to open ssh management sessions to different hosts. Furthermore, a novel testbed is proposed in context of 5G and usage of NFV elements within the virtualized environment which is expected in a real-life setup. The network packets are analysed directly within the virtual switch, without the use of a probe or an SDN element. Statistical characteristics are extracted from TCP and UDP packet flows and used to perform further analysis.

The next section shows the details of the experimental environment and the creation of the datasets used to train and to test the six supervised ML algorithms.

3 Experimental Setup and Dataset Creation

To simulate the east–west traffic within a virtualized NFV based network, the experimental environment is based on Oracle VirtualBox [24], which is installed on a single physical host with Ubuntu 18.04 Server. All elements are connected with Open vSwitch (OVS) [25, 26] that provides the network connectivity. The switch is connected to internet through the host in a bridge mode. All network packets flow through the OVS switch, the packets inside the environment, and the packets to and from internet. We are capturing the traffic directly on the OVS using Wireshark and tshark [27].

Mininet [28] is used as a network simulator. There are two different installations on two separate virtual machines, each with different network topology having 100 hosts, 20 switches and links among them and to the OVS. The hosts within the simulated networks are having private IP addresses and are able to communicate with each other. GRE tunnelling is used to link the two simulated Mininet networks. Some of the hosts within Mininet have NAT-ed IP addresses and are able to communicate to internet.

Ryu Controller [29] is used to control the simulated Mininet networks. It is installed and configured in a separate virtual machine.

VMs connected to the OVS are also used for traffic generation. Skype and Viber are installed onto them to simulate the VoIP traffic. When initiated, VoIP needs access to internet, but after that peer-to-peer communication can be observed within the OVS in a completely east–west direction. Script that starts ssh sessions is enabled on the VMs. Python script so start ssh sessions are from the Mininet hosts was developed, as well. The SSH sessions were started in intervals that followed Poisson distribution.

Distributed Internet Traffic Generator (D-ITG) [30] generates various TCP and UDP traffic among the hosts within Mininet. Different scripts are used to generate traffic at packet level, replicating appropriate stochastic processes for both IDT (Inter Departure Time) and PS (Packet Size) random variables.

Figure 1 is an overview of the experimental setup, showing the components symbolically.

We have made 50 different experiments to generate various traffic (using D-ITG, Skype, Viber, custom scripts) and to analyse it. The experiments were conducted in time intervals from 4 to 20 min in which VoIP calls lasted from 10 s to 10 min, following Poisson distribution. One dataset per experiment was generated. Different D-ITG scripts for different traffic simulation were used in every experiment. The scripts used different Mininet hosts and different paths in every try. The average number of captured packets was 1.262.375 and the average number of flows was 4090.

We have made specific classification of the traffic, using classes that are commonly used, based on experience from the traditional networks. As it will be shown in the results, the classification precision was calculated as an overall, but also for every

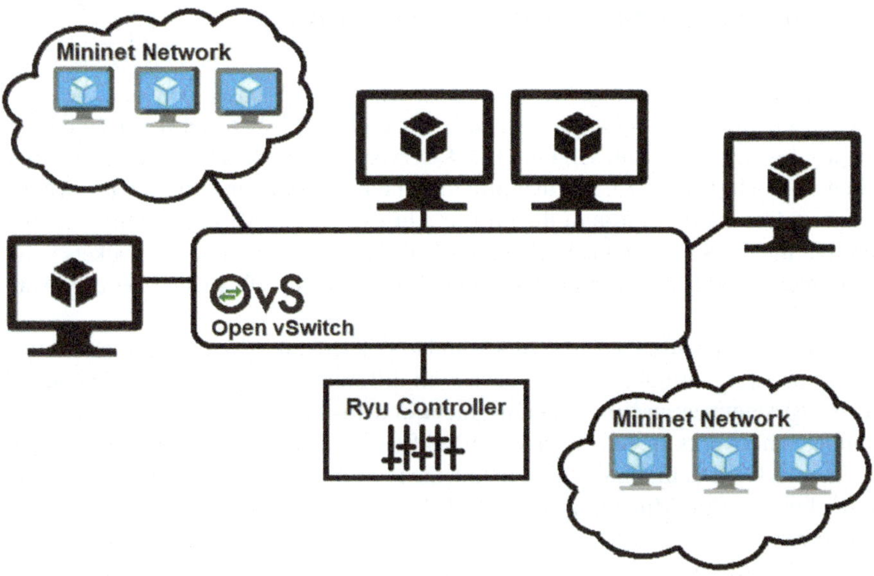

Fig. 1 Experimental environment

class independently, in order to calculate the macro-average precision in which every class contribution to the precision is treated equally (as the number of packets and flows varies for every class).

We used the following labels for the classes: DNS—for all the traffic used for name resolution, NETMGMT—all traffic used for hosts and network management, SSH—for the ssh sessions in the environment, WEB—for HTTP and HTTPS traffic, VOIP—for VoIP traffic, SVOIP—for encrypted VoIP.

From the generated Wireshark pcap files, UDP and TCP packet flows, and the classes used for ML training and latter for test precision and confirmation are identified using Argus [31]. Similar to [5], we define a flow as a bi-directional connection between two hosts. TCP flows are terminated either by flow time-out or by connection tear-down, whereas UDP flows are ended by flow time-out. When we observe the flows within the OVS, it can be seen that most of the traffic is east–west based, inside the virtual layout and between the hosts, but also the flows from the management generated by the hypervisor and the Ryu controller are detected. Because our focus is an NFV based environment, some of the flow features are not taken into consideration, such as the source and destination IP and MAC address, as well as the communication port that can vary inside the virtual environment.

To train and to test the supervised ML algorithms, we have used Weka [3, 32]. 2/3 of every dataset was used for training, while 1/3 was used for testing each of the algorithms. As not all the attributes have the same contribution to the classification, the AttributeSelectedClassifier with Ranker as an attribute ranking algorithm was used. InfoGainAttributeEval was used as an evaluator that determines the gain of information that the attributes carry. With this approach we are ranking the attributes that are used for the algorithms after which the information gain of every attribute is evaluated. This approach prevents a possible data leakage.

Based on experience from traditional networks and with careful observation of the gained datasets, we have selected the attributes given in Table 1, as features that characterize the flows. The payload is not used for reason of privacy within cloud environments and usage of different encryption methods that will make the payload irrelevant for the classification.

4 Results and Analysis

To create 50 datasets, we have performed as many experiments and all 6 supervised ML algorithms were tested. The performance of each algorithm is a combination of its precision and the time needed to make the classification. Because the time consumption is correlated to the performance of the machine where the analysis is performed, all classification tasks were performed on the same machine with a careful observation of all the processes on the machine that can influence the performance. A mean value of the 50 results was derived for all target metrics.

True Positive (TP), False Positive (FP), True Negative (TN) and False Negative (TN) rates are defined as:

Table 1 Flow attributes

	Abbrevation	Feature
1	Proto	Transaction protocol
2	Rate	Packets per second
3	Srate	Source packets per second
4	Drate	Destination packets per second
5	Sintpkt	Source interpacket arrival time
6	Dintpkt	Destination interpacket arrival time
7	Sjit	Source jitter
8	Djit	Destination jitter
9	Mdoffset	Mean of the data offset values of the packets in the flow
10	Smeansz	Mean of the flow packet size transmitted by the source
11	Dmeansz	Mean of the flow packet size transmitted by the destination
12	Smaxsz	Max packet size for source
13	Dmaxsz	Max packet size for destination
14	Sminsz	Min packet size for source
15	Dminsz	Min packet size for destination

- TP is number of instances that are truly identified of a class.
- FP is number of instances that are falsely identified of a class.
- TN is number of instances that are truly identified that are not of a class.
- FN is number of instances that are falsely identified that are not of a class.

The overall precision of the algorithms is calculated as the proportion between TP instances and all instances in the dataset [32]:

$$Precision = TP/(TP + FP) \tag{1}$$

Table 2 shows the average precision of the algorithms in all 50 experiments with the statistical standard deviation across the experiments, as a weighted average value.

Table 2 Algorithm precision

No.	ML algorithm	Precision
1	AdaBoost	0.744 ± 0.0292
2	BayesNet	0.9672 ± 0.0189
3	J48	0.9906 ± 0.0027
4	KNN	0.9172 ± 0.0438
5	NaiveBayes	0.8634 ± 0.0170
6	Decision tree	0.9914 ± 0.0033

It can be seen that overall Decision Tree algorithm has the best precision. It is followed by J48 and BayesNet. On the other hand, AdaBoost algorithm has the worst overall performance with the lowest precision of 74.4%.

To further explore the precision, we have calculated the micro average precision, which aggregates the contribution of all classes and calculates the average metric, as given by Eq. 2. The results are presented in Table 3.

$$Precision\_MIC = \frac{(TP1 + TP2 + \cdots + TPN)}{(TP1 + FP1 + TP2 + FP2 + \cdots + TPN + FPN)} \quad (2)$$

Not all classes have same or similar number of packets and flows and the data distribution is skewed. To avoid data balancing problem and to come to valid conclusions we are calculating the macro average precision, the recall and the F-1 score.

Macro average precision is an average of precisions of each class. This means that every class will weigh the same in the macro average precision. The following equation is used to calculate the macro average precision (Precision_MAC), where Pr1, Pr2 etc., denote the precision of the algorithm regarding the individual classes.

$$Precision\_MAC = \frac{Pr1 + Pr2 + \cdots PrN}{Count(Pr)} \quad (3)$$

These results are shown in Table 4. In this table the statistical standard deviation is calculated for the precision between classes.

If we evaluate Table 4, it becomes clear that the algorithms are not performing the same on all the classes. Decision Tree algorithm is the most constant with highest

Table 3 Algorithm micro average precision

No.	ML algorithm	Micro average precision
1	AdaBoost	0.8450 ± 0.0176
2	BayesNet	0.9954 ± 0.0027
3	J48	0.9984 ± 0.0006
4	KNN	0.9856 ± 0.0073
5	NaiveBayes	0.9752 ± 0.0027
6	Decision tree	0.9984 ± 0.0010

Table 4 Algorithm macro average precision

No.	ML algorithm	Macro average precision
1	AdaBoost	0.20335 ± 0.3064
2	BayesNet	0.8899 ± 0.1489
3	J48	0.9824 ± 0.0148
4	KNN	0.82735 ± 0.2202
5	NaiveBayes	0.78915 ± 0.2048
6	Decision tree	0.9848 ± 0.0107

macro average precision and the lowest standard deviation between classes, showing that it classifies all classes similarly. J48 is very close to Decision Tree, with over 98% precision. Opposite to them, AdaBoost algorithm shows very low macro average precision with high standard deviation, which means that it performs poorly on different classes. K-Nearest Neighbour algorithm is also underperforming, with just over 82% macro average precision. When we compare these results with the standard weighted precision in Table 2, we can see that the algorithms have the same order, but the macro precision of the lower end algorithms is worse, drawing the conclusion that AdaBoost and KNN have different precision for different classes.

To evaluate the impact of the false negative classified instances, Recall is used as a model metric. It is the proportion of true positive instances and total actual instances:

$$Recall = TP/(TP + FN) \tag{4}$$

We've used the Recall to calculate the F1-score of the supervised ML algorithms in our experiments. It is a metric that balances between the precision and the recall, so that false negative instances are taken into consideration. F1-score is calculated as a harmonic mean of the precision and the recall:

$$F1 - Score = 2 \cdot \frac{Precision \cdot Recall}{Precision + Recall} \tag{5}$$

Table 5 shows the F1-score values calculated for our experiments.

Decision Tree ML algorithm has the best F1-score, followed by J48, BayesNet, KNN, NaiveBayes and AdaBoost. The last one has F1-score of only 23.2% with very high standard deviation.

The tables are visually represented in Figs. 2, 3, 4 and 5.

The algorithm precision is only the first feature to determine the usability of the algorithm. The second important aspect is the time needed to perform the classification. If the time needed for classification is too high, the process will add latency to the network communication, thus making the benefit of the classification too costly. This is important especially in the protocols where latency can degrade the service, such as VoIP. Furthermore, this is also crucial in the 5G scenarios, where latency is one of the major concerns. Another point is that if the time spent by the algorithm is

Table 5 F-1 score

No.	ML algorithm	F1-score
1	AdaBoost	0.231575 ± 0.3356
2	BayesNet	0.913425 ± 0.1055
3	J48	0.975425 ± 0.0212
4	KNN	0.797425 ± 0.2295
5	NaiveBayes	0.782125 ± 0.1510
6	Decision tree	0.980475 ± 0.0152

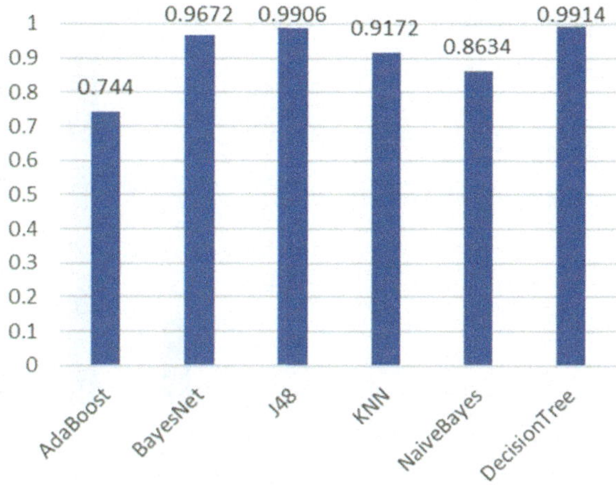

Fig. 2 Algorithm precision

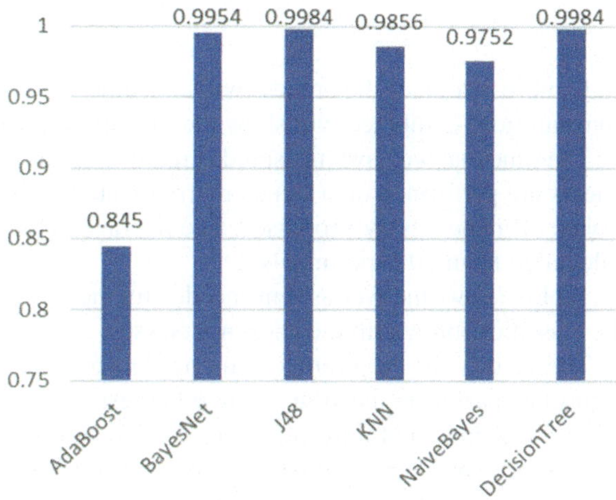

Fig. 3 Micro average precision

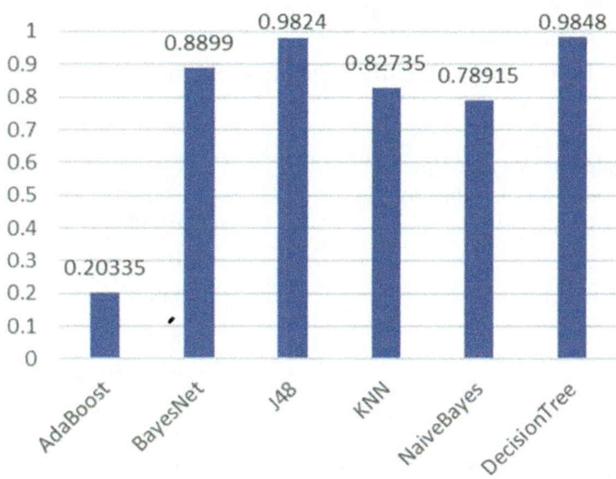

Fig. 4 Macro average precision

Fig. 5 F-1 score

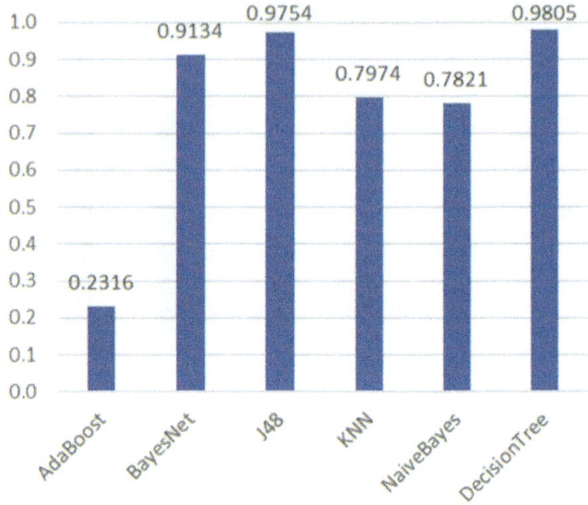

high, more resources the process will consume. The two metrics (precision and time consumption) combined will show the overall performance of the algorithms.

The time that we have measured is relative to our testbed environment. All experiments are performed on a same environment, where special care has been taken to isolate all unnecessary processes. The average value of the time consumption was calculated from 50 experiments.

Table 6 shows the average time needed for the six supervised algorithms to perform the classification within the chosen 8 classes.

The results for the average time needed for classification show that AdaBoost algorithm performs the best, with a highest speed. Decision Tree and BayesNet share the second and third place being 25% slower than AdaBoost. J48 has also a satisfactory speed. NaiveBayes is almost 9 times slower than AdaBoost and more than 6 times slower than Decision Tree KNN algorithm is the slowest. Decision Tree and AdaBoost spend only 5.9% of the time needed by KNN to perform the classification.

Figure 6 graphically represents the average time consumed by the algorithms.

Table 6 Average time needed for classification (in sec)

No.	ML algorithm	Average time in seconds
1	AdaBoost	0.012
2	BayesNet	0.016
3	J48	0.022
4	KNN	0.272
5	NaiveBayes	0.104
6	Decision tree	0.016

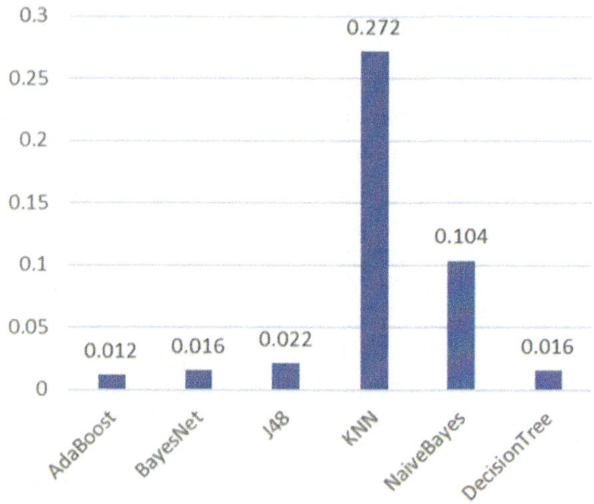

Fig. 6 Time needed for classification (in seconds)

To summarize, when we take a look at both precision and time needed for classification, Decision Tree supervised ML algorithm has the best overall performance. Although AdaBoost is the fastest algorithm, the classification precision is poor and unsteady across different classes, which makes this algorithm unreliable for our scenario. J48 has also high precision that is evenly distributed among classes, but it is slower than Decision Tree and BayesNet. Nevertheless, its speed is in the scale of Decision Tree and BayesNet, and it is also a valid choice. BayesNet has a high precision, but the macro average precision and the F1-score show that precision distribution among classes is not as good as Decision Tree and J48.

NaiveBayes is in the middle from both precision and time perspective, while KNN algorithm has about 83% macro average precision and 80% F1-score, but it is by far the slowest algorithm which makes it useful only in cases where time needed for classification has low importance.

5 Concluding Remarks and Future Research

The main goal of our paper is to compare the efficiency of six supervised ML algorithms for classification of network traffic that is moving in east–west direction within a scenario where NFV elements are placed.

The efficiency of the algorithms is explored from a point of precision of the algorithm but also form a point of time consumption needed to perform the classification. This is important from a virtualization point of view, where mixed cloud scenarios are a common practice, but also for the incoming penetration of 5G, where the network latency is of high importance.

Our experimental testbed is used to perform multiple experiments and to collect network traffic data from which IP flows are extracted. The statistical features of the flows are used as attributes for classification. Because attributes such as source and destination IP and MAC addressed and communication ports can vary inside a virtualized environment, they are not taken into consideration. Due to encryption and data privacy concerns, the payload of the data packets is also excluded from the datasets and it is not used for classification.

The environment that we use is not introducing any kind of network probes or SDN elements to perform the data collection, so that east–west traffic is completely unchanged. The traffic is completely intercepted within the virtual layer where it naturally resides. This has also an impact on the resource consumption, minimizing the additional latency that can be added to the network packets by redirecting or port replication used in traditional DPI.

The results have shown that Decision Tree algorithm has the best overall performance, from both classification precision and time consumption point of view. It has proved as a reliable classifier that is performing evenly across different classes. J48 and BayesNet are also performing well, with J48 having slightly better precision and BayesNet being faster. K-Nearest Neighbour and NaiveBayes have an average classification precision in a range of about 80%, but they are slow, especially KNN which is almost 20 times slower than Decision Tree and BayesNet. AdaBoost shows the worst performance with precision that varies a lot among different classes, which can be seen from the macro average precision and the F1-score.

The analysis in our paper can be used in practice within multiple systems that are built on top of cloud environments. NFV elements are now unavoidable part of such infrastructures. 5G infrastructure is relying onto these types of systems, but also connectivity to such systems is most likely to be done through 5G access technology. In those examples performing QoS, network and application security, data management, system and process monitoring and control is depending on valid network traffic classification that has to be precise and fast without taking considerable amount of system resources.

For future work we are planning to evaluate the impact of the number of classes on the classification results and the time consumption of the supervised ML algorithms by introducing large number of classes and reducing the classes. Another stream is to expand the experimental testbed to multiple hosts and distributed switches and to evaluate the network that is moving across multiple hosts.

References

1. Chiosi, M., et al.: Network Functions Virtualisation. Introductory White Paper. Cited 2019-10-10 Available at: https://portal.etsi.org/nfv/nfv_white_paper.pdf (2015)
2. Eiman, M.: Minimum Technical Performance Requirements for IMT-2020 Radio Interface(s). Presentation. Cited 2019-10-10. https://www.itu.int/en/ITU-R/study-groups/rsg5/rwp5d/imt-2020/Documents/S01-1_Requirements%20for%20IMT-2020_Rev.pdf (2018)

3. Frank, E., Hall, M.A., Witten, I.H.: Data Mining: Practical Machine Learning Tools and Techniques, Morgan Kaufmann, 4th edn. San Francisco, CA, USA (2016). ISBN: 0128042915 9780128042915
4. Vergara-Reyes, J., Martinez-Ordonez, M.C., Ordonezy, A., Rendon O.M.C.: IP traffic classification in NFV: a benchmarking of supervised machine learning algorithms. In: IEEE Colombian Conference on Communications and Computing (2017). https://doi.org/10.1109/ColComCon.2017.8088199
5. Alshammari, R., Nur Zincir-Heywood, A.: Identification of VoIP encrypted traffic using a machine learning approach. J. King Saud Univ. Comput. Inf. Sci. Arch. **27**(1), 77–92 (2015). https://doi.org/10.1016/j.jksuci.2014.03.013
6. Ma, B., Zhang, H., Guo, Y., Liu, Z., Zeng, Y.: A summary of traffic identification method depended on machine learning. In: Sensor Networks and Signal Processing (SNSP) 2018 International Conference, pp. 469–474 (2018). https://doi.org/10.1109/SNSP.2018.00094
7. Trivedi, U., Patel, M.: A fully automated deep packet inspection verification system with machine learning. In: IEEE International Conference on Advanced Networks and Telecommunications Systems (2016). https://doi.org/10.1109/ANTS.2016.7947802
8. Rezaei, S., Liu, X.: Deep learning for encrypted traffic classification: an overview. IEEE Commun. Mag. **57**(5), 76–81 (2019)
9. Shafiq, M., Yu, X., Laghari, A.A., Yao, L., Karn, N.K., Abdessamia, F.: Network traffic classification techniques and comparative analysis using machine learning algorithms. In: ICCC, pp. 2451–2455 (2016). https://doi.org/10.1109/CompComm.2016.7925139
10. Huang, U., Li, P., Gu, S.: Traffic scheduling for deep packet inspection in software-defined networks. Concurr. Comput. Pract. Exp. (2017). https://doi.org/10.1002/cpe.3967
11. Mousa, M., Bahaa-Eldin, A., Sobh, M.: Software defined networking concepts and challenges. In: 11th International Conference on Computer Engineering & Systems (ICCES), pp. 79–90 (2016). https://doi.org/10.1109/ICCES.2016.7821979
12. Polčák, L., et al.: High level policies in SDN. In: International Conference on E-Business and Telecommunications, pp. 39–57 (2016). https://doi.org/10.1007/978-3-319-30222-5_2
13. Arevalo Herrera, J., Camargo, J.E.: A survey on machine learning applications for software defined network security. In: Applied Cryptography and Network Security Workshops. ACNS 2019. Lecture Notes in Computer Science, vol. 11605. Springer, Cham. https://doi.org/10.1007/978-3-030-29729-9_4
14. Chowdhary, A., Huang, D., Alshamrani, A., Sabur, A., Kang, Kim, M.A., Velazquez, A.: SDFW: SDN-Based Stateful Distributed Firewall (2018). https://doi.org/10.13140/RG.2.2.11001.93281
15. Choudhury, S., Bhowal, A.: Comparative analysis of machine learning algorithms along with classifiers for network intrusion detection. In: ICSTM, Tamil Nadu, India, pp. 89–95 (2015). https://doi.org/10.1109/ICSTM.2015.7225395
16. Shafiq, M., Yu, X., Laghari, A.A., et al.: Wechat text and picture messages service flow traffic classification using machine learning technique. In: IEEE HPCC/SmartCity/DSS, vol. 58–62, pp. 58–62 (2016). https://doi.org/10.1109/HPCC-SmartCity-DSS.2016.0019
17. Reza, M., Sobouti, M.J., Raouf, S., Javidan, R.: Network traffic classification using machine learning techniques over software defined networks. International J. Netw. Comput. Appl. **8**, (2017). https://doi.org/10.14569/IJACSA.2017.080729
18. Karakus, M., Durresi, A.: Quality of service (QoS) in software defined networking (SDN): a survey. J. Netw. Comput. Appl. **80**, (2016). https://doi.org/10.1016/j.jnca.2016.12.019
19. Le, L., Sinh D., Lin, B. P., Tung, L.: Applying big data, machine learning, and SDN/NFV to 5G traffic clustering, forecasting, and management. In: 4th IEEE Conference on Network Softwarization and Workshops (NetSoft), Montreal, Canada (2018). https://doi.org/10.1109/NETSOFT.2018.8460129
20. Zander, S., Armitage, G.: Practical machine learning based multimedia traffic classification for distributed QOS management. In: IEEE LCN, Bonn, Germany, pp. 399–406 (2011). https://doi.org/10.1109/LCN.2011.6115322

21. Shu, J.H., et al.: Network traffic classification based on deep learning. In: First International Conference on Advanced Algorithms and Control Engineering, Journal of Physics: Conference Series, vol. 1087, p. 062021 (2018). https://doi.org/10.1088/1742-6596/1087/6/062021
22. Ma, W., Medina, C., Pan, D.: Traffic-aware placement of NFV middleboxes. In: IEEE GLOBECOM, San Diego, CA, USA, pp. 1–6 (2015). https://doi.org/10.1109/GLOCOM.2015.7417851
23. Bonfiglio, D., Mellia, M., Meo, M., Rossi, D., Tofanelli, P.: Revealing skype traffic: when randomness plays with you. ACM SIGCOMM Comput. Commun. Rev. **37**, 37–48 (2007). https://doi.org/10.1145/1282427.1282386
24. Oracle VirtualBox. Cited 2019-09-10. https://www.virtualbox.org (2019)
25. Bernal, M.V., Cerrato, I., Risso, F., Verbeiren, D.: Transparent optimization of inter-virtual network function communication in open vSwitch. In: IEEE Cloudnet, Pisa, Italy, pp. 76–82 (2016). https://doi.org/10.1109/CloudNet.2016.26
26. Linux Foundatrion, Open vSwitch Project. http://www.openvswitch.org (2016)
27. Wireshark. Cited 2019-09-10. https://www.wireshark.org/ (2006)
28. Team, M.: Mininet: an instant virtual network on your laptop (or other pc)-mininet. Cited 2019-09-12. http://mininet.org (2017)
29. Ryu Framework. Cited 2019-09-10. http://osrg.github.io/ryu/ (2019)
30. Botta, A., Dainotti, A., Pescapè, A.: A tool for the generation of realistic network workload for emerging networking scenarios. Comput. Netw. (Elsevier) **56**(15), 3531–3547 (2012). https://doi.org/10.1016/j.comnet.2012.02.019
31. Argus Quosient. Cited 2019-09-10. https://qosient.com/argus/ (2015)
32. Hall, M., Frank, E., Holmes, G., Pfahringer, B., Reutemann, P., Witten, I.H.: The WEKA data mining software: an update. SIGKDD Explor. Newsl. **11**(1), 10–18 (2009). https://doi.org/10.1145/1656274.1656278

Manipulation of URL Addresses Using Machine Learning to Provide Better Cyber Security

David Acev, Gorjan Nadzinski, Valentin Rakovic, and Aleksandar Risteski

Abstract In this new and modern reality, complex systems are no longer only reserved for the theoretical textbooks, research centres, or industrial plants. With computers and the Internet becoming indispensable for both homes and organizations, complex systems are now in the hands of regular civilian users, utilizing a combination of advanced concepts (such as control theory, artificial intelligence and machine learning, BigData, etc.) to deliver comfort, safety, robustness, and fast and easy communication to households, commerce, and industries alike. However, this sudden rise in complexity and opaqueness of the systems in regard to their everyday users prompts a serious rethink of the approach to securing IT systems. As a pinnacle example of complexity, the Internet offers many points of cyber vulnerability where regular users can be targeted. In this work we investigate the implementation of machine learning based approaches for detection of malicious URLs in order to improve cyber security. The proposed approaches show promising results in exposing dangerous links, and could be implemented in several points within a complex network, both locally and in different nodes within a network hierarchy, thus showing how data science and machine learning (two of the very driving forces in the rise and development of complex systems) stimulate the core of the cyber security. Moreover, the same proposed approaches are used for classification of URL addresses based on the content that each of the URLs provides, thus, exploiting the benefits of machine learning approaches even further.

D. Acev (✉) · G. Nadzinski · V. Rakovic · A. Risteski
Faculty of Electrical Engineering and Information Technologies, Ss. Cyril and Methodius University, Ruger Boskovic 18, 1000 Skopje, North Macedonia
e-mail: david_acev@hotmail.com

G. Nadzinski
e-mail: gorjan@feit.ukim.edu.mk

V. Rakovic
e-mail: valentin@feit.ukim.edu.mk

A. Risteski
e-mail: acerist@feit.ukim.edu.mk

P. Shi et al. (eds.), *Complex Systems: Spanning Control and Computational Cybernetics: Applications*, Studies in Systems, Decision and Control 415,
https://doi.org/10.1007/978-3-031-00978-5_22

Keywords Machine learning · URL filtering · Cyber security

1 Introduction

As a worldwide compilation of interconnected IP networks, the Internet presents the epitome of a modern complex system, one that is a complicated system exhibiting a simple behavior. As such, it is a junction of interdisciplinary concepts and architectural features which constantly produce new problems, challenges, and phenomena of scientific and technical nature [1].

The Internet is an integral part of human lives from social, economical, and technical point of view, and it is an ever evolving, multi-faceted, and diverse structure. Being an everyday tool and essence, the Internet is used and navigated by billions of users daily, who utilize its benefits without intrinsically knowing or understanding its highly complex background. This makes all careless Internet users, whether casual individuals or critical commercial or industrial systems, potential targets of damaging malicious attacks [2, 3]. Therefore, any approach attempting to study the relationships between the Internet's features and subsystems and how their interaction can be used to protect and secure communication, is an approach that can rightfully be classified as a study of a complex system [4, 5].

As a result, the topic of cyber security as a study of the protection of computer networks and systems in general from malicious attacks, damages, data leaks, and disruptions, is an extremely popular and fruitful subject of interest. Many such modern and advanced approaches exist, based on mathematical and statistical concepts [6], discrete-event modeling and simulation [7], adaptive learning and inference [8], artificial intelligence and machine learning [9], among others.

The evolution of Information and Communications Technology (ICT), and the Internet in general, introduces a massive storage of data and values, from trivial to vitally important. Thus, the form of thieves or someone who wants to realize malicious actions has shifted more to cyber crime. There are various techniques the tech-savvy criminals could use in order to steal someone else's identity, for instance: malware (virus), worm, trojan, spyware, spam, adware, etc.

Cyber crime increases rapidly and the systems used to impede it need to deliver a remarkable performance in order to keep the users well protected. Any online service requires certain security measures. Therefore, it is reasonable that certain safety measures are needed to be applied in the online world, in other words the enhanced way to undermine someone's personal data undoubtedly introduces the necessity of some kind of protection on all of the systems. The protection is called cyber security.

Cyber security encompasses many digital security topics, but most importantly we could classify them as: Information Security (INFOSEC), Information Assurance (IA), and Systems Security. There are different definitions of cyber security, where it is generally defined as a group of techniques and practices designed to protect data, or more precisely digital data, in other words data that is stored, transmitted or used on an information system.

There are different techniques that could boost the performance of the security network. One of them is Machine Learning (ML). New Artificial Intelligence (AI) tools are being developed to help so many different aspects of our everyday life to achieve their goals.

Machine Leaning is the practice of using algorithms to analyze data, then learn from that data and eventually make a determination or prediction from new data. It is well suited for problems where the existing solutions require a lot of hard-tuning or long lists of rules, by simplifying the code drastically and perform massively better than the traditional programming, or when changeable (fluctuating) environments are encountered [10].

With the use of well-structured ML algorithm, cyber security can be facilitated. In this paper, detailed machine learning approaches are scrutinized and their implementation in real cyber security tasks is presented. More concretely, it is investigated how certain URL address datasets can be classified based on the essence of the websites and the content they provide by applying machine learning algorithms [11].

2 Related Work

In [12], the cyber threat of phishing is encompassed, which is a major concern on the Internet today and many users suffer from it. It is argued that blacklisting is still the most common defence users have against such phishing websites, but is failing to deal with the increasing number of threats. A random forests machine learning approach was developed to classify the websites to good and bad so it will allow the system to block the malicious website from damaging a particular computer. However, there are various opinions on what the best approach is for certain features and algorithms. The objective of that paper is to evaluate the performance of the aforementioned machine learning algorithm using only a lexical dataset. The performance is benchmarked against other machine learning algorithms and additionally against those reported in the literature. According from the presented experiments the random forest algorithm performed best yielding an 86.9% accuracy.

Another interesting scenario is developed and presented in [13], in which malicious URLs are detected using multi-layer filtering model. It is known how harmful can malicious URLs be to every aspect of the digital processes. Detection of malicious website techniques includes black-list and white-list methodology and for that purpose, machine learning classification algorithms are used. It is argued that the black-list and white-list technology is useless if a particular URL is not in the list. In the paper, a multi-layer model for detecting malicious URL is proposed. The filter is able to explicitly determine whether the URL is benign or malicious by training the threshold of each layer filter when it reaches the threshold. Otherwise, the filter leaves the URL to the next layer.

Moreover, there is another sophisticated approach already researched, which is about URL classification using online incremental learning and it is presented in [14]. The challenge of large-scale topic classification of websites based on the minimal text available in the URLs is addressed. It is very interesting, but also very exigent approach, because of the sparsity of feature vectors that are derived from the URL text composition, and the typical asymmetry between the cardinality of train and test sets due to non-availability of sufficient sets of annotated URLs for training and very large test sets (for instance, in the case of large-scale focused crawling). The proposition elaborated in [14] is an online incremental learning algorithm which addresses those issues. It is such a novelty approach that requires steady reasoning methods.

As a support to the previous papers that we have mentioned, [15] is a further development of the URL classification for online security during web-surfing. That approach is stimulated by the fact that everyday users visit websites with suspicious background in their pursuit of information. Therefore, malicious websites are a significant threat to the Internet users. Malicious websites implant malwares into users' computers without their knowledge, through drive-by-downloads technology. A naive user could easily fall victim of such attack. That brings everyone to the core of the study to classify malicious websites from benign ones from their URL features. As in the previous paper that has been analyzed, the URL characteristic features are textual. In the paper it is argued that could be beneficial the textual representation to be converted to numeric values that truly represent the information of the original feature. The selection methods that are undertaken are: least absolute shrinkage and selection operator (LASSO) and Multi-objective Pareto Genetic algorithm (MOGA). Finally, the data consisting of selected features is used to train a support vector machine (SVM) classifier. A tenfold validation is used to estimate the performance of the approach. It is a promising approach that gives satisfactory results at the end.

The benefits of the machine learning, more precisely supervised machine learning methods, are argued in [16]. Similarly, the websites are distinguished to one another using some kind of machine learning reasoning and parsing in order to determine the nature of the URLs. It is a slightly different approach, because mainly encompasses the URL addresses regarding a certain country. The aim is to be able to automatically distinguish websites targeting a specific nation, using both the URL and the content of a website.

Similarly to the first approach that we have discussed in this section, [17] handles the issues with phishing threats using machine learning. The authors use supervised classification algorithm to determine whether a URL is malicious or not. In this paper it is argued to utilize the benefits of the naive Bayes algorithm which is a powerful approach for the classification tasks. Furthermore, it is pointed out that the detection of phishing websites is done using blacklists. There are many popular websites which host a list of blacklisted websites, for example PhisTank. The blacklisting techniques lack in two aspects, blacklists might not be exhaustive and do not detect a newly generated phishing website. At the end the examined algorithm gives desirable results that just enlarge the confidence towards that particular classification algorithm.

The last approach covered in this section is the deep learning approach for detecting malicious URLs. Deep learning is a subset of machine learning that uses neural networks in order to predict certain features from given dataset. In [18], the

URL classification is done with the help of Convolutional Neural Networks (CNN). The proposed model and the architecture in the paper addresses the problem statement. The taxonomy under which the URLs are segmented follows a hierarchical pattern. An ensemble nature of classifying the textual content into the respective branches is followed, depending on the unique confidence values for each branch.

This work performs the classification of URL addresses using supervised learning methods. The approach differs from the other mentioned in this section by the way the URL is parsed and manipulated before applying one of the models. The final results introduce a plethora of opportunities how the websites could be grouped.

3 ML-Based URL Detection

Malicious URL detection is a massive challenge in cyber security systems. As a result of the previously elaborated trends, state-of-art solutions for malicious URL detection require high sophistication and reliability [19]. There are different techniques how the cyber security can be enhanced. The approach described in this work is composed of parsing the URL addresses and latter applying supervised machine learning approaches [20].

The dataset used disposes of different labeled URLs and as it has been explained previously, certain patterns need to be found that classify one URL address as malicious or benign. ML to categorical or string values can introduce significant problems. Designing such ML model poses a very complex task [21–23]. One approach that can mitigate the aforementioned problems, is the utilization of an idiosyncratic approach. Some crucial patterns within the concept of the URL addresses definition are found to facilitate the way how the challenge is handled.

The approach described encompasses several steps. First, a dataset of total number of 60,000 URLs is being used, each referred as bad (malicious) or good (benign). In order to manipulate this dataset in a certain way, a separation of the URL addresses content is applied. Once the URLs are detached the algorithm is trained on each part of the URLs, in order to develop a robust ML model that reliably detect malicious URLs. The labels of the URLs are transformed in a numerical output, labeling bad as 1, and good as 0.

Before producing the results, a thorough cleaning of the dataset is provided. That signifies finding a reasonable way how the missing values will be handled, if any, to make sure that everything is represented the way it should be. Once the missing values are cleared, any digit that could be found in the corpus of the URL addresses is being substituted with a blank space. The following step includes substituting every character except the alphabetic letters and the '.' sign with a corresponding '/' character. The parsing of the URLs is done by splitting them out by the special character '/'. By the end of that process, the URLs parts are separated once again, however, this time they are being parsed by the '.' sign. The reasoning behind the second separation is due to the fact that the common parts of the URLs should also be removed from their content. The common parts of the URLs include: ‘com’, ‘www’, ‘php’, and ‘html’.

Another instance of how machine learning could stimulate cyber security is by detecting the type of the URL. The type of URLs is detected by the content the URLs provide. This example is similar to the previous one. Nonetheless the preferences are slightly different. The URL classification is important, because it can convey a valuable information for the website itself. By setting apart a particular URL from the rest or grouping that URL in a certain set, different options are provided. Therefore, in order to develop an algorithm for this case a different dataset is being used. This dataset is a little bit more convoluted than the previous. The technique of manipulating the data are the same. However, the output is different. The URLs that are being utilized belong to the groups of: arts (labeled as 0), business (labeled as 1), computers (labeled as 2), games (labeled as 3), and health (labeled as 4).

In order to validate the models, a cross validation is being utilized. The cross validation in this approach splits the data in 5 equal parts and each part is validated on the remaining data. For the final training of the models, 80% of the data is used for training and the remaining 20% of the data is used for validation. Three models of machine learning are used for training and classifying each URL that the model was not trained on. In order to separate the data into data for training and data for validation specially designed functions are being employed [24, 25].

In this work three ML models are used, that provide best results with respect to string manipulation: linear classifier, multinomial naive Bayes algorithm, and logistic regressor. All three models are based on supervised learning approach.

Linear classifiers explicitly work on data in the original input space. These classifiers give competitive performances on document data with nonlinear classifiers. An important benefit of linear classification algorithms is that the training and testing procedures are much more efficient. Thus, linear classification could be useful for large-scale applications [26].

The multinomial naive Bayes algorithm is extremely fast to implement and simple to interpret; each classifier is selected once the probability of that set is bigger than the probability of the other set. The classifier solely divides the group of data to the number of labels earlier presented in the model and compares the features which has already learned in order to make the classification as accurately as possible. Such model is called a generative model because it specifies the hypothetical random process that generates the data [27].

Logistic regression uses the logistic function in its core to model a binary variable. This regression can be implemented to classify samples using different types of data to perform the classification. Also, it can be used to assess what variables are important for classifying samples [28].

Most of the performance measures are calculated from the confusion matrix. The structure of a confusion matrix is shown in Table 1.

In order to determine the performance of the classification models, the following performance parameters are calculated: recall, precision, F1-score, and accuracy.

Better results can be provided when the parameters are adjusted inside of the models respectively to the necessities of the undergoing tasks. That is basically

Table 1 Confusion matrix

Confusion matrix	Predicted positive	Predicted negative
Actual positive	True positives (TP)	False negatives (FN)
Actual negative	False positives (FP)	True negatives (TN)

shown on the confusion matrix parameters, from there it can be concluded that if the performance of a certain model is improving or diminishing by experimenting with the model parameters in a particular way [29].

4 Performance Analysis

This section, analyses the performance of the three models (linear classifier, multinomial naive Bayes algorithm, and logistic regressor) elaborated in the previous section. For the implementation of the algorithms, Python programming language is used. Additionally, some specific libraries developed in Python (like: Pandas, NumPy, Scikit-learn, etc.) that facilitate the implementation of data science and artificial intelligence are incorporated [30, 31].

4.1 Malicious URL Detection

For the purpose of malicious URL detection, each model is scrutinized by measuring the model's accuracy as well as the corresponding confusion matrices.

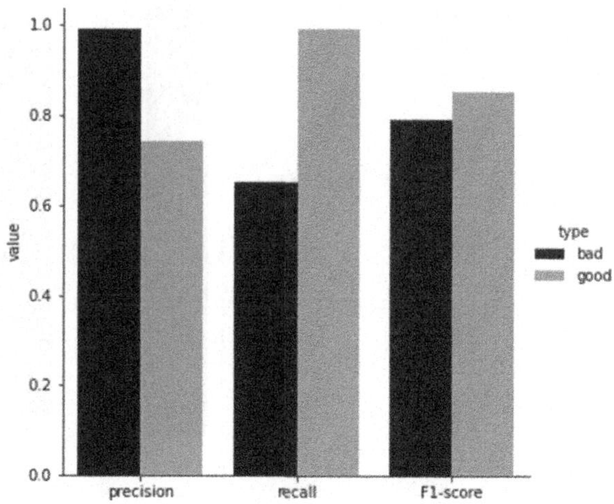

Fig. 1 Outcome of a linear classifier (malicious URL detection)

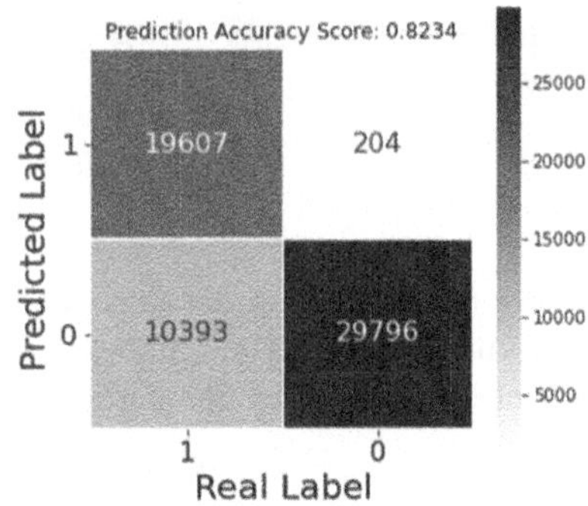

Fig. 2 Confusion matrix for a linear classifier (malicious URL detection)

Figure 1 depicts the accuracy of the linear classifier. As shown in Fig. 1, the performance of this model is not ideal.

The corresponding confusion matrix is shown in Fig. 2. The model provides good performances regarding the truly predicted labels. However, the total accuracy is lower compared to the truly predicted labels as a result of the low accuracy of the false predictions.

Figure 3 depicts the accuracy of the multinomial naive Bayes classifier. As it can be seen in Fig. 3, this model presents a massive improvement from the previous.

The corresponding confusion matrix is presented in the Fig. 4. The overall accuracy is solid 96.23% which is satisfactory. In this case, the URL labels are mainly predicted correctly. However, there is always a space for an additional improvement and the focus is to push this result couple of decimals upper.

Figure 5 demonstrates the performance of the logistic regression model.

In this situation it can be inferred that this model falls short in certain parts of the duty, but a remarkable result for the recall of the good URLs is achieved, which

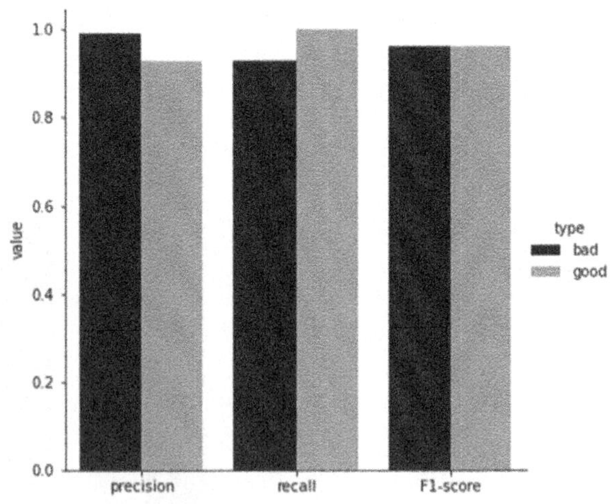

Fig. 3 Outcome of a multinomial naive Bayes classifier (malicious URL detection)

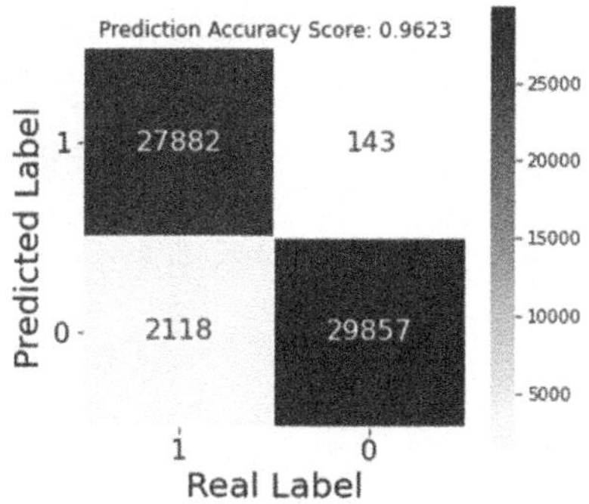

Fig. 4 Confusion matrix for a multinomial naive Bayes classifier (malicious URL detection)

Fig. 5 Outcome of a logistic regression (malicious URL detection)

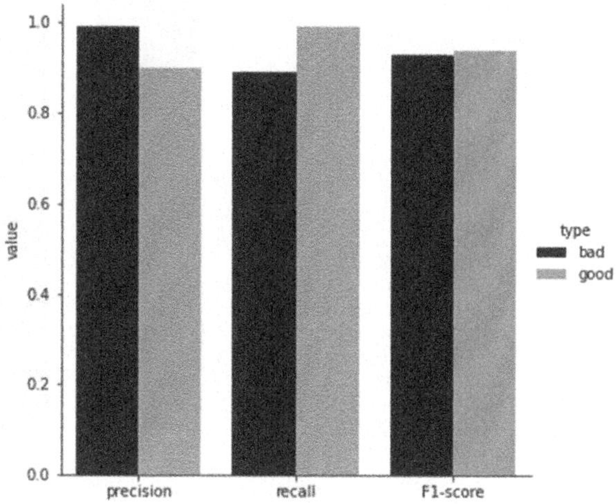

is 99%. All in all, this is a better model than the linear classifier, but apparently a worse model than the multinomial naive Bayes model for the data. The corresponding confusion matrix is shown in Fig. 6.

The models can be improved. Therefore, the results that were obtained in the preceding step do not have to be the final outcomes of the experiments. It can be experimented with the inner parameters of the models in order to aggrandize the performance of the algorithms. The parameters that can be adjusted are different on each model.

The particular model shown after adjustment is the multinomial naive Bayes algorithm, since the best results were obtained using that model. After a thorough investigation and research, the learning parameter of the model (the smoothing parameter) has been tuned so it can render even higher, more satisfying results at the end of the process.

The final results of the advanced algorithm are displayed in the Fig. 7. As shown on Fig. 7 the metrics are enhanced.

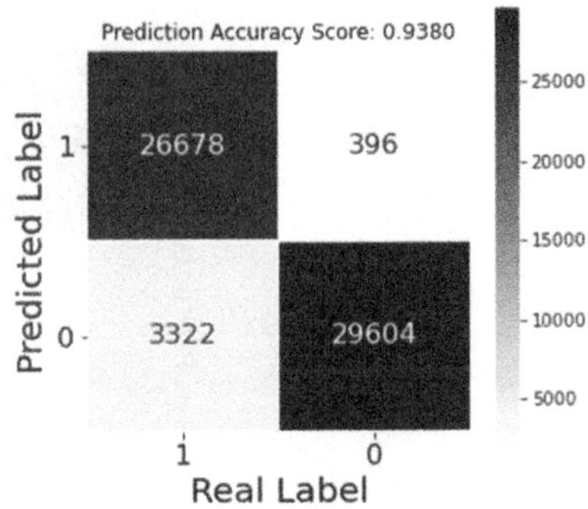

Fig. 6 Confusion matrix for a logistic regression (malicious URL detection)

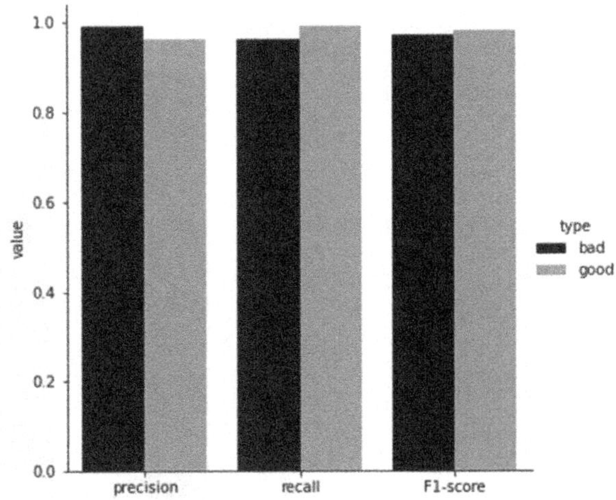

Fig. 7 Outcome of a multinomial naive Bayes classifier (after adjustment) (malicious URL detection)

As it can be seen from the confusion matrix (shown in Fig. 8) the accuracy of this model is 97.44%. It can be concluded that where all of the models slightly lack is the

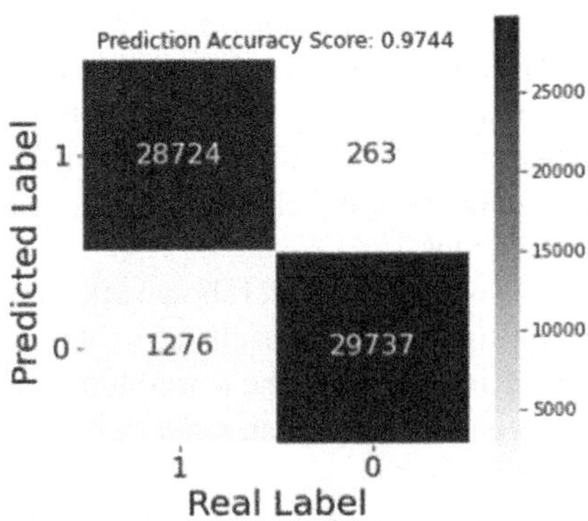

Fig. 8 Confusion matrix for a multinomial naive Bayes classifier (after adjustment) (malicious URL detection)

part where the model needs to predict bad (malicious) URL, however a malicious URL is not predicted at the end. Merely, looking at the last matrix, the model that has been optimized. The good URLs are usually classified as good and the bad are usually classified as bad, but still there are 1191 URLs from the dataset that were mistakenly predicted. That is the just a verification that the models, even well-designed, there is a chance that they can make mistakes.

The improvement of the ultimate results very much depends of the number of samples in the dataset and the core nature of the data. Nonetheless, the model that is pinpointed in this work provides satisfactory results and serves as a representable demonstration on how the machine learning is used in the cyber security.

4.2 URL Classification Based on the Content Provided

As it was discussed previously the models used in this task are the same as the models of malicious URL detection task. Nonetheless, the way how the data is organized is a little bit different, because this time there are five labels of the data. The final set comprises of 150,000 data samples each divided on five equal parts so each label has exactly 30,000 URLs attached to it.

Figure 9 depicts the accuracy of the linear classifier. It can be noted from the Fig. 9 that the performance is below the satisfactory level.

The corresponding confusion matrix is demonstrated in Fig. 10. As it can be seen from the confusion matrix, this is a pretty poor performance. The accuracy is down to 51.76%, and a different, more accurate model is needed to satisfy the task's requirements.

Figure 11 depicts the accuracy of the multinomial naive Bayes classifier. As it can be seen in Fig. 11, this model presents better results in every metric from the preceding case.

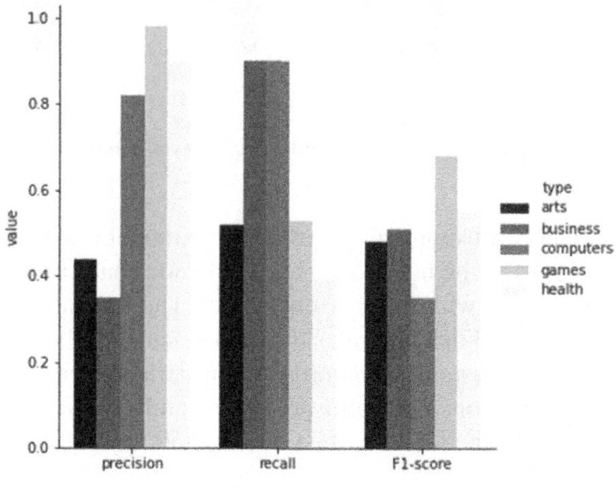

Fig. 9 Outcome of a linear classifier (URL classification)

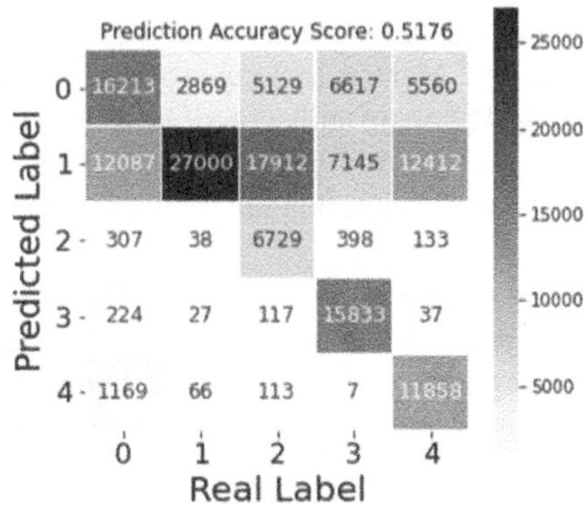

Fig. 10 Confusion matrix for a linear classifier (URL classification)

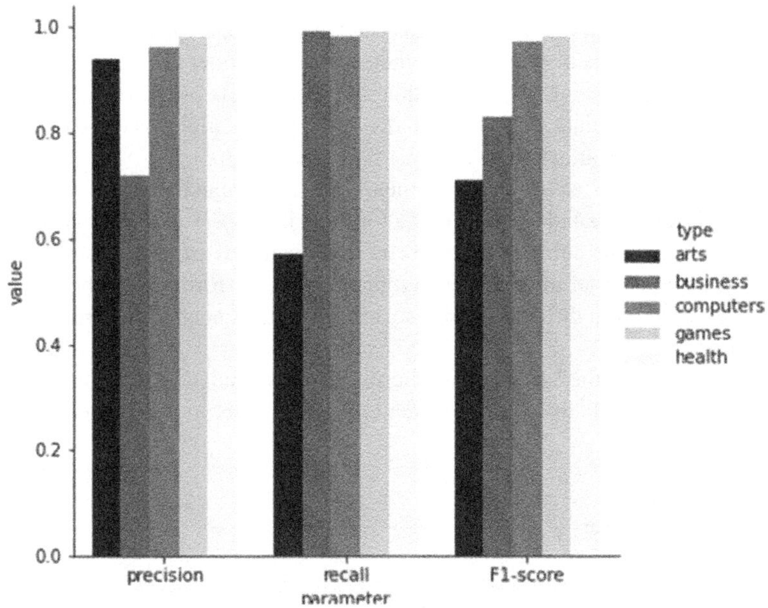

Fig. 11 Outcome of a multinomial naive Bayes classifier (URL classification)

The confusion matrix for the same model is shown in Fig. 12.

As it can be noticed fm the confusion matrix the results are massively better in comparison with the previous model. The final accuracy result is 90.16%.

Figure 13 demonstrates the performance of the logistic regression model.

The corresponding confusion matrix is shown in Fig. 14. As it can be seen from this regression, the results are significantly worse than the multinomial naive Bayes model and the final accuracy result is 50.59%.

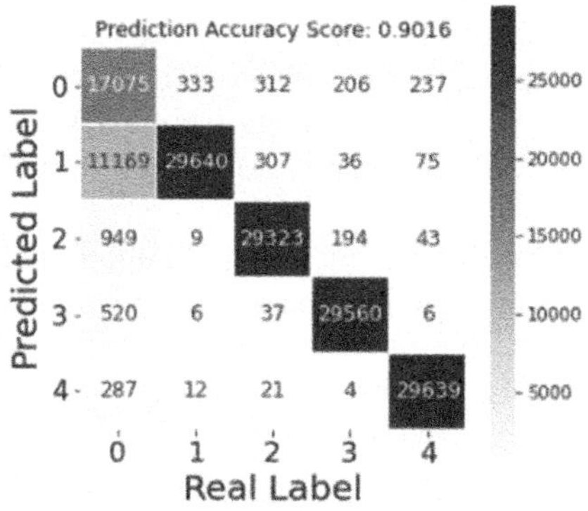

Fig. 12 Confusion matrix for a multinomial naive Bayes classifier (URL classification)

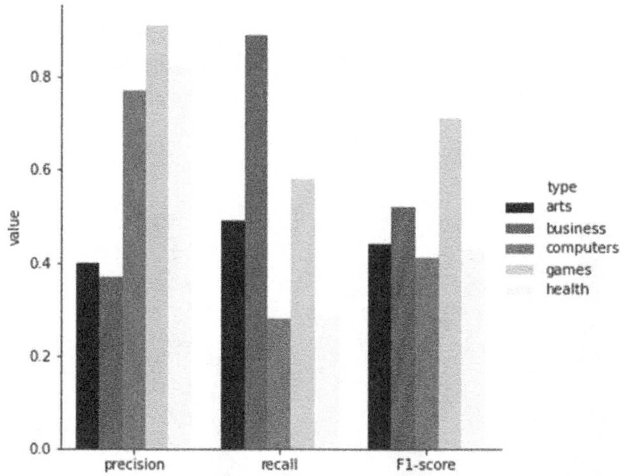

Fig. 13 Outcome of a logistic regression (URL classification)

Fig. 14 Confusion matrix for a logistic regression (URL classification)

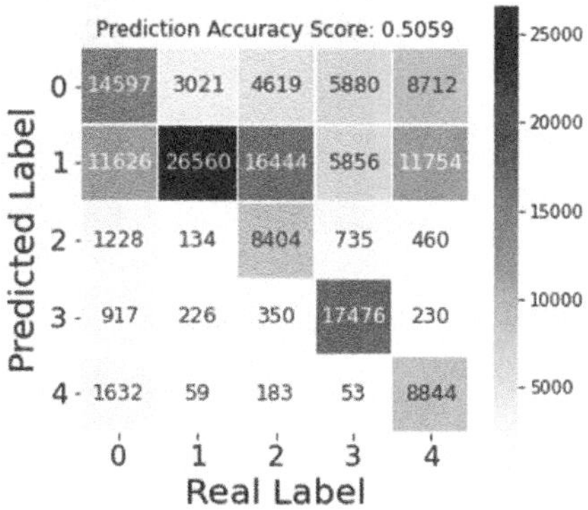

Having in mind all the results obtained with these prose approaches, it is needed to think of a way how the results can be improved, but at the same time have in mind the best results that could be produced using the available data. It could certainly be the case that from the available dataset the optimal results are not as high as we firstly planned to be.

After the investigation of the best model, the multinomial naive Bayes, and the tuning of its parameters (the smoothing parameter) the final performance is shown in Fig. 15.

The corresponding confusion matrix is shown in Fig. 16. It can undoubtedly be concluded that this example provides a decent percent of accuracy, and the new model designed after the adjustment gives accuracy level of 90.46%, which is a slight improvement from the first multinomial naive Bayes model.

It has been demonstrated how one approach differently affects two powerful dataset. In the first case both of the groups of labeled samples, the good and bad URLs, were separated with almost perfection, but in the second case, even though a relatively high percentage has been achieved is definitely not as perfect as the first one. The goal is to maximize the performance with the resources provided.

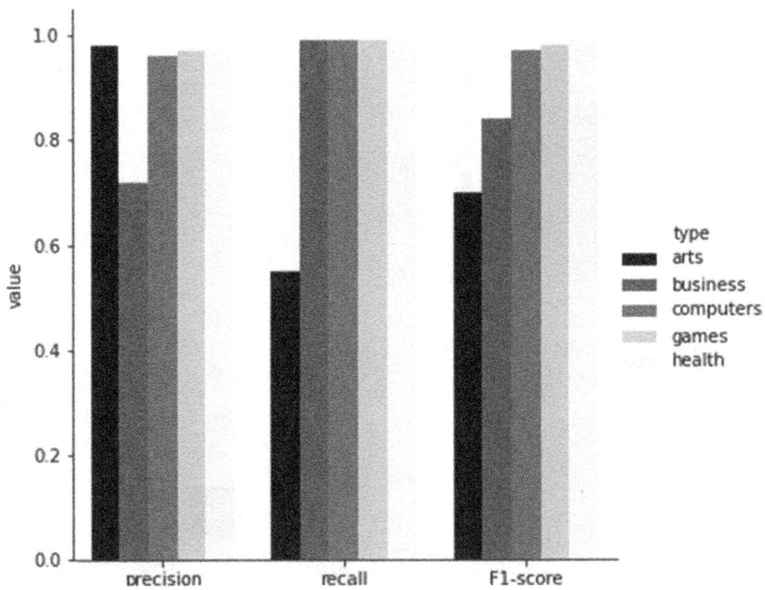

Fig. 15 Outcome of a multinomial naive Bayes classifier (after adjustment) (URL classification)

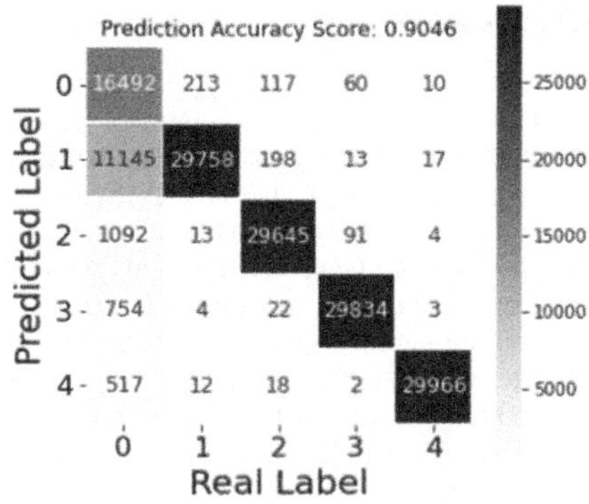

Fig. 16 Confusion matrix for a multinomial naive Bayes classifier (after adjustment) (URL classification)

5 Conclusion

All the machine learning tools are becoming vital in the field of cyber security. Nowadays it is almost unimaginable to think of cyber security protocols without implementing some kind of machine learning. It is definitely the way how the future is going to unfold. Constantly, there are different techniques that the malicious attackers develop to steal and manipulate the data, but so there are different techniques to stop them. There will possibly be some way the attackers might develop mighty algorithms to break into someone's system. However, that is why and when the artificial intelligence steps in. By implementing those algorithms, data science makes the algorithm learn by itself, so it will halt all the malicious algorithms trying to invade someone's computer.

Thus, there are definitely some certainties and some uncertainties about the future of the cyber security and artificial intelligence. It is certainly known that these years we are living in right now could be just the introduction of the data science on the big stage of cyber security. Those tools are more and more considered everywhere (in every field of the professional development). However, the real essence of the tools remains a partial enigma. It is true that we have the current models in play, but for sure they will evolve. What it is not certain is the direction and intensity they will evolve. The models that were designed in this paper are just one small and tiny example of the arsenal of the machine learning.

The advantages are enormous, but the models that will be developed should provide some security by themselves. What we try to say with that is what has been analyzed earlier, the attackers are smart, so smart to create an algorithm that can destroy a whole security system and can peek into someone's information. That implies they will also look for disadvantages in some of the models, because they would like to turn the models around and use the already designed machine learning algorithms to protect someone's system, as a potential threat to the customers. Therefore, the security machine learning algorithms need not only to provide security, but also to be secured themselves [32].

References

1. Park, K.: The internet as a complex system. The internet as a large-scale complex system (2005)
2. Pitts, V.: Cyber crimes: history of world's worst cyber attacks. Alpha Editions (2016)
3. Bendovschi, A.: Cyber-attacks–trends, patterns and security countermeasures. Proc. Econ. Financ. **28**, 24–31 (2015)
4. Bar-Yam, Y.: General features of complex systems. Encycl. Life Support Syst. **1**, 1–10 (2002)
5. Dimirovski, G.M.: Complex systems: relationships between control, communications and computing. In: Studies in Systems, Decision and Control. Springer (2016)
6. Meza, J., Campbell, S., Bailey, D.: Mathematical and statistical opportunities in cyber security (2009). arXiv:0904.1616
7. Ficco, M., Choraś, M., Kozik, R.: Simulation platform for cyber-security and vulnerability analysis of critical infrastructures. J. Comput. Sci. **22**, 179–186 (2017)
8. Chatterjee, S., Thekdi, S.: An iterative learning and inference approach to managing dynamic cyber vulnerabilities of complex systems. Reliab. Eng. Syst. Saf. 193 (2020)
9. Sarker, I.H., Kayes, A.S.M., Badsha, S., et al.: Cybersecurity data science: an overview from machine learning perspective. J. Big. Data. **7**, 41 (2020)
10. Sikos, L.F., Choo, K.-K.R.: Data science in cybersecurity and cyberthreat intelligence. Springer, Perth, WA, Australia (2017)
11. Sahoo, D., Liu, C., Hoi, S.C.H.: Malicious URL detection using machine learning: a survey. Singapore Management University, IEEE (Aug 2019)
12. Weedon, M., Tsaptsinos, D., Denholm-Price, J.: Random forest explorations for URL classification. Faculty of SEC, Kingston University, Kingston Upon Thames, United Kingdom, IEEE, London, UK (Oct 2017)
13. Kumar, R., Zhang, X., Tariq, H.A., Khan, R.U.: Malicious URL detection using multi-layer filterinag model. In: 14th International Computer Conference on Wavelet Active Media Technology and Information Processing (ICCWAMTIP). IEEE, Chengdu, China (Feb 2018)
14. Singh, N., Sandhawalia, H., Monet, N., Poirier, H., Coursimault, J.-M.: Large scale URL-based classification using online incremental learning. In: 11th International Conference on Machine Learning and Applications, IEEE, Boca Raton, FL, USA (Jan 2013)
15. Chakraborty, G., Lin, T.T.: A URL address aware classification of malicious websites for online security during web-surfing. In: IEEE International Conference on Advanced Networks and Telecommunications Systems (ANTS). IEEE, Bhubaneswar, India (June 2018)
16. Abdessamed, O., Zakaria, E.: Web site classification based on URL and content: Algerian vs. non-Algerian case. In:12th International Symposium on Programming and Systems (ISPS). IEEE, Algiers, Algeria (Sep 2015)
17. Kumar, J., Santhanavijayan, A., Janet, B., Rajendran, B., Bindhumadhava, B.S.: Phishing website classification and detection using machine learning. In: International Conference on Computer Communication and Informatics (ICCCI). IEEE, Coimbatore, India (June 2020)
18. Maladkar, K.: Content based hierarchical URL classification with convolutional neural networks. In: International Conference on Information Technology (ICIT). IEEE, Bhubaneswar, India (Mar 2020)
19. Sun, Y., Peng, M., Zhou, Y., Huang, Y., Mao, S.: Application of machine learning in wireless networks: key techniques and open issues. IEEE ResearchGate (Mar 2019)
20. VanderPlas, J.: Python data science handbook; essential tools for working with data. O'Reilly Media, Inc., 1005 Gravenstein Highway North; Sebastopol; CA 954722015, California, United States of America (Dec 2016)
21. Chio, C., Freeman, D.: Machine learning and security; protecting systems with data and algorithms. O'Reilly Media, Inc., 1005 Gravenstein Highway North; Sebastopol; CA 95472, California, United States of America (Feb 2018)
22. Jagannath, J., Polosky, N., Jagannath, A., Restuccia, F., Melodia, T.: Machine learning for wireless communications in the internet of things: a comprehensive survey. ResearchGate (Jan 2019)

23. Luo, F.-L.: Machine learning for future wireless communications. John Wiley & Sons Ltd., Silicon Valley, California, USA (2000). (Ph.D., IEEE Fellow)
24. Martin, K.: Cryptography: The Key to Digital Security, How it Works, and Why it Matters. W. W. Norton & Company (2020)
25. Sagduyu, Y.E., Shi, Y., Erpek, T., Headley, W., Flowers, B., Stantchev, G., Lu, Z.: When wireless security meets machine learning: motivation, challenges, and research directions. IEEE (Jan 2020)
26. Yuan, G.-X., Chia-Hua, H., Chih-Jen, L.: Recent advances of large-scale linear classification. Proc. IEEE **100**(9), 2584–2603 (2012)
27. Rish, I.: An empirical study of the naive Bayes classifier. In: IJCAI 2001 Workshop on Empirical Methods in Artificial Intelligence, vol. 3, no. 22 (2001)
28. Kleinbaum, D.G., et al.: Logistic regression. Springer, New York (2002)
29. Baykan, E., Henzinger, M., Marian, L., Weber, I.: Purely URL-Based Topic Classification. Fédérale de Lausanne Lausanne, Switzerland (Apr 2009)
30. Le, H., Pham, Q., Sahoo, D., Hoi, S.C.H.: URLNet: learning a URL representation with deep learning for malicious URL detection. Singapore Management University, IEEE (Mar 2018)
31. Poole, D.L., Mackworth, A.K.: Python code for artificial intelligence: foundations of computational agents, Version 0.7.3. (May 2018)
32. Nordrum, A.: Everything you need to know about 5G. IEEE Spectr. https://spectrum.ieee.org. (Jan 2017)

Modelling of Priority Buffering Systems Applicable for Commercial Mobile Networks

Strahil Panev and Pero Latkoski

Honoring Professor Georgi M. Dimirovski on his anniversary.

Abstract Software Defined Networking (SDN) is a popular technology paradigm that is embedded in the basic architecture of the 5th Generation (5G) of mobile networks. Today, OpenFlow is the most common protocol used on the southbound interface. OpenFlow switches generally follow two types of buffering mechanisms: (1) single buffer that is used to handle both control and user plane; (2) two different priority buffers, each serving the control and user plane packets. We try to examine and quantify the average packet loss rate of the two different buffer design principles, by developing an analytical proposal that incorporates Quasi-Birth–Death (QBD) processes. The proposed numerical model is also verified via extensive simulations in MATLAB. The obtained results clearly show that using priority buffering in the SDN switches increases the performance significantly, when compared to traditional shared buffering. When the probability of Packet-In messages is low, the arrival rate is increasing, and as the number of Mobile Nodes (MN) goes up, the priority buffering clearly outperforms the single buffering, in most of the scenarios by nearly 99% of lower packet losses. The obtained results can be used for predicting the average packet loss rate when designing OF-based mobile core networks.

S. Panev (✉)
Faculty of Computer Science, International Slavic University "G. R. Derzhavin", Sv. Nikole, North Macedonia
e-mail: strahil.panev@msu.edu.mk

P. Latkoski
Faculty of Electrical Engineering and Information Technologies, Ss. Cyril and Methodius University, Skopje, North Macedonia
e-mail: pero@feit.ukim.edu.mk

P. Shi et al. (eds.), *Complex Systems: Spanning Control and Computational Cybernetics: Applications*, Studies in Systems, Decision and Control 415,
https://doi.org/10.1007/978-3-031-00978-5_23

1 Introduction

Software Defined Networking (SDN) and Network Function Virtualization (NFV) are the key technologies used by mobile operators to cope with traditional challenges, such as high operational costs, vendor lock-in, slow time to market etc. SDN separates the control and user plane and brings the benefit of having open and easy programmability and application development, whereas NFV allows network operators to manage and expand their network capabilities on demand using virtual, software-based applications where physical boxes once stood in the network architecture [1]. Today, the most widely spread protocol for control-user plane communication is OpenFlow (OF) [2]. There is a typical sequence of events taking place if a packet flow cannot be matched into a flow table: (1) the packet is first sent to the controller, (2) the controller makes a decision regarding the network path of the packet and sends update instructions to all the involved switches, (3) all the subsequent packets are afterwards routed directly in the user plane without controller contact.

In existing Long-Term Evolution (LTE) networks, SDN is widely used in data centers and the mobile core network [3]. Initially, the concept of SDN was introduced in the Evolved Packet Core (EPC), by separating the Packet Data Network Gateway (PGW) and the Serving Gateway (SGW) into specific logical parts that belong to the control and user plane. It is important to note that SDN is embedded in the heart of the new 5G architecture since the very beginning, bringing many benefits for the newly defined 5G use cases. One of the greatest advantages of introducing SDN in mobile core networks is the reduction of the overall latency. In an SDN-based core network, in case of an MN (Mobile Node) handover procedure, specific OF messages are exchanged between the user and the control plane [4]. The so called "hard handover", includes a breakage of the existing user session followed by a set of OF signaling messages that include reconfiguration and management of network entities.

A buffer in the OF-switch is typically used to temporary store the incoming packets before the switch can actually serve them. The capacity of the output buffers and the buffer design in the OF-switch, have a high impact on the average switch service time. Regarding the buffer design, the most straightforward implementation today is the use of a single buffer that is shared by all ports. This is the cheapest and least complex solution that is widely popular in real implementations. However, there are proposals for other types of buffer designs, that usually include the use of two isolated separate buffers, one only for the data plane, and the other exclusively for the control plane.

The work in this article is an extension of our previous work, where our aim was to quantify the total handover delay experienced by MN and compare this delay when using a traditional single buffer switch with a two-buffer priority switch design. In this work, we extend the mathematical modelling of our previous work, with aim to quantify the packet loss rate of the two already created systems that have different buffer designs. The novelty of the work presented here lies in the applicability of mathematical modelling to shape a traditional buffering system, and a more advanced buffering mechanism that uses packet prioritization. The main contributions of this article, can be summarized in the following points:

(1) A mathematical model based on queuing theory is proposed. This analytical approach models the controller as M/M/1, and the switch as GI/M/1/K queue. The QBD processes are used to define the stationary behavior of the network, which later allows for performance metric calculation,
(2) Extensive simulations are performed to verify the numerical results. Key factors that impact the packet loss rate are analyzed, such as arrival rate, number of MNs, and controller-to-switch mean service rate ratio,
(3) The results obtained can be easily used by network designers when planning an SDN-based core network.

The rest of the article is organized in the following way: In Sect. 2, we review the past literature in the area of interest, and in Sect. 3 we define the two compared systems and the analytical model. In Sect. 4, we present the results and we comment in detail on the findings. Finally, in Sect. 5 we summarize the highlights of this research.

2 Previous Work

The very first analytical modelling of SDN networks was done by Jarchel et al. [5], where the authors propose an analytical model which was later verified via a hardware testbed. The model uses a M/M queue for the forwarding system at the switch, whereas a M/M-S queue is used to model the feedback system of the controller. Only a single switch and a single controller are modelled, which is the major drawback of the proposed mathematical approach. In [6], the authors use a Jackson network to mimic the behavior of the entities in an SDN network. They are interested in quantifying the average service time and the time distribution spent by the packets. The results determine the upper limit of the average service time at the switch, and the upper limit of the queue at the controller. The main disadvantage of the approach is the single node model, and the fact that the mathematical apparatus is not verified via simulations. Mahmood et al. [7] are the first to introduce multiple switches in the user plane in the analytical modeling. They extend their original work in [8] by introducing *Packet-In* messages. The main parameter of interest is the controller's packet sojourn time, whereas the propagation delay is not analyzed at all. In [9], the authors model the *Packet-In* messages, and they analyze in detail the impact of these messages to the average service time at the controller. M/H_2 queue is used for modelling of the switches, whereas the controller is modelled as M/M. The greatest shortcoming of this work is the fact that the authors do not verify their modelling by using simulations.

So far, all the described references take into consideration a single shared buffer that is used for both the user and control plane. The reason for this design choice in the analytical modeling, is the fact that it greatly simplifies the mathematical equations, however this is not accurate because of the implicit assumption of threating in the same way both the incoming packets and the packets that are returned by the controller. Moreover, if the OF-switch uses finite buffering [10], the Markovian properties of the buffer queues are not preserved. The latter is the main condition for obtaining product-form analysis [11]. A network with finite buffering can have its stationary behavior described only by using global-balance equations [12].

Yen et al. [13], use a two-stage tandem network by using the M/M/1 queue, however their proposed model does not clearly separate the control and user plane traffic. The authors in [14] perform experiments which prove that using M/M/1 queue for the OF-switches is not recommended, instead they consider the M/G/1 model and discuss that using this type of queue mimics much better the real SDN implementations. By applying similar thinking, in [15] the authors use MMPP/1/1 queue for both the switches and the controller, whereas the creators of [16] deploy the M/Geo/1 model. However, both do not take into consideration a distinction of the control and user plane packets.

Miao et al. [17] deploy preemptive priority queuing at the switches and model them by using infinite buffers. Best to our knowledge, this is the first article that deals with different buffer designs, by analyzing single shared buffering and comparing the results with a more advanced priority buffering mechanism. The results clearly show that the average service time of the switches and the overall network performance is significantly improved when deploying priority buffering, yet again at the cost of increased complexity. The authors extend their own modeling in [18], by assuming a bursty multimedia traffic, and by modelling the high priority queue as finite, and the low priority queue as infinite. The analytical modeling in [19, 20] model the low and high priority queues as finite by using the GI/M/1/K queue and by taking advantage of the QBD processes. The proposed model implements non-preemptive prioritization, which means that the packets in the queue with lower priority are served only in the case when there are no packets to be served in the high priority queue. The proposed mathematical approach is much more complex and product-form analysis cannot be used [21], instead the network is modelled by using global-balance equations [22].

In our proposal, we model the switches by using finite buffering by taking advantage of the GI/M/1/K queue, whereas the controller is modelled as M/M/1 queue, with infinite buffer capacity. The design implements two isolated priority queues, one for the user plane, the other for the control plane. We make use of the QBD processes approach to obtain the stationary state distribution of the network, which allows us to calculate the average packet loss rate.

3 Analytical Model

InIn our previous work [23], we proposed a mathematical approach to model the OF-based signaling messages exchanged within the procedure of "hard" handover. We compared two systems that use different OF-switch buffer designs, single shared buffer and two isolated buffers that incorporate prioritization. The aim was to quantify the handover delay experienced by the MN when moving from one switch to another within the SDN-based mobile core network. In this work, we are only interested in quantifying the average packet loss of the two systems.

Two types of messages are modelled: *Packet-In* and *Port-Status*. If an incoming packet cannot be matched against an existing flow table entry, then it is sent to the controller via a *Packet-In* message, whereas a *Port-Status* message is sent to the controller in case MN connects or disconnects to a port of the switch. The handover scenario of interest involves a single MN that disconnects from one switch (s_i), and due to its mobility, connects to another switch (s_j). The first event triggers a *Port-Status (off-port)* message that is sent the controller, informing it of MN's mobility, whereas the second event triggers a second *Port-Status* (*on-port*) message to indicate that the MN is now connected to the switch s_j. When the controller receives any of the *Port-Status* messages, it immediately sends a set of *Flow-mod* messages to update the flow tables in all the switches involved in the packet path. The generated sequence of signaling messages is shown in Fig. 1, where $^{(1)}$ notes the *off-port* signaling sequence, whereas $^{(2)}$ notes the *on-port* messaging sequence. The handover is modeled as follows: (1) the start of the handover is indicated by sending an *off-port* message by the source switch s_i ($Msg = Ps^{(1)}$); (2) the controller sends a set of *Flow-Mod* ($Msg = Fm^{(1)}$) messages to all the switches on the packet path; (3) the destination switch $s_j$sends an *on-port* message ($Msg = Ps^{(2)}$); (4) the controller sends a new set of *Flow-Mod* ($Msg = Fm^{(2)}$) to all involved switches to inform them about MN's

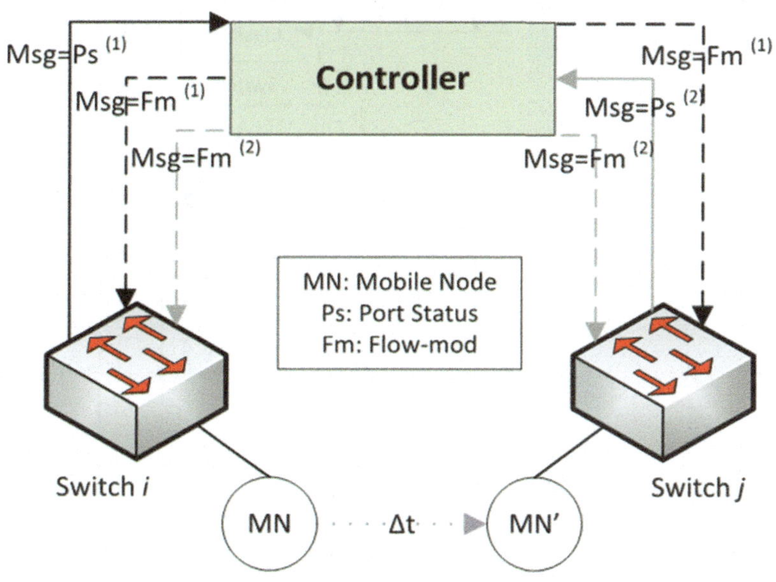

Fig. 1 Mobility triggered exchange of messages

mobility; (5) the handover procedure is finished when the *Flow-Mod* message is received at switch s_j and the respective flow table is updated.

Figure 2 shows the two system designs, Fig. 2a depicts the model called *Shared Finite Buffering (Model SFB)*, Fig. 2b gives the model called *Priority Finite Buffering (PFB)*. Model SFB deploys a switch with a single shared buffer without priority, where the traffic entering the switch is the sum of the traffic returned by the controller, and the external traffic of the incoming packets from the outside. This type of queue is described as GI/M/1/K queue (finite buffer), as the arrival rate follows a general distribution due to the mix of traffic that has independent arrival rates. Model SFB

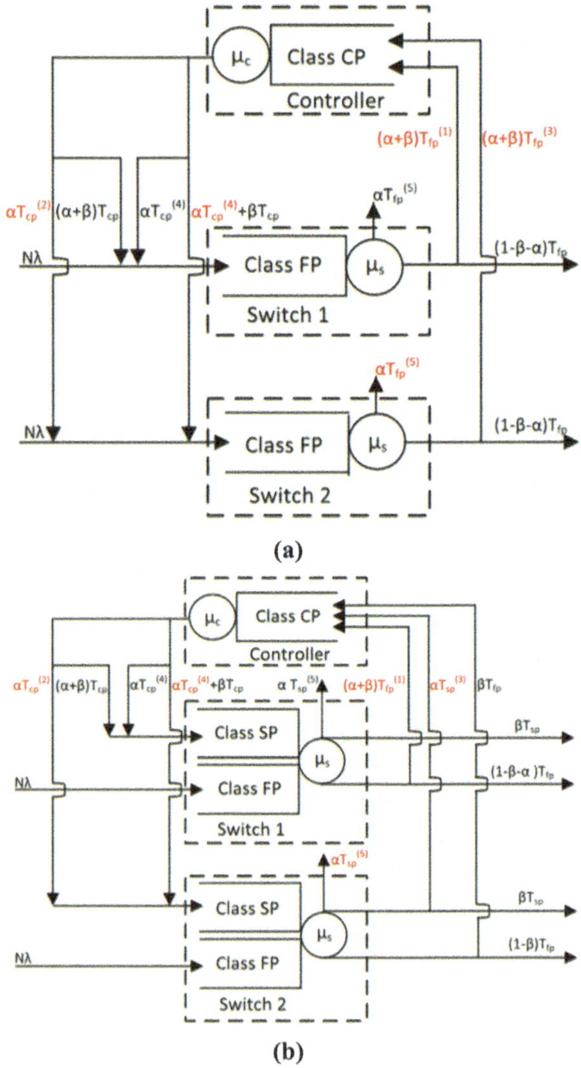

Fig. 2 Systems used for handover modelling at the switch: **a** Model SFB, **b** Model PFB

has two Classes: *Fast Path* (FP), used by the switch for processing all packets, and *Controller Path* (CP) for all the packets processed by the controller. The Class CP in this case is modelled as M/M/1 queue (infinite queue) and has a mean service rate of μ_c, and the Class FP is modelled with a mean service rate of μ_s. Model PFB incorporates priority queuing, and each switch has two classes: Class Slow Path (SP), that is used to receive the packets from the controller and to forward the *Port-Status* messages, and Class Fast Path (FP), where a packet is only processed in case there are no packets in the Class SP. μ_s is the mean service rate of the switch that is shared by both classes. We use non-preemptive priority queuing, meaning that Class FP is served only when Class SP has no packets. N is the number of MNs connected to a single switch, while K is the finite size of the buffer at the switch for Model SFB. For Model PFB the total queue capacity of the switch is K, Class SP has queue capacity of K_1, whereas Class FP has K_2, such that $K = K_1 + K_2$. Furthermore, we use β to indicate the probability of *Packet-In* messages sent to the controller when no match is found in the flow table of the switch, and α to indicate the probability of *Port-Status* messages, which are sent from the switch indicating the start of the handover procedure.

The basic idea is to design a system that will allow us to mimic a handover behavior. The relevant sequence of OF-messages exchanged due to handover signaling, is marked with red on Fig. 2 for Model SFB and for Model PFB. The sequence of events for Model SFB is designed as follows: (1) a *Port-Status (off-port)* message is received at switch 1, it is processed in the common Class FP at switch 1, and sent to the controller ($\alpha T_{fp}{}^{(1)}$,where T_{fp} is the throughput of Class FP); (2) the *Port-Status* message is then processed by the common Class CP, forwarded to switch 1, and an extra packet is generated to inform switch 2 about the *off-port* for the MN at switch 1 ($\alpha T_{cp}{}^{(2)}$, where T_{cp} is the throughput of Class CP); (3) we assume a totally synchronized system, meaning that immediately upon receiving the *off-port* message switch 2 knows that an *on-port* message is received at Class FP at switch 2, and this message is similarly forwarded from switch 2 to the controller ($\alpha T_{fp}{}^{(3)}$); (4) this packet is then processed by the controller and sent to inform both switch 1 and switch 2 about the change in the network ($\alpha T_{cp}{}^{(4)}$); (5) after the packet of interest is processed at switch 2 (and switch 1), it is dropped ($\alpha T_{fp}{}^{(5)}$). The only difference in the sequence of messages of model PFB compared with model SFB, is that when the packet is returned from the controller, it now enters a specialized buffer of Class SP, and is immediately processed and sent either back to the controller, or on the external output path.

The SFB system is architected as a continuous time Markov chain $\{n_{cp}(t), n_{fp1}(t), n_{fp2}(t)$, where $t > 0\}$, where $n_{cp}(t), n_{fp1}(t)$, and $n_{fp2}(t)$, represent the number of packets in the controller, switch 1, and switch 2, respectively. We take x, y, z to present a set of values for $n_{cp}(t), n_{fp1}(t)$, and $n_{fp2}(t)$, where $x \in Z_+$, $y \in Z_+^{(\leq K)}$, and $z \in Z_+^{(\leq K)}$. The allowed transitions of the Markov process are given in Table 1. The PFB system is similarly architected as a continuous time Markov chain $\{n_{cp}(t), n_{sp1}(t), n_{fp1}(t), n_{sp2}(t), n_{fp2}(t)$ where $t > 0\}$, where $n_{cp}(t), n_{sp1}(t), n_{fp1}(t), n_{sp2}(t)$, and $n_{fp2}(t)$ represent the number of packets in the controller, switch 1 (Class SP), switch 1 (Class FP), switch 2 (Class SP), and switch 2 (Class FP), respectively. We take v, w, x, y, z to

Table 1 Permissible transitions for Model SFB

	From	To	Rate
Packet arrives at switch 1	(x, y, z)	(x, y + 1, z)	$N\lambda$
Packet arrives at switch 2	(x, y, z)	(x, y, z + 1)	$N\lambda$
Packet forwarded from switch 1 to controller(1)	(x, y > 0, z)	(x + 1, y-1, z)	$(\alpha + \beta)\mu_s$
Packet serviced from controller to switch 1	(x > 0, y, z)	(x−1, y + 1, z)	μ_c
Packet serviced from controller to switch 2(2)	(x > 0, y, z)	(x−1, y, z + 1)	$\alpha\mu_c$
Packet forwarded from switch 2 to controller(3)	(x, y, z > 0)	(x + 1, y, z−1)	$(\alpha + \beta)\mu_s$
Packet serviced from controller to switch 1(4)	(x > 0, y, z)	(x−1, y + 1, z)	$\alpha\mu_c$
Packet serviced from controller to switch 2(4)	(x, y, z > 0)	(x−1, y, z + 1)	μ_c
Packet dropped at switch 1(5)	(x, y > 0, z)	(x, y−1, z)	$\alpha\mu_s$
Packet dropped at switch 2(5)	(x, y, z > 0)	(x, y, z−1)	$\alpha\mu_s$
Packet departs from switch 1	(x, y > 0, z)	(x, y−1, z)	$(1-\beta-\alpha)\mu_s$
Packet departs from switch 2	(x, y, z > 0)	(x, y, z−1)	$(1-\beta-\alpha)\mu_s$

present a set of values for $n_{cp}(t)$, $n_{sp1}(t)$, $n_{fp1}(t)$, $n_{sp2}(t)$, and $n_{fp2}(t)$, where $v \in Z_+$, $w \in Z_+^{(\leq K1)}$, $x \in Z_+^{(\leq K2)}$, $y \in Z_+^{(\leq K1)}$, $z \in Z_+^{(\leq K2)}$. The allowed transitions of the Markov process are given in Table 2.

In a small network operator (with ~20 Gbps total network throughput), the number of handover requests is around 40 pkt/sec on network level (data taken from a real mobile operator's statistics) for around 200 k simultaneously active users (with ongoing sessions). If we consider even a small throughput for a single MN of ~1Mbps (for MTU of 1500 Bytes, this is 83 pkt/sec), for the smallest value of $\beta = 0.01$ and one MN, this results in 0.83 pkt/s. So, for a single MN, when comparing the worst case of 0.83 pkt/s for the *Packet-In* messages, and an average of 0.0002 pkt/s (40 pk/s divided by 200 k users) for the *Port-Status* messages, it is obvious that $\alpha << \beta$ and $\alpha << 1$. Therefore, we consider $(\alpha + \beta)\mu_s \sim \beta\mu_s$, $(1-\beta-\alpha)\mu_s \sim (1-\beta)\mu_s$, and $(1 + \alpha)\mu_c \sim \mu_c$. We use these approximations when calculating the sub-matrices of the QBD process, and we use the permissible states in Tables 1 and 2 to define the sub-matrices of the transition rate generator matrix Q. The matrices are then input to Matrix-Analytic Methods (MAM) to compute the stationary distribution probabilities. The details of generating the submatrices are explained in [19, 23], and will not be repeated here.

The main performance metric of interest is the packet loss rate. With aim to quantify the average packet loss rate, we need to calculate the stationary distribution probability, π. For model SFB, the stationary distribution probability $\pi_{x,y,z}$ is to have x packets in the controller, y packets in Class FP/Switch 1, and z packets in Class FP/Switch 2. Similarly, for Model PFB, the stationary distribution probability $\pi_{v,w,x,y,z}$ is having v packets in the controller, w packets in Class SP/Switch 1, x packets in Class FP/Switch 1, y packets in Class SP/Switch 2, and z packets in Class FP/Switch 2. The throughput of each Class is the sum of probabilities that the Class has at least one packet to forward with the mean service rate of μ_s for the switch, and μ_c for the

Table 2 Permissible transitions for Model PFB

	From	To	Rate
Packet arrives at switch 1, Class FP	(v, w, x, y, z)	(v, w, x + 1, y, z)	$N\lambda$
Packet arrives at switch 2, Class FP	(v, w, x, y, z)	(v, w, x, y, z + 1)	$N\lambda$
Packet forwarded from switch 1 to controller, Class FP(1)	(v, 0, x > 0, y, z)	(v + 1, 0, x−1, y, z)	$(\alpha + \beta)\mu_s$
Packet serviced from controller to switch 1, Class SP	(v > 0, w, x, y, z)	(v−1, w + 1, x, y, z)	μ_c
Packet serviced from controller to switch 2, Class SP(2)	(v > 0, w, x, y, z)	(v−1, w, x, y + 1, z)	$\alpha\mu_c$
Packet forwarded from switch 2 to controller, Class SP(3)	(v, w, x, y > 0, z)	(v + 1, w, x, y−1, z)	μ_s
Packet forwarded from switch 2 to controller, Class FP	(v, w, x, 0, z > 0)	(v + 1, w, x, 0, z−1)	$\beta\mu_s$
Packet serviced from controller to switch 1, Class SP(4)	(v > 0, w, x, y, z)	(v−1, w + 1, x, y, z)	$\alpha\mu_c$
Packet serviced from controller to switch 2, Class SP(4)	(v > 0, w, x, y, z)	(v−1, w, x, y + 1, z)	$(1 + \alpha)\mu_c$
Packet dropped at switch 1, Class SP(5)	(v, w > 0, x, y, z)	(v, w−1, x, y, z)	$\alpha\mu_s$
Packet dropped at switch 2, Class SP(5)	(v, w, x, y > 0, z)	(v, w, x, y−1, z)	$\alpha\mu_s$
Packet departs from switch 1, Class FP	(v, 0, x > 0, y, z)	(v, 0, x−1, y, z)	$(1-\beta-\alpha)\mu_s$
Packet departs from switch 2, Class FP	(v, w, x, 0, z > 0)	(v, w, x, 0, z−1)	$(1-\beta)\mu_s$
Packet departs from switch 1, Class SP	(v, w > 0, x, y, z)	(v, w−1, x, y, z)	μ_s
Packet departs from switch 2, Class SP	(v, w, x, y > 0, z)	(v, w, x, y−1, z)	μ_s

controller. So, for Model SFB and Model PFB, the respective Class throughputs are given as:

$$T_{fp1}^{sfb} = \mu_s \sum_{x=0}^{\infty} \sum_{y=1}^{K} \sum_{z=0}^{K} \pi_{x,y,z}. \quad (1)$$

$$T_{fp2}^{sfb} = \mu_s \sum_{x=0}^{\infty} \sum_{y=0}^{K} \sum_{z=1}^{K} \pi_{x,y,z}, \quad (2)$$

$$T_{cp}^{sfb} = \mu_c \sum_{x=1}^{\infty} \sum_{y=0}^{K} \sum_{z=0}^{K} \pi_{x,y,z}. \quad (3)$$

$$T_{fp1}^{pfb} = \mu_s \sum_{v=0}^{\infty} \sum_{x=1}^{K2} \sum_{y=0}^{K1} \sum_{z=0}^{K2} \pi_{v,0,x,y,z}. \tag{4}$$

$$T_{sp1}^{pfb} = \mu_s \sum_{v=0}^{\infty} \sum_{w=1}^{K1} \sum_{x=0}^{K2} \sum_{y=0}^{K1} \sum_{z=0}^{K2} \pi_{v,w,x,y,z}, \tag{5}$$

$$T_{fp2}^{pfb} = \mu_s \sum_{v=0}^{\infty} \sum_{x=0}^{K2} \sum_{y=0}^{K1} \sum_{z=1}^{K2} \pi_{v,x,x,0,z}. \tag{6}$$

$$T_{sp2}^{pfb} = \mu_s \sum_{v=0}^{\infty} \sum_{w=0}^{K1} \sum_{x=0}^{K2} \sum_{y=1}^{K1} \sum_{z=0}^{K2} \pi_{v,w,x,y,z}, \tag{7}$$

$$T_{cp}^{pfb} = \mu_c \sum_{v=1}^{\infty} \sum_{w=0}^{K1} \sum_{x=0}^{K2} \sum_{y=0}^{K1} \sum_{z=0}^{K2} \pi_{v,w,x,y,z}, \tag{8}$$

where $T_{fp1}{}^{sfb}$, $T_{fp2}{}^{sfb}$, $T_{cp}{}^{sfb}$ is the throughput of Model SFB for Class FP/Switch 1, Class FP/Switch 2, and Class CP, respectively. Similarly, $T_{fp1}{}^{pfb}$, $T_{sp1}{}^{pfb}$, $T_{fp2}{}^{pfb}$, $T_{sp2}{}^{pfb}$ $T_{cp}{}^{pfb}$, is the throughput of Model PFB for Class FP/Switch 1, Class SP/Switch 1, Class FP/Switch 2, Class SP/Switch 2, and Class CP, respectively.

Next, we define the average packet loss rate. This parameter indicates the average number of packets that are blocked by any of the respective classes out of the total incoming packets. For Model SFB, for the Class FP/Switch 1 ($Pl_{fp1}{}^{sfb}$), and Class FP/Switch 2 ($Pl_{fp1}{}^{sfb}$), and for the Model PFB, for the Class FP/Switch 1 ($Pl_{fp1}{}^{pfb}$), Class SP/Switch 1 ($Pl_{sp1}{}^{pfb}$), Class FP/Switch 2 ($Pl_{fp2}{}^{pfb}$), Class SP/Switch 2 ($Pl_{sp1}{}^{pfb}$), the loss rates can be expressed as follows:

$$Pl_{fp1}^{sfb} = 1 - \frac{T_{fp1}^{sfb}}{N\lambda}, \tag{9}$$

$$Pl_{fp2}^{sfb} = 1 - \frac{T_{fp2}^{sfb}}{N\lambda}, \tag{10}$$

$$Pl_{fp1}^{pfb} = 1 - \frac{T_{fp1}^{pfb}}{N\lambda}, \tag{11}$$

$$Pl_{fp2}^{pfb} = 1 - \frac{T_{fp2}^{pfb}}{N\lambda}, \tag{12}$$

$$Pl_{sp1}^{pfb} = 1 - \frac{T_{sp1}^{pfb}}{T_{cp}^{pfb}}, \tag{13}$$

$$Pl_{sp1}^{pfb} = 1 - \frac{T_{sp2}^{pfb}}{T_{cp}^{pfb}}. \tag{14}$$

The total packet loss rate for Model SFB ($Pl_{tot}{}^{sfb}$) is the sum of the packet loss rates of the Class FP in both switches, whereas the total packet loss rate for Model PFB ($Pl_{tot}{}^{pfb}$) is the sum of the packet loss rates of the Class SP and Class FP in both switches, both are given with the following equations:

$$Pl_{tot}^{sfb} = Pl_{fp1}^{sfb} + Pl_{fp2}^{sfb}, \tag{15}$$

$$Pl_{tot}^{pfb} = Pl_{fp1}^{pfb} + Pl_{fp2}^{pfb} + Pl_{sp1}^{pfb} + Pl_{sp2}^{pfb}. \tag{16}$$

Finally, we define a new parameter called relative packet loss gain (Pl_{rel}) between the Model SFB and Model PFB. If Pl_{rel} has a positive value, it means that Model PFB has lower packet loss rate compared to Model SFB. The relative packet loss gain is calculated as:

$$Pl_{rel} = \frac{(Pl_{tot}^{sfb} - Pl_{tot}^{pfb})}{Pl_{tot}^{sfb}} \times 100\%. \tag{17}$$

4 Results

We now want to compare the packet loss rate in both proposed systems, Model SFB and Model PFB. The parameters are described in Table 3. The probability of *Packet-In* messages is in the range of 0.1 to 1, the controller-to-switch service rate, $r = \mu_c/\mu_s$, varies from 0.1 to 2, μ_s is set at 10,000 pkt/s. The arrival rate of the packets per MN is 400 pkt/s and 600 pkt/s, the maximum transmission unit (MTU) is 1500 Bytes, and the number of MNs varies in the range of 1 to 80. The numerical analysis is performed in MATLAB. The numerical results are afterwards compared with the simulation results obtained by using a simulator also developed in MATLAB. The

Table 3 Parameter Setting

Parameter	Value
Probability of Packet-In messages, β	0.1–1
Mean service time of the switch, μ_s(pkt/s)	10,000
Controller to switch mean service rate, r	0.1–2
Arrival rate, λ (pkt/s)	400, 600
Maximum Transmission Unit, MTU (Bytes)	1500
Number of mobile nodes per switch, N	1–80

Monte Carlo approach is used, and the simulations are executed 50 times (5 s each). We use MATLAB 2019a, a laptop with Core-i5-8350U 1.9 GHz with 16 GB RAM. At each simulation execution, the generated control traffic is stored in files, which are later post-processed for calculating the exact points in the graphs.

We first want to verify the numerical results for Model SFB and Model PFB, by comparing them with the simulation results obtained by our discrete event simulator. Figure 3 shows the validation results. β increases on the x-axis, whereas number of MNs is $N = 20$, and $r = 1$. Looking at the results, firstly we conclude that the simulation results are well in-line with the numerical values, the difference between the numerical and simulation results is between 0.7 and 2.6%. Figure 3a shows the packet loss rate dependency for Model SFB, whereas Fig. 3b depicts the same for Model PFB. For both graphs, as the probability of *Packet-In* messages increases, the packet loss rate also increases. As the arrival rate becomes higher, as expected, the packet loss rate is also going up. In Fig. 3a, for $\lambda = 400$, starting from $\beta = 0.5$, there

Fig. 3 Absolute packet loss rate validation for: **a** Model SFB, **b** Model PFB

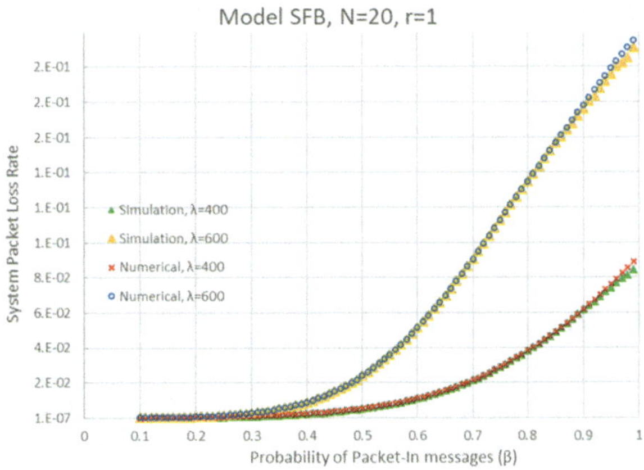

(a)

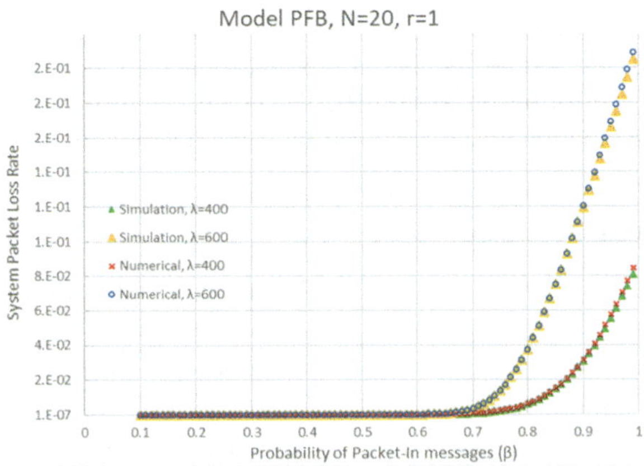

(b)

is more visible increase in the average packet loss rate. On the same figure, for λ = 600, the increase of *PL* is steeper, and it starts to be visible at lower value of β = 0.4. For Model PFB in Fig. 3b, a similar analysis can be done, as β increases, the *PL* also increases, and as the arrival rate is higher, the *PL* increases faster and has higher absolute values. However, when comparing Fig. 3a with Fig. 3b, we can conclude that the Model PFB is much more robust when the probability of *Packet-In* messages increases, meaning that the visible increase for λ = 400 starts at $\beta = 0.8$, and for λ = 600 starts at $\beta = 0.7$. To conclude, the model PFB offers lower average packet loss rate for lower values of β, and for $\beta = 1$, the packet loss rate has similar values for both models.

In Fig. 4 we analyze the relative *PL* gain, as defined in Eq. (17). Again, on the x-axis we vary β, whereas $r = 1$. The difference between Fig. 4a, b and c, is that number of MN changes with $N = 10$, $N = 20$, and $N = 40$, respectively. When analyzing Fig. 4a, we conclude that for lower values of β, for the case of λ = 600, Model PFB exhibits almost 100% reduction in the average packet loss rate, when compared to Model SFB. For values of β higher than 0.6, this *PL* reduction decreases. In Fig. 4a, for λ = 400, we see negative values for the packet loss gain, which basically means that for all β, Model SFB exhibits better results. Interestingly, as β increases for the lower curve in Fig. 4a, the improvement when compared to Model PFB decreases, but only up to $\beta = 0.65$, afterwards the *PL* of Model SFB again improves when comparing to Model PFB. The graph in Fig. 4b however, for $N = 20$, shows a different result. Again, for λ = 600, the reduction in *PL* of Model PFB is almost 100% for lower β, but the same applies also for λ = 400. The difference is that for λ = 400, after $\beta = 0.5$, there is a decrease in the relative *PL* gain, whereas for λ = 600, this decrease starts at $\beta = 0.6$. So, for Fig. 4b, Model PFB exhibits better results. Finally, we analyze Fig. 4c, where $N = 40$. For all β values, and both arrival rates, Model PFB shows better performance. The relative *PL* gain of almost 100% decreases at higher values of β, at β = 0.7 the gain tends to deteriorate. To conclude, Model PFB has lower average packet loss rate compared to Model SFB in almost all scenarios of interest. As the number of MNs increases, this improvement is highly dominant even for higher β values, and generally as the arrival rate becomes higher, Model PFB has superior performance when compared to Model SFB. The only case where Model SFB shows better results is when the arrival rate and the number of MNs is low, however for mobile networks, the very opposite is true.

Now in Fig. 5, we vary r on the x-axis, and we increase β in Fig. 5a, b, and c by using a fixed value of $\beta = 0.1$, $\beta = 0.5$, $\beta = 0.9$, respectively. By analyzing Fig. 5a, as r increases from 0.1 to 2, we notice almost linear *PL* gain, 99.8% for λ = 600, and around 99.76 for λ = 400. This means that Model PFB shows superior results for all r values and for both arrival rates of interest. In Fig. 5b, $\beta = 0.5$, we notice similar curve shapes, but now the *PL* gain is slightly decreased to 99.7%, and 98% for λ = 600, and λ = 400, respectively. Additionally, for λ = 400, for very small values of β of around 0.1, we notice a non-linear dependency, which promptly increases from 97.4% for $r = 0.1$, to 98% for $r = 0.2$. This means that if r is very small, and for lower arrival rates, Model PFB has slightly lower *PL* gain. Similar, but much more visible behavior can be seen on Fig. 5c, where $\beta = 0.9$. For values of $r = 0.18$ and

Fig. 4 Relative packet loss gain for: **a** N = 10, **b** N = 20, **c** N = 40. (r = 1)

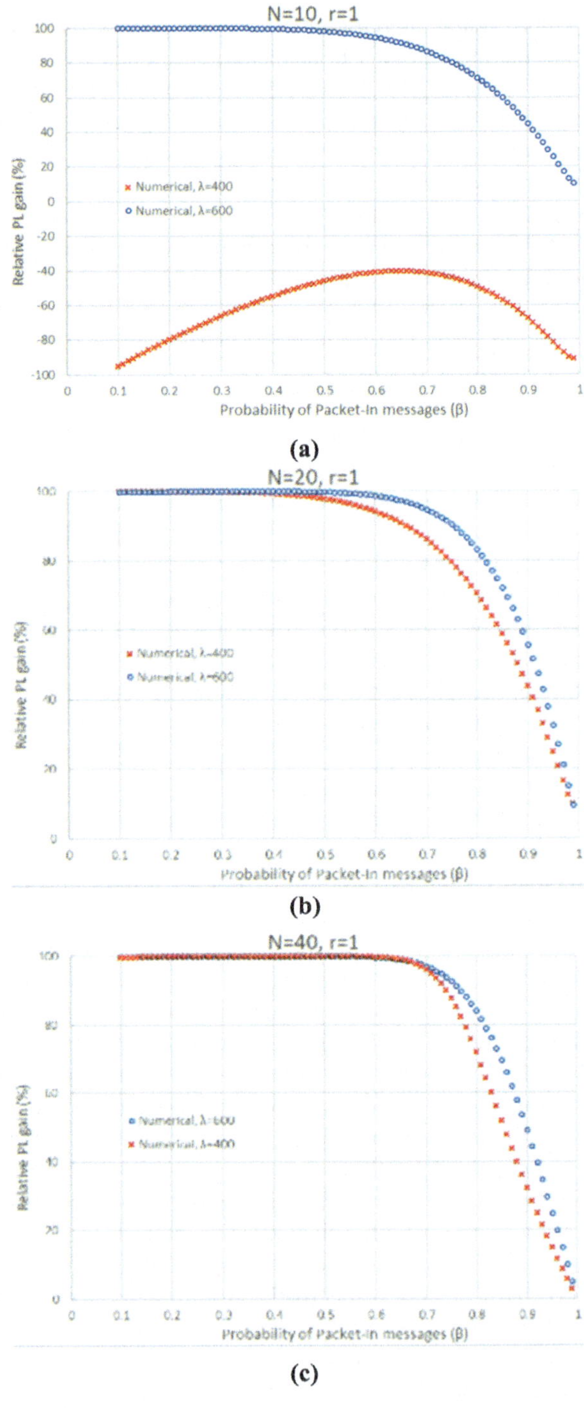

Fig. 5 Relative packet loss gain for: **a** β = 0.1, **b** β = 0.5, **c** β = 0.9 (N = 20)

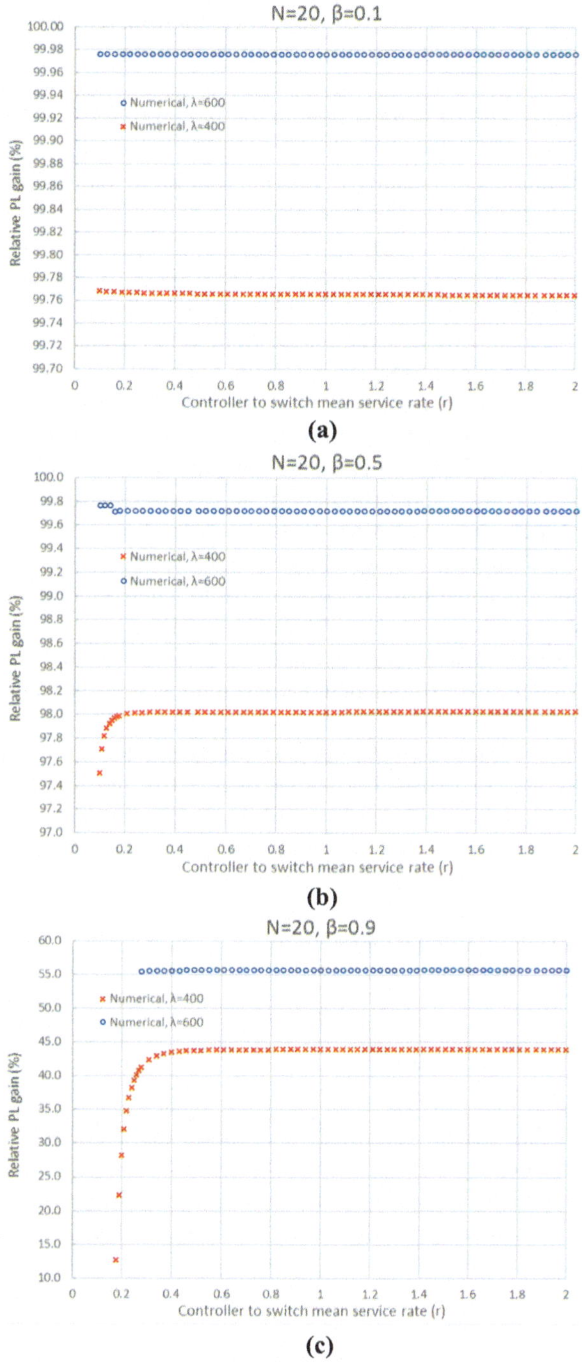

$\lambda = 400$, the relative *PL* gain is 12.5%, whereas for $r = 0.4$ is around 44%. For $\lambda =$ 600, such non-linear dependency is not seen, but it must be stressed that the *PL* gain in Fig. 5c, for both arrival rates, is significantly lower in comparison with Fig. 5a and b. The *PL* gain is now in the region of 44% to 56%, whereas previously it was above 97%. To conclude, Model PFB shows superior performance for all scenarios of interest, especially if β is not that high, and if the value of r is higher. The later only applies for the case of $\lambda = 400$ in Fig. 5b and c, where we saw a non-linear curve for the very small values of r. This generally means that increasing r does not have impactful performance boost, except for smaller arrival rates and the smallest r values, which is not typical for mobile networks.

Finally, we analyze the absolute relative packet loss rate in Fig. 6. The x-axis is the number of MNs (varies from 1 to 80), $r = 1$, whereas we increase β in Fig. 6a, b, and c, by using fixed value of $\beta = 0.1$, $\beta = 0.5$, $\beta = 0.9$, respectively. In Fig. 6a, $\beta = 0.1$, we notice that Model PFB shows superior performance when compared to Model SFB, especially if the number of MNs goes up. When the number of MNs is low, approximately up to $N = 10$, there is no significant difference in both Models, however, afterwards the difference increases as the number of MNs goes up. Similarly, in Fig. 6b, $\beta = 0.5$, we see similar superiority of the Model PFB, but this superiority is clearly visible when MN number is higher than 30. We also see significant average *PL* difference between the two models at $N = 80$, the same conclusion as in Fig. 6a. Figure 6c shows both arrival rate curves for $\beta = 0.9$. Again, Model PFB exhibits better performance, however now the absolute difference in *PL* for both models is much smaller, and this difference is visible starting at $N = 17$. To conclude, generally as β increases and the number of MNs increases, the relative *PL* also increases. For values of $\beta = 0.1$ and $\beta = 0.5$, the performance superiority of Model PFB is clearly visible especially for higher MN values, however if $\beta = 0.9$ the difference of both models in the values of *PL* tends to be smaller. As for mobile networks, high number of MNs is expected, and β is expected to be lower (in the region of 0.1 or 0.2), we conclude that the Model PFB is better suited for switch design in mobile core networks.

5 Conclusion

In this article we aimed to compare the average packet loss rate of two systems that deploy different OF-switch buffer designs, Model SFB that uses traditional shared buffering, and Model PFB that incorporates non-preemptive prioritization where the control packets are always prioritized. The proposed mathematical model heavily relies on queuing theory and the use of QBD processes. The numerical values are validated with the results obtained by using a discrete event simulator. The comparison error observed for the average packet loss rate is between 0.7% and 2.6%. The results reveal that when r and N are kept as constants, and β varies from 0.1 to 1, the packet loss rate increases as both β and the arrival rate increase. Furthermore, the Model PFB is superior as the sharp increase in the packet losses starts at higher

Fig. 6 Absolute packet loss gain for: **a** $\beta = 0.1$, **b** $\beta = 0.5$, **c** $\beta = 0.9$. ($r = 1$, $\lambda = 600$)

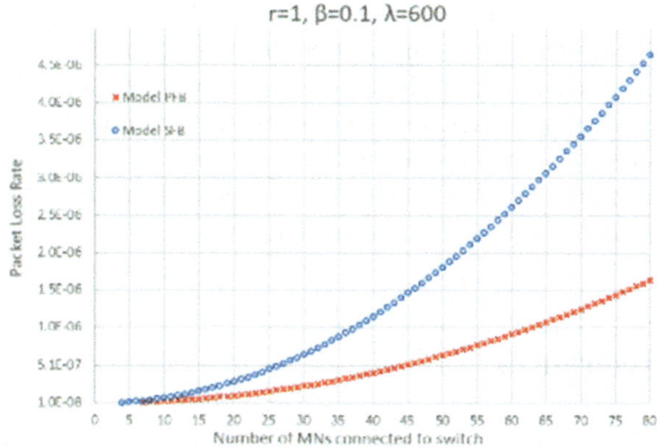

(a)

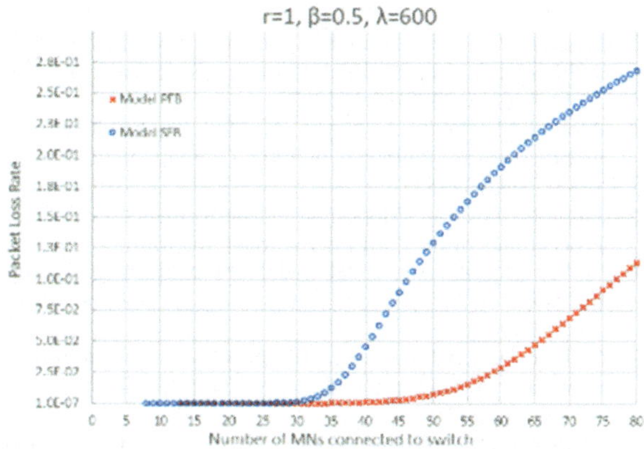

(b)

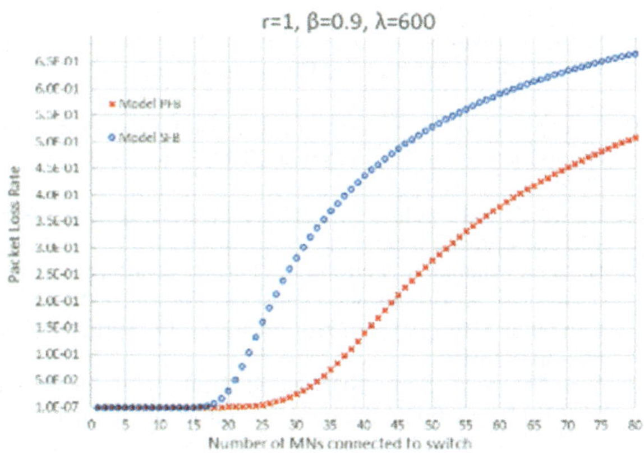

(c)

β values when compared to Model SFB. We also proved that varying the parameter r does not have an impactful performance influence for mobile networks where β is low, the number of MNs is high, and the arrival rates are high. Finally, as the number of MNs increases, the Model PFB again outperforms the Model SFB, however at very high β values, this difference tends to be smaller. For future work, we plan to develop a new analytical model that will incorporate more realistic SDN switches with detailed hardware considerations.

References

1. Cox, J.H., Cung, J., Donovan, S., et al.: Advancing software-defined networks: a survey. IEEE Access **5**, 25487–25526 (2017). https://doi.org/10.1109/JPROC.2014.2371999
2. Open Networking Foundation: OpenFlow switch specification version 1.3.1. Tech. Republic (2012)
3. Nguyen, V.G., Brunstrom, K., Grinnemo, J., et al.: SDN/NFV-based mobile packet core network architectures: a survey. IEEE Commun. Surv. Tutor. **19**(33), 1567–1602 (2017). https://doi.org/10.1109/COMST.2017.2690823
4. Alotabi, M., Helmy, A., Nayak, A.: Modeling handover signaling messages in OpenFlow-based mobile software-defined network. J. Comput. Netw. Commun. **2018** (2018). https://doi.org/10.1155/2018/1543531
5. Jarchel, M., et al.: Modeling and performance evaluation of an OpenFlow architecture. In: Proceedings of the 23rd International Teletraffic Congress, pp. 1–7 (2011)
6. Sarkar, C., et al.: Analytical model for OpenFlow-based software-defined network. In: Progress in Computer, Analytics and Networking, pp. 583–592 (2018). https://doi.org/10.1007/978-981-10-7871-2_56
7. Mahmood, K., Chilwan, A., Osterbo, O., et al.: Modelling of OpenFlow-based software-defined networks: the multiple node case. IET Netw. **4**(5), 278–284 (2015). https://doi.org/10.1049/iet-net.2014.0091
8. Mahmood, K., Chilwal, A., Osterbo, O., et al.: On the modeling of OpenFlow-based sdns: the single node case. Comput. Sci. Inf. Technol. (CS & IT) **4**, 207–214 (2014). https://doi.org/10.5121/csit.2014.41120
9. Shang, Z., Wolter, K.: Delay evaluation of OpenFlow network based on queueing model. In: 12th European Dependable Computing Conference (2016). https://doi.org/10.1016/j.comnet.2016.03.005
10. Mondal, A., Misra, S., Maity, I., et al.: Buffer size evaluation of open-flow systems in software-defined networks. IEEE Syst. J. **2018**, 1–8 (2018). https://doi.org/10.1109/JSYST.2018.2820745
11. Burke, P.J.: The output of a queuing system. Oper. Res. **4**(6), 699–704 (1956). https://doi.org/10.1287/opre.4.6.699
12. Busic, A., Gaujal, B., Perronin, F.: Perfect sampling of networks with finite and infinite capacity queues. In: Al-Begain, K., Fiems, D., Vincent, J.-M. (eds) ASMTA, Series Lecture Notes in Computer Science, vol. 7314, pp. 136–149. Springer (2012). https://doi.org/10.1007/978-3-642-30782-910
13. Yen, T.C., Su, C.S.: An SDN-based cloud computing architecture and its mathematical model. In: 2014 International conference on information science, electronics and electrical engineering (ISEEE), vol 3. IEEE, pp 1728–1731. https://doi.org/10.1109/InfoSEEE.2014.6946218
14. Javed, U., Iqbal, A., Saleh, S., et al.: A stochastic model for transit latency in OpenFlow SDNs. Comput. Netw. **113**, 218–229 (2017). https://doi.org/10.1016/j.comnet.2016.12.015
15. Lai, Y., Lai, C., Lai, A., et al.: Performance modeling and analysis of TCP connections over software defined networks. In: GLOBECOM 2017–2017 IEEE Global Communications Conference, pp. 1–6, IEEE (2017). https://doi.org/10.1109/GLOCOM.2017.8254078

16. Sood, K., Sood, S., Yu, S., et al.: Performance analysis of software defined network switch using M/Geo/1 model. IEEE Commun. Lett. **20**(12), 2522–2525 (2016). https://doi.org/10.1109/LCOMM.2016.2608894
17. Miao, W., Min, G., Wu, Y., et al.: Performance modelling of preemption-based packet scheduling for data plane in software defined networks. IEEE International Conference on Smart City/SocialCom/SustainCom (SmartCity) (2015). https://doi.org/10.1109/SmartCity.2015.48
18. Miao, W., Min, G., Wu, Y., et al.: Performance modelling and analysis of software-defined networking under Bursty multimedia traffic. ACM Trans. Multimed. Comput. Commun. Appl. (TOMM) **12**(5s), 77 (2016). https://doi.org/10.1145/2983637
19. Goto, Y., Masuyama, H., Ng, B., et al.: Queueing analysis of software defined network with realistic OpenFlow–based switch model. Comput. Netw. **164** (2019). https://doi.org/10.1109/MASCOTS.2016.30
20. Ogasawara, S., Takahashi, Y.: Performance analysis of traffic classification in an OpenFlow switch. In: Cloudification of the internet of things (CIoT). IEEE (2016). https://doi.org/10.1109/CIOT.2016.7872908
21. Schassberger, R.: Review of reversibility and stochastic networks. Perform. Eval. **1**(1), 90 (1981). ISBN: 9781107401150
22. Bolch, G., Greiner, S., Meer, H., et al.: Queueing networks and markov Chains, 2nd ed. John Wiley & Sons, Inc. Hoboken, New Jersey (2006). ISBN: 9780471791577
23. Panev, S., Latkoski, P.: Performance analysis of handover delay and buffer capacity in mobile OpenFlow-based networks. Int. J. Commun. Syst. **33**(Is.), 15 (2020). https://doi.org/10.1002/dac.4529

The manufacturer's authorised representative in the EU is Springer Nature Customer Service Centre GmbH, Europaplatz 3, 69115 Heidelberg, Germany. If you have any concerns regarding our products, please contact ProductSafety@springernature.com

Printed and bound by CPI Group (UK) Ltd, Croydon, CR0 4YY
28/07/2026
02179240-0003